Goldspink

SYMPOSIA OF THE
SOCIETY FOR EXPERIMENTAL BIOLOGY

NUMBER XLVII

PROCEEDINGS OF A MEETING
HELD AT THE UNIVERSITY OF BATH,
ENGLAND
28 SEPTEMBER – 1 OCTOBER 1992

SYMPOSIA OF THE
SOCIETY FOR EXPERIMENTAL BIOLOGY

SYMPOSIA OF THE
SOCIETY FOR EXPERIMENTAL BIOLOGY

NUMBER XLVII

CELL BEHAVIOUR:
ADHESION AND MOTILITY

EDITED BY
GARETH JONES, CAROLINE WIGLEY
AND RICHARD WARN

SOCIETY FOR EXPERIMENTAL BIOLOGY SYMPOSIA

SEB Symposia form a long-standing series of volumes first published in 1947. The series is annual and each publication is a collection of authoritative articles on an aspect of modern experimental Biology. The contributors are all invited and speak on specific topics within the chosen field. Meetings are held annually, in September, over a period of two or three days.

The aims of the Symposium series are to stimulate discussion and communication between scientists of all nationalities, to foster the development of, and research on, modern aspects of plant, animal and cell biology.

Published for the Society for Experimental Biology
by The Company of Biologists Limited
Department of Zoology, University of Cambridge
Downing Street, Cambridge CB2 3EJ

Typeset, Printed and Published by The Company of Biologists Limited,
Department of Zoology, University of Cambridge
Downing Street, Cambridge CB2 3EJ

© Society for Experimental Biology 1993
ISBN 0 948601 39 6
ISSN 0081–1386

A CIP Catalogue record for this book is available from The British Library

Cover photograph

Model of a crawling fibroblast showing the spatial relationships of the various actin assemblies in the lamellipodium, dorsal cortical microfilament sheath, stress fibers and perinuclear region. Further information can be found in the article by J. P. Heath and B. F. Holifield in this volume.

CONTENTS

Contents

PREFACE

This volume is the third in a series of published symposia dedicated to the memory of Michael Abercrombie. It follows the tradition of the previous volumes in attempting to produce a snapshot of recent developments in the field of cell motility. This most recent Symposium was held from 28th September - 1st October, 1992 at the University of Bath, and constituted a joint meeting of the Society for Experimental Biology and the British Society for Cell Biology.

The underlying theme of the Symposium and of this volume concerns the mechanisms governing cell motility considered at several levels. The papers published here are grouped into sections reflecting the different sessions of the Symposium. These develop the theme of cell motility from cell behaviour through to the molecular motors which effect motility. The first section entitled "Cell Behaviour" introduces the parameters thought to be integral to cell locomotion *in vitro* - contact responses and cytoskeletal changes - and describes a new method for the analysis of cell behaviour. The next session concentrates on cell-substratum adhesions, integrin-ligand binding and the roles of laminin and tenascin in particular. The activity of recently described motility factors on cell behaviour is the subject of the third section of this volume, which includes a discussion of potential effects on tumour cell metastasis. The two final sections are devoted to the cytoskeleton; firstly as the target for signalling pathways generated at the plasma membrane, and secondly as elements in the response of the cell to external stimuli. The wide range of cellular and molecular approaches taken by the authors to further our understanding of motile behaviour in cells is well illustrated in this volume.

The editors would like to thank the many people involved in the preparations for the Symposium and the publication of its proceedings. In addition to the financial support of the Society for Experimental Biology and the British Society for Cell Biology, the expenses of several contributors were met by the Royal Society, the British Council and the Yamanouchi Research Institute. Finally we would like to thank our session chairmen, Jim Weston, Sir Michael Stoker, Adam Curtis, John Lackie and Yuri Vasiliev and all the contributors to this timely review of a rapidly progressing field.

Gareth Jones, Richard Warn and Caroline Wigley

Printed in Great Britain © The Society of Experimental Biology 1993

CONTACT SIGNALLING AND CELL MOTILITY

D. GINGELL

Department of Anatomy and Developmental Biology, University College London, Gower Street, London WC1E 6BT, UK

Summary

There is now clear evidence that signals can be generated when cells make adhesions. This happens when contacts are made with the extracellular matrix, with other cells or with experimental substrata. It is shown that intracellular signals are triggered in a variety of situations by the aggregation of transmembrane glycoproteins. Such aggregation occurs at adhesion sites, and it is suggested that adhesion molecules which are capable of lateral diffusion become trapped at developing adhesion sites because their headgroups bind to arrays of external ligands. Signals generated by molecular clustering can influence cell spreading and motility via familiar transduction pathways.

Although it is not yet possible to reconstruct with confidence the entire sequence of events between initial glycoprotein clustering and motility, there is a lot of information which points to the identities of the molecular ingredients. I shall focus the discussion on the thin peripheral lamellar extensions produced by cell types ranging from amoebae to vertebrate nerve growth cones, and discuss the experimental evidence relating to their cytoskeletal architecture and dynamics. Finally I shall try to construct models of motility consistent with the experimental data, and suggest how lamellar motility may be modulated by signals generated by clustered adhesion molecules.

Introduction

The realization that intracellular signals can be generated by the binding of soluble ligands to receptor proteins of the cell membrane opened one of the most exciting and productive phases of cell biology. It brought into focus the significance of the lateral diffusion of membrane molecules which can activate signalling pathways. Much more recently it has become apparent that the engagement of transmembrane adhesion receptor proteins with ligands of the extracellular matrix, or those carried on other cells, can also trigger intracellular signalling mechanisms. A fresh impetus to thinking about the control of tissue cell motility has emerged from the realization that the actin-based motor can respond to the signals generated by adhesions. No longer are we constrained to view cell locomotion as being influenced purely by the mechanical effects of adhesive pathways, modulated by chemotactic signals from diffusible molecules. Instead we perceive moving cells "listening" to the structural texture of their surroundings, manifested as contact signals, and responding with cytoskeletal rearrangements which influence motility. This has important implications for unravelling the complexities of processes as diverse as path finding in neurogenesis and leucocyte migration.

Key words: cell motility, contact signalling, cell adhesion, extracellular matrix, glycoprotein, model.

I shall discuss contact-mediated signalling in a wide range of systems and then focus on the way in which such events may influence the cytomechanics of the thin peripheral lamellar extremities which play a prominent role in the motility of a variety of cell types. A plausible sequence linking the earliest formation of cell contacts made by the lamellae to the subsequent cytomechanical events will be put forward. The lamellipodium of fibroblasts appears to have a structural counterpart in a wide variety of cells, including nerve growth cones, the advancing edge of leucocytes, blood platelets and small amoebae. It appears to be a common basic element of motility in cells which exhibit crawling movement. Being ≈0.1 μm thick it lacks cytoplasmic inclusions and is devoid of microtubules and intermediate filaments. It is known to contain actin and the single-headed myosin-I which can associate with membranes. The membrane of the lamellipodium has adhesive proteins and, in the case of the nerve growth cone, abundant G-proteins which suggest signalling potential.

There is arguably no entirely new idea in what follows. What may be useful is in collecting several relatively new but well documented cellular functions and forming from them a chain of events. It may be helpful to outline the steps to avoid getting bogged down in the supporting details. The sequence is as follows. First, there is now good evidence that adhesions can generate cytoplasmic signals. It is shown that adhesion receptors which are free to diffuse laterally accumulate at contact sites, where engagement with appropriate bound ligands traps them. The receptors consequently cluster, and this generates local signals. In advancing lamellae, the cytoskeletal response probably involves protein phosphorylation, and this may control the activity of myosin-I molecules, which are thought to form mechanical links between contact sites and longitudinal actin filaments. The mechanochemical state of these links determines whether the lamella advances centrifugally, remains static or develops centripetally. Such considerations of the cytomechanics of lamellae build upon previous models put forward to explain nerve growth cone behaviour. Since the potential compass of the subject is very wide, and it would be cumbersome to approach it in the manner of a comprehensive review, I shall summarize only what seems to me to be among the most pertinent evidence. In the short time which has elapsed since these ideas were first put forward (Gingell and Owens 1992b) a number of important studies have appeared which add strength to the argument, and indeed remove contact-signalling from the realm of the arcane to general acceptance.

Adhesions send signals

A growing body of evidence makes it clear that contacts between cells and their surroundings can generate signals which elicit cellular responses. Such contact-dependent responses can apparently result from interactions with other cells, the extracellular matrix, or with experimental surfaces.

Perhaps the first clear evidence of contact signalling came from the highly imaginative but rather neglected experiments of Wright *et al.* (1983) illustrated in Fig. 1. Monocytes carry two distinctly different C3 complement receptors, CR1 (a member of the immunoglobulin superfamily) which binds to particles opsonized with C3b, and CR3

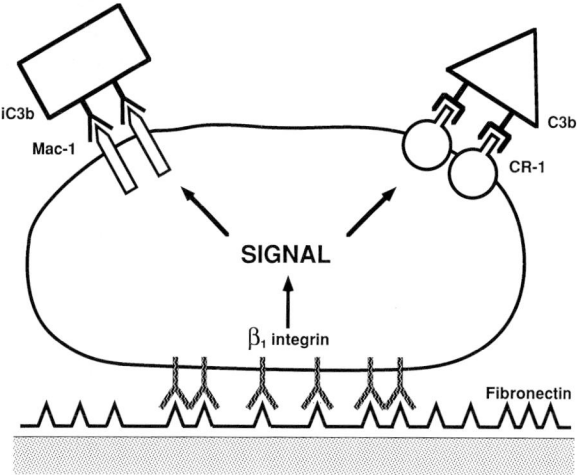

Fig. 1. Monocyte phagocytic response influenced by spreading on fibronectin.

(leucocyte β2 integrin CD11b/CD18, formerly called Mac-1) which binds to particles bearing iC3b. Both receptors normally exist at the cell surface in an inactive state. In the active state, appropriately opsonized particles are bound and endocytosed. No change in the number of surface receptors is involved in activation. Wright *et al.* found that monocytes, which were not otherwise activated by soluble factors, spread on albumin or gelatin coated surfaces and showed a minimal ability to internalize opsonized particles via the two types of C3 receptor. In sharp contrast, when monocytes were spread on surfaces coated with covalently bound fibronectin, the C3 receptors on the distant apical part of the cell became activated and were able to mediate endocytosis. This remarkable effect was reversed when the cells were re-spread on a surface lacking fibronectin. The authors concluded that the activation of C3 receptors on the apical surface of the cell occurs by a novel mechanism initiated by the interaction of substrate-bound fibronectin with membrane receptors in the basal part of the cell. In a following paper Bohnsack *et al.* (1985) reported a similar phenomenon in macrophages attached to laminin-coated glass. Since the cells spread more extensively on uncoated glass than on laminin, the authors were able to rule out a relation between degree of spreading and degree of phagocytosis. They aptly concluded that a "global change in phagocyte plasma membrane function" develops after contact with extracellular matrix proteins.

In neither of these studies was any suggestion made of cytoplasmic signalling, and adhesion was not discussed in terms of integrins, but viewed with hindsight, signals arising from integrin-ligand interactions are clearly the means by which communication occurs between the basal and apical cell membranes.

In a recent publication from the same laboratory Siu *et al.* (1991) discuss an analogous example of adhesive potentiation in the interaction between polymorphonuclear leucocytes (PMN) and endothelial cells. Initial adhesion is mediated by ELAM-1, expressed on cytokine-stimulated endothelial cells; this binds to specific carbohydrate

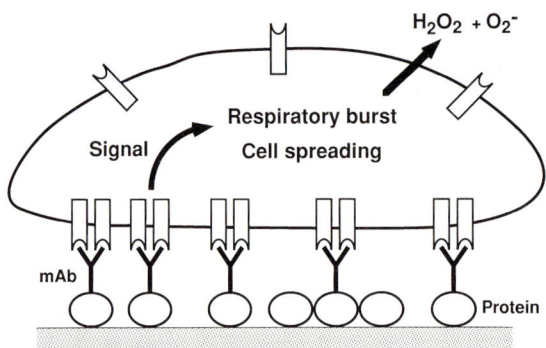

Fig. 2. Neutrophil spreading and respiratory burst in response to signals from immobilized antibodies to integrin chains.

residues on the leucocyte membrane. Binding causes the PMN to activate CR3, an event which increases the strength of adhesion between the two cells, since the activated CR3 interacts with uncharacterized carbohydrates on the endothelium. Studies with blocking mAbs showed that the two adhesion mechanisms operate in sequence, and suggest that the first triggers the second by means of intracellular signals, to give an adhesion cascade.

Signals generated by cell-substratum adhesion can influence the respiratory burst and actin organization in neutrophils (Berton *et al.* 1992). These authors measured the respiratory burst of cells which had spread on antibodies (Abs) bound indirectly to polystyrene (Fig. 2). Antibodies to the common $\beta2$ chain (CD18) as well as those to the three α chains (CD11a, b, c) of the leucocyte integrins LFA-1, p150,95 and CR3 (alias Mac-1) all promoted cell spreading. In contrast, the respiratory burst with H_2O_2 formation was triggered by antibodies to CD18, CD11a and CD11c but not the α_M (CD11b) chain of Mac-1, nor by bound Abs directed against a variety of other surface antigens. This result is particularly interesting since the binding of particles opsonized with iC3b to CR3 is reported not to stimulate the respiratory burst. The authors conclude that the contact- generated signals for spreading and for the respiratory burst are distinct, but they give only a brief discussion.

The nature of contact-induced signalling in T-cells has been elegantly demonstrated by work from Springer's laboratory (Dustin and Springer 1989; Springer 1990). On the T-cell surface LFA-1 is present in an inactive form. Activation can be achieved either by engagement of the TCR complex with the appropriate class of MHC (armed with a peptide fragment) or else by cross-linking the TCR complex with anti-CD3 followed by anti-Ig (see Fig. 3). Signals are generated by mechanisms apparently involving protein kinase-mediated phosphylation of the cytoplasmic tails of CD11a, leading to the association of actin filaments with LFA-1 molecules (Pardi *et al.* 1992). The activation of LFA-1 results in binding to ICAM-1 on target cells. Dustin and Springer coined the catchy phrase "adhesion servo motor" to describe the activation of a second adhesion mechanism due to signals originating from a different initial adhesion system. They emphasize that spontaneous reversion of LFA-1 to a low avidity state occurs within

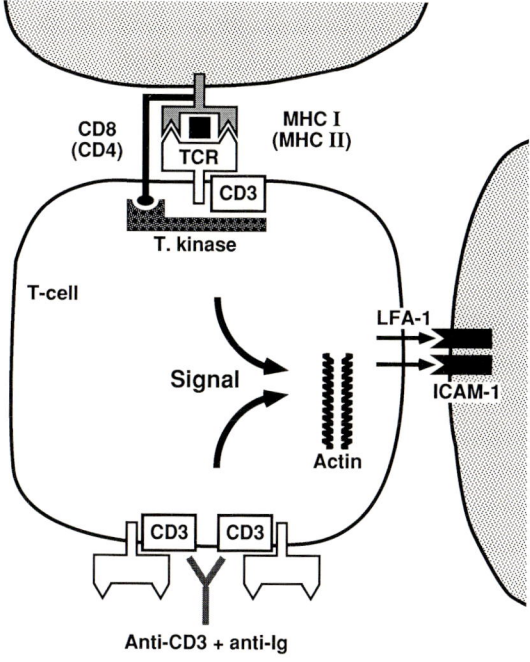

Fig. 3. Signals generated by engagement of the T cell receptor with MHC or by antibody-mediated cross-linking of the receptor complex activate the LFA-1 adhesion mechanism.

minutes, so that the system is well adapted for transient adhesions. Such reversibility clearly implicates a subtle feedback control mechanism.

An analogous activation process occurs in platelet adhesion (review : Andrews and Fox 1990). The platelet integrin IIb,IIIa (alias αIIb,β3) which is principally responsible for platelet-platelet aggregation by means of fibrinogen cross-bridging, is normally present in an inactive form at the cell surface. Platelet adhesion, for example, to Von Willebrand factors adsorbed on solid surfaces by means of the heterodimeric non-integrin receptor Ib,IX, may activate the fibrinogen receptor. A well documented activation mechanism involves the binding of thrombin to its receptor (CD36) on the platelet membrane. This stimulates an associated cytoplasmic G-protein, resulting in a signal cascade which in turn activates αIIb,β3 and confers the ability to bind soluble fibrinogen (review : Brass *et al.* 1991).

Contact-mediated signalling is clearly implicated in the spreading of HeLa cells on collagen. Chun and Jacobson (1992) have elegantly demonstrated that collagen receptor up-regulation, clustering and binding to the cytoskeleton accompany spreading. Receptor clustering released arachidonic acid, whose lipoxygenase products are central to the signalling pathway which results in cytoskeletal reorganization.

A further elegant demonstration of the fact that integrin-mediated adhesion on one aspect of a cell can influence adhesion at another has been recently provided by Curtis *et al.* (1992). BHK cells adhere but spread poorly on haemoglobin adsorbed to a solid substratum. However, when Dynabeads bearing fibronectin were allowed to come into contact with the cells, many were seen to spread avidly on the adsorbed haemoglobin and these cells were found to carry adherent beads on their upper surfaces, remote from the substratum. Although the signal pathway has not yet been analyzed, it seems clear that signals generated by the strictly localized binding of fibronectin to β_1 integrins on the apical surface of a cell stimulates adhesion at the basal surface, together with global cytoskeletal reorganization.

The engagement of integrins, as well as other types of receptor, can send signals which regulate cellular metabolism. A synergistic effect of fibronectin binding to the integrin VLA-5 (α_5,β_1) together with antibodies to CD3 was shown to cause proliferation in a lymphocyte sub-set by Matsuyama *et al.* (1989). Werb *et al.* (1989) showed that fibroblasts plated on intact fibronectin synthesize matrix-degrading metaloproteinases at a low level which increases markedly on addition of anti-α_5 antibody or large cell binding fragments of fibronectin. In mammary alveolar cells, the final state of differentiation requires a signal derived from the association of cellular integrins with the basement membrane. Streuli *et al.* (1991) found that antibodies against the integrin β_1 chain can block casein synthesis. During glandular involution, casein synthesis ceases at the time when the basement membrane is degraded by metaloproteinases, in which state contact signals can no longer be initiated. Damsky and Werb (1992) review evidence relating to fibroblast-ecm interactions which suggests that the association of α_5,β_1 integrin with the cell binding domain of fibronectin initiates a signal which results in metaloproteinase secretion and ecm remodelling. In contrast, simultaneous interaction with both the cell binding domain (via α_5,β_1) and heparin binding (via a transmembrane syndecan-like proteoglycan) inhibits proteinase release and promotes ecm assembly and focal contact formation. The engagement of heparin-binding receptors sends a signal which activates protein kinase C and is distinct from the integrin signal.

A further cooperative signal-generating interaction discussed by Damsky and Werb is the simultaneous association of fibroblast growth factor with both the extracellular part of a transmembrane proteoglycan and with the specific Ig superfamily receptor for the growth factor (Klagsbrun and Baird 1991). Similarly, the cooperative interaction between integrins and growth factors can lead to a signal cascade which regulates the Na^+/H^+ antiporter (Schwartz *et al.* 1991; Schwartz and Lechene, 1992).

Thus it is clear that engagement of a given type of receptor may send a signal resulting in a certain cellular response, whereas simultaneous cooperative engagement of that receptor plus another can generate a different signal-mediated response. This multiplicity may have a significant role in cell behaviour. The possible permutations from such duplex interactions greatly increases the spectrum of responses that can be generated for any given number of receptor types and readily suggest a basis for distinct patterns of cell behaviour in structurally complex environments.

Cell-to-cell adhesion provides a less well documented category of adhesion-mediated signalling responses. The first unequivocal evidence that cell-cell adhesion molecules can trigger transduction was obtained by Schurch *et al.* (1989) using rat PC12 adrenal tumour cells. Stimulation of the Ca^{2+}-independent immunoglobulin superfamily adhesion molecules L1 and NCAM with specific antibodies reduced cellular levels of phosphoinositides as well as lowering intracellular pH and increasing intracellular free Ca^{2+}. A crucial finding, though described only briefly, was that cell-cell adhesion produced similar changes in Ca^{2+} and pH.

Contact-mediated transduction events have been extensively studied by Doherty's group using the PC12 line (Doherty *et al.* 1991a,b; 1992). PC12 cells were co-cultured on 3T3 fibroblast monolayers, which carried transfected genes for N-cadherin and NCAM. In this situation the PC12 cells changed their phenotype from adrenal to typical neuronal cells, a transformation inhibited by Pertussis toxin (which inhibits G_i) and Ca^{2+} channel-blockers. No such changes occurred on control cultures of normal 3T3 cells. These results show that transmembrane signalling can be initiated by mutual cell contact.

Signalling events following contact of avian cells with adhesion molecules NCAM, L1 and N-cadherin bound to solid substrates were reported by Bixby and Jhabvala (1990). Neurite outgrowth was strongly promoted on these surfaces, in a manner dependent on protein kinase C function, thus apparently involving a signal pathway different from that activated by extracellular matrix molecules.

Adhesion receptors which diffuse laterally in the membrane concentrate at adhesion sites

Perhaps the earliest indication that adhesion-competent receptors can diffuse from the apical surfaces of spread cells and become specifically trapped at the basal membrane, due to ligands immobilized on the substratum, came from Michl *et al.* (1979a). Immune complexes specifically bound to DNP-derivatized polylysine adsorbed on glass presented their Fc domains towards settling macrophages (see Fig. 4). The authors used an attachment assay involving opsonized erythrocytes to monitor the disappearance of Fc receptors from the apical region of the attached macrophages. Similar experiments using coverslips coated with complement indicated that the distribution of the receptor for C3b could be independently modulated. In a further publication, it was suggested that receptors disappear from the upper surface of the spread cells by diffusion to the basal

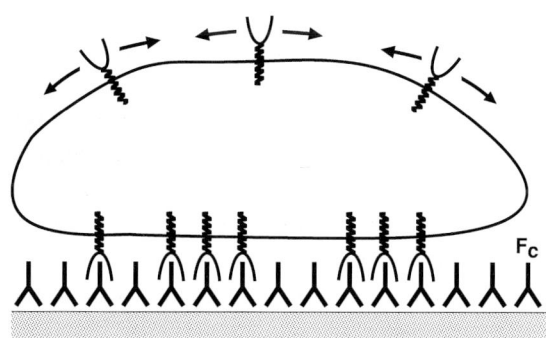

Fig. 4. Macrophage Fc receptors captured by diffusion into adhesion zone by immobilized orientated Ig.

membrane (Michl *et al.* 1979b). Similar conclusions were reached by Griffin and Mullinax (1981) who showed that lymphokine treatment increased C3b receptor mobility in the membranes of unstimulated macrophages and thereby promoted receptor redistribution.

Two later publications (Wright and Jong 1986; Bullock and Wright 1987) extended these studies to the receptor CR3 (the integrin Mac-1 or CD11b/CD18), as well as LFA-1 (CD11a/CD18) and P150,95 (CD11c/CD18), all of which show lateral diffusion and suffer individual specific capture by appropriate substratum-bound ligands (see Wright and Jong, op.cit. p.119).

The distribution of C3b receptors on neutrophils that had spread on glass or siliconized glass was also studied by Hafeman *et al.* (1982) using fluoresceinated monovalent anti-C3b Fab that does not cause patching. They observed clustered C3b receptors at the basal surface and also at the apical surface. The latter is perhaps unexpected, but may indicate a response to signals originating at the basal contact zone. This receives some support from the fact that no clustering was seen when cells were weakly attached to a phospholipid monolayer. Selective redistribution of receptors was shown by the failure of HLA molecules to respond to adhesion on a C3b coated surface.

An example which may involve non-specific trapping by lateral diffusion (though not discussed as such by the authors) can be seen in the results of Ursitti *et al.* (1991). Red cells attached to polylysine-treated surfaces were sheared off, leaving membranous fragments which showed strongly aggregated intramembranous particles, mostly dimers of band 3, which aggregate when freed from restraints imposed by the cytoskeleton. Persuasive evidence of receptor clustering accompanying cell spreading has recently been provided by Chun and Jacobson (1992). HeLa cells were allowed to spread on collagen and then fixed prior to labelling with anti-receptor antibody, followed by a rhodamine-labelled secondary antibody. The cells showed very pronounced receptor clustering at the basal surface contacting the substratum. Control cells in suspension showed much more diffuse labelling. Most interestingly, the receptors could also be clustered by exposure to primary antibody at 4°C prior to fixation and treatment with the secondary labelled antibody. This indicates that clustering occurs by passive lateral diffusion and antibody-mediated cross-linking of the membrane receptors.

The redistribution of receptors on lymphoma cells has been the subject of a clever novel experimental technique by Chan *et al.* (1991). The T-lymphocyte trans-membrane CD2 receptor binds to LFA-3 (Moingeon *et al.* 1989) in either its transmembrane or GPI-linked isoforms. Cells were allowed to spread on lipid membranes formed on a solid support. The lipids contained either GPI-linked LFA-3 (which was shown to diffuse laterally in the plane of the surface) or the transmembrane form (which was immobile). A measure of the density of bonds between CD2 and LFA-3 molecules in the adhesion zone was obtained by measuring the force required to shear cells from the surface. They concluded that CD2 receptors became more highly concentrated at the GPI/LFA-3 surface due to progressive mutual trapping. The essential point is that when *both* interacting components can diffuse the final density is not constrained to the mean value of the immobile partner. This study illustrates the most interesting point that ligand-receptor trapping may take place between cells on which both partners in the interaction are laterally mobile to give a local density of ligand-receptor pairs which exceeds that of either alone.

A number of studies, in which diffusible transmembrane adhesion molecules are seen to aggregate at sites of adhesion, also demonstrate that the cytoplasmic segments of adhesion molecules associate with the cytoskeleton. This leads to an important general question. Are the diffusible adhesion molecules sequestered by intracellular interactions with cytoskeletal elements which (for some reason) exist adjacent to substratum contacts, or are they first of all trapped as their extracellular headgroups engage immobilized arrays of multivalent ligands at contact points? If the latter is true, the local receptor concentration may effectively flip a switch which causes them to bind to the cytoskeleton and influence its organization. The second mechanism would avoid the difficulty of explaining why the cytoskeleton should be predisposed so as to trap adhesion molecules at sites adjacent to suitable arrays of binding ligands. Several studies have approached this question by cross-linking adhesion receptors with antibodies and then demonstrating the development of mechanical linkages between the clusters and the adjacent cytoskeleton. From this it is argued that fixed arrays of adhesive ligands would be expected to do the same by effectively cross-linking mobile receptors.

Evidence for the clustering and sequestration of proteoglycan adhesion receptors at the basal surface of epithelial cells as a result of contact with ligands of the basement membrane was published by Rapraeger *et al.* (1986). Subconfluent monolayers show apical labelling for proteoglycan, but this becomes restricted to the basal surface after confluence. The authors obtained evidence that the proteoglycan is anchored in the membrane, and speculate that it exhibits lateral diffusion. At the basal surface it is found to co-localize with F-actin. To determine whether the cytoskeleton traps the mobile receptors, or whether they are first trapped by adhesion to the substratum and only then are able to bind to the cytoskeleton, the authors cross-linked apical receptors at 4°C and followed their distribution at 37°C. The clusters rapidly become immobilized all over the cell surface by interaction with the actin cytoskeleton and become resistant to Triton extraction. Thus external cross-linking appears to cause receptors to associate with the cytoskeleton. These results correspond with the findings of Cody and Wicks (1986) who found that the antibody-induced clustering of laminin receptors resulted in their association with the detergent-insoluble cytoskeleton. They speculate that transmembrane laminin receptors cluster at cellular contacts with the extracellular matrix and that their cytoplasmic domains then bind to the cytoskeleton.

A further example (which will be referred to again later) is the linkage apparently induced by the adhesion of macroscopic beads to the surface of lamellae. This is deduced from the fact that the diffusion coefficients of some adherent particles can be much less than that of others and several orders of magnitude less than that of labelled membrane glycoproteins, obtained from FRAP measurements (Sheetz *et al.* 1989; Kucik *et al.* 1991).

A novel and informative approach from Yamada's laboratory to the question of whether it is the cytoskeleton or the substratum that traps adhesion molecules has been reported by La Flamme *et al.* (1992). Their study utilized cells which expressed chimeric receptors consisting of the extracellular and transmembrane domain of the interleukin-2 receptor combined with either the α_5 or β_1 intracellular integrin domain. Immunofluorescence studies showed that the β_1 chimera localized with the normal fibronectin receptor at focal contacts on a fibronectin surface. The α_5 chimera did not. This indicates that the intracellular domain alone of β_1 can, when activated, target the molecule to pre-formed focal adhesions, since the extracellular part cannot bind to fibronectin. They argue that the β_1 chimeras behave like ligand-occupied receptors in being constitutively switched to a state wherein they can bind to the cytoskeleton. This conclusion is supported by the fact that binding soluble fibronectin fragments to cells spread with focal contacts on vitronectin results in the trapping of fibronectin receptors at the focal contacts. Similarly, cells plated on fibronectin and exposed to GRGDS fragments (which bind to the α_v,β_3 vitronectin receptor but cannot bind to the substratum) show redistribution of the latter receptor to focal contacts. The overall conclusion from this study is that unoccupied normal heterodimeric receptors do not bind to the cytoskeleton, but that binding to appropriate ligands immobilized on a substratum effects a change in the intracellular segment of the β chain which allows association with the cytoskeleton. Thus diffusive trapping at adhesion sites drives focal contact assembly by causing integrins to associate with the adjacent cytoskeleton. A similar conclusion was reached in a recent study by Lu *et al.* (1992) regarding the trapping of non-integrin collagen receptors in the spreading of HeLa cells.

Synaptogenesis at the neuromuscular junction involves the lateral diffusion of acetylcholine receptor (AchR) molecules in post synaptic muscle cell membranes, in conjunction with some means of immobilizing them at the site of the developing synapse. The outstanding question here is, what initiates trapping the AchR molecules at this location? A cytoplasmic 43 kD protein is known to bind to the intracellular segment of the AchR complex and thus act to stabilize the growing receptor aggregates (review : Froehner 1991). However, this is unlikely to be the initial event (Stollberg and Fraser 1990) for basically the same reasons that localized binding of integrins to the cytoskeleton is unlikely to be the first event in the formation of focal contacts, namely it fails to account for the development of membrane-cytoplasmic specializations which coincide with sites of contact. One of the most promising candidates for an external ligand, apparently associated with nerve membrane (though not a transmembrane molecule), and with the basal lamina, has been named agrin (Nitkin *et al.* 1987) review : McMahan *et al.* 1992. However, the former authors' calculate that agrin at synapses may not be sufficiently abundant to directly bind and trap all the AchR. If so, agrin may initiate synaptogenesis by cross-linking and trapping either AchR or some other diffusible species, but that further catalytic events are involved in recreating further AchR molecules and stabilizing the assembly. This receives support from a recent report (Wallace *et al.* 1991) in which it was demonstrated that agrin induces tyrosine phosphorylation of the β subunit of the AchR. Although much remains to be understood, it does seem that trapping of laterally diffusible molecules by extracellular protein is responsible for initiating synaptogenesis, which may then bear interesting similarities to focal contact formation.

In view of the evidence now available, there can be no doubt that adhesion receptors can aggregate both at sites of cell-substratum and cell-cell adhesion, and there is very strong evidence that adhesive-trapping by extracellular ligands at the contacted surface is primarily responsible for the process of accumulation. However, it is less certain to what extent adhesion receptors become aggregated under the lamellar regions of actively locomoting cells. Several recent studies have addressed this question. In the case of nerve growth cones locomoting on polylsine-treated plastic, Letourneau and Shattuck (1989), using immunofluorescent methods, reported high concentrations of the adhesion molecule L1 (alias NgCAM) and also β_1 integrin on filopodia, but found uniform labelling of ACAM (alias N-cadherin) and NCAM, in agreement with Tsui *et al.* (1988). The 180 kD isoform of NCAM was less abundant on filopodia than elsewhere on the growth cone. Observations from Sheetz' laboratory on the lamellae of locomoting fish keratocytes using succinylated ConA (which being monovalent cannot cross-link its ligands) gave no indication of receptor concentration at the advancing edge (Kucik *et al.* 1991). In a careful quantitative study using DiI as a control with which to map the local distribution of membrane lipid in moving neutrophils, Pytowski *et al.* (1990) found no evidence that the receptors for iC3b and Fc are concentrated at the leading lamellae. Strong anterior immunofluorescence staining was shown to stem from a higher local concentration of membrane surface, presumably as folds. However, it may be significant that the cells were not locomoting on surfaces bearing iC3b or Ig and it would be informative to repeat the experiments using such surfaces. Letourneau and Shattuck (1989) recognized the problem of membrane folding or curvature but argued that the apparently uniform distribution of staining for some receptors effectively provides a control for those which do show spatial localization.

In experiments involving labelling and examination of the distribution of membrane receptors using fluorescent antibodies or immunogold derivatives on living moving cells, it is important to bear in mind the fact that probe access to contact sites may be restricted. This will lead to false negatives, and there is direct evidence from studies using total internal reflection fluorescence of labelled dextran molecules that high molecular weight substances cannot freely access the contact regions of actively locomoting fibroblasts (Gingell *et al.* 1985). Stationary fibroblasts with well formed focal contacts stand relatively proud of the surface between their adhesions, and no such restriction on diffusion was seen. A further point is that very small aggregates may be sufficient to send signals, as is the case in Mast cell triggering which is caused by small IgE-linked receptor oligomers (Menon *et al.* 1986a,b; Metzger *et al.* 1986).

It is well known that not all membrane-associated proteins are free to diffuse laterally. The various constraints on diffusion are discussed by Zhang *et al.* (1991) and Gumbiner (1991). Some are clearly under metabolic control and can exhibit phases of freedom or restraint at different times. This is evident from the work of Michl *et al.* (1979a,b) referred to earlier. They found that resident macrophages did not redistribute C3 adhesion receptors when the cells were spread on surfaces bearing the appropriate immobilized ligands, whereas after exposure to thioglycollate (or lymphokine treatment, described later by Griffin and Mullinax, 1981) receptors were laterally mobile. Reversible linkage to actin by means of the cytoplasmic segments of β integrins (Hynes 1992) and immunoglobulin superfamily receptors is now known to be responsible for limiting lateral diffusion. It has become increasingly clear that the engagement of membrane proteins by means of lectins, antibodies or in some cases solid particles in the submicron-micron range can cause receptors to link up with the cytoskeleton. This reduces the diffusion coefficient. For this reason labelled monovalent Fab fragments, with no second antibody, may be one of the more reliable diffusion markers. Holifield *et al.* (1990) conclude from their complex immunofluorescence labelling data that cross-linked membrane proteins can attach to cytoskeletal actin. The studies of Kucik *et al.* (1989) on keratocytes labelled with gold particles also show that receptor engagement and consequent cross-linking can lead to association with actin.

Unconstrained membrane proteins might be expected to exhibit the high lateral diffusion of GPI-linked proteins such as Thy-1 (Ishihara *et al.* 1987) which is more typical of membrane lipids than proteins. However, the picture may be more complicated, since both gold-labelled Thy-1 and gold particles bound electrostatically to membrane glycoproteins show diffusion which suggests high mobility within domains on a micron scale (de Brabander *et al.* 1991) possibly limited by cytoskeletal fences, like rabbits in small fields. Edidin and Stroynowski (1991) conclude that a lipid-linked Class-I related protein is freely diffusible whereas the closely related transmembrane molecule MHC-I is domain-restricted in its diffusion characteristics.

Any cellular control process which restricts the lateral diffusion of adhesion receptors will by the same token control the kinds of receptors that can aggregate at contacts, sending signals influencing motility and activating further adhesion mechanisms. A complex picture emerges wherein cellular response to contact signals depends on the nature of the substratum, requirements for the simultaneous or consecutive engagement of diverse adhesion mechanisms and the freedom of membrane receptors to diffuse. The latter may be under metabolic control.

Clustering of receptors generates signals

I have presented evidence which shows that diffusible membrane receptors can become trapped by adhesive interactions involving their external ligand-binding domains, and that such processes are primarily responsible for receptor accumulation at contact sites. The next step is to show that aggregation can generate intracellular signals, after which signalling pathways will be examined. It is inevitable that there will be a certain amount of overlap between these two sections.

Perhaps the most widely reported signal-mediated effect correlated with receptor aggregation is protein phosphorylation, which frequently accompanies binding of receptors to the cytoskeleton, yet in no case is it understood how these events are linked. During the formation of focal contacts, signals are initiated when β_1 integrins are clustered by engagement with extracellular ligands. How do these signals arise? Unlike growth factor receptors, integrins possess no intrinsic enzymic activity. Nor does there seem to be direct involvement with the class of receptors (with seven transmembrane helixes) which are typically linked to G-proteins (Manning and Brass 1991). Signals are also generated by β_1 integrins of tissue cells and by β_2 integrins of leucocytes, as well as β_3 integrins of platelets. Recent evidence suggests that the signalling machinery may be assembled in association with cytoskeletal molecules that bind to the cytoplasmic tails of integrin β subunits. In addition to α-actinin, talin and vinculin, fibroblast focal contacts contain tyrosine kinase pp125[FAK] (focal adhesion kinase : Horak *et al.* 1990) as well as protein kinase C. FAK is singular in being neither transmembrane nor membrane-binding and yet it appears to associate with the cytoskeleton at contact sites.

There is evidence that when fibroblasts spread on fibronectin, several cytoplasmic proteins, including FAK become phosphorylated, but integrins themselves do not. Ridley and Hall (1992) have reported that the cytoskeletal organization of focal contacts and the stress fibres associated with them is controlled by a small GTP binding protein named Rho, which may provide a basis for signal transduction.

Platelet activation is initiated by thrombin, whose receptor is linked on the cytoplasmic face of the membrane to a G-protein, as well as collagen which binds to the integrin α_2,β_1 (review : Brass *et al.* 1991). Signals from the activated G-protein complex sequentially activate the integrin α_{IIb},β_3 (inside-to-outside signalling) which is then able to bind soluble fibrinogen, thus leading to platelet cross-linking by this protein and thrombus formation. The engagement of the fibrinogen receptor initiates a second generation of signals (outside-to-inside) by a process which involves FAK and pp60[c-src] (Huang *et al.* 1991).

A temporal correlation between the clustering of integrin receptor molecules and tyrosine phosphorylation, suggestive of a signal generating system, has been demonstrated by Kornberg *et al.* (1991). Human carcinoma (KB) cells showed integrin clustering following incubation with primary anti-integrin and secondary fluorescent antibodies. While clustering was seen using antibodies specific for α_2, α_3, α_5, α_6 and β_1 subunits, tyrosine kinase activation only occurred after anti-α_3 or anti-β_1 antibodies. Since the α_3,β_1 integrin (VLA-3) can bind to fibronectin, collagen and laminin the authors suggest that kinase activation may occur naturally at focal contacts where integrins are clustered. Since integrins have no intrinsic kinase activity, it is argued that clustering of α_3,β_1 induces association with a kinase or phosphatase to initiate a signal cascade.

There is evidence that receptor clustering produces signals in the case of Mast cell activation by antibody. The FcεRI receptors which normally carry IgE must be aggregated by ligand-induced cross-linking in order to trigger the cell and cause the release of histamine from granules. (Menon *et al.* 1986 a,b; review : Metzger *et al.* 1986.) FRAP measurements show that the receptor complex is laterally mobile, with a diffusion coefficient, "typical" of proteins ($D = 10^{-10}$ cm^2/s). Anti-IgE or small oligomers of IgE adsorb and reduce the value of D by 2 orders of magnitude, presumably by inducing connections between the receptor and the cytoskeleton. Significantly, strong binding of monomeric IgE to the receptors does not induce degranulation and does not reduce their high lateral mobility. Direct evidence that receptor clustering occurs on binding of anti-IgE labelled with gold particles has been provided by Stump *et al.* (1989).

A meticulous study by Chun and Jacobson (1992) has shown the spreading of HeLa cells on collagen is accompanied by receptor clustering and the production of lipoxygenase metabolites of arachidonic acid. Their studies clearly show that spreading does not take place if arachidonic acid production and metabolism are inhibited, but that cross-linking of collagen receptors by externally added specific antibody does not require arachidonic acid metabolism. These results taken together very strongly suggest that receptor aggregation occurs by diffusive trapping at sites of cell contact with extracellular matrix, and that this aggregation is responsible for initiating a signal cascade involving the activation of phospholipase A_2 the production of arachidonic acid and ensuing lipoxygenase metabolites. Most significantly, these metabolites are shown to be essential for cell spreading, though how remains a mystery. This provides persuasive evidence for the diffusive-trapping-signalling hypothesis.

Transductive signalling pathways

It is hardly surprising to find that the transmission of signals from receptor aggregates at cell contacts involves some of the well-worn pathways familiar from other membrane signalling events. What is of particular interest is to discover how aggregation initiates signals and how cytoskeletal dynamics are subject to control by signals. Any models of how cell contact might influence motile processes, such as the formation and protrusion of lamellae, must take account of the ways in which control by signal systems is exercised.

The first evidence that cell-recognition can trigger transduction was provided by the experiments Schurch et al. (1989) alluded to earlier. They showed that both cell-cell contact and antibodies to the adhesion receptors NCAM and L1 (= NgCAM) lowered intracellular IP_2, IP_3 and pH_i, but increase Ca^{2+}_i by a Pertussis-toxin sensitive G-protein system (Fig. 5). This suggested that an inhibitory G_i-protein is stimulated by lateral

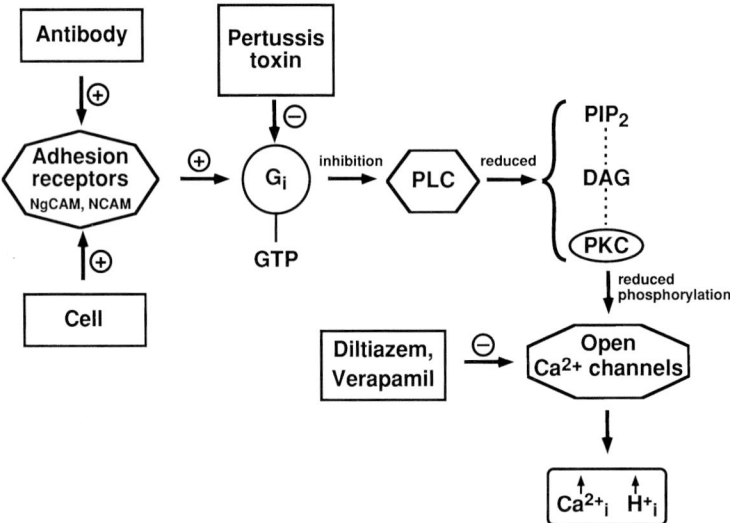

Fig. 5. Signal pathway (based on data from Schurch et al. 1989) triggered by engagement of PC12 cell adhesion receptors, resulting in increased intracellular Ca^{2+}_i concentration and reduced pH_i. External influences in rectangular boxes: activation (+), inhibition (−). Ca^{2+} channel opening inhibited by diltiazem or verapamil. PLC = phospholipase C; PKC = protein kinase C; DAG = diacylglycerol; PIP_2 = phosphatidyl inositol diphosphate.

association with an engaged (activated) adhesion receptor, and that subsequent inhibition of phospholipase-C lowers the rate of PIP_2 hydrolysis, reducing the levels of phosphatidylinositide intermediates and diacylglycerol, thus reducing PKC activity. They argue that the latter increases the calcium influx by controlling the degree of phosphylation of the H^+/Ca^{2+} antiporter.

Strittmater and Fishman (1991) in reviewing nerve growth cone transducers, discuss the evidence for signalling pathways activated by neurite outgrowth on laminin. The fact that added PKC can mimic contact with laminin, and that outgrowth is inhibited by PKC inhibitors and promoted by phorbol esters implicates PKC-mediated phosphylation in a pathway triggered by integrin aggregation. It is argued that a G_o-protein is stimulated by integrins which have bound to laminin (and have thus clustered). Phospholipase-C is stimulated in turn to activate the phosphoinositide pathway wherein diacylglycerol activates PKC. In contrast to Schurch *et al.* (1989) subsequent phosphorylation of calcium channels is thought to open them and thereby increase Ca^{2+}_i. These authors also provide evidence for an intracellular control loop involving a membrane-associated intracellular protein GAP43 which can also stimulate G_o when it binds calmodulin. Increased Ca^{2+}_i, or phosphorylation by PKC, cause calmodulin to be released, whereupon GAP43 cannot activate the G_o-protein.

Important results from Doherty's group (Doherty *et al.* 1991a,b; 1992) have added substantially to our knowledge of the signalling events which follow cell-cell adhesion based on homotypic NCAM interactions. The central feature of their results (Fig. 6) is the opening of Ca^{2+} channels (in common with Schurch *et al.* 1989 and Strittmater and Fishman 1991) which is indicated by a variety of inhibitor studies. They argue that increased Ca^{2+}_i activates a kinase (which is not PKC and is specifically inhibited by the kinase inhibitor K-252b) and that this is a necessary step prior to cell spreading. The control of channel opening is thought to depend on a Pertussis toxin-sensitive G-protein, which they say may act directly rather than via PKC. The likelihood that the signal pathways for CAM-induced and integrin-induced contact events may be different is also raised by Bixby and Jhabvala (1990). They show that PKC is activated early in neurite outgrowth on laminin, but that PKC is not part of the initial signal pathway involved in response to L1 or N-cadherin, in contrast to the conclusions reached by Schurch *et al.* (1989).

Although activation of the T-cell receptor (TCR) complex by MHC on another cell represents a very specialized example of contact-mediated signalling, it is of interest in the present context. Cross-linking the CD4/8-TCR-CD3 complex leads to signal generation, cell proliferation and activation of multiple secondary adhesion mechanisms including LFA-1 and several β_1 integrins (O'Rourke and Mescher 1990). The celebrated study of T-cell activation by Dustin and Springer (1989), showed that the LFA-1 adhesion mechanism could be activated by antibodies to the TCR, followed by a second cross-linking antibody, and also by phorbol esters. This indicates a role for PKC in the signalling pathway, but the fact that the kinase inhibitor staurosporine eliminates activation induced by PMA, while only partially blocking the activation caused by TCR cross-linking, points to two overlapping pathways of activation. This is reinforced by the fact that prior exposure of cells to dibutyryl cAMP inhibited stimulation of LFA-1 via the

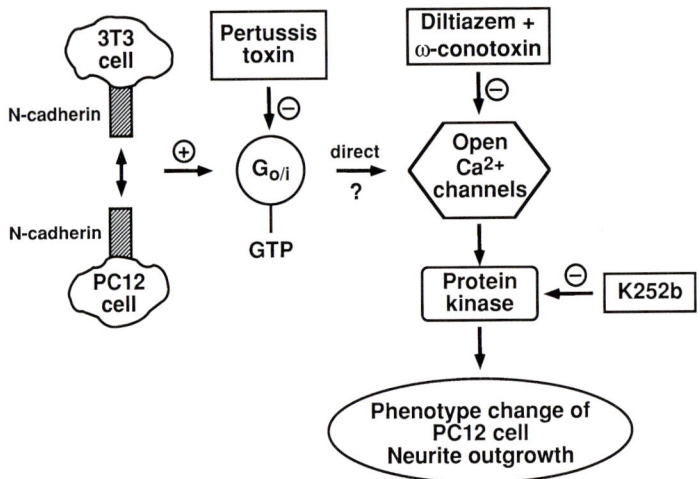

Fig. 6. Signal pathway (based on data of Doherty *et al.* 1991a) linking cell-cell adhesion to motile responses. External activators (+) or inhibitors (−) indicated by rectangular boxes. Homotypic adhesion between either N-cadherin or NCAM molecules stimulates phenotypic change and neurite extension of PC12 cells in Pertussis-toxin inhibitable manner, implicating G protein activity. Lack of sensitivity to wide-spectrum kinase inhibitors or PMA suggests no role for PKC. Motile response depends on opening of both L and N-type Ca^{2+} channels, since specific inhibitors of both types (diltiazem and ω-conotoxin, respectively) are needed to inhibit the cell response. Role for an unknown kinase late in signal path shown by sensitivity to specific inhibitor K252b which blocks motility.

TCR but had no effect on PMA-induced stimulation of LFA-1. Cross-linking the components of the TCR complex by its natural ligand, MHC, activates the associated lymphocyte specific tyrosine kinase p56[lck], but the relationship between the two kinases in LFA-1 activation is not clear. Additional effects of TCR stimulation include increased phosphoinositide turnover (indicating a mechanism for PKC activation via diacylglycerol), elevation of intracellular Ca^{2+} (in common with the results of Doherty and others discussed above) and reduction of cAMP levels.

The recent publication by Chun and Jacobson (1992) provides unambiguous evidence that when HeLa tumour cells spread on collagen, a carpet of GRGDS peptides, or the cells are treated with antibody to the β_1 integrin collagen receptor, arachidonic acid metabolism is activated. Their very carefully controlled experiments using fluorescent probes show that no measurable increase in Ca^{2+}_i or change in pH_i takes place as attached cells spread. In contrast, a series of studies using specific inhibitors shows that the earliest event detected is the activation of phospholipase A_2 which releases arachidonic acid from membrane lipid, and that cell spreading will not take place unless lipoxygenase metabolites of arachidonic acid are produced (Fig. 7). Arachidonic acid can also be released by the sequential action of phospholipase C (giving diacylglycerol) and DAG lipase. In view of these results it would be most instructive to investigate the possible role of lipoxygenase metabolites in other systems where PKC is activated by cell contacts.

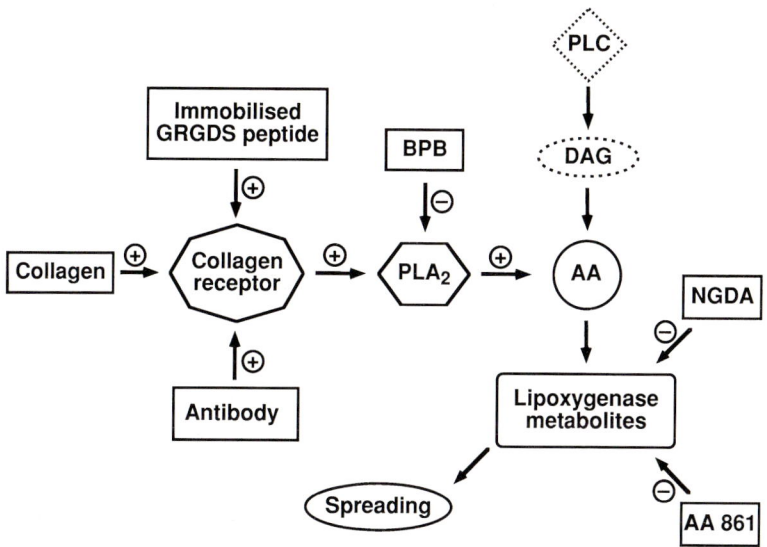

Fig. 7. Signal pathway (based on data of Chun and Jacobson 1992) leading from the engagement of HeLa cell integrin receptor for collagen to cell spreading. External influences shown in rectangular boxes: stimulatory (+), inhibitory (−). PLA2 = phospholipase A2; PLC = phospholipase C; DAG = diacylglycerol; AA = arachidonic acid; BPP = bromophenacylbromide; NGDA = nordihydroguaretic acid.

There is evidence that duplex signalling is sometimes needed in order to bring about a given cellular response (Damsky and Werb review 1992 : also see above). Matsuyama *et al.* (1989) described the stimulation of the lymphocyte proliferative response on surfaces bearing co-immobilized fibronectin and anti-CD3, whereas neither the antibody nor the protein alone were sufficient to do so. It would be interesting to know in such cases whether there is a requirement for two different simultaneous signals or whether a signal from the first adhesion mechanism potentiates the second, for example by increasing the lateral mobility of its receptors or by priming the second transduction mechanism in some way.

From the studies which have been outlined it is not possible to trace a single mechanism of signal transductions and indeed Doherty *et al.* (1991a) stress that the signal pathways for responses to integrins and to CAMS, appear to differ. However, a recurrent theme in several systems is the early activation of a G-protein. It is possible that this is the first event after receptor clustering. How a G-protein might be activated by receptor aggregation is not known, but parallels with G-protein activation by receptors which bind diffusible agonists are readily envisaged. The recent study by La Flamme *et al.* (1992) showed that the cytoplasmic domain of the β_1 integrin subunit can be activated by engagement of an extracellular ligand or constitutively activated in a chaemeric derivative. The activated $\beta1$ chain can then bind to the cytoskeleton, and it might be imagined that activation of a G-protein system by the active β_1 chain results in phosphorylation events leading directly to cytoskeletal association.

A quite different interpretation, which does not necessarily exclude the former, is that G-protein activation can also lead to phosphoinositide production, and these may exercise a direct control over actin via gelsolin. Hartwig (1992) has provided compelling evidence that long (≈ 1 μm) actin filaments in unactivated blood platelets are normally capped by gelsolin at their barbed ends. Spreading of platelets is accompanied by the release of gelsolin (itself activated by the transient rise in Ca^{2+}_i accompanying activation) which then causes multiple scission of the long filaments. This is soon followed by rapid actin polymerization. It is well known that calmodulin which is strongly bound in its Ca^{2+}-depleted state to the ends of actin filaments, can be displaced by phosphoinositides. It is interesting to speculate that a major function of the sequential activation of G-proteins and phospholipase-C in contact-mediated signalling during spreading and locomotion might be to generate phosphoinositides which then release calmodulin from F-actin and thereby promote cytoskeletal rearrangement.

Actin and myosin organization near the membrane

In order for a cell which is bounded by an essentially fluid membrane to locomote by crawling over a surface or through a network of extracellular matrix, it must make reversible adhesions with its surroundings and some elements of the cell's contractile actin-myosin motor must be capable of making mechanical connections to the outside world via the intracellular domains of its transmembrane adhesion molecules.

How are signals originating at adhesions likely to influence the motile apparatus? In several examples the contacts are known to be mechanically stabilized by the direct or indirect association of actin filaments with the clustered adhesion receptors. In focal contacts these associations are mediated by linking proteins. It is likely that the effects of contact signals in locomoting cells are relatively short ranged, if they are to control localized cytomechanical events. It is therefore necessary to outline what is known about the components of the cytoskeleton adjacent to the membrane, and in particular those which are candidates for making reversible mechanical linkages with the membrane.

It is now known that pseudopodia contain the m-RNA coding for β-actin (review : Singer 1992) and the polymerization of G-actin into F-actin is an early event in the extending pseudopodia of both leucocytes and small amoebae undergoing chemotaxis (reviews: Omann et al. 1987; Newell et al. 1988). Actin polymerization occurs during the initial transient retraction phase of the chemotactic response of Dictyostelium amoebae to cAMP, (Dharmawardhane et al. 1989). In the later pseudopodal extension phase there is evidence of extensive actin cross-linking by a 120 kD protein (Condeelis et al. 1988; Dharmawardhane et al. 1989).

An innovative kinetic analysis by Schwartz and Luna (1988) on isolated Dictyostelium membranes, using photoactivatable cross-linkage, showed that F-actin assembly is initiated on the cytoplasmic face at sites where pairs of transmembrane 17 KD glycoproteins (later shown by Chia et al. (1991) to be ponticulin, first described by Westehube and Luna (1987)) initiate the binding of actin, resulting in the assembly of F-actin filaments parallel to the membrane. This relationship is reminiscent of that of band 4.1 with F-actin in the red cell cytoskeleton, but Schwartz and Luna conclude that each additional actin monomer requires the recruitment of one more ponticulin molecule to the cluster, resulting in multiple bonding. These authors suggest that "extracellular factors that cluster membrane proteins may create sites for the formation of actin nuclei and thus trigger actin polymerization in the cell". In the leading lamellae of actively motile macrophages, unbranching actin filaments have been convincingly shown to lie parallel to the substratum in negatively stained aldehyde-detergent fixed whole mounts (Rinnerthaler *et al.* 1991; Small *et al.* 1993). The authors stress that deductions about filament organization based on the widely used critical point drying methods may be unreliable. Filament lengths of at least 0.5 μm are reported and a kinetic study of depolymerization in at least one type of lamella (Theriot and Mitchison 1991) make it unlikely that they extend unbroken through the whole length of the structure. A recent study by Hartwig (1992) has provided important insight into the cytoskeletal rearrangements that take place during platelet activation, which results in spreading and lamella formation.

Hartwig's results show that in resting platelets actin is present as filaments up to 1 μm long. These lie parallel to the membrane and each is bound at around six points to actin binding protein (ABP) which in turn is bound by a distinct site to membrane glycoprotein Ib,IX (and perhaps IIb, IIa : Fox *et al.* 1988). There is also on average one connection per filament to spectrin. In this state the barbed end of each filament is blocked by gelsolin. Activation somehow releases gelsolin and the transient rise in Ca^{2+}_i, activates gelsolin to fragment F-actin, resulting in numerous short membrane-attached pieces. Monomeric actin assembles on the exposed 'barbed' ends of F-actin within 30 sec. Hartwig demonstrates that chelation of Ca^{2+}, which prevents the transient Ca^{2+}_i rise, inhibits lamella formation.

The fact that gelsolin is known to be inhibited in its F-actin filament capping function by phosphoinositides (Yin *et al.* 1988; Hartwig 1992) is of particular interest in relation to activation and signalling, though this aspect is not emphasized by Hartwig. While this attractively detailed mechanism of platelet spreading is of great interest in relation to lamellar extension, some caution should be exercised before adopting it as a paradigm for the behaviour of lamellae in cells which can locomote. Nevertheless some of the processes in the neutrophil chemotaxis bear a remarkable resemblance to those in platelet spreading. The chemotactic peptide fMLP binds to a transmembrane receptor and activates a G-protein which activates the phosphatidylinositol pathway, giving phosphoinositides. Gundersen and Devreotes (1990) have shown that the α-2 subunit of a G-protein is phosphorylated within 40 sec of cAMP binding. This is accompanied by rapid actin polymerization (as well as myosin phosphorylation) and association of actin filaments with the plasma membrane (review : Omann *et al.* 1987). A recent study by Cano *et al.* (1991) has shown that the number of actin filaments doubles after fMLP stimulation, but their mean length is unchanged. Most of the newly polymerized actin is located in the leading lamella.

Our concept of the role played by myosin in cell locomotion has been transformed by the discovery of unconventional myosins (reviews : Korn and Hammer 1990; Pollard *et al.* 1991; Cheney and Mooseker 1992). *Dictyostelium* amoebae lacking myosin-II can locomote without major impediment (De Lozanne and Spudich 1987; Knecht and Loomis

1987) and produce a chemotactic response to cAMP, with the normal rise in F-actin concentration (Peters *et al.* 1988). The shockwaves produced by these remarkable results were quelled by the now famous experiments of Fukui *et al.* (1988, 1989) who showed that two types of myosin have different cellular locations corresponding to distinct functions. Myosin-II is found at the rear of locomoting amoebae, whereas myosin-I (one form of it) is concentrated in advancing pseudopodia and phagocytic cups.

The myosin-I family has some most interesting properties. Unlike muscle myosin, types MIA, MIB, MIC cannot self-associate into bipolar filaments. They can bind to membranes, perhaps to proteins as well as lipids, and are characteristically found in association with membranes. They have two distinct binding sites for actin, but only one is a Mg^{2+}-ATP-ase and the other is non-mechanochemical. The members of the myosin-I family exist in several genetic isoforms each with a distinct separately coded heavy chain which can be phosphorylated and this process increases the mechanochemical activity (Pollard *et al.* 1991). All can bind calmodulin at one or more sites and Ca^{2+} tends to dissociate calmodulin from myosin-I, but activates the Mg^{2+}-ATP-ase binding site.

Myata *et al.* (1989) provided evidence that myosin-I attachment to membranes takes place at a site independent of either actin-binding site. Phosphorylated myosin-IB bound to phospholipid membranes attached to glass has recently been shown to exert traction on actin filaments in the presence of ATP (Zot *et al.* 1992). There exists a requirement for phosphatidylserine, a component of the inner leaflet of plasma membranes, for efficient binding of myosin-I.

Dynamics of actin in lamellae: evidence from light microscopy

Having briefly examined some of the pertinent ultrastructural and biochemical features of actin and myosin which are currently thought to be important for crawling locomotion, I shall next discuss evidence from optical microscopy relating to the dynamics of actin during the formation of lamellae. This will lean heavily upon observations made on keratocytes and nerve growth cones. In the final section I shall attempt to put together a tentative scheme for the molecular basis of the control of lamellar motility by contact signals, which will incorporate the basic concepts developed earlier in this article.

Two major experimental approaches have provided valuable information about the behaviour of the cytoskeleton in lamellar protrusions. These are direct observation of labelled cytoplasmic components, employing fluorescent probes for actin, and the use of video-enhanced differential interference microscopy (DIC) to follow the motions of small particles attached to the dorsal surfaces lamellae.

The first direct evidence relating to the behaviour of actin in lamellae came from the photobleach-recovery experiment of Wang (1985). He calculated a low rearward rate of 0.8 μm/min. with respect to the edge of the cell. Unfortunately the rate of movement of the cell edge is not mentioned, but from the published results it would appear to be much less than 0.8 μm/min.

In a study of major importance, Forscher and Smith (1988) followed the dynamics of actin in growth cones of the sea slug *Aplysia*, using DIC and rhodamine-phalloidin staining. Since the latter technique necessitates cell permeabilization, it provides "snapshots" from which a temporal sequence must be put together. They found that after several minutes exposure to cytochalasin B or D, the fluorescent actin of the lamella had withdrawn centripetally from the edge of the cell, leaving an unstained peripheral margin which was also distinguishable under DIC, which made it possible to follow the process in real time. Most remarkably, after washing out the inhibitor of actin polymerization, recovery was accompanied by the development of actin fluorescence which began at the extreme edge of the lamella and broadened as it progressed in a centripetal direction. Recession of the actin margins occurred at ≈3.5 µm/min in cytochalasin. The cell margin in these experiments was virtually stationary before addition of the inhibitor, which then completely halted extension. The DIC images show that the lamella thickens and becomes invaded with organelles in zones where actin is depolymerized. The authors conclude that cytochalasin caps the barbed ends of the actin filaments (which are associated with the membrane at the extreme periphery of the cell), detaches them from the membrane (or near to their insertion) and that the filaments then pass at a uniform velocity towards the centre of the growth cone. In a following paper, Smith (1988) proposes a model based on these striking results. F-actin is assembled at the cell margin, and in non-locomoting growth cones it is moved centripetally by myosin-I molecules which are indirectly fixed to the substratum by association with adhesion receptors (op.cit Fig. 6a). F-actin is depolymerized somewhere towards the centre of the growth cone. The author presciently speculates on the possibility that signals arising from such adhesions may play a role in directing the cytomechanical events.

The concept of a centripetal flow by actin received further support from the DIC observations of Fisher *et al.* (1988) who recorded the motions of 0.5 µm aminated polystyrene beads on lamellae of 3T3 fibroblasts. Centripetal (=rearward or retrograde) particle movements at 12 µm/min were seen on the dorsal surface, reportedly regardless of whether the cell edge was static or advancing at up to 5 µm/min. Although the authors did not follow labelled actin, they state that bead motions corresponded with the rate of movement of cellular actin, based on the motion of small cytoplasmic inclusions seen under DIC.

Similar conclusions about the motion of particles attached to the dorsal surface of *Aplysia* growth cones were reached by Burmeister *et al.* (1991). Beads moved centripetally at 1-2 µm/min with respect to the substratum under conditions where the cell edge advanced at only 0.12 µm/min. Fisher *et al.* (1988) interpreted their results in terms of active contraction, based on myosin-I or II, resulting in the centripetal motion of actin away from the edge of the lamella, together with polymerization of G-actin at the edge, to give a continuous process.

Reviewing current progress in the same year, Mitchison and Kirchner (1988) concluded that "universal" retrograde actin flow probably reflects the activity of a basic motor for actin-dependent motility. They calculated that retrograde flow at 3 µm/min can occur on the basis of measured actin polymerization rates at a monomer concentration found in cells, though depolymerization may require some additional process such as filament scission. Mitchell and Kirchner put forward a model resembling that of Smith (1988).

An elegant study of actin dynamics in non-motile 3T3 fibroblasts was published several years later by Symons and Mitchison (1991). The cellular location of newly incorporated actin was shown using microinjected rhodamine-actin, and this was distinguished from normal actin polymerized *before* the microinjection event by permeabilizing the cell to give access to fluorescein-phalloidin. They also used Cap Z protein to block the barbed ends of actin filaments and thus inhibit polymerization. It was found that actin was first incorporated at the edge of the lamella and that it moves centripetally at ≈5 μm/min, in excellent agreement with Forscher and Smith (1988). Symons and Mitchison (1991) conjecture that incorporation of actin occurs predominantly at the extreme edges of the lamella because the barbed ends of the filaments are normally uncapped there, whereas barbed ends occurring elsewhere (the filaments are thought to be too short to traverse the entire lamella of several microns) are probably capped. They rationalize the apparent rate of F-actin depolymerization by assuming that it occurs throughout the lamella, rather than solely at the junction between the lamella and body of the growth cone.

In a later photobleach-recovery study, similar to that of Wang (1985), Okabe and Hirokawa (1991) reported that actin in mouse dorsal root growth cones and PC12 cells (the line used in signal transduction studies by Doherty's group, q.v.) is incorporated peripherally and moves centripetally at ≈1.6 μm/min with respect to the substratum, as the edge of the cell advances. However, not all bleached spots moved rearwards, and the distances that could be measured prior to recovery of fluorescence were small. Like others, they conclude that F-actin is assembled at the cell margin, but unlike others they admit that they are puzzled as to how this occurs, since the filaments apparently associate with the cell membrane at the margin.

There have been several further quantitative studies of the movement of particles attached to the dorsal surfaces of lamellae. Sheetz *et al.* (1989) examined mouse macrophages locomoting in the presence of 40 nm gold particles coated with Con A. Two distinct motions were discovered; some particles diffused randomly ($D \approx 3.6 \times 10^{-11}$ cm^2/sec) and others showed highly restricted diffusion ($D \approx 2 \times 10^{-12}$ cm^2/sec) together with cytochalasin-sensitive directional rearward transport (no rate given), reminiscent of the observations of Fisher *et al.* (1988). The value of 3.6×10^{-11} cm^2/sec is somewhat lower than that measured by photobleaching of labelled glycoproteins, but is 1000 times faster than gold sol particles in aqueous suspension. This suggests that the "free" diffusion rate measured on the membrane reflects attachment to the cell and is not significantly influenced by the viscosity of water. The results seem to indicate that particle diffusion on the membrane is always constrained to some degree, but that in one state the constraint is small and non-directional, whereas in the other it is much larger and vectorial.

Although the major thrust of Sheetz' paper is to discredit the membrane lipid-flow hypothesis of cell locomotion, its present significance is considerable. It strongly suggests that particles bound to surface glycoproteins can stimulate the cytoplasmic segments of the receptors to become associated with elements of the cytoskeleton such that rearward motion of the attached particles marks the natural motion of the cytoskeleton (presumably actin). This is in fact two separate points; rearward actin flow and adhesion-dependent receptor linkage to the cytoskeleton. The latter may be of wide significance, insofar as it hints that when adhesion receptors are engaged by the substratum, the receptors bind to the cytoskeleton. Precisely the same conclusion was reached, for very different reasons, by La Flamme *et al.* (1992) from their study of chimaeric receptors, discussed earlier.

The following year Sheetz *et al.* (1990) uncovered further complexity in their analysis of particle motions. They describe non-diffusive movements of the gold particles towards the edges of mouse neuronal growth cones, where they can become trapped. Retreat of the edge during locomotion carries these particles with it, indicating that the particles are less adherent to the substratum than to the cell. The authors argue that regardless of particle (or antigen) binding, some receptors are naturally trapped at the edges, though not by adhesion to the substratum. If this is so, its significance is unclear.

Very recently another study has reported that linkage of membrane adhesion molecules to rearwardly moving actin can be induced by particle binding. Forscher *et al.* (1992) reported that 240 nm polycationic polystyrene beads are transported rearwards on lamellae of *Aplysia* growth cones at ~5 μm/min. However, they too found the unexpected : charged beads could sometimes move according to the vector sum of the normal centripetal velocity plus a new non-directional flow of twice that speed, both sensitive to cytochalasin.

Several studies have appeared on the movement of fish keratocytes. These cells have large very rapidly moving lamellae which are quite distinctive. Perhaps the most surprising result is that of Theriot and Mitchison (1991). They injected a caged photoactivatable analogue of actin (resorufin) and followed the location of fluorescent spots. During rapid lamellar advance ($\approx$10 μm/min) the spots remained static with respect to the substratum. Fluorescence decay apparently occurred by a process of generalized depolymerization throughout the lamella. On the basis of these results it is suggested that actin polymerization must occur near the edges of the lamella and that cell advance is not dependent on the mutual of shearing of actin filaments, but may involve actin-myosin shearing. The major result that actin is immobile with respect to the substratum in advancing keratocytes is supported by Kucik *et al.* (1991) who used the new method of laser-optical trapping to apply forces to 40 nm gold particles placed on advancing lamellae. Beads which caused association of membrane proteins to the cytoskeleton remained almost static with respect to the substratum (0.08 μm/min directed centripetal motion) and showed a degree of random diffusion ($D\approx3.6 \times 10^{-11}$ cm²/sec) around tenfold lower than cell-attached particles which did not activate linkage to the cytoskeleton. At first sight, this satisfying corroboration sits rather awkwardly with an earlier study (Kucik *et al.* 1989) using the same cells, in which single gold particles were sometimes seen making directed anterior excursions up to twice the velocity of the advancing lamellae (both measured with respect to the substratum). It was also found that aggregates of particles executed directed centripetal movements at about the same speed as lamellar advance, $\approx$30 μm/min. Somewhat unexpectedly, it will be shown that the two types of particle transport described by Kucik *et al.* in these papers are in fact predicted by a model of lamellar dynamics.

Synthesis and models

At this stage it will be helpful to summarize what has gone before. Transmembrane glycoprotein adhesion receptors which are free to undergo lateral diffusion will accumulate at contact sites where their extra-cellular bonding domains engage appropriate ligands of the extracellular matrix, on other cells, or on experimental surfaces. Clustering of receptors somehow triggers transduction and generates intracellular signals that can influence a variety of cellular functions which include the activation of secondary adhesion mechanisms and cell movement. At some stage linkage of the cytoskeletal actin to the clustered adhesion receptors occurs, mediated in some cases by actin binding proteins.

While it is not yet possible to draw up a single scheme for the events surrounding contact-mediated signal transduction from the examples which have been studied, several features frequently occur. These include the activation of G-proteins, (which are abundant in nerve growth cone membranes) activation of phospholipases and protein kinases, the generation of phosphoinositides and arachidonic acid derivatives, and a rise in the concentration of intracellular Ca^{2+}. To this may be added the facts that kinase activation can give rise to phosphoinositides which are capable of releasing gelsolin from the barbed ends of actin filaments, and that the increase in intracellular Ca^{2+} may activate the released gelsolin, leading to actin filament scission and further polymerization.

What can be said of lamellar dynamics?

1. F-actin is present as 0.5-1.0 μm filaments parallel to the substratum, orientated with their barbed (more rapidly polymerizing) ends towards the margin of the cell.

2. Subunit addition occurs at the cell margin where the barbed ends of the actin filaments are close to (or inserted in) the membrane.

3. In most cells which have been studied the actin 'core' of the lamella moves backwards at several μm/min with respect to the substratum. This is apparently true whether the cell edge is advancing or stationary.

4. In keratocytes, the actin core is static with respect to the substratum, even in rapidly advancing lamellae.

5. Submicron particles attached to the upper surface of lamellae can either diffuse at random (sometimes within localized domains) or trigger an association with the cytoskeleton. In most cells, particles which have become linked to the cytoskeleton either remain static or are carried rearwards at several μm/min. with respect to the substratum. Which occurs is probably not correlated with whether the cell edge is advancing or static.

6. In the case of keratocytes, particles which link with the cytoskeleton either remain static with respect to the substratum or move *forwards*.

7. Lamellae of spreading or actively locomoting metazoan cells are formed by protrusion. In the case of *Dictyostelium* amoebae the lamella is formed by a distinct centripetal thinning process which proceeds at ≈2 μm/min (Gingell and Vince 1982; Owens *et al.* 1988; Gingell and Owens, 1992a,b).

Myosin, in the unconventional form myosin-I (A,B,C), is found in the advancing regions of crawling cells. It associates with membranes, and has two actin binding sites, one of which is a mechanochemical Mg^{2+}-ATP-ase, activated by a heavy chain kinase and potentiated by Ca^{2+} binding.

This list provides a yardstick for building models of lamellar motility. One must nevertheless be cautious, since it is drawn up from studies on various cell types, which clearly do not all work the same way, so the scope for spectacular errors in modelling is

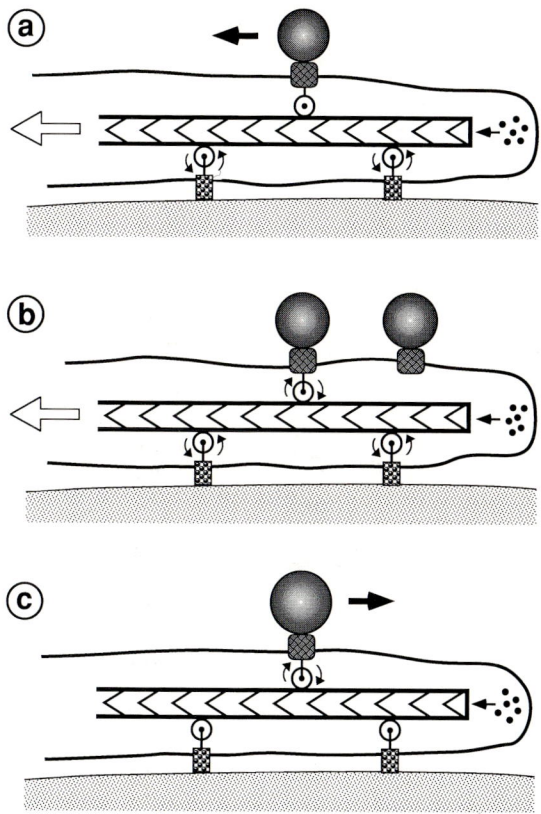

Fig. 8. Outline mechanics describing particle movements (large solid arrow) and actin translocation (large open arrow) in lamellae. F-actin core is shown as chevron block with 'barbed' end to the right where net polymerization from monomers (black spots) takes place. Myosin-I motors shown as rotating (active) or stationary (inactive) monocycles. Direction of rotation indicates sense of relative movement of myosin on actin. Lower myosins linked to substratum at adhesions in association with clustered membrane adhesion glycoproteins (hatched rectangles). Particles bound to clustered receptors on upper membrane may link to myosin-I. (a) Situation as in growth cone lamellae : inactive upper linkage results in rearward movement of particle, indirectly driven by lower myosins. (b) Similar, but active upper linkage results in little or no net particle displacement, like unlinked particle. (c) Situation in keratocyte. Lower links inactive. Particle translocated forward by active upper links. Actin core static. A clear prediction from all 3 cases (and also Fig. 9) is that whenever an adherent particle is translocated away from the cell margin, the core actin is translocated *at an equal or greater speed* in the same direction, with respect to the substratum. The converse is not necessarily true.

disconcertingly large. However, a clearly stated hypothesis has the dual advantages of stimulating discussion and of inviting experiments capable of disproving it.

Fig. 8. shows outline mechanical schemes for lamellar actin-myosin motors. Myosin-I is shown as a monocycle, bound either in a non-mechanochemically active or inactive state to actin. The direction of travel along the actin filament, towards the barbed end, is indicated by rotation. It is assumed that the engagement of membrane adhesion receptors with the substratum or with particles can cause myosin to form links between receptors and actin. Whether these are mechanochemically active may depend on the type of receptor and the degree of clustering as well as on signalling processes (shown in more detail in Fig. 10). Fig. 8a. shows the situation typified by growth cone lamellae, where the actin core is drawn back by myosin activity. Subunit addition occurs at the tip. An adherent particle linked passively to actin moves to the rear. In Fig. 8b the only difference is that a particle is linked actively to actin via the mechanochemical site of myosin. The result of two opposite motions is that the particle remains more or less static with respect to the substratum. Fig. 8c shows the situation in keratocytes, where links with the substratum are passive. The actin core is static with respect to the substratum, but an actively linked particle can move *forwards*. A passively linked particle (not shown) would not move.

Although the scheme shown in Fig. 8a corresponds with the available experimental data, it is counter-intuitive. What is the significance of the rearward translocation of actin? Does a rearwardly directed force exerted on the anterior filament-insertion zone promote actin polymerization there? Whatever the reason, the net rate of lamellar advance, involving anterior extension of the actin core by polymerization at the front, will be diminished by the apparently perverse activity of a motor which translocates the axial actin filaments in the opposite direction. Given the basic elements of membrane-associated myosin and an actin core, one might be tempted to design the system the other way around, reversing actin filament polarity, so that they would be driven forward.

Fig. 9 illustrates lamella formation in *Dictyostelium*. The accompanying sketch emphasizes that the cell periphery remains constant after spreading, but the lamella (dark) develops centripetally. If actin is moved by active engagement with fixed myosins, passively linked transmembrane molecules would be transported away from the cell edge. This would cause centripetal thinning, since the process is effectively like squeezing a toothpaste tube. It is possible that this process shares a common mechanism with the centripetal recovery process described by Forscher and Smith (1988). In nerve growth cones, after treatment with cytochalasin B and removal of the drug, the disorganized thickened lamella subsequently thins centripetally and the moving margin of the thinning zone corresponds with the inner limit of the redeveloping actin filaments.

Fig. 10 shows a possible but tentative molecular basis for a system like that of Fig. 8a. Adhesion receptors are shown clustered by binding to ligands on the substratum. Transductive elements (which may be G-proteins) associated with the membrane are activated by receptor clusters. Signals pass to adjacent myosin-I molecules, inducing mechanochemical activity. The core of cross-linked actin filaments is translocated to the rear. A cyclic process of polymerization occurs at the front and depolymerization occurs mainly at the base of the lamella. A passively-linked dorsal particle moves to the rear. Barbed ends of actin filaments in the lamella are capped (perhaps by gelsolin) but lamellar advance is associated with uncapping near the cell margin. In order for the actin core to be moved en bloc, its component filaments must be cross-linked. This might involve any one of several actin binding proteins. One possibility is that myosin-I

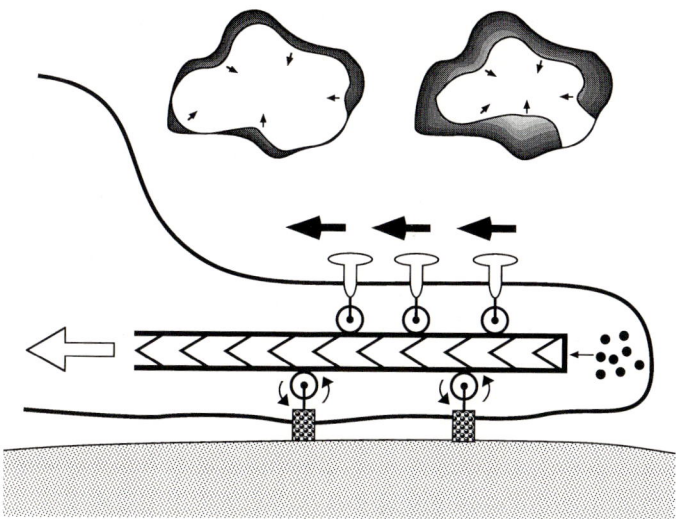

Fig. 9. Inserts show centripetal formation of lamella (dark) by *Dictyostelium*. Section through developing lamella shows predicted actin translocation. Inactive links to membrane glycoproteins drives them rearward : a vertical force component exerted on the upper membrane squeezes cytoplasm towards the centre of the cell and is responsible for centripetal progression of the inner margin of the lamella. Mechanically similar to situation in Fig. 8a.

molecules could assist this function if they attach with random orientation by their two binding sites. Regardless of their state of mechanochemical activity, the mechanism would lock-up just as if all links were inert.

There are several ways in which myosin-I might attach to the membrane. If clustered adhesion molecules nucleate short actin filaments, myosin might attach to the filaments. However, in this case it is necessary to explain why attachment is always via the non-mechanochemical site, in order to avoid the kind of lock-up previously mentioned. Alternatively, myosin may bind to some other intermediary linking protein which itself binds to the clustered receptors. In the diagram a simpler possibility is adopted. Myosin-I associated with phosphatidylinositol of the inner leaflet diffuses randomly until it becomes associated with clusters of adhesion molecules, perhaps by becoming immobilized in the clusters, to which additional linking proteins may then bind. Once it is immobilized, activation by signal-related phosphorylation and a localized signal-related increase in Ca^{2+} concentration, results in rearward translocation of actin.

There arises a problem which must not be fudged. In this model, adhesions at the lower membrane activate myosin-I and result in rearward motion of actin. Contacts at the upper membrane should activate the same mechanism and result in particle stasis, since they would have the rearward component due to bulk actin translocation should be counteracted by actin movement of myosins (associated with the particle) in the opposite direction, towards the barbed end of the filament. Hence it is necessary to explain why the upper links should be normally passive. One possible answer is that the generation of signals necessary for myosin activation depends on the size of the contact and consequent

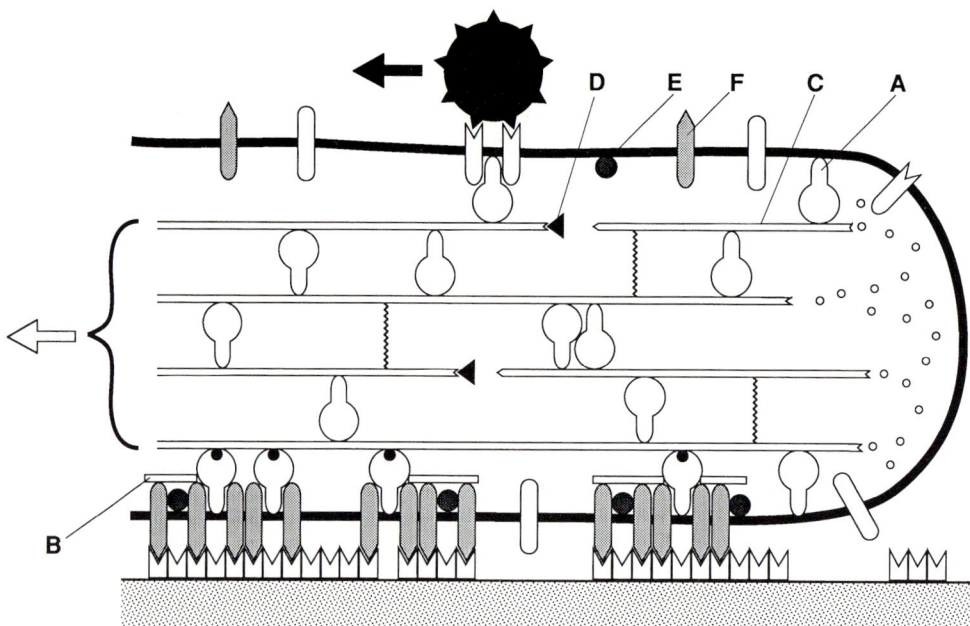

Fig. 10. Suggested mechanism for the case shown in Fig. 8a. Myosin-I may be linked to actin in a non-mechanochemically active state (A) : active myosins indicated by black spots. Myosins shown bound to membrane lipid, but their associations at adhesion sites may be stabilized by interaction with actin (or other protein : B) assembled on cytoplasmic domains of clustered adhesion receptors. Actin filaments (C) are capped at 'barbed' ends (D) except at edge of cell. Membrane-associated elements of transduction system (G-proteins/lipases/ kinases : E) may be activated at adhesion sites where adhesion receptors (F) diffuse and get trapped by immobilized ligands on the substratum.

size of the receptor cluster. If this is so, a study with a range of particle sizes would be illuminating. Alternatively, signal generation at the upper membrane may be limited by the engagement of key elements of the transductive system at the lower surface, if they are sufficiently scarce. In the case of *Dictyostelium*, it is necessary to postulate the existence of inert linkages at the upper membrane and active linkages below. This need raise no conceptual difficulty, since the intercalated membrane molecules involved in fixed linkage would be expected to move centripetally on both the upper and lower membranes, in contrast to the mechanochemically active lower linkages involving different transmembrane molecules, activated by receptor aggregation. Studies on the behaviour of lamellar actin as well as particle movement are clearly needed in this organism.

Finally, there is the question of the degree to which actin filaments are parallel, when the lamella is viewed not in side elevation but in plan. There is evidence that at least in some cells the majority of actin filaments form intersecting arrays (Rinnerthaler *et al.* 1991). In such cases it is less easy to explain the observed axial motions of particles and of actin, and further postulates are required (Gingell and Owens 1992b).

References

Andrews, R.K. and Fox, J.E.B. (1990). Platelet receptors in homeostasis. *Curr. Opin. Cell Biol.* **2**, 894-901.

Berton, G., Laudanna, C., Sorio, C. and Rossi, F. (1992). Generation of signals activating neutrophil functions by leukocyte integrins : LFA-1 and gp150/95, but not CR3 are able to stimulate the respiratory burst of human neutrophils. *J. Cell Biol.* **116**, 1007-1017.

Bixby, J.L. and Jhabvala, P. (1990). Extracellular matrix molecules and cell adhesion molecules induce neurites through different mechanisms. *J. Cell Biol.* **111**, 2725-2732.

Bohnsack, J.F., Kleinman, H.K., Takahashi, T., O'Shea, J.J. and Brown, E.J. (1985). Connective tissue proteins and phagocytic cell function. Laminin enhances complement and Fc-mediated phagocytosis by cultured human macrophages. *J. Exp. Med.* **161**, 912-923.

Brass, L.F., Manning, D.R. and Shattil, S.J. (1991). GTP-binding proteins and platelet activation. *Progr. Hemost. Thromb.* **10**, 127-174.

Bullock, W.E. and Wright, S.D. (1987). Role of the adherence-promoting receptors, CR3, LFA-1, and p150,95, in binding of *Histoplasma capsulatum* by human macrophages. *J. Exp. Med.* **165**, 195-210.

Burmeister, D.W., Rivas, R.J. and Goldberg, D.J. (1991). Substrate-bound factors stimulate engorgement of growth cone lamellipodia during neurite elongation. *Cell Motil. Cytoskel.* **19**, 25-258.

Cano, M.L., Lauffenburger, D.A. and Zigmond, S.H. (1991). Kinetic analysis of F-actin depolymerization in polymorphonuclear leukocyte lysates indicates that chemoattractant stimulation increases actin filament number without altering the filament length distribution. *J. Cell Biol.* **115**, 677-687.

Chan, P-Y., Lawrence, M.B., Dustin, M.L., Ferguson, L.M., Golan, D.E. and Springer, T.A. (1991). Influence of receptor lateral mobility on adhesion strengthening between membranes containing LFA-3 and CD2. *J. Cell Biol.* **115**, 245-255.

Cheny, R.E. and Mooseker, M.S. (1992). Unconventional myosins. *Curr. Opin. Cell Biol.* **4**, 27-35.

Chia, C.P., Hitt, A.L. and Luna, E.J. (1991). Direct binding of F-actin to ponticulin, an integral plasma membrane glycoprotein. *Cell Motil. Cytoskel.* **18**, 164-179.

Chun, J.-S. and Jacobson, B.S. (1992). Spreading of HeLa cells on a collagen substratum requires a second messenger formed by the lipoxygenase metabolism of arachidonic acid released by collagen receptor clustering. *Mol. Biol. Cell* **3**, 481-492.

Cody, R.L. and Wicks, M.S. (1986). Clustering of cell surface laminin enhances its association with cytoskeleton. *Exp. Cell Res.* **165**, 107-116.

Condeelis, J., Hall, A., Breswick, A., Warren, V., Hock, R., Bennett, H. and Ogihara, S. (1988). Actin polymerization and pseudopod extension during amoeboid chemotaxis. *Cell Motil. Cytoskel.* **10**, 77-90.

Curtis, A.S.G., McGrath, M. and Gasmi, L. (1992). Localized application of an activating signal to a cell. Experimental use of fibronectin bound to beads and the implications for mechanisms of adhesion. *J. Cell Sci.* **101**, 427-436.

Damsky, C.H. and Werb, Z. (1992). Signal transduction by integrin receptors for extracellular matrix : cooperative processing of extracellular information. *Curr. Opin. Cell Biol.* **4**, 772-781.

De Brabander, M., Nuydens, R., Ishihara, A., Holifield, B., Jacobson, K. and Geerts, H. (1991). Lateral diffusion and retrograde movements of individual cell surface components on single motile cells observed with nanovid microscopy. *J. Cell Biol.* **112**, 111-124.

De Lozanne, A. and Spudich, J.A. (1987). Disruption of the *Dictyostelium* myosin heavy chain gene by homologous recombination. *Science* **236**, 1086-1091.

Dharmawardhane, S., Warren, V., Hall, A.L. and Condeelis, J. (1989). Changes in the association of actin-binding proteins with the actin cytoskeleton during chemotactic stimulation of *Dictyostelium discoideum*. *Cell Motil. Cytoskel.* **13**, 57-63.

Doherty, P., Ashton, S.V., Moore, S.E. and Walsh, F.S. (1991a). Morphoregulatory activities of NCAM and N-Cadherin can be accounted for by G-protein-dependent activation of L- and N- type neuronal Ca^{2+} channels. *Cell* **67**, 21-33.

Doherty, P., Rowett, L.H., Moore, S.E., Mann, D.A. and Walsh, F.S. (1991b). Neurite outgrowth in response to transfected N-CAM and N-Cadherin reveals fundamental differences in neuronal responsiveness to CAMs. *Neuron* **6**, 247-258.

Doherty, P., Ashton, S.V., Skaper, S.D., Leon, A. and Walsh, F.S. (1992). Ganglioside modulation of neural cell adhesion molecule and N-cadherin-dependent neurite outgrowth. *J. Cell Biol.* **117**, 1093-1099.

Dustin, M.L. and Springer, T.L. (1989). T-cell receptor cross-linking transiently stimulates adhesiveness through LFA-1. *Nature* **341**, 619-624.

Edidin, M. and Stroynowski, I. (1991). Differences between the lateral organization of conventional and inositol phospholipid-anchored membrane proteins. A further definition of micrometer scale membrane domains. *J. Cell Biol.* **112**, 1143-1150.

Fisher, G.W., Conrad, P.A., DeBiasio, R.L. and Taylor, D.L. (1988). Centripetal transport of cytoplasm, actin, and the cell surface in lamellipodia of fibroblasts. *Cell Motil. Cytoskel.* **11**, 235-247.

Forscher, P. and Smith, S.J. (1988). Actions of cytochalasins on the organization of actin filaments and microtubules in a neuronal growth cone. *J. Cell Biol.* **107**, 1505-1516.

Forscher, P., Lin, C.H. and Thompson, C. (1992). Novel form of growth cone motility involving site-directed actin filament assembly. *Nature* **357**, 515-518.

Fox, J.E.B., Boyles, J.K., Berndt, M.C., Steffen, P.K. and Anderson, L.K. (1988). Identification of a membrane skeleton in platelets. *J. Cell Biol.* **106**, 1525-1538.

Froehner, S.C. (1991). The submembrane machinery for nicotinic acetylcholine receptor clustering. *J. Cell Biol.* **114**, 1-7.

Fukui, Y., Lynch, T.J., Brzeska, H. and Korn, E.D. (1988). Immunofluorescence localization of myosin I in *Dictyostelium*. *J. Cell Biol.* **107**, 471a.

Fukui, Y., Lynch, T.J., Brzeska, H. and Korn, E.D. (1989). Myosin I is located at the leading edges of locomoting *Dictyostelium* amoebae. *Nature* **341**, 328-331.

Gingell, D. and Owens, N.F. (1992a). Adhesion of *Dictyostelium* amoebae to deposited Langmuir-Blodgett monolayers and derivatized solid surfaces. A study of the conditions which trigger a contact-mediated cytoplasmic response. *Biofouling* **5**, 205-226.

Gingell, D. and Owens, N. (1992b). How do cells sense and respond to adhesive contacts? Diffusion-trapping of laterally mobile membrane proteins at maturing adhesions may initiate signals leading to local cytoskeletal assembly response and lamella formation. *J. Cell Sci.* **101**, 255-266.

Gingell, D. and Vince, S.M. (1982). Substratum wettability and charge influence the spreading of *Dictyostelium* amoebae and the formation of ultrathin cytoplasmic lamellae. *J. Cell Sci.* **54**, 255-285.

Gingell, D., Todd, I. and Bailey, J. (1985). Topography of cell-glass apposition revealed by total internal reflection fluorescence of volume markers. *J. Cell Biol.* **100**, 1334-1338.

Griffin, F.M. and Mullinax, P.J. (1981). Augmentation of macrophage complement receptor function in vitro. lll. C3b receptors that promote phagocytosis migrate within the plane of the macrophage plasma membrane. *J. Exp. Med.* **154**, 291-305.

Gumbiner, B.M. (1991). Membrane glycoprotein choreography. *Curr. Biol.* **1**, 271-273.

Gundersen, R. and Devreotes, P.N. (1990). In vivo receptor mediated phosphylation of a G-protein in *Dictyostelium*. *Science* **248**, 591-593.

Hafeman, D.G., Smith, L.M., Fearon, D.T. and McConnell, H.M. (1982). Lipid monolayer-coated solid surfaces do not perturb the lateral motion and distribution of C3b receptors on neutrophils. *J. Cell Biol.* **94**, 224-227.

Hartwig, J. (1992). Mechanisms for actin rearrangements mediating platelet activation. *J. Cell Biol.* **118**, 1421-1442.

Holifield, B.F., Ishihara, A. and Jacobson, K. (1990). Comparative behaviour of membrane protein-antibody complexes on motile fibroblasts: Implications for a mechanism of capping. *J. Cell Biol.* **111**, 2499-2512.

Horak, I.D., Corcoran, M.L., Thompson, P.A., Wahl, L.M. and Bolen, J.B. (1990). Expression of p60[fyn] in human platelets. *Oncogene* **5**, 597-602.

Huang, M.M., Bolen, J.B., Barnwell, J.W., Shattil, S.J. and Brugge, J.S. (1991). Membrane glycoprotein IV (CD36) is physically associated with the fyn, lyn and yes protein-tyrosine kinases in human platelets. *Proc. Natl. Acad. Sci. USA* **88**, 7844-7848.

Hynes, R.O. (1992). Integrins: versatility, modulation, and signaling in cell adhesion. *Cell* **69**, 11-25.

Ishihara, A., Hou, Y. and Jacobson, K. (1987). The Thy-1 antigen exhibits rapid lateral diffusion in the plasma membrane of rodent lymphoid cells and fibroblasts. *Proc. Natl. Acad. Sci. USA* **84**, 1290-1293.

Klagsbrun, M. and Baird, A. (1991). A dual receptor system is required for basic fibroblast growth factor activity. *Cell* **67**, 229-231.

Knecht, D.A. and Loomis, W.F. (1987). Antisense RNA inactivation of myosin heavy chain gene expression in *Dictyostelium discoideum. Science* **236**, 1081-1086.

Korn, E.D. and Hammer, J.A. (1990). Myosin I. *Curr. Opin. Cell Biol.* **2**, 57-61.

Kornberg, L.J., Earp, H.S., Turner, C.E., Prockop, C. and Juliano, R.L. (1991). Signal transduction by integrins : increased protein tyrosine phosphorylation caused by clustering of β_1 integrins. *Proc. Natl. Acad. Sci. U.S.A.* **88**, 8392-8396.

Kucik, D.F., Elson, E.L. and Sheetz, M.P. (1989). Forward transport of glycoproteins on leading lamellipodia in locomoting cells. *Nature* **340**, 315-317.

Kucik, D.F., Kuo, S.C., Elson, E.L. and Sheetz, M.P. (1991). Preferential attachment of membrane glycoproteins to the cytoskeleton at the leading edge of lamella. *J. Cell Biol.* **114**, 1029-1036.

La Flamme, S.E., Akiyama, S.K. and Yamada, K.M. (1992). Regulation of fibronectin receptor distribution. *J. Cell Biol.* **117**, 437-447.

Letourneau, P.C. and Shattuck, T.A. (1989). Distribution and possible interactions of actin-associated proteins and cell adhesion molecules of nerve growth cones. *Development* **105**, 505-519.

Lu, M.L., McCarron, R.J. and Jacobson, B.S. (1992). Initiation of HeLa cell adhesion to collagen is dependent upon collagen receptor upregulation, segregation to the basal plasma membrane, clustering and binding to the cytoskeleton. *J. Cell Sci.* **101**, 873-883.

Manning, D.R. and Brass, L.F. (1991). The role of GTP-binding proteins in platelet activation. *Thromb. Haemost.* **66**, 393-399.

Matsuyama, T., Yamada, A.K.J., Yamada, K.M., Akiyama, S.K., Schlossman, S.F. and Morimoto, C. (1989). Activation of CD4 cells by fibronectin and anti-CD3 antibody. A synergestic effect mediated by the VLA-5 fibronectin receptor complex. *J. Exp. Med.* **170**, 1133-1148.

McMahan, U.J., Horton, S.E., Werle, M.J., Honig, L.S., Kroger, S., Ruegg, M.A. and Escher, G. (1992). Agrin isoforms and their role in synaptogenesis. *Curr. Opin. Cell Biol.* **4**, 869-874.

Menon, A.K., Holowka, D., Webb, W.W., and Baird, B. (1986a). Clustering, mobility and triggering activily of small oligomers of immunoglobulin E on rat basophilic leukemia cells. *J. Cell Biol.* **102**, 534-540.

Menon, A.K., Holowka, D., Webb, W.W., and Baird, B. (1986b). Cross-linking of receptor-bound IgE to aggregetes larger than dimers leads to rapid immobilization. *J. Cell Biol.* **102**, 541-550.

Metzger, H., Alcaraz, G., Holman, R., Kinet, J.-P., Pribluda, V. and Quarto, R. (1986). The receptor with high affinity for immunoglobulin E. *Annu. Rev. Immunol.* **4**, 419-470.

Michl, J., Pieczonka, M.M., Unkeless, J.C. and Silverstein, S.C. (1979a). Effects of immobilized immune complexes on F_c and complement-receptor function in resident and thioglycollate-elicited mouse peritoneal macrophages. *J. Exp. Med.* **150**, 607-621.

Michl, J., Unkeless, J.C., Pieczonka, M.M. and Silverstein, S.C. (1979b). Mechanism of macrophage Fc receptor modulation by immobilized antigen-antibody complexes. *J. Cell Biol.* **83**, 295a.

Mitchison, T. and Kirschner, M. (1988). Cytoskeletal dynamics and nerve growth. *Neuron* **1**, 761-772.

Miyata, H., Bowers, B. and Korn, E.D. (1989). Plasma membrane association of *Acanthamoeba* myosin I. *J. Cell Biol.* **109**, 1519-1528.

Moingeon, P., Chang, H-C., Wallner, B.P., Stebbins, C., Frey, A.Z. and Reinherz, E.L. (1989). CD2-mediated adhesion facilitates T lymphocyte antigen recognition function. *Nature* **339**, 312-314.

Newell, P.C., Europe-Finner, G.N., Small, N.V. and Liu, G. (1988). Inositol phosphates, G proteins and *ras* genes involved in chemotactic signal transduction of *Dictyostelium. J. Cell Sci.* **89**, 123-127.

Nitkin, R.M., Smith, M.A., Magill, C., Fallon, J.R., Yao, Y-M.M., Wallace, B.G. and McMahan, U.J. (1987). Identification of agrin, a synaptic organizing protein from *Torpedo* electric organ. *J. Cell. Biol.* **105**, 2471-2478.

Okabe, S. and Hirokawa, N. (1991). Actin dynamics in growth cones. *J. Neuroscience* **11**, 1918-1929.

Omann, G.M., Allen, R.A., Bokoch, G.M., Painter, R. G., Traynor, A. E. and Sklar, L.A. (1987). Signal transduction and cytoskeletal activation in the neutrophil. *Physiol. Rev.* **67**, 285-322.

O'Rourke, A.M. and Mescher, M.F. (1990). T-cell receptor-activated adhesion systems. *Curr. Opin. Cell Biol.* **2**, 888-893.

Owens, N.F., Gingell, D. and Bailey, J. (1988). Contact-mediated triggering of lamella formation by *Dictyostelium* amoebae on solid surfaces. *J. Cell Sci.* **91**, 367-377.

Pardi, R., Inverardi, L., Rugarli, C. and Bender, J.R. (1992). Antigen-receptor complex stimulation triggers protein kinase C-dependent CD11a/CD18-cytoskeleton association in T-lymphocytes. *J. Cell Biol.* **116**, 1211-1220.

Peters, D.J.M., Knecht, D.A., Loomis, W.F., DeLozanne, A., Spudich, J. and Van Haastert, P.J.M.

(1988). Signal transduction, chemotaxis and cell aggregation in *Dictyostelium discoideum* cells without myosin heavy chain. *Develop. Biol.* **128**, 158-163.

Pollard, T.D., Doberstein, S.K. and Zot, H.G. (1991). Myosin-I. *Annu. Rev. Physiol.* **53**, 653-681.

Pytowski, B., Maxfield, F.R. and Michl, J. (1990). Fc and C3bi receptors and the differentiation antigen BH2-Ag are randomly distributed in the plasma membrane of locomoting neutrophils. *J. Cell Biol.* **110**, 661-668.

Rapraeger, A., Jalkanen, M. and Bernfield, M. (1986). Cell surface proteoglycan associates with the cytoskeleton at the basolateral cell surface of mouse mammary epithelial cells. *J. Cell Biol.* **103**, 2683-2693.

Ridley, A.J. and Hall, A. (1992). The small GTP-binding protein rho regulates the assembly of focal adhesions and actin stress fibres in response to growth factors. *Cell* **70**, 389-399.

Rinnerthaler, G., Herzog, M., Klappacher, M., Kunka, H. and Small, J.V. (1991). Leading edge movement and ultrastructure in mouse macrophages. *J. Struct. Biol.* **106**, 1-16.

Schurch, V., Lohse, M.J. and Schachner, M. (1989). Neural cell adhesion molecules influence second messenger systems. *Neuron* **3**, 13-20.

Schwartz, M.A. and Lechene, C. (1992). Adhesion is required for protein kinase C-dependent activation of the Na^+/H^+ antiporter by PDGF. *Proc. Natl. Acad. Sci. USA.* **89**, 6138-6141.

Schwartz, M.A. and Luna, E.J. (1988). How actin binds and assembles into plasma membranes from *Dictyostelium discoideum. J. Cell Biol.* **107**, 201-209.

Schwartz, M.A., Lechene, C. and Ingber, D.E. (1991). Insoluble fibronectin activates the Na^+/H^+ antiporter by clustering and immobilizing integrin α_5/β_1, independent of cell shape. *Proc. Natl. Acad. Sci. USA.* **88**, 7849-7853.

Sheetz, M.P., Turney, S., Qian, H. and Elson, E.L. (1989). Nanometre-level analysis demonstrates that lipid flow does not drive membrane glycoprotein movements. *Nature* **340**, 284-288.

Sheetz, M.P., Baumrind, N.L., Wayne, D.B. and Pearlman, A.L. (1990). Concentration of membrane antigens by forward transport and trapping in neuronal growth cones. *Cell* **61**, 231-241.

Singer, R.H. (1992). The cytoskeleton and on RNA localization. *Curr. Opin. Cell Biol.* **4**, 15-19.

Siu, K. Lo., Lee, S., Ramos, R.A., Lobb, R., Rosa, M., Chi-Rosso, G. and Wright, S.D. (1991). Endothelial-leukocyte adhesion molecule 1 stimulates the adhesive activity of leukocyte integrin CR3 (CD11b/DC18, Mac-1, α_m/β_2) on human neutrophils. *J. Exp. Med.* **173**, 1493-1500.

Small, J.V., Rohlfs, A. and Herzog, M. (1993). Actin and Cell Movement (this volume).

Smith, S.J. (1988). Neuronal cytomechanics : the actin-based motility of growth cones. *Science* **242**, 708-715.

Springer, T.A. (1990). Adhesion receptors of the immune system. *Nature* **346**, 425-434.

Stollberg, J. and Fraser, S.E. (1990). Acetylcholine receptor clustering is triggered by a change in the density of a nonreceptor molecule. *J. Cell Biol.* **111**, 2029-2039.

Streuli, C.H., Bailey, N. and Bissell, M.J. (1991). Control of mammary epithelial differentiation : basement membrane induces tissue-specific gene expression in the absence of cell-cell interaction and morphological polarity. *J. Cell Biol.* **115**, 1383-1395.

Strittmater, S.M. and Fishman, M.C. (1991). The neuronal growth cone as a specialized transduction system. *Bioessays* **13**, 127-134.

Stump, R.F., Pfeiffer, J.R., Schneebeck, M.C., Seegrave, J.C. and Oliver, J.M. (1989). Mapping gold-labelled receptors on cell surfaces by backscattered electron imaging and digital image analysis: studies of the IgE receptor on Mast cells. *Amer. J. Anat.* **185**, 128-141.

Symons, M.H. and Mitchison, T.J. (1991). Control of actin polymerization in live and permeabilized fibroblasts. *J. Cell Biol.* **114**, 503-513.

Theriot, J.A. and Mitchison, T.J. (1991). Actin microfilament dynamics in locomoting cells. *Nature* **352**, 126-131.

Tsui, H-C. T., Schubert, D. and Klein, W.L. (1988). Molecular basis of growth cone adhesion : anchoring of adheron-containing filaments at adhesive loci. *J. Cell Biol.* **106**, 2095-2108.

Ursitti, J.A., Pumplin, D.W., Wade, J.B. and Bloch, R.J. (1991). Ultrastructure of the human erythrocyte cytoskeleton and its attachment to the membrane. *Cell Motil. Cytoskel.* **19**, 227-243.

Wallace, B.G., Qu, Z. and Huganir, R.L. (1991). Agrin induces phosphorylation of the nicotinic acetylcholine receptor. *Neuron* **6**, 869-878.

Wang, Y.-L. (1985). Exchange of actin subunits at the leading edge of living fibroblasts: possible role of treadmilling. *J. Cell Biol.* **101**, 597-602.

Werb, Z., Tremble, P.M., Behrendsten, O., Crowley, E. and Damsky, C.H. (1989). Signal transduction through the fibronectin receptor induces collagenase and stromelysin gene expression. *J. Cell Biol.* **109**, 877-889.

Wright, S.C., Craigmyle, L.S. and Silverstein, S.C. (1983). Fibronectin and serum amyloid P component stimulate C3b- and C3bi-mediated phagocytosis in cultured human monocytes. *J. Exp. Med.* **158**, 1338-1343.

Wright, S.D. and Jong, M.T.C. (1986). Adhesion-promoting receptors on human macrophages recognize *Escherichia coli* by binding to lipopolysaccharide. *J. Exp. Med.* **164**, 1876-1888.

Wuestehube, L.J. and Luna, E.J. (1987). F-actin binds to the cytoplasmic surface of ponticulin, a 17kD integral glycoprotein from *Dictyostelium* dissociated plasma membranes. *J. Cell Biol.* **105**, 1741-1751.

Yin, H.L., Iida, K. and Janmey, P.A. (1988). Identification of a polyphosphoinositide-modulated domain in gelsolin which binds to the sides of actin filaments. *J. Cell Biol.* **106**, 805-812.

Zhang, F., Crise, B.S., Hou, Y., Rose, J.K., Bothwell, A. and Jacobson, K. (1991). Lateral diffusion of membrane-spanning and glycosylphosphatidylinositol-linked proteins: towards establishing rules governing the lateral mobility of membrane proteins. *J. Cell Biol.* **115**, 75-84.

Zot, H.G., Doberstein, S.K. and Pollard, T.D. (1992). Myosin-I moves actin filaments on a phospholipid substrate : implications for membrane targeting. *J. Cell Biol.* **116**, 367-376.

Printed in Great Britain © The Society of Experimental Biology 1993   35

ON THE MECHANISMS OF CORTICAL ACTIN FLOW AND ITS ROLE IN CYTOSKELETAL ORGANISATION OF FIBROBLASTS

JULIAN P. HEATH and BRUCE F. HOLIFIELD

Departments of Pediatrics (CNRC) and Cell Biology, Baylor College of Medicine, Houston, TX 77030, USA

Summary

In this review we discuss the organization of F-actin in motile fibroblasts in relation to the phenomenon of cortical actin flow; we review some of the mechanisms proposed to drive this process and then relate some of our new findings on the interactions of cortical flow and substratum adhesions in fibroblasts. It is our thesis that the centripetal flow of F-actin through the lamellipodium and leading lamella is the major determining factor organising the polarized stress fiber system and associated cell-substratum adhesions in crawling fibroblastic cells.

The broad flattened region of cytoplasm anterior to the nucleus, called the leading lamella, is fringed with much thinner structures, the lamellipodia, which are the primary protrusive organelles of motile cells. Lamellipodia are filled with a criss-crossed network of actin filaments interspersed with small bundles or ribs. In the leading lamella, the actin cytoskeleton is largely confined to two cortical layers beneath the dorsal and ventral cell surfaces. The ventral cortex is engaged in cell adhesion; the dorsal cortex is made up of a circumferentially orientated sheet of actin filaments and bundles which we term the dorsal cortical microfilament sheath (DCMS). Stress fibers insert into the ventral adhesions and pass back through the lamella rising up to meet the DCMS.

Cortical flow appears to be a constitutive process in most types of cells when they become motile. In fibroblasts a continuous centripetal flux of structure is seen flowing through the lamellipodium and into the more central regions of the lamella at approximately 0.1 μm per second. Some of the structures engaged in the flux pass back and merge into the DCMS which is also moving rearward at a slower rate of 1 to 5 μm per minute.

We find that the formation of a stress fibre is a direct consequence of cortical flow. Initially, stress fibre formation involves the establishment of a focal adhesion between an F-actin bundle in the lamellipodium and the substratum. Subsequently, a fibre grows centripetally from the adhesion and elongates coordinately with the rearward flowing cortex. Cortical flow is restrained locally at the distal end of the nascent fibre leading to indentation and folding of the sheet of filaments. This fold develops into an arc.

We demonstrate that mechanical linkages exist between the lamellipodium and the DCMS. Cytochalasin induces a sudden and massive centripetal collapse of the DCMS which drags the lamellipodia with it. This mechanical linkage suggests that myosin-II-based contraction in the lamella may be responsible for centripetal flow in the lamellipodium by pulling lamellipodial microfilaments rearwards to be incorporated into the rearward flowing DCMS.

Key words: cortical actin, cell motility, microscopy.

These findings indicate that there is centripetal flux of material from the tip of the lamellipodium back to the nucleus which involves three types of F-actin assemblies having dissimilar microfilament organizations. Portions of the actin cortex become anchored by adhesions to the substratum and continued flow of the cortex drives elongation of stress fibers from these adhesions. We conclude that the development of an organised actin cytoskeleton in the leading lamella is to a large part a consequence of the interaction of cortical flow and cell-substratum contact.

Introduction

"There is more to cell movement than an arrangement of the classical muscle proteins" Michael Abercrombie (1973).

With these words, Abercrombie closed his remarks about the status of research on cell locomotion in 1973 at a CIBA Foundation meeting. Now, some twenty years on, we know and understand a good deal more about the molecular biology and the structural interactions of cytoskeletal proteins. Armed with this information, the way is now open to extend and fully develop the pioneering holistic approaches of Abercrombie and his contemporaries into a deeper understanding of how differentiated tissue cells integrate the properties of the cytoskeletal proteins to produce movement.

In migratory tissue cells, such as neurones and fibroblasts, the actin cytoskeleton plays many roles. Among these are the protrusion of lamellipodia by the growth of actin filaments at the cell margin, the development of contractility and matrix structuring via focal adhesions and their associated stress fibers, and the regulation of the topography of the plasma membrane and its receptors through cortical networks of actin filaments.

It is understandable that structurally less complicated cells such as polymorphonuclear leucocytes, keratocytes and the amoeboid cells of invertebrates and Protists should be popular subjects for the study of cell motility. These cells often have only a single actin-rich lamellipodium that drags the perinuclear region along behind it; stress fibers and higher order assemblies of F-actin are reduced or absent. This simplified form undoubtedly accounts for their high motility, around 1 μm per second, which is up to fifty fold higher than that of fibroblasts and epithelial cells. In particular, the fish or amphibian epidermal keratocyte has given new insights into actin dynamics; and use of genetically-manipulated and mutant cells of Protists such the slime mould has provided information of the possible roles of myosins I and II in actin cortical organization and in motility.

But the fact remains that fibroblasts do have a higher order of cytoskeletal organization than say keratocytes and therefore it must be recognised that in some areas of cytoskeletal function fibroblasts employ somewhat different mechanisms for the various tasks that they fulfill in tissue morphogenesis and homeostasis. Determining how these differing levels of cytoskeletal organization are controlled is a major challenge faced by students of tissue cell locomotion.

Cortical actin flow

The superficial layers of a eukaryotic cell contain a network of cytoskeletal proteins

LAMELLIPODIAL ZONE   LAMELLAR ZONE   PERINUCLEAR ZONE

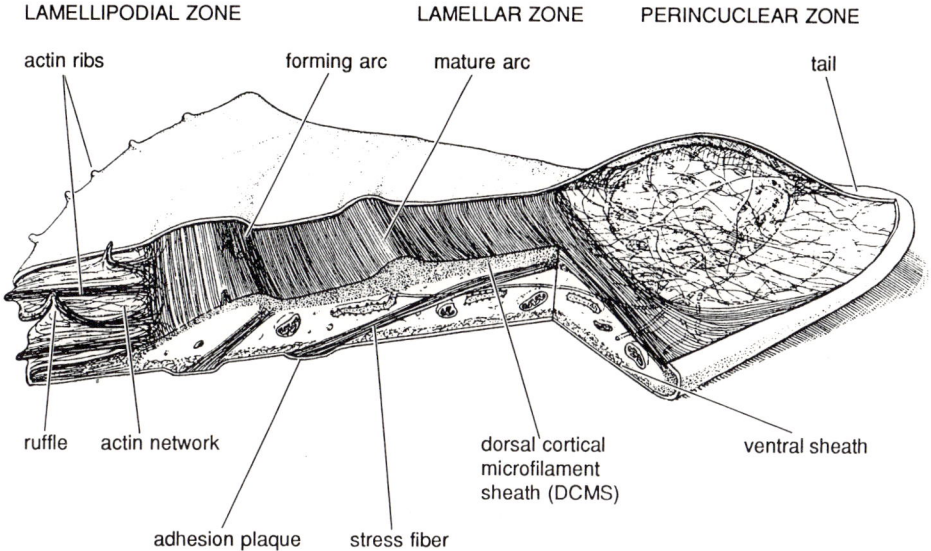

Fig. 1. Model of a crawling fibroblast showing the spatial relationships of the various actin assemblies in the lamellipodium, dorsal cortical microfilament sheath, stress fibers and perinuclear region. Note the clear separation of the ventral from the dorsal actin cortex into which the distal ends of stress fibres insert.

that forms a visco-elastic gel or cortex that helps in the maintenance of cell shape and form. The term cortical flow describes the bulk movement of this cortex, and the flow seems to be a constitutive process of most, if not all, motile cells (Abercrombie, 1980; Bray and White, 1988). The flow may be subtle and short range such as in the lamellipodia of neutrophils or neuronal growth cones, or dramatic and long range such as the sweeping of arcs through the large leading lamellae of fibroblasts. Significantly, cortical flow is seen in motile cells, in those that are moving and those that happen to have momentarily become stationary, but is absent or much reduced in non-motile quiescent cells; how the activation of the actin cytoskeleton relates to cell cycle and growth control is another major issue to be addressed by cell motility researchers in the near future.

For the past three years, we have been using using video microscopy and correlated electron microscopy to examine spatial and functional relationships between cortical F-actin structures of respreading and crawling cells of the IMR-90 human fibroblast line. These cells have large lamellipodia and broad lamellae that show some of the best examples of cortical actin flow. The centripetal movement of actin arcs was first described in these cells by Soranno and Bell (1982).

In an earlier review we discussed membrane and actin flow in fibroblasts (Heath and Holifield, 1991a). Here we focus mainly on the cytoskeleton and include some of our own recent findings on cortical flow and stress fiber formation. In Figure 1 we present a model of the fibroblast cytoskeleton in order to define some terms that we employ to describe the architecture of the fibroblast cytoskeleton, and to to provide a frame of reference for the discussion.

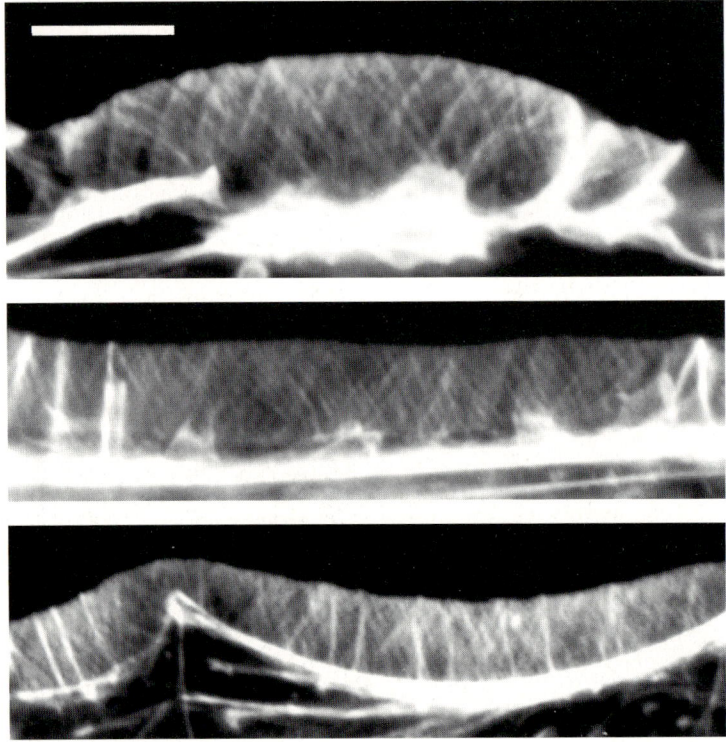

Fig. 2. Fluorescent-phalloidin images showing three examples of actin organisation in lamellipodia. Note the criss-cross pattern of fine actin fibers and larger fibers, or ribs, stretching across the lamellipodium. Scale bar = 10 μm.

Structure of the fibroblast lamellipodium

Lamellipodia are the most visible and hence the most familar of the locomotory organelles of crawling cells, and with few exceptions, all lamellipodia have a similar structure. In fibroblasts, lamellipodia are the thin, veil-like processes that commonly extend from the margins of the lamella, but may occur from less active regions of the cell and from the cell centre. Using phalloidin one can show a characteristic criss-cross pattern of F-actin filaments. This pattern is broken up by thicker bundles of F-actin called ribs or microspikes (Figure 2). These F-actin ribs arise suddenly within the criss-cross network and then can display a complicated pattern of extension and retraction, lateral motion and fusion of adjacent ribs as described by Fisher *et al.* (1988) and Izzard (1988).

Electron microscopy of fibroblastic cells shows that lamellipodia contain a network of F-actin filaments predominantly oriented with their "barbed" or fast growing ends towards the distal margin of the lamellipodium (Small *et al*, 1978; Hoglund *et al*, 1980; Small, 1981; see also Small, this volume). These microfilament bundles appear to meet the lamellipodial margin at electron dense nodes which could contain nucleation sites for the filaments. These nodes are enriched in the cytoskeletal protein talin (DePasquale and Izzard, 1991) and the transmembrane protein integrin (Memmo and Izzard, 1991); the

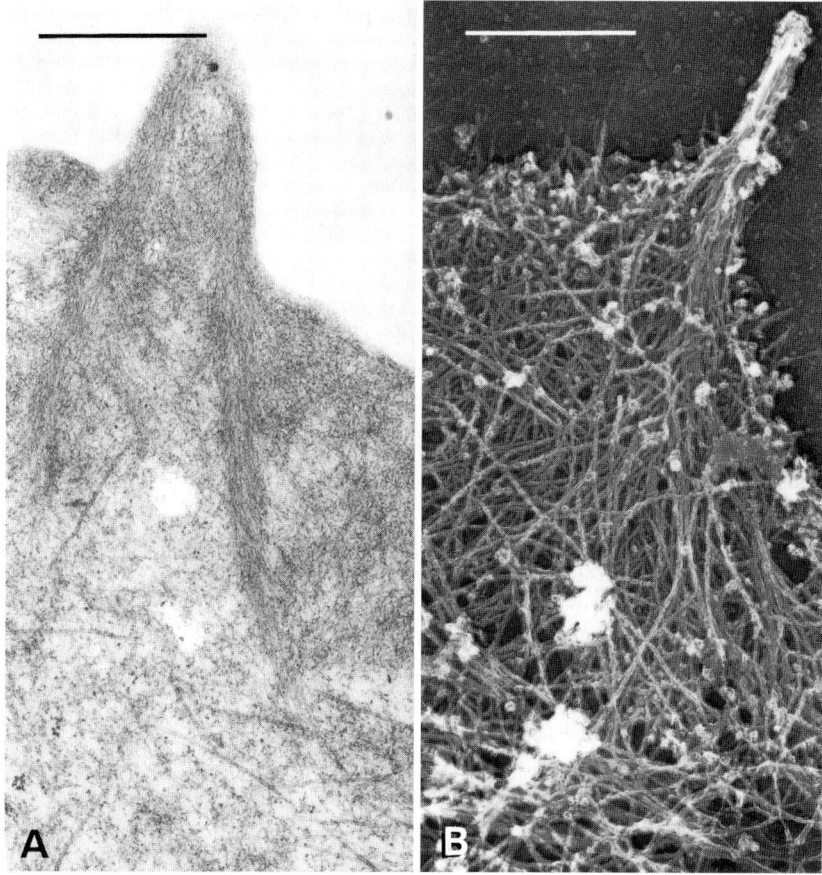

Fig. 3A and 3B. Electron microscopy of lamellipodial actin networks and ribs. (A) Conventional glutaraldehyde, osmium, ethanol, and plastic processing reveals a fine fibrillar network with more densely-packed filaments in the ribs. (B) By contrast, cryo fixation and deep etching after detergent extraction and glutaraldehyde fixation reveals filaments more clearly. (Fig. 3B from Heath *et al*, 1991c). Scale bars (a) 1 μm (b) 0.5 μm.

nodes may help anchor the actin filaments and regulate subunit addition. The lamellipodium contains several other cytoskeletal proteins including filamin and alpha-actinin that could be involved in filament bundling, but the dynamic interactions of these proteins in the F-actin network remains to be resolved.

Figures 3A and 3B show actin filaments and F-actin ribs in cells prepared by two different techniques for electron microscopy. Conventional processing reveals a feltwork of filamentous material (Figure 3A), but deep etch replicas (Figure 3B) clearly show that some bundles of filaments span the breadth of the lamellipodium as suggested by the fluorescence images in Figure 2. The presence of a criss-cross network of actin filaments is also shown in the beautiful negative-stain electron micrographs of fibroblast and macrophage lamellipodia by Small (1981) and Small *et al*. (1992).

Actin flow in the lamellipodium

In fibroblasts lamellipodia continually extend, lift up, and then fall back and dissolve into the dorsal surface of the lamella. This is the familiar process called ruffling. But a more interesting feature of lamellipodia is the prominent rearward flux of structure best seen in time-lapse video recordings, but conspicuous enough to be seen by careful direct observation of living cells on the microscope. In differential interference contrast optics (DIC) the flow takes the form of a mottled pattern of densities that moves rearwards at 0.1-0.2 μm/sec and occurs both along F-actin ribs and between them from the tip of the lamellipodium to its base (Fisher *et al*, 1988). The flow appears to be always directly rearwards, but curiously the motion of F-actin ribs can be orthogonal to the main direction of flow.

The flux could be a sign of treadmilling of monomers through actin filaments. Wang (1985) microinjected cells with fluorescent G-actin and showed that a photobleached spot moved rearwards at 0.013 μm/sec. This result is consistent with polymerization of actin filaments at the lamellipodial margin and depolymerization at the base which would produce the rearward flux, or treadmilling, of actin monomers through stationary filaments.

This possibility has been strengthened by the EM demonstration of the addition of microinjected biotinylated actin to the anterior ends of microspikes in fibroblasts (Okabe and Hirokawa, 1989). More recently Symons and Mitchison examined lamellipodial actin assembly using microinjected rhodamine actin (Symons and Mitchison, 1991). Their data were consistent with a flux from tip to base at a rate of >4.5 μm per min. Furthermore, incorporation of microinjected actin into the lamellipodium was inhibited by CapZ, a barbed-end capping protein.

An alternate explanation for lamellipodial flow is that not only are subunits moving rearward but the filament network itself is continuously transported rearwards. Such a model of F-actin flow in lamellipodia was developed by Fisher *et al* (1988) and Forscher and Smith (1988) and has been discussed in detail in reviews by Smith (1988), Mitchison and Kirschner (1988), Adams and Pollard (1989), Small (1988,1989), Heath and Holifield (1991a, 1991b) and Heath (1992). A key piece of evidence for the model comes from the work of Forscher and Smith (1988) on cytochalasin B-treated *Aplysia* neuronal growth cones. The drug caused a rapid retreat of the whole F-actin network from the cell margin; this would not be expected of a stationary actin network since cytochalasin B blocks polymerization at the fast-growing anteriorly sited ends of actin filaments and so promotes depolymerization of the network from the slow-growing ends of the filaments outwards to the cell margin. Forscher and Smith proposed that retreat of the F-actin away from the cell margin is due to a centripetal pull on the filaments by a myosin motor.

But what happens to the filaments at the base of the lamellipodium in a fibroblast? Two possibilities are that either the F-actin is rapidly broken to monomers or oligomers by depolymerization in concert with actin-fragmenting proteins such as gelsolin, or it remains in a filamentous form and is redistributed elsewhere; we think a substantial portion goes on into the lamella.

Where is the motor for lamellipodial flow?

What is the motor for the F-actin flow and where is it located in fibroblasts? In view of the absence of myosin II from lamellipodia, new models focus on the role of myosin I in rearward lamellipodial transport. Myosin I may lie in a stable submembraneous layer in the lamellipodium where it can drive F-actin rearwards (Mitchison and Kirschner, 1988; Smith, 1988; Gingell and Owens, 1992). An alternate possibility is that the motor is located either on some stable elements within the lamellipodium, such as the F-actin ribs. We favour a third option which is that the motor, in part at least, resides in the more central regions of the cytoskeleton in particlar within the dorsal cortical F-actin sheath of the lamella, where myosin II is found (Heath, 1983a). The role of myosin I molecules in the lamellipodium could be to bind to membrane complexes and transport them anteriorly (Small, 1988; Adams and Pollard, 1989); in this regard, the occasional anterior transport of particles, against the more obvious centripetal direction of lamellipodial flow, has been observed on highly motile fish keratocytes (Kucik *et al*, 1989).

Filament dynamics in lamellipodia

A recent development in the field comes from the work of Theriot and Mitchison (1991). They injected fish keratocytes with a photoactivatable probe, resorufin-conjugated G-actin. A fluorescent bar activated at the leading edge remained stationary as the cell crawled forwards, but its intensity decreased rapidly. Theriot and Mitchison propose that lamellipodium of keratocytes is filled not by a uniform network of long filaments but by large numbers of much shorter actin filaments that are continuously treadmilling subunits. The filaments are nucleated at the cell margin and then released to fill the lamellipodium. This model, the nucleation-release model, would lead to a uniform density of short actin filaments in the lamellipodium of fish keratocytes (but see Small, this volume). Fluorescent images of phalloidin-stained fibroblasts do sometimes show a uniform intensity across the lamellipodium (Figure 2). However, Symons and Mitchison (1991) described a gradient of filament density in the lamellipodium of fibroblasts. The reasons for the apparent differences in filament densities and dynamics between different lamellipodia are not yet resolved. Actin filaments in various cell types exhibit the same criss-cross arrangement (Heath and Holifield, 1991b; Small *et al.*, 1992). Whether or not gradients of filament density reflect different stages in the protrusion and retraction cycle of lamellipodia, and how the type and pattern of cell-substratum adhesions affect filament dynamics, are issues that require combining high resolution structural and cell behavioral approaches with measurements of molecular dynamics of actin and its cross-linking proteins. We summarize some of the possible ways that F-actin could be organised in the fibroblast lamellipodium in the schematic in Figure 4.

Interactions of lamellipodia with the substratum

What types of interaction occur between the actin filaments and membrane-associated adhesion molecules and how would these interactions influence the operation of the motor for lamellipodial flow? For the motor to do any work, it must be anchored so that forces propelling the F-actin rearward are not diffused by forward motion of the motor.

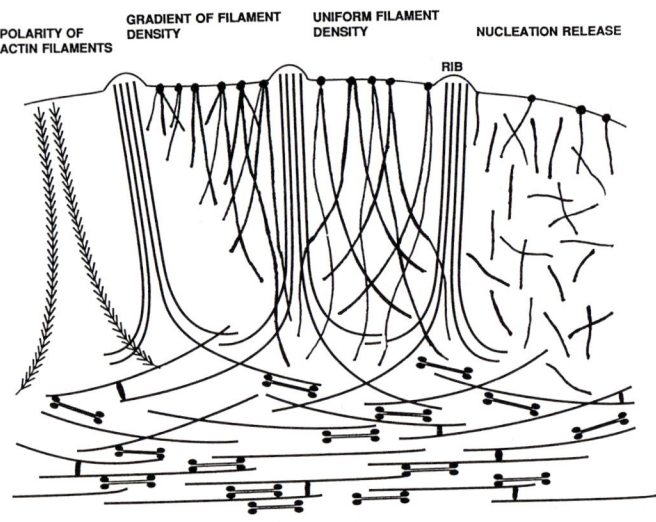

POLARITY OF ACTIN FILAMENTS | GRADIENT OF FILAMENT DENSITY | UNIFORM FILAMENT DENSITY | NUCLEATION RELEASE | RIB

DORSAL SHEATH FILAMENTS WITH MYOSIN AND FILAMENT CROSS-LINKERS

Fig. 4. Schematic showing three models for the organisation of actin filaments in a lamellipodium. The actin ribs spanning the breath of the lamellopodium are comprised of long actin filaments interspersed with bundling proteins. In between the ribs, actin filaments arise from nucleation sites at the cell margin. In a steady state, involving nucleation at the front and depolymerization at the rear, the filaments will span the lamellipodium giving rise to a uniform density of filaments; when new filaments appear and/or existing ones are clipped, a gradient in filament length and density will be seen. If the filaments are released from the cell margin shortly after nucleation, as in the nucleation release model, the lamellipodium will contain a fairly uniform density of short filaments. At the base of the lamellipodium, the actin filaments splay out and merge into the dorsal sheath where at least part of the motor for lamellipodial flow may reside.

Transmembrane linkages to the substratum could stabilise the motor. But it is difficult to conceive that stable adhesive contacts are formed if the filaments are moving backwards and/or disassembling. For this reason the concept of a molecular clutch has been introduced (Mitchison and Kirschner, 1988); the clutch allows slippage between actin filaments and any associated substratum-based transmembrane complexes along the length of the lamellipodium. The concept of such a clutch mechanism would be supported if it were demonstrated that there was a difference in rates of the rearward actin flow dependant on the rate of protrusion and on the location of adhesions.

But there is a problem for this mechanism in the case of the fibroblast lamellipodium which is typically adherent to the substratum in a very limited number of sites. During protrusion, the major portion of a lamellipodium often projects anteriorly some distance above the substratum. This is clearly seen by interference reflection light microscopy and by electron microscopy, and is indicated by the fact that particles can attach and move back on both surfaces of the fibroblast lamellipodium (Harris and Dunn, 1972). In a similar fashion, fibroblasts moving through a three dimension lattice of gelled collagen fibers extend lamellipodia into the open spaces of the gel (Heath and Peachey, 1989) and so extension clearly does not require adhesion of lamellipodia to the gel fibres. One

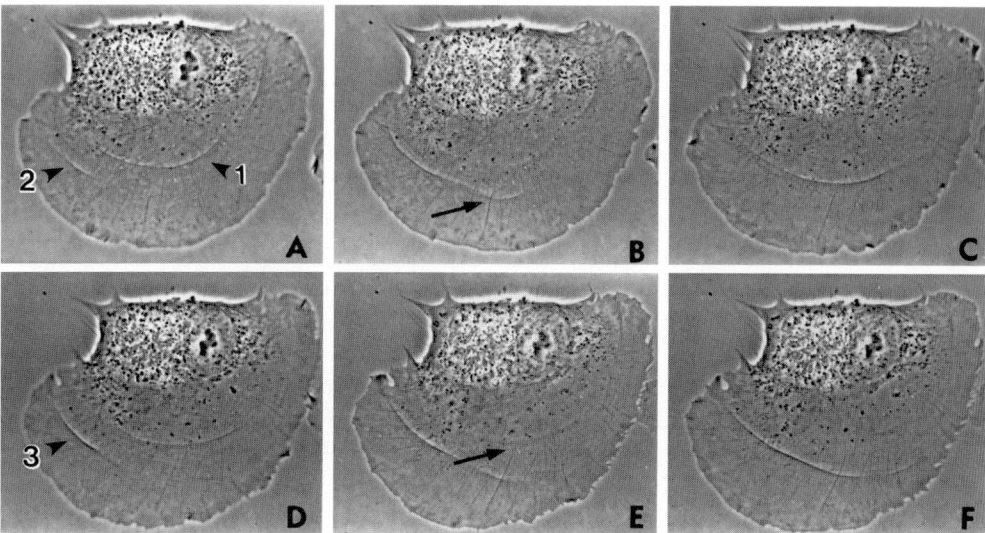

Fig. 5. Sequence of phase-contrast images showing the formation and centripetal movement of three arcs in the leading lamella. The cell is shown at 1 minute intervals. Arcs form behind the lamellipodium, move inwards and fade away near the nucleus. Arcs 2 and 3 elongate towards the right within 1 minute after their initial appearance (arrowheads). Note that the end of a stress fiber is associated with arc 2 and grows inwards as the arc moves (arrows). Scale bar = 10 μm.

answer to the problem of how to develop a transmembrane linkage between a motile sheet of actin filaments and a stationary substratum is to accept that some of the F-actin, possibly the ribs, within the lamellipodium does not engage in centripetal flow.

Keratocytes, on the other hand, have managed to couple actin filament dynamics to a very efficient adhesion, and importantly deadhesion, mechanism so that the forces involved in filament flux are totally directed towards traction. Clearly then lamellipodial extension would be a balance between the rate of filament polymerisation and the action of a molecular clutch that transmits force to the substratum. It is interesting to note that if the forces responsible for the 0.2 μm per min rearward flux in a fibroblast lamellipodium were efficiently harnessed to drive protrusion, fibroblasts would indeed begin to compete with their faster counterparts.

Structure of the leading lamella of motile fibroblasts

The rapid centripetal flux of cytoskeletal material in crawling cells is not confined to lamellipodia. This fact was noted by Abercrombie *et al.* (1970) and others many years ago. There is a rearward flux of F-actin in the form of arc-shaped structures in the large fan-shaped lamellae of crawling cells (Heath and Dunn, 1978; Soranno and Bell, 1982; Wang, 1984; Heath, 1983a, 1983b; Holifield *et al.*, 1990; Holifield and Jacobson, 1991). The formation, movement and disappearance of arcs is illustrated in Figure 5. These structures serve as conspicuous markers for the more general rearward flow that occurs throughout the dorsal lamellar cortex; in the absence of arcs, the flux of actin is best seen

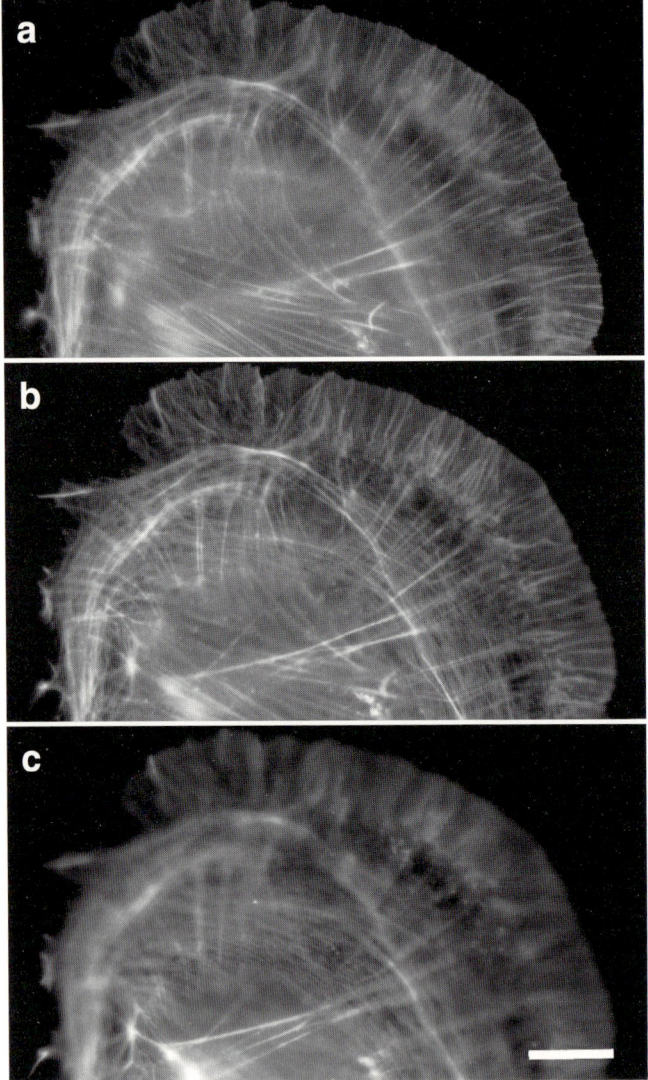

Fig. 6. The organisation of F-actin in a fibroblast as seen in three focal planes. The cell was fixed with 0.2% glutaraldehyde and stained with rhodamine-phalloidin. The plane closest to the substratum (a) shows the lamellipodium and ventral stress fibres; rising up through the cell (b) the dorsal cortical sheath and an arc are imaged; in the highest plane (c) the rear of the dorsal sheath and the supranuclear actin are seen. Scale bar = 10 μm.

by time-lapse recording and appears, as described by Abercrombie and colleagues (Abercrombie, *et al.*, 1970), as "... a series of indefinite shadows chasing each other steadily backwards ..." from the base of the lamellipodium to near the cell center.

To understand cortical actin flow in the lamella requires an appreciation of the three-dimensional organization of the actin cytoskeleton (at this point the reader is referred back to Figure 1). In Figure 6 we show three focal planes from a cell stained with fluorescein-phalloidin that were imaged using a 1.4 N.A. objective. The fibroblast leading lamella inclines approximately 4 degrees as it passes back from the base of the lamellipodium where it is about 300 nm thick to the perinuclear region where it is 2-3 microns in thickness (Fig. 6c). This change in thickness precludes capturing the entire volume of the cell in reasonable focus by any high resolution (high numerical aperture) optical microscopy technique; consequently, examination of multiple focal planes is necessary. Note in particular how incomplete a picture of the lamellar F-actin organization is given by a single focal plane near the substratum (Figure 6a).

Arcs are particularly large bundles of F-actin situated among much finer circumferential fibers which form a continous sheet rising up from the base of the lamellipodium and passing back to near the cell apex over the nucleus. Electron microscopy of sections cut parallel to the substratum show this sheet to consist of densely packed microfilaments (Figure 7). Because this sheet encloses the bulk of the cytoplasm and has a characteristic ultrastructural organization, it has been described as a dorsal cortical microfilament sheath (DCMS) to distinguish it from the less ordered microfilament assemblies present in the cortex of amoeboid cells such as leukocytes.

Immunofluorescence localisation shows that the DCMS contains the bulk of a fibroblast's complement of contractile proteins. It is notable that alpha-actinin, myosin and caldesmon, proteins involved in organizing or regulating muscle contraction, are found in discrete spots distributed throughout the DCMS (Figure 8). This pattern is also reflected in the distribution of electron-dense bodies similar in appearance to the dense bodies of smooth muscle (see Figure 7). The sarcomere-like organisation suggests that a sliding filament mechanism could regulate contraction in the DCMS.

The lamella also contains stress fibers, the long radial F-actin bundles that insert into cell-substratum adhesions at the ventral cell surface and rise obliquely toward the nucleus. Note that the central end of a stress fiber may terminate at an arc (Figures 9, 10, 11). The peripheral end of a stress fiber invariably terminates in a focal adhesion to the substratum. Focal adhesions are specialized sites of attachment of F-actin to the plasma membrane. A number of cytoskeletal proteins are concentrated there, as well as integral membrane protein receptors for some extracellular matrix proteins (reviewed in Burridge *et al.*, 1988).

Mechanism of cortical flow in the leading lamella

If the mechanism for generating actin flow in the lamellipodium is still unclear, then the movement of the dorsal cortical F-actin in the lamella may be simpler to understand. The sliding of actomyosin filaments within the circumferential bundles would lead to a constriction, drawing the DCMS towards the cell center. Rearward movement of punctate accumulations of myosin-II has been documented (McKenna *et al*, 1989). There is a striking parallel in the behavior of the contractile ring, a band of actin filaments that develops at the site of cleavage during cytokinesis. The association of myosin with these filaments results in a potentially contractile structure which apparently possesses some mechanical connections to the plasma membrane. Interestingly, cross-linked membrane proteins on mitotic cells are transported to the cleavage furrow as the contractile ring forms (Koppel *et al*, 1982), in a process strikingly similar to the transport of aggregated membrane proteins with moving arcs in fibroblasts (Heath, 1983b). It has been argued that a release of cortical tension at the poles of a dividing cell could induce contractile ring formation (White and Borisy, 1983). Whether or not an analogous release of tension

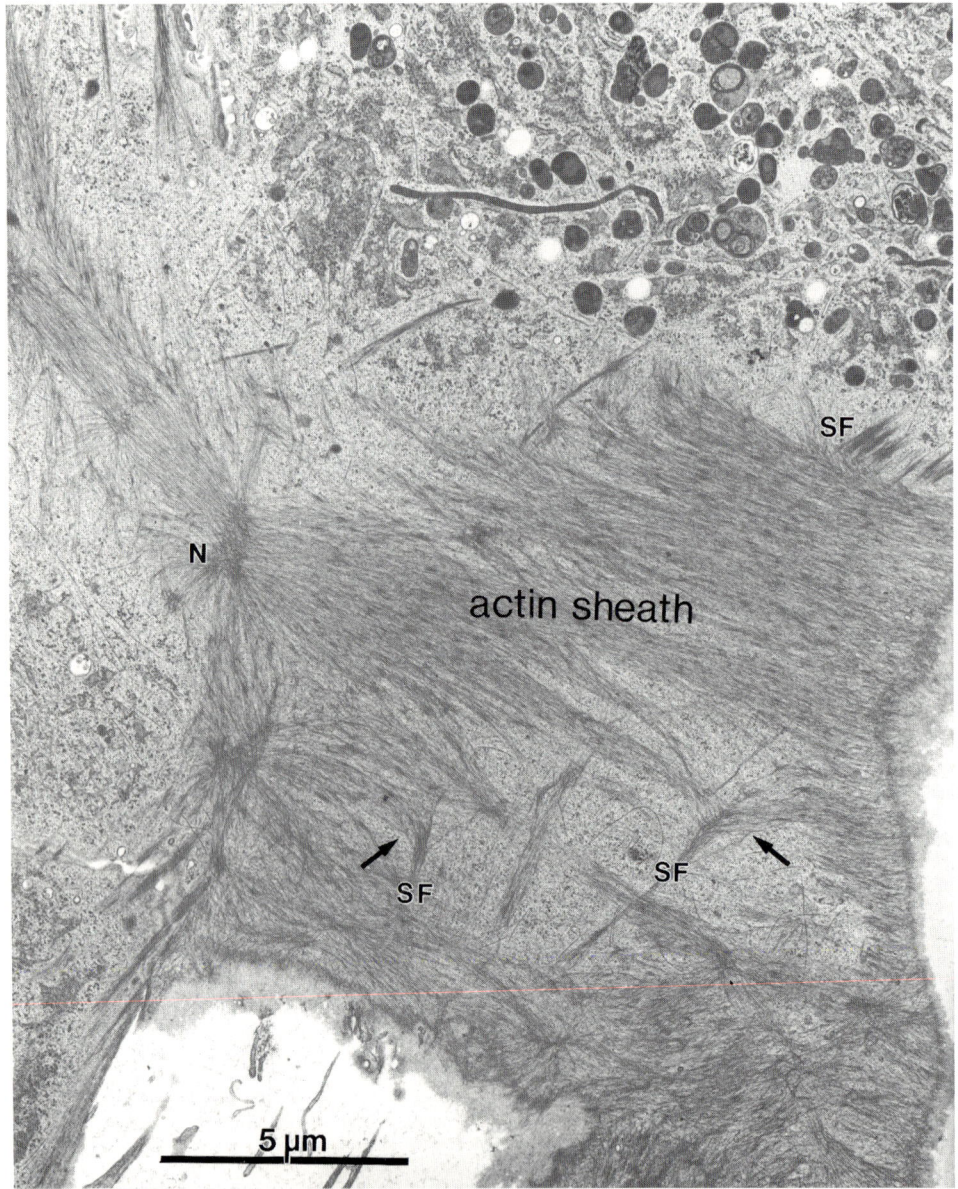

Fig. 7. Horizontal section through the leading lamella of an IMR fibroblast. The circumferential arrangement of the dorsal actin sheath is seen. Note the many electron densities within the sheath that reflect the sarcomeric-like arrangement of the cortical cytoskeleton. Some stress fibers (sf) insert into the sheath (arrows). On the left the sheath breaks up into a polygonal network (N). Scale bar = 5 μm.

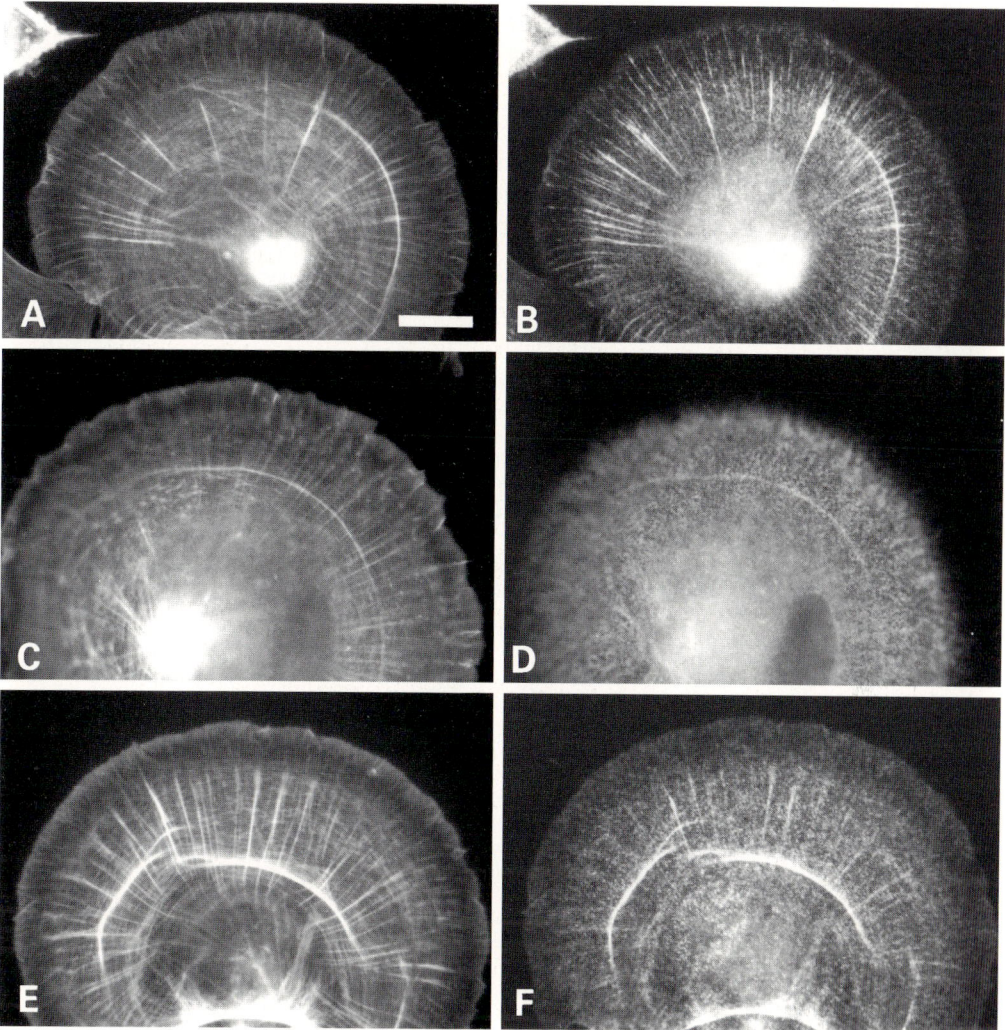

Fig. 8 A-F. Organisation of actin, α-actinin, myosin II and caldesmon in the leading lamella. Paired images of cells stained with phalloidin to visualize F-actin (A,C,E) and antibodies for (B) α-actinin; (D) myosin II; (F) caldesmon. Alpha-actinin and caldesmon are located in both the DCMS and the lamellipodia. Myosin II is present in the DCMS but absent from the lamellipodium. Scale bar = 10 μm.

at the margin of a fibroblast is important to the mechanism in lamellar cortical flow is still an open question (Bray and White, 1988).

A relation between stress fibers and cortical flow

To many people, the most characteristic feature of the actin cytoskeleton of cultured fibroblasts and epithelial cells is the array of large, mature stress fibres with their characteristic sarcomere-like distribution of many of the muscle proteins. For a fibroblast,

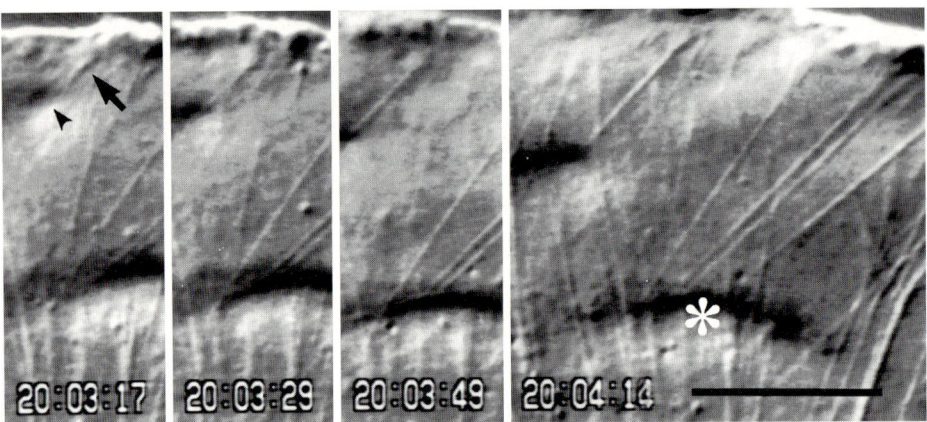

Fig. 9. Sequence of images from a video-DIC recording showing the elongation of a stress fibre. In the first frame, the newly-formed stress fibre extends from a focal adhesion (arrow) and inserts into a small dimple in the DCMS. The stress fibre eleongates cenripetally, and as the dimple moves inwards with the surrounding cortical flow, it gradually takes on the appearance of a mature arc. Further back in the lamella, several older stress fibres insert into a larger arc and elongate as the arc moves rearwards (asterisk). Scale bar = 10 μm.

the development of a well organized array of stress fibers is perhaps the sign that the cytoskeleton has reached the degree of order that is required to fulfill what undoubtedly is the primary function of fibroblasts: tractional structuring of the extracellular matrix. This is the application of tension to an extracellular matrix, for example collagen, resulting in its dramatic reorganization (Harris *et al*, 1981). This is potentially a powerful morphogenetic mechanism in embryogenesis which could guide cell movments, but can be a problem when, as is often the case in wound healing, the tractional forces result in the build up of excessive scar tissue. Because the specialized focal adhesions at the ends of stress fibers can mediate connections of the actin cytoskeleton to the extracellular matrix, it may be that the contractile force generated through stress fibers is the route by which fibroblasts accomplish their work of remodelling their physical environment. One has to recognise that stress fibers can also generate cell movement. If the cell body instead of the extracellular matrix fibrils yields when stress fibers contract, then the cell body will be drawn forward towards the anterior focal adhesions and the cell will have translocated. Whether stress fibers are necessary for translocation or not is still a moot point.

A stress fiber is typically aligned along the axis of cell movement, with the peripheral end marking the direction of spreading and the central end indicating the direction of cortical flow. That this geometry provides a clue to the process of stress fibre formation was realized by Dunn who proposed the continuous contraction model of cytoskeletal function in locomotion (Dunn, 1980). He envisaged that a cytoskeletal meshwork in the lamella was continuously contracting rearwards to a site where it was broken down and recycled to the front of the cell. Formation of a focal adhesion could anchor a portion of the contracting meshwork, preventing or slowing its rearward movement, and lead to rapid formation of a stress fiber extending between the adhesion and the perinuclear region (Heath and Dunn, 1978).

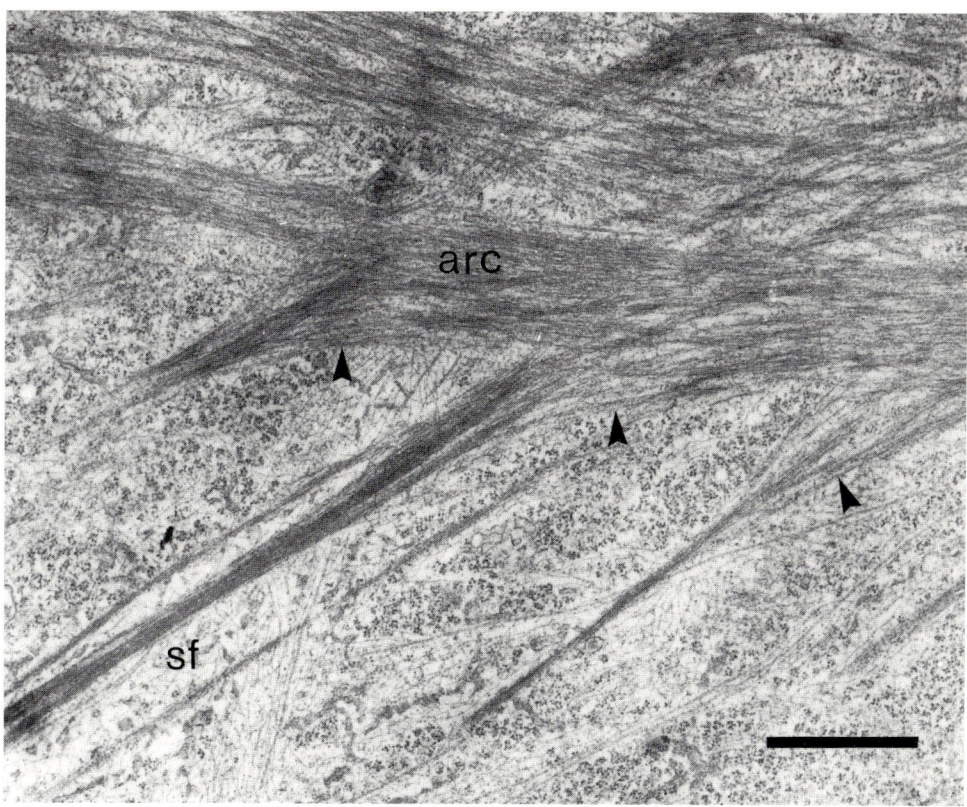

Fig. 10. Intersection of stress fibers and the dorsal cortical sheath at the site of an arc. Note how actin filaments pass directly from the stress fibers and realign as they insert into the sheath (arrows). Scale bar = 1 μm.

We have found that the formation of stress fibers is indeed related to cortical flow (Heath and Holifield, 1992). Izzard and Lochner (1980) first drew attention to the growth of a stress fibre inward from a newly formed focal adhesion. DePasquale and Izzard (1987) documented a sequence of events in chick fibroblasts leading to formation of the focal adhesion from its cytoplasmic precursor structure, the lamellipodial F-actin rib. We have confirmed this same sequence of events in human fibroblasts and have followed the subsequent centripetal elongation of the stress fiber (Heath and Holifield, 1992). Briefly, the first indication of stress fiber formation is the deposition of a focal adhesion in association with the F-actin rib. The anterior portion of the rib then separates from the focal adhesion. The posterior portion of the rib remains in place as a part of the adhesion and appears to have a pre-existing connection to the DCMS. As the DCMS moves rearward a small narrow fiber extends centripetally from the focal adhesion and lengthens in concert with the flow of the dorsal lamellar cortex. Because the forming stress fiber is anchored at one end and inserts into the moving DCMS at the other (Figures 9, 10, 11 and 12), the stress fiber comes under tension. Eventually it resists movement of the sheath, producing a small fold or dimple in the dorsal cortex (see Figure 11). These small folds often develop into much larger arcs. Thus, arcs form in the DCMS as a result of tension

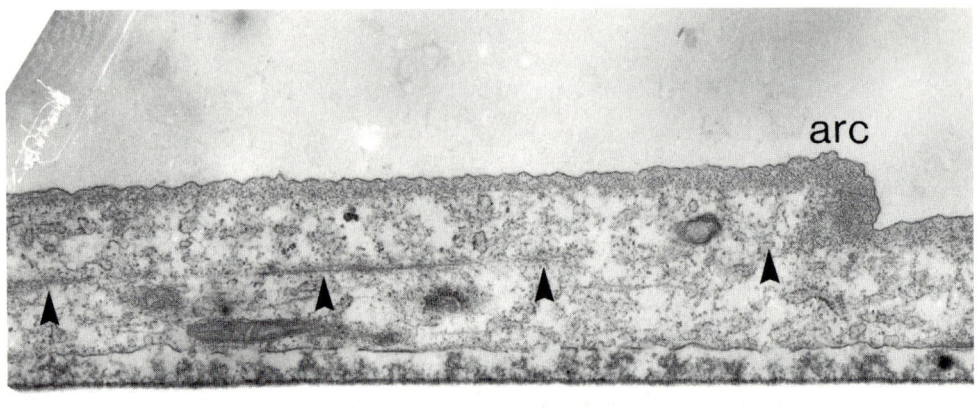

Fig. 11. Vertical section through leading lamella of an IMR fibroblast. The dorsal cortical actin microfilament sheath is seen in cross section. On the left is an arc, which is an S-shaped fold of the DCMS. Note the segments of stress fibers (arrows) that rise up through the lamella to meet the sheath close to the arc. Scale bar = 1 μm.

transmitted by stress fibers between stationary cell-substratum adhesions and the rearward-contracting dorsal F-actin cortex; this tension could be generated by actomyosin contraction either in the DCMS, the stress fiber or both. The schematic in Figure 13 summarises this sequence of steps.

Cytochalasin B causes collapse of the DCMS and reveals interconnection of F-actin assemblies

The interconnections of the three principle microfilament assemblies in a motile fibroblast, the lamellipodia, dorsal cortex and the stress fibers, are more clearly revealed by the response to cytochalasin-B. This drug caps the fast-growing end of actin polymers, thereby blocking further monomer addition, and reduces the viscosity of preformed F-actin gels apparently by severing filaments. We have shown that cytochalasin B causes a rapid and systematic collapse of the dorsal F-actin sheath which is useful in assaying its interaction with other cytoplasmic structures (Holifield and Heath, 1990). This response is superficially similar to the effect of cytochalasin on *Aplysia* growth cones (Forscher and Smith, 1988), but differs with respect to the organization of the collapsing microfilament assembly and so also with respect to the way myosin might be involved.

When exposed to cytochalasin (Figure 14) the lamellipodium rapidly ceases to extend and the cell margin retreats back to the edge of DCMS. Within 1 minute, the entire DCMS collapses towards the cell center. In essence, the DCMS is removed from the still spread lamella, leaving behind an actin-deficient dorsal cortex. The time to onset of this response is not uniform in the cell population; the rates of DCMS collapse may vary an order of magnitude (2-20 μm per min). The reason for this variability is evident when cells are fixed and stained to visualize F-actin. Cells that respond slowly or incompletely show numerous radial stress fibers persisting in association with the DCMS and the substratum adhesions. Collapse of the sheath ceases or slows at the point where the stress fibers rise

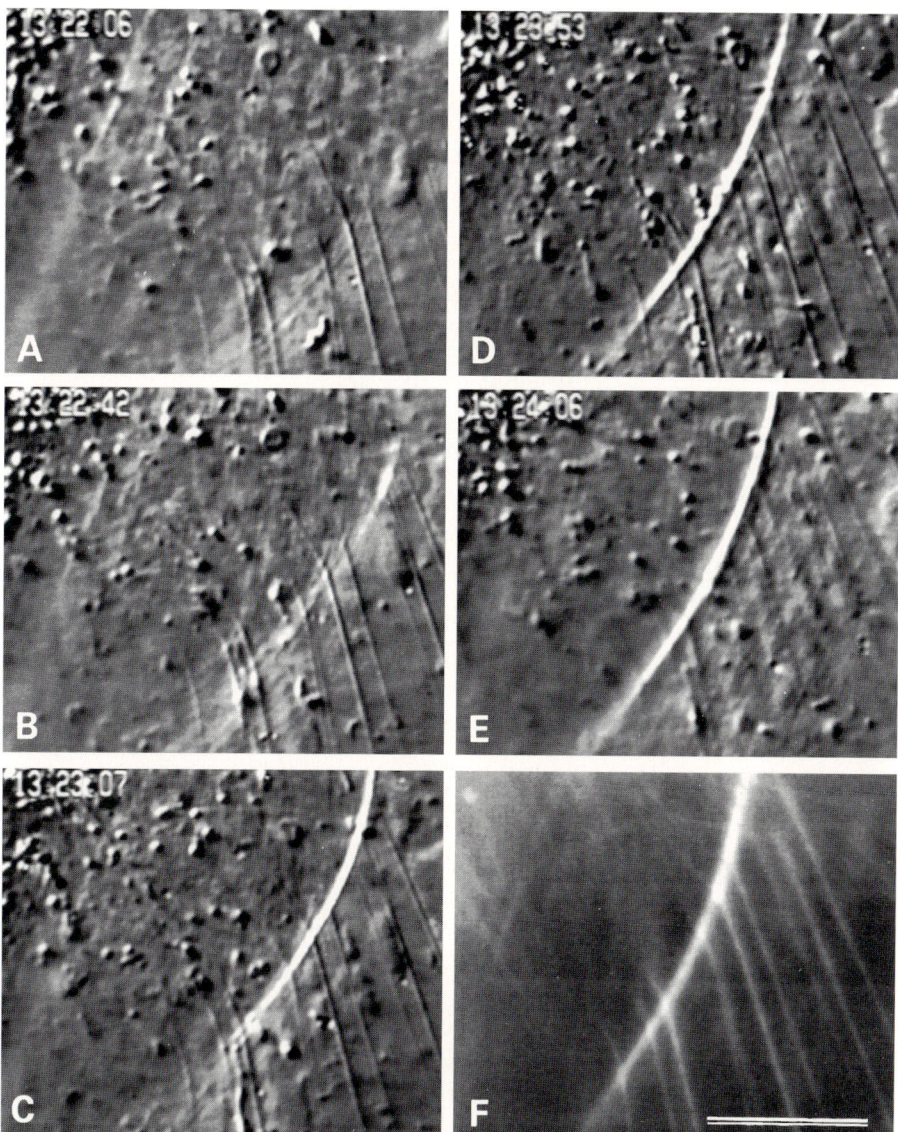

Fig. 12. Images from a video-DIC recording showing an arc and stress fibers. The ends of several stress fibres are seen associated with fine transverse striations which are microfilament bundles in the dorsal cortex. The stress fibres elongate towards the upper left of the field in concert with the centripetal flow of the cortex. Cortical flow is restrained at the tips of the stress fibers, leading to an indentation. Folding of the DCMS occurs as trailing and unrestrained cortical bundles, the striations visible at a slightly higher focal plane in E, overtake the tips of the stress fibres and accumulate there. The cell was fixed after E and stained with rhodamine-phalloidin to visualize F-actin, panel F. Scale bar = 10 μm.

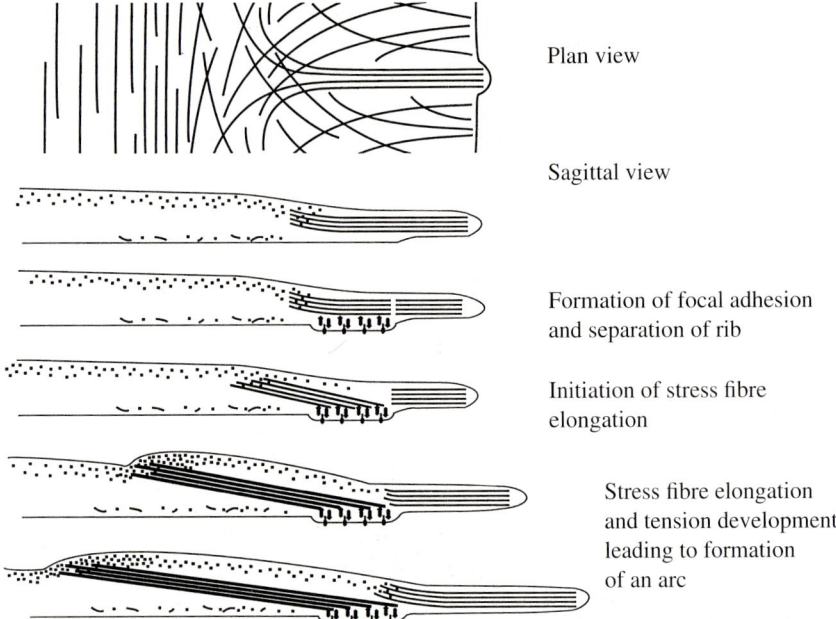

Plan view

Sagittal view

Formation of focal adhesion
and separation of rib

Initiation of stress fibre
elongation

Stress fibre elongation
and tension development
leading to formation
of an arc

Fig. 13. A model for stress fiber formation in crawling fibroblasts. The schematic shows our interpretation of the steps that lead from the interaction of focal contacts with the margin of the dorsal sheath and the subsequent centripetal elongation of stress fibers in concert with cortical flow in the DCMS. See text for further details.

from their focal adhesions to meet it, and this site is predictable in living cells by differential focusing with high-numerical aperture DIC optics; the DCMS appears to remain staked out by way of tethers to the stress fibres and the substratum. Most stress fibers eventually do break and their central segments are then carried centripetally as the DCMS collapse resumes.

From these observations we conclude that the DCMS can interact with stress fibers not only at their tips, but also along their lengths. The stress fibers appear to be under considerable tension since their interruption allows a rapid recoil of the DCMS. This recoil is probably partly elastic and partly contractile, given that the DCMS is is formed from a circumferential network of actomyosin bundles.

In cells that show a very rapid collapse of the DCMS, the lamellipodium does not have a chance to disassemble before the sheath starts to retract; in these cases a dramatic event occurs that confirms our conclusion, drawn from observations of stress fiber formation, that the lamellipodial microfilament array possesses mechanical connections to the DCMS. As Figure 15 shows, the lamellipodium goes along with the collapsing DCMS. These connections are likely to be formed by a subset of microfilaments that do not depolymerize at the base of the lamellipodium but instead pass deeper into the DCMS. Thus, normal contraction and rearward flow of the DCMS can exert a pull on the cross-linked lamellipodial microfilament network. We suggest this pull, driven by myosin II-based contraction, plays a role in the rearward flow of F-actin structures within the lamellipodium.

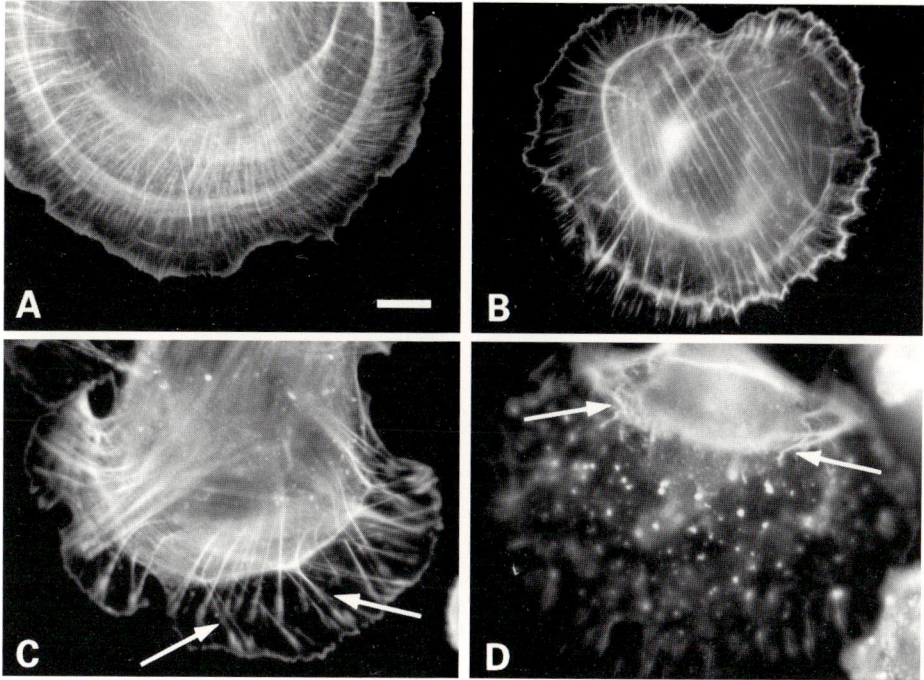

Fig. 14. Cytochalasin B treatment produces collapse of the sheath and reveals the interconnections of actin assemblies in fibroblasts. Cells stained with rhodamine-phalloidin. (A) shows the typical relationship of the dorsal F-actin cortex and stress fibers prior to treatment with 5 μM cytochalasin B; (B) 30 seconds after treatment; (C) 1 minute after treatment; (D) and 5 minutes after treatment. Note how the dorsal cortex F-actin cortex rapidly collapses inwards. The mechanical link between the cortex and underlying stress fibres is shown by the fact that the fibers are first strained (arrows in C indicate bending of stress fibers) and eventually severed (arrows in D) by the collapse of the sheath. Scale bar = 10 μm.

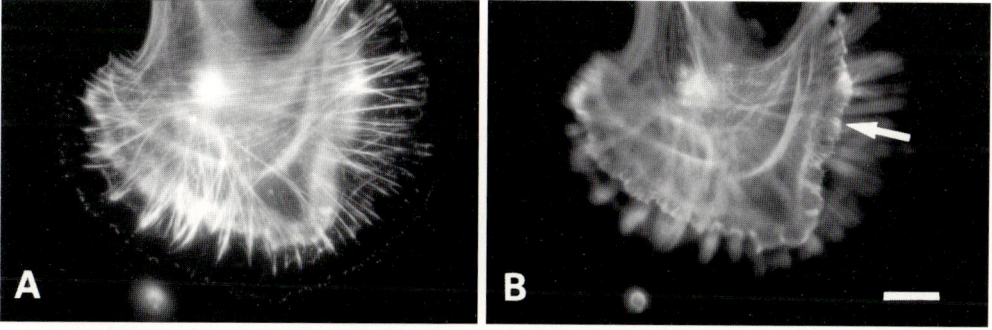

Fig. 15. Integrity of the lamellipodium and the cortical sheath is revealed by CB. The fibroblast was stained with rhodamine-phalloidin after CB treatment and is shown at two focal planes ventral (a) and dorsal (b) In cells such as this one that respond rapidly to cytochalasin treatment, the margin of the collapsing dorsal cortex carries the lamellipodium inwards, indicating mechanical continuity between the lamellipodial microfilament assembly and the dorsal cortical microfilament sheath. The lamellipodium is recognizable at the higher focal plane (b) by its ruffled appearance and the persistence of ribs (arrow). Scale bar = 10 μm.

Conclusions

As we have indicated in our model of the cytoskeletal organisation of a crawling fibroblast (Figure 1), the three major types of actin assemblies found in the lamellipodium and leading lamella are structurally linked and functionally interdependent. The flux of actin through the lamellipodium continues into the dorsal cortex of the lamella and seems to be responsible in part for the building of the dorsal cortex. The semi-circular arrangement of actin and myosin II filaments drives the cortical flow of the sheath back to the nucleus. The association of focal adhesions at the base of the lamellipodium leads to the development and centripetal elongation of stress fibers in concert with the flow. And the large actin arcs seen in the dorsal cortex form as a result of interaction of stress fibers and the cortical sheath.

As some of the other contributors to this Symposium have shown, we are now starting to understand the details of the regulation of actin filament assembly and dynamics in the lamellipodium and in stress fibers. But the existence of a continuous cortical flow of the cytoskeleton demands that there must also be carefully regulated mechanisms for disassembly of the cytoskeleton in order that the constituent molecules can recycle back to the leading edge of the cell.

The interdependence of the different actin assemblies in the mechanism that drives fibroblast crawling is a powerful argument for the need for integrated molecular and cellular approaches to resolve the mechanisms of cell locomotion.

The authors gratefully acknowledge the inspiration and guidance they received while working in the laboratories of the late Michael Abercrombie and Graham Dunn (for JPH) and while associated with Albert Harris (for BFH). Without the wisdom of these individuals and their insightful scientific approaches, the field of cell locomotory behaviour would not have the prominence that it proudly bears today.

We are grateful to Keith Burridge (University of North Carolina) for providing the rabbit antibody to alpha-actinin and to Joseph Bryan (Baylor College of Medicine) for the rabbit antibody to caldesmon.

The new work presented here is supported by a grant from the National Science Foundation to JPH (DCB 8820262).

References

Abercrombie, M. (1973). Locomotion of tissue cells. *CIBA Foundation Series* 14, Elsevier North Holland.

Abercrombie, M. (1980). The crawling movement of metazoan cells. *Proc. R. Soc. Lond.* B. **207**, 129-147.

Abercrombie, M., J.E.M. Heaysman and S.M. Pegrum (1970). The locomotion of fibroblasts in culture. III Movements of particles on the dorsal surface of the leading lamella. *Exp. Cell Res.* **62**, 389-398.

Adams, R.D. and Pollard, T.D. (1989). Membrane-bound myosin-1 provides new mechanisms in cell motility. *Cell Motility and the Cytoskeleton* **14**, 178-182.

Bray, D. and White, J.G. (1988). Cortical flow in animal cells. *Science* **239**, 883-888.

Burridge, K., Fath, K., Kelly, T., Nuckolls, G and Turner, C. (1988). Focal adhesions, transmemberane junctions between the extracellular matrix and the cytoskeleton. *Ann Rev Cell Biol* **4**, 487-525.

DePasquale, J.A. and Izzard, C.S. (1987). Evidence for an actin-containing cytoplasmic precursor of the focal contact and the timing of incorporation of vinculin at the focal contact. *J. Cell Biology* **105**, 2803-2808.

DePasquale, J.A. and C.S. Izzard (1991). Accumulation of talin in nodes at the edge of the lamellipodium and separate incorporation into adhesion plaques at focal contacts in fibroblasts. *J. Cell Biol.* **113**, 1351-1360.

Dunn, G.A. (1980). Mechanisms of fibroblast locomotion. In, Cell Adhesion and Motility. eds. Curtis, A.S.G. and Pitts, J.D. *Brit. Soc. Cell Biol. Symp.* **3**, 409-423.

Fisher, G. Conrad, P.A., DeBasio, R.L. and Taylor, D.L. (1988). Centripetal transport of cytoplasm, actin and the cell surface in lamellipodia of fibroblasts. *Cell Motil. Cytoskel.* **11**, 235-247.

Forscher, P. and Smith, S.J. (1988). Actions of cytochalasins on the organization of actin filaments and microtubules in a neuronal growth cone. *J Cell Biology* **107**, 1505-1516.

Gingell, D. and Owens, N. (1992). How do cells sense and respond to adhesive contacts? Diffusion trapping of laterally mobile membrane proteins at maturing adhesions may initiate signals leading to local cytoskeletal assembly response and lamella formation. *J. Cell Science* **101**, 255-266.

Harris, A. K. and Dunn, G.A. (1972). Centripetal transport of attached particles on both surfaces of moving fibroblasts. *Exp. Cell Res.* **73**, 519-523.

Harris, A.K., D. Stopak and P. Wild (1981). Fibroblast traction as a mechanism for collagen morphogenesis. *Nature* **290**, 249-251.

Heath, J.P. (1983a). Behaviour and structure of the leading lamella in moving fibroblasts. I. Occurrence and centripetal movement of arc-shaped microfilament bundles beneath the dorsal cell surface. *J. Cell Sci.* **60**, 331-354.

Heath, J.P. (1983b). Direct evidence for microfilament-mediated capping of surface receptors on crawling fibroblasts. *Nature (Lond.).* **302**, 532-534.

Heath, J.P. (1992). A worm's eye view of cell motility. *Current Biology* **2**, 301-303.

Heath, J.P. and Dunn, G.A. (1978). Cell to substratum contacts of chick fibroblasts and their relation to the microfilament system. A correlated interference-reflexion and high-voltage electron-microscope study. *J. Cell Sci.* **29**, 197-212.

Heath, J.P. and Holifield, B.F. (1991a). Cell Locomotion, New research tests old ideas about membrane and cytoskeletal flow. *Cell motil. Cytoskel.* **18**, (4). 1-11.

Heath, J.P. and Holifield, B.F. (1991b). Actin alone in lamellipodia. *Nature* **352**, 107-108.

Heath, J.P., Turner, D. and Holifield, B.F. (1991c). Correlative light microscopy, electron microscopy, and cytochemistry of cytoskeletal structures in motile fibroblasts. Proc 49 Ann Meeting EMSA pp168-169, San Francisco Press.

Heath, J.P. and Holifield, B.F. (1992). Lamellipodia, stress fibers and the actin cortex of motile cells, their structure and dynamics interaction studies by video DIC microscopy and correlated electron microscopy. Proc 50 Ann Meeting EMSA. pp 428-429, San Francisco Press.

Heath, J.P. and Peachey, L.D. (1989). Morphology of fibroblasts in collagen gels, A study using 400 keV electron microscopy and computer graphics. *Cell Motil. Cytoskel.* **14**, 382-392.

Hoglund, A-S, Karlsson, R., Arro, E., Frederiksson, B-A. and Lindberg, U. (1980). Visualization of the peripheral weave of microfilaments in glia cells. *J. Muscle Res. and Cell Motil.* **1**, 127-146.

Holifield, B.F. and Heath, J.P. (1990). Cytochalasin treatment reveals the interconnections between actin assemblies in fibroblasts. *J. Cell Biology.* **111**, 288a.

Holifield, B.F., Ishihara, A., and Jacobson, K. (1990). Comparative behavior of membrane protein-antibody complexes on motile fibroblasts, Implications for a mechanism of capping. *J. Cell Biology* **111**, 2499-2512.

Holifield, B.F. and K. Jacobson (1991). Mapping trajectories of Pgp-1 membrane protein patches on surfaces of motile fibroblasts reveals a distinct boundary separating capping on the lamella and forward transport on the retracting tail. *J. Cell Sci.* **98**, 191-203.

Izzard, C.S. (1988). A precursor of the focal contact in cultured fibroblasts. *Cell Motil. Cytosk.* **10**, 137-142.

Izzard, C.S. and Lochner, L.R. (1980). Formation of cell to substrate contacts during fibroblast motility, An interference reflexion study. *J Cell Sci* **42**, 81.

Koppel, D.E., Oliver, J.M. and Berlin, R.D. (1982). Surface functions during mitosis. III. Quantitative analysis of ligand-receptor movment into the cleavage furrow, diffusion versus flow. *J. Cell Biology* **93**, 950-960.

Kucik, D.F., Elson, E.L. and Sheetz, M.P. (1989). Forward transport of glycoproteins on leading lamellipodia in locomoting cells. *Nature* **340**, 315-317.

McKenna, N.M., Wang, Y-L. and Konkel, M.E. (1989). Formation and movment of myosin-containing structrues in living fibroblasts. *J. Cell Biology* **109**, 1163-1172.

Memmo, L.M. and Izzard, C.S. Distribution of the integrin b1 subunit on lamellipodia of fibroblasts. (1991). *J Cell Biology* **115**, 392a.

Mitchison, T. and Kirschner, M. (1988). Cytoskeletal dynamics and nerve growth. *Neuron* **1**, 761-772

Okabe, S. and Hirokawa, N. (1989). Incorporation and turnover of biotin-labeled actin microinjected into fibroblastic cells, an immunoelectron microscopic study. *J Cell Biology* **109**, 1581-1595.

Small, J.V., Isenberg, G. and Celis, J.E. (1978). Polarity of actin at the leading edge of cultured cells. *Nature* **272**, 638-639.

Small, J.V. (1981). Organization of actin in the leading edge of cultured cells, influence of osmium tetroxide and dehydration on the ultrastructure of actin meshworks. *J Cell Biol.* **90**, 222-235

Small, J.V. (1988). The actin cytoskeleton. *Electron Microscopy Reviews* **1**, 155-174.

Small, J.V. (1989). Microfilament-based motility in non-muscle cells. *Current Opinion in Cell Biology,* **1**, 75-79

Small, J.V., Rohlfs, A. and Herzog, M. (1992). Actin filament dynamics in locomoting fish keratinocytes. Proceedings EMBO Meeting, Alpbach, Austria 1992.

Smith, S.J. (1988). Neuronal cytomechanics, the actin-based motility of growth cones. *Science* **242**, 708-715.

Soranno, T. and Bell, E. (1982). Cytostructural dynamics of spreading and translocating cells. *J. Cell Biol.* **95**, 127-136.

Symons, M.H. and Mitchison, T.J. (1991). Control of actin polymerization in live and permeabilized fibroblasts. *J Cell Biology* **114**, 503-513.

Theriot, J. and Mitchison, T. (1991). Actin microfilament dynamics in locomoting cells. *Nature* **352**, 126-131.

Wang, Y.-L. (1984). Reorganisation of actin filament bundles in living fibroblasts. *J. Cell Biology* **99**, 1478-1485.

Wang, Y.-L. (1985). Exchange of actin subunits at the leading edge of living fibroblasts, possible role of treadmilling. *J. Cell Biol.* **101**, 597-602.

White, J.G. and Borisy, G.G. (1983). On the mechanisms of cytokinesis in animal cells. *J Theor. Biology* **101**, 289.

Printed in Great Britain © The Society of Experimental Biology 1993   57

ACTIN AND CELL MOVEMENT

J. V. SMALL, A. ROHLFS and M. HERZOG

Institute of Molecular Biology of the Austrian Academy of Sciences, Billrothstr. 11, 5020 Salzburg, Austria

Summary

The primary locomotory organelle of non-muscle, eukaryotic cells is the lamellipodium, a thin layer of cytoplasm that exhibits active protrusive activity. Earlier studies have implicated actin polymerization in the formation of lamellipodia, whereby actin monomers insert at the front and dissociate at the rear, in a treadmilling fashion. However, other models based on gel swelling and a breakdown of actin networks at the site of protrusion of the lamellipodium have also been proposed. By employing videomicroscopy and electron microscopy of the same cells, in this case mouse macrophages, it could be shown that lamellipodial protrusion is directly linked with the growth of dense actin meshworks. The gel swelling and cortical breakdown models are not supported by this data.

Using rapidly locomoting fish keratocytes, Theriot and Mitchison (*Nature* **352**, 126-131, 1991) recently obtained results that they interpreted as supporting a form of actin filament dynamics different from treadmilling. In their new "nucleation release model" the actin filaments in the lamellipodium are predicted as being very short (less than 0.5 µm) and randomly organized. We have now investigated the ultrastructure of the keratocyte cytoskeleton. Our results show that the actin filaments in these cells are very long and organized in dense and regular, more or less orthogonal networks. A gradient of filament density across the rear part of the lamellipodium suggests that the filaments are graded in length. These data support a treadmilling type model for the keratocyte.

Introduction

In the more hardy days of time-lapse micro-cinematography it was Abercrombie and his colleagues who were the first to draw close attention to the phenomenon of substrate-associated cell movement of metazoan cells (for review see Heath and Holifield, 1991a). Studying mainly fibroblasts, they could show, around 1970, that the advance of a cell over a substrate was driven primarily by the protrusive activity of a thin layer of cytoplasm at the cell front, the lamellipodium. Subsequent work by others was to reveal the beauty and complexity of the cytoskeleton (Lazarides and Weber, 1974) and in particular the more or less exclusive presence of actin (and its associated proteins) in the lamellipodium (for review see Small, 1988). These and other findings, including the observed inhibition of cell movement by the actin antagonist cytochalasin B (Spooner et al., 1971) implicated actin as the major cytoskeletal element involved in the locomotion process. But the mechanisms involved were by no means clear.

Key words: actin, cell movement, lamellipodium.

With hindsight it may be said that the use of fibroblasts may have been an unfortunate choice for the first locomotion studies. Apart from an actin- rich lamellipodium, these cells contain an extensive system of actin bundles (see Fig. 4a), or stress fibres, that are involved in tight substrate anchorage at so-called "focal contacts" (Burridge, 1986). The presence of these strapped-down stress fibres explains the jerky "creeping and crawling" (Trinkaus, 1984) of fibroblasts which arises from a tug-of-war between the advancing lamellipodium and the focal contact sites. Other cells, such as leucocytes (e.g. Keller *et al.*, 1989), keratocytes (Cooper and Schliwa, 1986) and amoeba (Fukui *et al.*, 1989) lack stress fibres but have well developed lamellipodia and move much faster than fibroblasts. Nevertheless, the aesthetic appeal of stress fibres led to early suggestions that they are involved in cell movement (e.g. Abercrombie, 1980): however, they would appear to hinder rather than support it (Badley *et al.*, 1980).

That the lamellipodium is the primary locomotory organelle of migrating cells was most dramatically shown by experiments with fish keratocytes. Lamellipodial fragments separated from these cells by microdissection were shown to locomote for a considerable time as autonomous units at the speed of the parent cell (Euteneuer and Schliwa, 1984). Studies of cell movement are, in consequence, now focussed on the mechanism underlying the protrusive activity of the lamellipodium. To form a basis for speculation about how lamellipodia move, we have concentrated efforts on defining their structural organization. In this paper we summarise earlier and more recent studies on different cell types, including macrophages and keratocytes and discuss plausible models of protrusion.

Visualizing the ultrastructure of the lamellipodium

To gain information about the ultrastructure of the lamellipodium a number of techniques have been applied (for more details see Small, 1988). In their very early studies Spooner *et al.* (1971) recognised the presence of actin filaments in thin plastic sections of lamellipodia, but embedding methods, involving osmium tetroxide post-fixation and dehydration in organic solvents yield poorly structured images of this region of the cell. Fortunately, the lamellipodium is so thin (around 0.1 μm - 0.2 μm: Abercrombie *et al.*, 1971) that it is readily accessible to electron microscope analysis in whole cells or cytoskeletons. The chosen method of preparation (critical point drying, freeze drying, negative staining etc.) can, however, produce markedly different levels of structural preservation in cell or cytoskeleton whole mounts. We shall not dwell on these differences here since they have been discussed elsewhere (Small, 1988). For the present it suffices to say that the most ordered images of the lamellipodium have so for been obtained by the adoption of extraction buffers that arrest locomotion rapidly without change in cell shape (Small *et al.*, 1982; Rinnerthaler *et al.*, 1988, 1991) and the use of negative staining for image contrasting. Improvements in filament order may be achieved by the added use of phalloidin to stabilise aldehyde-fixed actin filaments against slight distortions introduced by drying in the negative stain.

Using these methods, it could be shown that the lamellipodium of fibroblasts, glial and neuroblastoma cells is composed of arrays of actin filaments organised in dense, diagonal

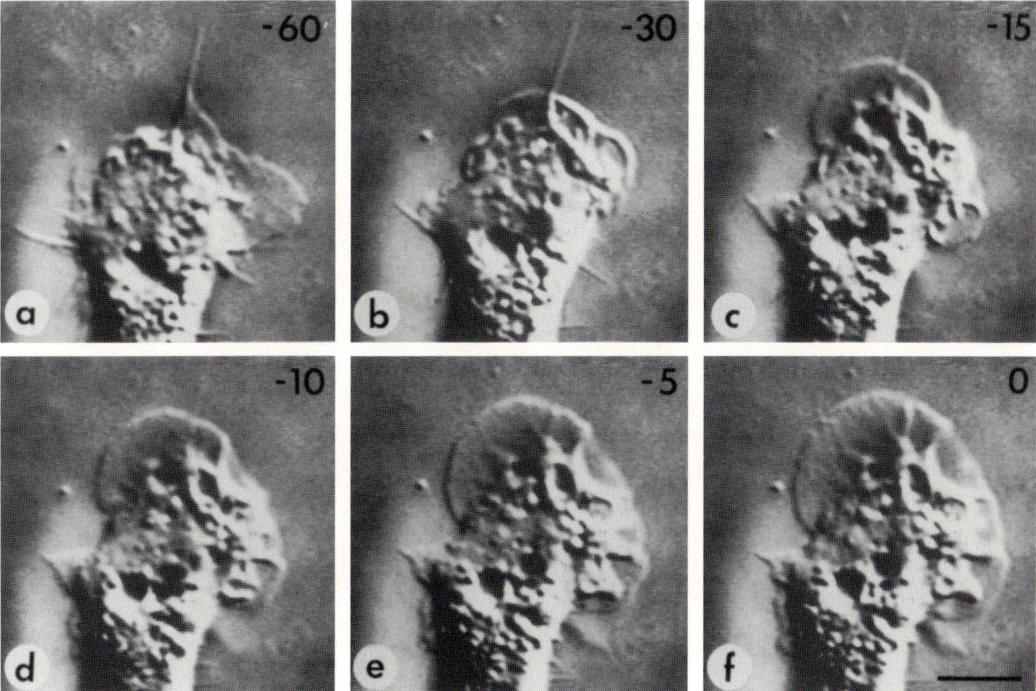

Fig. 1. a - f. Video series of a macrophage growing on an electron microscope support film. A lamellipodium is protruded at a rate of 4-5 μm/min over the last 30 seconds prior to fixation and extraction at O (f). The ultrastructure of the same lamellipodium is shown in Fig. 2. Bar, 5 μm. Reproduced from Rinnerthaler *et al.* (1991), with permission.

meshworks or in radially-arranged filament bundles (Small and Celis, 1978; Isenberg and Small, 1978; Höglund *et al.*, 1980; Small, 1981, 1988). The existence of filament arrays, apparently in transition between the meshworks and the bundles (see e.g. Small *et al.*, 1982; Small, 1981, 1988) suggested that these two organizational forms were readily interchangeable. It was also apparent that the actin filaments, at least in the transitional arrays were long, extending across most of the width of the lamellipodium. We shall return to the problem of filament length in a later section.

Correlating ultrastructure and movement

Abercrombie *et al.* (1970) showed that the motile activity of the lamellipodium in fibroblasts is highly variable, so that two parts of the same lamellipodium, only 6 μm apart, may be moving in completely different ways. In order to relate lamellipodium structure with movement it was therefore necessary to perform structural analysis on cells whose locomotion history, up to the time of fixation for electron microscopy, was known. This was achieved in a recent study on mouse macrophages (Rinnerthaler *et al.*, 1991).

Figure 1 shows a video series of the lamellipodium of a macrophage moving on an electron microscope support film for a period of one minute prior to arrest and fixation, at "O" (f). Over the terminal half of this period, the lamellipodium protruded at a relatively

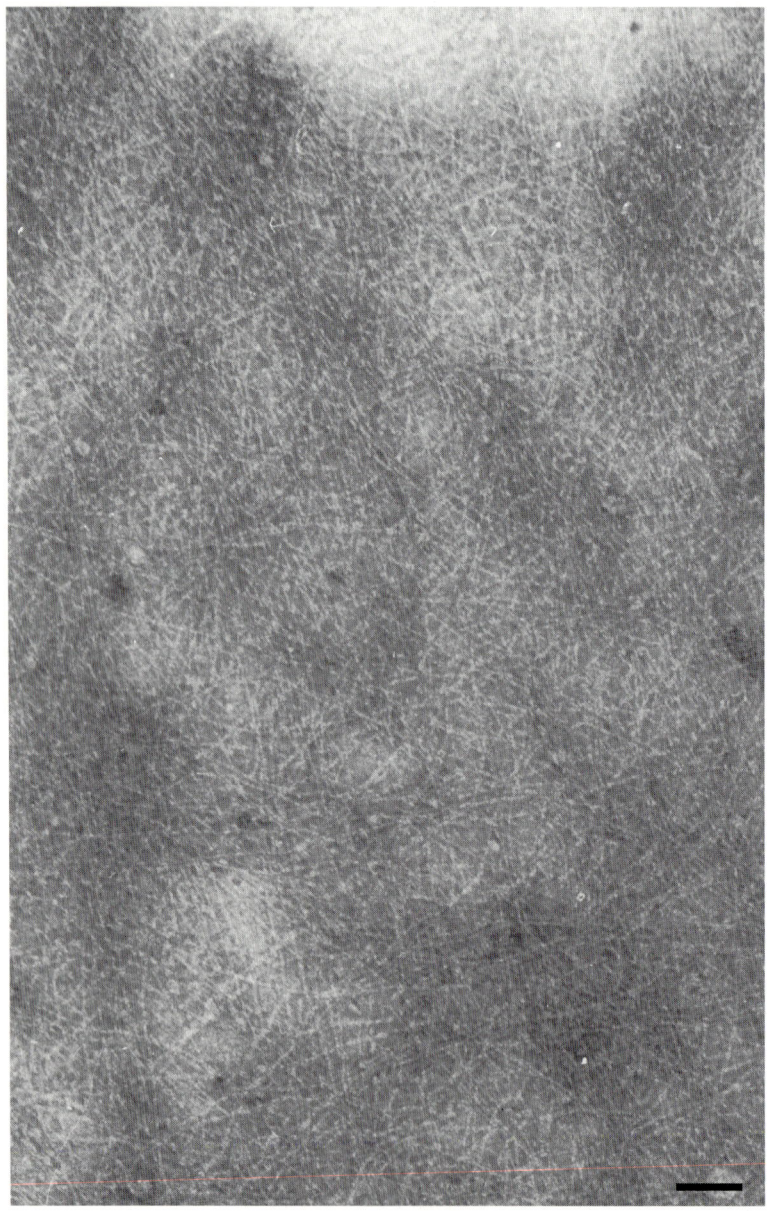

Fig. 2. Front part of the lamellipodium shown in Fig. 1f after fixation and negative staining. A dense meshwork of actin filaments is evident. Cell front is at top. Modified from Rinnerthaler *et al.* (1991). Bar, 0.2 μm.

constant rate of 4 - 5 μm/min. Electron microscopy of the same cell showed that the rapidly protruding regions contained densely-packed meshworks of actin filaments (Fig. 2). Analysis of lamellipodia showing different motile behaviour showed that protrusion was generally associated with dense meshwork formation. Behind cell edges that were relatively stationary, dense meshworks were absent and, instead, a loose bundle of actin filaments, lying parallel to the cell edge was commonly observed (Rinnerthaler *et al.*, 1991).

These studies were the first to demonstrate in a direct way the association of actin filament meshwork formation with protrusion. They also provided evidence against such models of cell locomotion based on a swelling of preformed actin networks (e.g. Condeelis, 1992) and "toothpaste models" that invoke a localized breakdown of actin filaments at the protruding site and cortical contraction elsewhere in the cell (Zigmond, 1989).

Since radial bundles of actin filaments are not a general feature of lamellipodia they are not essential for the protrusion process. Rather, they seem to be involved in sensing the surroundings (see e.g. Bray, 1982) and may take part in the initiation of focal contacts (De Pasquale and Izzard, 1987). As rigid structures, these bundles can also be presumed to play an important role in stabilising the lamellipodium, but the signals involved in their formation and dynamic movements are only now coming under closer scrutiny (Forscher *et al.*, 1992).

The keratocyte model

Epidermal fish and amphibian keratocytes are among the most rapidly moving metazoan cells (Goodrich, 1924; Bereiter-Hahn *et al.*, 1981; Euteneuer and Schliwa, 1984) and have recently attracted increased attention from investigators of cell motility (Kucik *et al.*, 1989, 1990, 1991; Theriot and Mitchison, 1991, 1992). Although keratocytes are obtainable only in explant cultures for several hours at a time, the remarkable locomotory behaviour of these cells more than offsets the efforts required to produce sufficient quantities of them for analysis.

A striking feature of keratocytes is their shape (Fig. 3): Goodrich (1924) likened them to a "canoe" or "fan" and we shall use his term "fan cells" to describe them. Fig. 3 shows a video series of a moving fan cell derived from trout epidermis. When unobstructed, these cells move on glass at velocities of up to 10 µ/min. And this motion is achieved, remarkably, without a significant change in overall cell shape (Fig. 3a-c)! As well as illustrating this point the video series was chosen to show that the lamellipodium exerts enough traction to cause gross distortions in cell shape when the nucleus becomes tethered by debris on the substrate. Observation of many cells reveals that when they exhibit the common fan-like shape, with the nucleus comfortably seated at the rear, they are moving rapidly and persistently in the direction of the convex front. This information is particularly relevant to the ultrastructural studies (below, and Small, Rohlfs and Herzog, in preparation), since the fan-like cell shape can be used as an index of persistent forward movement. The complications that arise with other cells, such as fibroblasts or macrophages, that show constant fluctuations and rearward ruffling of the lamellipodium, are not met with the keratocyte.

Intact keratocyte cytoskeletons may be produced by extraction with Triton X-100 alone or in combination with aldehyde; details of the conditions used will be given elsewhere (Small, Rohlfs and Herzog, in preparation). When stained with rhodamine-conjugated phalloidin, the keratocyte lamellipodium appears as a broad, more or less

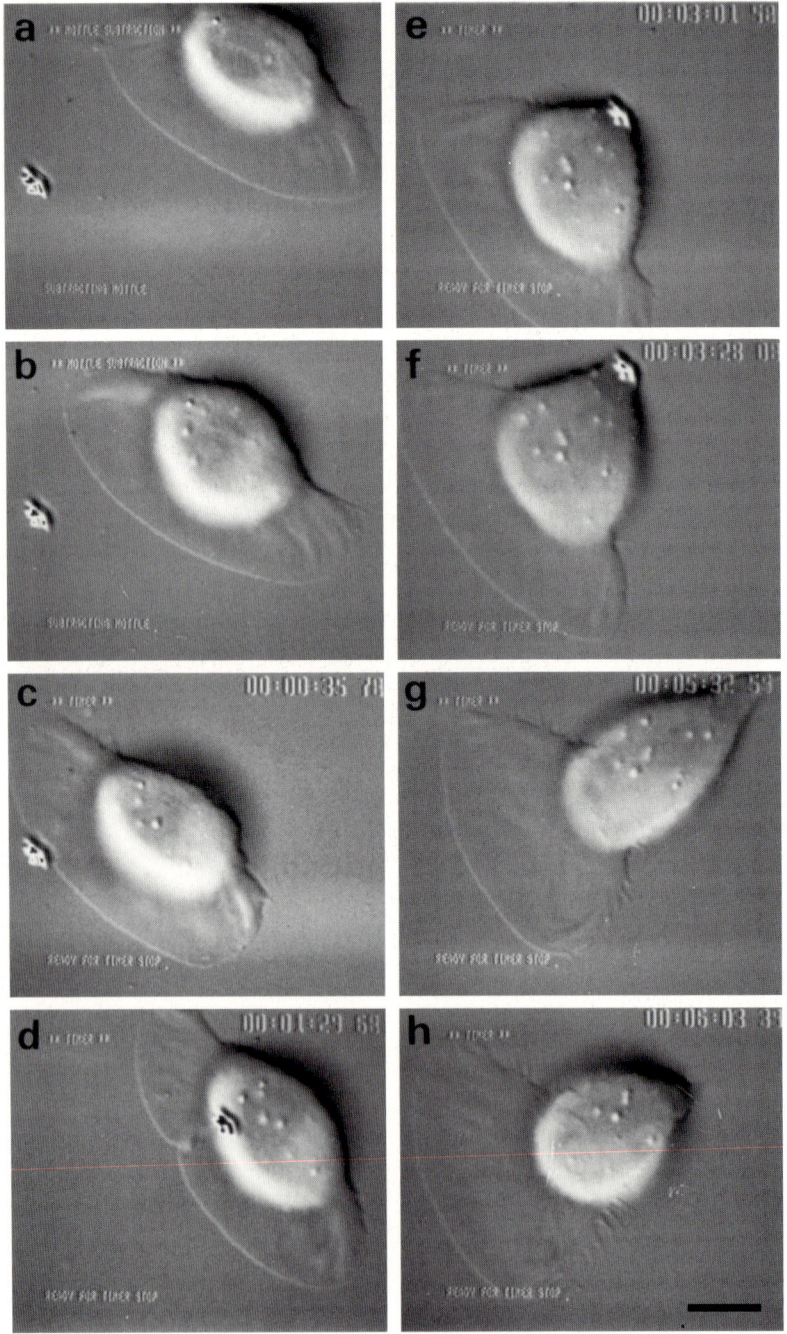

Fig. 3. a - h. Video series of a migrating trout keratocyte "fan cell". a - c. Time between frames, 45 sec: cell is moving at around 7.5 μm/min. d - f. Obstruction on substrate tethers rear part of nuclear region, but lamellipodium still advances. h. Obstruction is released and nucleus recoils towards lamellipodium. Time in top right corner (c - h) is in minutes and seconds. Bar, 10 μm.

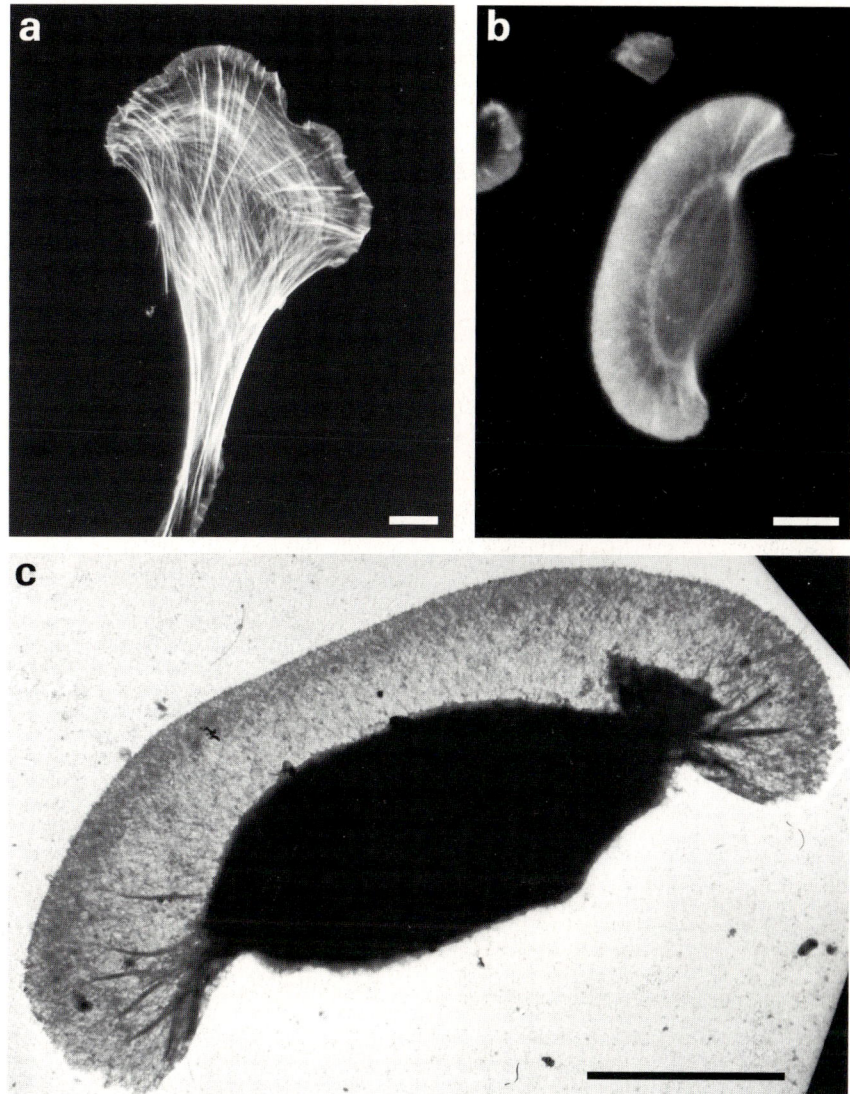

Fig. 4. a, b. Comparison of the actin cytoskeletons of a chick heart fibroblast (a) and a trout keratocyte (b), as observed after staining with rhodamine-conjugated phalloidin. The lamellipodium dominates the keratocyte and is homogenously stained; it lacks the radial bundles seen in the narrower fibroblast lamellipodium. c. Whole mount electron micrograph of detergent-extracted, fixed and negatively-stained keratocyte. Bar, 10 μm.

homogeneously-stained band of up to around 10 um in width (Fig. 4b; Euteneuer and Schliwa, 1984; Cooper and Schliwa, 1986; Theriot and Mitchison, 1991; Heath and Holifield, 1991b). No stress fibres or radial bundles are found stretching from the front to the rear as in fibroblasts (Fig. 4a) but laterally-oriented bundles beneath or behind the nuclear region may sometimes be observed (e.g. Heath and Holifield, 1991b). Also, radially-oriented ruffles at the lateral edges of the fan-shaped lamellipodium are not uncommon (Fig. 3a, b; Fig. 4b).

Keratocytes migrate equally well on plastic support films, facilitating analysis of their

Fig. 5. Electron micrograph of anterior edge of a trout keratocyte cytoskeleton, showing orthogonal arrangement of actin filaments. The filaments subtend an angle of around 45° with the cell front. Negatively stained with sodium silicotungstate. Bar, 0.2 μm.

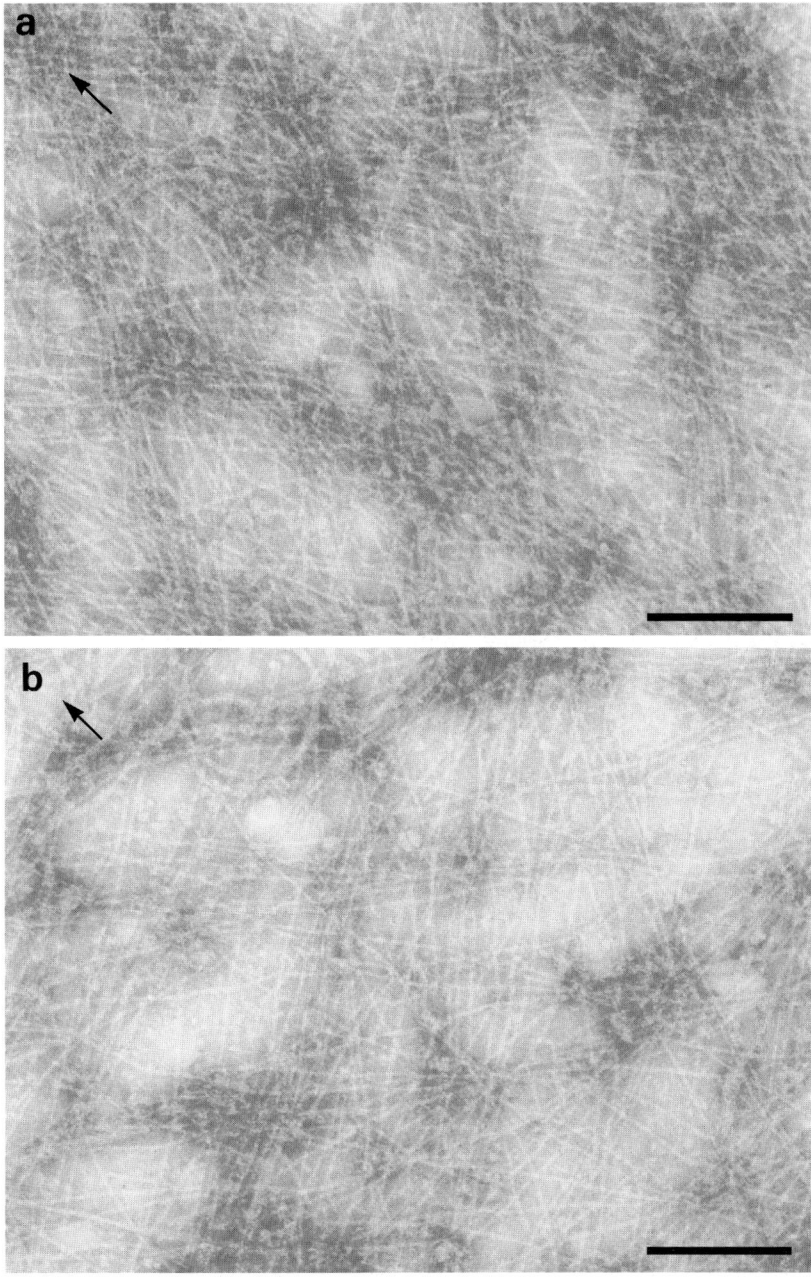

Fig. 6. Electron micrograph of deeper regions in a keratocyte cytoskeleton, respectively 6 μm (a) and 7.5 μm (b) from the front edge of the lamellipodium. The orthogonal arrangement of filaments is still approximately preserved and the filaments can be seen to be very long, extending beyond the borders of the micrographs. There is an obvious decrease in filament density between a and b. Arrows point to and are perpendicular to the cell front. Bars, 0.2 μm.

cytoskeleton by electron microscopy (Figs 4c, 5 and 6). At low magnification (Fig. 4c) the lamellipodium appears relatively homogeneous in structure and this arises from a dense, homogeneous organization of actin filaments, as would be expected from the results of the phalloidin staining. Closer inspection shows that the actin filaments are predominantly organised in a criss-cross network such that they subtend an angle of around 45° with the cell front (Fig. 5). Accordingly, the lamellipodium appears to be composed, to a first approximation, of two roughly orthogonal arrays of actin filaments. This orthogonality is even more obvious in deeper regions of the lamellipodium, a few microns away from the anterior edge, where the filament density is lower (Fig. 6). In these regions it is also possible to see that the filaments are extremely long, in fact many microns in length. The striking orthogonal arrangement of actin in these cells compares with regions sometimes found in fibroblasts (see e.g. Small, 1981, Fig. 2b), but is much more uniform and extensive. Indeed, already published micrographs of phalloidin-stained keratocytes show a criss-cross pattern of fibres throughout the lamellipodium (Cooper and Schliwa, 1986; Heath and Holifield, 1991b), consistent with our electron microscope observations.

As already indicated, the actin filament density was observed to decrease towards the rear of the lamellipodium. Counts of the number of filaments bisected by a 1 µm scale line drawn parallel to the cell front showed that the number of filaments was constant over the anterior 3 to 5 µm of the lamellipodium and then declined towards the nuclear region (Small, Rohlfs and Herzog, in preparation). Figures 6a and 6b show two areas on the same lamellipodium, respectively 6 µm and 7.5 µm from the cell front and illustrate the decay in filament density in this rearward half. We conclude from these findings that the minimum length of filaments in the keratocyte corresponds to that needed to span the region of constant filament density (that is 3-5 times 1.4 µm, taking the subtended angle as around 45°). Behind this zone of constant filament density there is then a graded fall off in filament length, giving rise to the observed density decrease.

Actin polymerization, cytoskeletal flow and protrusion

By labelling fibroblasts and neuroblastoma cells with the S-1 subfragment of myosin, it could be shown that the filaments in the bundles and meshworks of lamellipodia were all similarly polarized, with their fast growing barbed ends oriented towards the leading membrane edge (Small et al., 1978; Small and Celis, 1978). (This same polarity has now been demonstrated for actin filaments in the keratocyte; Small, Rohlfs and Herzog, in preparation). Taken together with the noteworthy findings that actin polymerization was associated with the acrosomal reaction of echinoderm sperm (Tilney et al., 1973), it was suggested that the polarised growth of actin drives, or at least accompanies lamellipodia protrusion (Small et al., 1978). In subsequent work (Wang, 1985; Forscher and Smith, 1988; Okabe and Hirokawa, 1989; Symons and Mitchison, 1991) on living cells and cytoskeletons, it has been shown, indeed, that actin monomers are exclusively incorporated at the cell front and, further that there is a centripetal flux, or treadmilling, of actin filaments through the lamellipodium at rates of up to 5 µm/min or more (Symons and Mitchison, 1991; Forscher and Smith, 1988). Significantly this flux also occurs in stationary lamellipodia (Forscher and Smith, 1988). The question arises as to how this rearward flux is related to forward protrusion.

Mitchison and Kirschner (1988) have made the reasonable proposal that there may be a variable coupling or "molecular clutch" between the actin cytoskeleton and the receptor-substrate complexes at the ventral membrane: in stationary lamella the clutch is uncoupled whereas in rapidly protruding lamella it is fully engaged. This idea fits especially well to the speedy keratocyte in the light of the recent results of Theriot and Mitchison (1991). From the microinjection of a photoactivatable actin probe, these authors have shown that the actin cytoskeleton in keratocytes remains stationary, relative to the substrate, as cells move rapidly forward. There is therefore no slippage of the actin cytoskeleton relative to the substrate, the clutch is fully engaged and the cells move at the maximum possible velocity, corresponding to the rate of retrograde cytoskeletal flow.

On the basis of other data, however, Theriot and Mitchison (1991, 1992) did not favour such a model of locomotion for the keratocyte. Their reservations were based on measurements of the phalloidin staining intensity across the keratocyte lamellipodium which suggested that the actin filament density was constant from the front to the rear. To reconcile this finding with the decay in intensity of photoactivated actin fluorescence in microinjected cells they proposed a "nucleation release model" of locomotion that differs in principle from the treadmilling model proposed for other cells. In this model, actin filaments are nucleated at the cell front but then dissociate from the leading membrane and continue to polymerise, randomly oriented in the lamellipodium: filament lengths in the nucleation release model are estimated as less than 0.5 μm (Theriot and Mitchison, 1991).

Our studies of the ultrastructure of the keratocyte lamellipodium (here and Small, Rohlfs and Herzog, in preparation) are inconsistent with the nucleation release model. First we show that the filaments in the lamellipodium are highly ordered and second that they are very long. Further, our data on the decay of filament density towards the rear of the lamellipodium are in full accord with the observed decay in fluorescence of the actin probe in Theriot and Mitchison's microinjection experiments. Their error lies simply in attempting to estimate filament density from phalloidin staining: at the high levels of filament density in the keratocyte the fluorescence intensity of their images did not give a reliable measure of actual filament content.

A model of protrusion

Mitchison and Kirschner (1988) and Smith (1988) have already proposed models to explain the phenomenon of centripetal actin flux in lamellipodia and we have already discussed the idea of variable coupling that could explain how this rearward flow is converted into protrusion. An important feature of these models is the integration of myosin I, which Fukui et al. (1989) have shown is specifically localized in the lamellipodia of Dictyostelium amoebae. We only wish to go one step further with these models and suggest that it is myosin I that facilitates the polymerization of actin by creating a gap at the actin filament-membrane interface for actin monomer insertion to take place (Fig. 7). Actin polymerization would then be coupled to the step-wise

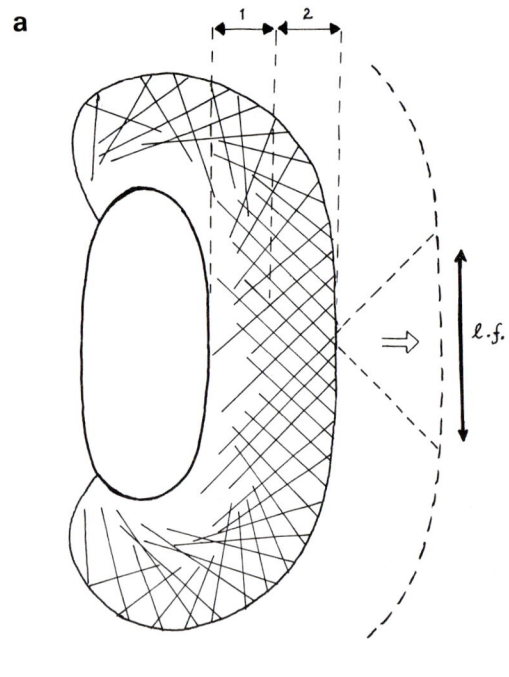

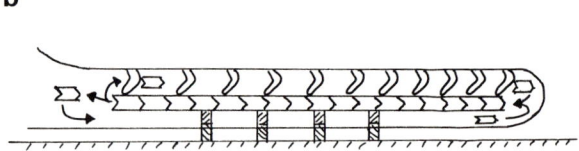

Fig. 7. a) Schematic illustration of actin filament organization in the keratocyte lamellipodium. In zone 2 there is a constant density of filaments and in zone 1 a graded decline in filament density towards the nucleus. In this zone the breakdown of filaments takes place. Polymerization of actin occurs exclusively at the cell front. Since the formed actin filaments remain stationary, relative to the substrate (Theriot and Mitchison, 1991) there is a lateral flow (l.f.) of filament ends (indicated by divergence of dotted lines) along the leading membrane, as the cell moves forward. b) Model of protrusion in which actin polymerization is coupled to the movement of myosin-I. Actin filaments are rendered stationary relative to the substrate via transient, transmembrane linkages (hatched bars). Membrane-bound myosin-I molecules (knees) move along actin filaments at the same rate as actin polymerises. Actin monomers (chevrons) may be transported anteriorly by myosin-I (upper part of scheme) or otherwise (lower part). Myosin-I is required to create a gap between the actin filament ends and the membrane to allow insertion of actin monomers.

movement of myosin along the filament. An intriguing possibility, suggested by the presence of a second actin-binding site on Acanthamoeba myosin I (see Pollard et al., 1991) is that myosin I could actually transport actin monomers (or complexes with other proteins) to feed the polymerization process, but there is yet no further in vitro data that might support this idea.

In the model shown in Fig. 7, actin polymerization occurs exclusively at the front edge of the lamellipodium, consistent with earlier work by others on fibroblasts (discussed above). The only free barbed filament ends exist at this site. At a rate of movement of 10 μm/min, around 50 subunits per second must be added to the growing filament ends: in the test tube this would require an actin monomer concentration of around 4.5 μM or 0.2 mg/ml. We shall not enter into discussion about the possible source of actin monomers since this subject has been treated in detail elsewhere (e.g. Zigmond, 1989; Stossel, 1989; Cooper, 1991; Weeds and Way, 1991). We shall only note that *in vivo* information is so far very scanty about the localization of actin-sequestering complexes with respect to the moving parts of a cell.

We have noted that the filament density is constant over the first 3-5 μm of the lamellipodium, and conclude that all filaments span this zone: taking into account their oblique arrangement, the minimum length of filaments would then range from around 4 μm - 7 μm. We propose that this anterior zone may be stabilised by actin cross-linking proteins like filamin and alpha-actinin. Filaments extending behind this zone to the perinuclear region may be up to 15 μm long. Finally, the geometry of actin in the keratocyte has interesting consequences as concerns the flow of filaments in the lamellipodium as the cell moves forward. Since the actin cytoskeleton remains stationary relative to the substrate (Theriot and Mitchison, 1991) the growing filament ends must move laterally along the anterior edge of the lamellipodium as the cell moves forward. It is possible that this lateral flow contributes to the formation of lateral ruffles commonly seen in migrating cells. According to this scheme, new filaments must arise from the lateral edges of the lamellipodium to feed the lateral flow.

Conclusions

The results of ultrastructural studies of slow and rapidly migrating cells support a unified treadmilling-like model of lamellipodium protrusion. Information on how the polymerization and depolymerizatiion of actin are regulated *in vivo* as well as on the involvement of other components, such as myosin I in motility is currently very fragmentary. We await progress in this direction with interest.

Acknowledgements

We thank Prof. M. Schliwa and Dr. U. Euteneuer for helpful suggestions about culturing fish keratocytes. Special thanks go to Fischerei Bayrhammer (Adnet, Salzburg) for generously providing free trout for the duration of this project. We also thank Mr. A. Weber for photographic assistance and Mrs. G. Mc Coy for typing.

References

Abercrombie, M. (1980). The crawling movement of metazoan cells. *Proc. R. Soc. Lond.* B **207**, 129-147.

Abercrombie, M., Heaysman, J.E.M. and Pegrum, S.M. (1970). The locomotion of fibroblasts in culture. I. Movements of the leading edge. *Expl. Cell Res.* **59**, 393-398.

Abercrombie, M., Heaysman, J.E.M. and Pegrum, S.M. (1971). The locomotion of fibroblasts in culture. IV. Electron microscopy of the leading lamella. *Expl. Cell Res.* **67**, 359-367.

Badley, R.A., Couchman, J.K. and Rees, D.A. (1980). Comparison of the cell cytoskeleton in migrating and stationary chick fibroblasts. *J. Muscle Res. Cell Motil.* **1**, 5-14.

Bereiter-Hahn, J., Strohmeier, R., Kunzenbacher, I., Beck, K. and Vöth, M. (1981). Locomotion of *Xenopus* epidermal cells in primary culture. *J. Cell Sci.* **52**, 289-311.

Bray, D. (1982). Filopodial contraction and growth cone guidance. In *Cell Behaviour: A tribute to Michael Abercrombie*, (ed. R. Bellairs, A. Curtis and G. Dunn) pp. 299-317. Cambridge Univ. Press.

Burridge, K. (1986). Substrate adhesions in normal and transformed fibroblasts: organization and regulation of cytoskeletal membrane and extracellular matrix components at focal contacts. *Cancer Rev.* **4**, 18-78.

Condeelis, J. (1992). Are all pseudopods created equal? *Cell Motil. Cytoskeleton* **22**, 1-6.

Cooper, M.S. and Schliwa, M. (1986). Motility of cultured fish epidermal cells in the presence and absence of direct current electric fields. *J. Cell Biol.* **102**, 1384-1399.

Cooper, J.A. (1991). The role of actin polymerization in cell motility. *Ann. Rev. Physiol.* **53**, 585-605.

De Pasquale, J.A. and Izzard, C.S. (1987). Evidence for an actin-containing precursor of the focal contact and the timing of incorporation of vinculin at the focal contact. *J. Cell Biol.* **105**, 2803- 2809.

Euteneuer, U. and Schliwa, M. (1984). Persistent, directional motility of cells and cytoplasmic fragments in the absence of microtubules. *Nature* **310**, 58-61.

Forscher, P. and Smith, S.J. (1988). Actions of cytochalasins on the organization of actin filaments and microtubules in a neuronal growth cone. *J. Cell Biol.* **107**, 1505-1516.

Forscher, P., Lin, C.H. and Thompson, C. (1992). Novel form of growth cone motility involving site-directed actin filament assembly. *Nature* **357**, 515-518.

Fukui, Y., Lynch, T.J., Brzeska, H. and Korn, E.D. (1989). Myosin I is located at the leading edges of locomoting *Dictyostelium* amoebae. *Nature* **341**, 328-331.

Goodrich, H.B. (1924). Cell behaviour in tissue cultures. *Biol. Bull. (Woods Hole)* **46**, 252-262.

Heath, J.P. and Holified, B.F. (1991a). Cell locomotion : new research tests old ideas on membrane and cytoskeletal flow. *Cell Motil. Cytoskeleton* **18**, 245-257.

Heath, J.P. and Holifield, B. (1991b). Actin alone in lamellipodia. *Nature* **352**, 107-108.

Höglund, A.S., Karlsson, R., Arro, E., Fredriksson, B.F. and Lindberg, U. (1980). Visualization of the peripheral weave of microfilaments in glia cells. *J. Muscle Res. Cell Motil.* **1**, 127-146.

Isenberg, G. and Small, J.V. (1978). Filamentous actin, 100 A filaments and microtubules in neuroblastoma cells. *Cytobiologie* **16**, 326-344.

Keller, H.U., Niggli, V. and Zimmermann, A. (1989). Diacylglycerols and PMA induce actin polymerization and distinct shape changes in lymphocytes: relation to fluid pinocytosis and locomotion. *J. Cell Sci.* **93**, 457-465.

Kucik, D.F., Elson, E.L. and Sheetz, M.P. (1989). Forward transport of glycoproteins on leading lamellipodia in locomoting cells. *Nature* **340**, 315-317.

Kucik, D.F., Elson, E.L. and Sheetz, M.P. (1990). Cell migration does not produce membrane flow. *J. Cell Biol.* **111**, 1617-1622.

Kucik, D.F., Kuo, S.C., Elson, E.L. and Sheetz, M.P. (1991). Preferential attachment of membrane glycoproteins to the cytoskeleton at the leading edge of lamella. *J. Cell Biol.* **114**, 1029-1036.

Lazarides, E. and Weber, K. (1974). Actin antibody: the specific visualization of actin filaments in nonmuscle cells. *Proc. Natl. Acad. Sci. U.S.A.* **71**, 2268-2272.

Mitchison, T. and Kirschner, M. (1988). Cytoskeletal dynamics and nerve growth. *Neuron* **1**, 761-772.

Okabe, S. and Hirokawa, N. (1989). Incorporation and turnover of biotin-labeled actin microinjected into fibroblastic cells: an immunoelectron microscopic study. *J. Cell Biol.* **109**, 1581-1595.

Pollard, T.D., Doberstein, S.K. and Zot, H.G. (1991). Myosin-I. *Ann. Rev. Physiol.* **53**, 653-681.

Rinnerthaler, G., Geiger, B. and Small, J.V. (1988). Contact formation during fibroblast locomotion: involvement of membrane ruffles and microtubules. *J. Cell Biol.* **106**, 747-760.

Rinnerthaler, B., Herzog, M., Klappacher, M., Kunka, H. and Small, J.V. (1991). Leading edge movement and ultrastructure in mouse macrophages. *J. Struct. Biol.* **106**, 1-16.

Small, J.V. (1988). The actin cytoskeleton. *Electron Microsc. Rev.* **1**, 155-174.

Small, J.V. and Celis, J.E. (1978). Filament arrangements in negatively stained cultured cells: the organization of actin. *Cytobiologie* **16**, 308-325.

Small, J.V., Isenberg, G. and Celis, J.E. (1978). Polarity of actin at the leading edge of cultured cells. *Nature* **272**, 638-639.

Small, J.V. (1981). Organization of actin in the leading edge of cultured cells; influence of osmium tetroxide and dehydration on the ultrastructure of actin meshworks. *J. Cell Biol.* **91**, 695-705.

Small, J.V., Rinnerthaler, G. and Hinssen, H. (1982). Organization of actin meshworks in cultured cells: the leading edge. *Cold Spring Harb. Symp., quant. Biol.* **46**, 599-611.

Smith, S.J. (1988). Neuronal cytomechanics: the actin-based motility of growth cones. *Science* **242**, 708-715.

Spooner, B.S., Yamada, K.M. and Wessels, N.K. (1971). Microfilaments and cell locomotion. *J. Cell Biol.* **49**, 595-613.

Stossel, T.P. (1989). From signal to pseudopod. *J. Biol. Chem.* **264**, 18261-18264.

Symons, M.H. and Mitchison, T.J. (1991). Control of actin polymerization in live and permeabilized fibroblasts. *J. Cell Biol.* **114**, 503-513.

Theriot, J.A. and Mitchison, T.J. (1991). Actin microfilament dynamics in locomoting cells. *Nature* **352**, 126-131.

Theriot, J.A. and Mitchison, T.J. (1992). The nucleation-release model of actin filament dynamics in cell motility. *Trends Cell Biol.* **2**, 219-222.

Tilney, L.G., Hatano, S., Ishikawa, H. and Mooseker, M.S. (1973). The polymerization of actin: its role in the generation of the acrosomal process of certain echinoderm sperm. *J. Cell Biol.* **59**, 109-126.

Trinkaus, J.P. (1984). *Cells into Organs.* Prentice Hall, N.J.

Wang, Y.-L. (1985). Exchange of actin subunits at the leading edge of living fibroblasts: possible role of treadmilling. *J. Cell Biol.* **101**, 597-602.

Weeds, A. and Way, M. (1991). Is thymosin-β_4 the missing link? *Curr. Biol.* **1**, 307-308.

Zigmond, S.H. (1989). Cell locomotion and chemotaxis. *Curr. Opin. Cell Biol.* **1**, 80-86.

Printed in Great Britain © The Society of Experimental Biology 1993   73

THE FISH EPIDERMAL KERATOCYTE AS A MODEL SYSTEM FOR THE STUDY OF CELL LOCOMOTION

JULIET LEE, AKIRA ISHIHARA and KEN JACOBSON

Department of Cell Biology and Anatomy, University of North Carolina at Chapel Hill, NC 27599, USA

Summary

Keratocytes provide an excellent system for the study of locomotion because their simple shape allows the description of basic principles that govern their movement. The graded radial extension (GRE) model is a kinematic description of keratocyte locomotion which relates cell shape to movement. It predicts the circumferential motion of morphological features within the plane of the moving cell. The detection of circumferential motion allows the GRE model to be distinguished from the process of spreading and parallel extension. The circumferential motion of morphological features and the curvature of lines "photomarked" into macromolecular assemblies can be explained in terms of the regulation of actin filament dynamics. The GRE model can thus relate molecular scale events to the locomotion of a whole cell. The principles of the GRE model may operate in other cell types especially since different modes of locomotion appear to be part of the same phenomenon.

Introduction

One of the major goals in the study of cell locomotion is to elucidate the mechanism(s) fundamental to the locomotion of all cells. However, since moving cells display a variety of shapes and modes of locomotion, it is difficult to determine if the same principles underlie the locomotion of a variety of cell types or if a different set of rules applies to each.

Since early morphological observations of moving cells (Trinkaus, 1984; Lackie, 1986) research into cell locomotion has focused increasingly on finding a common molecular basis of locomotion. For the past several decades the fibroblast has been the most extensively used cell type for both morphological and biochemical investigations of locomotion (Abercrombie *et al.*, 1970a; 1970b; 1977; Abercrombie, 1982). Leukocytes have also been the subject of much research, particularly the chemotactic response of these cells because of its clinical importance in the functioning of the immune system (Lackie and Wilkinson, 1984; Matthes and Gruler, 1988; Zigmond, 1989; Schmid-Schonbein, 1990). Another intensively studied system is the motile amoebae of the slime mould Dictyostelium discoidum (Spudich, 1989) which has been used to explore the genetic basis of motility. However, despite the increasingly detailed information about locomoting cells at the molecular level we still do not know how cells move. Part of the problem is the difficulty in relating the vast intricacies of molecular behaviour to the

Key words: keratocyte, fish epidermis, model system, cell locomotion.

movement of a whole cell. Another problem is that the cell types used in most studies display complex shapes and modes of locomotion which may be further complicated by dramatic ruffling and blebbing activity of the dorsal cell surface. In order to deduce some of the fundamental principles that may be operative in a variety of cells it is essential to develop a simple concept of how molecular scale events lead to the locomotion of a whole cell. In this regard, the simple shape and rapid gliding motion of keratocytes provides some distinct advantages over the study of more irregularly shaped cells.

Keratocytes in vivo form part of the surface epithelium of fish and amphibians. Their rapid, persistent locomotion is thought to play an important role in the wound healing response (Lash, 1955; Krawczyk, 1971; Radice, 1980). In addition the behaviour of keratocytes in vivo and in vitro is very similar. Although keratocytes have been used in a variety of studies their use has been relatively small compared with the above cell types. However, over the past ten years studies of keratocyte locomotion have increased and it is rapidly becoming a focus of research into cell locomotion. Here we will discuss reasons for the increasing popularity of keratocytes as an experimental system. Emphasis will be placed on the importance of keratocyte motility in revealing general principles of locomotion and their possible application to other cell types.

Characteristics of keratocyte locomotion

The most striking feature of moving keratocytes in vitro is their simple shape and gliding mode of locomotion. Each motile cell consists almost entirely of one flat, semicircular or ellipsoidal lamella. In strict morphological terms the keratocyte lamella is more like a lamellipodium, being devoid of organelles and only a few tenths of a micron in thickness. Cellular organelles are located peripheral to the nucleus which is situated at the rear of the cell. This facilitates observation of particles on the cell surface (Kucik *et al.*, 1989; 1990; Sheetz *et al.*, 1989) without the interference of the complex surface structures seen in fibroblastic cells. Keratocytes can move rapidly at a constant speed with virtually no changes in size or shape. They are the fastest known eukaryotic cells moving at top speeds of about 1 μm/sec. Such rapid movement facilitates the detection of a systematic "flow" of particles that is associated with keratocyte locomotion as opposed random motion. Similarly, in fluorescence recovery after photobleaching (FRAP) experiments, rapid locomotion allows observation of the photobleached mark before diffusional recovery erases it. The highly persistent mode of keratocyte movement is particularly useful in studying the regulation of locomotion so that the effects of various treatments can be distinguished from random changes in cell shape or direction of locomotion. Despite the many attributes that make keratocytes excellent models for the study of locomotion they lack the degree of molecular characterisation that exists for fibroblasts. As a result there are virtually no molecular based tools with which to investigate keratocyte locomotion.

The lamellae of keratocytes is composed almost entirely of a dense orthogonal meshwork of f-actin filaments (Bereiter-Hahn *et al.*, 1981; Euteneuer and Schliwa, 1986; Cooper and Schliwa, 1986; Heath and Holifield, 1991). Actin binding proteins, α-actinin (Bereiter-Hahn *et al.*, 1981) and myosin II (Fig.2d) have been found associated with this meshwork. In addition keratocytes contain other types of filamentous structures such as microtubules and intermediate filaments. These tend to be restricted to the perinuclear region with very few filaments penetrating the lamella. The locomotion of keratocytes has been shown to be soley dependent on the presence of the f-actin meshwork within the lamella. Depolymerisation of actin filaments with cytochalasin completely disrupts keratocyte locomotion (Cooper and Schliwa, 1988). However, depolymerisation of microtubules with nocadozole or colcemid had no effect on keratocyte locomotion. Furthermore, small anucleate lamellar fragments of keratocyte lamellae which are devoid of microtubules could locomote with the characteristic "fan" shape and rapid gliding manner indistinguishable from normal keratocytes (Euteneuer and Schliwa, 1984; 1986).

A model for the locomotion of keratocytes

The shape of the keratocyte lamella appears to be a consequence of the forces which drive locomotion. This is suggested by the close association between the semicircular shape of these cells and the "gliding" mode of locomotion. The graded radial extension (GRE) model has been proposed (Lee *et al.*, 1993) to describe keratocyte locomotion. The model is a kinematic description of the direction and rate of lamellar extension (and retraction) consistent with the maintenance of a semicircular shape and a constant velocity of the whole cell (Fig.1a and d). It is proposed that extension of the front and retraction of the rear of the cell occurs perpendicular to the cell edge. In addition a graded distribution of extension and retraction rates is arranged along the cell margin such that cell shape and size are maintained. A simpler alternative to the GRE model is provided by the parallel extension model (Fig.1b) in which all regions along the cell margin are proposed to advance with the same speed and direction. However, graded radial extension is distinctly different from spreading. In contrast to the GRE model, spreading involves the equal extension of the cell margin in all directions and does not lead to cell movement (Fig.1c). The GRE model predicts that regions on extending or retracting edges will move along the circumference of the cell during locomotion (Fig.1d). Regions on the right side of the lamella are predicted to move in a clockwise direction along the cell margin whereas regions on the left side are expected to move anticlockwise. In addition regions within the plane of the lamella are predicted to behave in the same way. This movement which we term circumferential motion is the resultant motion arising from either graded radial extension or retraction. Within the front half of the cell circumferential motion leads to an increase in distance between two points on the edge whereas this distance decreases in retracting regions at the rear. Thus expansion of the lamella at the front is counterbalanced by its compression at the rear. Circumferential motion is not predicted by the parallel extension model or by the process of spreading and is therefore a specific test for the GRE model.

Evidence for the GRE model

Circumferential motion of subcellular structures

The GRE model of locomotion is supported by evidence obtained at both

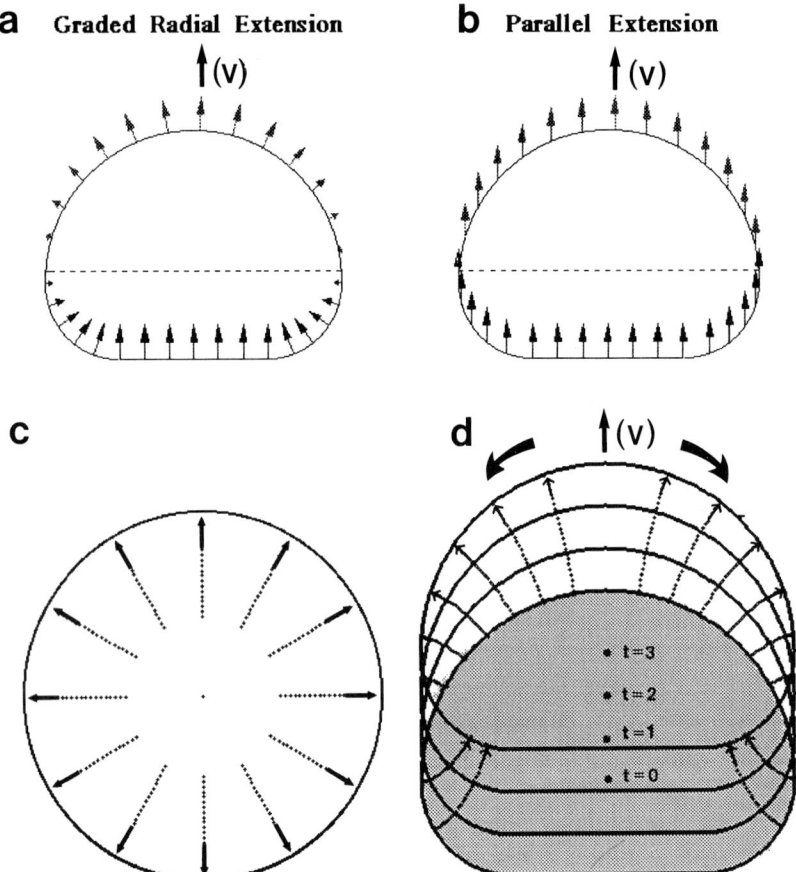

Fig. 1. (a) An idealized shape of a motile keratocyte moving with a constant velocity (**V**) according to the GRE model. The speed and direction of lamellar extension (or retraction) is indicated by the length and direction of arrows, respectively. The same position dependence of extension and retraction rates holds for regions within the plane of the lamella. Thus the velocity of cell movement (**V**) is the net result of a specific organization of extension and retraction rates within the lamella. (b) An idealized keratocyte moving according to the parallel extension model. The speed and direction (arrows) of lamellar extension (or retraction) occurs parallel to the velocity (**V**) of the whole cell. In contrast to (a) cell movement results from all regions of the cell having the same velocity. (c) Diagrammatic representation of a spreading cell. Extension is indicated (arrows) as occurring at the same rate in all directions, perpendicular to the cell margin. (d) Simulation of graded radial extension (dotted arrows) at the front of the cell and graded radial retraction (dotted arrows) at the rear. Small regions on the edge of an idealised keratocyte at t = 0 (stippled area) are shown to extend perpendicular to each successive cell outline. The rate of extension is greatest near to the cell apex and decreases towards the rear. Regions at the rear are shown to retract perpendicular to each cell outline. The amount of lamellar extension is counterbalanced exactly by lamellar retraction. Thus cell size, shape and velocity of locomotion (**V**) remain constant. Regions on the right side extending of the edge undergo circumferential motion in a clockwise direction (curved arrow) while those on the left move in an anticlockwise direction (curved arrow) with respect to the plane of the cell. During extension each region traces a curved trajectory with respect to the substratum. The advance of the cell's centroid is marked by a dot at each time point (t = 0, 1, 2, 3).

morphological and molecular levels (Lee *et al.*, 1993). Morphological evidence is obtained from observations of lamellar ridges, the formation of curved retraction fibers and the dynamics of close cell-substratum contact regions (Lee *et al.*, in preparation).

Lamellar ridges are shown by scanning electron microscopy to be shallow folds in the dorsal surface of the lamella (Fig.2a). These ridges represent cytoplasmic thickenings within the lamella as shown by the increased fluorescence of a non-specific cytoplasmic stain in these regions (Fig. 2b and c). However, there is no specialisation of actin filament organisation (Fig. 2d) within lamellar ridges. In addition a uniform distribution of myosin II staining (Fig. 2e) does not suggest any localised actomyosin contraction in these ridges. The formation of lamellar ridges appears to result from the periodic arrangement of regions of close cell-substratum contact along the cell margin (Fig.5). Differential cell adhesiveness along the cell edge thus appears to cause pleating of the lamella to form ridges.

During keratocyte locomotion ridges on the right side of the cell rotate in a clockwise direction, those on the left in an anticlockwise direction as predicted by the GRE model (Fig.3a-f). During elongation ridges also become broader and bifurcate to form two ridges (Fig.3e and f). Such broadening is also predicted by the GRE model (Fig.1d). The distal tips of most ridges elongate perpendicular to and at the same rate as the extending edge. This elongation is not immediately apparent in sequential images of moving keratocytes. However, if successive images of ridges are overlain it can be seen that they elongate in a graded radial manner with respect to the substratum in accord with the GRE model (Fig.4). This mode of elongation leads to the circumferential motion of lamellar ridges with respect to the cell's frame of reference (Fig.3). Elongation and broadening of lamellar ridges at the extending cell margin is counterbalanced by compression and shortening of lamellar ridges at the rear and is consistent with the predictions of the GRE model. The biological significance of this contrasting behaviour is that it provides means for maintaining surface area of the cell. The fact that lamellar ridges, especially those at the rear are composed of many smaller "micro-folds" suggests that they could provide a reservoir of surface material for rapid extension at the front of the cell. Indeed, the retraction induced spreading (RIS) model of Chen (1979; 1981a; 1981b) suggests that folds on the dorsal surface of fibroblasts, generated by retraction of the rear, provide surface area for lamellar extension. Thus the behaviour of lamellar ridges appears to be the "polar-coordinate" equivalent of the RIS model.

Morphological evidence for graded radial retraction is provided by the formation of curved retraction fibers at the rear edge of a moving keratocyte (Lee *et al.*, 1993). These fibers form from the elongation of small regions of the lamella that remain attached to the substratum as the cell advances. Since these fibers remain stationary with respect to the substratum they, in effect trace the course of the retracting cell margin at specific regions along its edge. The length and curvature of these fibres is very similar to those simulated by applying the rules of graded radial retraction to points along the rear cell margin.

The dynamics of close cell-substratum contact regions

In common with other rapidly moving cells (Kolega *et al.*, 1982) the ventral surface of fish keratocytes is occupied by varying degrees of close contacts (Fig.5a). These are

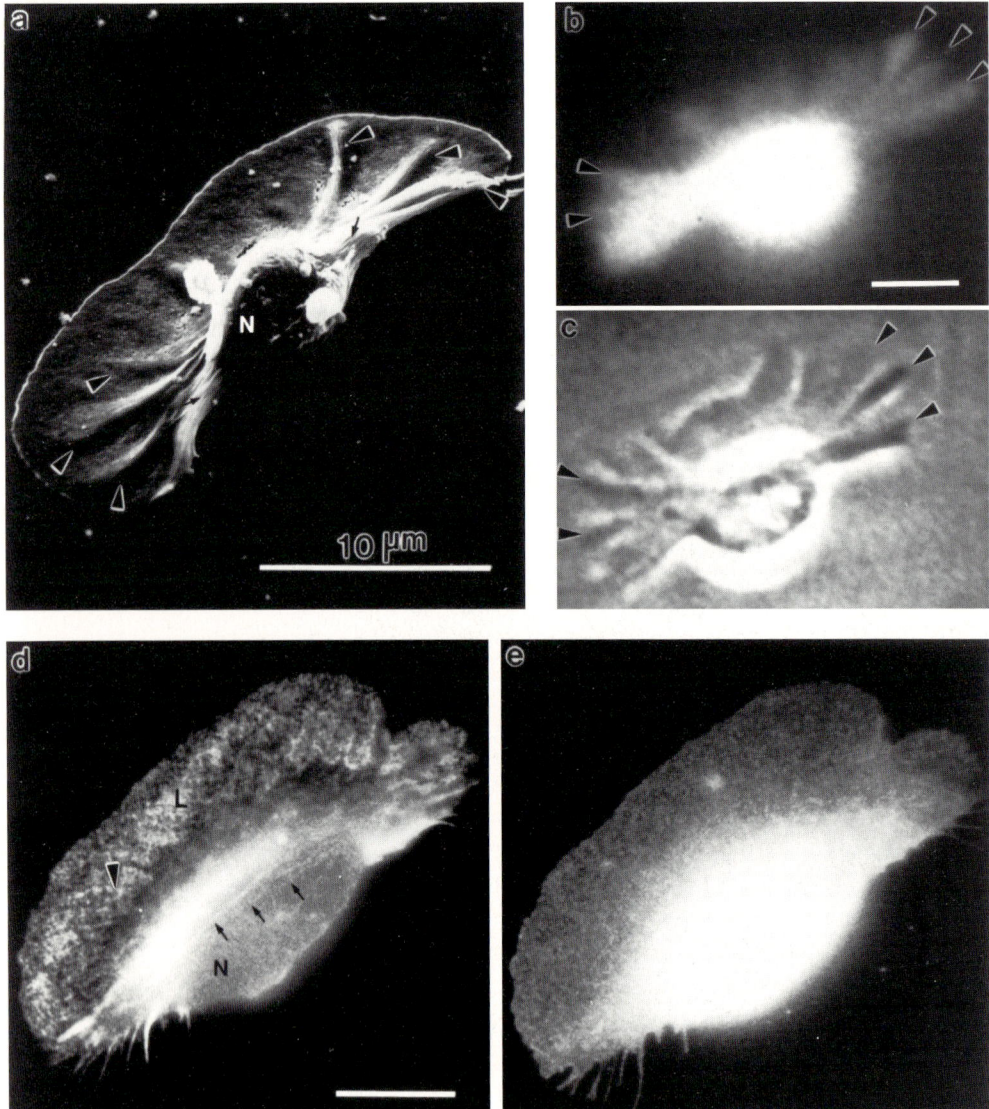

Fig. 2. (a) Scanning electron microscopy of a single keratocyte displaying a typical motile morphology. Six lamellar ridges are arranged radially (arrowheads). They are broader and flatter at their distal edges and appear to consist of many smaller "microfolds" (small arrows). N = nucleus. (b-c) Paired, fluorescence (b) and phase contrast (c) video images of a moving keratocyte stained with a bulk cytoplasmic marker, thiozole orange. A radial arrangement of cytoplasmic thickenings (arrowheads) appear as bright "bars" which correspond to the phase dense ridges in (c). The phase image was acquired 5 secs after (b). Bar = 10 μm. (d) Rhodamine phalloidin staining of actin filaments shows a fine orthogonal meshwork (arrowhead) of actin filaments within the keratocyte lamella (L). A parallel array of actin filaments (small arrows) span the nuclear region (N). Bar = 10 μm. (e) Indirect immunofluoresence of myosin II staining of the cell in (d) shows a uniform fluorescence throughout the lamella. Brighter staining corresponds to the increased thickness of the nuclear region.

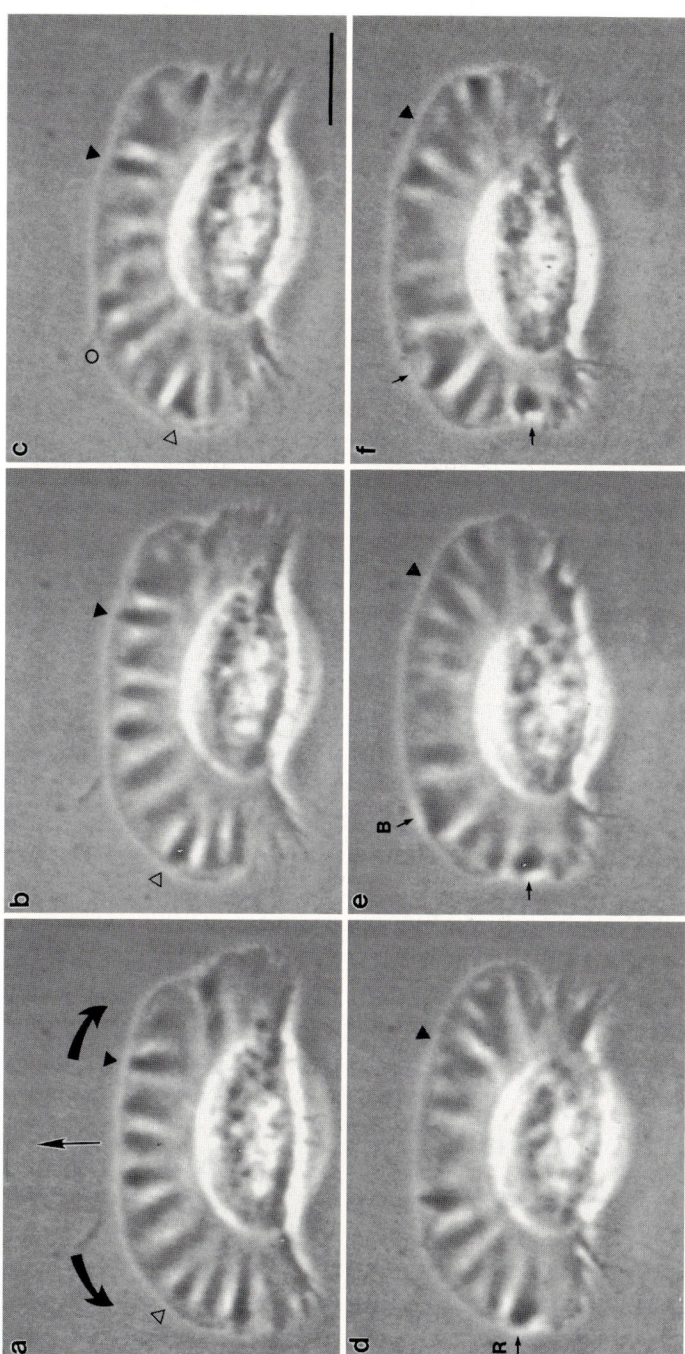

Fig. 3. (a-f) A series of video enhanced phase contrast images at 12 sec intervals, of a keratocyte moving at 16.6 μm/min in the direction of the straight arrow. Circumferential motion of lamellar ridges (curved arrows) is seen on the left (open triangle) and on the right (closed triangle) sides of the extending lamella. The distal tip of a ridge may broaden before bifurcating (**B**). In retracting regions ridges shorten before disappearing (**R**). Bar = 10 μm.

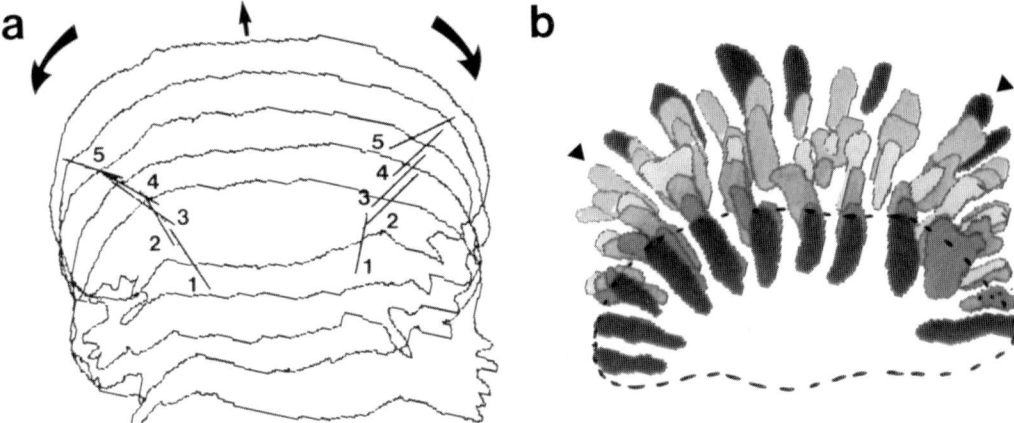

Fig. 4. (a) The change in position of the lamellar ridges (straight lines) in Fig.3c (open circle, closed triangle) is shown with respect to the substratum together with a cell outline, every 12 secs during 48 secs of locomotion. Successive ridge positions (1-5) follow a curved trajectory in the direction indicated (curved arrows) with respect to the substratum. Note the similarity between these trajectories and those predicted by the GRE model in Fig.1d. (b) Diagram showing the graded elongation at the tips of lamellar ridges perpendicular to the cell margin during the sequence shown in Fig.3. All the lamellar ridges in Fig.3a are shown as black objects radially arranged within the cell margin (dashed line). Elongated regions of each ridge are shown after 6 secs of locomotion. Successive time intervals are represented by different shades of gray. Ridge bifurcations are also included. Elongation of ridges is greatest close to the cell apex and decreases in a graded manner towards the left and right edges of the cell. No elongation occurs at the rear of the cell. The elongation of ridges (triangles) in similar positions to those in (a) trace a curved trajectory with respect to the substratum.

arranged in a narrow band of alternating close and more distant contacts along the cell margin. Interference reflection microscope (IRM) observations of moving keratocytes shows these regions to be highly dynamic, exhibiting circumferential motion and bifurcations as displayed by lamellar ridges (Lee *et al.*, in preparation). As yet, little is known about the molecular composition of close contacts in contrast to focal contacts (Burridge *et al.*, 1988). However, they appear to have an important influence on the morphology of the lamella. A comparison between phase contrast and IRM images of fixed motile keratocytes shows that lamellar ridges are flanked by two regions of close contact (Fig.5b and c). In cells that have no lamellar ridges an alternating pattern of contacts cannot be seen (Fig.5d and e). Instead, these cells have a narrow rim of close contact along the edge of the cell margin while the remainder of the ventral surface is occupied by an intermediate region of close contact. The behaviour of lamellar ridges thus appears to be a higher order phenomenon arising from the dynamics of close contact regions which is presumably linked in some way to the f-actin meshwork within the lamella.

The dynamics of molecular assemblies

At the molecular level two pieces of evidence for the GRE model are provided by the curvature of lines "photomarked" into components of the cell surface and cytoskeletal

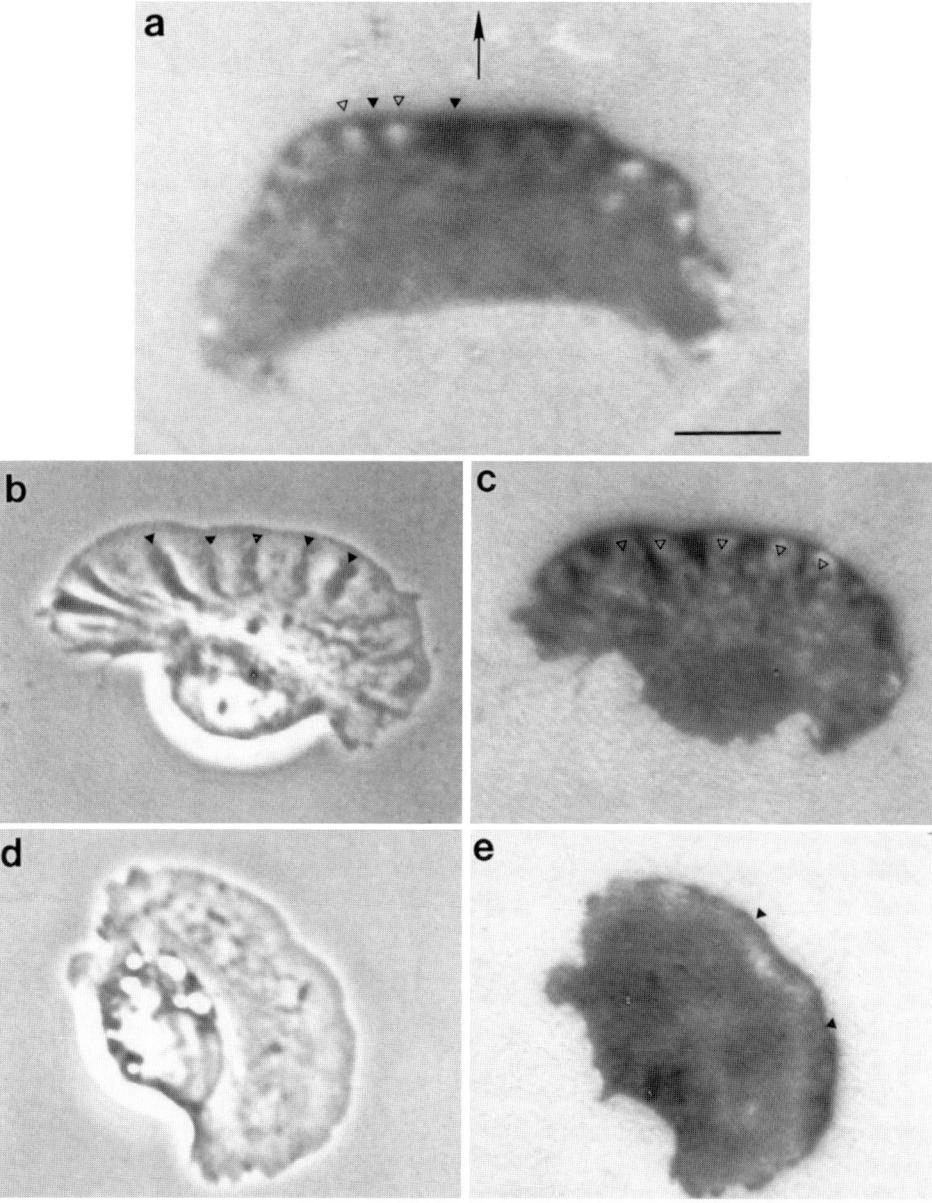

Fig. 5. (a) Interference reflection image of a keratocyte moving in the direction of the arrow. An alternating pattern of close (triangle) and more distant contacts (open triangle) are arranged in a narrow band along the entire length of the cell margin. The remainder of the ventral cell surface is occupied by a gray intermediate region of close contact. (b-e) Paired phase contrast and IRM images of fixed keratocytes. (b) Lamellar ridges (triangle) correspond to regions of more distant contact (open triangle) in (c). (d) A phase contrast image of a keratocyte with a lamella devoid of ridges. This corresponds to a simple IRM contact pattern (e) consisting of a rim of close contact along the cell margin (triangle) and a gray region of intermediate contact elsewhere. Bar = 10 μm.

assemblies (Lee *et al.*, 1993). The surface of moving keratocytes was stained with a rhodamine labelled, monovalent, snail (*Helix aspersa*) lectin. No patching or capping was observed as evidenced by a uniform staining of the lamella which did not change during locomotion. A photobleached line made across the extending lamella becomes progressively curved as predicted by the GRE model (Fig.6). The photobleached line is generally displaced with respect to the substratum in the direction of cell movement. However, a velocity component perpendicular to the direction of cell movement causes the line to curve. Thus rate and direction of line movement matches the rate of lamellar extension perpendicular to the cell edge. This result agrees with the finding that the membrane lipid of polymorphonuclear leukocytes moves as a unit with the extending edge (Lee *et al.*, 1990). FRAP measurements of lectin stained keratocytes give a value of ~1 × 10^{-9} cm^2/sec for the diffusion coefficient of the lectin receptor complexes. This value is closer to that of membrane lipids (Jacobson, 1980; Jacobson *et al.*, 1987) than membrane proteins suggesting that the collective behaviour of snail lectin-receptor complexes does actually reflect the bulk movement of the cell surface.

The second piece of molecular evidence for the GRE model is provided by experiments where caged fluorescent actin is photoactivated in a line across the extending lamella of a moving keratocyte (Fig.7). Initially, the photoactivated line remains stationary as the cell advances due to actin polymerisation predominately at the extending cell margin (Theriot and Mitchison, 1991) However, the photoactivated line becomes progressively curved beginning at the point where the line enters retracting regions of the lamella. Simulations of photoactivated lines suggest that such curvature can be accounted for by a graded radial retraction throughout the actin filament meshwork within the rear of the lamella.

The role of actin filament dynamics in keratocyte locomotion

A recent study of actin filament dynamics in keratocytes has shown that actin filament polymerisation predominately at the leading edge is coupled directly to cell movement (Theriot and Mitchison, 1991). Actin filament turnover is extremely rapid throughout the lamella, having a half life of only 23 secs. These findings are consistent with the idea that f-actin within the keratocyte lamella consists of numerous short (0.5 µm) filaments. In addition they do not appear to be arranged with any particular orientation (Fig. 8). This raises the question of how can such a homogeneous actin meshwork be involved in both extension and retraction of the keratocyte lamella.

Since actin polymerisation occurs predominately at the leading edge, it is likely that this is also graded along the cell margin. This in turn is suggested to be involved in driving the passive graded radial motion of the cell surface, including the snail lectin receptors. Similarly, the circumferential motion of phase dense ridges may be thought of as resulting from a graded radial polymerisation of actin at the extending edge which is coupled in some way to the dynamics of close cell-substratum contacts. Whereas lamellar extension involves the actin filaments at the cell edge, retraction involves a graded radial retraction throughout the rear of the cell. However, the rate of actin depolymerisation appears to be constant throughout the lamella (Theriot and Mitchison, 1991) as does the distribution of myosin II staining (Fig.2e). This suggests a uniform actomyosin like

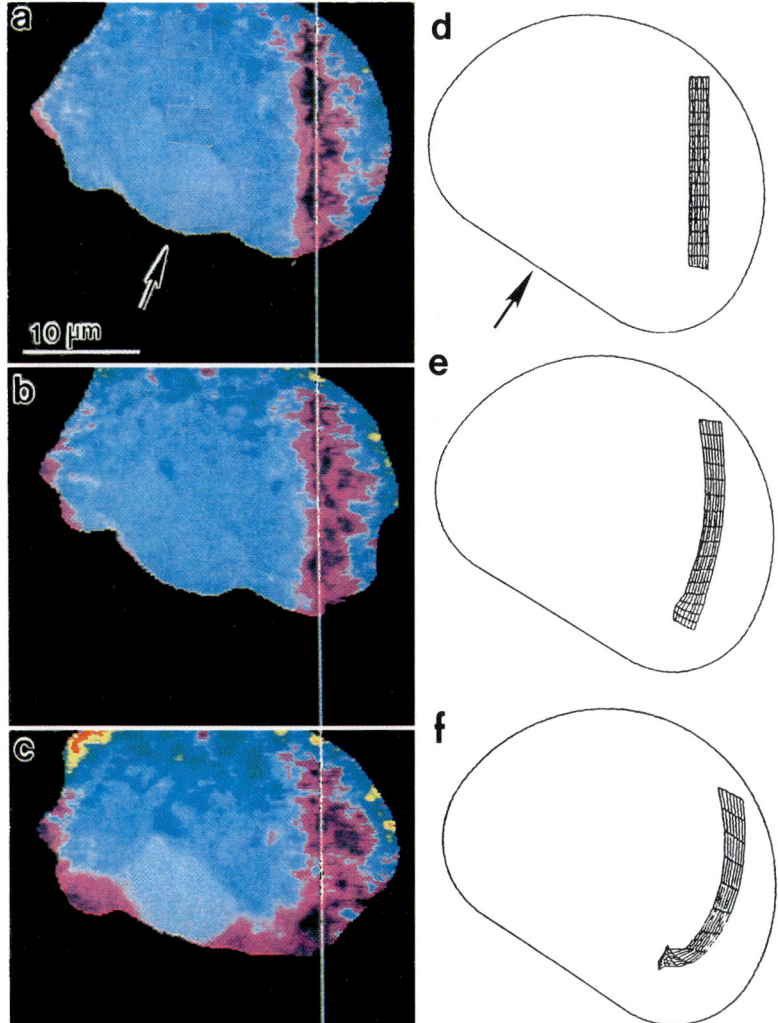

Fig. 6. (a-c) A sequence of ratio images at 200 msecs, 2 secs and 6 secs after photobleaching, showing the progressive curvature of a photobleached line across the extending lamella of a moving keratocyte, stained with a rhodamine labeled lectin. The vertical line in (a) passes through the long axis of the bleached line and represents a stationary marker with respect to the substratum. The slight line curvature in (b) is more pronounced in (c). Each postbleach image was aligned with the prebleach image prior to producing a ratio image, to remove any artifacts due to cell motion. (d-f) The simulated behavior of a photobleached line according to the GRE model reproduces the behavior of the observed photobleached line. The line was simulated by applying the GRE model to points on the vertices of a bar shaped grid across the extending lamella of a simulated cell moving at the same velocity (arrow) as in (a). The simulated line is equivalent in size and orientation to the bleached line in (a).

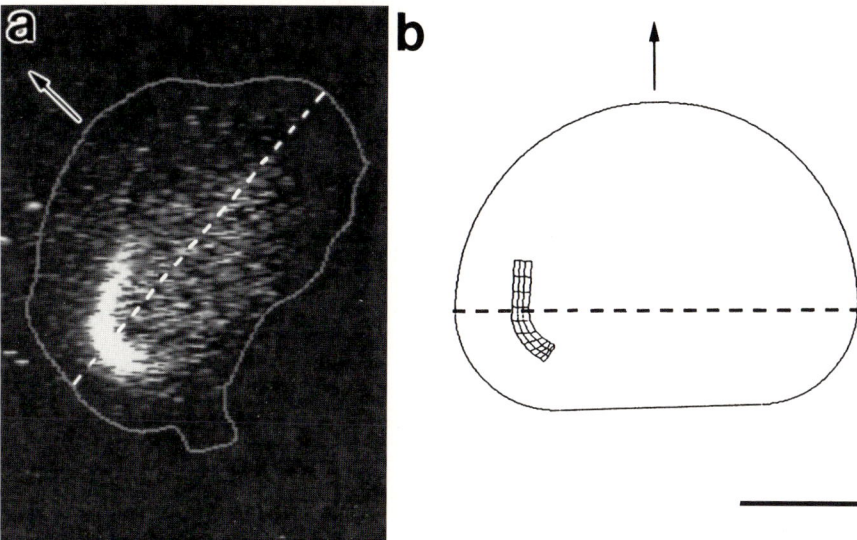

Fig. 7. (a) A fluorescent image obtained at 3 min after photoactivation of a fluorescent line across the extending lamella of a keratocyte, cooled to 15°C, moving at speed 3.40 μm/min (straight arrow). A cell outline from a corresponding phase image is superimposed on this image. The dotted straight line represents a hypothetical 'equator' which separates net extending (front) from net retracting (rear) region of the lamella. The photoactivated line is curved in retracting region only. Note the similarity between this behavior and that of a simulated actin bar. (b) A simulated photoactivated actin bar is curved in the retracting region. Simulation of the photoactivated actin bar is made by assuming that it is stationary with respect to the substratum except in net retracting regions. The rules for graded radial retraction are then applied to points at the vertices of a bar shaped grid within retracting regions of the lamella. Cell size and speed are similar to that of the cell in (a). Bar = 10 μm.

contraction throughout the keratocyte lamella. In this case a graded radial retraction could occur only if the contractile force were opposed by a graded radial extension at the rear edge otherwise it would retract in all directions at the same rate. Keratocyte movement may thus be viewed as the result of a balance between an actin based contractility and a protrusive force generated by actin polymerisation at the edge. Regions of net lamellar extension or retraction would be determined by the magnitudes of these opposing forces.

The control of keratocyte locomotion

The fact that keratocytes can maintain their shape and persistent gliding mode of locomotion implies a strict regulation of lamellar extension and retraction rates along the cell margin. The smooth outline of the extending lamella suggests that the control of actin polymerisation rate is highly localised at each point along the edge. Factors such as intracellular calcium concentration and actin binding proteins which are known to regulate actin polymerisation could provide this localised control (Brundage *et al.*, 1991; Cunningham *et al.*, 1991 and 1992; Kolega *et al.*, 1991). In keratocytes an influx of calcium ions and the operation of a Na$^+$/K$^+$ pump has been shown to be essential for the maintenance of cell shape and locomotion (Cooper and Schliwa, 1988; Strohmeiher and Bereiter-Hahn, 1984; Mittal and Bereiter-Hahn, 1985; Bereiter-Hahn and Voth, 1988). However, for keratocyte locomotion to occur actin poymerisation rates must also be

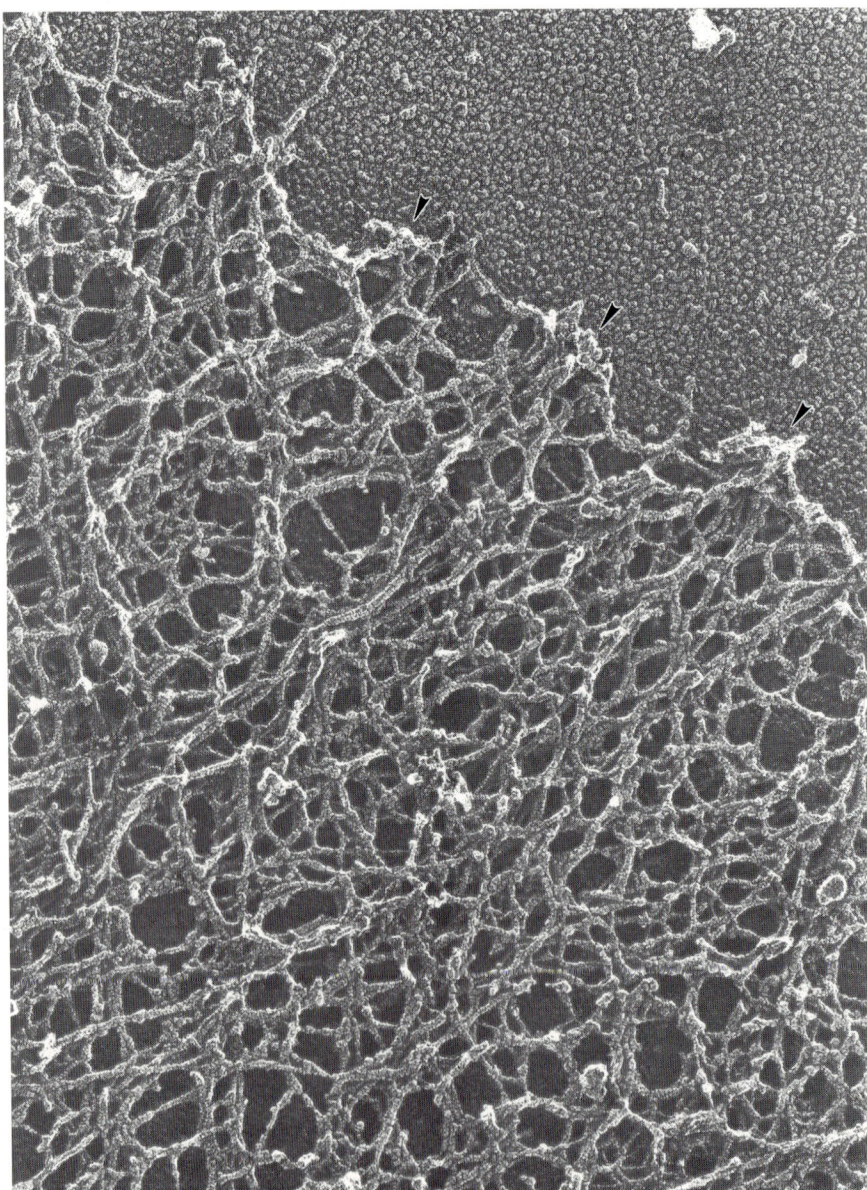

Fig. 8. A freeze-etched detergent extracted cytoskeleton of a keratocyte just behind the leading lamella. A dense orthogonal meshwork of actin filaments can be seen, which appears to be composed of numerous crosslinked actin filaments of random orientation. Regions where the leading edge is attached to the substratum are indicated (arrowheads). Magnification = 130,000. This electron micrograph was reproduced with the permission of J. Heuser.

organised on a larger scale along the cell margin. This implies a global regulation of actin polymerisation at the level of the whole cell. Increased mechanical tension in the keratocyte lamella has been shown to reduce the rate of actin polymerisation in keratocytes (Kolega, 1986) and other cells (Vasiliev, 1982). Furthermore, lamellar tension may regulate stretch activated calcium ion channels which can influence the rate of actin polymerisation by altering the concentration of cytoplasmic calcium ions (Pender and McCulloch, 1991). Thus interactions between local and global controls of actin polymerisation could provide a continual maintenance of graded radial extension and retraction.

Generality of the GRE model

The simplicity of keratocyte shape and mode of locomotion seems to be a unique feature of epithelial cells from lower vertebrates. Therefore, how much of what is learnt from keratocytes can be applied to other cells types? The answer to this question depends both on if the GRE model can be used to describe the locomotion of other cell types and whether this is really a single phenomenon. Generally cells that are shaped like keratocytes also tend to display rapid gliding modes of locomotion. For example, rat bladder carcinoma cells (Tchao, 1982) and some amoebae (Harris, personal communication) are remarkably similar in shape and movement to keratocytes. Perhaps the most unusual resemblance to keratocytes is shown by the amoeboid sperm of the parasitic nematode *Ascaris suum* (Roberts and King, 1991). These cells have semicircular pseudopods and can glide rapidly across a glass substratum. The motility of these cells is based entirely upon the dynamics of a system of non-actin containing microfilaments. The behaviour of this microfilament system appears to be strikingly similar to that of keratocytes. Forward cell movement is coupled to the elongation of tips of microfilaments that are perpendicular to the cell margin. In addition elongation of microfilaments appears to be graded along the cell margin in accordance with the GRE model. Thus the amoeboid sperm provide an interesting example of how the same principles of locomotion are conserved in a microfilament system that does not contain actin.

Further indications that the GRE model is operative in other cell types is shown by the circumferential motion of certain morphological features. For example, behavior analogous to circumferential motion is shown by cytoplasmic swellings moving along the extending edge of frog epithelial cells (Bereiter-Hahn *et al.*, 1981). In advancing chicken growth cones, microspikes move along the cell circumference (Bray and Chapman, 1985). In addition the frequency of microspike formation is greatest at the apex of the growth cone but gradually decreases, along the perimeter at increasing distances either side of the apex, in agreement with the rate of lamellar extension predicted by the GRE model. The formation of curved retraction fibers in moving macrophages is indicative of a graded radial retraction (Gudima *et al.*, 1988).

It is not so difficult to see how the GRE model might be applied to semicircular or ellipsoid cells (Fig.9a and b) especially when there is evidence for circumferential motion. However, the question remains of how these principles might describe the discontinuous locomotion of more irregular shaped cells such as fibroblasts or

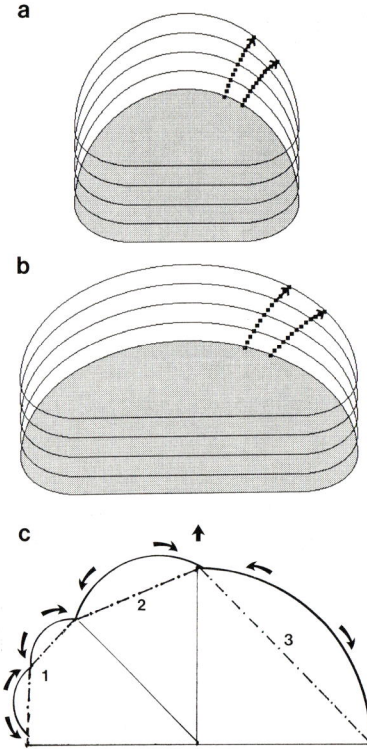

Fig. 9. (a) Application of the GRE model to a semi-circular cell and (b) an ellipsoid cell. Arrows show circumferential motion of two points on the extending cell margin. The rate of circumferential motion depends on cell size and shape. The rate of circumferential motion is greater for an ellipsoidal cell than for a semicircular cell advancing with the same velocity. (c) An illustration of how the GRE model may be applied to an irregularly shaped cell whose lamella is composed of smaller semicircular or convex extensions numbered in order of increasing size. Circumferential motion is shown to occur (curved arrows) independently on each lamellar "subunit".

leukocytes. One possibility is that in these cells the GRE model is operative simultaneously over a range of size scales (Fig.9c). This is consistent with the observation that the lamellae of many irregular shaped cells are composed of smaller semicircular or convex extensions. However, this possibility may be hard to verify due to the difficulty in detecting circumferential motion on very small lamellar "subunits". Indirect evidence for the operation of the GRE model in irregularly shaped cells is given by observations that suggest the locomotion of different cell types is part of the same phenomenon. For example cells of the same type such as fish keratocytes can assume different shapes and modes of locomotion in addition to their typical gliding motion (Fig. 10). Some keratocytes can display a remarkable resemblance to fibroblasts both in shape and movement (Fig. 10a). Conversely different cell types may display the same mode of locomotion. As discussed earlier a number of simple-shaped cell types can resemble keratocytes. However, it is of interest to find that even fibroblasts which are typically irregularly shaped can bear some resemblance to a keratocyte. This usually occurs within

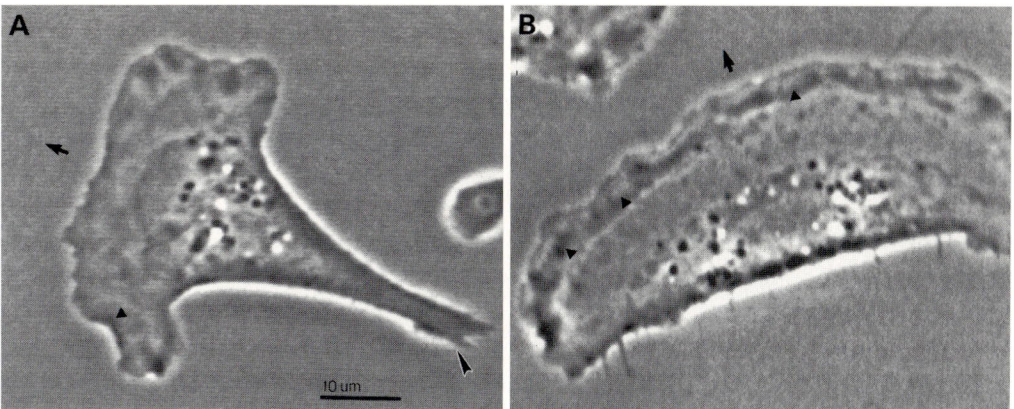

Fig. 10. Phase contrast images of keratocytes with irregular shapes moving in the direction indicated (arrow). (a) This keratocyte is shaped like a fibroblast with a ruffling lamella (triangle) and a tail at the rear (arrowhead). It also moves in the same discontinuous manner as fibroblasts. (b) A large ellipsoidal keratocyte with an irregular leading edge. Ruffles (triangles) can be seen arranged in a rim along the cell margin.

the first 30 mins after plating during which time locomotion is most rapid. Later, as focal contacts develop cell shape and mode of locomotion becomes characteristically fibroblastic together with a decrease in locomotion speed.

Although the principles for keratocyte locomotion have not yet been applied directly to cells with complex shapes, molecular based studies have revealed many similarities between different locomoting cell types. For example, the locomotion of most cell types is dependent on a system of actin microfilaments in which actin addition occurs predominately at the leading edge (Symmons and Mitchison, 1991). In fast moving cells such as leukocytes actin filament turnover is generally very rapid (Cano *et al.*, 1991). Furthermore, in motile mouse macrophages actin polymerisation appears to occur perpendicular to the cell margin (Rinnerthaler *et al.*, 1991). There is also some evidence that moving cells share common mechanisms to regulate actin filament dynamics such as intracellular calcium ions and lamellar tension. However, despite these similarities it appears that the organisation of lamellar extension and retraction rates along the cell margin is important in determining cell shape and mode of locomotion. Thus in keratocyte movement, a highly symmetrical organisation of lamellar extension and retraction rates leads to a rapid, synchronous advance of the whole cell. This contrasts with the more randomised organisation of lamellar extension and retraction rates along the cell margin of a fibroblast. Thus various modes of locomotion may be envisaged as forming part of a continuum in which motile keratocytes represent a very efficient mode of movement while fibroblasts represent the less efficient, discontinuous locomotors.

Although some evidence suggests that the locomotion of different cell types is a range of related phenomena much still needs to be learnt about how these differences arise. In relating molecular dynamics to the movement of whole cells, the GRE model provides a new conceptual basis on which to further build our understanding of cell locomotion.

This work was supported by a fellowship from the Human Frontier Science Program Organisation awarded to J.L. and by the NIH (K.J.). We thank J. Theriot for supplying photoactivation data and J. Heuser for a freeze-etch electronmicrograph.

References

Abercrombie, M. (1982). The crawling movement of metazoan cells. In *Cell behaviour* (ed. Bellairs, R., Curtis, A. and Dunn, G.), pp. 19-48. Cambridge: Cambridge University Press.

Abercrombie, M., Dunn, G. A. and Heath, J. P. (1977). The shape and movement of fibroblasts in culture. In *Cell and Tissue Interactions* (ed. Lash, J.W. and Burger, M.M.), pp. 57-70. New York: Raven Press.

Abercrombie, M., Heaysman, J. M. and Pegrum, S. M. (1970a). The locomotion of fibroblasts in culture. I. Movements of the leading edge. *Expt. Cell Res.* **59**, 393-398.

Abercrombie, M., Heaysman, J. M. and Pegrum, S. M. (1970b). The locomotion of fibroblasts in culture. III. Movements of particles on the dorsal surface of the leading lamella. *Exp. Cell Res.* **62**, 389-398.

Bereiter-Hahn, J., Strohmeier, R., Kunzenbacher, I., Beck, K. and Voth, M. (1981). Locomotion of Xenopus epidermis cells in primary culture. *J. Cell Sci.* **52**, 289-311.

Bereiter-Hahn, J. and Voth, M. (1988). Ionic control of locomotion and shape of epithelial cells: II. Role of monovalent cations. *Cell Motility and the Cytoskeleton* **10**, 528-536.

Bray, D. and Chapman, K. (1985). Analysis of microspike movements on the neuronal growth cone. *J. Neuroscience* **5**, 3204-3213.

Brundage, R. A., Fogarty, K. E., Tuft, R. A. and Fay, F. S. (1991). Calcium gradients underlying polarisation and chemotaxis of eosinophils. *Science* **254**, 703-706.

Burridge, K., Fath, K., Kelly, T., Nuckolls, G. and Turner, C. (1988). Focal adhesion: transmembrane junction between the extracellular matrix and the cytoskeleton. *Ann. Rev. Cell Biol.* **4**, 487-525.

Cano, M. L., Lauffenburger, D. A. and Zigmond, S. H. (1991). Kinetic analysis of F-actin depolymerization in polymorphonuclear leukocytes lysates indicates that chemoattractant stimulation increases actin filament number without altering the filament length distribution. *J. Cell Biol.* **115**, 677-687.

Chen, W.-T. (1979). Induction of spreading during fibroblast movement. *J. Cell Biol.* **81**, 684-691.

Chen, W.-T. (1981a). Surface changes during retraction-induced spreading of fibroblasts. *J. Cell Sci.* **49**, 1-13.

Chen, W.-T. (1981b). Mechanism of retraction of the trailing edge during fibroblast movement. *J. Cell Biol.* **90**, 187-200.

Cooper, M. S. and Schliwa, M. (1986). Motility of cultured fish epidermal cells in the presence and absence of direct current electric fields. *J. Cell Biol.* **102**, 1384-1399.

Cooper, M. S. and Schliwa, M. (1988). Ca-Channels and amoeboid cell movement. In *Signal transduction in cytoplasmic organization and cell motility*, pp. 271-278. New York: Alan R. Liss, Inc.

Cunningham, C. C., Gorlin, J. B., Kwiatkowski, D. J., Hartwig, J. H., Janmey, P. A., Byers, H. R. and Stossel, T. P. (1992). Actin binding protein requirement for cortical stability and efficient locomotion. *Science* **25**, 325-327.

Cunningham, C. C., Stossel, T. P. and Kwiatkowski, D. J. (1991). Enhanced motility in NIH 3T3 fibroblasts that overexpress gelsolin. *Science* **251**, 233-1236.

Euteneur, U. and Schliwa, M. (1984). Persistent, directional motility of cells and cytoplasmic fragments in the absence of microtubles. *Nature (London)* **310**, 58-61.

Euteneur, U. and Schliwa, M. (1986). The function of microtubules in directional cell movement. *Ann. N. Y. Acad. Sci.* **466**, 867-886.

Gudima, G. O., Vorobjev, T. A. and Chentsov, Y. S. (1988). Centriolar location during blood cell spreading and motion *in vitro*: an ultrastructural analysis. *J. Cell Sci.* **89**, 225-241.

Heath, J. and Holifield, B. (1991). Cell locomotion: Actin alone in lamellipodia. *Nature (London)* **352**, 107-108.

Jacobson, K. (1980). Fluorescence recovery after photobleaching: lateral mobility of lipids and proteins in model membranes and on single cell surfaces. in *Laser in biology and medicine* (eds. Hillkamp, F., Pratesi, R., and Sacchi, C. A.), pp. 271-288. New York: Plenum Publishing Corp.

Jacobson, K., Ishihara, A. and Inman, R. (1987). Lateral diffusion of proteins in membranes. *Ann. Rev. Physiol.* **49**, 163-175.

Kolega, J. (1986). Effects of mechanical tension on protrusive activity and microfilament and intermediate filament organization in an epidermal epithelium moving in culture. *J. Cell Biol.* **102**, 1400-1411.

Kolega, J., Shure, M. S., Chen, W. T. and Young, N. D. (1982). Rapid cellular translocation is related to close contacts formed between various cultured cells and their substrata. *J. Cell Sci.* **54**, 23-34.

Kolega, J., Janson, L.W. and Lansing Taylor, D. (1991). The role of solation-contraction coupling in regulating stress fiber dynamics in non muscle cells. *J. Cell Biol.* **114**, 993-1003.

Krawczyk, W. S. (1971). A pattern of epidermal cell migration during wound healing. *J. Cell Biol.* **49**, 247-263.

Kucik, D. F., Elson, E. L. and Sheetz, M. P. (1990). Cell migration does not produce membrane flow. *J. Cell Biol.* **111**, 1617-1622.

Kucik, D. F., Elson, E. L. and Sheetz, M. P. (1989). Forward transport of glycoproteins on leading lamellipodia in locomoting cells. *Nature (London)* **340**, 315-317.

Lackie, J. M. (1986). *Cell movement and cell behaviour.* London: Allen & Unwin.

Lackie, J. M. and Wilkinson, P. C. (1984). Adhesion and locomotion of netrophil leucocytes on 2-D substrata and in 3-D matrices. in *White cell mechanics: basic science and clinical aspects*, pp. 237-254, New York: Alan R. Liss, Inc.

Lash, J. W. (1955). Studies on wound closure in urodeles. *J. exp. Zool.* **128**, 13-28.

Lee, J., Gustafsson, M., Magnusson, K.-E. and Jacobson, K. (1990). The direction of membrane lipid flow in locomoting polymorphonuclear leukocytes. *Science* **247**, 1229-1233.

Lee, J., Ishihara, A., Theriot, J. A. and Jacobson, K. (1993). Principles of locomotion for simple-shaped cells. *Nature* **362**, 167-171.

Matthes, Th. and Gruler, H. (1988). Analysis of cell locomotion: contact guidance of human polymorphonuclear leukocytes. *Eur. Biophys. J.* **15**, 343-357.

Mittal, A. K. and Bereiter-Hahn, J. (1985). Ionic control of locomotion and shape of epithelial cells: I. Role of calcium influx. *Cell Motility* **5**, 123-136.

Pender, N. and McCulloch, C. A. G. (1991). Quantitation of actin polymerization in two human fibroblast sub-types responding to mechanical stretching. *J. Cell Sci.* **100**, 187-193.

Radice, G. P. (1980). Locomotion and and cell-substratum contacts of Xenopus epidermal cells *in vitro* and *in situ. J. Cell Sci.* **44**, 201-223.

Rinnerthaler, G., Herzog, M., Klappacher, M., Kunka, H. and Small, J. V. (1991). Leading edge movement and ultrastructure in mouse macrophages. *J. Structural Biol.* **106**, 1-16.

Roberts, T. M. and King, K. L. (1991). Centripetal flow and directed reassembly of the major sperm protein (MSP) cytoskeleton in the amoeboid sperm of the Nematode, Ascaris suum. *Cell Motil. and the Cytoskeleton* **20**, 228-241.

Schmid-Schonbein, G. W. (1990). Leukocyte biophysics: an invited review. *Cell Biophysics* **17**, 107-135.

Sheetz, M. P., Turney, S., Qian, H. and Elson, E. L. (1989). Nanometer-level analysis demonstrates that lipid flow does not drive membrane glycoprotein movements. *Nature (London)* **340**, 284-288.

Spudich, J. A. (1989). In pursuit of myosin function. *Cell regulation* **1**, 1-11.

Strohmeiher, R. and Bereiter-Hahn, J. (1984). Control of cell shape and locomotion by external calcium. *Exp. Cell Res.* **154**, 412-420.

Symons, M. H. and Mitchison, T. J. (1991). Control of actin polymerization in live and permeabilized fibroblasts. *J. Cell Biol.* **114**, 503-513.

Tchao, R. (1982). Novel forms of epithelial cell motility on collagen and on glass surfaces. *Cell Motility* **4**, 33-341.

Theriot, J. A. and Mitchison, T. J. (1991). Actin microfilament dynamics in locomoting cells. *Nature (London)* **352**, 126-131.

Trinkaus, J. P. (1984). *Cells into organs. The forces that shape the embryo.* 2nd ed. New Jersey: Prentice Hall Inc.

Vasiliev, J. M. (1982). Spreading of nontransformed and transformed cells. *Biochim. Biophys. Acta.* **780**, 21-65.

Zigmond, S.H. (1989). Cell locomotion and chemotaxis. *Current Opinion in Cell Biology* **1**, 80-86.

Printed in Great Britain © *The Society of Experimental Biology 1993*   91

PHASE-SHIFTING INTERFERENCE MICROSCOPY APPLIED TO THE ANALYSIS OF CELL BEHAVIOUR

G. A. DUNN and D. ZICHA

MRC Muscle and Cell Motility Unit, King's College London, 26-29 Drury Lane, London WC2B 5RL, UK

Summary

The theory of phase-shifting interferometry is not new but it is only recently, with the advent of solid-state detector arrays and fast image processors, that it has become a practical imaging technique. In conjunction with transmission interference microscopy, phase-shifting presents a new way of introducing contrast into the images of transparent microscopic objects such as cultured cells. An earlier paper from our laboratory has emphasised the advantages of transmission interference microscopy over phase contrast or differential interference contrast microscopy for the computerised analysis of cell behaviour. Phase-shifting greatly improves the accuracy, long-term stability and range of application of this technique but it has not previously, to our knowledge, been combined with transmission interference microscopy for the study of cultured cells. The resulting image is especially well suited to quantitative analysis by computer since it is a direct representation of the distribution of non-aqueous cellular material in the specimen. The image is not degraded by uneven illumination; by heterogeneous sensitivity of the detector array; or by differential absorption of light in the optics or specimen. Our main purpose in developing the method is to obtain sequences of images of the motile behaviour of cells in culture for analysis by computer. This type of analysis is potentially a powerful tool for studying the motile responses of cells and the operation and control of their locomotory machinery. Not only can the method be used for studying cell translocation and the dynamics of intracellular movement of non-aqueous material, but it is now possible to study in detail the time course of growth in individual cultured cells.

Introduction

In bright-field light microscopy or transmission electron microscopy of stained specimens the intensity at each point in the image is directly related to the amount of absorptive material at the corresponding part of the specimen. It is this correspondence between the image and the distribution of material in the specimen that makes these images particularly easy to interpret, not only by human observers but also by computer. However, conventional methods for introducing contrast into images of transparent microscopical objects such as living cells in culture do not have this property. With phase contrast or differential interference contrast (DIC) microscopy, the relationship between

Key words: phase-shifting interferometry, cell behaviour, interference microscopy.

image brightness and the distribution of material in the specimen is complex and gives rise to artefacts such as the phase halo of phase contrast and the direction-dependent enhancement of edges with DIC. Human observers can learn to allow for these defects to a limited extent but it is far more difficult to teach computers to do this. With phase contrast microscopy, for example, elaborate image processing rarely succeeds in extracting even the most basic information such as the location of the boundary of a cell in culture.

Brown and Dunn (1989) showed that images obtained by the neglected method of transmission interference microscopy could be interpreted much more easily by computer. In this case there is a direct and simple relationship between the image and the distribution of material in the specimen. The microscope can be adjusted so that image brightness is almost linearly related to the optical path difference (*OPD*) introduced by the specimen over a limited range of about one third of a wavelength. For most cellular materials, an *OPD* of one third of a wavelength of green light corresponds closely to an areal density of non-aqueous matter of one picogram for each square micrometre of the specimen (Davies *et al.*, 1954). Thus the image can be interpreted directly and automatically as the distribution of dry matter in the specimen provided that this never exceeds an areal density of 1 pg μm^{-2}. Well spread fibroblasts in culture rarely exceed this areal density but it is a serious limitation when studying larger and/or less well spread cells.

In this paper we show how transmission microinterferometry can be adapted for phase-shifting which overcomes many of the residual problems of recording cell behaviour using microinterferometry. The techniques of phase-stepping and phase-shifting interferometry have only recently been introduced (Creath, 1988) but the theory of these methods is much older. In the image from any interferometer, the intensity at each point depends on three generally unknown quantities: the intensity of illumination; the modulation of the interference fringes and the phase difference between object and reference beams. It has long been appreciated that three intensity measurements taken at three different phase settings of the reference beam are necessary and sufficient to solve for these three unknowns but it is only since the introduction of solid state detector arrays and fast image processors that it is practical to perform this calculation on a point-by-point basis for a whole image. In phase-stepping or phase-shifting interferometry, three or more images are captured with known phase changes of the reference beam and an image representing any of the three unknowns may then be calculated. The image of greatest interest is usually the one representing the phase difference introduced by the specimen. After calculating this image, no further calibration is necessary and the intensity at each point in the image is an exact linear coding of the phase difference over the range of a whole wavelength.

Phase-stepping and phase-shifting interferometry were developed mainly for optical testing in industry but phase-shifting has now been incorporated into commercial reflection-mode microscopes, such as the Zygo Maxim-3D Laser Interferometric Microscope, for measuring surface profiles to an accuracy of better than one nanometre. Here we show that, when incorporated into a transmission interference microscope, phase-shifting can yield an almost ideal image which directly displays the distribution of material in the specimen. It has four main advantages over conventional interference microscopy. Firstly, the accuracy of measuring *OPD* is greatly improved, with no need to calibrate the images, since spatial and temporal variations in illumination or fringe contrast are automatically removed during calculation of the final image. Secondly, any drift in the phase setting of the microscope is easily compensated by adjusting the intensity of the final image to a constant background level. Thirdly, the range over which image brightness is linearly related to *OPD* is extended threefold to a whole wavelength. For green light, this corresponds to an areal density of dry mass of 3 pg μm^{-2} which is rarely exceeded even by large or rounded cells. Fourthly, the distribution of *OPD* in much thicker objects may usually be determined by further image processing; techniques have been devised to do this with conventional microinterferometry but these depend on interpolation between fringes and have an application only to specimens with gentle variations in phase such as sea-urchin eggs (Bereiter-Hahn, 1985).

Prerequisites for quantitative interference microscopy

In this paper we will briefly describe the modifications, ancillary equipment and software needed to implement phase-shifting on a Zeiss (Oberkochen) Universal microscope equipped with Jamin-Lebedeff interference optics. The chief disadvantage of the Jamin-Lebedeff system, which achieves isolation of object and reference beams by orthogonal polarisation, is the presence of a blurred and astigmatic secondary image or ghost image which is shifted laterally from the primary image by a distance equivalent to about one third of the field width of the particular objective in use. This severely restricts the use of the microscope on extended objects but is not too limiting for the study of isolated cells. In any case, for quantitative work with any form of transmission interference microscope, it is usually necessary that some area of the image plane is free from objects in order to serve as a reference background. Unfortunately, the Jamin-Lebedeff optics are no longer manufactured by Zeiss but many sets are still in good condition and much of what follows applies equally to any type of transmission interference microscope; only the modifications to the microscope itself may need to be changed.

It is important for conventional interference microscopy that the culture chamber should have very good optical properties and good dimensional precision and stability. While these criteria can be relaxed a little in the case of phase-shifting interferometry, they are nevertheless important if the full advantages of the method are to be realised. We use bacteria counting chambers of the Helber type which can be obtained without the usual rulings (Z3 special unruled) from the manufacturer (Weber Scientific International Ltd, 40 Udney Park Road, Teddington, Middx. TW11 9BG). These have a uniform and precise chamber depth of 20 μm when used with a rigid coverslip such as a No. 3 or a haemocytometer coverslip.

The recently introduced solid-state photodetector arrays are a major improvement over conventional video cameras for quantitative microinterferometry since they have a much lower geometrical distortion and better photometric properties. Miniature video cameras incorporating charge-coupled device (CCD) arrays of 756×581 picture elements (pixels) are now readily available (Sony XC-77CE). These have a sensitivity of 0.5 Lux and can be adapted for Peltier cooling in order to reduce image noise. For photometric work, the AGC should be switched off and the gamma set to 1.0 which is the factory presetting. In our system, the video signal from a Peltier-cooled XC-77CE is fed to a DT2867 (Data Translation Inc) frame grabber and processor board mounted in a PC AT compatible host computer. This board is well suited for phase-shifting since it has three frame-store buffers and can perform real-time frame averaging and two-frame arithmetic operations. It is also well matched to the camera having an image array of 768×512 square pixels. The histogram function provided in the basic software that comes with the board is indispensable for initially adjusting the microscope, camera and board controls so that the range of image intensities encountered during phase-shifting occupies the full range of grey levels without clipping.

Phase-shifting interferometry

Theory of phase-shifting interferometry

In the interference microscope used with monochromatic light the intensity, I, at each point in the image is given by the interference equation:

$$I = I_0\{1 + \gamma\cos(\phi + \phi_0)\} \tag{1}$$

where ϕ_0 is the background phase difference in radians between object and reference beams and this can be set to a known value by adjusting the microscope. The remaining three quantities on the r.h.s. are unknown: I_0, the dc intensity; γ, the modulation or contrast of the interference fringes and ϕ, the additional phase difference introduced by the specimen. The meaning of all these quantities is illustrated in Figure 1 which is a plot of I against ϕ over the range of 0 to 2π radians. In this figure, ϕ_0 is set to $7\pi/6$ radians which corresponds to its optimal setting for the DRIMAS method of microinterferometry

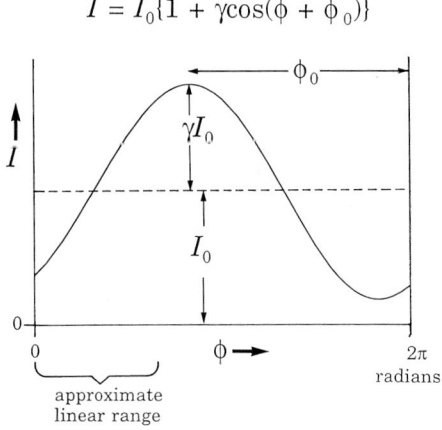

$$I = I_0\{1 + \gamma\cos(\phi + \phi_0)\}$$

Fig. 1. A plot of I against ϕ over the range of 0 to 2π radians. In this figure, ϕ_0 is set to $7\pi/6$ radians which corresponds to its optimal setting for the DRIMAS method of microinterferometry (see text).

(Brown and Dunn, 1989) and it can be seen that I is an approximately linearly increasing function of ø over the limited range of 0 to $2\pi/3$ radians. If it is known *a priori* that ø lies within this limited range, then the intensity of the image can be calibrated to yield the distribution of ø. The main advantages of phase-stepping are that it allows ø to be determined over the full range of 0 to 2π radians, with no need for calibration, and it is independent of any variation in I_0 or γ, either with time or across the image.

In the phase-stepping method three images are obtained by incrementing $ø_0$ in steps of $\pi/2$ starting at $\pi/4$:

$$I_1 = I_0\{1 + \gamma\cos(ø + \pi/4)\} \tag{2}$$

$$I_2 = I_0\{1 + \gamma\cos(ø + 3\pi/4)\} \tag{3}$$

$$I_3 = I_0\{1 + \gamma\cos(ø + 5\pi/4)\} \tag{4}$$

These equations can be solved for any of the three unknowns and the solution for ø, which is usually the only one required, can be obtained from:

$$\tan(ø) = (I_3 - I_2)/(I_1 - I_2) \tag{5}$$

However, using the standard arctangent function to solve this equation gives ø modulo π, i.e. as ø increases the values obtained "wrap round" so that they are always within the range $\pm\pi/2$. To obtain ø modulo 2π requires that we consider the signs of the numerator and denominator on the r.h.s. of Equation 5. This is done automatically by the two-argument arctangent function available in many computer languages. The final solution may be presented as a new digital image in which the grey level value, GV, is a strictly linear function of ø over the range 0 to 2π radians.

The OPD introduced by the specimen can be expressed as $(n + ø/2\pi)\lambda$ where n is an integer and λ is the wavelength of the light used. Thus, in the phase-stepped image, OPDs of 0.5, 1.5 or 2.5 wavelengths, for example, all give the same grey level because they all give a phase difference of π radians. This *phase ambiguity* is illustrated in Figure 2a which is a plot of GV against OPD over the range of 0 to 2 wavelengths. In contrast to Figure 2b, which is a similar plot for a final image obtained by the DRIMAS method, the image intensity is now a strictly linear function of ø over the range of 0 to 2π and so it is a sawtooth function of OPD instead of a truncated sinusoidal function. In the case of the sawtooth function, there is no ambiguity within the range of one wavelength and the phase ambiguity is easily resolved to give an extended range since the phase boundary usually appears as a sharp discontinuity in the image.

Phase-shifting or integrating-bucket interferometry is simply a variant of phase-stepping interferometry in which $ø_0$ is increased uniformly and continuously during the acquisition of the three images instead of being incremented between images. It has the advantage that the phase-modulating mechanism does not need to be repeatedly stopped and started which wastes time and can introduce vibration. The three images are now obtained by integrating the intensity over three time intervals or buckets during which $ø_0$ increases respectively from 0 to $\pi/2$, from $\pi/2$ to π and from π to $3\pi/2$ radians. The only effect of integrating over these quarter-wavelength buckets instead of stepping is that the fringe contrast or modulation, γ is reduced by 10%. The calculation of ø is unchanged.

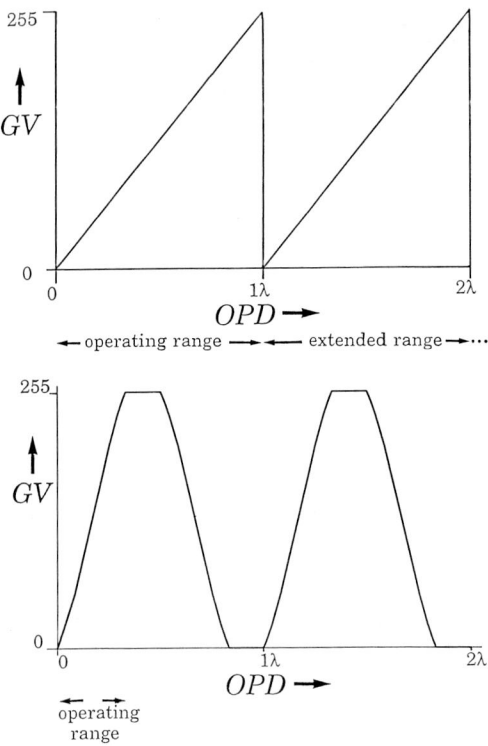

Fig. 2. A plot of the grey level value, *GV*, against *OPD* over the range 0 to 2 wavelengths for the phase-stepping method (2a) and the DRIMAS method (2b).

Very many variants of phase-stepping and phase-shifting have been proposed using different numbers of steps or of integration buckets and different step increments or integration intervals (Creath, 1988).

Implementing phase-shifting

A convenient way of implementing phase-shifting in the Jamin-Lebedeff microscope is Sénarmont compensation which consists of rotating the analyser in the presence of a specially oriented 1/4 wave retardation plate. The phase is shifted at a uniform rate through 2π radians for each half rotation of the analyser. We coupled a synchronous motor/gearbox combination to the main drive shaft of a Zeiss analyser after disassembling the control knob and vernier scale (Figure 3). One rotation of this drive shaft rotates the analyser 1/4 turn to give a phase shift of π radians. The position of the top lens of the analyser may need very careful adjustment (after loosening its three clamping screws) to ensure that the image does not move laterally as the analyser rotates. Care must also be taken to ensure that the sensitivity of the CCD camera does not depend on the analyser's orientation by avoiding beam splitters and anisotropic optical elements in coupling it to the microscope.

There are two considerations in choosing the speed of rotation of the motorised

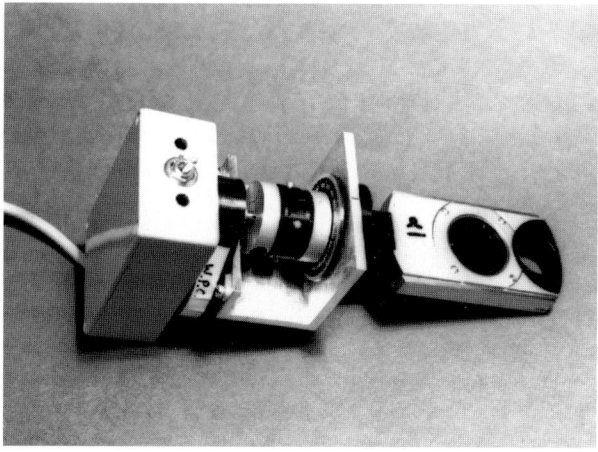

Fig. 3. A synchronous motor/gearbox combination coupled to the main drive shaft of a Zeiss analyser after disassembling the control knob and vernier scale.

analyser. Firstly, the time taken for the analyser to sweep through $\pi/2$ radians or one bucket should correspond to a whole number of video frames in order that the same number of frames occupy the same relative positions within each bucket. Video frames are captured at a rate of 25 per second and, with our system, the number of frames per bucket is 750 divided by the speed of the motor/gearbox in rpm. It is therefore necessary to select the speed so that this number is an integer. Secondly, the choice of speed should allow the integration of a sufficient number of frames to reduce image noise to an acceptable level while giving a sufficiently short exposure (total capture time) to freeze the motion of interest. The motor/gearbox we most commonly use runs at 50 rpm which gives 15 frames per bucket. It is not essential to capture and integrate all frames provided that the same number of frames in the same relative positions in each bucket are captured. We capture the first 11 frames in each bucket while the motor is running continuously. This gives a total capture time of 1.64 s and leaves sufficient time within each bucket for the necessary intermediate operations on the DT2867 board.

Before the beginning of each exposure, the host computer opens an automatic shutter, which protects the cells from unnecessary illumination between exposures, and starts the synchronous motor. The motor is allowed to run until the image background has just passed its maximum intensity before frame capturing begins. Assuming that the background occupies the greater part of the image, the background intensity can be determined as the modal value of grey level obtained from the DT2867 histogram function. This procedure ensures that ϕ_0 is set approximately to zero before the exposure begins even though the microscope phase setting may be drifting. It is not strictly necessary since any drift can be compensated later but it does facilitate monitoring of the images during recording. The position of the three buckets is illustrated in Figure 4 which shows how the intensity of the background changes as ϕ_0 is shifted. The synchronous motor is stopped at the end of the exposure and the shutter is closed.

There are three image buffers on the DT2867 board; two are 8-bit buffers (256 grey levels) and one is a 16-bit buffer (64k grey levels). During an exposure, the first 11 video frames in each bucket are averaged in the 16-bit buffer. The remaining 4 video frames of the first bucket allow plenty of time to transfer the image to an 8-bit buffer. In the case of the second bucket, however, three operations are performed during this final 4-frame period: the 16-bit buffer is copied to the second 8-bit buffer; the contents of the first 8-bit buffer are subtracted from the 16-bit buffer and the result is copied to the first 8-bit buffer which now contains the image difference $I_2 - I_1$. The last two operations are performed again during the third bucket using the second instead of the first 8-bit buffer which leaves $I_3 - I_2$ in this buffer. The time between exposures is used to transfer the two image differences from the 8-bit buffers into computer memory; to calculate the arctangent function using a pretabulated 2D-array of values and to store the final image on disc. The look-up table for arctangents is constructed so that the grey levels in the range 0-255 in the final image code for ø in the range 0 to 2π radians.

Figure 5 shows three images of a chick heart fibroblast captured during the three buckets and the final image calculated from this information. It can be seen that variations in background intensity, debris contaminating the camera faceplate, and vertical striping due to an electronic artefact have all been automatically removed from the final image. A lower magnification than this would normally be used to record a sequence of images for analysing cell behaviour. With our present system, the maximum rate of recording is 12 exposures per minute which generates data at the rate of about 300 Mbytes per hour. Sequences of images are conveniently stored in a single file using the TIFF 5.0 image format (Aldus/Microsoft Technical Memorandum: 8/8/88) which is compatible with a wide range of image processing and display software. Data compression of the whole file can reduce the large storage space required and we typically achieve a compression ratio of 5:1 using the SPLINT technique which is based on Huffman coding (Rikitake and Wakatabe, 1989, The National Public Domain Software Archive at Lancaster: micros/ibmpc/dos/e/e107).

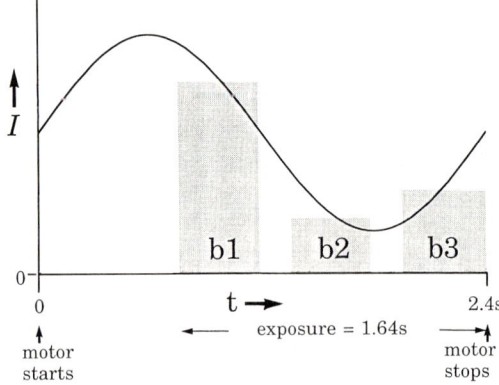

Fig. 4. A plot showing how the intensity of the background changes as the analyser rotates and illustrating the positions of the three integration buckets.

Comparison with other microscopy modes

Figure 6 shows three images of the same cell as in Figure 5 captured in the frame store using three other modes of image contrast: phase contrast, dark field and DIC. Note that the margin of the cell is generally much more prominent in these images than in the phase-shifted image of Figure 5d. Although often useful for visually discerning the margins of thinly spread cells, this edge enhancement can be misleading and the true density of material in the marginal regions is faithfully represented only in the phase-shifted image. But the main advantages of phase-shifting become apparent only when we try to analyse the images. Figure 7 shows the result of thresholding, in which all grey levels below a certain threshold level are set to zero, applied to all four modes of microscopy. In each case we have tried to achieve the optimal threshold level for enhancing the cell margin. This produces good results only in the case of the phase-stepped image shown in 7d. The residual noise surrounding the cell is mainly disconnected from it and can easily be removed by further image processing. In all the three other modes, large regions of the interior of the cell are below threshold and would need to be filled in order to achieve automatic recognition of the cell shape. The thresholded phase contrast image in 7a shows the cell's margin quite well but this is grossly exaggerated by the phase halo in the thicker parts of the margin whereas it is broken in the thinner parts. Sophisticated image processing is needed to retrieve even a

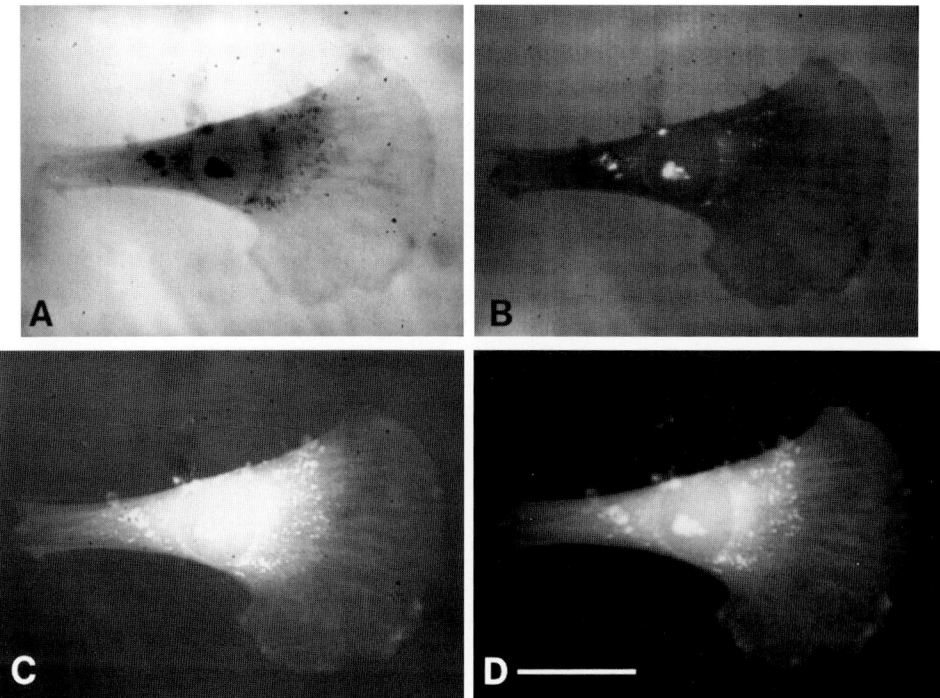

Fig. 5. Three images of a chick heart fibroblast captured during the three buckets (5a, 5b, 5c) and the final image calculated from this information (5d). Scale bar = 20 μm.

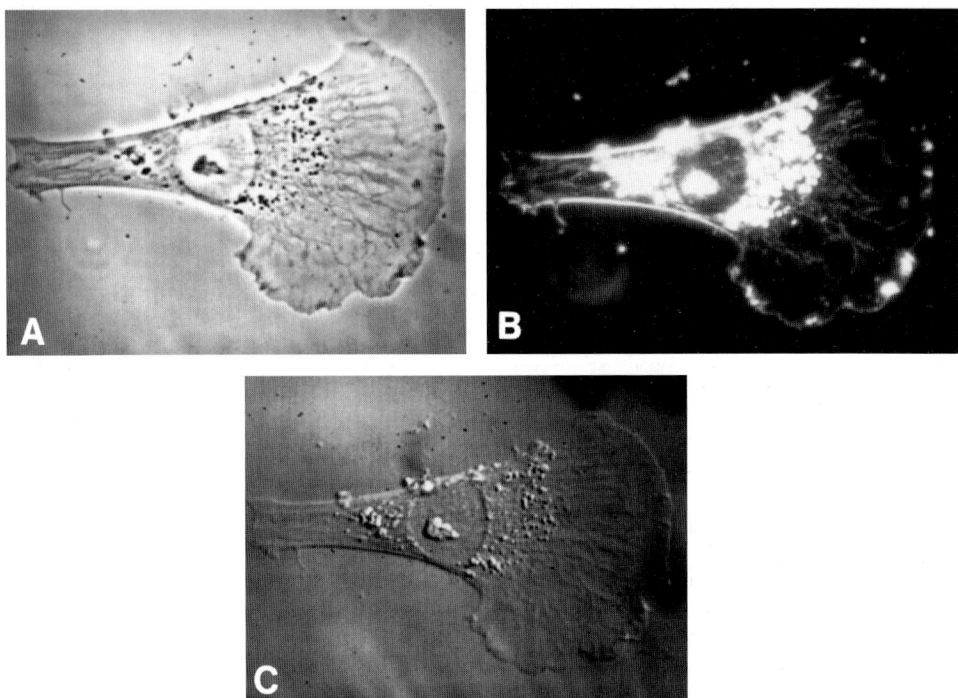

Fig. 6. Three images of the same cell as in Figure 5 captured in the frame store using three other modes of image contrast: phase contrast (6a), dark field (6b) and DIC (6c).

crude and unreliable approximation to the cell's true shape. The thresholded dark field image in 7b is rather better but the margin is still so broken that the same remarks apply. In the case of DIC it is quite impractical to retrieve the cell shape by thresholding as shown in 7c although it may be possible to achieve quite good results by integrating intensity along the shear direction.

Further image processing

An unusual feature of phase-shifted images is that the grey level scale can be regarded as a circular measure, with 0 and 255 being adjacent values, since the range of available grey level values represents the full range of phase angles. It is therefore possible, without loss of information about relative phases, to "rotate" the grey level values of all the pixels of an image using modular arithmetic. This consists of incrementing the values equally and taking the results modulo 256 (i.e. taking the remainders after dividing each result by 256) to give new values within the available range.

Removing spatial variations in background level

Spatial variations in background grey level due to non-uniform phase differences introduced by the microscope or specimen chamber may be dealt with by various techniques of *background subtraction* and returning the results modulo 256. Such

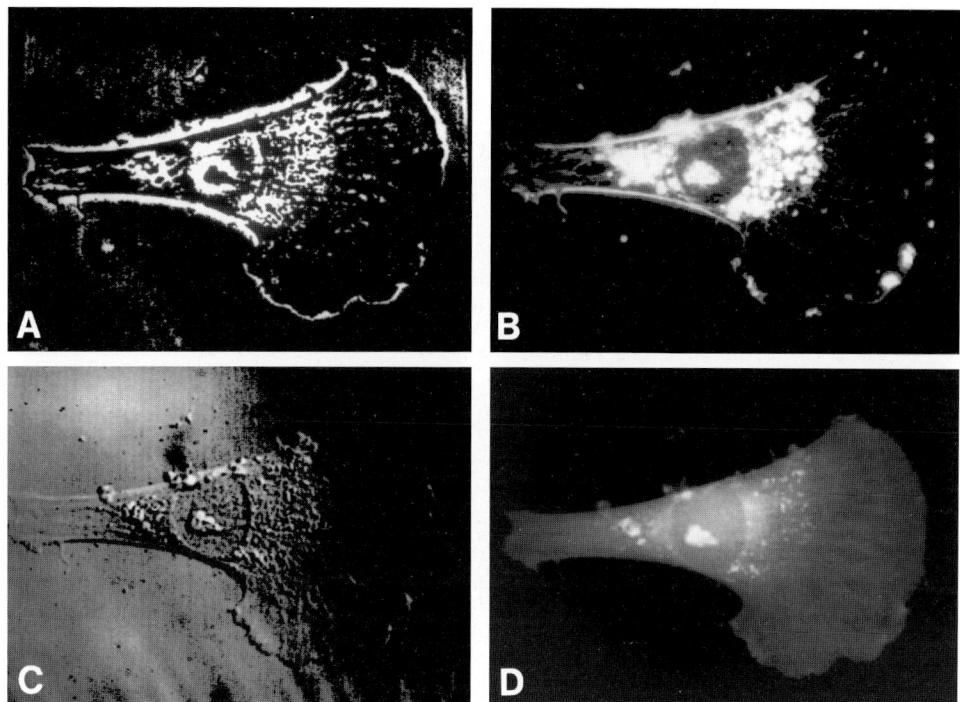

Fig. 7. The effect of thresholding, in which all grey levels below a certain threshold level are set to zero, applied to the phase contrast (7a), dark field (7b) and DIC (7c) images of Figure 6 and to the phase-shifted image of Figure 5d (7d).

variations must be removed before quantitative evaluation of the image but they are better avoided than remedied and the techniques will not be described here.

Removing temporal variations in background level

Adjusting the timing of exposures only crudely compensates for phase drift in the microscope (see '*Implementing phase-shifting*'). Residual fluctuations in the phase setting may be compensated by rotating the grey level values of each image so that the modal value of grey level is the same for all images. If the background occupies the greater part of the total image area and is reasonably uniform, the modal value is indicated by a large and sharply defined peak in the histogram of grey levels as shown in Figure 8. This value, GV^*, serves as the reference standard of zero phase. Prior to quantitative evaluation of an image, the grey level values should be rotated so that GV^* becomes zero. This method will compensate phase error to the nearest grey level but, if greater accuracy is required and the image quality is very good, GV^* may be estimated to a fraction of a grey level using curve-fitting techniques. After adjusting the image for the integer part of GV^*, a fractional adjustment may be made by adding 1 to the grey level values of a randomly selected fraction of the pixels.

After compensating the image, the background level will coincide with the phase boundary and image noise will have the effect that many of the background pixels will have grey levels at or close to 255 while the rest will have grey levels at or close to zero. Before quantitative analysis, this background noise must be removed by further processing as described in the next section. For displaying a sequence of images in the form of a movie without further processing, GV^* may be set to a constant level above zero to avoid the phase boundary as in Figure 8. Alternatively, a look-up table may be constructed for displaying the images in pseudocolour with no sudden discontinuity at the phase boundary.

Image segmentation

After setting GV^* to zero, the background noise may be removed by image segmentation. This serves to separate and to identify the non-background objects prior to quantitative analysis. Segmentation is essential if several cells or other objects occupy the image plane and the dry mass distribution of each must be analysed separately. Ideally, the grey levels of all pixels identified as belonging to the background or to contaminating objects should be given grey level values of zero while the objects of interest should be left unchanged. This is achieved firstly by thresholding the image so that very high grey level values and one or two of the lowest grey levels are given values of zero. The aim of this thresholding is to reduce the background noise sufficiently to leave it sparse and disconnected while not significantly eroding the margins of even thinly spread cells. The residual noise may then be removed by one of several algorithms available for scanning the image and identifying objects on the basis of connectivity of the non-zero pixels. Any cell in the image is then identifiable as an object with certain attributes, such as a large area, that distinguishes it from residual background noise and debris. The grey levels of pixels belonging to non-cellular objects may then be set to zero.

Resolving the phase ambiguity

If the *OPD* in any region of the specimen exceeds one wavelength, rotating the grey

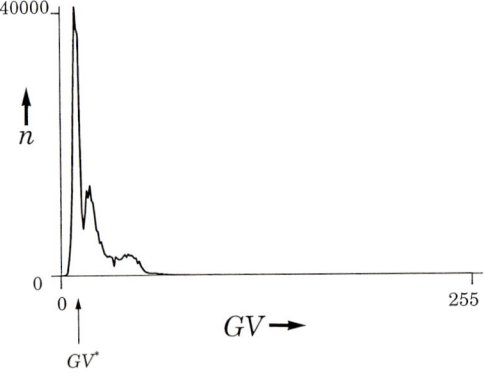

Fig. 8. A histogram of the number of pixels, n, with grey level values GV in the image of a fibroblast. Even though this cell occupies the larger part of the image area, the background grey level value, GV^*, is still represented by the tallest peak because the background is more uniform than the cell. The second peak represents the cell's lamellar regions and the dense perinuclear region gives rise to the hump on the right.

level values of its phase-shifted image will never prevent the phase boundary from appearing. If it is known that there are no sharp transitions in *OPD* within the specimen, the phase boundary is easily recognised as a sharp transition in grey level values. Resolving the phase ambiguity consists in transforming the image from one in which grey levels code for phase to one in which they code for *OPD*. The general strategy is to decrement or increment the grey level values of selected pixels by 256 until no two neighbouring pixels have a difference in grey levels of greater than 128. This requires extending the available range of grey levels or compressing their values into the available range. Since this technique is very rarely required for isolated cells in tissue culture, it will not be described further here.

Quantitative image analysis

Quantifying dry mass density

Quantitation of cellular material is based on the fact that the areal dry mass density in pg μm^{-2} is given by OPD/χ where the *OPD* is measured in μm and χ is measured in $\mu m^3 pg^{-1}$ and is 100 times the value of the *specific refraction increment* of the cellular material. Davies *et al.* (1954) state that χ is so similar for different cellular materials that the assumption of an average value of 0.18 will never result in an error greater than +/− 10% and usually far less. In our grey level coding, a difference in grey levels of 1 corresponds to a phase difference of $2\pi/256$. Provided there is no phase ambiguity, this corresponds to an *OPD* of 1/256 of a wavelength which, for green Hg light, is 0.546/256 μm and represents a difference in dry mass density of 0.0118 pg μm^{-2}. In an image that is fully corrected, regions with a grey level value of *GV* therefore signify a dry mass density of 0.0118*GV* pg μm^{-2}. The dry mass corresponding to a single pixel whose projected area in specimen space is A μm^2 is given by 0.0118*GV/A* pg which, with our system using a 40× objective, is only 740*GV* ag (ag=10^{-18} g).

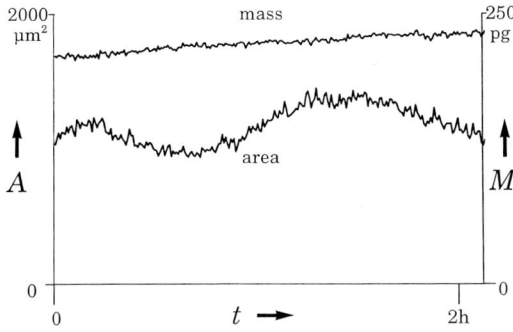

Fig. 9. The spread area and total dry mass of a chick heart fibroblast obtained from each of 256 phase-stepped images obtained over a total period 127 min. The error in relative mass between consecutive images is generally less than 2%, even though this cell is ruffling vigorously. It is clear that accurate estimates of the growth rate of an individual cell can be made over periods of 2 h or less.

Analysing dry mass distribution

In our laboratory, the analysis of these images usually starts by taking moments (Brown and Dunn, 1989; Dunn and Brown, 1990). These are two sequences of numbers, the areal moments and the mass moments, that can together describe the size, location, orientation, shape and mass distribution of a cell. Areal and mass moments of order zero give respectively the total spread area of a cell and its total mass. The position of the geometric centroid or of the true mass centroid may be obtained from the first order areal and mass moments. Moments of successively higher orders describe the cell's shape and mass distribution to any required degree of accuracy.

Figures 9 and 10 illustrate the consistency that may be achieved in determining the total mass and mass centroid location of a cell from a sequence of phase-shifted images. In Figure 9, the short-term fluctuations in area, caused mainly by the vigorous ruffling of the cell, are of the order of 10% whereas the short-term fluctuations in mass, caused probably by measurement error, are generally less than 2%. The increase in mass throughout the 2 h period is clearly discernible and, by filtering out the error noise, it appears that the method will allow accurate estimates of growth rate over periods as short as 1 h. In Figure 10, judging from the persistence in direction between consecutive displacements, it appears that translocations of the cell's centroid as small as 0.1 μm can be reliably detected and measured.

The phase-shifted images are also suitable for another form of analysis which is more appropriate for studying the dynamics of intracellular motility. Brown and Dunn (1989) showed that an estimate of the pattern of minimal intracellular flow required to account for a given change in dry mass distribution may be obtained by solving the Poisson field equation using *finite element analysis*. This analysis can result in images of a cell showing the velocity and direction of the minimal flow; the magnitude and direction of the mass flux; and the distribution of the kinetic energy of this minimal flow. Integrating kinetic energy over the whole cell gives a single measure of the non-steady state intracellular activity. This may prove to be as fundamental as measures of translocation in assessing

Fig. 10. A trajectory of the successive positions of the mass centroid of a chick heart fibroblast at 30 s intervals during a period of slow locomotion. Judging from the spacing of points in regions of persistent motility, it is clear that translocations as small as 0.1 μm can be reliably detected which is well below the resolution limit of the 40× objective used. Scale bar = 1 μm.

the function of motor proteins. The improved stability of phase-shifting is not a significant advantage for this form of analysis but the unrestricted range of *OPD* offered by phase-shifting will allow a much wider application to larger cells or other biological systems.

Conclusions

Since the final images acquired by phase-shifting are remarkably clean and easily interpreted in terms of cellular material, the method may find a wide application for qualitative microscopy if commercial instruments become available*. This simplicity of interpretation is essential for automatic image analysis by computer. Automatic image interpretation is especially important for the analysis of cell behaviour since large quantities of data must be gathered in order to estimate accurately even the most basic motile parameters such as speed or persistence of translocation of the cells. For quantifying dry mass, the sensitivity in terms of the smallest detectable change in dry mass of the specimen can be estimated from the fact that, with our system using a 40× objective, a change in brightness of one grey level over an area of one pixel corresponds to a change in mass of 7.4×10^{-16} g or 740 ag. This is the mass of approximately 900 molecules of myosin although it must be admitted that this estimate does not take account of noise and resolution limits of the microscope optics. Nevertheless, a sensitivity of this order or even better might well be achievable under ideal conditions using an immersion objective. The improved long-term stability introduced by phase-shifting opens up a new field of study since it enables the total dry mass of cells to be measured sufficiently accurately for studying the growth of individual cells.

The chief disadvantages of phase-shifting microscopy compared with most other forms of microscopy arise because the images are obtained by digital processing. Digital processing is much slower than the optical processing and/or analogue video signal processing of other contrasting modes and generally requires more expensive equipment. Nevertheless, the slow rate of image acquisition is not a serious disadvantage for the time-lapse recording of slowly moving objects such as fibroblasts in culture. If shorter exposures are required, the DT2867 is capable of capturing three successive video frames into the three buffers without doing any processing on board and a system could be designed with the absolute minimum exposure time of 0.12 s. For other applications in the future, it may eventually be possible to construct a Jamin-Lebedeff phase-shifted system that would operate at video rate by splitting the light path above the objective into three paths with different phase shifts and using three matched detector arrays. Some of the many forms of transmission interference microscope (see Krug *et al.*, 1964) may be more appropriate for combining with phase-shifting than Jamin-Lebedeff optics and we are currently experimenting with the Horn interference microscope which was

*Note added in proof:

Since submitting this paper, we have been informed by Mr Barry Leete of Carl Zeiss that they are currently developing in Jena a software package called Jenamap for implementing phase-stepping on the Jenapol interphako shearing-interference microscope. More information will be available shortly.

manufactured by Leitz and has the advantage of giving no ghost image. It is to be hoped that microscope manufacturers will be stimulated by the success of commercial reflection-mode, phase-shifting microscopes to consider introducing a new transmission-mode instrument for cell biology.

References

Bereiter-Hahn, J. (1985). Computer assisted microscope interferometry by image analysis of living cells. In *Advances in Microscopy. Progress in Clinical and Biological Research*, vol. 196 (ed. R. R. Cowden and F. W. Harrison), pp. 27-44. New York: Alan R. Liss.

Brown, A. F. and Dunn, G. A. (1989). Microinterferometry of the movement of dry matter in fibroblasts. *J. Cell Sci.* **92**, 379-389.

Creath, K. (1988) Phase-Measurement Interferometry Techniques. In *Progress in Optics*, vol. XXVI (ed. E. Wolf), pp. 349-393. Amsterdam: North Holland.

Davies, H. G., Wilkins, M. H. F., Chayen, J. and La Cour, L. F. (1954). The use of the interference microscope to determine dry mass in living cells and as a quantitative cytochemical method. *J. Microsc. Sci.*, **95**, 271-304.

Dunn, G.A. and Brown, A.F. (1990). Quantifying cellular shape using moment invariants. In *Biological Motion. Lecture Notes in Biomathematics*, vol. 89 (ed. W. Alt and G. Hoffman), pp. 10-34. Berlin: Springer-Verlag.

Krug, W., Rienitz, J. and Schulz, G. (1964). *Contributions to Interference Microscopy*. London: Hilger Watts Ltd, translated by J. Home Dickson from *Beitrage zur Interferenzmikroskopie*, Berlin: Akademie-Verlag.

Printed in Great Britain © The Society of Experimental Biology 1993   107

INTEGRINS: A REVIEW OF THEIR STRUCTURE AND MECHANISMS OF LIGAND BINDING

DANNY S. TUCKWELL, SUSAN A. WESTON and MARTIN J. HUMPHRIES

Department of Biochemistry and Molecular Biology, University of Manchester, Stopford Building, Oxford Road, Manchester, M13 9PT, U.K.

Summary

Adhesive interactions between cells and between cells and extracellular matrices play key roles in determining spatiotemporal positioning, influencing site-specific gene expression, and dictating proliferation rate. In addition, aberrant adhesion contributes to various aspects of disease pathology. These phenotypic effects of adhesion are mediated initially by the recognition of adhesive components of the extracellular matrix by membrane-intercalated receptor molecules and ultimately by the transduction of chemical and physical signals to the cell interior. Cell-cell and cell-matrix interactions are highly complex, since they involve the interfacing of surface membrane structures with each other or with three-dimensional aggregates of glycoproteins and proteoglycans, and it is this complexity that provides the necessary versatility for cells to react appropriately to either gross or subtle changes in their environment. Reagents with the ability to modulate adhesion could have many types of use: They could be employed to dissect the role of cell migration in development, provide insight into how adhesion might regulate gene expression and cell phenotype, and they could have widespread therapeutic applications in the treatment of thrombosis, inflammation and cancer. The quest to develop such reagents has necessitated the elucidation of the mechanisms of cell adhesion, and in particular the identification of the molecules involved and their modes of interaction. This article reviews the state of this quest; in particular, the molecular basis of ligand binding by integrin receptors.

Integrins

The integrin family

The integrins are a large group of transmembrane glycoproteins that bind to adhesive macromolecules on cell surfaces and in extracellular matrices. They are the major class of adhesion receptor and make a dominant contribution to the process of cell adhesion (for recent reviews, see Ruoslahti and Pierschbacher, 1987; Akiyama *et al.*, 1990; Hemler *et al.*, 1990; Hynes, 1992). Each receptor is a dimer of an unrelated α and β subunit, with at present 14 α subunits (α1-8, αM, αL, αX, αV, αIIb, and αIEL) and 8 β subunits (β1-8) having been identified in vertebrates. The family is still growing, but slowly.

In many cases, members of the integrin family were originally identified as antigens

Key words: integrin, ligand binding, cell-cell interaction, cell-matrix interaction.

recognised by anti-cell surface monoclonal antibodies (reviewed in Hynes, 1987). Two anti-β1 antibodies, CSAT and JG22, had the ability to block the attachment of avian cells to fibronectin (Greve and Gottlieb, 1982; Neff *et al.*, 1982; Chapman, 1984; Decker *et al.*, 1984), while other antibodies, now known to be directed against the α subunits complexing with β1 and β2, were useful reagents in studying the changes in T cell surface proteins occurring after activation and in distinguishing the various sub-classes of human and mouse leukocytes (Hemler *et al.*, 1990; Springer, 1990). In a complementary approach, ligand affinity chromatography was used to isolate adhesion receptors for molecules such as fibronectin, vitronectin, collagens and von Willebrand factor (Pytela *et al.*, 1985a, b; Giancotti *et al.*, 1986; Patel and Lodish, 1986; Cardarelli and Pierschbacher, 1987). In most cases, these studies identified heterodimeric proteins of subunit molecular weights 115,000-160,000. With subsequent cloning and sequencing of the cDNAs encoding these molecules came the realisation that they were related and the coining of the term "integrin" to describe the family (Tamkun *et al.*, 1986; Hynes, 1987). In the last five years, the techniques used for the initial characterisation of the integrins, namely the use of inhibitory antibodies and/or affinity chromatography, have been subsequently used for the isolation of novel integrins or the identification of novel integrin-ligand combinations (for example, Wayner and Carter, 1987; Mould *et al.*, 1990).

Although integins are αβ dimers, not all of the possible combinations of subunits exist. In addition, the range of dimers that can be formed depends on the particular subunit, with α1, α2, α3 and α5 only forming dimers with β1, and αIIb only with β3, while αV can interact with β1, β3, β5, β6 and β8 (combinations given in full in Humphries, 1990; Hynes, 1992). Ligand specificity also varies from integrin to integrin and includes both extracellular matrix and cell surface proteins, with some integrins only appearing to bind to a limited number of molecules, for example, α5β1, which binds to fibronectin only, while others are very promiscuous, for example, αIIbβ3, which can bind fibrinogen, fibronectin, von Willebrand factor, vitronectin and thrombospondin (Plow *et al.*, 1985). Generally, it appears that β1 and β3 integrins predominantly mediate cell-matrix interactions, with β1 integrins binding to connective tissue proteins such as collagens, laminin and fibronectin, and β3 integrins binding to vascular-associated molecules such as fibrinogen, vitronectin and von Willebrand factor, while β2 integrins participate in cell-cell interactions by binding to ligands such as complement protein C3bi, ICAM-1 and ICAM-2 (Humphries, 1990).

The tissue distribution of integrins is also very variable, with β1 integrins being almost universally expressed (for example, Zutter and Santoro, 1990) and β2 integrins being restricted to leukocytes (Springer, 1990). In addition, integrin expression can also be subject to developmental regulation (Bronner-Fraser *et al.*, 1992; Song *et al.*, 1992; Wadsworth *et al.*, 1992). Integrins have a widespread evolutionary distribution: they are well characterised in mammals and other vertebrates, but have also been demonstrated in *Drosophila* (Bogaert *et al.*, 1987; Leptin *et al.*, 1987), and are likely to be present in *Candida albicans* and *Caenorhabditis elegans* (Marcantonio and Hynes, 1988; Gustafson *et al.*, 1991). There is also indirect evidence that integrins are found in plant cells (Schindler *et al.*, 1989).

Integrins function as important components in a wide range of biological events. The binding of fibrinogen and von Willebrand factor by αIIbβ3 is instrumental in platelet aggregation and therefore blood clotting (Phillips *et al.*, 1991) since in patients with the αIIbβ3 disorder known as Glanzmann's thrombasthemia, platelet aggregation does not occur, and this leads to uncontrolled bleeding (Loftus *et al.*, 1990). Integrin-extracellular matrix interactions are important in cell migration, and peptides which block these interactions have been shown to inhibit platelet aggregation and experimental metastasis *in vivo* (Humphries *et al.*, 1986b). In addition, integrins appear to be involved in the ordering of the extracellular matrix, since anti-β1 antibodies are reported to inhibit the assembly of fibronectin by cells in culture (Fogerty *et al.*, 1990). Finally, via integrins, the extracellular matrix can alter gene expression (Werb *et al.*, 1989; Streuli *et al.*, 1991).

Integrin structure

The majority of integrin subunits are composed of a large extracellular domain, a transmembrane segment and a short cytoplasmic domain. Although there are no crystallographic data available for integrins, analysis of primary sequence data suggest that both the α and β subunits contain a number of identifiable domains (Fig. 1).

α-subunits

EF-hand domains: The N-terminal portion of integrin α subunits is made up of 7 repeating modules. These modules contain a sequence that shows homology to the well characterised cation-binding EF-hand domain (Kretsinger and Nockolds, 1973; Strynadka and James, 1989). In practice, the homology is very poor towards the N-terminal end and therefore this similarity is only really apparent in domains 4-7 (5-7 in some subunits). The presence of the EF-hand sequences correlates well with the known divalent cation-dependent function of the integrins (Gailit and Ruoslahti, 1988; Kirchhofer *et al.*, 1991) and the biophysical demonstration of multiple cation binding sites in these proteins (Rivas and Gonzales-Rodriguez, 1991; Gulino *et al.*, 1992). The EF-hand consensus broadly specifies a 13-residue sequence with coordinating side-chain carboxyls or hydroxyls and backbone carbonyls at positions 1, 3, 5, 7, 9 and 12 (Strynadka and James, 1989). In the integrin sequences, however, the important coordination site at position 12 is absent. Although this differentiates between the

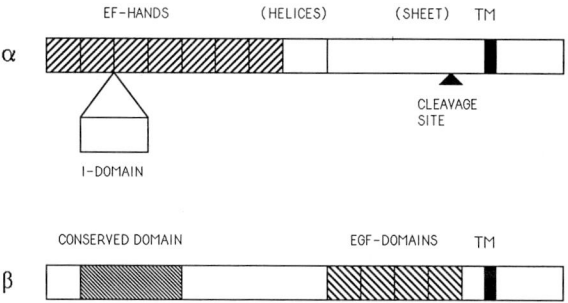

Fig. 1. Diagram to show the domains comprising the integrin α and β subunits. Putative domains are shown in brackets; details are provided in the text.

classical EF-hand proteins, such as parvalbumin and calmodulin, and the integrins, a functional cation binding site resembling that of the integrins has been found in an *E. coli* galactose-binding protein (GBP; Vyas *et al.*, 1987; Edwards *et al.*, 1988). A variety of evidence, reviewed below, suggests that the integrin EF-hand is an important binding site for integrin ligands. In view of the potential importance of these domains in integrin function, we have carried out a homology modelling study of the integrin EF-hand sequences. We were able to demonstrate their suitability as cation binding sites and to generate potential structures for them (Fig. 2; Tuckwell *et al.*, 1992).

I-Domains: Certain α subunits, α1, α2, αL, αM and αX, contain a distinct domain between EF-hand domains 2 and 3, known as the inserted- or I-domain (Corbi *et al.*, 1988; Pytela, 1988; Larson *et al.*, 1989). The role of the integrin I-domain is at present unknown although it is homologous in sequence to the A-domain of von Willebrand factor. This module is found in other proteins such as cartilage matrix protein, factor B and collagen VI where it has been implicated as an active site for the binding of collagen and complement molecules (Mole *et al.*, 1984; Bentley, 1986; Argraves *et al.*, 1987; Girma *et al.*, 1987). Since the I-domain is generally found in those subunits known to participate in binding to collagenous or complement proteins (Humphries, 1990; Hynes, 1992), it is conceivable, though unproven, that they might support the same function in integrins. Alternatively, they might simply serve as structural units.

Protease cleavage site: The region between the EF-hand domains and the transmembrane region does not contain any homology to known structural domains. However, the α3, α5, α6, α7, α8, αIIb and αV subunits are subject to a post-translational cleavage at a conserved site within this region to give two chains which remain covalently linked via a disulphide bond. On the basis of electron microscopical studies and protein secondary structure predictions, Nermut *et al.* (1988) have proposed that the region

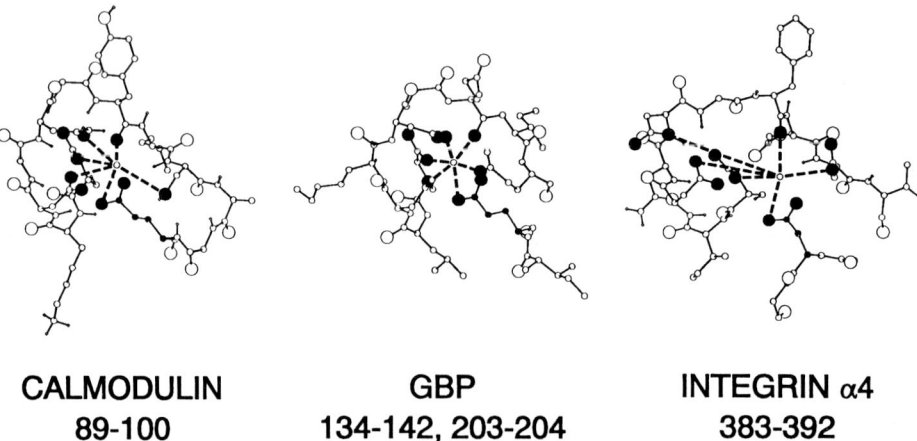

CALMODULIN   GBP   INTEGRIN α4
89-100   134-142, 203-204   383-392

Fig. 2. Diagrams of EF-hands from calmodulin and GBP and the homology modelled structure for integrin α4. Coordinating residues are shaded and the coordination to cation is shown by dashed lines. Experimental details are given in Tuckwell *et al.* (1992).

between the EF-hands and the cleavage site is composed of a short α helix-rich region followed by 12 β strands. The presence of the I-domain or the cleavage site appears to be mutually exclusive and corresponds with the division of the integrin α subunits into two broad groups on the basis of percentage amino acid similarity (Hynes, 1992). The α4 subunit does not have an I-domain and, although it possesses a cleavage site, this is at a different location to that in the other cleaved integrins, and the disulphide bond which keeps the cleaved chains together in the other integrins is not present in α4 (Takada *et al.*, 1989). Interestingly, this intermediate position between the two α-subunits groups fits with sequence comparisons (Hughes, 1992; Hynes, 1992).

Intracellular domains: The α subunit intracellular domain is approximately 30 amino acids long and, at least in some subunits, is subject to alternative splicing (α3, Tamura *et al.*, 1991; α6, Cooper *et al.*, 1991; Hogervorst *et al.*, 1991); this may have an important role in integrin function (see below). In addition, differential splicing of the extracellular domain has been reported, perhaps also modulating integrin function, with identification of an optional domain between EF-hand domains 3 and 4 of *Drosophila* αPS2 (Brown *et al.*, 1989) and of an alternatively spliced exon between the protease cleavage site and the transmembrane domain of αIIb (Bray *et al.*, 1990).

β subunits

β subunits are approximately 800 residues long, although β4 is longer because of its unusually large cytoplasmic tail (Fig. 1; Hogervorst *et al.*, 1990; Suzuki and Naitoh, 1990; Moyle *et al.*, 1991). The extracellular, N-terminal region contains a segment of about 200 residues which is highly conserved between subunits (Moyle *et al.*, 1991). This conservation, the experimental cross-linking of integrin ligands to this region (D'Souza *et al.*, 1988; Smith and Cheresh, 1988), and the presence in this area of mutations which lead to a loss of integrin function (Loftus *et al.*, 1990; Bajt *et al.*, 1992), make it likely that this region is somehow involved in the ligand binding process. The β subunit may also contain a divalent cation binding site (Rivas and Gonzales-Rodriguez, 1991) which could be within this region (Loftus *et al.*, 1990). Interestingly, although the β subunit is involved in the ligand binding process, ligands have been demonstrated to bind to the EF-hand containing region of the α subunit without any apparent requirement for other sequences (D'Souza *et al.*, 1991; Gulino *et al.*, 1992). The exact role of the β subunit in ligand binding is therefore unclear. The other well characterised section of the integrin β subunit is a cysteine-rich region in the C-terminal part of the extracellular domain, which is made up of four EGF-like domains (Nermut *et al.*, 1988; Yuan *et al.*, 1990).

Alternative splicing of β-subunit cytoplasmic domains has been observed for β1 (Altruda *et al.*, 1990; Languino and Ruoslahti, 1992), β3 (van Kuppevelt *et al.*, 1989) and β4 (Suzuki and Naitoh, 1990; Tamura *et al.*, 1990). In addition to splicing, proteolysis of the cytoplasmic domain also occurs in the case of β4 (Hemler *et al.*, 1989; Giancotti *et al.*, 1992). Alternative splicing of the β-subunit extracellular domains may also occur, since a putative splice variant of β7 in which most of the EGF-domains would not be present has been reported (Erle *et al.*, 1991). Finally, there are four fibronectin type III repeats in the cytoplasmic region of integrin β4 (Suzuki and Naitoh, 1990).

The integrin heterodimer

Association of α and β subunits to form an integrin heterodimer is divalent cation-dependent (Carrell *et al.*, 1985; Weisel *et al.*, 1992), and appears to be mediated predominantly by the extracellular domains of the molecules (Hayashi *et al.*, 1990; Frachet *et al.*, 1992). Electron microscopy studies have shown the heterodimer to have a globular head supported by two stalks which extend into the plasma membrane (Carrell *et al.*, 1985; Kelly *et al.*, 1987; Nermut *et al.*, 1988; Weisel *et al.*, 1992). Nermut *et al.* (1988) proposed that the α subunit contributes its EF-hand domains and the putative helical region to the head, and the β sheet region to one stalk. All of the N-terminal region of the β subunit up to the EGF-domains was proposed to form the rest of the head structure, with the EGF-repeats forming the β subunit stalk. This model for the integrin heterodimer implies that the head region is made up of the N-terminal regions, is the site of subunit contact and also contains the putative ligand binding domains. This is supported by data from a variety of approaches: first, electron microscopy studies have demonstrated ligand binding to the head region of the integrin (Weisel *et al.*, 1992); second, the N-terminal parts of α4 and αIIb are able to interact with intact β1 and β3, respectively, to form complexes that will support ligand binding (Pulido *et al.*, 1991; Lam, 1992); and third, an antibody directed against the central region of the β3 extracellular domain is reported to inhibit ligand binding to αIIbβ3 (Ramsamooj *et al.*, 1991). Some modification of the model may, however, be required, since cross-linking studies have shown that, at least under some circumstances, regions of the α and β subunits proposed to be in the stalk regions are proximal to the ligand binding site (Calvete *et al.*, 1992). Integrins may also require additional components for function since the disialoganglioside GD2 has been shown to be instrumental in human melanoma cell adhesion to fibronectin and vitronectin, and associates with the integrin molecule in a calcium-dependent fashion (Cheresh, 1989). The function of the ganglioside is, however, not yet known.

Modulation of integrin function: integrin activation states

Integrin-ligand binding is dependent on the presence of divalent cations, typically calcium or magnesium (Gailit and Ruoslahti, 1988; Kirchhofer *et al.*, 1991). There are, however, considerable differences in the effects that particular cations have on the function of individual integrins, for example, the αVβ1 interaction with fibronectin is supported by magnesium but not by calcium (Kirchhofer *et al.*, 1991) as is that of α2β1 with collagen (Santoro, 1986), while the interaction of αVβ3 with vitronectin is supported principally by calcium (Kirchhofer *et al.*, 1991). Different heterodimers therefore have different cation specificities, but different ligands for the same integrin can also require different cations, as in the case of α3β1, the interaction of which with collagen was found to be relatively insensitive to cation type, while its interaction with fibronectin was supported by magnesium but not calcium (Elices *et al.*, 1991).

So far, no clear relationship has emerged between integrin type, cation requirement and the ligand bound. It does, however, seem likely that cation specificities could at least be partly explained by subtle differences in the amino acid sequences and/or conformation of the α subunit EF-hand cation binding domains, since this has been demonstrated for the cation binding site of GBP, which resembles that of the integrins (Falke *et al.*, 1991).

In classical EF-hand proteins, cation binding leads to a conformational change enabling binding to a target protein, as in calmodulin (Strynadka and James, 1989), or regulation of its own function, as in calpain. Cation binding by the integrins also appears to be associated with conformational changes as an epitope recognised by a monoclonal antibody 24 has been described whose expression is induced in αLβ2 by binding of magnesium ions but not calcium ions. The expression of the 24 epitope coincides with the acquisition by the integrin of the ability to bind ligand (Dransfield and Hogg, 1989; Dransfield *et al.*, 1992a, b). Other epitopes, for example, those recognised by PMI-1 (Frelinger *et al.*, 1988), LIBS (Frelinger *et al.*, 1990), and PAC-1 (Shattil *et al.*, 1985), have also been characterised which report the conversion of αIIbβ3 from a resting to an activated, ligand-binding state. These data suggest that integrins can exist in active and inactive states, with activation (ability to bind ligand) being dependent on cations and mediated by a conformational change. This hypothesis is supported by the description of a number of antibodies with the ability to bind to and activate integrins. Interestingly, these antibodies can function by binding either to the α subunit (Keizer *et al.*, 1988) or to the β subunit (Caixia *et al.*, 1991; Arroyo *et al.*, 1992; Kovach *et al.*, 1992). Although probably physiologically irrelevant, manganese ions can substitute for calcium or magnesium ions in supporting ligand binding, and are in fact generally found to induce a greater level of binding (Dransfield *et al.*, 1990; Elices *et al.*, 1991). The mechanism by which manganese ions produce these enhanced effects and its relationship to integrin activation states are not yet understood.

The induction of the 24 epitope is sensitive, not only to cations, but also to reduced temperature and metabolic inhibitors (Dransfield and Hogg, 1989). This implies that integrin-ligand binding, and therefore activation state, can be controlled from within the cell. This level of control is observed in leukocyte and platelet activation, where ligand affinities can be increased without altering receptor numbers (Shimizu *et al.*, 1990a; Phillips *et al.*, 1991; Postigo *et al.*, 1991). It also appears that integrins can be activated by their ligands (Frelinger *et al.*, 1988; Du *et al.*, 1991), suggesting that activation state can be controlled by a number of factors.

Other integrin states have also been defined by the induction of the NKI-L16 epitope on αLβ2 (Keizer *et al.*, 1988). Absence of this epitope corresponds to a cation-free state and it is proposed that this corresponds to dispersed heterodimers, and that induction of the epitope, by calcium ions, leads to receptor clustering, a further integrin state (Figdor *et al.*, 1990). It has been suggested that clustering of receptors is required for stable cell-ligand interactions (Detmers *et al.*, 1987). Four states for αLβ2 have therefore been suggested by Dransfield *et al.* (1990); inactive, clustered, calcium-bound, inactive receptors, dispersed, magnesium-bound active receptors, and clustered, magnesium-bound active receptors, to which either intermediate state can convert. The actual details of integrin activation are, however, likely to be even more complex: An anti-β1 integrin Mab (TASC) has been developed which promotes adhesion to laminin and collagen while inhibiting adhesion to vitronectin (Neugebauer and Reichardt, 1991). Also, αIIbβ3 can be activated either by antibodies which promote its interaction with fibrinogen only or by antibodies which enable it to bind to fibrinogen and fibronectin (Frelinger *et al.*, 1991). Together, these observations suggest that multiple activation states may exist.

The ability of integrins to exist in active and inactive states is of considerable importance as it allows integrin-ligand binding to occur only when required, for example, the restriction of platelet activation and subsequent integrin-mediated aggregation to the site of vascular damage (Phillips *et al.*, 1991), or the control of T-cell activation and therefore of the integrin-mediated interactions of T-cells and antigen-presenting cells (Shimizu *et al.*, 1990b). Integrins can also be inactivated (for example, Adams and Watt, 1990) and it appears that a dynamic alternation between active and inactive states is likely to be central to integrin function (Dransfield *et al.*, 1992a). Not all integrins appear to require activation as some exist in a constitutively active state (for example, α2β1; Staatz *et al.*, 1991; Grzesiak *et al.*, 1992). However, even for constitutively active integrins, further subtle differences in activation state may explain the reported differences in ligand specificities for the same integrin from different sources, for example, the recognition of laminin by endothelial cell but not platelet α2β1 (Staatz *et al.*, 1991) or the currently unique recognition of laminin by α6β4 (Lee *et al.*, 1992). This phenomenon may also account for differences in the observed cation specificities of the same integrins in different systems (Dransfield *et al.*, 1990).

Integrin cytoplasmic domains

The cytoplasmic domains of the integrins represent potential sites for the modulation of integrin-ligand binding via intracellular mechanisms, as in activation, and for the relaying of ligand-binding events to cellular effector systems, for example, as occurs in migration or integrin-mediated changes in gene expression. The cytoplasmic domain of the β subunit is required for the localization of β1 integrins to focal contacts and for ligand binding (Hayashi *et al.*, 1990). A region close to the transmembrane domain appears to be important for this property, at least for localization of β1 integrins to focal contacts on fibronectin (Marcantonio *et al.*, 1990; Reszka *et al.*, 1992), and two other sections within the cytoplasmic domain also may be involved (Reszka *et al.*, 1992). These domains are probably important in other β subunits since a region close to the transmembrane region of the β2 subunit was found to be critical for the binding of αLβ2 to ICAM-1 (Hibbs *et al.*, 1991). The cytoplasmic domain of the β subunit also appears to be involved in integrin activation since expression of a mutant β3 subunit lacking almost all of the cytoplasmic domain generated a constitutively active form of αIIbβ3 (O'Toole *et al.*, 1991).

The cytoplasmic domain of the β subunit has been shown to interact with the actin cytoskeleton, consistent with its role in the formation of focal contacts. This association has been observed by immunohistochemistry (Larjava *et al.*, 1990), and there is biochemical evidence for the interaction of the cytoplasmic domain with talin (Horwitz *et al.*, 1986) and α-actinin (Otey *et al.*, 1990). In addition, the distinctive β4 cytoplasmic domain may interact with cytokeratins (intermediate filament cytoskeleton), a suggestion prompted by the localization of α6β4 in hemidesmosomes (Stepp *et al.*, 1990). The role of the β subunit cytoplasmic domain in integrin-ligand binding suggests that it may be involved in the transduction of signals between intracellular pathways and the extracellular domains of the integrin. In this respect, it is interesting that an alternatively spliced variant of the β1 cytoplasmic domain encodes a region with homology to a phosphotyrosine binding domain (Languino and Ruoslahti, 1992). Splice variants of other β subunits have also been reported (see above) as has tissue-specific proteolysis of the β4 cytoplasmic domain (Giancotti *et al.*, 1992). These may also be important in the modulation of integrin function.

The role of the cytoplasmic domain of the α subunit is less clear, but it may serve a modulatory role in regulating the association of the β subunit cytoplasmic domain with the cytoskeleton. Expression of C-terminal truncated mutants of αL had no effect on binding of αLβ2 to ICAM-1 (Hibbs *et al.*, 1991), whereas truncation of the cytoplasmic domain of αIIb increased the affinity of αIIbβ3 for ligands (O'Toole *et al.*, 1991). Different results were obtained by Chan *et al.* (1992) who studied chimeras between the extracellular domain of α2 and the intracellular domain of α2, α4 or α5. While cells containing any of the constructs bound to both collagen and laminin, the functional response to ligand binding differed. Cells containing the α2- and α5-cytoplasmic domain constructs supported collagen gel contraction but migrated poorly on collagen or laminin, whereas cells with the α4-cytoplasmic construct behaved in the opposite manner. It therefore appears that, at least under some circumstances, the cytoplasmic domain may control ligand binding and/or link ligand binding to an intracellular response. Alternative splicing of the α subunit cytoplasmic domains has also been observed (see above) and, although evidence is lacking, this is also likely to be important in the regulation of integrin function. It is not known if the α subunit cytoplasmic domain associates with the cytoskeleton. A 60 kD protein has been identified which binds to a conserved sequence near the transmembrane region of the α subunit, although it is not clear if it is involved in integrin function or if it has a role in integrin biosynthesis (Rojiani *et al.*, 1991). In a recent study using chimeras between the α5 or β1 cytoplasmic domains and the extracellular portion of the gp55 interleukin-2 receptor (LaFlamme *et al.*, 1992), it was observed that the β1 chimera, but not the α5 chimera, was competent to assemble at focal contacts. This ligand-independent association with the cytoskeleton suggest that the β1 cytoplasmic domain, when present without its α subunit partner, is constitutively active, and therefore that the α subunit domain may normally prevent cytoskeletal association. When the receptor is activated, either as a result of intracellular signalling or through ligand binding, it is conceivable that there is a reorganisation of the intracellular segments of the integrin leading to cytoskeletal interaction. In general, then, the cytoplasmic domains of the integrins appear to be involved in controlling integrin activation state and transmitting the ligand-binding event to intracellular components. It seems likely that the functions of the domains have some degree of subunit-specific function and this may account for the differences between the reports cited.

Integrins and signal transduction

The communication between integrin-ligand binding and the variety of cellular responses which ensue suggests that the integrins may be associated with second messenger-type signalling pathways. A number of second messengers used by receptors to transmit signals to intracellular effectors have now been characterised. In general, receptor-ligand binding can lead to the activation of GTP-binding and -hydrolysing proteins (G proteins) which can in turn activate, or inhibit, adenylate cyclase. The cyclic AMP produced in the case of activation can interact with protein kinase A thus leading to protein phosphorylation. G protein activation can also stimulate phospholipase C with the production of inositol phosphates and diacyl glycerol, potentially leading to an increase in intracellular calcium, and the production of active lipid metabolites or protein kinase C activation, respectively. Receptor activation can also lead to its own phosphorlyation or to phosphorylation of other proteins by tyrosine kinases, thus potentially affecting receptor function.

Integrins appear to be able to mobilise most, if not all, of these second messenger systems. There is evidence that G proteins are involved in cell spreading (Symons and Mitchison, 1992) and that spreading also requires a metabolite of diacylglycerol (Chun and Jacobson, 1992). The phosphorylation of specific proteins including the focal adhesion kinase, FAK, can also occur in response to ligand binding (Golden et al., 1990; Kornberg et al., 1991), although protein kinase C was found not to be involved in the spreading of HeLa cells on collagen (Chun and Jacobson, 1992), and phorbol esters, which are believed to activate protein kinase C, inhibited cell spreading of Xenopus fibroblasts (Symons and Mitchison, 1992). In addition to these second messengers, integrins are also able to produce a specific intracellular pH rise of 0.1-0.2 pH units on ligand binding (Schwartz et al., 1991). This may be significant since intracellular pH can act as a regulator of cell growth and differentiation.

In addition to the second messengers that integrins can generate, similar systems appear to be involved in the regulation of the integrins themselves, although these two aspects can overlap, as in cell spreading. Phorbol esters stimulate integrin-mediated adhesion in T-lymphocytes, macrophages and platelets (Dustin and Springer, 1989; van Kooyk et al., 1989; Shaw et al., 1990) and the activation of platelet fibrinogen receptors has been found to involve G proteins and phosphoinositides (Shattil et al., 1987). In addition, the degree of $\alpha M\beta 2$-mediated adhesion of leukocytes was found to correlate well with $\beta 2$ phosphorylation (though not with that of the α subunit; Valmu et al., 1991), and phosphorylation of $\alpha 6$, but not $\beta 1$, was observed in phorbol ester-activated macrophage adhesion to laminin (Shaw et al., 1990).

In some cases, the mechanisms by which the integrins regulate their own activity may have been elucidated. A lipid factor, IMF-1 (integrin modulating factor-1), which is produced by activated leukocytes and binds to $\alpha M\beta 2$ thereby increasing its ability to bind to its ligand C3bi has been isolated (Hermanowski-Vosatka et al., 1992). This factor may be a product of phospholipase activation. In addition, IAP (integrin associated protein), a potential effector protein for integrin-mediated events, has recently been characterised (Brown et al., 1990). Although not an integrin itself, monoclonal antibodies directed against this protein inhibit integrin-mediated phagocytosis by leukocytes. The protein appears to be a membrane channel or transport protein and its function may be under integrin control (E. Brown, personal communication).

Mechanisms of integrin-ligand binding

Strategies

A great deal of research effort has been concentrated on elucidating the mechanisms of adhesion at the molecular level. Specifically, this has entailed pinpointing the active sites in both sets of molecules that mediate ligand-receptor binding, and providing a three-dimensional description of the interactions. In recent years, the sub-domains of extracellular matrix macromolecules that confer adhesive activity have been mapped by progressive truncation. Large fragments of fibronectin, laminin and the interstitial collagens have been isolated and shown to retain similar levels of cell-binding activity to their parent molecules. These fragments have subsequently been dissected further to the point that in a number of instances adhesive activity can surprisingly be reproduced in the form of synthetic peptides. As described below, the ability to pinpoint adhesive activity to short, easily synthesised sequences has revolutionised both our view of how adhesion occurs *in vivo* and our ability to generate agents with therapeutic potential. However, adhesive peptides generally exhibit lower activity than their parent molecules and for some molecules it has not proven possible to describe their adhesive activity in the form of a linear sequence (including molecules which are known to have widely differing structures from those of most extracellular matrix glycoproteins such as collagens and members of the immunoglobulin superfamily). In the remainder of this article, we will assess the importance of established adhesive recognition sequences to the molecules from which they are derived, review their possible mechanisms of action and debate whether all adhesive events require such sequences. For this purpose, we will compare and contrast two classes of adhesive molecule, the non-collagenous and collagenous extracellular matrix glycoproteins.

Aspartate-containing peptide motifs

The RGDS sequence of fibronectin

The first cell adhesive sequence to be characterised was the Arg-Gly-Asp-Ser (RGDS) tetrapeptide found in the central cell-binding domain (CCBD) of the extracellular matrix glycoprotein fibronectin. A series of progressively shorter fragments from this region, culminating in a 108 amino acid, 11.5-kDa molecule, were found to be adhesive, although significantly less active than intact fibronectin (Pierschbacher *et al.*, 1981; Akiyama *et al.*, 1985). The 11.5-kDa fragment was reproduced synthetically in the form of a series of overlapping peptides, and ultimately, further peptide analyses identified the minimal active sequence as RGDS (Pierschbacher and Ruoslahti, 1984a; Yamada and Kennedy, 1984). As discussed below, other sites located outside the 11.5-kDa fragment are now known to function in combination with RGDS to generate the full adhesive activity of the CCBD and to determine the difference in activity between 11.5-kDa and fibronectin.

Although the fact that a simple tetrapeptide could account for a process as complex as adhesion was initally surprising, the specificity of RGDS peptides has since been established by testing a series of peptide homologues containing substitutions, deletions or inversions in each position. These studies revealed an absolute requirement for the first three residues, but a degree of flexibility in those amino acids tolerated in the fourth position (Pierschbacher and Ruoslahti, 1984a, b, 1987; Yamada and Kennedy, 1987). The functional specificity of RGD peptides has since been confirmed by the development by a number of pharmaceutical companies of non-proteinogenic analogues mimicking the peptide. These analogues retain the essential pharmacophore of RGD peptides (including appropriately spaced guanidino and carboxyl groups), yet contain very rigid hydrocarbon backbones. Refinement of the structure of these lead compounds should result in the production of therapeutically active anti-adhesive agents in the next few years.

In subsequent studies, other approaches have been used to demonstrate conclusively that the RGDS sequence satisfies all of the other experimental criteria for an adhesive active site (Table 1). Specifically, (i) monoclonal antibodies that bind close to the RGDS site perturb cell adhesion *in vitro* (Pierschbacher *et al.*, 1981; Nagai *et al.*, 1991), (ii) the inhibitory activity of RGDS peptides in cell adhesion assays can be overcome by an excess of intact fibronectin, i.e. it is competitive (Yamada and Kennedy, 1984; Akiyama and Yamada, 1985), (iii) RGDS peptide elutes the integrin $\alpha5\beta1$ from an affinity column derivatised with a CCBD proteolytic fragment (Pytela *et al.*, 1985a) and (iv) fibronectin-β-galactosidase fusion proteins mutated at the RGDS site exhibit greatly reduced adhesive activity (Obara *et al.*, 1988).

RGD sequences in other adhesion proteins

The core RGD tripeptide is found in many other adhesive glycoproteins and RGD-containing peptides have been found to block adhesion to collagens, laminin, fibrinogen, von Willebrand factor, vitronectin, tenascin, nidogen/entactin, thrombospondin, complement component iC3b, bone sialoprotein and osteopontin (reviewed in Humphries, 1990). Since these results could be explained by an indirect effect of the peptide, efforts have been made to test whether all RGD sequences satisfy the criteria in Table 1. For some proteins, RGD sequences are completely conserved

Table 1. *Experimental criteria for assessing the activity of adhesive recognition sequences*

- The sequence should be conserved between species at the same site in the protein

- The sequence should map to proteolytic or chemically-derived fragments that retain the cell adhesive activity of the parent protein

- The sequence may display adhesive activity when reproduced in the form of a synthetic peptide

- If a peptide containing the sequence is active, it should be a competitive inhibitor of the adhesive function of the parent protein and should competitively inhibit the binding of the protein to cells

- The peptide and its parent molecule should be recognised by the same receptor molecule

- Antibodies either raised against the peptide or that bind close to the peptide within the parent molecule should perturb the adhesive function of the sequence

- Mutagenesis of the sequence within the parent molecule should result in a loss of adhesive activity (for example human, mouse, pig and rat osteopontin), but in other cases the tripeptide is altered and presumably non-functional (for example the murine laminin A chain and chicken tenascin sequences). Since large peptides containing the flanking sequences of RGD sites do not retain the specificity found in the intact protein, it is not possible to examine the specificity of receptor binding to the peptide. However, RGD peptides have been shown to disrupt the binding of fibronectin, fibrinogen, thrombospondin and von Willebrand factor to αIIbβ3 and vitronectin to αVβ3 (Haverstick *et al.*, 1985; Plow *et al.*, 1985; Pytela *et al.*, 1985b). Although the adhesive activity of vitronectin, fibrinogen and tenascin has been localised to defined proteolytic fragments (Suzuki *et al.*, 1984; Friedlander *et al.*, 1988; Cheresh *et al.*, 1989), this truncation has not been pursued iteratively to the RGD tripeptide. A monoclonal antibody directed against the RGD site in the fibrinogen α-subunit blocks binding of fibrinogen to αVβ3, but not to αIIbβ3. The value of investigating the effects of mutation on putative active RGD sequences has been emphasised by two recent studies. First, alteration of RGD to RGE or RAD in von Willebrand factor resulted in a significant reduction in adhesive activity (Beacham *et al.*, 1992), and a similar mutation of RGD to AAA within complement component C3 was found to have no effect on αMβ2 binding (Taniguchi-Sidle and Isenman, 1992). Together with the other data described above, this suggests that the RGD in von Willebrand factor is probably functional, whereas that in C3 is not.

Other aspartate-containing adhesive recognition sequences

The successful strategy used for the identification of the fibronectin RGDS active site has now been applied to other proteins that are recognised by integrin receptors. As a result, four other short peptide sequences have been identified. Interestingly, three of these sequences possess a critical aspartate residue as a common functional element. The sequences are LDV, derived from the IIICS region of fibronectin and bound by α4β1, QAGDV, from the γ-chain of fibrinogen and bound by αIIbβ3, DGEA, from the α1 chain of type I collagen recognised by α2β1, and KQNCLSSRASFRGCVRNLRLSR, from the C-terminal end of the A chain of laminin which interacts with α3β1.

The IIICS region of fibronectin is a 120-amino acid-long, cell type-specific adhesive domain. Two active sites, represented by the 25- and 20-mer peptides CS1 and CS5, have been identified (Humphries *et al.*, 1986a, 1987). The minimal active sequence within CS5 is an RGD-like tetrapeptide, while CS1, which dominates the adhesive activity of the IIICS, has the tripeptide Leu-Asp-Val (LDV) as its minimal active site (Komoriya *et al.*, 1991). LDV shares many of the functional properties of RGD; substitution of E for D in either peptide greatly reduces activity and both peptides are competitive inhibitors of each other's function.

Both CS1 and CS5 bind to the integrin α4β1 (Wayner *et al.*, 1989; Guan and Hynes, 1990; Mould *et al.*, 1990, 1991), a receptor that is also reported to recognise site(s) in the adjacent HepII heparin-binding domain (Wayner *et al.*, 1989). A sequence related to LDV has been reported to account for this recognition (Mould and Humphries, 1991). Interestingly, this region is homologous to CS1 and contains as its minimal active sequence the pentapeptide IDA(PS). The LDV and IDA sites are completely conserved in human, rat, bovine and chicken fibronectins. However, to date, neither antibody blocking nor mutagenesis experiments have been performed for either site.

The QAGDV cell-binding site is located at the extreme COOH-terminus of the γ-chain of fibrinogen. A variety of techniques including antibody blocking and, most recently, electron microscopic visualisation of ligand-receptor complexes, suggest that this particular subunit dominates the adhesive activity of fibrinogen. As for other glycoproteins, the γ-chain active site was originally identified by chemical and proteolytic cleavage, and narrowed down to its minimal sequence through the use of synthetic peptides (Kloczewiak *et al.*, 1982, 1984). Although the pentapeptide QAGDV is probably the minimal active site, this is much less active than a dodecapeptide, HHLGGAKQAGDV (Kloczewiak *et al.*, 1984). This peptide competitively blocks the RGD-dependent binding of fibronectin and von Willebrand factor to platelets, indicating a possible shared mechanism of interaction of these sequences with the cell surface (Kloczewiak *et al.*, 1984), a finding reminiscent of the similar relationship between RGD and LDV described above.

The latest integrin-binding peptide to be identified, and the first to lack an aspartate residue, is a 22-mer termed GD-6 (primary sequence KQNCLSSRA-SFRGCVRNLRLSR). This peptide, which was examined initially because of its high content of hydrophilic amino acids, supports cell adhesion in a manner which is blocked by anti-α3 and anti-β1 antibodies, binds α3β1 in affinity chromatography experiments, displaces α3β1 from laminin affinity columns and blocks adhesion to laminin (Gehlsen *et al.*, 1992b). In addition, polyclonal antisera that are directed against the peptide at least partially cross-react with intact laminin and again block its adhesive activity. The adhesive activity that can be demonstrated for GD-6 suggests that it should be possible to observe similar levels of adhesion in proteolytic fragments from the extreme C-terminus of the laminin A chain. This part of the laminin molecule does not contain a contribution from either B chain and therefore, although mutagenesis of laminin is usually hampered by the trimeric structure of the molecule, in this case it should be possible to test the functionality of GD-6 within the context of its parent protein using this approach.

In addition to those recognition sequences that contain critical aspartate residues, a number of other peptides have been reported to mediate non-integrin-dependent cell adhesion (reviewed by Humphries, 1992). With few exceptions, however, these peptides are not as well characterised as the integrin-binding sequences described above and it is possible that the precedent set by RGD sequences may have been slightly misleading. Caution should therefore be employed in extrapolating the significance of these studies to too many other adhesion systems. There is no reason to suppose that all receptor-ligand interactions will involve the recognition of short peptides and, for those sequences that have been characterised, synthetic mimics are generally less active than the molecules from which they are derived. One explanation for this could be a strict dependence of ligand-receptor binding on the three-dimensional conformation of the active site peptide, a possibility supported by data from studies that have tested the activities of RGD analogues differing in their flanking residues or that are restricted in their conformation through cyclisation (Pierschbacher and Ruoslahti, 1987; Aumailley *et al.*, 1991). Alternatively, adhesive proteins may possess synergistic domains that act in concert with aspartate sequences to generate the full affinity and specificity of receptor binding. Fibronectin, thrombospondin, tenascin/cytotactin and von Willebrand factor have each been shown to possess cell adhesion domains distinct from their RGD sites. In the case of fibronectin, extensive deletion mutagenesis and mapping and functional testing of a range of monoclonal antibodies has pinpointed a region that contributes to the activity of the CCBD (Obara *et al.*, 1988; Aota *et al.*, 1991; Nagai *et al.*, 1991). This site cannot be described by a short, linear array of amino acids and probably relies on precise folding of non-contiguous regions of the polypeptide backbone for full activity. A further possibility is that the use of short peptide sequences may only be a feature of a subset of adhesive macromolecules and that others employ more conventional binding mechanisms involving amino acid residues separated in terms of sequence if not space. Candidates for this type of binding would be the immunoglobulin-based ligands such as ICAM-1 and -2 and VCAM-1. Current evidence from mutagenesis and antibody blocking suggests that one face of the ICAM-1 molecule is bound by its integrin receptor $\alpha L\beta 2$ rather than a limited number of short peptide sequences (Staunton *et al.*, 1990).

Integrin-collagen interactions

The structural feature common to all collagens is the presence of at least one stretch of polypeptide containing a repeating Gly-X-Y triplet motif. All collagens are trimeric molecules and the domains containing these triplets assemble into a triple helical coil, with the Gly hydrogen side-chains being directed towards the centre of the helix. Owing to the unique structure of the collagen triple helix, it would not be anticipated that integrins would employ a similar mechanism for binding as that used for non-collagenous glycoproteins. Determination of the sites within collagens that interact with integrins offers, therefore, an important insight into how cells adhere to extracellular matrices.

Although integrins do indeed bind to the collagen triple helix, there appears to be considerable diversity both in terms of the integrin receptors that are active and the ligand specificity they exhibit. Members of the $\beta 1$ and $\beta 3$ integrin subfamilies have been reported to bind various collagen types (reviewed by Hynes, 1992). In the case of the $\beta 1$ subfamily, $\alpha 1\beta 1$, $\alpha 2\beta 1$ and $\alpha 3\beta 1$ are active. However, it should be noted that these three receptors also interact with laminin and that $\alpha 3\beta 1$ binds to fibronectin. A single member of the $\beta 3$ subfamily, $\alpha V\beta 3$, binds collagens and like the $\beta 1$ integrins, it also has multiple ligands which include vitronectin, fibrinogen and fibronectin (reviewed by Humphries, 1990).

$\alpha 2\beta 1$-collagen interactions in platelet aggregation

Fibrillar collagen is one of the most thrombogenic vessel wall macromolecules. Following vascular injury, blood platelets rapidly adhere to exposed collagen and a cascade of events leads to the eventual formation of a hemostatic platelet plug (Baumgartner 1977; Hawiger, 1987). Although several platelet surface molecules have been proposed to mediate platelet-collagen adhesion, a key contribution is made by the glycoprotein Ia-IIa, now known as $\alpha 2\beta 1$ (Santoro, 1986; Wayner and Carter, 1987; Santoro *et al.*, 1988; Kunicki *et al.*, 1988; Coller *et al.*, 1989).

Platelets adhere to a variety of collagens including native types I, II, III, IV, and VI

(Staatz *et.al.*,1989; Wayner and Carter, 1987). This finding suggests that platelets recognise something common to all collagens and implies that this may be one or more sites within the triple helix. However, further investigations (Morton *et al.*, 1989; Staatz *et al.*, 1990a) have revealed that both intact platelets and liposomes containing the purified $\alpha2\beta1$ receptor exhibit similar attachment to both native and heat-denatured type I collagen in a Mg2+-dependent manner. This observation raised the possibility that such adhesion could be independent of triple helical conformation and permitted similar approaches to be employed to those used for non-collagenous glycoproteins to narrow down the sites involved.

As would be predicted from the finding that heat-denatured collagen type I supported platelet adhesion, purified and separated $\alpha1(I)$ and $\alpha2(I)$ chains were found to exhibit similar activity (Staatz *et al.*, 1990b). Following this, chemical cleavage with cyanogen bromide was used to prepare fragments of the $\alpha1(I)$ chain. Only one fragment, $\alpha1(I)$-CB3, supported Mg2+-dependent platelet adhesion. This activity was blocked by the anti-$\alpha2$ monoclonal antibody P1H5. Next, through the use of an overlapping series of synthetic peptides, a minimal recognition sequence responsible for $\alpha2\beta1$ binding has been identified. This is the tetrapeptide Asp-Gly-Glu-Ala (DGEA) corresponding to residues 435-438 of the $\alpha1(I)$ sequence. DGEA-containing peptides were able to specifically inhibit platelet adhesion to collagen, but had little or no inhibitory effect on either $\alpha5\beta1$-mediated adhesion of platelets to fibronectin or $\alpha6\beta1$-mediated platelet adhesion to laminin. In contrast, DGEA-containing peptides were also found to inhibit adhesion of T47D breast adenocarcinoma cells to both collagen and laminin. This process is known to be mediated by $\alpha2\beta1$ and suggests, first, that DGEA may serve as a recognition site for cells other than platelets and second, that blockage of the collagen-binding site in $\alpha2\beta1$ may prevent receptor interaction with other ligands.

Interestingly, structure-function analyses of the DGEA sequence have revealed some similarities with other integrin-binding peptides. Specifically, replacement of the aspartate residue with an alanine abrogates activity, suggesting that this sequence may function in a similar manner to RGD, LDV, etc. (see below). However, although DGEA was identified by progressive truncation of the type I collagen molecule, there are still many experimental criteria that remain to be tested before it can be fully established as an adhesive recognition sequence. First, it has not been shown that $\alpha2\beta1$ is able to bind the DGEA-containing $\alpha1(I)$ CB3 fragment when immobilised on an affinity matrix. If this binding were to occur, then elution of bound receptor with a peptide containing the DGEA sequence could be tested. Second, antibodies directed against a DGEA-containing synthetic peptide and reacting with the native collagen molecule should be able to inhibit $\alpha2\beta1$-mediated adhesion. Third, mutagenesis of the DGEA sequence within type I collagen should affect its adhesive activity. Although this experiment is complicated by the heterotrimeric structure of type I collagen, it would be a definitive approach to confirm whether this recognition determinant is active in the parent molecule.

Interactions of $\alpha1\beta1$ and $\alpha2\beta1$ with basement membrane collagen type IV

Type IV collagen is the major collagenous constituent of basement membranes and most frequently consists of two α1(IV) and one α2(IV) chains (Brazel *et al.*, 1988). In the extracellular matrix, type IV collagen forms a complex non-fibrillar network, through cross-linking of its N- and C-termini (Timpl *et al.*, 1981). This network acts as a mechanically stable support for cells of the basement membrane zone (Timpl and Dziadek, 1986). Collagen type IV displays a broad spectrum of reactivity. There are a substantial number of reports in the literature which show, *in vitro*, that type IV collagen is a highly specific substrate which interacts with endothelial (Palotie *et al.*, 1983b), epidermal (Murray *et al.*, 1979) and mesenchymal cells (Aumailley and Timpl, 1986) and certain transformed and metastatic tumor cells derived from these lineages (Dennis *et al.*, 1982; Engvall *et al.*, 1982; Vlodavsky *et al.*, 1982; Palotie *et al.*, 1983a; Aumailley and Timpl, 1986).

Consistent with the β1 integrin subfamily interacting with components of the ECM, a number of reports (Marcantonio and Hynes, 1988; Kramer and Marks, 1989; Hemler *et al.*, 1990) have identified members of this family as being active in interactions with type IV collagen. Aumailley and Timpl (1986) demonstrated that cell binding sites within type IV collagen are conformation-dependent and have been ascribed to the triple helical domain of the molecule. This contrasts the situation described above for type I collagen where similar binding to both the native and denatured molecule was reported.

Vandenberg and colleagues (Vandenberg *et al.*, 1991) have recently identified a major cell binding site within type IV collagen. Following cyanogen-bromide fragmentation, under conditions which conserved the triple helical conformation via interchain disulphide bonds, a trimeric fragment, designated as CB3, was found to contain a major cell binding site. Cell attachment assays, performed in the presence of either native type IV collagen or its trimeric CB3 fragment showed similar cell binding activities. A polyclonal antibody raised against the CB3 fragment was capable of totally inhibiting cell attachment to the CB3 fragment and also inhibited attachment to the parent type IV collagen by 70-80%. Subsequent digestion of CB3 with trypsin produced four subfragments. The longest of these fragments, designated fragment 1, a trimer of 150 amino acids which comprised almost the entire triple helical part of CB3, was found to display similar attachment capacity to that observed with CB3 itself. By contrast, fragments 2, 3, and 4, which were truncated versions of fragment 1, were of low activity suggesting that sequences at both the N- and C-termini of fragment 1 were essential for cell binding. Although it is possible that the CB3 fragment only retains activity because of its inter-chain disulphide bonds and that the anti-CB3 antibody cross-reacts with other parts of the helix, these studies suggest that integrin recognition of type IV collagen is localised to a specific region of the triple helix.

The cell surface receptors responsible for binding to type IV collagen were isolated by affinity chromatography on Sepharose-immobilised CB3. Two β1 integrin species were identified, namely α1β1 and α2β1, defined on the basis of N-terminal sequencing and immunoblotting. However, it remains to be demonstrated if these integrin receptors also interect with tryptic fragment 1, the minimal triple-helical subfragment of CB3 which supports cell attachment.

RGD-dependent integrin-collagen interactions.

As a ubiquitous component of the extracellular matrix, collagen type VI has been implicated as one of the major matrix proteins involved in cell binding (Carter, 1982a, b; Von der Mark *et al.*, 1984; Bruns *et al.*, 1986). In the extracellular matrix, collagen type VI, composed of α1, α2 and α3 chains, is assembled into its characteristic microfibrillar structure made up of disulphide-linked, dumb-bell shaped monomers consisting of two globular domains and a 100 nm triple helix. Aumailley *et al.* (1989) have described the cell attachment properties of collagen type VI with particular reference to the RGD-dependent binding of cells to the α2(VI) and α3(VI) chains. They demonstrated that a number of cell types, including fibroblasts and tumor cells, were able to attach and spread on intact and pepsin-solubilised collagen type VI. The structural requirements for cellular interactions with collagen type VI were ascribed to the triple helical domain of the molecule. Cell attachment and spreading were promoted by this domain in both its intact and denatured forms, albeit via different receptors. Native type VI collagen was bound by α1β1 and α2β1, while the denatured molecule interacted with αVβ3. This latter finding is consistent with the preferential recognition of open, flexible RGD sequences by αVβ3. Figure 3 is a representation of the possible conformation of RGD sequences within the context of a triple helix. Even though there would be three such sequences in a homotrimeric molecule, the inter-sidechain distance between any pair of guanidino and carboxyl groups is generally less than that thought to represent an active RGD pharmacophore (approx. 14 Å). This suggests that helical RGD would not efficiently fold into an integrin-binding conformation.

Adhesive activity was attributed to the α3(VI) and α2(VI) segments but not the α1(VI) segment. Furthermore, RGD, but not RGE, containing synthetic peptides completely inhibited cell attachment to the individual chains within a concentration range of 10-100 μM. This finding was consistent with those already reported for fibronectin, vitronectin and collagen type I (Pierschbacher and Ruoslahti, 1984b;

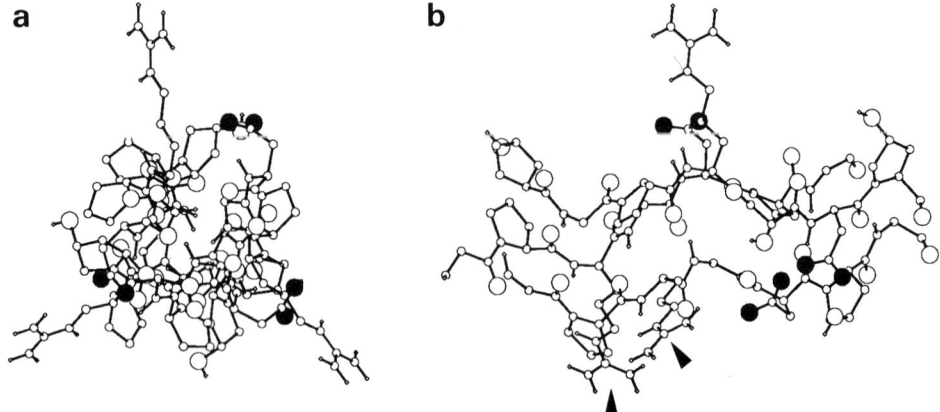

Fig. 3. Representation of the structure of an RGD peptide within a collagen triple helix shown (a) along the axis of the helix and (b) at right angles to the helix. The arginine side-chains are marked with arrow heads and carboxyl oxygens of aspartate side-chains are shaded.

Ginsberg *et al.*, 1985; Hayman *et al.*, 1985; Yamada and Kennedy, 1987). There are a total of 11 RGD sequences in the triple helical domains of the three chains of type VI collagen, with the active vitronectin (RGDV) and fibronectin (RGDS) cell binding sequences being present in the α2(VI) and/or α3(VI) chains. In addition, α3(VI) shows a cluster of three RGD sequences within 20 amino acids which may explain the high attachment efficiency to this fragment compared to α2(VI) and α1(VI). In contrast, binding of cells to triple helical collagen type VI was resistant to RGD peptide inhibition. This observation agrees with previous reports which showed that cell attachment to intact collagen type VI was RGD-independent (Wayner and Carter, 1987). Several explanations were offered in interpreting this cryptic activity of RGD sequences within collagen type VI. These included the possibility that recognition of triple helical type VI could involve at least two synergistic binding sites, only one of which is RGD-dependent or alternatively, that the RGD-dependent sites within the α2(VI) and α3(VI) chains may not be available in the triple helical structure. Since both chains induce significant cell spreading which involves cytoskeletal rearrangement and is a phenomenon usually considered to reflect relevant cell-matrix interactions (Woods and Couchman, 1988), it was suggested that cell binding to collagen type VI chains involves receptors linking cytoskeletal to extracellular ligands. Integrins are considered to be favourable target receptors for such interactions.

Additional cryptic RGD recognition sites have recently been described (Davis, 1992). In these studies the adherence of human melanoma cells to denatured versus native type I collagen was found to be mediated through different integrins. There are several published reports showing that β1 integrins bind to intact collagen type I in a manner which is RGD-independent (Wayner and Carter, 1987; Gullberg *et al.*, 1989; Kramer and Marks, 1989; Hall *et al.*, 1990; Staatz *et al.*, 1991). By contrast, αVβ3 has been shown to interact with denatured type I collagen via an RGD-dependent mechanism while exhibiting only weak binding to native type I collagen. Results of affinity chromatography experiments showed that α2β1 bound to native collagen I-Sepharose while showing minimal binding to denatured collagen I-Sepharose. This is in direct contrast with what was observed for αVβ3 which bound to denatured collagen I-sepharose and could be eluted with RGD. However, low but detectable levels of αVβ3 were also observed in eluates from the native collagen I column. These data were further confirmed by immunoprecipitation experiments which employed anti-integrin antibodies as well as cell adhesion and spreading assays. Without exception, binding of αvβ3 was found to be restricted to denatured collagen type I and the interaction was again demonstrated to be RGD-dependent. Given this information, Davis (1992) came to the conclusion that the ability of melanoma cells to recognise native versus denatured type I collagen via distinct integrin receptors could have an important functional role. For example, damaged ECM may present new adhesive signals to cells which are in some way masked when the matrix is intact. This hypothesis is supported by recent evidence associating increased αVβ3 expression with the progression of melanoma cells to a more malignant and metastatic state (Albelda *et al.*, 1990; Gehlsen *et al.*, 1992a). It is further suggested that such cells may cause local damage to the ECM which through the release of proteases could expose RGD sites in matrix proteins, such as collagen, with which αVβ3 could interact. Such a possibility underpins the possible importance of this integrin in wound repair as well as tumor invasion and metastasis.

Structural basis of ligand-receptor binding

Having identified short peptide sequences within adhesive ligands that are involved in receptor binding, a logical progression has been to identify the sites on the receptor molecules interacting with these sequences. Chemical cross-linking studies have localized binding sites for both the RGD and HHLGGAKQAGDV peptides on αVβ3 and αIIbβ3 (D'Souza et al., 1988, 1990; Smith and Cheresh, 1988, 1990). In the case of the RGD peptides, binding sites have been found on both α and β subunits. In the α subunits, the sites were close to EF-hand repeats (D'Souza et al., 1990; Smith and Cheresh, 1990). However, interestingly, one β subunit binding site was also homologous to an EF-hand structure, although the sequence did not fit the consensus as well as the sites in the α subunits (D'Souza et al., 1988; Smith and Cheresh, 1988). An additional β subunit binding site has been identified using a peptide-based approach. This site encompasses residues 211-222 (Charo et al., 1991). Cross-linking with HHLGGAKQAGDV peptide also revealed a site in αIIb, again at an EF-hand. The results of cross-linking studies have recently been supported by analysis of a mutant αIIbβ3 integrin derived from patients with the Cam variant of the bleeding disorder Glanzmann's thrombasthenia (Loftus et al., 1990). The mutant receptor was found to lack the ability to bind fibrinogen and to interact with divalent cations. This phenotype was caused by a point mutation in the β3 subunit that converted one of the aspartate residues in the putative EF-hand consensus identified in cross-linking studies to tyrosine. Furthermore, a peptide termed B12, derived from the sequence of the HHLGGAKQAGDV binding site on αIIb, has recently been shown to bind to fibrinogen directly and to block both fibrinogen-αIIbβ3 binding and platelet aggregation (D'Souza et al., 1991). Taken together, these results are now consistent with a hypothesis that interaction with receptor-bound cations may be a common mechanism for ligand binding to integrins.

Comparison of the sequence of the α subunit divalent cation-binding sites with the EF-hand consensus reveals an interesting difference: The amino acid occupying the important -z coordination position is always a small hydrophobic residue in integrin α subunits, whereas the consensus for a functional EF-hand is either aspartate or glutamate. The essential structural feature common to the best characterised adhesive recognition signals (RGD, LDV, QAGDV and DGEA) is the presence of a critical aspartate residue. This suggests that the aspartate residue in adhesive recognition signals may play a functional role in cell adhesion by providing an alternative coordination group for integrins to chelate divalent cation. The conformation of the EF-hand repeats in different integrin subunits combined with the conformation of the different aspartate-containing active sites in adhesion proteins may then be important in determining the adhesive specificity of various receptors. For receptors that bind multiple extracellular matrix molecules, it will be interesting to determine whether a ligand-binding site is shared, albeit with different affinities of interaction, or whether adjacent sites are employed (perhaps adjacent EF-hands). In biochemical studies, two pairs of peptides have now set a precedent for aspartate-containing sequences binding to the same or mutually exclusive binding sites on an integrin. QAGDV and RGDS (from fibrinogen) compete for the binding of each other to αIIbβ3 (Lam *et al.*, 1987; Santoro and Lawing, 1987; D'Souza *et al.*, 1990) and LDV and REDV (from fibronectin) have a similar relationship for binding to α4β1 (Mould *et al.*, 1991).

As discussed above, aspartate-containing active site peptides cannot always account either structurally or energetically for the entire receptor-ligand binding event. It is therefore important to determine the functional role of such peptides. Analyses of the secondary and tertiary structures of active RGD sites in fibronectin, snake venom disintegrins (both by nmr), foot-and-mouth disease virus coat protein and γ-crystallin (both by X-ray crystallography), have shown them to be located in a highly flexible loop structure (Adler *et al.*, 1991; Baron *et al.*, 1992). Thus, rather than the polypeptide flanking RGD sites contributing to a rigid optimised conformation, it appears that active site flexibility may be important for integrin binding. A flexible structure may provide a rapid on-rate for the interaction of RGD with an integrin. RGD, and perhaps other aspartate-based peptides, might then be used to initiate receptor contact. Subsequent to initial binding there may be receptor-induced shifts in the peptide into an optimised conformation followed by engagement of other regions of the ligand. In this way, the energetic requirement of integrin-ligand binding could be satisfied. As yet, there are no kinetic data for the interaction of parental and mutant RGD proteins with integrins, but such studies may yield valuable data in the future. In addition to a possible role in determining the kinetics of integrin-ligand interactions, as discussed earlier, aspartate-containing peptide sites may also have important roles in regulating the activation state of integrins.

Grateful acknowledgment is made to The Wellcome Trust and the Wigan and District Cancer Research Committee for financial support. We thank Joe Sheridan for help with generating the images in Fig. 3.

References

Adams, J. C. and Watt, F. M. (1990). Changes in keratinocyte adhesion during terminal differentiation: reduction in fibronectin binding precedes α5β1 integrin loss from the cell surface. *Cell (Cambridge, Mass.)* **63**, 425-435.

Adler, M., Lazarus, R. A., Dennis, M. S. and Wagner, G. (1991). Solution structure of kistrin, a potent platelet aggregation inhibitor and GP IIb-IIIa antagonist. *Science* **253**, 445-448.

Akiyama, S.K. and Yamada, K.M. (1985). Synthetic peptides competitively inhibit both direct binding to fibroblasts and functional biological assays for the purified cell-binding domain of fibronectin. *J. Biol. Chem.* **260**, 10402-10405.

Akiyama, S. K., Nagata, K. and Yamada, K. M. (1990). Cell surface receptors for extracellular matrix components. *Biochim. Biophys. Acta* **1031**, 91-110.

Albelda, S. M., Mette, S. A., Elder, D. E., Stewart, R., Damjanovich, L., Herlyn, M. and Buck, C. A. (1990). Integrin distribution in malignant melanoma: association of the β3 subunit with tumor progression. *Cancer Res.* **50**, 6757-64.

Altruda, F., Cervella, P., Tarone, G., Botta, C., Balzac, F., Stefanuto, G. and Silengo, L. (1990). A human integrin β1 subunit with a unique cytoplasmic domain generated by alternative mRNA processing. *Gene* **95**, 261-266.

Aota, S., Nagai, T. and Yamada, K. M. (1991). Characterization of regions of fibronectin besides the arginine-glycine-aspartic acid sequence required for adhesive function of the cell-binding domain using site-directed mutagenesis. *J. Biol. Chem.* **266**, 15938-15943.

Argraves, W. S., Deak, F., Sparks, K. J., Kiss, I. and Goetinck, P. F. (1987). Structural features of cartilage matrix protein deduced from cDNA. *Proc. Natl. Acad. Sci. USA* **84**, 464-468.

Arroyo, A. G., Sanchez-Mateos, P., Campanero, M. R., Martin-Padura, I., Dejana, E. and Sanchez-Madrid, F. (1992). Regulation of the VLA integrin-ligand interactions through the β1 subunit. *J. Cell Biol.* **117**, 659-670.

Aumailley, M. and Timpl, R. (1986). Attachment of cells to basement collagen type IV. *J. Cell Biol.* **103**, 1569-1575.

Aumailley, M., Mann, K., Von der Mark, H. and Timpl, R. (1989). Cell attachment properties of collagen type VI and Arg-Gly-Asp dependent binding to its α2(VI) and α3(VI) chains. *Exp. Cell Res.* **181**, 462-474.

Aumailley, M., Timpl, R. and Risau, W. (1991). Differences in laminin fragment interactions of normal and transformed endothelial cells. *Exp. Cell Res.* **196**, 177-183.

Bajt, M. L., Ginsberg, M. H., Frelinger, A. L., III, Berndt, M. C. and Loftus, J. C. (1992). A spontaneous mutation of integrin αIIbβ3 (platelet glycoprotein IIb-IIIa) helps define a ligand binding site. *J. Biol. Chem.* **267**, 3789-3794.

Baron, M., Main, A. L., Driscoll, P. C., Mardon, H. J., Boyd, J. and Campbell, I. D. (1992). ^{1}H NMR assignment and secondary structure of the cell adhesion type III module of fibronectin. *Biochemistry* **31**, 2068-2073.

Baumgartner, H. R. (1977). Platelet interaction with collagen fibrils in flowing blood. I. Reaction of human platelets with α-chymotrypsin-digested subendothelium. *Thromb. Haemostas.* **37**, 1-16.

Beacham, D. A., Wise, R. J., Turci, S. M. and Handin, R. I. (1992). Selective inactivation of the Arg-Gly-Asp-Ser (RGDS) binding site in von Willebrand factor by site-directed mutagenesis. *J. Biol. Chem.* **267**, 3409-3415.

Bentley, D. R. (1986). Primary structure of human complement component 2. *Biochem. J.* **239**, 339-345.

Bogaert, T., Brown, N. and Wilcox, M. (1987). The Drosophila PS2 antigen is an invertebrate integrin that, like the fibronectin receptor, becomes localized to muscle attachments. *Cell (Cambridge, Mass.)* **51**, 929-940.

Bray, P. F., Leung, C. S.-I. and Shuman, M. A. (1990). Human platelets and megakaryocytes contain alternatively spliced glycoprotein IIb mRNAs. *J. Biol. Chem.* **265**, 9587-9590.

Brazel, D., Pollner, R., Oberbaumer, I. and Kühn, K. (1988). Human basement membrane collagen (type IV). The amino acid sequence of the α2(IV) chain and its comparison with the α1(IV) chain reveals deletions in the α1(IV) chain. *Eur. J. Biochem.* **172**, 35-42.

Bronner-Fraser, M., Artinger, M., Muschler, J. and Horwitz, A. F. (1992). Developmentally regulated expression of α6 integrin in avian embryos. *Development (Cambridge, UK)* **115**, 197-211.

Brown, E., Hooper, L., Ho, T. and Gresham, H. (1990). Integrin-associated protein: a 50-kD plasma membrane antigen physically and functionally associated with integrins. *J. Cell Biol.* **111**, 2785-2794.

Brown, N. H., King, D. L., Wilcox, M. and Kafatos, F. C. (1989). Developmentally regulated alternative splicing of Drosophila integrin PS2 α transcripts. *Cell (Cambridge, Mass.)* **59**, 185-195.

Bruns, R. R., Press, W., Engvall, E., Timpl, R. and Gross, J. (1986) Type VI collagen in extracellular, 100-nm periodic filaments and fibrils: identification by immunoelectron microscopy. *J. Cell Biol.* **103**, 393-404.

Caixia, S., Stewart, S., Wayner, E., Carter, W. and Wilkins, J. (1991). Antibodies to different members of the β1 (CD29) integrins induce homotypic and heterotypic cellular aggregation. *Cell. Immunol.* **138**, 216-228.

Calvete, J. J., Schäfer, W., Mann, K., Henschen, A. and González-Rodríguez, J. (1992). Localization of the cross-linking sites of RGD and KQAGDV peptides to the isolated fibrinogen receptor, the human platelet integrin glycoprotein IIb/IIIa. Influence of peptide length. *Eur. J. Biochem.* **206**, 759-765.

Cardarelli, P. M. and Pierschbacher, M. D. (1987). Identification of fibronectin receptors on T lymphocytes. *J. Cell Biol.* **105**, 499-506.

Carrell, N. A., Fitzgerald, L. A., Steiner, B., Erickson, H. P. and Phillips, D. R. (1985). Structure of human platelet membrane glycoproteins IIb and IIIa as determined by electron microscopy. *J. Biol. Chem.* **260**, 1743-1749.

Carter, W. G. (1982a). The cooperative role of the transformation-sensitive glycoproteins, GP140 and fibronectin, in cell attachment and spreading. *J. Biol. Chem.* **257**, 3249-3257.

Carter, W. G. (1982b). Transformation-dependent alterations in glycoproteins of the extracellular matrix of human fibroblasts. Characterization of GP250 and the collagen-like GP140. *J. Biol. Chem.* **257**, 13805-13815.

Chan, B. M. C., Kassner, P. D., Schiro, J. A., Byers, H. R., Kupper, T. S. and Hemler, M. E. (1992). Distinct cellular functions mediated by different VLA integrin α subunit cytoplasmic domains. *Cell (Cambridge, Mass.)* **68**, 1051-1060.

Chapman, A. E. (1984). Characterization of a 140kD cell surface glycoprotein involved in myoblast adhesion. *J. Cell. Biochem.* **25**, 109-121.

Charo, I. F., Nannizzi, L., Phillips, D. R., Hsu, M. A. and Scarborough, R. M. (1991). Inhibition of fibrinogen binding to GP IIb-IIIa by a GP IIIa peptide. *J. Biol. Chem.* **266**, 1415-1421.

Cheresh, D. A. (1989). Human melanoma cell attachment involves an Arg-Gly-Asp-directed adhesion receptor and the disialoganglioside GD2. *Prog. Clin. Biol. Res.* **288**, 3-24.

Cheresh, D.A., Berliner, S., Vicente, V. and Ruggeri, Z. (1989). Recognition of distinct adhesive sites on fibrinogen by related integrins on platelets and endothelial cells. *Cell* **58**, 945-953.

Chun, J.-S. and Jacobson, B. S. (1992). Spreading of HeLa cells on a collagen substratum requires a second messenger formed by the lipoxygenase metabolism of arachidonic acid released by collagen receptor clustering. *Mol. Biol. Cell* **3**, 481-492.

Coller, B.S., Beer, J.H., Scudder, L.E. and Steinberg, M.H. (1989). Collagen-platelet interactions: evidence for a direct interaction of collagen with platelet GPIa/IIa and an indirect interaction with platelet GPIIb/IIIa mediated by adhesion proteins. *Blood* **74**, 182-192.

Cooper, H. M., Tamura, R. N. and Quaranta, V. (1991). The major laminin receptor of mouse embryonic stem cells is a novel isoform of the α6β1 integrin. *J. Cell Biol.* **115**, 843-850.

Corbi, A. L., Kishimoto, T. K., Miller, L. J. and Springer, T. A. (1988). The human leukocyte adhesion glycoprotein Mac-1 (complement receptor type III, CD11b) alpha subunit. Cloning, primary structure, and relation to the integrins, von Willebrand factor and factor B. *J. Biol. Chem.* **263**, 12403-12411.

Davis, G. E. (1992). Affinity of integrins for damaged extracellular matrix: αVβ3 binds to denatured collagen type I through RGD sites. *Biochem. Biophys. Res. Commun.* **182**, 1025-1031.

Decker, C., Gregg, R., Duggan, K., Stubbs, J. and Horwitz, A. (1984). Adhesive multiplicity in the interaction of embryonic fibroblasts and myoblasts with extracellular matrices. *J. Cell Biol.* **99**, 1398-1404.

Dennis, J., Waller, C., Timpl, R. and Schirrmacher, V. (1982). Surface sialic acid reduces attachment of metastatic tumor cells to collagen type IV and fibronectin. *Nature* **300**, 274-276.

Detmers, P. A., Wright, S. D., Olsen, E., Kimball, B. and Cohn, Z. A. (1987). Aggregation of complement receptors on human neutrophils in the absence of ligand. *J. Cell Biol.* **105**, 1137-1145.

Dransfield, I. and Hogg, N. (1989). Regulated expression of magnesium binding epitope on leukocyte integrin α subunits. *Embo J.* **8**, 3759-3765.

Dransfield, I., Buckle, A. M. and Hogg, N. (1990). Early events of the immune response mediated by leukocyte integrins. *Immunol. Rev.* **114**, 29-44.

Dransfield, I., Cabanas, C., Barrett, J. and Hogg, N. (1992a). Interaction of leukocyte integrins with ligand is necessary but not sufficient for function. *J. Cell Biol.* **116**, 1527-1535.

Dransfield, I., Cabanas, C., Craig, A. and Hogg, N. (1992b). Divalent cation regulation of the function of the leukocyte integrin LFA-1. *J. Cell Biol.* **116**, 219-226.

D'Souza, S. E., Ginsberg, M. H., Burke, T. A., Lam, S. C. T. and Plow, E. F. (1988). Localization of an Arg-Gly-Asp recognition site within an integrin adhesion receptor. *Science* **242**, 91-93.

D'Souza, S. E., Ginsberg, M. H., Burke, T. A. and Plow, E. F. (1990). The ligand binding site of the platelet integrin receptor GPIIb-IIIa is proximal to the second calcium binding domain of its α subunit. *J. Biol. Chem.* **265**, 3440-3446.

D'Souza, S. E., Ginsberg, M. H., Matsueda, G. R. and Plow, E. F. (1991). A discrete sequence in a platelet integrin is involved in ligand recognition. *Nature (London)* **350**, 66-68.

Du, X., Plow, E. F., Frelinger, A. L., III, O'Toole, T. E., Loftus, J. C. and Ginsberg, M. H. (1991). Ligands "activate" integrin αIIbβ3 (platelet GPIIb-IIIa). *Cell (Cambridge, Mass.)* **65**, 409-416.

Dustin, M. L. and Springer, T. A. (1989). T-cell receptor cross-linking transiently stimulates adhesiveness through LFA-1. *Nature (London)* **341**, 619-624.

Edwards, J. G., Hameed, H. and Campbell, G. (1988). Induction of fibroblast spreading by manganese(2+): a possible role for unusual binding sites for divalent cations in receptors for proteins containing Arg-Gly-Asp. *J. Cell Sci.* **89**, 507-513.

Elices, M. J., Urry, L. A. and Hemler, M. E. (1991). Receptor functions for the integrin VLA-3:

fibronectin, collagen, and laminin binding are differentially influenced by Arg-Gly-Asp peptide and by divalent cations. *J. Cell Biol.* **112**, 169-181.

Engvall, E., Bell, M. L., Carlsson, N. L. K., Miller, E. J. and Ruoslahti, E. (1982). Nonhelical, fibronectin-binding basement membrane collagen from endodermal cell culture. *Cell* **29**, 475-482.

Erle, D. J., Rüegg, C., Sheppard, D. and Pytela, R. (1991). Complete amino acid sequence of an integrin β subunit (β7) identified in leukocytes. *J. Biol. Chem.* **266**, 11009-11016.

Falke, J. J., Snyder, E. E., Thatcher, K. C. and Voertler, C. S. (1991). Quantitating and engineering the ion specificity of an EF-hand-like Ca^{2+} binding site. *Biochemistry* **30**, 8690-8697.

Figdor, C. G., Van Kooyk, Y. and Keizer, G. D. (1990). On the mode of action of LFA-1. *Immunol. Today* **11**, 277-280.

Fogerty, F. J., Akiyama, S. K., Yamada, K. M. and Mosher, D. F. (1990). Inhibition of binding of fibronectin to matrix assembly sites by anti-integrin (α5β1) antibodies. *J. Cell Biol.* **111**, 699-708.

Frachet, P., Duperray, A., Delachanal, E. and Marguerie, G. (1992). Role of the transmembrane and cytoplasmic domains in the assembly and surface exposure of the platelet integrin GPIIb/IIIa. *Biochemistry* **31**, 2408-2415.

Frelinger, A. L., Du, X., Plow, E. F. and Ginsberg, M. H. (1991). Monoclonal antibodies to ligand-occupied conformers of integrin αIIbβ3 (glycoprotein IIb-IIIa) alter receptor affinity, specificity, and function. *J. Biol. Chem.* **266**, 17106-17111.

Frelinger, A. L., Isaac, C., Plow, E. F., Smith, M. A., Jones, R., Lam, S. C. T. and Ginsberg, M. H. (1990). Selective inhibition of integrin function by antibodies specific for ligand-occupied receptor conformers. *J. Biol. Chem.* **265**, 6346-6352.

Frelinger, A. L., Lam, S. C. T., Plow, E. F., Smith, M. A., Loftus, J. C. and Ginsberg, M. H. (1988). Occupancy of an adhesive glycoprotein receptor modulates expression of an antigenic site involved in cell adhesion. *J. Biol. Chem.* **263**, 12397-12402.

Friedlander, D.R., Hoffman, S. and Edelman, G.M. (1988). Functional mapping of cytotactin: Proteolytic fragments active in cell-substrate adhesion. *J. Cell Biol.* **107**, 2329-2340.

Gailit, J. and Ruoslahti, E. (1988). Regulation of the fibronectin receptor affinity by divalent cations. *J. Biol. Chem.* **263**, 12927-12932.

Gehlsen, K. R., Davis, G. E. and Sriramarao, P. (1992a). Integrin expression in human melanoma cells with differing invasive and metastatic properties. *Clin. Exp. Metastasis* **10**, 111-20.

Gehlsen, K. R., Sriramarao, P., Furcht, L. T. and Skubitz, A. P. N. (1992b). A synthetic peptide derived from the carboxy terminus of the laminin A chain represents a binding site for the α3β1 integrin. *J. Cell Biol.* **117**, 449-459.

Giancotti, F. G., Comoglio, P. M. and Tarone, G. (1986). A 135,000 molecular weight plasma membrane glycoprotein involved in fibronectin-mediated cell adhesion. Immunofluorescence localization in normal and RSV-transformed fibroblasts. *Exp. Cell Res.* **163**, 47-62.

Giancotti, F. G., Stepp, M. A., Suzuki, S., Engvall, E. and Ruoslahti, E. (1992). Proteolytic processing of endogenous and recombinant β4 integrin subunit. *J. Cell Biol.* **118**, 951-959.

Ginsberg, M. H., Pierschbacher, M. D., Ruoslahti, E., Marguerie, G. and Plow, E. (1985). Inhibition of fibronectin binding to platelets by proteolytic fragments and synthetic peptides which support fibroblast adhesion. *J. Biol. Chem.* **260**, 3931-3936.

Girma, J.-P., Meyer, C. L., Verweij, C. L., Pannekoek, H. and Sixma, J. J. (1987). Structure-function relationship of human von Willebrand factor. *Blood* **70**, 605-611.

Golden, A., Brugge, J. S. and Shattil, S. J. (1990). Role of platelet membrane glycoprotein IIb-IIIa in agonist-induced tyrosine phosphorylation of platelet proteins. *J. Cell Biol.* **111**, 3117-3127.

Greve, J. M. and Gottlieb, D. I. (1982). Monoclonal antibodies which alter the morphology of cultured chick myogenic cells. *J. Cell. Biochem.* **18**, 221-230.

Grzesiak, J. J., Davis, G. E., Kirchhofer, D. and Pierschbacher, M. D. (1992). Regulation of α2β1-mediated fibroblast migration on type I collagen by shifts in the concentrations of extracellular magnesium and calcium. *J. Cell Biol.* **117**, 1109-1117.

Guan, J. L. and Hynes, R. O. (1990). Lymphoid cells recognize an alternatively spliced segment of fibronectin via the integrin receptor α4β1. *Cell (Cambridge, Mass.)* **60**, 53-61.

Gulino, D., Boudignon, C., Zhang, L., Concord, E., Rabiet, M. J. and Marguerie, G. (1992). Calcium-binding properties of the platelet glycoprotein IIb ligand-interacting domain. *J. Biol. Chem.* **267**, 1001-1007.

Gullberg, D., Terracio, L., Borg, T.K. and Rubin, K. (1989) Identification of integrin-like matrix receptors with affinity for interstitial collagens. *J. Biol. Chem.* **264**, 12686-12694.

Gustafson, K. S., Vercellotti, G. M., Bendel, C. M. and Hostetter, M. K. (1991). Molecular mimicry in Candida albicans. Role of an integrin analog in adhesion of the yeast to human endothelium. *J. Clin. Invest.* **87**, 1896-1902.

Hall, D. E., Reichardt, L. F., Crowley, E., Holley, B., Moezzi, H., Sonnenberg, A. and Damsky, C. H. (1990). The α1/β1 and α6/β1 integrin heterodimers mediate cell attachment to distinct sites on laminin. *J. Cell Biol.* **110**, 2175-2184.

Haverstick, D.M., Cowan, J.F., Yamada, K.M. and Santoro, S.A. (1985). Inhibition of platelet adhesion to fibronectin, fibrinogen, and von Willebrand factor substrates by a synthetic tetrapeptide derived from the cell-binding domain of fibronectin. *Blood* **66**, 946-952.

Hawiger, J. (1987). Formation and regulation of platelet and fibrin hemostatic plug. *Human Pathol.* **18**, 111-122.

Hayashi, Y., Haimovich, B., Reszka, A., Boettiger, D. and Horwitz, A. (1990). Expression and function of chicken integrin β1 subunit and its cytoplasmic domain mutants in mouse NIH 3T3 cells. *J. Cell Biol.* **110**, 175-184.

Hayman, E. G., Pierschbacher, M. D. and Ruoslahti, E. (1985). Detachment of cells from culture substrate by soluble fibronectin peptides. *J. Cell Biol.* **100**, 1948-1954.

Hemler, M. E., Crouse, C. and Sonnenberg, A. (1989). Association of the VLA α6 subunit with a novel protein. A possible alternative to the common VLA β1 subunit on certain cell lines. *J. Biol. Chem.* **264**, 6529-6535.

Hemler, M. E., Elices, M. J., Parker, C. and Takada, Y. (1990). Structure of the integrin VLA-4 and its cell-cell and cell-matrix adhesion functions. *Immunol. Rev.* **114**, 45-65.

Hermanowski-Vosatka, A., Van Strijp, J. A. G., Swiggard, W. J. and Wright, S. D. (1992). Integrin modulating factor-1: a lipid that alters the function of leukocyte integrins. *Cell (Cambridge, Mass.)* **68**, 341-352.

Hibbs, M. L., Xu, H., Stacker, S. A. and Springer, T. A. (1991). Regulation of adhesion to ICAM-1 by the cytoplasmic domain of LFA-1 integrin β subunit. *Science* **251**, 1611-1613.

Hogervorst, F., Kuikman, I., Van Kessel, A. G. and Sonnenberg, A. (1991). Molecular cloning of the human α6 integrin subunit. Alternative splicing of α6 mRNA and chromosomal localization of the α6 and β4 genes. *Eur. J. Biochem.* **199**, 425-433.

Hogervorst, F., Kuikman, I., Von dem Borne, A. E. G. K. and Sonnenberg, A. (1990). Cloning and sequence analysis of beta-4 cDNA: an integrin subunit that contains a unique 118 kd cytoplasmic domain. *Embo J.* **9**, 765-770.

Horwitz, A., Duggan, K., Buck, C., Beckerle, M. C. and Burridge, K. (1986). Interaction of plasma membrane fibronectin receptor with talin - a transmembrane linkage. *Nature (London)* **320**, 531-533.

Hughes, A. L. (1992). Coevolution of the vertebrate integrin α- and β-chain genes. *Mol. Biol. Evol.* **9**, 216-234.

Humphries, M. J. (1990). The molecular basis and specificity of integrin-ligand interactions. *J. Cell Sci.* **97**, 585-592.

Humphries, M. J. (1992). Peptide sequences in matrix proteins recognized by adhesion receptors. In *Molecular and Cellular Aspects of Basement Membranes* (ed. D. Rohrbach and R. Timpl). Academic Press, Orlando, FL (in press).

Humphries, M.J., Akiyama, S.K., Komoriya, A., Olden, K. and Yamada, K.M. (1986a). Identification of an alternatively spliced site in human plasma fibronectin that mediates cell type-specific adhesion. *J. Cell Biol.* **103**, 2637-2647.

Humphries, M.J., Komoriya, A., Akiyama, S.K., Olden, K. and Yamada, K.M. (1987). Identification of two distinct regions of the type III connecting segment of human plasma fibronectin that promote cell type-specific adhesion. *J. Biol. Chem.* **262**, 6886-6892.

Humphries, M.J., Olden, K. and Yamada, K.M. (1986b). A synthetic peptide from fibronectin inhibits experimental metastasis of murine melanoma cells. *Science* **233**, 467-470.

Hynes, R. O. (1987). Integrins: a family of cell surface receptors. *Cell (Cambridge, Mass.)* **48**, 549-554.

Hynes, R. O. (1992). Integrins: versatility, modulation, and signaling in cell adhesion. *Cell* (Cambridge, Mass.) **69**, 11-25.

Keizer, G. D., Visser, W., Vliem, M. and Figdor, C. G. (1988). A monoclonal antibody (NKI-L16) directed against a unique epitope on the α-chain of human leukocyte function-associated antigen 1 induces homotypic cell-cell interactions. *J. Immunol.* **140**, 1393-1400.

Kelly, T., Molony, L. and Burridge, K. (1987). Purification of two smooth muscle glycoproteins related to integrin. Distribution in cultured chicken embryo fibroblasts. *J. Biol. Chem.* **262**, 17189-17199.

Kirchhofer, D., Grzesiak, J. and Pierschbacher, M. D. (1991). Calcium as a potential physiological regulator of integrin-mediated cell adhesion. *J. Biol. Chem.* **266**, 4471-4477.

Kloczewiak, M., Timmons, S. and Hawiger, J. (1982). Localization of a site interacting with human platelet receptor on carboxy-terminal segment of human fibrinogen γ chain. *Biochem. Biophys. Res. Commun.* **107**, 181-187.

Kloczewiak, M., Timmons, S., Lukas, T. J. and Hawiger, J. (1984). Platelet receptor recognition site on human fibrinogen. Synthesis and structure-function relationship of peptides corresponding to the carboxy-terminal segment of the γ chain. *Biochemistry* **23**, 1767-1774.

Komoriya, A., Green, L. J., Mervic, M., Yamada, S. S., Yamada, K. M. and Humphries, M. J. (1991). The minimal essential sequence for a major cell type-specific adhesion site (CS1) within the alternatively spliced type III connecting segment domain of fibronectin is leucine-aspartic acid-valine. *J. Biol. Chem.* **266**, 15075-15079.

Kornberg, L. J., Earp, H. S., Turner, C. E., Prockop, C. and Juliano, R. L. (1991). Signal transduction by integrins: increased protein tyrosine phosphorylation caused by clustering of β1 integrins. *Proc. Natl. Acad. Sci. U. S. A.* **88**, 8392-8396.

Kovach, N. L., Carlos, T. M., Yee, E. and Harlan, J. M. (1992). A monoclonal antibody to β1 integrin (CD29) stimulates VLA-dependent adherence of leukocytes to human umbilical vein endothelial cells and matrix components. *J. Cell Biol.* **116**, 499-509.

Kramer, R. H. and Marks, N. (1989). Identification of integrin collagen receptors on human melanoma cells. *J. Biol. Chem.* **264**, 4684-4688.

Kretsinger, R. H. and Nockolds, C. E. (1973). Carp muscle calcium-binding protein II: Structure determination and general description. *J. Biol. Chem.* **248**, 3313-3326.

Kunicki, T. J., Nugent, D. J., Staats, S. J., Orchekowski, R. P., Wayner, E. A. and Carter, W. G. (1988). The human fibroblast class II extracellular matrix receptor mediates platelet adhesion to collagen and is identical to the platelet glycoprotein Ia-IIa complex. *J. Biol. Chem.* **263**, 4516-4519.

LaFlamme, S. E., Akiyama, S. K. and Yamada, K. M. (1992). Regulation of fibronectin receptor distribution. *J. Cell Biol.* **117**, 437-447.

Lam, S. C. T. (1992). Isolation and characterization of a chymotryptic fragment of platelet glycoprotein IIb-IIIa retaining Arg-Gly-Asp binding activity. *J. Biol. Chem.* **267**, 5649-5655.

Lam, S. C. T., Plow, E. F., Smith, M. A., Andrieux, A., Ryckwaert, J. J., Marguerie, G. and Ginsberg, M. H. (1987). Evidence that arginyl-glycyl-aspartate peptides and fibrinogen γ chain peptides share a common binding site on platelets. *J. Biol. Chem.* **262**, 947-950.

Languino, L. R. and Ruoslahti, E. (1992). An alternative form of the integrin β1 subunit with a variant cytoplasmic domain. *J. Biol. Chem.* **267**, 7116-7120.

Larjava, H., Peltonen, J., Akiyama, S. K., Yamada, S. S., Gralnick, H. R., Uitto, J. and Yamada, K. M. (1990). Novel function for β1 integrins in keratinocyte cell-cell interactions. *J. Cell Biol.* **110**, 803-815.

Larson, R. S., Corbi, A. L., Berman, L. and Springer, T. (1989). Primary structure of the leukocyte function-associated molecule-1 α subunit: an integrin with an embedded domain defining a protein superfamily. *J. Cell Biol.* **108**, 703-712.

Lee, E. C., Lotz, M. M., Steele, G. D., Jr. and Mercurio, A. M. (1992). The integrin α6β4 is a laminin receptor. *J. Cell Biol.* **117**, 671-678.

Leptin, M., Aebersold, R. and Wilcox, M. (1987). Drosophila position-specific antigens resemble the vertebrate fibronectin-receptor family. *Embo J.* **6**, 1037-1043.

Loftus, J. C., O'Toole, T. E., Plow, E. F., Glass, A., Frelinger, A. L., III and Ginsberg, M. H. (1990). A β3 integrin mutation abolishes ligand binding and alters divalent cation-dependent conformation. *Science* **249**, 915-918.

Marcantonio, E. E. and Hynes, R. O. (1988). Antibodies to the conserved cytoplasmic domain of the integrin β1 subunit react with proteins in vertebrates, invertebrates, and fungi. *J. Cell Biol.* **106**, 1765-1772.

Marcantonio, E. E., Guan, J. L., Trevithick, J. E. and Hynes, R. O. (1990). Mapping of the functional determinants of the integrin β1 cytoplasmic domain by site-directed mutagenesis. *Cell Regul.* **1**, 597-604.

Mole, J. E., Anderson, J. K., Davison, E. A. and Woods, D. E. (1984) Complete primary structure for the zymogen of complement factor B. *J. Biol. Chem.* **259**, 3407-3412.

Morton, L. F., Peachey, A. R. and Barnes, M. J. (1989). Platelet-reactive sites in collagens type I and type III. *Biochem. J.* **258**, 157-163.

Mould, A. P. and Humphries, M. J. (1991). Identification of a novel recognition sequence for the integrin α4β1 in the carboxy-terminal heparin-binding domain of fibronectin. *Embo J.* **10**, 4089-4095.

Mould, A. P., Komoriya, A., Yamada, K. M. and Humphries, M. J. (1991). The CS5 peptide is a second site in the IIICS region of fibronectin recognized by the integrin α4β1. Inhibition of α4β1 function by RGD peptide homologs. *J. Biol. Chem.* **266**, 3579-3585.

Mould, A. P., Wheldon, L. A., Komoriya, A., Wayner, E. A., Yamada, K. M. and Humphries, M. J. (1990). Affinity chromatographic isolation of the melanoma adhesion receptor for the IIICS region of fibronectin and its identification as the integrin α4β1. *J. Biol. Chem.* **265**, 4020-4024.

Moyle, M., Napier, M. A. and McLean, J. W. (1991). Cloning and expression of a divergent integrin subunit β8. *J. Biol. Chem.* **266**, 19650-19658.

Murray, J. C., Stingl, G., Kleinman, H. K., Martin, G. R. and Katz, S. I. (1979). Epidermal cells adhere preferentially to type IV (basement membrane) collagen. *J. Cell Biol.* **80**, 197-212.

Nagai, T., Yamakawa, N., Aota, S., Yamada, S. S., Akiyama, S. K., Olden, K. and Yamada, K. M. (1991). Monoclonal antibody characterization of two distant sites required for function of the central cell-binding domain of fibronectin in cell adhesion, cell migration, and matrix assembly. *J. Cell Biol.* **114**, 1295-1305.

Neff, N. T., Lowrey, C., Decker, C., Tovar, A., Damsky, C., Buck, C. and Horwitz, A. F. (1982). A monoclonal antibody detaches embryonic skeletal muscle from extracellular matrices. *J. Cell Biol.* **95**, 654-666.

Nermut, M. V., Green, N. M., Eason, P., Yamada, S. S. and Yamada, K. M. (1988). Electron microscopy and structural model of human fibronectin receptor. *Embo J.* **7**, 4093-4099.

Neugebauer, K. M. and Reichardt, L. F. (1991). Cell-surface regulation of β1-integrin activity on developing retinal neurons. *Nature (London)* **350**, 68-71.

Obara, M., Kang, M.S. and Yamada, K.M. (1988). Site-directed mutagenesis of the cell-binding domain of human fibronectin: Separable, synergistic sites mediate adhesive function. *Cell* **53**, 649-657.

Otey, C. A., Pavalko, F. M. and Burridge, K. (1990). An interaction between α-actinin and the β1 integrin subunit in vitro. *J. Cell Biol.* **111**, 721-729.

O'Toole, T. E., Mandelman, D., Forsyth, J., Shattil, S. J., Plow, E. F. and Ginsberg, M. H. (1991). Modulation of the affinity of integrin αIIbβ3 (GPIIb-IIIa) by the cytoplasmic domain of αIIb. *Science* **254**, 845-847.

Palotie, A., Peltonen, L., Risteli, L. and Risteli, J. (1983a) Effects of the structural components of basement membranes on the attachment of tetracarcinoma-derived endodermal cells. *Exp. Cell Res.* **144**, 31-37.

Palotie, A., Tryggvason, K., Peltonen, L. and Seppa, H. (1983b). Components of subendothelial aorta basement membrane. Immunohistochemical localization and the role in cell attachment. *Lab. Invest.* **49**, 362-370.

Patel, V. P. and Lodish, H. F. (1986). The fibronectin receptor on mammalian erythroid precursor cells: characterization and developmental regulation. *J. Cell Biol.* **102**, 449-456.

Phillips, D. R., Charo, I. F. and Scarborough, R. M. (1991). GPIIb-IIIa: the responsive integrin. *Cell (Cambridge, Mass.)* **65**, 359-362.

Pierschbacher, M.D. and Ruoslahti, E. (1984a). The cell attachment activity of fibronectin can be duplicated by small synthetic fragments of the molecule. *Nature* **309**, 30-33.

Pierschbacher, M.D. and Ruoslahti, E. (1984b). Variants of the cell recognition site of fibronectin that retain attachment-promoting activity. *Proc. Natl. Acad. Sci. USA* **81**, 5985-5988.

Pierschbacher, M. D. and Ruoslahti, E. (1987). Influence of stereochemistry of the sequence Arg-Gly-Asp-Xaa on binding specificity in cell adhesion. *J. Biol. Chem.* **262**, 17294-17298.

Pierschbacher, M.D., Hayman, E.G. and Ruoslahti, E. (1981). Location of the cell-attachment site in fibronectin with monoclonal antibodies and proteolytic fragments of the molecule. *Cell* **26**, 259-267.

Plow, E.F., McEver, R.P., Coller, B.S., Woods, V.L. and Marguerie, G.A. (1985). Related binding mechanisms for fibrinogen, fibronectin, von Willebrand factor, and thrombospondin on thrombin-stimulated human platelets. *Blood* **66**, 724-727.

Postigo, A. A., Pulido, R., Campanero, M. R., Acevedo, A., Garcia-Pardo, A., Corbi, A., Sanchez-Madrid, F. and De Landazuri, M. O. (1991). Differential expression of VLA-4 integrin by resident and peripheral blood B lymphocytes. Acquisition of functionally active α4β1-fibronectin receptors upon B cell activation. *Eur. J. Immunol.* **21**, 2437-2445.

Pulido, R., Elices, M. J., Campanero, M. R., Osborn, L., Schiffer, S., Garcia-Pardo, A., Lobb, R., Hemler, M. E. and Sanchez-Madrid, F. (1991). Functional evidence for three distinct and independently inhibitable adhesion activities mediated by the human integrin VLA-4. Correlation with distinct α4 epitopes. *J. Biol. Chem.* **266**, 10241-10245.

Pytela, R. (1988). Amino acid sequence of the murine Mac-1 α chain reveals homology with the integrin family and an additional domain related to von Willebrand factor. *Embo J.* **7**, 1371-1378.

Pytela, R., Pierschbacher, M. D. and Ruoslahti, E. (1985a). A 125/115-kDa cell surface receptor specific for vitronectin interacts with the arginine-glycine-aspartic acid adhesion sequence derived from fibronectin. *Proc. Natl. Acad. Sci. U. S. A.* **82**, 5766-5770.

Pytela, R., Pierschbacher, M. D. and Ruoslahti, E. (1985b). Identification and isolation of a 140 kd cell surface glycoprotein with properties expected of a fibronectin receptor. *Cell (Cambridge, Mass.)* **40**, 191-198.

Ramsamooj, P., Lively, M. O. and Hantgan, R. R. (1991). Evidence that the central region of glycoprotein IIIa participates in integrin receptor function. *Biochem. J.* **276**, 725-732.

Reszka, A. A., Hayashi, Y. and Horwitz, A. F. (1992). Identification of amino acid sequences in the integrin β1 cytoplasmic domain implicated in cytoskeletal association. *J. Cell Biol.* **117**, 1321-1330.

Rivas, G. A. and González-Rodríguez, J. (1991). Calcium binding to human platelet integrin GPIIb/IIIa and to its constituent glycoproteins. *Biochem. J.* **276**, 35-40.

Rojiani, M., V, Finlay, B. B., Gray, V. and Dedhar, S. (1991). In vitro interaction of a polypeptide homologous to human Ro/SS-A antigen (calreticulin) with a highly conserved amino acid sequence in the cytoplasmic domain of integrin α subunits. *Biochemistry* **30**, 9859-9866.

Ruoslahti, E. and Pierschbacher, M. D. (1987). New perspectives in cell adhesion: RGD and integrins. *Science* **238**, 491-497.

Santoro, S. A. (1986). Identification of a 160,000 dalton platelet membrane protein that mediates the initial divalent cation-dependent adhesion of platelets to collagen. *Cell* **46**, 913-920.

Santoro, S. A. and Lawing, W. J., Jr. (1987). Competition for related but nonidentical binding sites on the glycoprotein IIb-IIIa complex by peptides derived from platelet adhesive proteins. *Cell (Cambridge, Mass.)* **48**, 867-873.

Santoro, S. A., Rajpara, S. M., Staatz, W. D. and Woods, V. L. (1988). Isolation and characterization of a platelet surface collagen binding complex related to VLA-2. *Biochem. Biophys. Res. Commun.* **153**, 217-223.

Schindler, M., Meiners, S. and Cheresh, D. A. (1989). RGD-dependent linkage between plant cell wall and plasma membrane: consequences for growth. *J. Cell Biol.* **108**, 1955-1965.

Schwartz, M. A., Lechene, C. and Ingber, D. E. (1991). Insoluble fibronectin activates the sodium/hydrogen ion antiporter by clustering and immobilizing integrin α5β1, independent of cell shape. *Proc. Natl. Acad. Sci. U. S. A.* **88**, 7849-7853.

Shattil, S. J. and Brass, L. F. (1987). Induction of the fibrinogen receptor on human platelets by intracellular mediators. *J. Biol. Chem.* **262**, 992-1000.

Shattil, S. J., Hoxie, J. A., Cunningham, M. and Brass, L. F. (1985). Changes in the platelet membrane glycoprotein IIb-IIIa complex during platelet activation. *J. Biol. Chem.* **260**, 11107-11114.

Shaw, L. M., Messier, J. M. and Mercurio, A. M. (1990). The activation dependent adhesion of macrophages to laminin involves cytoskeletal anchoring and phosphorylation of the α6β1 integrin. *J. Cell Biol.* **110**, 2167-2174.

Shimizu, Y., Van Seventer, G. A., Horgan, K. J. and Shaw, S. (1990a). Regulated expression and binding of three VLA (β1) integrin receptors on T cells. *Nature (London)* **345**, 250-253.

Shimizu, Y., Van Seventer, G. A., Horgan, K. J. and Shaw, S. (1990b). Roles of adhesion molecules in T-cell recognition: Fundamental similarities between four integrins on resting human T cells (LFA-1, VLA-4, VLA-5, VLA-6) in expression, binding, and costimulation. *Immunol. Rev.* **114**, 109-143.

Smith, J. W. and Cheresh, D. A. (1988). The Arg-Gly-Asp binding domain of the vitronectin receptor. Photoaffinity cross-linking implicates amino acid residues 61-203 of the β subunit. *J. Biol. Chem.* **263**, 18726-18731.

Smith, J. W. and Cheresh, D. A. (1990). Integrin (αVβ3)-ligand interaction. Identification of a heterodimeric RGD binding site on the vitronectin receptor. *J. Biol. Chem.* **265**, 2168-2172.

Song, W. K., Wang, W., Foster, R. F., Bielser, D. A. and Kaufman, S. J. (1992). H36-α7 is a novel integrin alpha chain that is developmentally regulated during skeletal myogenesis. *J. Cell Biol.* **117**, 643-657.

Springer, T. A. (1990). Adhesion receptors of the immune system. *Nature (London)* **346**, 425-434.

Staatz, W. D., Fok, K. F., Zutter, M. M., Adams, S. P., Rodriguez, B. A. and Santoro, S. A. (1991). Identification of a tetrapeptide recognition sequence for the $\alpha2\beta1$ integrin in collagen. *J. Biol. Chem.* **266**, 7363-7367.

Staatz, W. D., Peters, K. J. and Santoro, S. A. (1990a). Divalent cation-dependent structure in the platelet membrane glycoprotein Ia-IIa (VLA-2) complex. *Biochem. Biophys. Res. Commun.* **168**, 107-113.

Staatz, W. D., Rajpara, S. M., Wayner, E. A., Carter, W. G. and Santoro, S. A. (1989). The membrane glycoprotein Ia-IIa (VLA-2) complex mediates the magnesium-dependent adhesion of platelets to collagen. *J. Cell Biol.* **108**, 1917-1924.

Staatz, W. D., Walsh, J. J., Pexton, T. and Santoro, S. A. (1990b). The $\alpha2\beta1$ integrin cell surface collagen receptor binds to the $\alpha1(I)$-CB3 peptide of collagen. *J. Biol. Chem.* **265**, 4778-4781.

Staunton, D. E., Dustin, M. L., Erickson, H. P. and Springer, T. A. (1990). The arrangement of the immunoglobulin-like domains of ICAM-1 and the binding sites for LFA-1 and rhinovirus. *Cell (Cambridge, Mass.)* **61**, 243-254.

Stepp, M. A., Spurr-Michaud, S., Tisdale, A., Elwell, J. and Gipson, I. K. (1990). The $\alpha6\beta4$ integrin heterodimer is a component of hemidesmosomes. *Proc. Natl. Acad. Sci. U. S. A.* **87**, 8970-8974.

Streuli, C. H., Bailey, N. and Bissell, M. J. (1991). Control of mammary epithelial differentiation: basement membrane induces tissue-specific gene expression in the absence of cell-cell interaction and morphological polarity. *J. Cell Biol.* **115**, 1383-1395.

Strynadka, N. and James, M. (1989). Crystal structures of the helix-loop-helix calcium-binding proteins. *Annu. Rev. Biochem.* **58**, 951-998.

Suzuki, S. and Naitoh, Y. (1990). Amino acid sequence of a novel integrin $\beta4$ subunit and primary expression of the mRNA in epithelial cells. *Embo J.* **9**, 757-763.

Suzuki, S., Pierschbacher, M.D., Hayman, E.G., Nguyen, K., Ohlgren, Y. and Ruoslahti, E. (1984). Domain structure of vitronectin. Alignment of active sites. *J. Biol. Chem.* **259**, 15307-15314.

Symons, M. H. and Mitchison, T. J. (1992). A GTPase controls cell-substrate adhesion in *Xenopus* XTC fibroblasts. *J. Cell Biol.* **118**, 1235-1244.

Takada, Y. and Hemler, M. E. (1989). The primary structure of the VLA-2/collagen receptor $\alpha2$ subunit (platelet GPIa): homology to other integrins and the presence of a possible collagen-binding domain. *J. Cell Biol.* **109**, 397-407.

Tamkun, J. W., DeSimone, D. W., Fonda, D., Patel, R. S., Buck, C., Horwitz, A. F. and Hynes, R. O. (1986). Structure of integrin, a glycoprotein involved in the transmembrane linkage between fibronectin and actin. *Cell (Cambridge, Mass.)* **46**, 271-282.

Tamura, R. N., Cooper, H. M., Collo, G. and Quaranta, V. (1991) Cell-type specific integrin variants with alternative α chain cytoplasmic domains. *Proc. Natl. Acad. Sci. USA* **88**, 10183-10187.

Tamura, R. N., Rozzo, C., Starr, L., Chambers, J., Reichardt, L. F., Cooper, H. M. and Quaranta, V. (1990). Epithelial integrin $\alpha6\beta4$: complete primary structure of $\alpha6$ and variant forms of $\beta4$. *J. Cell Biol.* **111**, 1593-1604.

Taniguchi-Sidle, A. and Isenman, D. E. (1992). Mutagenesis of the Arg-Gly-Asp triplet in human complement component C3 does not abolish binding of iC3b to the leukocyte integrin complement receptor type III (CR3, CD11b/CD18). *J. Biol. Chem.* **267**, 635-643.

Timpl, R. and Dziadek, M. (1986). Structure, development and molecular pathology of basement membranes. *Int. Rev. Exp. Pathol.* **29**, 1-112.

Timpl, R., Wiedemann, H., Van Delden, V., Furthmayr, H. and Kühn, K. (1981). A network model for the organization of type IV collagen molecules in basement membranes. *Eur. J. Biochem.* **120**, 203-211.

Tuckwell, D. S., Brass, A. and Humphries, M. J. (1992). Homology modeling of integrin EF-hands. Evidence for widespread use of a conserved cation-binding site. *Biochem. J.* **285**, 325-331.

Valmu, L., Autero, M., Siljander, P., Patarroyo, M. and Gahmberg, C. G. (1991). Phosphorylation of the β-subunit of CD11/CD18 integrins by protein kinase C correlates with leukocyte adhesion. *Eur. J. Immunol.* **21**, 2857-2862.

Vandenberg, P., Kern, A., Ries, A., Luckenbill-Edds, L., Mann, K. and Kühn, K. (1991). Characterization of a type IV collagen major cell binding site with affinity to the $\alpha1\beta1$ and $\alpha2\beta1$ integrins. *J. Cell Biol.* **113**, 1475-1483.

Van Kooyk, Y., Van de Wiel-Van Kemenade, P., Weder, P., Kuijpers, T. W. and Figdor, C. G. (1989). Enhancement of LFA-1-mediated cell adhesion by triggering through CD2 and CD3 on T lymphocytes. *Nature (London)* **342**, 811-813.

Van Kuppevelt, T. H. M. S. M., Languino, L. R., Gailit, J. O., Suzuki, S. and Ruoslahti, E. (1989). An alternative cytoplasmic domain of the integrin β3 subunit. *Proc. Natl. Acad. Sci. U. S. A.* **86**, 5415-5418.

Vlodavsky, I., Levi, A., Lax, I., Fuks, Z. and Schlessinger, J. (1982). Induction of cell attachment and morphological differentiation in a pheochromocytoma cell line and emryonal sensory cells by the extracellular matrix. *Dev. Biol.* **93**,285-300.

Von der Mark, H., Aumailley, M., Wick, G., Fleischmajer, R. and Timpl, R. (1984). Immunochemistry, genuine size and tissue localization of collagen VI. *Eur. J. Biochem.* **142**, 493-502.

Vyas, N. K., Vyas, M. N. and Quiocho, F. A. (1987). A novel calcium binding site in the galactose-binding protein of bacterial transport and chemotaxis. *Nature* **327**, 635-638.

Wadsworth, S., Halvorson, M. J. and Coligan, J. E. (1992). Developmentally regulated expression of the β4 integrin on immature mouse thymocytes. *J. Immunol.* **149**, 421-428.

Wayner, E. A. and Carter, W. G. (1987). Identification of multiple cell adhesion receptors for collagen and fibronectin in human fibrosarcoma cells possessing unique α and common β subunits. *J. Cell Biol.* **105**, 1873-1884.

Wayner, E. A., Garcia-Pardo, A., Humphries, M. J., McDonald, J. A. and Carter, W. G. (1989). Identification and characterization of the T lymphocyte adhesion receptor for an alternative cell attachment domain (CS-1) in plasma fibronectin. *J. Cell Biol.* **109**, 1321-1330.

Weisel, J. W., Nagaswami, C., Vilaire, G. and Bennett, J. S. (1992). Examination of the platelet membrane glycoprotein IIb-IIIa complex and its interaction with fibrinogen and other ligands by electron microscopy. *J. Biol. Chem.* **267**, 16637-16643.

Werb, Z., Tremble, P. M., Behrendtsen, O., Crowley, E. and Damsky, C. H. (1989). Signal transduction through the fibronectin receptor induces collagenase and stromelysin gene expression. *J. Cell Biol.* **109**, 877-889.

Woods, A. and Couchman, J. R. (1988). Focal adhesions and cell-matrix interactions. *Collagen Relat. Res.* **8**, 155-82.

Yamada, K. M. and Kennedy, D. W. (1984). Dualistic nature of adhesive protein function: fibronectin and its biologically active peptide fragments can autoinhibit fibronectin function. *J. Cell Biol.* **99**, 29-36.

Yamada, K. M. and Kennedy, D. W. (1987). Peptide inhibitors of fibronectin, laminin, and other adhesion molecules: Unique and shared features. *J. Cell. Physiol.* **130**, 21-28.

Yuan, Q., Jiang, W.-M., Krissansen, G. W. and Watson, J. D. (1990). Cloning and sequence analysis of a novel β2-related integrin transcript from T lymphocytes: Homology of integrin cysteine-rich repeats to domain III of laminin B chains. *Int. Immunol.* **2**, 1097-1108.

Zutter, M. M. and Santoro, S. A. (1990). Widespread histologic distribution of the α2β1 integrin cell-surface collagen receptor. *Am. J. Pathol.* **137**, 113-120.

Printed in Great Britain © The Society of Experimental Biology 1993   137

DEVELOPMENTAL REGULATION OF BASEMENT MEMBRANE DEPOSITION

DAVID EDGAR, JULIE CARTER, SARAH RUNSWICK and PATRICIA YBOT

Department of Human Anatomy and Cell Biology, P.O. Box 147, Liverpool L69 3BX, UK

Summary

The expression and localisation of the basement membrane glycoprotein laminin in tissues are two factors that determine to what extent it can influence cell behaviour during development. In order to begin to unravel the potential functions of laminin in vivo, the mechanisms regulating its expression, structure and stabilization in the basement membranes of developing tissues are discussed.

Introduction

Tissue development during embryogenesis and cell homeostasis in mature tissues is regulated by both cell-autonomous and epigenetic mechanisms. Epigenetic influences include interactions of cells with soluble molecules such as growth factors and hormones, or with molecules fixed in the immediate environment, either on the membranes of other cells or within the extracellular matrix. Over the last decade it has become clear that the extracellular matrix protein laminin - the major non-collagenous glycoprotein of basement membranes - can profoundly influence the adhesion, survival and differentiation of cells *in vitro*. The ability of cells to respond to laminin depends on their expression of functional laminin receptors (Reichardt and Tomaselli, 1991; Mecham, 1991; Elices *et al.*, 1991), and there is now abundant evidence that such receptors include members of the integrin superfamily, the expression and/or activity of which is itself epigenetically regulated by cell-cell interactions (Cohen *et al.*, 1989; de-Curtis *et al.*, 1991). Clearly, however, cells must have access to laminin if they are to respond to it. In contrast to the wealth of information currently available on the integrins, there is comparatively little discussion of the factors modulating laminin expression and deposition into the extracellular matrix of developing tissues - the subject of this chapter.

Laminin variants and the regulation of their expression

The laminin first isolated from the Engelbreth-Holm-Swarm (EHS) mouse sarcoma tumour (Timpl *et al.*, 1979) consisted of three covalently-bound polypeptide chain subunits of 400 kDa (A) 220 kDa (B1) and 210 kDa (B2), which together constitute the large cruciform molecule (Engel *et al.*, 1981). However, observations on the expression of laminin subunits in the early embryo showed that the B chains appeared before any A

Key words: laminin, nidogen, basement membrane, extracellular matrix, development, nerve, nervous system, gene expression.

chain could be detected (Cooper and Macqueen, 1983), and similar observations have been made in tissues later during development (Klein *et al.*, 1988). Correspondingly, numerous cultured cell lines synthesized and secreted B subunits in the absence of A chain (for examples see Lander *et al.*, 1985; Edgar *et al.*, 1988). Furthermore, the presence of subunits differing from the A chain but nevertheless covalently associated with the laminin B chains was reported in laminin from several tissues and cell lines (Ohno *et al.*, 1983; Engvall *et al.*, 1990; Paulsson and Saladin, 1989; Tokida *et al.*, 1990). Consequently it is now apparent that several variants of laminin exist which differ in their subunit composition and are expressed in a time- and tissue-specific manner. Presently three A chain variants have been described (A, A′ and M or merosin) and one B1 chain variant (S). (See Paulsson, 1992 for a recent review on this subject). Members of either the A or B1 classes of chain variants appear to be mutually exclusive in the laminin expressed in any one location (Sanes *et al.*, 1990; Engvall *et al.*, 1990). It should be noted that at the time of writing, the existence of other laminin variants is becoming apparent (Carter *et al.*, 1991; Rousselle *et al.*, 1991). The molecules kalinin and epiligrin found in association with the hemidesmosomes and anchoring filaments of the epidermis comprise subunits that display sequence homology to known laminin subunits, and which may be found covalently bound to the classical laminin B subunits (unpublished observations).

The pattern of laminin expression in developing tissues appears to be subject to two distinct mechanisms: a coordinated regulation of expression of subunits present in a given tissue (see Laurie *et al.*, 1989; Senior *et al.*, 1988; Kücherer-Ehret *et al.*, 1990), paralleling the need for rapid basement membrane construction during development, followed by a reduced need in maturity due to the low turn-over rate of basement membranes and their constituents in the adult (Trier *et al.*, 1990). Similar pleiotropic regulation of subunit expression has been observed in cultured mouse F9 teratocarcinoma cells, induced to differentiate with cAMP and retinoic acid (Cooper *et al.*, 1981; Kleinman *et al.*, 1987). Presumably this reflects the presence of the cAMP and retinoic acid upstream response elements reported for those subunit genes which have been analysed (Kallunki *et al.*, 1991; Vasios *et al.*, 1989). In addition, laminin expression has been reported to be regulated by factors including TGFβ and interferon (Maheshwari *et al.*, 1991; Vollberg *et al.*, 1991). However, the mechanisms responsible for the differential regulation of laminin subunit expression seen during the differentiation of cell types and tissues mentioned above (Klein *et al.*, 1988; Klein *et al.*, 1990) are largely unexplored. In this connection, however, it should be noted that it was reported recently that treatment of endothelial cells with steroids resulted in an enhancement of expression and incorporation of the A subunit into laminin, while the A′ subunit was down-regulated (Tokida *et al.*, 1990).

Laminin assembly and secretion from cells

Metabolic labelling and immunoprecipitation studies on a variety of cell types have shown that the assembly of laminin subunits occurs very soon after their synthesis in the rough endoplasmic reticulum, and that assembly proceeds via the formation of a B1-B2 dimer, followed by addition of the A chain (Peters *et al.*, 1985; Morita *et al.*, 1985;

Tokida *et al.,* 1990). These studies are consistent with dissociation and reconstitution experiments in vitro, showing that the B1 and B2 subunits have a greater propensity to assemble via a coiled-coil double helix than the A chain which will, however, recombine readily with the B chain dimer to form a coiled-coil triple helix (Hunter *et al.,* 1990). The kinetics of A chain association with B chain dimers and subsequent rapid laminin secretion, coupled with the observation that A chains alone may be secreted from PYS-2 teratocarcinoma cells (Cooper *et al.,* 1981), has led to the hypothesis that the A chain may function as a "secretory peptide" (Peters *et al.,* 1985). Whether or not an A chain variant subunit is necessary for secretion of laminin molecules lacking "authentic" A chains is not known.

The consequences of such structural variations on the ability of laminin to interact with other matrix molecules or cells remain largely unexplored, although it has recently been reported that the substitution of an S subunit for B1 (as found in the laminin localized at the neuromuscular junction) may stabilize the growth of motorneuron neurites (Hunter *et al.,* 1989). Furthermore, the laminin isolated from human placenta which has an M subunit substituted for the A chain has been reported to interact with cellular integrin receptors that differ from those binding EHS laminin (Brown and Goodman, 1991). Thus it is likely that laminin structural variants will display a variety of different cellular responses.

Incorporation of laminin into basement membranes

The availability of anti-laminin antibodies showed that laminin is a ubiquitous component of basement membranes, the well-defined, thin, sheet-like structures of the extracellular matrix that surround either individual cells or groups of cells and hence demarcate boundaries within the tissues (Timpl *et al.,* 1979). Although essentially confined to basement membranes in the adult, a much more wide-spread distribution has been reported during development; thus laminin immunoreactivity has been observed both on cell membranes (Liesi, 1985; Halfter and Song-Fua, 1987; Schiff and Rosenbluth, 1986; Cohen *et al.,* 1987; McLoon *et al.,* 1988; Kücherer-Ehret *et al.,* 1990) and in the interstitial extracellular matrix (Ekblom *et al.,* 1980; Ekblom *et al.,* 1990; Dziadek and Mitrangas, 1989; Kücherer-Ehret *et al.,* 1990) during embryogenesis and in the neonatal period. The presence of laminin outside basement membranes explains its high solubility and sensitivity to degradation during this time (Kücherer-Ehret *et al.,* 1990; Dziadek and Mitrangas, 1989). The biological significance of the above observations are three-fold: the laminin present during development may be rapidly broken down, permitting a remodelling of the extracellular matrix, and more potential cell binding sites on extra-basement membrane laminin may be exposed than would be the case when it is firmly embedded in a basement membrane (see Kücherer-Ehret *et al.,* 1990 for discussion). Furthermore, many cell types can contact laminin, irrespective of whether or not they are juxtaposed to a basement membrane. It is therefore likely that the roles of laminin during development are more transient, diverse and widespread than they are in the adult.

The reason for the developmental change in distribution of laminin is not known, no differences in subunit composition being detected during the time when laminin shifts to becoming confined to the basement membranes, and no differences being seen between the subunit composition of soluble and basement-membrane associated laminin during development (Kücherer-Ehret *et al.*, 1990). However, the fact that the extra-basement membrane laminin is found at a time when laminin gene expression is high, as reflected by levels of mRNA coding for individual subunits (Laurie *et al.*, 1989; Senior *et al.*, 1988; Kücherer-Ehret *et al.*, 1990), indicates that rate of synthesis exceeds that with which laminin may be stably incorporated into the basement membranes (Kücherer-Ehret *et al.*, 1990). The rate-limiting step in laminin incorporation into basement membranes, and hence the rate limiting step for basement membrane assembly itself is unknown.

In addition to laminin, basement membranes comprise several other major components including collagen type IV, heparan sulphate proteoglycan and nidogen/entactin, together with numerous minor components that have a more restricted distribution. The tissue-chimeric experiments of Simon-Assmann and co-workers have demonstrated that both epithelial cells and cells from the underlying mesenchyme contribute distinct components to the basement membrane which separates them (Simo *et al.*, 1991; Simon-Assmann *et al.*, 1989). Thus, while the laminin of gut basement membrane may derive either from the epithelium or mesenchyme, heparan sulphate proteoglycan and collagen type IV are derived from epithelium and mesenchyme, respectively (Simo *et al.*, 1991; Simon-Assmann *et al.*, 1989), and nidogen/entactin is almost exclusively mesenchymal in origin (Simon-Assmann and Edgar, unpublished observations; see also Dong and Chung, 1991).

The capacities of these components for self-assembly and their abilities to bind one another contribute to basement membrane assembly (Timpl, 1989; Paulsson, 1992), although the precise structure of individual basement membranes in vivo and the mechanisms of their assembly and turnover are not well elucidated (Yurchenco, 1990). Indeed, recent work has shown that tumour cells can produce an extracellular matrix containing basement membranes which lack both nidogen/entactin and collagen type IV. Such structures are relatively unstable, relying on cation-dependent non-covalent interactions between laminin molecules for their stability (Yurchenco *et al.*, 1992).

The ability of collagen type IV and laminin molecules to self-associate into three-dimensional reticular structures presents the possibility for basement membrane stabilization via interactions between the two structures (Yurchenco *et al.*, 1992). Although the laminin isolated from many cell types is often non-covalently but tightly associated with the protein entactin/nidogen (Cooper *et al.*, 1981; Lander *et al.*, 1985; Edgar *et al.*, 1988) and with heparan sulphate proteoglycan (Lander *et al.*, 1985), the binding of laminin to collagen type IV has been controversial until recently. It is now clear, however, that this binding is not direct, but rather is mediated by entactin/nidogen, which has distinct high-affinity binding sites for both laminin and collagen type IV (Fox *et al.*, 1991). Given the necessity of stoichiometric amounts of entactin/nidogen for laminin binding to collagen type IV, then both over-expression and under-expression of entactin/nidogen will preclude stabilization of basement membranes by binding of the laminin network to that of collagen type IV.

Finally, in mature tissues the individual components of basement membranes are often very difficult to extract and are found covalently linked together (Paulsson, 1992), indicating a further stabilization of structure. The mechanism of this stabilization is currently under investigation, and recent observations have shown that nidogen/entactin is a good substrate for tissue transglutaminases (Aeschlimann and Paulsson, 1991), pointing to an extracellular enzymic mechanism of basement membrane stabilization via intermolecular covalent bonds in the adult.

Laminin expression in a specific developing tissue: The nervous system

Previous work on laminin has shown that it promotes the proliferation, migration, differentiation and survival of vertebrate neural cells in vitro (see Mercurio, 1990 and Edgar, 1991 for reviews). Furthermore, the transitory appearance of laminin immunoreactivity in the developing central nervous system at the time of migration and/or initial axon elongation (Cohen *et al.,* 1987; Liesi, 1985), and the ability of laminin-containing basement membranes of the peripheral nervous system to support regeneration of lesioned axons in the adult (Ide *et al.,* 1983), indicates that laminin may well exert a role in nervous system development and regeneration in vivo analogous to its activities demonstrated in vitro. In addition to laminin, it is however clear that cell-cell interactions involving cell adhesion molecules and neurotrophic factors are required for optimal regeneration (Hausmann *et al.,* 1989; Bixby and Jhabvala, 1990; Kawaja and Gage, 1991).

Reports describing laminin immunoreactivity in neurons of the central nervous system (Hagg *et al.,* 1989; Suzuki *et al.,* 1990) and mRNAB1 in retinal ganglion cells (Sarthy and Fu, 1990) remain controversial (Jucker *et al.,* 1992). It is, however, well-established that laminin is expressed by at least some neuroepithelial and glial cells during embryogenesis. In the adult, laminin expression is confined to those areas which retain some capacity for regeneration including the peripheral nervous system and the olfactory system of the CNS (Liesi, 1985). Although astrocytes of the mature CNS show no expression in situ, after lesion, cultured or reactive astrocytes do display laminin immunoreactivity (Liesi *et al.,* 1984).

The expression and deposition of laminin subunits have mainly been analysed in the peripheral nervous system. In the sciatic nerve, both schwann cells and perineurial cells have been shown to express laminin (Jaakkola *et al.,* 1989; Sanes *et al.,* 1990). These cells express the laminin B chains, and although no evidence for A chain synthesis by cultured schwann cells was originally seen (Cornbrookes *et al.,* 1983), it could recently be shown that the endoneurial basement membrane contains the alternative A chain, merosin, whereas the perineurial extracellular matrix displayed (very weak) A chain immunoreactivity. It is presently not clear if the merosin-specific chain is only expressed by the schwann cells in vivo, or alternatively if the merosin of the endoneurial basement membrane is contributed by some other cell type in the nerve.

The crucial roles and interactions mediated by the basement membrane of peripheral nerve have been demonstrated in a series of experiments carried out by Bunge and co-workers (Bunge *et al.,* 1990). Although isolated schwann cells express laminin, axon-schwann cell contact is necessary for basement membrane formation, which is itself necessary for myelination of the axon by the schwann cell (Eldridge *et al.,* 1987; Eldridge *et al.,* 1989). These workers went on to show that optimal synthesis of collagen type IV together with fibroblasts and/or the interstitial extracellular matrix were normally necessary for basement membrane formation. However, addition of exogenous laminin to cultured schwann cells was able to stimulate the appearance of basement membrane-like structures (Eldridge *et al.,* 1987; Bunge *et al.,* 1990). While the molecular bases of these observations remain unknown, they indicate that basement membrane deposition depends upon the provision of the appropriate cellular and extracellular matrix environment, and that the stoichiometry of basement membrane components is critical for their deposition.

Acknowledgements

The work carried out in our laboratory is supported by the Wellcome Trust, the Anatomical Society of Great Britain and Ireland, and the Commission of the European Community.

References

Aeschlimann, D. and Paulsson, M. (1991). Cross-linking of laminin-nidogen complexes by tissue transglutaminase. A novel mechanism for basement membrane stabilization. *J.Biol.Chem.* **266**, 15308-15317.

Bixby, J.L. and Jhabvala, P. (1990). Extracellular matrix molecules and cell adhesion molecules induce neurites through different mechanisms. *J.Cell Biol.* **111**, 2725-2732.

Brown, J.C. and Goodman, S.L. (1991). Different cellular receptors for human placental laminin and murine EHS laminin. *FEBS Lett.* **282**, 5-8.

Bunge, M.B., Clark, M.B., Dean, A.C., Eldridge, C.F. and Bunge, R.P. (1990). Schwann cell function depends upon axonal signals and basal lamina components. *Ann.N.Y.Acad.Sci.* **580**, 281-287.

Carter, W.G., Ryan, M.C. and Gahr, P.J. (1991). Epiligrin, a new cell adhesion ligand for integrin alpha 3 beta 1 in epithelial basement membranes. *Cell* **65**, 599-610.

Cohen, J., Burne, J.F., McKinlay, C. and Winter, J. (1987). The role of laminin and the laminin/fibronectin receptor complex in the outgrowth of retinal ganglion cell axons. *Devl.Biol.* **122**, 407-418.

Cohen, J., Nurcombe, V., Jeffrey, P. and Edgar, D. (1989). Developmental loss of functional laminin receptors on retinal ganglion cells is regulated by their target tissue, the optic tectum. *Development* **107**, 381-387.

Cooper, A.R., Kurkinen, M., Taylor, A. and Hogan, B.L.M. (1981). Studies on the biosynthesis of laminin by murine parietal endoderm cells. *Eur.J.Biochem.* **119**, 189-197.

Cooper, A.R. and MacQueen, H.A. (1983). Subunits of laminin are differentially synthesized in mouse eggs and early embryos. *Devl.Biol.* **96**, 467-471.

Cornbrookes C.J., Carey, D.J., McDonald, J.A., Timpl, R., and Bunge, R.P. (1983). In vivo and in vitro observations on laminin production by Schwann cells. *Proc.Natl.Acad.Sci.USA* **80**, 3850-3854.

de-Curtis, I., Quaranta, V., Tamura, R.N. and Reichardt, L.F. (1991). Laminin receptors in the retina: sequence analysis of the chick integrin alpha 6 subunit. Evidence for transcriptional and posttranslational regulation. *J.Cell Biol.* **113**, 405-416.

Dong, L.-J. and Chung, A.E. (1991). The expression of the genes for entactin, laminin A, laminin B1 and laminin B2 in murine lens morphogenesis and eye development. *Differentiation* **48**, 157-172.

Dziadek, M. and Mitrangas, K. (1989). Differences in the solubility and susceptibility to proteolytic degradation of basement membrane components in adult and embryonic mouse tissues. *Am.J.Anat.* **184**, 298-310.

Edgar, D. (1991). The expression and distribution of laminin in the developing nervous system. In: *Nerve Cell Biology* (ed. D.Bray, N.Holder, R.Keynes, A.Lumsden and V.H.Perry) pp.9-12. Cambridge,U.K.: The Company of Biologists, Ltd.

Edgar, D., Timpl, R. and Thoenen, H. (1988). Structural requirements for the stimulation of neurite outgrowth by two variants of laminin and their inhibition by antibodies. *J.Cell Biol.* **106**, 1299-1306.

Ekblom, M., Klein, G., Mugrauer, G., Fecker, L., Deutzmann, R., Timpl, R. and Ekblom, P. (1990). Transient and locally restricted expression of laminin A chain mRNA by developing epithelial cells during kidney organogenesis. *Cell* **60**, 337-346.

Ekblom, P., Alitalo, K., Vaheri, A., Timpl, R. and Saxen, L. (1980). Induction of a basement membrane glycoprotein in embryonic kidney: possible role of laminin in morphogenesis. *Proc.Natl.Acad.Sci.USA* **77**, 485-489.

Eldridge, C.F., Bunge, M.B., Bunge, R.P. and Wood, P.M. (1987). Differentiation of axon-related schwann cells in vitro. I. Ascorbic acid regulates basal lamina assembly and myelin formation. *J.Cell Biol.* **105**, 1023-1034.

Eldridge, C.F., Bunge, M.B., Bunge, R., Rigamonti, L., Procacci, P. and Ledda, M. (1989). Differentiation of axon-related schwann cells in vitro: II. Control of myelin formation by basal lamina. *J.Neurosci.* **9**, 625-638.

Elices, M.J., Urry, L.A. and Hemler, M.E. (1991). Receptor functions for the integrin VLA-3: fibronectin, collagen, and laminin binding are differentially influenced by Arg-Gly-Asp peptide and by divalent cations. *J.Cell Biol.* **112**, 169-181.

Engel, J., Odermatt, E., Engel, A., Madri, J.A., Furthmayr, H., Rohed, H. and Timpl, R. (1981). Shapes, domain organisations and flexibility of laminin and fibronectin, two multifunctional proteins of the extracellular matrix. *J.Mol.Biol.* **150**, 97-120.

Engvall, E., Earwicker, D., Haaparanta, T., Ruoslahti, E. and Sanes, J.R. (1990). Distribution and isolation of four laminin variants; tissue restricted distribution of heterotrimers assembled from five different subunits. *Cell Regul.* **1**, 731-740.

Fox, J.W., Mayer, U., Nischt, R., Aumailley, M., Reinhardt, D., Wiedemann, H., Mann, K., Timpl, R., Krieg, T., Engel, J. and Chu, M.-L. (1991). Recombinant nidogen consists of three globular domains and mediates binding of laminin to collagen type IV. *EMBO (Eur.Mol.Biol.Org.) J.* **10**, 3137-3146.

Hagg, T., Muir, D., Engvall, E., Varon, S. and Manthorpe, M. (1989). Laminin-like antigen in rat CNS neurons: distribution and changes upon brain injury and nerve growth factor treatment. *Neuron* **3**, 721-732.

Halfter, W. and Song-Fua, C. (1987). Immunohistochemical localization of laminin, neural cell adhesion molecule, collagen type IV and T-61 antigen in the embryonic retina of the japanese quail by in vivo injection of antibodies. *Cell Tissue Res.* **249**, 487-496.

Hausmann, B., Sievers, J., Hermanns, J. and Berry, M. (1989). Regeneration of axons from the adult rat optic nerve: Influence of fetal brain grafts, laminin, and artificial basement membrane. *J.Comp.Neurol.* **281**, 447-466.

Hunter, D.D., Porter, B.E., Bulock, J.W., Adams, S.P., Merlie, J.P. and Sanes, J.R. (1989). Primary sequence of a motor neuron-selective adhesive site in the synaptic basal lamina protein S-laminin. *Cell* **59**, 905-913.

Hunter, I., Schultheis, T., Bruch, M., Beck, K. and Engel, J. (1990). Evidence for a specific mechanism of laminin assembly. *Eur.J.Biochem.* **188**, 205-211.

Ide, C., Tohyama, K., Yokota, R., Nitatori, T. and Onodera, S. (1983). Schwann cell basal lamina and nerve regeneration. *Brain Res.* **288**, 61-75.

Jaakkola, S., Peltonen, J. and Uitto, J.J. (1989). Perineurial cells coexpress genes encoding interstitial collagens and basement membrane zone components. *J.Cell Biol.* **108**, 1157-1163.

Jucker, M., Bialobok, P., Hagg, T. and Ingram, D. (1992).Laminin immunohistochemistry is dependent on method of tissue fixation. *Brain Res.* **586**, 166-170.

Kallunki, T., Ikonen, J., Chow, L.T., Kallunki, P. and Tryggvason, K. (1991). Structure of the human laminin B2 chain gene reveals extensive divergence from the laminin B1 chain gene. *J.Biol.Chem.* **266**, 221-228.

Kawaja, M. and Gage, F.H. (1991). Reactive astrocytes are substrates for the growth of adult CNS axons in the presence of elevated levels of nerve growth factor. *Neuron* **7**, 1019-1030.

Klein, G., Ekblom, M., Fecker, L., Timpl, R. and Ekblom, P. (1990). Differential expression of laminin A and B chains during development of embryonic mouse organs. *Development* **110**, 823-837.

Klein, G., Langegger, M., Timpl, R. and Ekblom, P. (1988). Role of laminin A chain in the development of epithelial cell polarity. *Cell* **55**, 331-341.

Kleinman, H.K., Ebihara, I., Killen, P.D., Sasaki, M., Cannon, F.B., Yamada, Y. and Martin, G.R. (1987). Genes for basement membrane proteins are coordinately expressed in differentiating F9 cells but not in normal adult murine tissues. *Devl.Biol.* **112**, 373-378.

Kücherer-Ehret, A., Pottgiesser, J., Kreutzberg, G.W., Thoenen, H. and Edger, D. (1990). Developmental loss of laminin from the interstitial extracellular matrix correlates with decreased laminin gene expression. *Development* **110**, 1285-1293.

Lander, A.D., Fujii, D.K. and Reichardt, L.F. (1985). Purification of a factor that promotes neurite outgrowth: isolation of laminin and associated molecules. *J.Cell Biol.* **101**, 898-913.

Laurie, G.W., Horikoshi, S., Killen, P.B., Segui-Real, B. and Yamada, Y. (1989). In situ hybridization reveals temporal and spatial changes in cell expression of mRNA for a laminin receptor, laminin, and basement membrane type iv collagen. *J.Cell Biol.* **109**, 1351-1362.

Liesi, P. (1985). Laminin-immunoreactive glia distinguish regenerative adult CNS systems from non-regenerative ones. *EMBO (Eur.Mol.Biol.Org.) J.* **4**, 2505-2511.

Liesi, P., Kaakkola, S., Dahl, D. and Vaheri, A. (1984). Laminin is induced in astrocytes of adult brain by injury. *EMBO (Eur.Mol.Biol.Org.) J.* **3**, 683-686.

Maheshwari, R.K., Kedar, V.P., Coon, H.C. and Bhartiya, D. (1991). Regulation of laminin expression by interferon. *J.Interferon Res.* **11**, 75-80.

McLoon, S.C., McLoon, L.K., Palm, S.L. and Furcht, L.T. (1988). Transient expression of laminin in the optic nerve of the developing rat. *J.Neurosci.* **8**, 1981-1990.

Mecham, R.P. (1991). Laminin receptors. *Annu.Rev.Cell Biol.* **7**, 71-91.

Mercurio, A.M. (1990). Laminin: multiple forms, multiple receptors. *Curr.Opin.Cell Biol.* **2**, 845-849.

Morita, A., Sugimoto, E. and Kitagawa, Y. (1985). Post-translational assembly and glycosylation of laminin subunits in parietal endoderm-like F9 cells. *Biochem.J.* **229**, 259-264.

Ohno, M., Martinez-Hernandez, A., Ohno, N. and Kefalides, N.A. (1983). Isolation of laminin from human placental basement membranes:amnion, chorion and chorionic microvessels. *Biochem.Biophys.Res.Commun.* **112**, 1091-1098.

Paulsson, M. (1992).Basement membrane proteins: structure, assembly and cellular interactions. *Critical Reviews Biochem.Mol.Biol.* **27**, 93-127.

Paulsson, M. and Saladin, K. (1989). Mouse heart laminin. Purification of the native protein and structural comparison with Engelbreth-Holm-Swarm tumor laminin. *J.Biol.Chem.* **264**, 18726-18732.

Peters, B.P., Hartle, R.J., Krzesick, R.F., Kroll, T.G., Perini, F.C., Balun, J.E., Goldstein, I.J. and Ruddon, R.W. (1985). The biosynthesis, processing and secretion of laminin by human choriocarcinoma cells. *J.Biol.Chem.* **260**, 14732-14742.

Reichardt, L.F. and Tomaselli, K.J. (1991). Extracellular matrix molecules and their receptors: Functions in neural development. *Annu.Rev.Neurosci.* **14**, 531-570.

Rousselle, P., Lunstrum, G.P., Keene, D.R. and Burgeson, R.E. (1991). Kalinin: An epithelium-specific basement membrane adhesion molecule that is a component of anchoring filaments. *J.Cell Biol.* **114**, 567-576.

Sanes, J.R., Engvall, E., Butkowski, R. and Hunter, D.D. (1990). Molecular heterogeneity of basal laminae: Isoforms of laminin and collagen IV at the neuromuscular junction and elsewhere. *J.Cell Biol.* **111**, 1685-1699.

Sarthy, P.V. and Fu, M. (1990). Localization of laminin B1 mRNA in retinal ganglion cells by in situ hybridization. *J.Cell Biol.* **110**, 2099-2108.

Schiff, R. and Rosenbluth, J. (1986). Ultrastructural localization of laminin in rat sensory ganglia. *J.Histochem.Cytochem.* **34**, 1691-1699.

Senior, P.V., Critchley, D.R., Beck, F., Walker, R.A. and Varley, J.M. (1988). The localization of laminin mRNA and protein in the postimplantation embryo and placenta of the mouse: An in situ hybridization and immunocytochemical study. *Development* **104**, 431-446.

Simo, P., Simon-Assmann, P., Bouziges, F., Leberquier, C., Kedinger, M., Ekblom, P. and Sorokin, L. (1991). Changes in the expression of laminin during intestinal development. *Development* **112**, 477-487.

Simon-Assmann, P., Bouziges, F., Vigny, M. and Kedinger, M. (1989). Origin and deposition of basement membrane heparan sulfate proteoglycan in the developing intestine. *J.Cell Biol.* **109**, 1837-1848.

Suzuki, H., Yamamoto, T., Yamamoto, H., Konno, H., Iwasaki, Y., Ohara, Y. and Terunuma, H. (1990). Intraneuronal laminin-like immunoreactivity in the human central nervous system. *Brain Res.* **520**, 324-329.

Timpl, R. (1989). Structure and biological activity of basement membrane proteins. *Eur.J.Biochem.* **180**, 487-502.

Timpl, R., Rohde, H., Robey, P.G., Rennerd, S.I., Foidart, J.-M. and Martin, G.R. (1979). Laminin - a glycoprotein from basement membranes. *J.Biol.Chem.* **254**, 9933-9937.

Tokida, Y., Aratani, Y., Morita, A. and Kitagawa, Y. (1990). Production of two variant laminin forms by endothelial cells and shift of their relative levels by angiostatic steroids. *J.Biol.Chem.* **265**, 18123-18129.

Trier, J.S., Allan, C.H., Abrahamson, D.R. and Hagen, S.J. (1990). Epithelial basement membrane of mouse jejunum. Evidence for laminin turnover along the entire crypt-villus axis. *J.Clin.Invest.* **86**, 87-95.

Vasios, G.W., Gold, J.D., Petkovich, M., Chambon, P. and Gudas, L.J. (1989). A retinoic acid-responsive element is present in the 5′ flanking region of the laminin B1 gene. *Proc.Natl.Acad.Sci.USA* **86**, 9099-9103.

Vollberg, T.M.S., George, M.D. and Jetten, A.M. (1991). Induction of extracellular matrix gene expression in normal human keratinocytes by transforming growth factor beta is altered by cellular differentiation. *Exp.Cell Res.* **193**, 93-100.

Yurchenco, P.D. (1990). Assembly of basement membranes. *Ann.N.Y.Acad.Sci.* **580**, 195-213.

Yurchenco, P.D., Cheng, Y.S. and Colognato, H. (1992). Laminin forms an independent network in basement membranes. *J.Cell Biol.* **117**, 1119-1133.

Printed in Great Britain © The Society of Experimental Biology 1993 147

LAMININ OLIGOSACCHARIDES PLAY A PIVOTAL ROLE IN CELL SPREADING

MARVIN L. TANZER*, MARTIN S. GINIGER and S. CHANDRASEKARAN

Department of Biostructure and Function, School of Dental Medicine University of Connecticut Health Center, Farmington, CT 06030-3705

Summary

The basement membrane glycoprotein laminin promotes cell adhesion, spreading and neurite outgrowth. We can uncouple cell adhesion and spreading (or neurite outgrowth) when unglycosylated laminin is used as a substratum. Mouse melanoma cells, B16F1 line, readily attach to unglycosylated laminin but fail to spread once adherent. Spreading can be restored by titration with glycosylated laminin or with laminin glycopeptides. When the laminin substratum is absent in the test chambers, the cells do not adhere when either intact laminin or its glycopeptides are then added. Analyses show that these added substances are recoverable from the culture medium and do not bind to the chamber surfaces. Use of selective inhibitors which interfere with carbohydrate processing yields several glycoforms of laminin which we have isolated and examined for their ability to support cell adhesion and spreading. Laminin which is enriched in high mannose oligosaccharides is much more effective in promoting cell spreading than laminin which is enriched in hybrid oligosaccharides. These results are consistent with earlier studies which showed that ConA, which primarily recognizes mannose residues, could also uncouple cell adhesion and spreading. Although mono- and disaccharides failed to restore cell spreading, we have found that addition of various mannose oligosaccharides to adherent cells effectively reestablishes their spreading behavior. The extent of cell spreading which is achieved by the added saccharides is related to their amount, their duration of addition, and their molecular structures. The composite results indicate that: 1) melanoma cells recognize and adhere to laminin protein domains; 2) subsequent cell spreading is dependent upon recognition of laminin oligosaccharides, either in solution or bound to laminin; 3) mannose-containing oligosaccharides, *per se*, suffice to promote spreading of laminin-adherent melanoma cells.

Introduction

The in vitro model of cell interactions with substrata has provided insights into the molecular determinants of the substrata and into the nature of the cell surface receptors. It is generally assumed that cell spreading, which sequentially follows cell adhesion, relies upon the same determinants which are implicated in the adhesion process. In other words, integrins and their counterpart ligands within the substrata (e.g., fibronectin, laminin) are

*Author for correspondence.

Key words: glycosyl pathway inhibitors, mannosyl oligosaccharides, cell receptors.

thought to be necessary and to suffice for both cell adhesion and spreading (Hynes 1992, Ruoslahti 1991). Recent data suggest that reciprocal lectin-mediated interactions between the cell surface and laminin also play a role in cell adhesion (Bouzon *et al.*, 1990, Chammas *et al.*, 1991, Lotan and Raz 1988). We have developed an experimental model in which cell adhesion and spreading can be uncoupled by use of unglycosylated laminin as a substratum (Dean *et al.*, 1990). Spreading is restored in this system by addition of suitable glycosyl groups, either as components of the substratum or in the fluid surrounding the cells (Chandrasekaran *et al.*, 1991). These results evoke other examples of carbohydrate-mediated cellular responses by, e.g., selectins (Stoolman 1992), NCAMs (Rutishauser *et al.*, 1988), and hepatic cell lectins (Ashwell and Harford 1982). The informational carbohydrates which act as recognition signals in these examples are as structurally specific as the peptide determinants recognized by receptors such as the integrins. It has already been shown that there are biological processes in which both peptide and carbohydrate determinants are essential for progression of a response (Stoolman 1992); it is this situation which has been uncovered in our experimental system.

Analysis of cell spreading

Initially, using EHS tumor laminin as a substratum, we found that preincubation of the laminin surface with the lectins WGA or GSA-I prevented cell adhesion while ConA permitted full adhesion but halted cell spreading (and neurite outgrowth) (Dean *et al.*, 1988). Several possible explanations could account for this latter result including non-specific steric hinderance by the lectin. To critically test whether laminin glycosyl groups were, in fact, implicated in cell spreading we used unglycosylated laminin as a substratum. Production of this form of laminin depended upon addition of optimal amounts of tunicamycin to cultures of murine cells which constituitively produce laminin. A serious complication of this approach is that the unglycosylated laminin is not secreted by the cells and must be purified from cell lysates. By using sequential immunoaffinity chromatography and lectin affinity chromatography we were able to isolate unglycosylated laminin free of its glycosylated counterpart as well as free of detectable cellular proteins (Dean *et al.*, 1990). A comprehensive analysis of the molecular structure of unglycosylated laminin indicated that, except for lack of carbohydrates, it had the same structure as glycosylated laminin (Table 1). No differences could be found at the level of primary (i.e., subunits), secondary and tertiary protein

Table 1. *Summary of the molecular properties of unglycosylated laminin*

Molecular feature	Criterion
Cruciform shape	Electron microscopy
Secondary structure	Circular dichroism
Molecular size	Gel electrophoresis
Subunit composition (A,B1,B2)	Immunoblots
Absent carbohydrate	(1) Metabolic labelling; (2) Lectin blots

Table 2. *Restoration of cell spreading by laminin glycopeptides*

Additive	Relative concentration	Spreading
None	N/A	–
Pronase digest	1×	–
Pronase digest	3×	+
Pronase digest	5×	++
Intact laminin	1×	++

Melanoma cells were presented with an unglycosylated laminin substratum, with or without an additive in the medium. Relative concentrations are expressed with regard to the amount of intact, glycosylated laminin shown in the last line of the table. A pronase digest of glycosylated laminin was incrementally added as indicated. Control studies, using the same amounts of a pronase digest of unglycosylated laminin, failed to elicit a spreading response (Chandrasekaran *et al.*, 1991).

structure. Both forms of laminin adsorbed equally well to the surfaces of the plastic wells used in the cell adhesion and spreading assays. The test cells, either murine B16F1 melanoma cells or rat PC12 pheochromacytoma cells, adhered equally well to the two forms of laminin. However, the cells adherent to unglycosylated laminin failed to spread or to project neurite extensions, respectively. An example of adherent versus adherent/spread melanoma cells is shown in Figure 1. This figure dramatically demonstrates the difference in cellular responsiveness, when cells are presented with an unglycosylated substratum plus or minus a mannose-rich oligosaccharide in the medium.

Depletion of the carbohydrates from laminin by use of tunicamycin could have produced subtle effects upon laminin structure which may not have been detected by molecular analyses. In order to determine that it was the absent carbohydrates, *per se*, which were responsible for the altered cellular response, we replenished those carbohydrates (Figs. 1 and 2). Replenishement was carried out in two ways, by blending the two types of laminin in varying proportions or by adding digests of the two types of laminin to the assay system. In the former case, there was a progressive response of cell spreading as the proportion of glycosylated laminin increased (Fig. 2). In the latter case, there was a progressive response of cell spreading as greater amounts of glycopeptides were added (Table 2). Control studies indicated that the glycopeptides did not adhere to the test chamber but were in solution throughout the incubation, obviating the notion that they must be bound to be effective.

Glycosylation inhibitors

Another approach to defining which category of laminin carbohydrates might be responsible for triggering cell spreading was to use selective inhibitors of the glycosylation pathway. Inhibitors are available which interfere with several major steps in the maturation of N-linked glycosyl groups (Elbein 1987). The results of one such experiment, in which PC12 cells were measured for neurite outgrowth, is shown in Table 3. The data indicate that those laminin glycoforms which are enriched in high mannose oligosaccharides are more effective in promoting cell spreading than laminin which is enriched in hybrid oligosaccharides, containing fewer mannose residues.

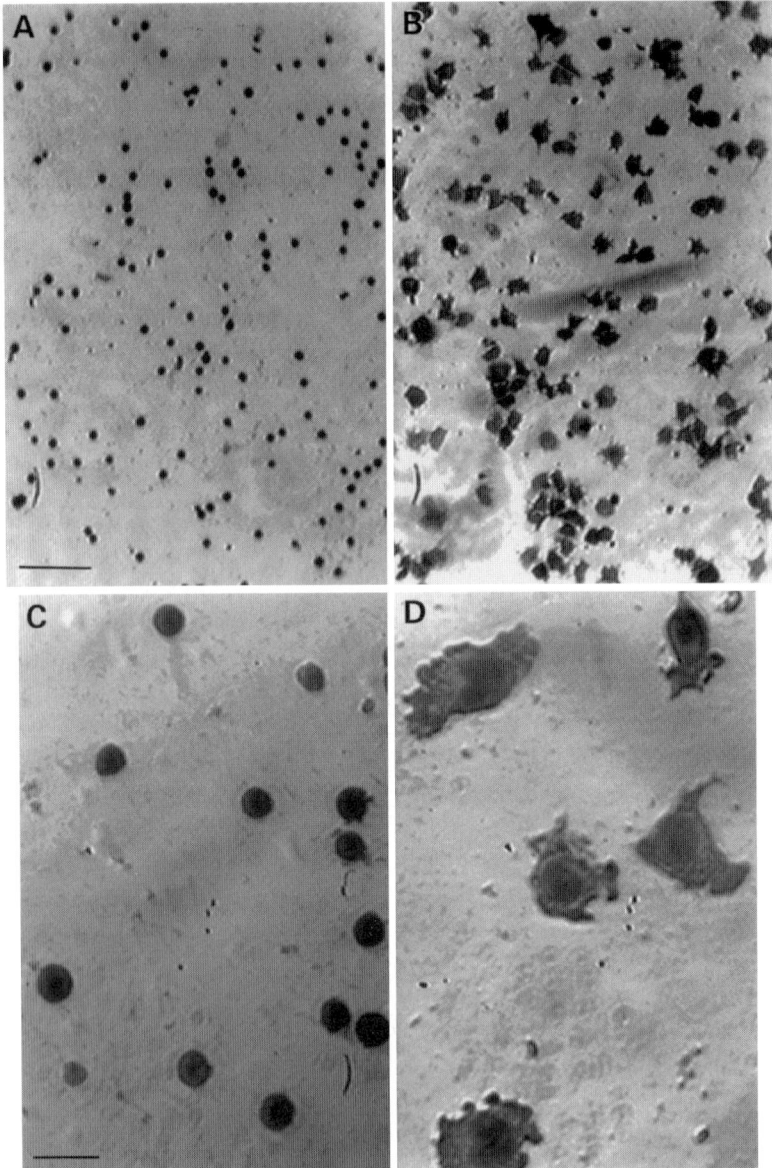

Fig. 1. Adherent melanoma cells (panels A and C) compared to adherent/spread melanoma cells (panels B and D). Cells were added to test wells containing an unglycosylated laminin substratum, in the presence (B,D) or absence (A,C) of a mannose-rich oligosaccharide. A,B, bar=100 µm; C,D, bar=20 µm.

High mannose structures

This observation prompted us to examine whether other, specific carbohydrates would substitute for the laminin-derived carbohydrates. Although mono- and disaccharides failed to restore cell spreading (Chandrasekaran *et al.*, 1991), addition of various

Table 3. *Neurite outgrowth on different laminin glycoforms*

Laminin substrate	3 µg/well	6 µg/well
Complex and high mannose oligosaccharides	49±6%	48±4%
High mannose oligosaccharides	47±3%	55±5%
Hybrid oligosaccharides	21±6%	23±3%
Without oligosaccharides	4±2%	2±2%

PC12 cells were presented with several different laminin glycoforms as substrata. The laminin glycoforms were obtained from cultures of murine cells incubated in the presence of selective inhibitors of the glycosylation pathway. The laminins were isolated and characterized (Chandrasekaran *et al.*, 1991) prior to their use in the neurite outgrowth assays. Laminin containing complex and high mannose oligosaccharides is murine EHS tumor laminin (Arumugham *et al.*, 1986, Fujiwara *et al.*, 1988, Knibbs *et al.*, 1989). Laminin containing high mannose oliogosaccharides is from castanospermine-treated cells while laminin containing hybrid oligosaccharides is from swainsonine-treated cells. Laminin without oligosaccharides is from tunicamycin-treated cells.

Table 4. *Restoration of cell spreading by oligosaccharides and polysaccharides*

Additive	Amount	Cell area, μm^2±SEM
None	0	162±4.1
Amylopectin	10 µg	136±3.2
Pullulan	10 µg	149±3.1
Laminarin	10 µg	191±4.1
Chitin	10 µg	190±4.8
Glycogen	10 µg	181±4.9
Mannan	10 µg	500±19
Man3	2 nmol	246±9.0
Man6	2 nmol	614±16
Man9	2 nmol	884±48

Melanoma cells were presented with unglycosylated laminin as a substratum, with or without an additive in the medium. Cell areas were measured essentially by the method of Jones *et al.* (1986) except that a video image of the microsopic field was captured into a computer and stored for analysis. The image was digitized and all non-contacting cells were selected for automatic measurement of their areas. At least one hundred cells were analyzed per sample well and duplicate or triplicate wells were measured. A mixture of oligosaccharides was used in the case of laminarin and chitin. 10 µg of mannan produced maximal spreading; therefore the other heterogeneous substances were compared at equivalent mass amounts. Man3, 6 or 9 refer to a related family of oligosaccharides, all based upon a chitobiose core and differing by their content of mannose units. These substances have precise molecular weights and were compared on a molar basis.

mannose-containing structures to adherent cells effectively reestablished their spreading behavior (Giniger *et al.*, unpublished). Thus, yeast mannan and the large mannose oligosaccharides Man6 and Man9 are effective, whereas a variety of other polysaccharides and the small mannose oligosaccharide Man3 are ineffective (Table 4). Titration of the response with the effective substances shows a series of related saturation curves. However, the level of saturation which is achieved varies with the specific titrant. The maximal effect is obtained with an oligosaccharide which contains nine mannoses while a less than maximal effect is found with an oligosaccharide which contains six

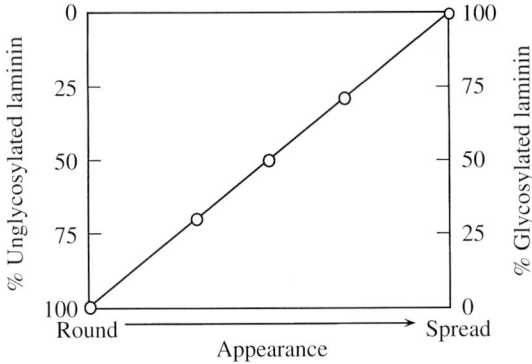

Fig. 2. Restoration of cell spreading by glycosylated laminin. Summary of the experimental protocol and results of an experiment in which melanoma cells were presented with blended substrata containing different proportions of glycosylated and unglycosylated laminin. (Adapted from Chandrasekaran *et al.*, 1991).

mannoses. An oligosaccharide which contains three mannoses has virtually no effect. The maximally responding cells show an increase of average cell area over five-fold greater than the control adherent, spherical cells. Intermediate responses are obtained using yeast mannan or various laminin glycoforms; the increases in cell area promoted by these substances are two to three-fold greater than the control cell area.

Putative receptors

The titration data imply that a receptor-mediated response is occuring, as there are both time-dependent and concentration-dependent responses, each reaching a plateau. The differences in extent of the cellular spreading response may be due to efficacy in occupying a receptor site, as suggested by the responses to Man3, Man6 and Man9. The differences may also be due to the structural forms of the mannose oligosaccharides, the major determinants being number of mannoses, type of branching, specific array of linkages and anomeric configuration. Such features are important in the response of mammalian hepatic Gal/GalNAc receptors (Lee 1989). Model building studies, based upon NMR-derived conformations (Wooten *et al.*, 1990), may provide insight into which structural features are important for receptor-mediated recognition of mannose oligosaccharides. Concurrently, we are attempting to isolate and characterize the putative cell surface receptors from B16 murine melanoma cells. We have previously speculated that the cellular carbohydrate recognition sites might either reside within an integrin or be entirely separate (Tanzer *et al.*, 1991). If the former case is found, it would imply that integrins may have at least two distinct binding sites, one for amino acid/peptide recognition and one for carbohydrate recognition. Integrin binding properties are modulated by cells (Hynes 1992) and, conceivably, carbohydrate ligands might influence the specificity and affinity of a given integrin receptor.

Membrane-bound animal cell lectins are ubiquitous and, although specific for individual saccharides, have a wide range of carbohydrate recognition sites (Varki 1992).

A mannose-binding lectin has been detected in macrophages and has been well-characterized (Lennartz *et al.*, 1987). It may be present in liver endothelial cells but has not been described in other cell types. In contrast, the mannose-6-phosphate lectins are virtually ubiquitous and are primarily implicated in intracellular routing of newly synthesized lysososmal enzymes (Varki 1992). However, one form of these lectins serves as a cell surface receptor for extracellular mannose-6-phosphate. Conceivably, the melanoma cells and PC12 cells, both of cancer origin, may express variants of either or both of these mannose-specific lectins.

Laminin high mannose oligosaccharides

Interestingly, high mannose oligosaccharides have been detected only on one laminin-derived proteolytic fragment, E7 (Fujiwara *et al.*, 1988). The location of that fragment has not been determined but it most likely arises from the long arm of laminin. Compositional analyses show that E7 contains 72% carbohydrate, has a mass of 29 kDa, is acidic and lacks Cys, Met and Tyr (Fujiwara *et al.*, 1988, Ott *et al.*, 1982). Calculation, based upon these values, suggests that E7 contains a minimum of 74 amino acids. Examination of the distribution of N-linked consensus sites for oligosaccharides in a model of the long arm of laminin (Beck *et al.*, 1990) suggests that one location best fits these particular molecular and compositional criteria. That locus is about 150 amino acids from the beginning of the long arm, towards the globular end of that arm. Presumably, E7 is derived by elastase cleavage of all three of the laminin chains although there are no definitive data concerning this point. We are currently trying to pinpoint the precise site(s) of mannose attachment and should be able to identify those specific Asn residues which are substituted by mannose-containing oligosaccharides. Several such oligosaccharides are estimated to be present in the E7 fragment (Fujiwara *et al.*, 1988).

Conclusions

We have developed a model system in which cell adhesion and cell spreading are uncoupled from each other. Cells adhere equally well to laminin lacking N-linked oligosaccharides as to fully glycosylated laminin, but cells adherent to unglycosylated laminin fail to spread on that substratum. Spreading is readily restored by addition of suitable carbohydrates, either as part of the substratum or in solution, with mannose-rich oliogsaccharides being the most effective substances. The extent of cell spreading is related to structural features of the effective oligosaccharides and the kinetics of response suggest that a receptor-mediated process is occurring.

Several immediate questions evolve from the current information. Which part(s) of the laminin molecule participate in the cell adhesion and cell spreading responses? Can oligosaccharides themselves suffice for both spreading and adhesion? How do the cells recognize the mannose-rich oligosaccharides? Are both cell surface integrins and putative carbohydrate receptors mediating adhesion and spreading? How general is the role of carbohydrates in promoting cell spreading? These questions and related ones potentially will be answered as greater attention becomes focused on the role of carbohydrates in mediating cellular responses.

This work was supported by NIH grants AR 17220 and DE 00157.

References

Arumugham, R. G., Hsieh, T.-C., Tanzer, M. L. and Laine, R. A. (1986). Structures of the asparagine-linked sugar chains of laminin. *Biochim. Biophys. Acta.* **883**, 112-126.

Ashwell, G. and Harford, J. (1982). Carbohydrate-specific receptors of the liver. *A. Rev. Biochem.* **51**, 531-554.

Beck, K., Hunter, I. and Engel, J. (1990). Structure and function of laminin: anatomy of a multidomain glycoprotein. *FASEB J.* **4**, 148-160.

Bouzon, M., Dussert, C., Lissitzky, J.-C. and Martin, P.-M. (1990). Spreading of B16 F1 cells on laminin and its proteolytic fragments P1 and E8: involvement of laminin carbohydrate chains. *Expl Cell Res.* **190**, 47-56.

Chammas, R., Veiga, S. S., Line, S., Potochajk, P. and Brentani, R. R. (1991). Asn-linked oligosaccharide dependent interaction between laminin and gp 120/140. *J. biol. Chem.* **266**, 3349-3355.

Chandrasekaran, S., Dean, J. W., Giniger, M. S. and Tanzer, M. L. (1991). Laminin carbohydrates are implicated in cell signalling. *J. Cell Biochem.* **46**, 115-124.

Dean, J. W., Chandrasekaran, S. and Tanzer, M. L. (1988). Lectins inhibit cell binding and spreading on a laminin substrate. *Biochem. Biophys. Res. Commun.* **156**, 411-416.

Dean, J. W., Chandrasekaran, S. and Tanzer, M. L. (1990). A biological role of the carbohydrate moieties of laminin. *J. biol. Chem.* **265**, 12553-12562.

Elbein, A. D. (1987). Inhibitors of the biosynthesis and processing of N-linked oligosaccharide chains. *A. Rev. Biochem.* **56**, 497-534.

Fujiwara, S., Shinkai, H., Deutzmann, R., Paulsson, M. and Timpl, R. (1988). Structure and distribution of N-linked oligosaccharide chains on various domains of mouse tumor laminin. *Biochem. J.* **252**, 453-461.

Hynes, R. O. (1992). Integrins: versatility, modulation, and signaling in cell adhesion. *Cell* **69**, 11-25.

Jones, G. E., Arumugham, R. and Tanzer, M. L. (1986). Fibronectin glycosylation modulates fibroblast adhesion and spreading. *J. Cell Biol.* **103**, 1663-1670.

Knibbs, R. N., Perini, F. and Goldstein, I. J. (1989). Structure of the major concanavalin A reactive oligosaccharides of the extracellular matrix component laminin. *Biochemistry* **28**, 6379-6392.

Lee, Y. C. (1989). Binding modes of mammalian hepatic Gal/GalNAc receptors. In *Carbohydrate Recognition in Cellular Function,* (ed. G. Bock and S. Harnett), pp. 80-95. Chichester: Wiley.

Lennartz, M. R., Wileman, T. E. and Stahl, P. D. (1987). Isolation and characterization of a mannose-specific receptor from rabbit alveolar macrophages. *Biochem. J.* **245**, 705-711.

Lotan, R. and Raz, A. (1988). Endogenous lectins as mediators of tumor cell adhesion. *J. Cell Biochem.* **37**, 107-117.

Ott, U., Odermatt, E., Engel, J., Furthmayr, H. and Timpl, R. (1982). Protease resistance and conformation of laminin. *Eur. J. Biochem.* **123**, 63-72.

Ruoslahti, E. (1991). Integrins. *J. clin. Invest.* **87**, 1-5.

Rutishauser, U., Acheson, A., Hall, A. K., Mann, D. M. and Sunshine, J. (1988). The neural cell adhesion molecule (NCAM) as a regulator of cell-cell interactions. *Science* **240**, 53-57.

Stoolman, L. M. (1992). Selectins (LEC-CAMs): Lectin-like receptors involved in lymphocyte recirculation and leukocyte recruitment. In *Cell Surface Carbohydrates and Cell Development,* (ed. M. Fukuda), pp. 71-97. Boca Raton: CRC Press.

Tanzer, M. L., Dean, J. W. and Chandrasekaran, S. (1991). Cell signalling: a role for laminin carboydrates. *Trends Glycosci. Glycotech.* **3**, 302-314.

Varki, A. (1992). Role of oligosaccharides in the intracellular and intercellular trafficking of mammalian glycoproteins. In *Cell Surface Carbohydrates and Cell Development,* (ed. M. Fukuda), pp.25-69. Boca Raton: CRC Press.

Wooten, E. W., Bazzo, R., Edge, C. J., Zamze, S., Dwek, R. A. and Rademacher, T. W. (1990). Primary sequence dependence of conformation in oligomannose oligosaccharides. *Eur. Biophys. J.* **18**, 139-148.

Printed in Great Britain © The Society of Experimental Biology 1993   155

TENASCIN FUNCTION AND REGULATION OF EXPRESSION

NEVENKA VRUČINIĆ-FILIPI and RUTH CHIQUET-EHRISMANN

Friedrich Miescher Institute, P.O. Box 2543, CH-4002 Basel, Switzerland

Summary

Tenascin is a large hexameric extracellular matrix protein. Each subunit consists of a linear array of the following structural domains: an N-terminal cysteine-rich region and heptad repeats both involved in the oligomerization, followed by EGF-like repeats, fibronectin type III domains and a C-terminal globular domain homologous to fibrinogens. Depending on the species, the number of EGF-like repeats as well as of the fibronectin type III repeats differs. A model proposing a new nomenclature for the numbering of the fibronectin type III repeats is presented. It enables the assignement of corresponding repeats from different species and shows that for human tenascin several more alternatively spliced fibronectin type III repeats have been described than for chicken or mouse tenascin. Tenascin is generally classified as an anti-adhesive protein, since many cells do not adhere to tenascin or if they adhere they do not spread. The addition of tenascin can also lead to the loss of focal contacts in well spread cells. On the other hand growth cones adhere well to tenascin and a tenascin substrate can support neurite extension. Using recombinant fragments of tenascin we could identify an adhesive as well as an anti-adhesive region within one tenascin subunit. The major interest in tenascin was prompted by its interesting tissue distribution. Tenascin is often transiently expressed during organogenesis, tissue remodeling, cell migration and the development of the nervous system. We thus investigated possibilities to regulate tenascin expression in primary chick embryo fibroblast cultures. We found that among the growth factors tested, namely, PDGF, acidic and basic FGF, EGF and TGF-β, only the latter was able to induce tenascin production by the cells. Furthermore, tenascin synthesis could be stimulated by fetal calf serum. Fractionation of the serum showed that tenascin-inducing activity could be found both in the fraction binding to heparin-Sepharose, as well as in the non-binding fraction. Interestingly the fraction eluted with 0.5M NaCl contained an activity that specifically induced the synthesis of tenascin, but not of fibronectin, whereas all other materials tested affected both tenascin and fibronectin in parallel. Since the specific induction of tenascin may be of physiological importance it will be interesting to identify the active component in this serum fraction.

Introduction

Tenascin is an extracellular matrix protein with a dynamically changing tissue distribution. It is transiently expressed in many developing organs such as connective tissues and the mesenchyme of epithelial organs and it is often reexpressed in the stroma of malignant epithelial tumors. It is furthermore present in the central and peripheral

Key words: tenascin, extracellular matrix, fibronectin, growth factors.

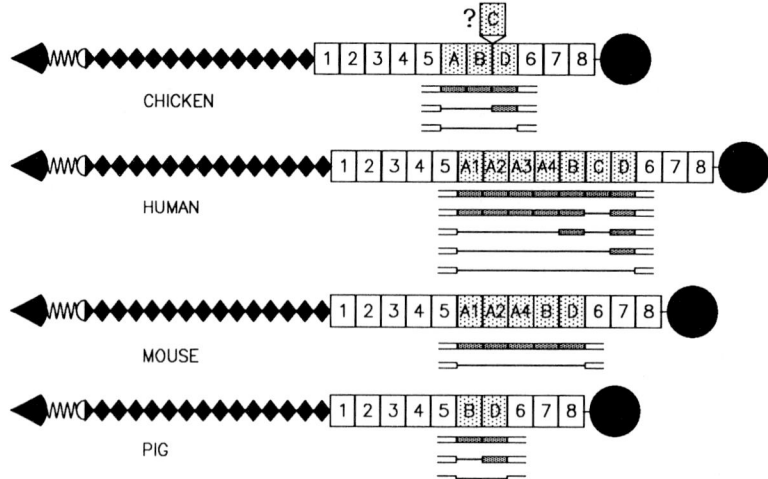

Fig. 1. Model of one subunit of chicken, human, mouse, and pig tenascin. The subunits consist of an N-terminal part involved in the oligomerization to trimers (heptad repeats; waved line) and hexamers (central domain; filled sector), followed by tenascin-type EGF-like repeats (filled diamonds), fibronectin type III repeats (rectangles) and a domain homologous to the globular part of β- and γ-fibrinogen. Alternatively spliced extra repeats are stippled and letter coded. Below each tenascin subunit the published splicing variants are shown. Repeat C in chicken tenascin has so far only been identified on the gene level and it remains to be determined, whether it is expressed as part of a tenascin protein.

nervous system. The distinctive and highly regulated expression of tenascin has provoked interest in trying to identify possible functions of tenascin in cell-cell and cell-substratum adhesion, cell migration, growth and cell differentiation during morphogenesis (for reviews see Erickson and Bourdon, 1989; Chiquet-Ehrismann, 1990; Chiquet et al., 1991). We have shown previously that the three characterized chicken tenascin variants can differ both in function as well as tissue distribution. The largest tenascin form is more sensitive to proteolysis than the smallest one (Chiquet et al., 1991) whereas the smallest tenascin variant binds more strongly to fibronectin (Chiquet-Ehrismann et al., 1991). In chicken gizzard we found that tendon contains the smallest tenascin variant exclusively, whereas smooth muscle expresses intermediate sized tenascin and the stroma underlying

Fig. 2. Cell adhesion to recombinant tenascin fragments. Tissue culture plates were coated with drops of the following protein solutions: purified fibronectin and tenascin, or β-galactosidase-tenascin fusion proteins as indicated in the figure. The cell-binding fragment contains the fibronectin type III repeats 6-8 and the fibrinogen globe, whereas the anti-adhesive fragment contains the EGF-like repeats and the first $1\frac{1}{2}$ fibronectin type III repeats. Details of the experimental procedure can be found in Spring et al. (1989). The cells in the middle panel were fixed and stained 10 min after plating, whereas all other cells were fixed after 60 min. Cells adhere and spread on fibronectin, but do not adhere efficiently on tenascin. One tenascin fragment promotes the rapid adhesion of the cells within 10 min, but no spreading even after 60 min, whereas the other tenascin fragment does not allow any cell adhesion even after 60 min when the cells already adhere to the surrounding area blocked with bovine serum albumin.

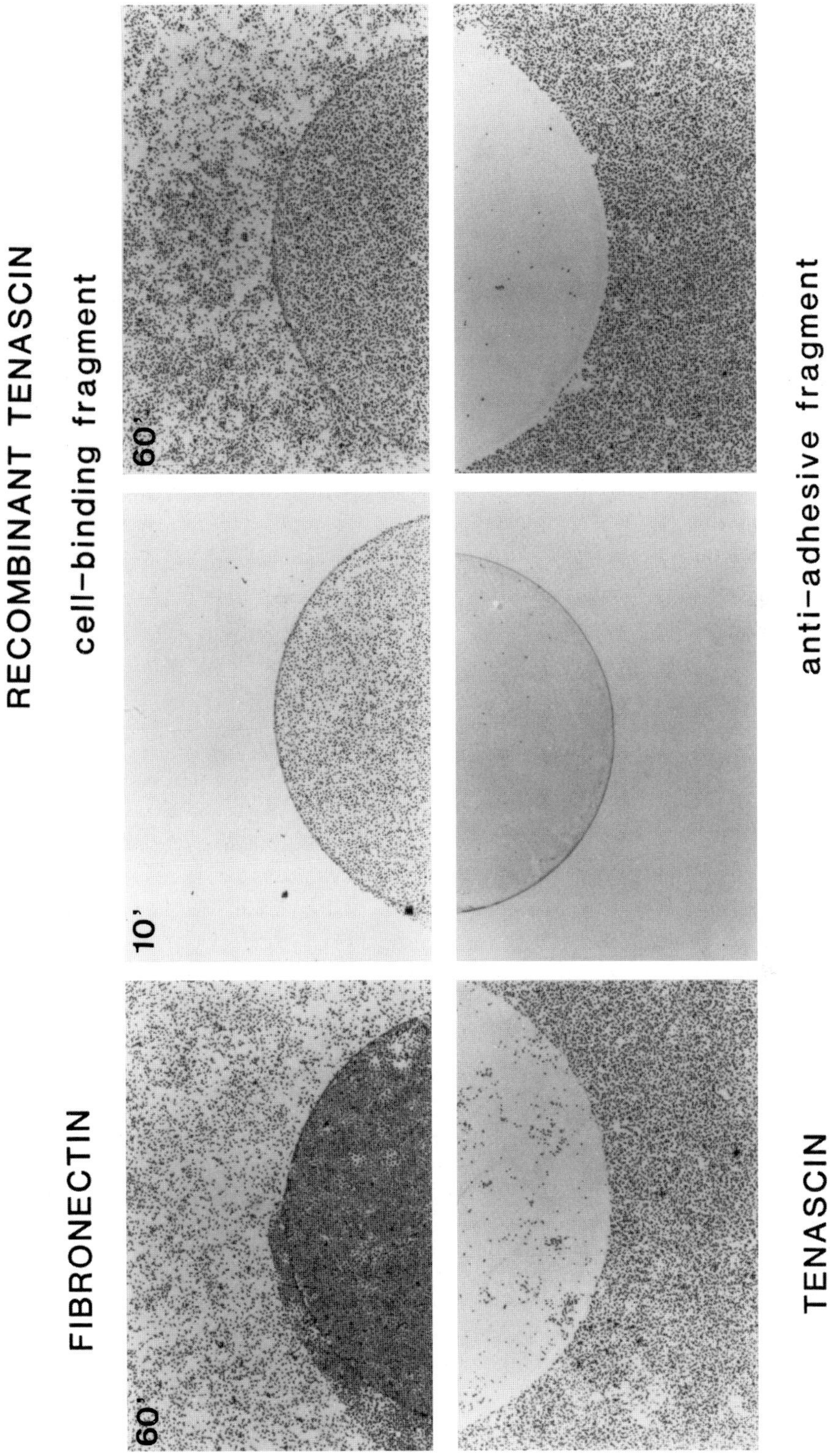
RECOMBINANT TENASCIN
cell-binding fragment
anti-adhesive fragment
FIBRONECTIN
TENASCIN
60'
10'
60'

whereas smooth muscle expresses intermediate sized tenascin and the stroma underlying the villous epithelium contains the largest form (Chiquet-Ehrismann *et al.*, 1991).

In this manuscript we will first review the structure and the postulated function of tenascin. We will then report on our experiments regarding possible factors involved in the regulation of tenascin expression.

Tenascin structure

Tenascin is a disulfide-linked hexameric protein with subunit molecular weights in the range of 200-300 kD. Different subunits are generated by alternative splicing of one common primary transcript. The model shown in figure 1 reveals the domain structure of chicken, human, mouse and pig tenascin (cf. Spring *et al.*, 1989; Siri *et al.*, 1991; Saga *et al.*, 1991; Nishi *et al.*, 1991). The most prominent domains are the tenascin-type EGF-like repeats, the fibronectin type III repeats as well as the fibrinogen domain. When the fibronectin type III repeats in tenascins from different species are aligned according to their highest similarity it becomes clear that human tenascin contains more extra-repeats subject to alternative splicing than the tenascin of any other species. In order to simplify the comparison between tenascins of different species we propose to adopt the nomenclature presented in figure 1. The constant repeats found in all variants in all species are numbered (1 to 8). In the alternatively spliced region the repeats are letter coded (A to D). In human and mouse tenascin four or three repeats, respectively, show highest homology to chicken repeat A and are therefore called A1 to A4, since they seem to have arisen by duplication of an ancestral A repeat. The human extra-repeat C has so far not been described in any other species. We have however recently found that also in the chicken tenascin gene an exon corresponding to repeat C is present between the exons for repeats B and D. Figure 1 also includes the different splicing variants reported in the respective species.

Tenascin function

When we started to study cell adhesion to tenascin substrates, we noted that in contrast to fibronectin, tenascin was a very poor adhesion molecule and the addition of tenascin to the cell culture medium inhibited fibroblast adhesion to fibronectin (Chiquet-Ehrismann *et al.*, 1988). In another study it was shown that cells were able to adhere to tenascin at 4°C, but in contrast to fibronectin this adhesion became weaker when the cells were warmed up to 37°C (Lotz *et al.*, 1989). Because of these studies tenascin has been classified as an anti-adhesive or adhesion-modulating extracellular matrix protein (Chiquet-Ehrismann, 1991; Sage and Bornstein, 1991). Contrary to these negative effects on cell adhesion, Wehrle and Chiquet (1990) demonstrated that tenascin promoted neurite outgrowth. They showed that even though neuronal cell bodies did not adhere to tenascin, growthcones were able to attach and spread and this leads to the generation of neurites. Using monoclonal antibodies to tenascin as well as tenascin fragments expressed as fusion proteins in E. coli several labs attempted to identify the sites within the tenascin molecule responsible for the adhesive as well as anti-adhesive activities of

fold induction

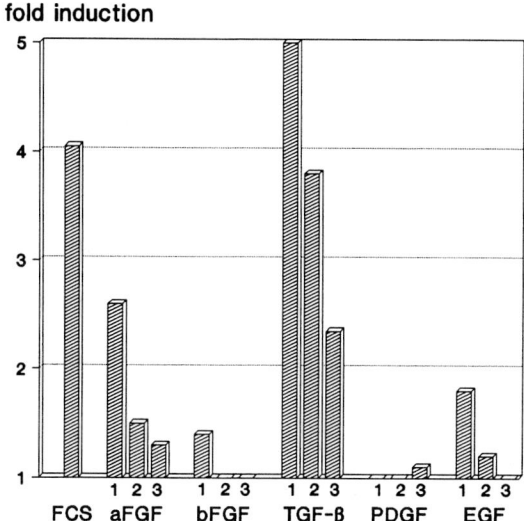

Fig. 3. Induction of tenascin secretion by growth factors. We added each of the indicated growth factors at 100 ng/ml (1), 10 ng/ml (2) and 1 ng/ml (3) to fibroblasts grown at 0.3% FCS and tested for their tenascin inducing activity using the method described in (Pearson *et al.* 1988). As a comparison the induction with 10% FCS is shown.

tenascin. We found that a monoclonal antibody reacting with repeat 7 neutralized the anti-adhesive effect of tenascin thus implying a cell interaction site in that area of the tenascin molecule (Chiquet-Ehrismann *et al.*, 1988). In a different study a monoclonal antibody presumably reacting with repeat 6 was reported to neutralize the neurite outgrowth promoting activity of tenascin (Lochter *et al.*, 1991). Judging from these two studies cell interaction sites seem to be present in the distal fibronectin type III domains following the alternatively spliced extra repeats. In experiments using recombinant fragments of tenascin the following two results were obtained. As shown in figure 2, we found that cells adhere to recombinant tenascin fragments containing the distal fibronectin type III repeats, whereas the fragments containing the EGF-like repeats were anti-adhesive for the same cells (cf. Spring *et al.*, 1989). Assaying for the downregulation of focal adhesion in aortic endothelial cells it was reported that a recombinant tenascin fragment encompassing the seven human extra repeats was active and monoclonal antibodies to the extra repeats of tenascin were able to neutralize the downregulation of focal contacts by the addition of purified tenascin (Murphy-Ullrich *et al.*, 1991). Thus so far two regions within the tenascin molecule have been postulated to contribute to its anti-adhesive effects, namely the region of the EGF-like repeats and the extra repeats. It remains to be determined which cellular receptors interact with the different domains of tenascin and how the anti-adhesive and adhesive signals are transmitted to the cells.

Regulation of tenascin expression

We have shown previously that tenascin mRNA and protein accumulation is induced either by fetal calf serum (FCS) or the addition of TGF-β to chick embryo fibroblasts

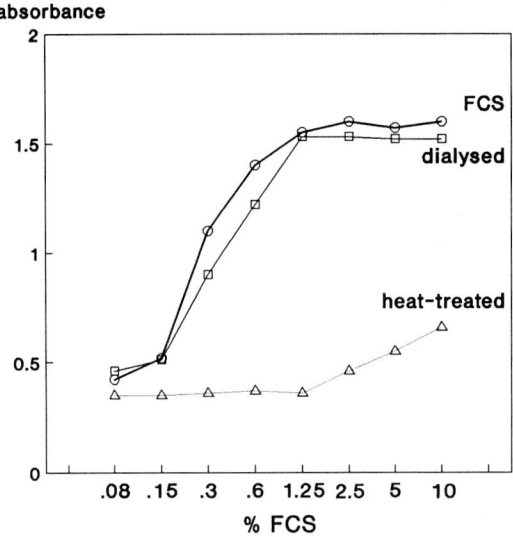

Fig. 4. Induction of tenascin secretion by fetal calf serum. Increasing concentrations of untreated FCS (○), dialysed FCS (□), or FCS treated for 3 min at 100°C (△) were added to fibroblasts grown at 0.3% FCS and tested for their effect on tenascin secretion. Whereas untreated and dialysed FCS showed equal activity in inducing tenascin, heat-treatment destroyed the activity almost completely. Quantitation of tenascin accumulation in the medium was performed by an ELISA assay as described by Pearson *et al.* (1988).

cultured at low FCS concentration (Pearson *et al.*, 1988). With the exception of TGF-β none of the factors affecting tenascin expression have been identified. We therefore analyzed known and commercially available growth factors for their potential to affect tenascin secretion. As shown in figure 3, none of the newly tested growth factors, acidic and basic fibroblast growth factor (aFGF and bFGF), platelet-derived growth factor (PDGF), and epidermal growth factor (EGF) had a dramatic effect on tenascin expression. Only aFGF showed a significant induction at the highest concentration tested. Since the active concentration of 100 ng/ml is rather high we do not think that aFGF could be the factor responsible for the serum induction observed. In order to test for the nature of the serum factors involved, we investigated whether boiling or dialysis of FCS was able to destroy the inducing activity. As shown in figure 4, the activity was non-dialysable and heat labile. We thus concluded that the activity may be a protein with a molecular weight larger than the cutoff size of 3.5 kD of the dialysis membrane used. We thus decided to fractionate the FCS in an attempt to further characterize the serum factor(s) involved. We loaded FCS on a heparin column, collected the flow through fraction and the following eluates. The chromatogram and a protein gel resolving the proteins in these preparations are presented in figure 5A. After dialysis of all fractions they were tested for their activity in inducing the secretion of tenascin as well as fibronectin in fibroblast cultures (figure 5B). Interestingly we found tenascin inducing activity in several of the fractions tested, namely the heparin non-binding flowthrough fraction as well as in the first three salt steps. No activity remained after elution with 0.65M NaCl. Since FGFs are known to be eluted from heparin-Sepharose between 1-2M NaCl they cannot be

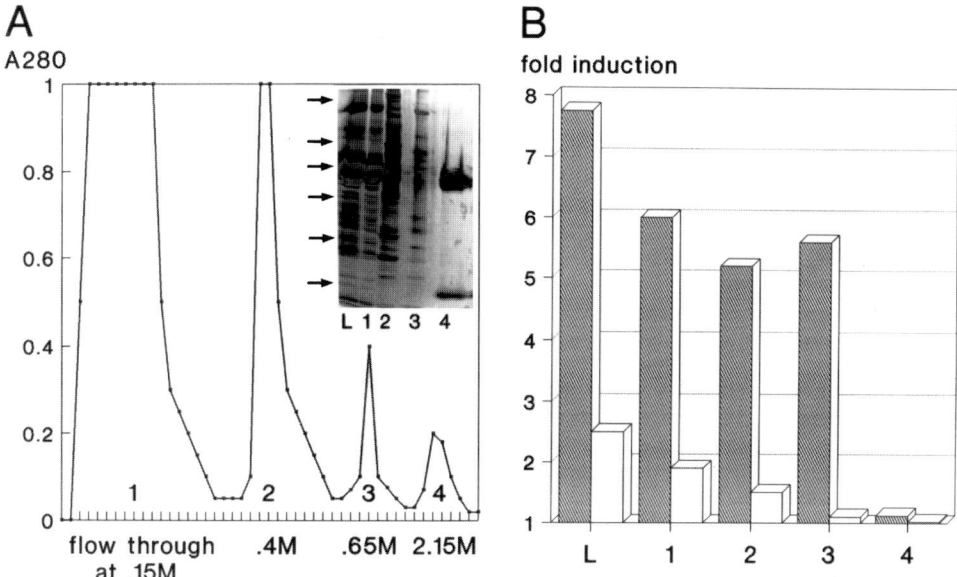

Fig. 5. Fractionation of fetal calf serum and identification of a fraction inducing tenascin, but not fibronectin secretion. A: Chromatogram of the fractionation of FCS over heparin-Sepharose and 10% SDS-Page of the loaded FCS (L) and the fractions collected of the salt steps at the indicated NaCl concentrations (1-4). The molecular weights markers migrating at 200 kD, 97 kD, 66 kD, 45 kD, 31 kD, 21.5 kD are indicated from top to bottom to the left of the gel. B: Induction of tenascin (hatched bars) and fibronectin (white bars) secretion by the addition of the fractions shown in figure 5A to fibroblast cultures. The unfractionated FCS and the fractions 1 and 2 induced both, fibronectin as well as tenascin, whereas fraction 3 stimulated tenascin accumulation only and fraction 4 was ineffective for both proteins analysed.

responsible for the major tenascin-inducing activity by FCS. Most interestingly, the fraction eluted with 0.65M NaCl specifically induced the accumulation of tenascin, but not fibronectin, whereas all other fractions were stimulating both proteins.

Because of the opposing effects of tenascin and fibronectin on cell adhesion we believe that independent regulation of these two proteins may have important physiological consequences. It is conceivable that a relative increase of tenascin over fibronectin in a given tissue may lead to changes in cell shape with its possible consequences (cf. Ben-Ze'ev, 1991). It will therefore be interesting to try to identify the factor specifically leading to an increased accumulation of tenascin and not of fibronectin.

References

Ben-Ze'ev, A. (1991). Animal cell shape changes and gene expression. *BioEssays* **13**, 207-211.

Chiquet, M., Vručinić-Filipi, N., Schenk, S., Beck, K. and Chiquet-Ehrismann, R. (1991). Isolation of chick tenascin variants and fragments. *Eur. J. Biochem.* **199**, 379-388

Chiquet, M., Wehrle-Haller, B. and Koch, M. (1991). Tenascin (cytotactin): an extracellular matrix protein involved in morphogenesis of the nervous system. *The Neurosciences* **3**, 341-350.

Chiquet-Ehrismann, R., Kalla, P., Pearson, C.A., Beck, K. and Chiquet, M. (1988). Tenascin interferes with fibronectin action. *Cell* **53**, 383-390.

Chiquet-Ehrismann, R. (1990). What distinguishes tenascin from fibronectin. *FASEB J.* **4**, 2598-2604.

Chiquet-Ehrismann, R. (1991). Anti-adhesive molecules of the extracellular matrix. *Curr. Opin. Cell Biol.* **3**, 800-804.

Chiquet-Ehrismann, R., Matsuoka, Y., Hofer, U., Spring, J., Bernasconi, C. and Chiquet, M. (1991). Tenascin variants: differential binding to fibronectin and distinct distribution in cell cultures and tissues. *Cell Reg.* **2**, 927-938

Erickson, H.P. and Bourdon, M.A. (1989). Tenascin: an extracellular matrix protein prominent in specialized embryonic tissues and tumors. *Ann. Rev. Cell Biol.* **5**, 71-92.

Lochter, A., Vaughan, L., Kaplony, A., Prochiantz, A., Schachner, M. and Faissner, A. (1991). J1/tenascin in substrate-bound and soluble form displays contrary effects on neurite outgrowth. *J. Cell Biol.* **113**, 1159-1171.

Lotz, M.M., Burdsal, C.A., Erickson, H.P. and McClay, D.R. (1989). Cell adhesion to fibronectin and tenascin: quantitative measurements of initial binding and subsequent strengthening response. *J. Cell Biol.* **109**, 1795-1805.

Murphy-Ullrich, J.E., Lightner, V.A., Aukhil, I., Yan, Y.Z. Erickson, H.P. and Höök, M. (1991). Focal adhesion integrity is downregulated by the alternatively spliced domain of human tenascin. *J. Cell Biol.* **115**, 1127-1136.

Nishi, T., Weinstein, J., Gillespie, W.M. and Paulson, J.C. (1991). Complete primary structure of porcine tenascin. *Eur. J. Biochem.* **202**, 643-648.

Pearson, C.A., Pearson, D., Shibahara, S., Hofsteenge, J. and Chiquet-Ehrismann, R. (1988). Tenascin: cDNA cloning and induction by TGF-β. *EMBO J.* **7**, 2677-2981.

Saga, Y., Tsukamoto, T., Jing, N., Kusakabe, M. and Sakakura, T. (1991). Murine tenascin: cDNA cloning, structure and temporal expression of isoforms. *Gene* **104**, 177-185.

Sage, E.H. and Bornstein, P. (1991). Extracellular proteins that modulate cell-matrix interactions. *J. Biol. Chem.* **266**, 14831-14834.

Siri, A., Carnemolla, B., Saginati, M., Leprini, A., Casari, G., Baralle, F. and Zardi, L. (1991). Human tenascin: primary structure, pre-mRNA splicing patterns and localization of the epitopes recognized by two monoclonal antibodies. *Nucleic Acid. Res.* **19**, 525-531.

Spring, J., Beck, K. and Chiquet-Ehrismann, R. (1989). Two contrary functions of tenascin: dissection of the active sites by recombinant tenascin fragments. *Cell* **59**, 325-334.

Wehrle, B. and Chiquet, M. (1990). Tenascin is accumulated along peripheral nerves and allows neurite outgrowth in vitro. *Development* **110**, 401-415.

Printed in Great Britain © The Society of Experimental Biology 1993   163

HEPATOCYTE GROWTH FACTOR/SCATTER FACTOR (HGF/SF), THE c-*met* RECEPTOR AND THE BEHAVIOUR OF EPITHELIAL CELLS

ERMANNO GHERARDI*, MELANIE SHARPE, KAREN LANE,
ANDRES SIRULNIK and MICHAEL STOKER

ICRF Cell Interactions Laboratory, Cambridge University Medical School, MRC Centre,
Cambridge, UK

Summary

Hepatocyte growth factor/scatter factor (HGF/SF) is a multifunctional protein produced by fibroblasts and other mesenchymal cells and active on epithelial and endothelial cells. The factor shares the basic domain organization of plasminogen and other blood proteases and is produced as a single-chain high molecular weight precursor which is subsequently cleaved to produce a biologically active heterodimer.

HGF/SF acts on target cells through binding to a cell surface, high-affinity receptor with tyrosine kinase activity encoded by the c-*met* proto-oncogene. Transduction of the HGF/SF signal in epithelial cells by the *met* receptor leads either to: (i) cell dissociation, with loss of adhesion and junctional communication, (ii) cell division or (iii) differentiation and morphogenesis. These responses depend on target cells and culture conditions.

There is preliminary evidence for a role for HGF/SF *in vivo*, suggesting that the factor may be involved in the early stages of embryo development and in liver regeneration. A role in cancer growth or dissemination has been proposed but remains to be clarified.

I. Introduction

Growth, movement and differentiation

The behaviour of the specialised cells of complex organisms results from regulation of their growth, movement and differentiation. However, the order and the mechanisms by which these three processes are regulated vary considerably.

In the initial stages of embryo development, for example, a rapid succession of cell divisions takes place during which blastomeres do not change appreciably in shape or position. When the formation of the blastula is completed, the rate of cell division decreases and cells begin to show extensive movement and differentiation; the result of these processes is the formation of the three germ layers: the ectoderm, endoderm and mesoderm. Further differentiation of the primitive ectoderm is associated with complex

*Address for correspondence: ICRF Cell Interactions Laboratory, Cambridge University Medical School, MRC Centre, Hills Road, Cambridge CB2 2QH, UK.

Key words: HGF/SF, c-*met* receptor, epithelial growth, movement and differentiation.

cell movements and leads to the separation of three subpopulations of cells: the epidermal ectoderm, the neural ectoderm and the neural crest cells. The latter cells, soon after their appearance, begin a remarkable process of migration across the embryo and resume growth and differentiation only when they have reached their destination [Le Douarin, 1982].

The reason for outlining these processes is to illustrate the fact that, in the embryo, regulation of cell growth, movement and differentiation can be distinct as illustrated by stages of "growth alone" (cleavage stage) or "movement alone" (neural crest cell migration), or can be tightly coordinated, as during gastrulation or neurulation.

Adult life offers several other examples of separate and concerted regulation of growth, movement and differentiation. For example, the repair of skin wounds requires that keratinocytes (cells which, in the normal skin, divide and terminally differentiate but do not show appreciable cell movement) migrate into the wounded area before they re-enter cell division and differentiation. Liver regeneration after partial hepatectomy, on the other hand, is an example of a much more complex set of events in which both cell growth and movement have to be re-initiated and coordinated in at least three cell types: hepatocytes, biliary epithelial cells and endothelial cells. The result of this process is the astonishing ability of the liver to reconstitute a substantial proportion of the organ within a few weeks of partial hepatectomy. Other tissues however, such as neural tissue and skeletal muscle, cannot resume growth in adult life, as established by Bizzozero a century ago [Bizzozero, 1894]. Thus, both in the embryo and in adult organisms, cell growth, movement and differentiation appear to be regulated separately or coordinately in different cell types and at different stages of life.

Paracrine regulation of cell behaviour

It has been known for a considerable period of time that the behaviour of cells of complex organisms is regulated by local extracellular signals [Grobstein, 1967]. In the embryo this process is responsible for the phenomenon of *induction* [reviewed in Jessel and Melton, 1992]. For example, equatorial cells of a 32-cell stage *Xenopus* embryo differentiate into mesoderm as a result of signals from the adjacent vegetal cells and similarly, the ectoderm differentiates into neural tissue because of signals from the dorsal mesoderm, as first described in the classic experiments of Spemann and Mangold [1924].

It has been known for a long time that mesoderm-derived mesenchymal signals are especially important in paracrine regulation of epithelial cells both in development and adult life [Kratochwil, 1983; Sharpe and Ferguson, 1988]. Only recently, however, the molecular nature of such interactions has begun to emerge.

While it remains unclear whether mesenchymal and epithelial cells can communicate by gap junctions [Fentiman *et al.*, 1976; Pitts and Burk, 1976], there are at least three other well established forms of communication between these cells. The first one is mediated by surface molecules on mesenchymal cells which, upon contact with epithelial cells, induce epithelial growth or morphogenesis. Studies with monoclonal antibodies have identified surface gangliosides [Sariola *et al.*, 1988] or proteins [Hirai *et al*, 1992] with such activity and it is to be expected that other molecules with similar functions will be found. The second form of communication between mesenchymal and epithelial cells is mediated by the extracellular matrix. Mesenchymal cells produce a complex matrix which makes contact with epithelial cells through a class of receptors known as integrins. The integrins were initially discovered and characterised as binding sites for substrate adhesion molecules but recent work has clearly established that they are true receptors with the ability to transduce a number of signals including mitogenic ones [reviewed in Damsky and Werb, 1992]. The third form of communication between mesenchymal and epithelial cells is mediated by soluble factors. Fibroblasts and other mesenchymal cells are an important source of soluble proteins which affect epithelial behaviour and this chapter is devoted to one such protein, HGF/SF. This factor affects epithelial growth, movement and morphogenesis *in vitro* and may be involved in the regulation of epithelial behaviour *in vivo*.

It should be noted at this point that, although HGF/SF was discovered for its activities on epithelial cells, it is now clear that the target specificity of HGF/SF extends outside the epithelial lineage. There is clear evidence for an effect of HGF/SF on endothelium [see Rosen *et al*, this volume] and there are recent reports implying an effect of HGF/SF on hematopoietic progenitor cells [Kmiecik *et al.*, 1992] and neutrophils [Jiang *et al.*, 1992]. Moreover, from the expression patterns of HGF/SF and its receptor it can be predicted that the final list of target cells may be even more numerous. In spite of this multiplicity of targets, this chapter will be confined to the effects of HGF/SF on epithelium, the cell type which allowed the initial isolation of HGF/SF and probably a major target of HGF/SF activity *in vivo*.

II. Properties of HGF/SF

Discovery of HGF/SF as an epithelial scatter factor and as a hepatocyte growth factor

In 1984 Stoker observed that when human mammary epithelial cells were cultured in the presence of medium conditioned by the MRC5 strain of human lung embryo fibroblasts, the majority of the colonies appeared as loose aggregates of cells lacking the typical features of epithelial colonies [Stoker, 1984]. Stoker and Perryman [1985] subsequently found that the Madin Darby canine kidney (MDCK) epithelial cell line represented a sensitive target for the epithelial scatter activity present in the MRC5 conditioned medium. Since then, this cell line has provided the basis for a reliable biological assay for HGF/SF which has been used for the purification and characterization of the factor [Stoker *et al.*, 1987; Gherardi *et al.*, 1989].

Parallel and independent studies on growth regulation of hepatocytes in culture had led, at the same time, to the identification of a protein which induced DNA synthesis and cell division in short term, primary cultures of rat hepatocytes [Nakamura *et al.*, 1986, 1987, 1989; Gohda *et al.*, 1988; Miyazawa *et al*, 1989; Zarnegar and Michalopoulos, 1989; Selden and Hodgson, 1989]. The factor had been given various names (hepatocyte growth factor, hepatopoietin A and hepatotropin) but it was realised quite early on that the different names referred to the same molecule. What was not realised for some time was the fact that the hepatocyte growth factor and the fibroblast-derived epithelial scatter factor were the same molecule. This only became apparent when partial [Gherardi and Stoker, 1990; Weidner *et al.*, 1990] and later complete [Weidner *et al.*, 1991] sequences of the epithelial scatter factor became available. The factor is now designated by us and others as hepatocyte growth factor/scatter factor (HGF/SF) [Gherardi and Stoker, 1991].

The HGF/SF protein

HGF/SF is a heterodimer of a large (A or α, M_r ~ 57,000) and a small (B or β, M_r ~ 30,000) protein subunit held together by a single interchain disulphide bond [Nakamura *et al.*, 1987; Gohda *et al*, 1988; Gherardi *et al*, 1989; Zarnegar and Michalopoulos, 1989]. The factor exhibits strong heparin-binding activity [Rosen *et al*, 1989], a feature suggesting that proteoglycans may be invoved in the binding and/or presentation of the factor to target cells *in vivo*. Sequencing of full length cDNA clones for human [Nakamura *et al*, 1989; Miyazawa *et al*, 1989], rat [Tashiro *et al*, 1990] and mouse HGF/SF [Sharpe, Lane and Gherardi, unpublished results] has shown that HGF/SF is synthesised as a single chain precursor protein which is subsequently cleaved to generate the two-chain heterodimer [Nakamura *et al.*, 1989; Miyazawa *et al*, 1989]. The cleavage step is required for biological activity [Gak *et al*, 1992; Hartmann *et al*, 1992; Lokker *et al.*, 1992]. At least two proteases are now known to cleave the HGF/SF precursor into the mature factor. The first is urokinase [Naldini *et al*, 1992], one of the enzymes involved in plasminogen activation, the second is a novel serine protease recently isolated from fetal calf serum [Shimomura *et al*, 1992].

HGF/SF is a protein of 728 amino acids which shows extensive sequence polymorphism among different isolates [reviewed in Gherardi *et al*, 1993] and a 30-40% overall sequence identity with plasminogen, a pro-enzyme of 791 amino acids which is converted into the active protease plasmin by cleavage at residues Arg_{561}-Val_{562}. The A subunit of plasminogen contains 5 kringles, a protein module defined by six conserved cysteine residues and found in several other proteins such as tissue-type and urokinase-type plasminogen activators, factor XII and prothrombin. The A subunit of HGF-SF contains 4 kringles (Fig. 1). The B subunit of HGF/SF resembles the catalytic subunit of plasmin and other serine proteases (Fig. 1) but in HGF/SF the His and Ser residues of the catalytic site are replaced by Gln and Tyr [Nakamura *et al*, 1989; Miyazawa *et al.*, 1989]. These substitutions account for lack of protease activity in HGF/SF [Rosen *et al.*, 1990b; Coffer *et al.*, 1991], consistent with our inability to abolish activity with inhibitors of serine proteases [Gherardi *et al*, 1989].

Two major variants of human HGF/SF have been described. One is 723 amino acids in length [Seki *et al*, 1990; Rubin *et al.*, 1991; Weidner *et al.*, 1991], lacks 5 amino acids in kringle 1 and has full biological activity. The other is a truncated form of HGF/SF which extends from the N-terminus until the end of kringle 2 (Miyazawa *et al*, 1991; Chan *et al.*, 1991, Hartmann *et al.*, 1992). The two-kringle variant of HGF/SF lacks mitogenic activity [Miyazawa *et al*, 1991] but seems to retains some motogenic activity [Hartmann *et al*, 1992]. Interestingly, this variant appears to compete for the binding of full length

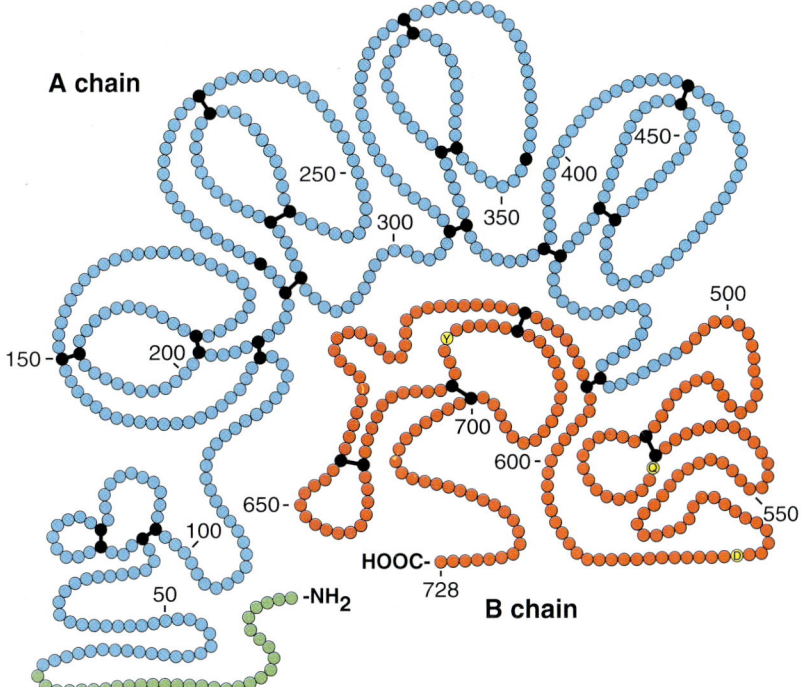

A chain

250 -

300

350

400

450-

200

150 -

100

50

500

550

650 -

700

600 -

HOOC-

728

-NH₂

B chain

Fig. 1. Proposed primary structure of mouse HGF/SF precursor. The model is based on the conservation of critical cysteine residues in HGF/SF and plasminogen. In the latter protein, the pattern of disulphide bonds has largely been established experimentally. The A and the B chains of mature HGF/SF are shown as blue and red circles respectively, except for cysteine residues which are shown as black circles. A 30 residue long signal sequence is shown as green circles. The three residues corresponding to the catalytic triad of the serine protease are shown in yellow.

HGF/SF to its receptor and could, therefore, function as a natural inhibitor of the factor [Chan *et al.*, 1991; Lokker *et al.*, 1992; Okigaki *et al.*, 1992].

III. The receptor for HGF/SF

Studies in two different laboratories have established that the receptor for HGF/SF is the protein encoded by the *c-met* proto-oncogene [Bottaro *et al*, 1991; Naldini *et al.*, 1991a], an oncogene originally identified in a human osteogenic sarcoma cell line (MNNG/HOS) transformed *in vitro* with N-methyl-N′-nitronitrosoguanidine [Cooper *et al*, 1984]. The sequence of c-*met*-proto-oncogene cDNAs predicts a transmembrane protein with a large N-terminal extracellular domain and a cytoplasmic tyrosine kinase domain [Park *et al.*, 1987; Chan *et al.*, 1988; Iyer *et al*, 1990]. The mature receptor consists of two subunits [Chan *et al.*, 1988; Giordano *et al.*, 1989] and recent work suggests that the larger subunit (the so-called β subunit) is involved in HGF/SF binding [Naldini *et al*, 1991b].

Given that the predominant responses to HGF/SF *in vitro* are mitogenicity and motogenicity, it was unclear for some time whether the *met* receptor mediates both responses. Recent experiments of Weidner and colleagues [1993] have clearly established that this is the case. When a chimaeric receptor composed of the cytoplasmic portion of *met* and the extracellular portion of the high affinity *trk* receptor for nerve growth factor (NGF) was transfected in MDCK cells, (which lack expression of the endogenous *trk*), NGF could induce both a motogenic response and a mitogenic response in the transfectants [Weidner *et al*, 1993]. Thus the cytoplasmic domain of the *met* receptor can initiate both responses and the "choice" between them is probably determined by additional signals (as discussed below) and the complement of kinases, kinase substrates and phosphatases expressed in different target cells.

IV. Sources of HGF/SF

The first systematic study of sources and targets of HGF/SF was carried out using the MDCK scatter assay as a measure of production of HGF/SF and by studying the motility response of a variety of cells to MRC5-derived HGF/SF [Stoker et al., 1987]. The results of this study led to two main conclusions.

The first is that the factor acts in a paracrine manner. Producer cells are certain fibroblasts and these cells do not respond to the factor itself. Conversely, most epithelia (including skin, kidney, and breast) are sensitive to the factor in one or more motility assays but do not produce it [Stoker *et al.*, 1987]. Subsequent studies by other groups have extended the list of producer and responder cells and confirmed the paracrine mechanism [Rosen *et al.*, 1989, 1990b] and further support has also emerged from *in situ* experiments. Sonnenberg *et al* [1993] have recently reported that in the mouse embryo c-*met* transcripts are found in a variety of epithelia whereas HGF/SF transcripts are found in the adjacent mesenchyme. Similarly, several studies on expression of HGF/SF in adult rat liver have revealed that non parenchymal cells express the factor [Noji *et al.*, 1990]. Thus the initial suggestion of paracrine activity of HGF/SF [Stoker *et al.*, 1987] has been extensively confirmed and examples of autocrine stimulation of epithelial movement or growth by HGF/SF should be regarded, so far, as exceptions [Adams *et al.*, 1991]. The possibility that an autocrine mechanism may play a role in certain epithelial tumours will be addressed below.

The second conclusion which emerged from the early work on cell sources and specificity is that, among producer cells, those of embryonic origin show the highest expression of HGF/SF [Stoker *et al.*, 1987]. When embryo and post-natal fibroblasts were compared it became clear that embryonic cells consistently produced higher levels of HGF/SF. This led us to investigate a role of HGF/SF in the development of the chick embryo and the initial results of this work have raised the possibility that HGF/SF may have neural-inducing activity [Stern *et al*, 1990]. Other studies are now in progress in our and several other laboratories to investigate further the role of HGF/SF and *met* in embryogenesis.

V. Activities of HGF/SF on epithelia *in vitro*

Effects on the motility, growth and differentiation of normal epithelia

The motility response of a number of target epithelia to HGF/SF has been extensively studied by time lapse cinematography [Stoker and Perryman, 1985; Stoker *et al*, 1987] and the MDCK cell line will be used here as an example of the motility response of epithelial cells to HGF/SF.

Within minutes of the addition of the factor to a single cell suspension of MDCK cells, the cells exhibit extensive membrane ruffling and begin to protrude numerous pseudopodia which extend and retract very rapidly. The morphology of the cells changes from rounded to multipolar and the cells show extensive local movement [Stoker and Perryman, 1985]. If the factor is added to preformed epithelial colonies, the colonies show a rapid expansion and separate into single cells after 4 to 6 hours. After separation, the cells show the same changes in morphology and motility observed with individual cells and remain unable to form stable cell-cell interactions (Fig. 2B). In the presence of low concentrations of factor, however, the cells eventually re-associate but the colonies remain expanded (Fig. 2C).

A response similar to that of MDCK cells is observed with a variety of epithelia but with some, such as normal human breast, colony dispersion is less pronounced [Stoker and Perryman, 1985; Stoker *et al.*, 1987]. There are only two reported cases of normal epithelia failing to show a motility response to HGF/SF: human fetal keratinocytes [Stoker *et al.*, 1987] and human primary melanocytes [Matsumoto *et al.*, 1991b] (Table 1). There is evidence from another study, however, that the migration of normal melanocytes in a Boyden chamber is enhanced by HGF/SF [Halaban *et al.*, 1992]. Thus all normal epithelia that have been studied with the single exception of the fetal keratinocytes show a motility response to HGF/SF. It should be noted, however, that data are not yet available on the motility response to HGF/SF of certain epithelia, most notably liver epithelia (Table 1).

When the identity of the epithelial scatter factor and hepatocyte growth factor was established, a contrast emerged between the broad epithelial specificity of the motogenic action of HGF/SF and the apparent narrow target specificity of its mitogenic action (the hepatocyte). This contrast has now been solved by a number of recent studies which indicate that the mitogenic action of HGF/SF is as broad as its motogenic action (Table 1). Thus, in addition to rat hepatocytes, which were used as the primary target for the identification of HGF/SF as a hepatocyte growth factor, biliary epithelium, renal, lung and skin epithelia all respond to HGF/SF with increased DNA synthesis and/or growth. There are few exceptions to this rule. The first one, and the most extensively studied, is the MDCK cell line which shows a motility but not a growth response to HGF/SF and a similar result was obtained in early studies on cultures of human breast epithelium [Stoker and Perryman, 1985]. The only other case of lack of growth stimulation by HGF/SF is that of keratinocyte cultures in low Ca^{++}. Interestingly, however, the growth of the same cells in high Ca^{++} medium was enhanced by the factor [Matsumoto *et al.,* 1991a]. The lack of effect of HGF/SF on the growth of MDCK cells is of special significance since it initially led to the identification of HGF/SF as a motility factor. The growth of these cells is not stimulated by HGF/SF in standard cultures on plastic but the factor stimulates their growth in collagen gels (Table 1). Thus, transduction of the HGF/SF and a collagen signal by the *met* receptor and, presumably, by an integrin receptor leads to growth stimulation in MDCK cells while the HGF/SF signal alone, in these cells is not sufficient to elicit growth. It is of interest that MDCK cells cultured in collagen gels respond to HGF/SF not only with growth stimulation but also with the formation of branched tubules, a result which has been attributed to a morphogenetic activity of HGF/SF on the epithelium [Montesano *et al,* 1991]. This response, like the growth response of MDCK cells to HGF/SF, occurs in collagen gels and requires an additional signal probably mediated by an integrin receptor. Whether other epithelia in collagen culture respond to HGF/SF in a similar manner, i.e. with differentiation and morphogenesis, remains to be seen since similar studies have not been carried out in other cell types (Table 1).

Epithelial tumours

The motility stimulation induced by HGF/SF in normal epithelia led to an early interest in the possibility that HGF/SF could be involved in the spreading of epithelial tumours [Stoker *et al.,* 1987]. This interest was strengthened by the discovery that the HGF/SF receptor is a member of the tyrosine kinase family of membrane receptors, whose activation commonly leads to transformation and tumorigenicity [Ullrich and Schlessinger, 1990]. Unfortunately, the role of HGF/SF and *met* in cancer is far from clear.

The earliest study focused on the motility response to HGF/SF of a number of breast carcinoma lines and showed no effect [Stoker *et al.,* 1987] (Table 2). The growth response of these cell lines to HGF/SF was not studied at the time but the growth of a different breast carcinoma line (B5/589) was later found to be enhanced by the factor

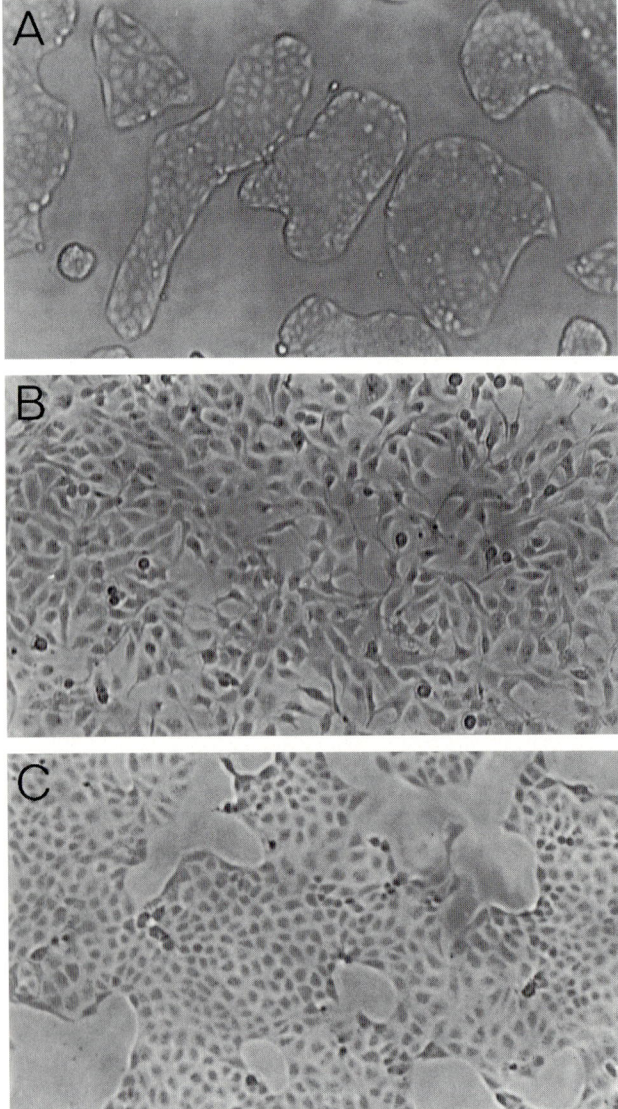

Fig. 2. Effects of HGF/SF on the appearance of MDCK cells after 3 days in culture. Cells were plated at a density of $10^4/cm^2$ in 10% FCS and, after 8 hours, mouse recombinant HGF/SF was added to cultures B and C at concentrations of 3×10^{-10} M and 3×10^{-11} M. A control culture with no exogenous HGF/SF is shown in A. After three days in culture in the presence of 3×10^{-10} M HGF/SF the cells remained dissociated (panel B). At the lower concentration of factor (3×10^{-11} M HGF/SF) the cells re-associated although the colonies remained expanded (panel C).

[Rubin *et al*, 1991]. Other studies had, meanwhile, produced evidence that the motility and invasiveness of carcinoma lines from other tissues was enhanced by HGF/SF [Weidner *et al.,* 1990; Rosen *et al*, 1990a] (Table 2).

In the same year, however, HGF/SF was re-isolated and partially sequenced as a "tumour cytotoxic factor" [Higashio et al., 1990]. Although it soon became clear that the effect of HGF/SF on several tumour lines was growth inhibition rather than cytotoxicity, the report by Higashio and colleagues and several subsequent reports prompted a re-assessment of the role of HGF/SF in tumour growth. A number of these studies are summarised in Table 2. The data are incomplete because, as often happens, a number of studies have addressed the question of the growth response but not of the motility response or *vice versa* and there is only one report of an effect of HGF/SF on the differentiation *in vitro* of target tumour lines [Tsarfaty *et al*, 1992]. Two conclusions appear, nevertheless, legitimate: first, HGF/SF enhances the motility of a variety of epithelial tumour lines with the possible exception of breast carcinoma and melanoma lines (Table 2). Unfortunately little information is available on the motility of liver, stomach and colon carcinoma lines which would be of interest given the high level of expression of *met* in epithelial tumours derived from these organs. Second, the growth response is variable and shows no correlation with the motility response (Table 2). Growth inhibition seems to be consistent among the liver carcinoma lines but not among carcinomas from other tissues or in the melanoma lines which have been studied (Table 2).

Although it is not easy to rationalise the current data one possibility is that the effect of HGF/SF on tumour cell lines *in vitro* may depend upon the level of differentiation of the target cells. Poorly differentiated lines may respond with growth and/or motility stimulation whereas more differentiated lines may respond with further differentiation and a reduced rate of growth. This interpretation is supported by the results of an early study [Weidner *et al.,* 1990] and future work will establish if it has general validity.

We have alluded earlier to the possibility that certain carcinoma lines, unlike normal epithelia, might produce HGF/SF and respond to the factor in an autocrine manner. A number of carcinoma lines have been studied for expression of HGF/SF but only four have been reported so far which produce the factor [Rygaard *et al.,* 1992; Yoshinaga *et al*, 1992]. Among these, only one expresses the *met* receptor [Rygaard *et al.,* 1992] and the possibility of an autocrine response to HGF/SF in epithelial tumours currently appears, at best, a rare event. Thus the question of HGF/SF and tumours remains an open one. More work on tumour cell lines in culture might help in clarifying the issue but it is likely that more definite answers to this problem will come from the study of HGF/SF or *met* transgene animals and from studies on the effect of HGF/SF and receptor antagonists in experimental and human cancer. Work along these lines is well underway in a number of laboratories.

VI. Perspectives and conclusions

The notion that polypeptide growth factors regulate exclusively (or mainly) cell growth can no longer be maintained. These factors exert a variety of other effects on

Table 1. *Effects of HGF/SF on the behaviour of normal epithelia in vitro*

	Target cell	Motility	Growth	Differentiation	References
Breast	Human adult, primary	↑ (time lapse)	— (cell count)	n.r.	Stoker and Perryman (1985)
	Mouse adult, Comma 1D	↑ (time lapse, wound)	n.r.	,,	Stoker et al. (1987)
	Mouse fetal, NMG	↑ (,, ,,)	,,	,,	,,
Lung	Monkey adult, 4MBr-5	n.r.	↑ (DNA synthesis)	,,	Rubin et al. (1991)
	Mink fetal, Mv1Lu	,,	↑ (cell count)	,,	Tajima et al. (1992)
Kidney	Dog adult, MDCK	↑ (time lapse, wound, colony morphology, Boyden chamber)	— (cell count) ↑ (cell count) (in collagen gel)	↑ (tubule formation in collagen gel)	Stoker and Perryman (1985) Stoker et al. (1987) Gherardi et al. (1989) Montesano et al. (1991)
	,, ,, MDCK (*)	↑ (colony morphology, Boyden chamber)	↑ (soft agar cloning)	n.r.	Uehara and Kitamura (1992)
	Rat adult, primary	n.r.	↑ (DNA synthesis and cell count)	,,	Kan et al. (1991)
	Monkey adult, BS-C1	↑ (time lapse, wound)	n.r.	,,	Stoker et al. (1987)
Liver	Rat adult hepatocytes, primary	n.r.	↑ (DNA synthesis)	,,	Nakamura et al. (1987) Godha et al. (1988) Zarnegar and Michalopoulos (1989)
	Rat fetal hepatocytes, primary	,,	↑ ,, ,,	,,	Fabregat et al. (1992)
	Human adult hepatocytes, primary	,,	↑ ,, ,,	,,	Strain et al. (1991)
	Human adult biliary epithelium, primary	,,	↑ (DNA synthesis and cell count)	,,	Joplin et al. (1992)

Table 1. *Continued*

Target cell	Motility	Growth	Differentiation	References
Skin				
Keratinocytes Human fetal, primary	— (time lapse)	n.r.	n.r.	Stoker et al. (1987)
,, postnatal, primary	↑ ,, ,,	,,	,,	,,
Human adult, primary	↑ (colony morphology)	↓ (DNA synthesis and cell count) (low Ca^{++})	,,	Matsumoto et al. (1991a)
		↑ (DNA synthesis and cell count) (high Ca^{++})	,,	,,
Mouse postnatal, MK	n.r.	↑ (DNA synthesis)	,,	Rubin et al. (1991)
,, ,, PAM212	↑ (colony morphology)	↑ (cell count)	,,	Tajima et al. (1992)
Melanocytes Human adult, primary	— (colony morphology)	↑ (DNA synthesis and cell count)	,,	Matsumoto et al. (1991b)
,, postnatal, primary	n.r.	↑ (DNA synthesis)	,,	Rubin et al. (1991)
,, ,, ,,	↑ (Boyden chamber)	↑ (,, ,,)	,,	Halaban et al. (1992)

↑, — and ↓ mean stimulation, no change and inhibition; n.r. means that the effect was not investigated or reported. Motility stimulation assessed by colony morphology means that the colony expands and/or dissociates (scatters) in the presence of HGF/SF as shown initially by Stoker and Perryman [1985] for human breast epithelium and the MDCK cell line (see Fig. 2B for an example). A brief description of the other motility assays can be found in Stoker and Gherardi [1991]. The growth effect of HGF/SF was assessed generally by measuring DNA synthesis or by cell counts except in a few studies in which it was measured by staining the cultures with crystal violet and measuring the amount of eluted dye photometrically. This method correlates well with cell numbers and, therefore, the data from these studies are also expressed in the tables as "cell counts". Effects of HGF/SF on target cells have been evaluated by adding exogenous factor (natural or recombinant) to the cultures except in the case of cell lines marked with an asterisk (*) which have been transfected with a full length cDNA for HGF/SF.

Table 2. *Effects of HGF/SF on the behaviour of epithelial tumour lines in vitro*

	Target cell line	Motility	Growth	Differentiation	References
Breast	ZR751	— (time lapse)	n.r.	n.r.	Stoker et al. (1987)
	MDA157	— "	"	"	"
	CAMA1	— "	"	"	"
	T47D	— "	"	"	"
	BT474	— "	"	"	"
	BT20	— "	"	"	"
	MCF7	— "	"	"	"
	B5/589	n.r.	↑ (DNA synthesis)	"	Rubin et al. (1991)
Lung	LXF 289	↑ (collagen invasion)	n.r.	n.r.	Weidner et al. (1990)
	A549	↑ "	"	"	"
	LX1	— "			
	Lu99	↑ (colony morphology)	↑ (cell count)	"	Tajima et al. (1992)
Pancreas	HS766T	↑ (collagen invasion)	n.r.	"	Weidner et al. (1990)
	Capan 1	↑ "	"	"	"
	DAN-G	↑ "	"	"	"
	Capan 2	— "	"	"	"
Bladder	EJ28	↑ (collagen invasion)	"	"	"
	RT112	↑ "	"	"	"
	RT4	— "	"	"	"
Liver	HepG2	— (time lapse) ↑ (colony morphology)	↓ (DNA synthesis and cell count)	"	Stoker et al. (1987) Tajima et al. (1991 and 1992)
	Hep3B	n.r.	↓ (cell count)	"	Shiota et al. (1992)
	SKHep1	"	→ "	"	"
	TON6	"	→ "	"	"
	Ha22T	"	→ "	"	"
	HuH7	"	→ "	"	"
	Focus	"	→ "	"	"
	Fao	"	→ "	"	"
	Fao (*)	↓ (colony morphology)	↓ (DNA synthesis and tumorigenicity)	"	"

Table 2. *Continued*

	Target cell line	Motility	Growth	Differentiation	References
Stomach	MKN-74	↑ (colony morphology, Boyden chamber)	↑ (DNA synthesis, cell count)	n.r.	Shibamoto *et al.* (1992)
Colon	SW480	n.r.	n.r.	↑ (lumen formation)	Tsarfaty *et al.* (1992)
	HT29	,,	,,	↑ ,,	,,
Skin (carcinomas)	KB	↑ (colony morphology)	↓ (DNA synthesis and cell count)	n.r.	Higashio *et al.* (1990)
	A431	↑ ,,	↑ (cell count)	,,	Tajima *et al.* (1991 and 1992)
	B6/F1	n.r.	↓ (DNA synthesis and cell count)	,,	Tajima *et al.* (1992)
(melanomas)	NEL M1	,,	↑ (DNA synthesis)	,,	Tajima *et al.* (1991 and 1992)
	WW94	,,	,,	,,	Kan *et al.* (1991)
	YUSAC2	,,	—	,,	Halaban *et al.* (1992)
	YUZA26	— (Boyden chamber)	↑	,,	,,
	MeWO	— ,,	→ ↑	,,	,,
Other	A253	↑ (colony morphology and migration)	n.r.	,,	Rosen *et al.* (1990a)
	Fa Du	↑ ,,	,,	,,	,,
	SQCC/Y1	↑ ,,	,,	,,	,,
	CE81	— ,,	,,	,,	,,

For footnotes see Table 1.

cells, two of which, regulation of cell movement and differentiation, are as important as regulation of cell division in determining the fate and behaviour of responder cells [Sporn and Roberts, 1988; Gherardi, 1991; Cross and Dexter, 1991]. The complex response of target cells to HGF/SF therefore does not put this factor in a class of its own. Other growth factors, such as PDGF, EGF, acidic and basic FGF and TGFβ, have equally broad and complex activities on the growth, movement and differentiation of their target cells. At the moment, the distinctive feature of HGF/SF appears to be the target specificity (epithelia and endothelia) and a paracrine action which identifies the factor as an effector of mesenchymal-epithelial and endothelial interactions. The discovery of new target cells outside the epithelial and endothelial lineages [Kmiecik *et al*, 1992; Jiang *et al.*, 1992] may change this emphasis but, at present, this conclusion seems justified.

There are, however, important aspects of the biology of HGF/SF and the *met* receptor which need to be emphasised and will undoubtedly attract further study. Epithelial cells, like fibroblasts, have extensive contacts with substrate adhesion molecules and it can be predicted that the early response to HGF/SF may induce changes in the phosphorylation of integrins, talin and vinculin in areas of focal contacts as occurs in fibroblasts in response to other growth factors [Ullrich and Schlessinger, 1990].

Unlike fibroblasts, however, epithelial cells also have a complex pattern of cell-cell interactions mediated by cell adhesion molecules and junctional systems. Clearly the signal transduced by the HGF/SF receptor leads, at least *in vitro*, to the loss of such cell-cell contacts but the effects of HGF/SF on cell adhesion molecules and junctional systems have not yet been thoroughly investigated. Recent progress in the isolation and characterization of epithelial cell adhesion molecules [Edelman and Crossin, 1991] and the protein constituents of epithelial junctions [Schwarz *et al.*, 1990] have now set the scene for these studies.

Let us now return to the question, which we outlined at the beginning of this review, of separate control of movement or growth or differentiation. Although there remain candidate molecules for separate signals, such as the motility factors AMF and MSF [Liotta *et al.*, 1986; Grey *et al*, 1989], the main strategy for regulation does not appear to be different signals for different responses. Indeed most growth factors can initiate multiple responses and the way in which a cell chooses between different responses to the same signal is not obvious.

The effects of HGF/SF on MDCK cells in the absence or presence of collagen suggest a possible strategy. The strategy relies on the combination and integration of different extracellular signals, such as a growth factor signal and a signal from the extracellular matrix or two signals from two different growth factors. Downstream, signal processing by the array of kinases and phosphatases available to the cell (which depend on its state of differentiation) probably offers a further mechanism for directing the response. While this is just a working hypothesis, HGF/SF and *met* constitute an ideal system for testing it and provide us with further insights on regulation of growth, movement and differentiation in epithelial cells in the years to come.

We are grateful to Sue Fletcher for typing Tables 1 and 2.

References

Adams, J. C., Furlong, R. A., and Watt, F. M. (1991) Production of scatter factor by ndk, a strain of epithelial cells, and inhibition of scatter factor activity by suramin. *J. Cell Sci.* **98**, 385-394.

Bizzozero, G. (1894) Accrescimento e rigenerazione nell' organismo. *Arch. Scienze Med.* **18**, 245-287.

Bottaro, D.P., Rubin, J.S., Faletto, D.L., Chan, A.M.-L., Kmiecik, T.E., Vande Woude, G.F., and Aaronson S.A. (1991) Identification of the hepatocyte growth factor receptor as the c-*met* proto-oncogene product. *Science* **251**, 802-804.

Chan, A.M-L., King, H.V.S., Deakin, E.A., Tempest, P.R., Hilkens, J., Kroezen, V., Edwards, D.R., Wills, A.J., Brookes, P., and Cooper, C.S. (1988) Characterization of the mouse *met* protooncogene. *Oncogene* **2**, 593-599.

Chan, A.M.-L, Rubin, J.S., Bottaro, D.P., Hirschfield, D.W., Chedid, M., and Aaronson, S.A. (1991) Identification of a competitive HGF antagonist encoded by an alternative transcript. *Science* **254**, 1382-1385.

Coffer, A., Fellows, J., Young, S., Pappin, D., and Rahman, D. (1991) Purification and characterization of biologically active scatter factor from ras-transformed NIH 3T3 conditioned medium. *Biochem. J.* **278**, 35-41.

Cooper, C.S., Park., Blair, D.G., Oskarsson, M.K., Tainsky, M.A., Eader, L.A., and Vande Woude, G.F. (1984) Molecular cloning of a new transforming gene from a chemically-transformed human cell line. *Nature* **311**, 29-33.

Cross, M., and Dexter, M. (1991) Growth factors and development, transformation and tumorigenesis. *Cell* **64**, 271-280.

Damsky, C.H. and Werb, Z. (1992) Signal transduction by integrin receptors for extracellular matrix: cooperative processing of extracellular information. *Curr. Opin. Cell Biol.* **4**, 772-781.

Edelman, G.M. and Crossin, K.L. (1991) Cell adhesion molecules: implications for a molecular histology. *Annu. Rev. Biochem.* **60**, 155-190

Fabregat, I., de Juan, C., Nakamura, T. and Benito, M. (1992) Growth stimulation of rat fetal hepatocytes in response to hepatocyte growth factor: modulation of c-myc and c-fos expression. *Biochem. Biophys. Res. Comm.* **189**, 684-690.

Fentiman, I., Taylor-Papadimitriou, J., and Stoker, M. (1976) Selective contact-dependent cell communication. *Nature* **264**, 760-762.

Gak, E., Taylor, W. G., Chan, A., and Rubin, J. S. (1992) Processing of hepatocyte growth-factor to the heterodimeric form is required for biological activity. *Fed. Eur. Biochem. Soc. Lett.* **311**, 17-21.

Gherardi, E., and Stoker, M. (1990) Hepatocytes and scatter factor. *Nature* **346**, 228.

Gherardi, E., and Stoker, M. (1991) Hepatocyte growth factor/scatter factor: mitogen, motogen, and met. *Cancer Cells* **3**, 227-32.

Gherardi, E. (1991) Growth factors and cell movement. *Eur J. Cancer* **27**, 403-405.

Gherardi, E., Gray, J., Stoker, M., Perryman, M., and Furlong, R. (1989) Purification of scatter factor, a fibroblast-derived basic protein that modulates epithelial interactions and movement. *Proc. Natl. Acad. Sci. U.S.A.* **86**, 5844-5848.

Gherardi, E., Sharpe, M., and Lane, K. (1993) Properties and structure-function relationship of HGF-SF. In: *Hepatocyte Growth Factor - Scatter Factor (HGF-SF) and the c-*Met *Receptor.* Goldberg I.D. (ed) Birkhäuser Verlag, Basel, pp.31-48.

Giordano, S., Ponzetto, C., Di Renzo, M.F., Cooper, C.S., and Comoglio, P.M. (1989) Tyrosine kinase receptor indistinguishable from the c-*met* protein. *Nature* **339**, 155-156.

Gohda, E., Tsubouchi, H., Nakayama, H., Hirono, S., Sakiyama, O., Takahashi, K., Miyazaki, H., Hashimoto, S., and Daikuhara, Y. (1988) Purification and partial characterization of hepatocyte growth factor from plasma of a patient with fulminant hepatic failure. *J. Clin. Invest.* **81**, 414-419.

Grey, A-M., Schor, A.M., Rushton, G., Ellis, I., and Schor, S.L. (1989) Purification of the migration stimulating factor produced by fetal and breast cancer patient fibroblasts. *Proc. Natl. Acad. Sci. U.S.A.* **86**, 2438-2442.

Grobstein, C. (1967) Mechanism of organogenetic tissue interaction. *Natl Cancer Inst. Monogr.* **26**, 279-299.

Halaban, R., Rubin, J.S., Funasaka, Y., Cobb, M., Boulton, T., Faletto, D., Rosen, E., Chan, A.,

Yoko, K., White, W., Cook, C., and Moellmann, G. (1992) *Met* and hepatocyte growth factor/scatter factor signal transduction in normal melanocytes and melanoma cells. *Oncogene.* **7**, 2195-2206.

Hartmann, G., Naldini, L., Weidner, K.M., Sachs, M., Vigna, E., Comoglio, P.M. and Birchmeier, W. (1992) A functional domain in the heavy chain of scatter factor/hepatocyte growth factor binds the c-*met* receptor, induces cell dissociation but not mitogenesis. *Proc Natl. Acad. Sci. U.S.A.* **89**, 11574-11578.

Higashio, K., Shima, N., Goto, M., Itagaki, Y., Nagao, M., Yasuda, H., and Morinaga, T. (1990) Identity of a tumor cytotoxic factor from human fibroblasts and hepatocyte growth factor. *Biochem. Biophys. Res. Comm.* **170**, 397-404.

Hirai, Y., Takebe, K., Takashina, M., Kobayashi, S. and Takeichi, M. (1992) Epimorphin: a mesenchymal protein essential for epithelial morphogenesis. *Cell* **69**, 471-481.

Iyer, A., Kmiecik, T.E., Park, M., Daar, I., Blair, D., Dunn, J., Sutrave, P., Ihle, J.N., Bodescot, M., and Vande Woude, G.F. (1990) Structure, tissue-specific expression and transforming activity of the mouse *met* protooncogene. *Cell Growth Diff.* **1**, 87-95.

Jessel, T.M., and Melton, D.A. (1992) Diffusible factors in vertebrate embryonic development. *Cell* **68**, 257-70

Jiang, W., Puntis, M., Nakamura, T., and Hallett, M. B. (1992) Neutrophil priming by hepatocyte growth-factor, a novel cytokine. *Immunology* **77**, 147-149.

Joplin, R., Hishida, T., Tsubouchi, H., Daikuhara, Y., Ayres, R., Neuberger, J. M., and Strain, A.J. (1992) Human Intrahepatic Biliary Epithelial Cells Proliferate In Vitro in Response to Human Hepatocyte Growth Factor. *J. Clin Invest.* **90**, 1284-1289.

Kan, M., Zhang G.H., Zarnegar, R., Michalopoulos, G., Myoken, Y., McKeehan, W.L., and Stevens, J.L. (1991) Hepatocyte growth-factor hepatopoietin-a stimulates the growth of rat-kidney proximal tubule epithelial cells (rpte), rat non parenchymal liver cells, human melanoma cells, mouse keratinocytes and stimulates anchorage-independent growth of sv40-transformed rpte. *Biochem. Biophys. Res. Comm.* **174**, 331-337.

Kmiecik, T. E., Keller, J. R., Rosen, E., and Vandewoude, G. F. (1992) Hepatocyte growth-factor is a synergistic factor for the growth of hematopoietic progenitor cells. *Blood* **80**, 2454-2457.

Kratochwil, K. (1983) Embryonic induction. In: *Cell Interactions and development. Molecular mechanisms.* Yamada K.M. (ed). John Wiley & Sons, New York, pp. 100-122

Le Douarin, N. (1982) The neural crest. Cambridge University Press, Cambridge.

Liotta, L.A., Mandler, R., Murano, G., Katz, D.A., Gordon, R.K., Chiang, P.K., and Schiffmann, E. (1986) Tumor cell autocrine motility factor. *Proc. Natl. Acad. Sci. U.S.A.* **83**, 3302-3306.

Lokker, N. A., Mark, M. R., Luis, E. A., Bennett, G. L., Robbins, K. A., Baker, J. B., and Godowski, P. J. (1992) Structure-function analysis of hepatocyte growth-factor - identification of variants that lack mitogenic activity yet retain high-affinity receptor-binding. *Eur. Mol. Biol. Org. J.* **11**, 2503-2510.

Matsumoto, K., Hashimoto, K., Yoshikawa, K., and Nakamura, T., (1991a) Marked stimulation of growth and motility of human keratinocytes by hepatocyte growth factor. *Exp. Cell Res.* **196**, 114-120.

Matsumoto, K., Tajima, H., and Nakamura, T. (1991b) Hepatocyte growth factor is a potent stimulator of human melanocyte DNA synthesis and growth. *Biochem. Biophys. Res. Comm.* **176**, 45-51.

Miyazawa, K., Kitamura, A., Naka, D., and Kitamura, A. (1991) An alternatively processed mRNA generated from human hepatocyte growth factor gene. *Eur J Biochem* **197**, 15-22.

Miyazawa, K., Tsubouchi, H., Naka, D., Takahashi, K., Okigaki, M., Arakaki, N., Nakayama, H., Hirono, S., Sakiyama, O., Takahashi, K., Godha, E., Daikuhara, Y., and Kitamura, N. (1989) Molecular cloning and sequence anlalysis of cDNA for human hepatocyte growth factor. *Biochem. Biophys. Res. Comm.* **163**, 967-973.

Montesano, R., Matsumoto, K., Nakamura, T., and Orci, L. (1991) Identification of a fibroblast-derived epithelial morphogen as hepatocyte growth factor. *Cell.* **67**, 901-908.

Nakamura, T., Nawa, K., Ichihara, A., Kaise, N., and Nishino, T. (1987) Purification and subunit structure of hepatocyte growth factor from rat platelets. *Fed. Eur. Biochem. Soc. Lett.* **224**, 311-316.

Nakamura, T., Nishizawa, T., Hagiya, M., Seki, T., Shimonishi, M., Sugimura, A., Tashiro, K., and Shimizu, S. (1989) Molecular cloning and expression of human hepatocyte growth factor. *Nature* **342**, 440-443.

Nakamura, T., Teramoto, H., and Ichihara, A. (1986) Purification and characterization of a growth factor from rat platelets for mature parenchymal hepatocytes in primary cultures. *Proc. Natl. Acad. Sci. U.S.A.* **86**, 6489-6493.

Naldini, L., Vigna, E., Narsimhan, R., Gaudino, G., Zarnegar, R., Michalopoulos, G.K., and Comoglio, P.M. (1991a) Hepatocyte growth factor (HGF) stimulates the tyrosine kinase activity of the receptor encoded by the proto-oncogene c-*MET*. *Oncogene* **6**, 501-504.

Naldini, L., Weidner, K. M., Vigna, E., Gaudino, G., Bardelli, A., Ponzetto, C., Narsimhan, R. P., Hartmann, G., Zarnegar, R., Michalopoulos, G. K., Birchmeier, W. and Comoglio, P.M. (1991b) Scatter factor and hepatocyte growth factor are indistinguishable ligands for the MET receptor. *Eur. Mol. Biol. Org. J.* **10**, 2867-78.

Naldini, L., Tamagnone, L., Vigna, E., Sachs, M., Hartmann, G., Birchmeier, W., Daikuhara, Y., Tsubouchi, H., Blasi, F., and Comoglio, P. M. (1992) Extracellular proteolytic cleavage by urokinase is required for activation of hepatocyte growth-factor scatter factor. *Eur. Mol. Biol. Org. J.* **11**, 4825-4833.

Noji, S., Tashiro, M., Koyama, E., Nohno, T., Ohyama, K., Taniguchi, S., and Nakamura, T. (1990) Expression of hepatocyte growth factor gene in endothelial and Kupffer cells of damaged rat livers, as revealed by *in situ* hybridization. *Biochem. Biophys. Res. Comm.* **173**, 42-47.

Okigaki, M., Komada, M. Uehara, Y., Miyazawa, K., and Kitamura, N. (1992) Functional characterization of human hepatocyte growth factor mutants obtained by deletion of structural domains. *Biochemistry* **31**, 9555-9561.

Park, M., Dean, M., Kaul, K., Braun, M.J., Gonda, M.A., and Vande Woude, G. (1987) Sequence of MET protooncogene cDNA has features characteristic of the tyrosine kinase family of growth-factor receptors. *Proc. Natl. Acad. Sci. U.S.A.* **84**, 6379-6383.

Pitts., J.D., and Burk, R.R. (1976) Specificity of junctional communication between animal cells. *Nature* **264**, 762-764.

Rosen, E. M., Goldberg, I. D., Kacinski, B. M., Buckholz, T., and Vinter, D. W. (1989) Smooth muscle releases an epithelial cell scatter factor which binds to heparin. *In Vitro Cell Dev. Biol.* **25**, 163-73.

Rosen, E.M., Meromsky, L., Setter, E., Vinter, D.W., and Goldberg, I.D. (1990a) Smooth muscle-derived factor stimulates mobility of human tumour cells. Invas. *Metastasis* **10**, 49-64.

Rosen, E.M., Meromsky, L., Setter, E., Vinter, D.W., and Goldberg, I.D. (1990b) Purified scatter factor stimulates epithelial and vascular endothelial cell migration. *Proc. Soc. Exp. Biol. Med.* **195**, 34-43.

Rubin, J.S., Chan, A. M.-L., Bottaro, D.P., Burgess, W.H., Taylor, W.G., Cech, A.C., Hirschfield, D.W., Wong, J., Miki, T., Finch, P.W., and Aaronson, S.A. (1991) A broad spectrum human lung fibroblast-derived mitogen is a variant of hepatocyte growth factor. *Proc. Natl. Acad. Sci. U.S.A.* **88**, 415-419.

Rygaard, K., Nakamura, T., and Spangthomsen, M. (1992) Expression of the protooncogenes c-met and c-kit and their ligands, hepatocyte growth factor-scatter factor and stem cell factor, in sclc cell lines and xenografts. *Brit. J. Cancer* **67**, 37-46.

Sariola, H., Aufderheide, E., Bernhard, H., Henke-Fahle, S. Dippold, W and Ekblom, P. (1988) Antibodies to cell surface ganglioside GD3 perturb inductive epithelial-mesenchymal interactions. *Cell* **54**, 235-245.

Schwarz, M.A., Owaribe, K., Kartenbeck, J., and Franke, W.W. (1990) Desmosomes and hemidesmosomes, constitutive molecular components. *Annu. Rev. Cell Biol* **6**, 461-491.

Seki, T., Ihara, I., Sugimura, A., Shimonishi, M., Nishizawa, T., Asami, O., Hagiya, M., Nakamura, T., and Shimizu, S. (1990) Isolation and expression of cDNA for different forms of hepatocyte growth factor from human leukocytes. *Biochem. Biophys. Res. Comm.* **172**, 321-327.

Selden, C., and Hodgson, H. (1989) Further characterization of hepatotropin, a high molecular weight hepatotropic factor in rat serum. *J. Hepatol.* **9**, 167-176.

Sharpe, P.M. and Ferguson, W.J. (1988) Mesenchymal influences on epithelial differentiation in developing systems. *J. Cell Sci. Suppl.* **10**, 195-230.

Shibamoto, S., Hayakawa, M., Hori, T., Oku, N., Miyazawa, K., Kitamura, N., and Ito, F. (1992) Hepatocyte growth-factor and transforming growth-factor-beta stimulate both cell-growth and migration of human gastric adenocarcinoma cells. *Cell Struct. Funct.* **17**, 185-190.

Shimomura, T., Ochiai, M., Kondo, J., and Morimoto, Y. (1992) A novel protease obtained from fbs-containing culture supernatant, that processes single chain form hepatocyte growth-factor to 2 chain form in serum-free culture. *Cytotechnology* **8**, 219-229.

Shiota, G., Rhoads, D. B.,Wang, T. C., Nakammura, T., and Schmidt, E.V. (1992) Hepatocyte growth factor inhibits growth of hepatocellular carcinoma cells. *Proc. Natl. Acad. Sci. U. S. A.* **89**, 373-377.

Sonnenberg, E. Weidner, K.M., and Birchemeir, C. (1993) Expression of the met receptor and its ligand, HGF/SF during mouse embryogenesis. In Hepatocyte Growth Factor - "Scatter Factor (HGF-SF) and the C-*Met* Receptor" (ed. I.D. Goldberg) *Birkhäuser Verlag, Basel*, pp.381-394.

Spemann, H., and Mangold, H. (1924) Über induktion von embryonanlagen durch implantation artfremder organisatoren. *Wilh. Roux Arch. EntMech. Organ.* **100**, 599-638.

Sporn, M.B. and Roberts, A.B. (1988) Peptide growth factors are multifunctional. *Nature* **332**, 217-219.

Stern, C.D., Ireland, G.W., Herrick, S.E., Gherardi, E., Gray, J., Perryman, M., and Stoker, M. (1990) Epithelial scatter factor and development of the chick embryonic axis. *Development* **110**, 1271-1284

Stoker, M. (1984) Junctional competence in clones of mammary epithelial cells, and modulation by conditioned medium. *J. Cell. Physiol.* **121**, 174-83.

Stoker, M. and Gherardi, E. (1991) Regulation of cell movement: the motogenic cytokines. *Biochim. Biophys. Acta (Cancer Reviews)* **1072**, 81-102.

Stoker, M., and Perryman, M. (1985) An epithelial scatter factor released by embryo fibroblasts. *J. Cell Sci.* **77**, 209-223.

Stoker, M., Gherardi, E., Perryman, M., and Gray, J. (1987) Scatter factor is a fibroblast-derived modulator of epithelial cell mobility. *Nature* **327**, 239-42.

Strain, A.J., Ismail, T., Tsubouchi, H., Arakaki, N., Hishida, T., Kitamura, N., Daikuhara, Y. and McMaster, P. (1991) Native and recombinant human hepatocyte growth factors are highly potent promoters of DNA synthesis in both human and rat hepatocytes. *J. Clin. Invest.* **87**, 1853-1857.

Tajima, H., Matsumoto, K., and Nakamura, T. (1991) Hepatocyte growth factor has potent anti-proliferative activity in various tumor cell lines. *Fed. Eur. biochem. Soc. Lett* **291**, 229-232.

Tajima, H. Matsumoto, K., and Nakamura, T. (1992) Regulation of cell growth and motility by hepatocyte growth factor and receptor expression in various cell species. *Exp. Cell Res.* **202**, 423-431.

Tashiro, K., Hagiya, M., Nishizawa, T., Seki, T., Shimonishi, M., Shimizu, S., and Nakamura, T. (1990) Deduced primary structure of rat hepatocyte growth factor and expression of the mRNA in rat tissues. *Proc. Natl. Acad. Sci. U.S.A.* **87**, 3200-3204.

Tsarfaty, I., Resau, J.H., Rulong, S., Keydar, I., Faletto, D.L. and Vande Woude, G.F. (1992) The *met* proto-oncogene receptor and lumen formation. *Science* **257**, 1258-1261.

Uehara, Y., and Kitamura, N. (1992) Expression of a human hepatocyte growth factor/scatter factor cDNA in MDCK epithelial cells influences cell morphology, motility and anchorage-independent growth. *J. Cell Biol.* **117**, 889-894.

Ullrich, A. and Schlessinger, J. (1990) Signal transduction by receptors with tyrosine kinase activity. *Cell* **61**, 203-212.

Weidner, K. M., Arakaki, N., Hartmann, G., Vandekerckhove, J., Weingart, S., Rieder, H., Fonatsch, C., Tsubouchi, H., Hishida, T., Daikuhara, Y., and Birchmeier W. (1991) Evidence for the identity of human scatter factor and human hepatocyte growth factor. *Proc. Natl. Acad. Sci. U.S.A.* **88**, 7001-7005.

Weidner, M. K., Behrens, J., Vandekerckhove, J. and Birchmeier, W. (1990) Scatter factor: Molecular characteristics and effect on the invasiveness of epithelial cells. *J. Cell Biol.* **111**, 2097-2108.

Weidner, M.K., Sachs, M., and Birchmeier, W. (1993) The Met receptor tyrosine kinase transduces motility, proliferation and morphogenic signals of scatter factor/hepatocyte growth factor in epithelial cells. *J. Cell Biol.* **121**, 145-154.

Yoshinaga, Y., Fujita, S., Gotoh, M., Nakamura, T., Kikuchi, M., and Hirohashi, S. (1992) Human lung-cancer cell-line producing hepatocyte growth-factor scatter factor. *Jpn. J. Cancer Res.* **83**, 1257-1261.

Zarnegar, R., and Michalopoulos, G. (1989) Purification and biological characterization of human hepatopoietin A, a polypeptide growth factor for hepatocytes. *Cancer Res.* **49**, 3314-3320.

Printed in Great Britain © The Society of Experimental Biology 1993   183

INVOLVEMENT OF CELL MOTILITY IN TUMOR PROGRESSION

BRIGITTE BOYER, ANA-MARIA VALLÉS, GORDON C. TUCKER, ANNIE DELOUVÉE and JEAN PAUL THIERY

Laboratoire de Physiopathologie du Développement URA CNRS 1337 Ecole Normale Supérieure, 46 rue d'Ulm 75230 Paris Cedex 05, France

Summary

Since one crucial step in tumor progression consists of the acquisition of invasive and metastatic properties, it is important to analyze the mechanisms used by cancer cells to disperse. Among the possible mechanisms of cell dispersion, cell motility appears as a central phenomenon that still needs to be understood at the molecular level. Our experimental approach to the contribution of cell motility in carcinoma cell dissemination is based on the study of the NBT-II rat bladder carcinoma cell line. This epithelial cell line gives rise to isolated, actively migrating, fibroblast-like cells in response to specific stimuli (collagens and acidic fibroblast growth factor [aFGF]). Analysis of the scattering response indicates that the different stimuli can synergize, leading to increased motility and invasiveness. NBT-II cells have two types of response to aFGF: they can either proliferate or scatter. In addition, the two reponses are mutually exclusive, suggesting that the cell status can dictate whether or not tumor cells will disperse after exposure to a scatter factor. Finally, recent studies on the involvement of epithelial-specific cadherins in the process of aFGF-induced cell scattering indicate that a sustained expression of E-cadherin is not sufficient to protect cells from dispersing. In conclusion, our experimental model offers the opportunity to dissect the molecular events leading to tumor cell dissemination.

Introduction

The mechanisms whereby cells acquire motility remain poorly understood. Cell motility is, however, a phenomenon which contributes to several physiological processes and diseases, such as embryonic morphogenesis and adult wound healing or tumor progression (Trinkaus, 1976; Zetter, 1990). In certain types of epithelia, the acquisition of motility is correlated with dramatic changes in the state of differentiation since migrating cells no longer express epithelial characteristics and acquire mesenchymal properties. Such profound changes are observed during embryogenesis in the conversion of epithelial to mesenchymal cells (i.e., the dissociation of some cells from cohesive epithelial sheets and their transformation into elongated, fibroblast-like, motile cells). It has been also suggested that similar processes of epithelium-to-mesenchyme transition (EMT) might occur during invasion and metastasis of carcinoma cells.

It is becoming evident that at least three components are capable of inducing tumor cell locomotion: the extracellular matrix (ECM), motility factors and growth factors.

Key words: cell motility, invasion, metastasis, carcinoma, growth factor, acidic fibroblast growth factor, collagen, extracellular matrix.

Collagens, fibronectin and laminin (Mensing *et al.*, 1985; Mc Carthy *et al.*, 1985; Mc Carthy *et al.*, 1986; Aznavoorian *et al.*, 1990) are extracellular matrix molecules which can serve adhesive as well as signal-transducing functions. The question as to how the binding of ECM components to their specific receptors on the cell surface (termed integrins) can transduce intracellular signals has recently been approached. Agonist-induced activation of platelet aggregation leads to a wave of tyrosine phosphorylations involving several platelet proteins (reviewed by Kroll and Schafer, 1989). Treatments which interfere with the function of IIb/IIa, a major platelet integrin, block the phosphorylation of these proteins, suggesting that integrins can regulate phosphorylation events (Golden *et al.*, 1990). Ligand-induced activation of integrins in carcinoma cells (Kornberg *et al.*, 1991) and fibroblasts (Guan *et al.*, 1991) also induce phosphorylation events. The major molecular species to be phosphorylated is a 120 kD protein, located in focal contacts where it co-localizes with β-1 integrins. This protein appears to be identical to a protein of similar size found to be phosphorylated by growth factors or oncogenes belonging to the tyrosine kinase signalling pathway (for review, see Hynes, 1992).

The second group of inducers of cell motility are composed of motility factors which were first isolated on the basis of their biological effect on cell locomotion. However, it has recently been found that some of them have more complex effects: scatter factor that disperses colonies of epithelial cells *in vitro* (Weidner *et al.*, 1990) is identical to hepatocyte growth factor (Naldini *et al.*, 1991) and binds to the *c-met* protooncogenic receptor which is a transmembrane tyrosine kinase receptor. Autocrine motility factor (AMF) first isolated from the conditioned medium of human melanoma cells (Liotta *et al.*, 1986), was found to bind to a 78 kD cell surface glycoprotein that displays significant homology with the human tumor suppressor p53 (Nabi *et al.*, 1990). Migration-stimulating factor (MSF) was first purified from foetal fibroblasts or from fibroblasts of breast cancer patients (Grey *et al.*, 1989). However its specific receptors remain to be characterized.

The third group of inducers of cell motility belong to the large family of cytokines and growth factors. Although first described as mitogenic agents, cytokines have also been found to induce the acquisition of a motile phenotype thereby promoting the dispersion of cancer cells. Insulin-like growth factor-I (IGF-I) is chemotactic for human melanoma cells (Stracke *et al.*, 1989) while IGF-II has been reported to induce motility of human rhabdomyosarcoma cells *in vitro* (El Badry *et al.*, 1990). Epidermal growth factor (EGF) stimulates the non-directed movement of keratinocytes and induces membrane ruffling and filopodia extension in carcinoma cells (Chinkers *et al.*, 1981), whereas platelet-derived growth factor (PDGF) and transforming growth factor-β (TGF-β) are chemotactic for fibroblasts and other cell types (Seppa *et al.*, 1982; Grotendorst, 1984; Postlethwhaite *et al.*, 1987; Wahl *et al.*, 1987).

Since each group of inducers of cell motility have complex and pleiotropic effects, it is conceivable that in certain situations factors belonging to two different groups could act synergistically. In that respect, an overwhelming body of evidence has demonstrated the role of growth factors on the remodelling of extracellular matrix (for review, see Nathan and Sporn, 1991). EGF/TGF-α and fibroblast growth factor (FGF) are potent inducers of metalloproteinases which can degrade ECM components (Gavrilovic *et al.*, 1990;

Jouanneau *et al.*, 1991). Conversely, TGF-β regulates the level of ECM degradation by reducing the synthesis of enzymes involved in ECM proteolysis (Moses *et al.*, 1990; Overall *et al.*, 1991) or by increasing the expression of plasminogen activator-1 (Reilly and McFall, 1991) or of tissue inhibitor of metalloproteinases (TIMP) (Lotz and Guerne, 1991; Overall *et al.*, 1991).

TGF-β also enhances gene expression of several extracellular matrix elements, including collagens, fibronectin, thrombospondin, osteopontin and proteoglycans (Kahari *et al.*, 1991; Shönherr *et al.*, 1991; Ura *et al.*, 1991).

Moreover, cytokines regulate the expression of receptors specific for ECM components. Nerve growth factor (NGF), which promotes the neural differentiation of the PC12 pheochromocytoma cell line, induces the expression of a laminin-binding integrin in these cells (Rossino *et al.*, 1990). EGF and PDGF activate the synthesis of β1-integrins in a fibroblastic cell line (Bellas *et al.*, 1991). TGF-β increases the expression of several integrins in various cell lines (Heino *et al.*, 1989; Ignotz *et al.*, 1989). Interleukin-1 increases β1-integrin expression in human osteosarcoma cells (Dedhar, 1989). It should be stressed that increased expression of ECM components or of integrins could have diverse effects on cell migration, depending on the type of ECM proteins or of integrins being overexpressed. Indeed, the type and the specificity of the interactions between ECM and cells dictate the cellular response (attachment, spreading, motility...) to a given local environment.

On the other hand, the antagonist effects of TGF-β1 and MSF on fibroblast migration have been recently reported (Ellis *et al.*, 1992) and could be implicated in dermal wound healing.

The epithelium-to-mesenchyme transition of a rat bladder carcinoma cell line

We have used the rat bladder carcinoma-derived cell line NBT-II (Toyoshima *et al.*, 1971) to investigate the mechanisms whereby cancer cells of epithelial origin can reversibly convert from an epithelial to a fibroblastic phenotype (Boyer *et al.*, 1989). Our studies have focussed on the role of ECM components and soluble factors on epithelial cell scattering.

1. The epithelium-to-mesenchyme transition induced by collagens

Since it is well established that during the invasive and metastatic processes, cancer cells have to migrate through connective tissues, we have carried out a series of experiments to study the interactions of the NBT-II cell line with different components of the extracellular matrix (e.g., fibronectins, laminin, collagens of different types) (Tucker *et al.*, 1990). We have observed that all ECM components tested permitted the attachment and spreading of NBT-II cells. To analyse the capacity of ECM proteins to promote cell dispersion, we have studied cell scattering from aggregates of NBT-II cells deposited on different substrates. All ECM components tested promoted aggregate spreading and the subsequent formation of a monocellular halo around the aggregate. In contrast, the appearance of isolated cells was observed only on collagens (types I, III, IV and V) in a dose-dependent manner (Figure 1). While all four collagens permitted translocation,

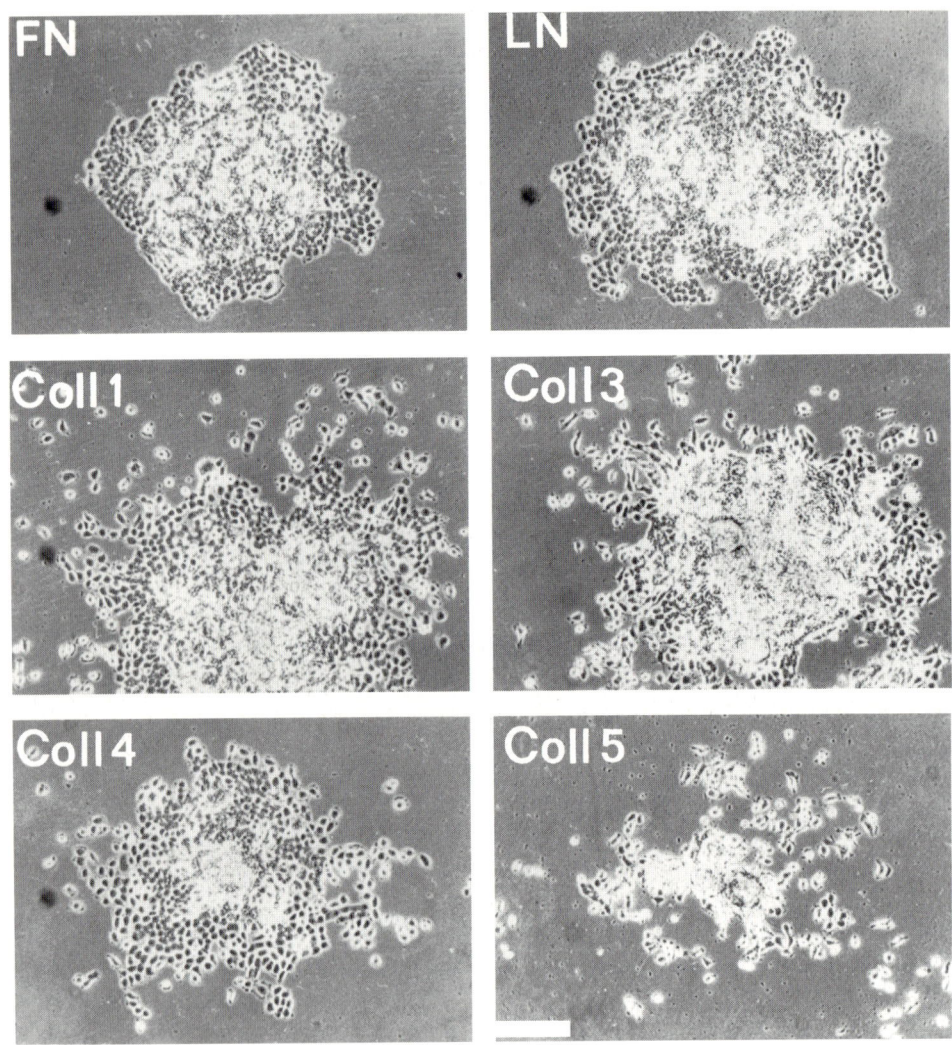

Fig. 1. Collagen-mediated dispersion of NBT-II cells. NBT-II cell aggregates were deposited on Petri dishes coated with different ECM components (FN= fibronectin; LN=laminin; Coll= collagen). Aggregate spreading and emergence of isolated cells emigrating from the aggregate were observed by light microscopy. Note that FN and LN have no scattering effect. Among all types of collagens tested, collagen type IV has the least pronounced effect. Bar, 100 μm.

types I and III maintained their effects for longer periods of time and thus were more efficient. Furthermore, the interaction between collagens and NBT-II cells required a native helical structure since different fragments of collagen type I or denatured collagen could not mimic the original scattering response observed with fibrillar collagen. When NBT-II cells were cultured on or included in collagen gels, the scattering behaviour induced by collagens gave rise to various patterns of cell invasion, with single cells, small groups of cells or "indian files" infiltrating the gel. All the effects of collagens were reproduced with fragments of NBT-II tumors obtained by injection of NBT-II cells into nude mice, suggesting that *in vivo*, interactions with collagens might favor cell infiltration into the connective tissue surrounding the tumor.

2. Acidic FGF promotes the dispersion of NBT-II cells

The ability of exogenous soluble factors to promote cell scattering was also investigated. NBT-II cells on exposure to aFGF exhibited the same morphological changes as on collagens; they became progressively elongated and detached from one another, migrating as individual cells (Figure 2) (Boyer *et al.*, 1989; Vallés *et al.*, 1990a). A prominent feature of the EMT is the disruption of the extended cell contacts and intercellular junctions interconnecting epithelial cells. To study the possible mechanisms contributing to NBT-II cell dissociation we analysed, by immunofluorescence, the distribution of desmosome-specific proteins desmoplakins (DP) I and II and their fate during the conversion (Boyer *et al.*, 1989). Disappearance of DP cortical staining was first observed after 5 h of culture in the presence of aFGF. After 8 h of culture, most cells were devoid of DP cortical staining. Decrease of cortical immunoreactivity correlated

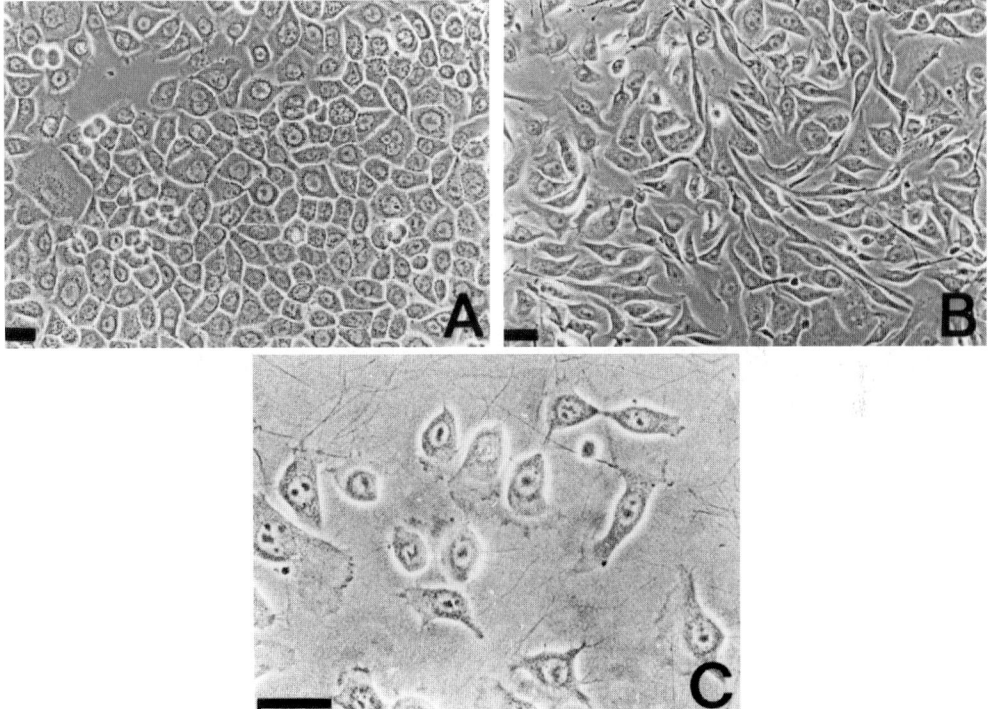

Fig. 2. Morphological changes induced by aFGF and collagen type 1. In A, NBT-II cells cultured in standard medium formed a cobblestone-like monolayer of polygonal cells typical of an epithelium-derived cell line. In B, cells exposed to 20 ng/ml aFGF displayed a spindle-shaped morphology typical of a cell line of fibroblastic origin. In C, NBT-II cells deposited on collagen 1 fibers exhibited morphological changes similar to those observed in B. Bar, 20 μm.

with the appearance of intracellular immunofluorescent dots, suggesting that the modifications in DP immunostaining were related to the rapid internalization of desmosomes occuring after aFGF stimulation. Biochemical analyses further revealed that during the first hours of aFGF induction, the soluble pool of DP molecules transiently increased while the insoluble pool of DP molecules, cross-linked to cytoskeletal proteins, decreased. These results suggested that solubilization of DP might correspond to an early step in desmosome destabilization. We also demonstrated that phosphorylation events were required in the pathway leading to desmosome internalization as indicated by the blocking effect of 6-dimethylaminopurine, a specific kinase inhibitor. Using immuofluorescence microscopy, we also showed that the changes in cell morphology and behavior were accompanied by the reorganization of the cytokeratin and the actin filament systems and by the appearance of vimentin intermediate filaments (Figure 3).

To understand the mechanisms of action of aFGF on cell scattering, we sought to characterize aFGF receptors on NBT-II cells (Vallés et al., 1990a). Specific high-affinity receptors (Kd=25pM) were identified. Cross-linking experiments revealed the existence of three distinct aFGF-binding species migrating at apparent molecular weights of 190, 150 and 130 kD. Most interesting was the finding that basic FGF, while being a mitogen for NBT-II cells similar to aFGF was unable to induce EMT. Binding experiments demonstrated that NBT-II cells were equipped with low-affinity receptors for basic FGF. The differential effects observed with acidic and basic FGFs suggested that the signalling pathway leading to EMT could be distinct from that leading to cell proliferation.

To explore further the dual effect of aFGF, we investigated the response of NBT-II cells to this cytokine as a function of the environmental conditions of culture (Vallés et al., 1990b). We observed that at low cell density, aFGF promoted EMT, characterized by cell dissociation, morphological changes toward a fibroblast-like phenotype and acquisition of cell motility. Under these conditions, NBT-II cells were unresponsive to the growth-promoting effect of aFGF. At high cell density, aFGF was a potent mitogenic agent but its scattering effect was essentially abolished (Figure 4). A slight decrease in the binding of aFGF to its specific receptors was observed at high cell density but was not sufficient to explain the modifications of the biological response. Furthermore, NBT-II cells located at the edge of artificial wounds mimicked the behavior of subconfluent cells, while cells located in the center of colonies had a response similar to that of confluent cultures. These results suggested that the cellular response to a growth factor might depend on the localization of the responding cell within a cell colony.

It is important to understand how, after interacting with growth factors, cells are directed to enter cell division or to undergo EMT. One attractive hypothesis could be that the dual function of aFGF results from the presence or absence of cellular effector molecules specifically determining the choice between the two pathways. This hypothesis is currently under investigation.

3. Acidic FGF potentiates cell dispersion induced by collagens

The possible cooperation between aFGF and collagens was studied on aggregates of NBT-II cells (Tucker et al., 1991). In the presence of aFGF, no peripheral single cell dispersion occurred on laminin or fibronectin. In sharp contrast, aFGF dramatically

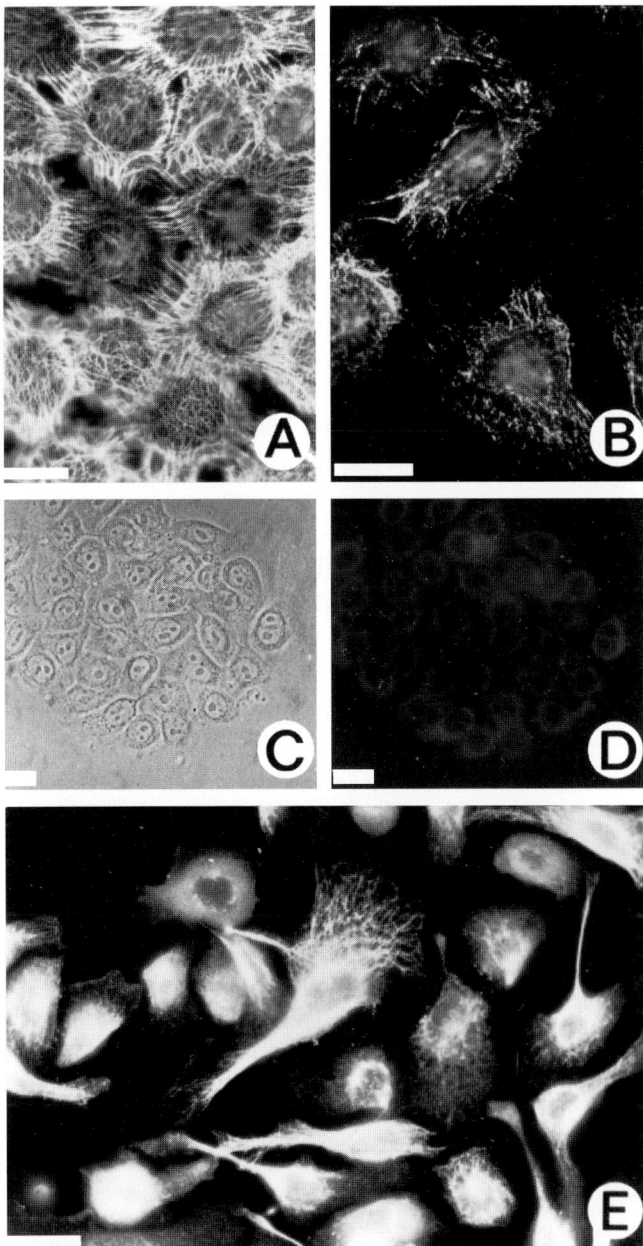

Fig. 3. Immunofluorescence microscopy of NBT-II cells cultured in the presence or absence of aFGF. A, C, D: cell culture was done under standard conditions. B, E: cells were exposed to 20 ng/ml aFGF. A, B: immunostaining with anti-cytokeratin monoclonal antibody. D, E: immunostaining with anti-vimentin monoclonal antibody. C: phase-contrast micrograph of D. Note that cells cultured in the presence of aFGF display a disorganized pattern of cytokeratin immunolabelling and express vimentin filaments which are not revealed in epithelial-like cells. Bar, 10 μm.

A B

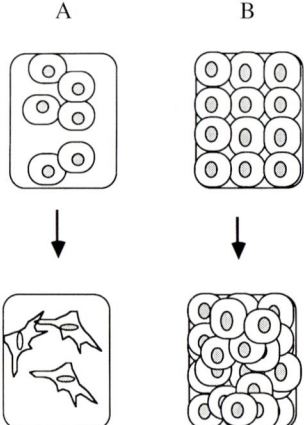

Fig. 4. Schematic representation of the dual effect of aFGF. At subconfluency (A), aFGF induces cell dispersion but not cell proliferation. When cells reach confluency (B), aFGF is mitogenic but has no scattering effect.

accelerated collagen-mediated NBT-II individual cell dispersion and locomotion in a reversible way (Figure 5). As a marker of cell dissociation, we studied desmosome distribution in aggregate cultures: desmosomes were present in aggregates formed in suspension even in the presence of aFGF, whereas their internalization occurred after cell contact with a permissive substratum (collagens but not laminin or fibronectin). These observations suggested that scatter effects induced by soluble factors are dependent on the composition of the extracellular matrix. Furthermore, the synergistic effects of aFGF and collagens might be due to a direct action of aFGF on collagen-specific cellular integrins. In that respect, we have recently shown that aFGF induces a rapid upregulation of $\alpha2\beta1$-integrin expression in NBT-II cells.

We then sought to evaluate the effects of extracellular matrix and growth factors on tumor cell invasion under conditions which resemble the *in vivo* situation. We therefore developed an *in vitro* invasion test specifically adapted for NBT-II cells derived from urinary bladder. A fragment of urinary bladder of adult rat was confronted in organotypic culture with an aggregate of NBT-II cells. After a period of culture of 7 days in the absence or presence of aFGF, explants were fixed, sectioned and stained. NBT-II cells were able to replace the normal urothelium but were unable to penetrate the underlying stroma. In contrast, when aFGF was added in the culture medium, cells invaded the stroma and the underlying muscular layer. These results demonstrated unambiguously that aFGF, whilst being a scatter factor for cells cultured under standard conditions could also be considered as an invasion-promoting agent.

4. Acidic FGF-induced regulation of cell adhesion

Among all the possible mechanisms of carcinoma cell invasion, disruption of intercellular connecting systems has received much attention. Cell-adhesion molecules (CAMs) are cell surface glycoproteins involved in processes of cell-cell recognition and

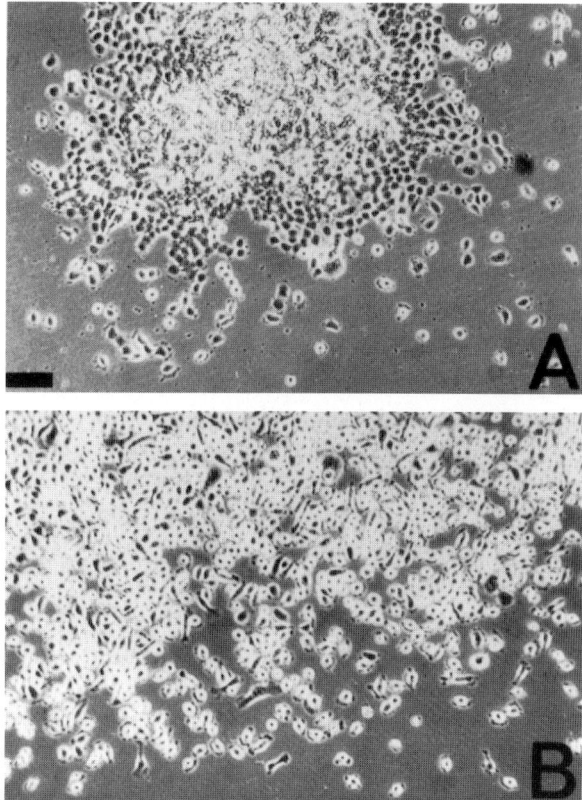

Fig. 5. Synergistic effects of aFGF and collagen type 1. NBT-II cell aggregates were deposited on a Petri dish coated with collagen I and cultured in the absence (A) or presence (B) of 20 ng/ml aFGF. Cell dispersion was observed the next day by light microscopy. In A, the aggregate has not entirely collapsed, while in B, it is completely decompacted and has generated numerous isolated cells. Bar, 50 μm.

interaction during embryogenesis. Since the pioneering experiments of Steinberg and colleagues (Steinberg *et al.*, 1973) and Takeichi's group (Takeichi *et al.*, 1979) evidence has accumulated to suggest that calcium-dependent CAMs, also termed cadherins, play a key role in these cell-cell interactions. Cadherins form a family of cell surface glycoproteins (reviewed by Takeichi, 1988). Among them, E-cadherin, which is expressed as early as the blastula stage, becomes developmentally restricted to epithelial tissues. Several lines of evidence have indicated that E-cadherin is a master molecule for the maintenance of epithelial integrity and polarized function (Gumbiner and Simons, 1986; McNeill *et al.*, 1990) and conversely, that the suppression of E-cadherin expression or function might trigger the dispersion of epithelial (Behrens *et al.*, 1985) and carcinoma cells (Behrens *et al*, 1989; Takeichi, 1991). Since E-cadherin-mediated intercellular binding regulates the cohesiveness of epithelial tissues and since the study of E-cadherin in human tumor cell lines has revealed differences in the levels of expression of these molecules depending on the metastatic potential (Shimoyama *et al.*, 1989), it is tempting

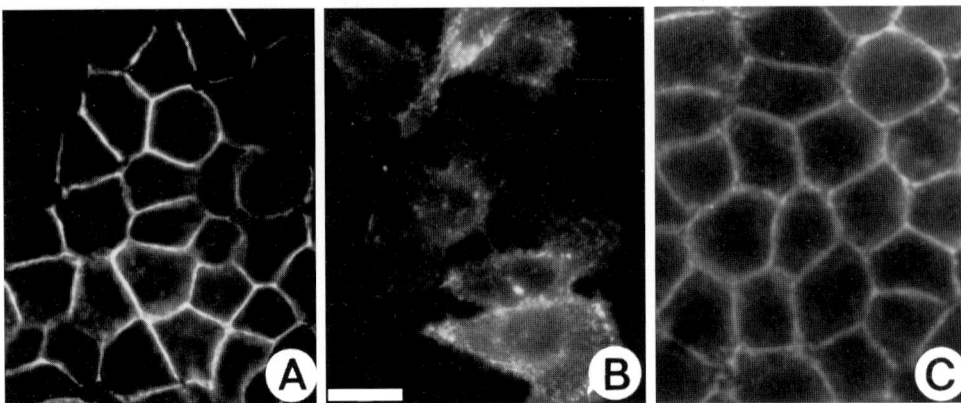

Fig. 6. Immunostaining of NBT-II cells with anti-E-cadherin antibody. A: standard conditions of culture. B: sparse culture of fibroblast-like NBT-II cells exposed to 20 ng/ml aFGF. C: confluent culture of fibroblast-like NBT-II cells exposed to 20 ng/ml aFGF. Note that E-cadherin is concentrated in regions of apposition between epithelial cells and is not present in free cell edges (A) while fibroblast-like cells exhibit a punctate immunolabelling covering the entire cell surface (B). Bar, 10 μm.

to speculate that the modulation of E-cadherin is an early step in carcinoma cell dissemination, i.e., in invasion and metastasis of tumor cells derived from epithelial tissues. In that respect, Behrens and colleagues have shown that epithelial cells acquire invasive properties when E-cadherin cell-surface expression is experimentally abrogated (Behrens *et al.*, 1989). However, the scatter factor that disperses several carcinoma and normal epithelial cell lines does not affect E-cadherin expression in a normal epithelial cell line (Weidner *et al.*, 1990). In this context, it was interesting to determine the role of E-cadherin in the aFGF-induced dispersion of NBT-II cells.

Epithelial NBT-II cells expressed E-cadherin that was not downregulated during the aFGF-induced transition toward the migratory phenotype (Fig. 6). However, in isolated elongated cells emerging upon exposure to aFGF, E-cadherin molecules were no longer concentrated at basolateral membranes, and were instead redistribued over the entire cell surface. These modifications were observed very early in the course of EMT and could serve to lower the strengh of intercellular adhesion. More strikingly, formation of intercellular contacts beween fibroblastic cells promoted the redistribution of E-cadherin at regions of cell apposition but did not abrogate cell motility (Boyer *et al.*, 1992). The persistence of motility might be due to the fact that E-cadherin molecules were not properly retargeted to their specific sites at lateral membranes (Boller *et al.*, 1985) or not correctly linked to the cytoskeleton, or that the establishment of other types of intercellular junctions was necessary to strengthen the intercellular adhesion and induce a stationary state.

It has been reported that high levels of E-cadherin expression in carcinomas are correlated with a well-differentiated epithelial phenotype and a low invasive potential while low levels of E-cadherin are associated with a less differentiated phenotype and increased invasiveness (Shimoyama *et al.*, 1989). Low levels of E-cadherin might thus account for the responsiveness of cancer cells to scattering factors. To test this hypothesis, NBT-II cells were transfected with an expression vector encoding mouse E-cadherin. Overexpression of E-cadherin in stables clones derived from transfected cells did not protect them from the scattering effect of aFGF. It thus seems reasonable to conclude that high levels of E-cadherin are not sufficient to induce a non-motile, non-invasive phenotype.

In the past few years, research on cancer invasion and metastasis has benefited from the striking advances in our knowledge of the molecular mechanisms that are involved in the acquisition of cell motility. Model systems specifically designed to study the role of molecules known to be involved in the modulation of cell motility will probably increase our understanding of the pathogenesis of malignancy.

This work was supported by grants from the CNRS, the Ligue contre le Cancer (Comité National and Comité de Paris), the Association pour la Recherche sur le Cancer, the National Cancer Institute of the NIH (1 R01 CA 49417-01A2).

References

Aznavoorian, S.A. Stracke, M.L., Krutzsch, H., Schiffman, E., and Liotta, L.A. (1990). Signal transduction for chemotaxis and haptotosis by matrix molecules in tumor cells. *J. Cell Biol.* **110**, 1427-1438.

Behrens, J., Birchmeier, W., Goodman, S.L., and Imhof, B.A. (1985). Dissociation of Madin Darby canine kidney epithelial cells by the monoclonal antibody anti-Arc-1: mechanistic aspects and identification of the antigen as a component related to uvomorulin. *J. Cell Biol.* **101**, 1307-1315.

Behrens, J., Mareel, M. M., Van Roy, F.M., and Birchmeier, W. (1989). Dissecting tumor cell invasion: epithelial cells acquire invasive properties after the loss of uvomorulin-mediated cell-cell adhesion. *J. Cell Biol.* **108**, 2435-2447.

Bellas, R. E., Bendori, R. and Farmer, S. R. (1991). Epidermal growth factor activation of vinculin and β_1-integrin gene transcription in quiescent Swiss 3T3 cells. *J. Biol. Chem.* **266**, 12008-12014.

Boller, K., Vestweber, D., and Kemler, R. (1985). Cell-adhesion molecule uvomorulin is localized in the intermediate junctions of adult intestinal epithelial cells. *J. Cell Biol.* **100**, 327-332.

Boyer, B., Tucker, G.C., Vallés, A.M., Franke, W.W., and Thiery, J.P. (1989). Rearrangements of desmosomal and cytoskeletal proteins during the transition from epithelial to fibroblastoid organization in cultured rat bladder carcinoma cells. *J. Cell Biol.* **109**, 1495-1509.

Boyer, B., Dufour, S., and Thiery, J.P. (1992). E-cadherin expression during the acidic FGF-induced dispersion of a rat bladder carcinoma cell line. *Exp. Cell Res.* **201**, 347-357.

Chinkers, M., Mckanna, J.A. and Cohen, S. (1981) . Rapid rounding of human epidermoid carcinoma cells A-431 induced by epidermal growth factor. *J. Cell Biol.* **88**, 422-429.

Dedhar, S. (1989). Regulation of expression of the cell adhesion receptors, integrins, by recombinant human interleukin-1β in human osteosarcoma cells: inhibition of cell proliferation and stimulation of alkaline phosphatase activity. *J. Cell Physiol.* **138**, 291-299.

El Badry, O.M., Minniti, C., Kohn, E.C., Houghton, P.J., Daughaday, W.H. and Helman, L.J. (1990) Insulin-like growth factor II acts as an autocrine motility factor in human rhabdomyosarcoma cells. *Cell Growth and Diff.* **1**, 325-331.

Ellis, I., Grey, A.M., Schor, A.M. and Schor, S.L. (1992). Antagonistic effects of TGF β1 and MSF on fibrobast migration and hyaluronic acid synthesis: possible implications for dermal wound healing. *J. Cell Sci.* **102**, 447-456.

Gavrilovic, J., Moens, G., Thiery, J. P. and Jouanneau, J. (1990). Expression of transfected transforming growth factor alpha induces a motile fibroblastic-like phenotype with extracellular matrix-degrading potential in a rat bladder carcinoma cell line. *Cell Regulation* **1**, 1003-1014.

Golden, A., Brugge, J.S., and Shattil, S.J. (1990). Role of platelet membrane glycoprotein IIb-IIIa in agonist-induced tyrosine phosphorylation of platelets. *J. Cell Biol*. **111**, 3117-3127.

Grey, A.M., Schor, A.M., Rushton, G., Ellis, I., and Schor, S.L. (1989). Purification of the migration stimulating factor produced by foetal and breast cancer patient fibroblasts. *Proc. Natl. Acad. Sci. USA* **86**, 2438-2442.

Grotendorst, G. R. (1984). Alteration of the chemotactic response of NIH/3T3 cells to PDGF by growth factors, transformation, and tumor promotors. *Cell* **36**, 279-285.

Guan, J.L., Trevithick, J.E., and Hynes, R.O. (1991). Fibronectin/integrin interaction induces tyrosine posphorylation of a 120-kda protein. *Cell Regulation* **2**, 951-964.

Gumbiner, B. and Simons, K. (1986) A functional assay for proteins involved in establishing identification of a uvomorulin-like polypeptide. *J. Cell Biol*. **102**, 457-468.

Heino, J., Ignotz, R. A., Hemler, M. E., Crouse, C. and Massagué, J. (1989). Regulation of cell adhesion receptors by transforming growth factor-β. *J. Biol. Chem*. **264**, 380-388.

Hynes, R.O. (1992). Integrins: versatility, modulation, and signaling in cell adhesion. *Cell* **69**, 11-25.

Ignotz, R. A., Heino, J. and Massagué, J. (1989). Regulation of cell adhesion receptors by transforming growth factor-β. *J. Biol. Chem*. **264**, 389-392.

Jouanneau, J., Gavrilovic, J., Caruelle, D., Jaye, M., Moens, G., Caruelle, J. P. and Thiery, J. P. (1991). Secreted or nonsecreted forms of acidic fibroblast growth factor produced by transfected epithelial cells influence cell morphology, motility, and invasive potential. *Proc. Natl. Acad. Sci. U.S.A*. **88**, 2893-2897.

Kahari, V. M., Larjava, H. and Uitto, J. (1991). Differential regulation of extracellular matrix proteoglycan (PG) gene expression. *J. Biol. Chem*. **266**, 10608-10615.

Kornberg, L.J., Earp, H.S., Turner, C.E., Prockop, C., and Juliano, R. (1991). Signal transduction by integrins: increased protein tyrosine phosphrylation caused by clustering of β1 integrins. *Proc. Natl. Acad. Sci. USA* **88**, 8392-8396.

Kroll, M.H. and Schafer, A.I. (1989). Biochemical mechanisms of platelet activation. *Blood* **74**, 1181-1195.

Liotta, L.A., Mandler, R., Murano, G., Katz, D.A., Gordon, R.K., Chang, P.K., and Schiffman, E. (1986). Tumor cell autocrine motility factor. *Proc. Natl. Acad. Sci. USA* **83**, 3302-3306.

Lotz, M. and Guerne, P. A. (1991). Interleukin-6 induces the synthesis of tissue inhibitor of metalloproteinase-1/erythroid potentiating activity (TIMP-1/EPA). *J. Biol. Chem*. **266**, 2017-2020.

McCarthy, J.B., Basara, M.L., Palm, SL., Sas, D.F., and Furcht, L.T. (1985). The role of cell adhesion proteins laminin and fibronectin in the movement of malignant and metastatic cells. *Cancer Met. Rev*. **4**, 125-152.

McCarthy, J.B., Hager, S.J., and Furcht, L.T.F. (1986). Human fibronectin contains distinct adhesion- and motility promoting domains for metastatic cells. *J. Cell Biol*. **102**, 179-188

McNeill, H., Ozawa, M., Kemler, R. and Nelson, W.J. (1990) Novel function of the cell adhesion molecule uvomorulin as an inducer of cell surface polarity. *Cell* **62**, 309-316.

Mensing, H., Albini, A., Krieg, T., Pontz, B.F., and Muller, P.K. (1985). Enhanced chemotaxis of tumor-derived and virus-transformed cells to fibronectin and fibroblast conditioned medium. *Int. J. Cancer* **33**, 43-48.

Moses, H. L., Yang, E. Y. and Pietenpol, J. A. (1990). TGF-β stimulation and inhibition of cell proliferation: new mechanistic insights. *Cell* **63**, 245-247.

Nabi, I.R., Watanabe, H., and Raz, A. (1990). Identification of B16-F1 melanoma autocrine motility-like factor receptor. *Cancer Res*. **50**, 409-414.

Naldini, L., Weidner, K.M., Vigna, E., Gaudino, G., Bardelli, A., Ponzetto, C., Narsimhan, R.P., hartmann, G., Zarnegar, R., Michalopoulos, G.K. et al. (1991). Scatter factor and hepatocyte growth factor are indistinguishable ligands for the MET receptor. *EMBO J*. **10**, 2867-2878.

Nathan, C. and Sporn, M. (1991). Cytokines in context. *J. Cell Biol*. **113**, 981-986.

Overall, C. M., Wrana, J. L. and Sodek, J. (1991). Transcriptional and post-transcriptional regulation of 72-kDa gelatinase/Type IV collagenase by transforming growth factor-β1 in human fibroblasts. *J. Biol. Chem*. **266**, 14064-14071.

Postlethwaite, A. E., Keski-Oja, J., Moses, H. L. and Kang, A. H. (1987). Stimulation of the chemotactic migration of human fibroblasts by TGF-β. *J. Exp. Med*. **165**: 251-256.

Reilly, C. F. and McFall, R. C. (1991). Platelet-derived growth factor and transforming growth factor-β regulate plasminogen activator inhibitor-1 synthesis in vascular smooth muscle cells. *J. Biol. Chem*. **266**, 9419-9427.

Rossino, P., Gavazzi, I., Timpl, R., Aumailley, M., Abbadini, M., Giancotti, F., Silengo, L., Marchisio, P. C. and Tarone, G. (1990). Nerve growth factor induces increased expression of a laminin-binding integrin in rat pheochromocytoma PC12 cells. *Exp. Cell Res.* **189**, 100-108.

Schönherr, E., Järveläinen, H. T., Sandell, L. J. and Wight, T. N. (1991). Effects of platelet-derived growth factor and transforming growth factor-β1 on the synthesis of a large versican-like chondroitin sulfate proteoglycan by arterial smooth muscle cells. *J. Biol. Chem.* **266**, 17640-17647.

Seppä, H., Grotendorst, G., Seppä, Schiffmann, E. and Martin, G.R. (1982) Platelet derived growth factor is chemotactc for fibroblasts. *J. Cell Biol.* **92**, 584-588.

Shimoyama, Y., Hirohashi, S., Hirano, S., Noguchi, M., Shimasato, Y., Takeichi, M., and Abe, O. (1989). Cadherin cell-adhesion molecules in human epithelial tissues and carcinomas. *Cancer Res.* **49**, 2128-2133.

Steinberg, M.S., Amstrong, P.B., and Granger, R.E. (1973). On the recovery of adhesiveness by trypsin-dissociated cells. *J. Membr. Biol.* **13**, 97-128.

Stracke, M.L., Engel, J.D., Wilson, L.W., Rechler, M.M., Liotta, L.A. and Schiffmann, E. (1989). The type I insulin-like growth factor is a motility receptor in human melanoma cells. *J. Biol. Chem.* **264**, 21544-21549.

Takeichi, M. (1988). The cadherins: cell-cell adhesion molecules controlling animal morphogenesis. *Development* **102**, 639-655.

Takeichi, M. (1991) Cadherin cell adhesion receptors as a morphogenetic regulator. *Science* **251**, 1451-1455.

Takeichi, M., Azaki, H.S., Tokunaga, K., and Okada, T.S. (1979). Experimental manipulation of cell surface to affect cellular recognition mechanisms. *Dev. Biol.* **70**, 195-205.

Toyoshima, K., Ito, N., Hiasa, Y., Kamamoto, Y., and Makiura, S. (1971). Tissue culture of urinary bladder tumor induced in a rat by N-butyl-N-(4-hydroxybutyl)nitrosamine: establishment of a cell line, Nara Bladder Tumor II. *J. Natl. Cancer Inst.* **47**, 979-985.

Trinkaus, J.P. (1976). On the mechanisms of metazoan cell movements. In *The cell surace in animal embryogensis and development* (eds. G. Poste and G.I. Nicolson), pp. 226-329. Amsterdam: North-Holland.

Tucker, G.C., Boyer, B., Gavrilovic, J., Emonard, H., and Thiery, J.P. (1990). Collagen-mediated dispersion of NBT-II rat bladder carcinoma cells. *Cancer Res.* **50,** 129-137.

Tucker, G.C., Boyer, B., Vallés, A.-M., and Thiery, J.P. (1991). Combined effects of extracellular matrix and growth factors on NBT-II rat bladder carcinoma cell dispersion. *J. Cell Sci.* **100**, 371-380.

Ura, H., Obara, T., Yokota, K., Shibata, Y., Okamura, K. and Namiki, M. (1991). Effects of transforming growth factor-β released from gastric carcinoma cells on the contraction of collagen-matrix gels containing fibroblasts. *Cancer Res.* **51**, 3550-3554.

Vallés, A.M., Boyer, B., Badet, J., Tucker, G.C., Barritault, D., and Thiery, J.P. (1990a). Acidic fibroblast growth factor is a modulator of epithelial plasticity in a rat bladder carcinoma cell line. *Proc. Natl. Acad. Sci. USA* **87**, 1124-1128.

Vallés, A. M., Tucker, G. C., Thiery, J. P. and Boyer, B. (1990b). Alternative patterns of mitogenesis and cell scattering induced by acidic FGF as a function of cell density in a rat bladder carcinoma cell line. *Cell Regulation* **1**, 975-988.

Wahl, L. M., Roberts, A. B. and Sporn, M. B. (1987). Transforming growth factor type β induces monocyte chemotaxis and growth factor production. *Proc. Natl. Acad. Sci. U.S.A* **84:** 5788-5792.

Weidner, K.M., Behrens, J., Vandekerckohove, J., and Birchmeier, W. (1990). Scatter factor: molecular characteristics and effect on the invasiveness of epithelial cells. *J. Cell Biol.* **111**, 2097-2108.

Zetter, B.R. (1990). The cellular basis of site specific tumor metastasis. *NEJM* **322**, 605-612.

Printed in Great Britain © The Society of Experimental Biology 1993   197

THE ROLE OF AUTOTAXIN AND OTHER MOTILITY STIMULATING FACTORS IN THE REGULATION OF TUMOR CELL MOTILITY

MARY STRACKE*, LANCE A. LIOTTA and ELLIOTT SCHIFFMANN

Laboratory of Pathology, National Cancer Institute, National Institutes of Health, Bldg 10/Rm 2A33, Bethesda, MD 20892, USA

Summary

Active cellular motility is required for tumor cell penetration of the basement membrane and the interstitial stroma during the transition from *in situ* to invasive carcinoma. Multiple factors, both autocrine and paracrine in origin, appear to influence this motile response. Recently, a potent new cytokine with molecular mass 120 kDa has been purified to homogeneity from a human melanoma cell line (A2058). This new protein, termed autotaxin (ATX), is a basic glycoprotein with pI ~ 7.7. ATX is active in the picomolar range, stimulating pertussis toxin sensitive chemotactic and chemokinetic responses by the same cell line that produces it. Sequence information, obtained on 11 purified tryptic peptides (114 residues), confirmed that the protein is unique with no significant homology to growth factors or previously described motility factors. It is hypothesized that an autocrine motility factor, such as ATX, could play a role in the initiation of the metastatic cascade by stimulating tumor cells to move away from the primary tumor. Other motility stimulating factors, such as components of the extracellular matrix or growth factors, could then influence both the time course and the localization of tumor cell spread.

Introduction

The process of metastasis consists of a series of linked, sequential steps involving multiple host-tumor interactions (Fidler and Hart, 1982; Liotta *et al.*, 1983; Schirrmacher, 1985). Creation of a metastatic nidus requires a cell or group of cells to leave the primary tumor, invade locally, enter the circulation, arrest at a distant vascular bed, extravasate into the target organ, and proliferate as a secondary colony (Fig. 1). Tumor cell locomotion has been shown to play an important role in several steps of this metastatic cascade (Haemmerli *et al.*, 1982; Strauli and Haemmerli, 1984; Liotta and Schiffmann, 1988).

Although it has now been well established that active cellular motility is a necessary

Abbreviations: AMF, autocrine motility factor; ATX, autotaxin; FGF, fibroblast growth factor; HGF-SF, hepatocyte growth factor-scatter factor; IGF, insulin-like growth factor; IL, interleukin; PDGF, platelet-derived growth factor; TGF, transforming growth factor.
    *Author for correspondence.

Key words: motility factor, growth factor, extracellular matrix.

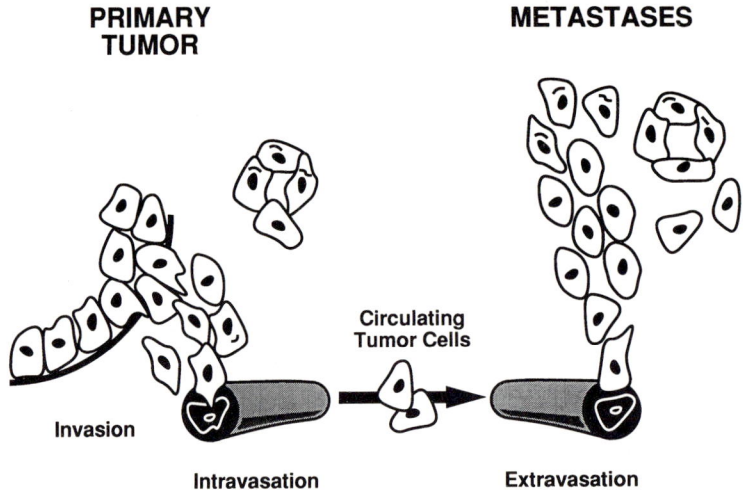

Fig. 1. Metastatic dissemination of tumor cells. Individual tumor cells dissociate from the primary tumor mass, move through the stroma to reach and intravasate into a conduit such as a blood vessel, are carried to a distant site where they extravasate and proliferate as a metastatic colony.

component of metastasis, the regulation of this motility is still poorly understood. Under normal physiological conditions, cell motility is tightly controlled. However, tumor cell motility may be aberrantly regulated or even autoregulated. Tumor cells can respond in a motile fashion to a variety of agents, including extracellular matrix components, host-derived motility and growth factors, and tumor-secreted or autocrine factors. Many of these same agents stimulate motility in embryonic or other physiologically motile cells. However, the tumor cells appear to respond in a motile fashion to multiple diverse stimuli. This gives the cells great flexibility in adapting to a variety of microenvironments. This also makes the regulation of tumor cell motility an inherently complicated issue.

Autocrine motility factors

In early studies, Hayashi and colleagues described a 70 kDa chemotactic factor derived from extracts of rat hepatoma cell tumors grown subcutaneously in Donryu rats (Hayashi *et al.*, 1970). This factor was chemotactic for several tumor cell lines including the cells of origin. Later, it was observed that cultured human melanoma cells (the A2058 cell line) secreted an attractant material into serum-free media. This factor stimulated a motility response that was both chemokinetic (random) and chemotactic (directed toward a positive concentration gradient) in nature. Because this property accorded with the early dissemination of single cells from the primary tumor, the autocrine motility factor (AMF) hypothesis was proposed (Fig. 2). Cells in the primary tumor are presumed to secrete AMF until the concentration rises enough to stimulate motility via receptors on the responding cells. Previous work by Todaro (Todaro *et al.*, 1980), Anzano (Anzano *et al.*, 1983) and their colleagues had demonstrated the presence of specific autocrine growth

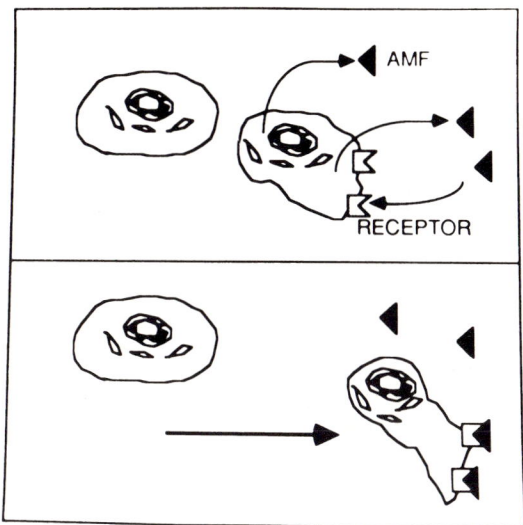

Fig. 2. The autocrine motility factor hypothesis. The tumor cells synthesize and secrete a motility factor which can then activate the same or adjacent tumor cells, presumably through a cell surface receptor.

factors for tumor cells, thereby setting a precedent for autocrine control of neoplastic cell behavior.

A family of tumor cell-derived motility-inducing cytokines has now been reported. In addition to the 70 kDa factor produced by rat hepatoma cells (Hayashi *et al.*, 1970) and the 60 kDa factor produced by human melanoma cells (A2058), these include a 53 kDa factor from rat mammary adenocarcinoma (Atnip *et al.*, 1987), a 64 kDa protein from murine melanoma (Siletti *et al.*, 1991), a >10 kDa protein from rat and human malignant glioma cells (Ohnishi *et al.*, 1990), a <30 kDa factor from rat prostatic adenocarcinoma cells (Evans *et al.*, 1991), a 150 - 200 kDa protein from *ras*-transfected NIH 3T3 cells (Seiki *et al.*, 1991), and a newly purified 120 kDa glycoprotein from A2058 melanoma cells (Stracke *et al.*, 1992). These autocrine motility factors are not specific for a given type of cancer cell and have a wide spectrum of activity on many types of cancer cells (Hayashi *et al.*, 1970; Kohn *et al.*, 1990; Evans *et al.*, 1991; Seiki *et al.*, 1991). However, these cytokines have little effect on leukocytes (Ozaki *et al.*, 1971; Liotta *et al.*, 1986) and less stimulatory effect on NIH-3T3 cells than PDGF (Liotta *et al.*, 1986). In addition, growth factors (El-Badry *et al.*, 1990) or glycosaminoglycans (Turley *et al.*, 1991) may be secreted by certain tumor cells that induce a motile response by the same cells and therefore function as relatively specific autocrine motility factors.

The 60 kDa cytokine produced by the A2058 cell line is a protein that stimulates pseudopodial extrusion to initiate cell locomotion (Guirguis *et al.*, 1987). This response is sensitive to pertussis toxin (Stracke *et al.*, 1987), to inhibitors of methylation (Liotta *et al.*, 1986), and to pharmacologic agents that act on microtubules or microfilaments, such as cis-tubulozole, taxol, and cytochalasin B or D (Stracke *et al.*, 1993). Although a G protein might be involved in the signal transduction, a variety of agents that affect

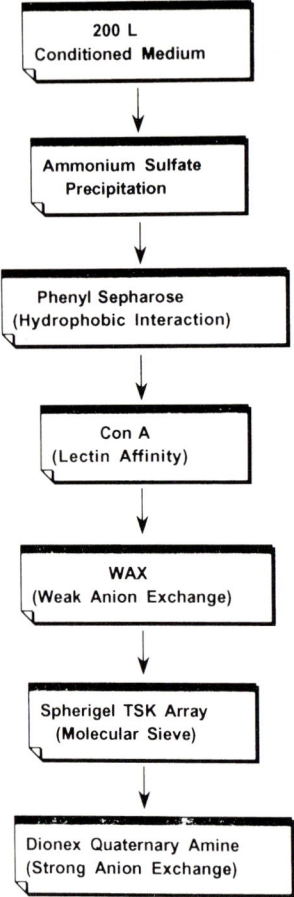

Fig. 3. Purification of ATX. This is a schematic presentation of the chromatographic steps in the process of purification. The details can be found in Stracke *et al.* (1992).

adenylate cyclase have no effect on AMF-stimulated motility, indicating that cAMP is not the necessary second messenger (Stracke *et al.*, 1987). Recently, a possible cell surface receptor to a murine melanoma autocrine motility factor has been isolated (Siletti *et al.*, 1991). Antibodies produced against this 78,000 kDa glycoprotein stimulated motility (Nabi *et al.*, 1990) and enhanced metastatic ability in high metastatic variants (Watanabe *et al.*, 1991).

Autotaxin, a novel autocrine motility factor

We have recently isolated a potent new cytokine with molecular mass 120 kDa from the conditioned medium of A20587 cells. Utilizing sequential chromatographic methods (Fig. 3), we have purified this factor to homogeneity (Fig. 4) (Stracke *et al.*, 1992a). This new cytokine, termed autotaxin (ATX), is a basic glycoprotein with pI ~ 7.7 (Fig. 5). ATX is active in the picomolar range (Fig. 6) stimulating both chemotactic and

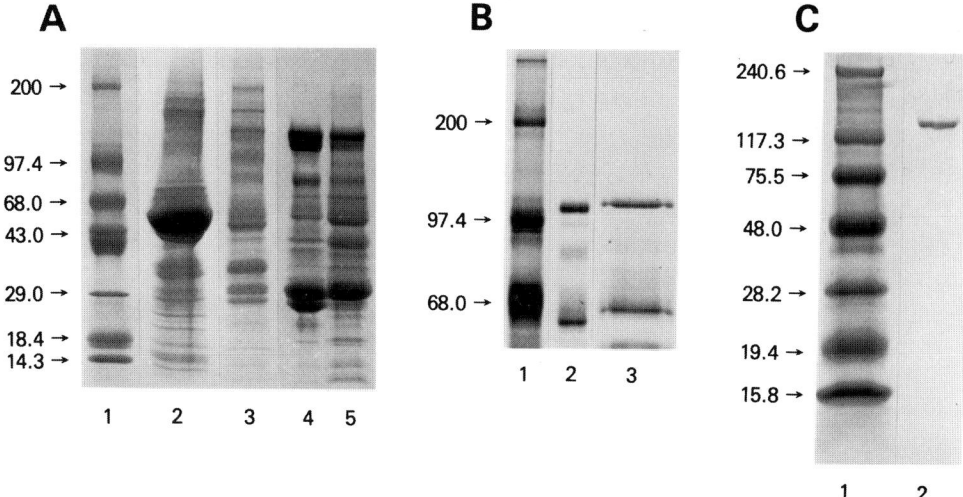

Fig. 4. Electrophoretic analysis of the different steps of ATX purification. The activity peak from each chromatographic fractionation was pooled, concentrated and analyzed by SDS-polyacrylamide gel electrophoresis. Molecular weight standards are in Lane 1 for each panel. (A) 8-16% gradient gel of the first three purification steps, run under non-reducing conditions. Lane 2 is an aliquot of the pooled activity peak eluted from the phenyl sepharose fractionation. Lane 3 is an aliquot of the pooled activity peak eluted from the Con A affinity purification. Lanes 4 and 5 show the "peak" and "shoulder" of activity fractionated by weak anion exchange chromatography. (B) 7% gel of the activity peak fractionated by molecular sieve exclusion chromatography. Lanes 2 and 3 show the protein separation pattern of the total pooled activity peak when the gel was run under non-reducing and reducing conditions, respectively. (C) 8-16% gradient gel of the final strong anionic exchange chromatographic separation, run under non-reducing conditions. Lane 2 comprises ~1% of the total pooled activity peak eluted from the column.

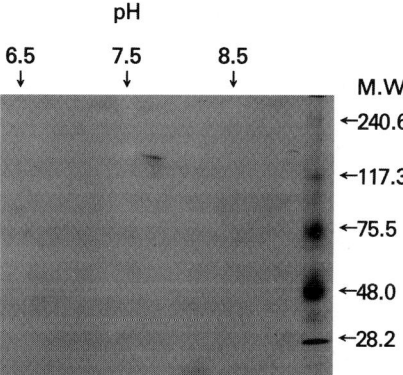

Fig. 5. 2-dimensional gel electrophoresis of ATX. Purified ATX was subjected to non-equilibrium isoelectric focusing (5 hr at 500v), then applied to a 7.5% SDS-polyacrylamide gel for the second dimension. The pH separation which resulted is shown at the top. Molecular weight standards for the second dimension are shown on the right. This analysis revealed a single component with pI = 7.7 ± 0.2 and M_r = 120,000.

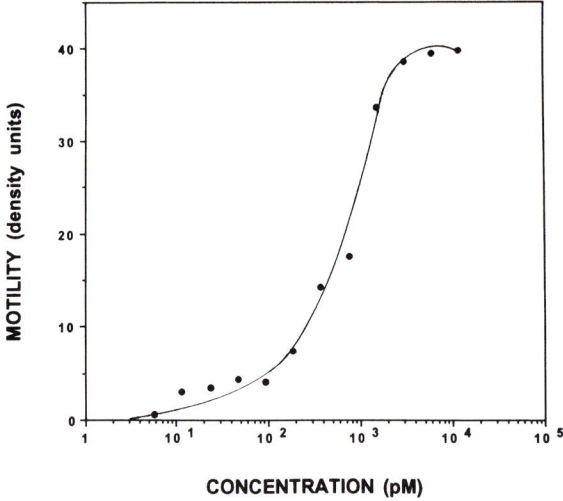

CONCENTRATION (pM)

Fig. 6. Dilution curve of ATX. Purified ATX was serially diluted and tested for motility-stimulating activity. The result, with unstimulated background motility subtracted out, shows that activity is half-maximal at ~500 pM ATX.

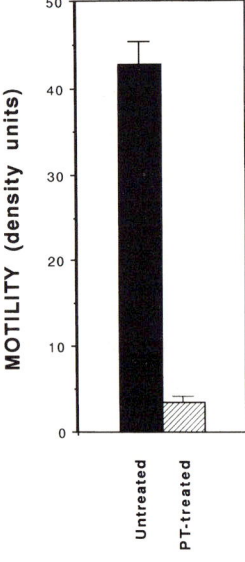

Fig. 7. Effect of pertussis toxin on ATX-stimulated motility. A2058 cells were pre-treated for 1 hr prior to the start of the motility assay with 0.5 μg/ml PT in 0.1% BSA-DMEM or with 0.1% BSA-DMEM alone (for untreated control). The motility activity stimulated by purified ATX was then assessed for the two treatment groups. The result, expressed as cells/HPF ± S.E.M. with unstimulated background motility subtracted out, reveals profound inhibition of PT-treated cells (hatched) compared to untreated cells (solid). PT had no effect on cell viability. S.E.M.s were < 10%.

Table 1. *Peptide sequences for autotaxin*

Name	Amino acid sequence
ATX-18	W-H-V-A-A-N
ATX-19	P-X-L-D-V-Y-K
ATX-20	Y-P-A-F-K
ATX-24	Q-A-E-V-S
ATX-29	P-E-E-V-T-X-P-N-Y-L
ATX-47	Y-D-V-P-W-N-E-T-I
ATX-48	S-P-P-F-E-N-I-N-L-Y
ATX-100	G-G-Q-P-L-W-I-T-A-T-K
ATX-101	V-N-S-M-Q-T-V-F-V-G-Y-G-P-T-F-K
ATX-102	D-l-E-H-L-T-S-L-D-F-F-R
ATX-103	T-E-F-L-S-N-Y-L-T-N-V-D-D-I-T-L-V-P-G-T-L-G-R

chemokinetic responses in the same A2058 cells. When cells are pre-treated with pertussis toxin, this motile response is abolished (Fig. 7). ATX may therefore act through a G protein-linked cell surface receptor. Sequence information, obtained on 11 purified tryptic peptides, confirmed that the protein is unique with no significant homology to growth factors or previously described motility factors (Table 1). ATX, therefore, appears to represent a new member of the AMF family.

The role of the extracellular matrix in tumor cell motility

Components of the extracellular matrix are ubiquitous molecules that the tumor cell encounters repeatedly during the process of metastasis. In particular, basement membrane barriers are traversed by the tumor cell when it enters or leaves blood vessels as well as when it invades nerves and muscles. It is a general observation for all types of carcinomas that invasion is associated with defects in the basement membrane (Barsky *et al.*, 1983). This implies that aggressive tumor cells may interact with the extracellular matrix in a manner fundamentally different from normal cells. Several components of the extracellular matrix have been found to stimulate locomotion in tumor cells. These include vitronectin (Basara *et al.*, 1985), fibronectin (McCarthy and Furcht, 1984; Mensing *et al.*, 1984; McCarthy *et al.,* 1986; Aznavoorian *et al.*, 1990 a and b; Makabe *et al.*, 1990), laminin (McCarthy, 1983; McCarthy and Furcht, 1984; Situ *et al.*, 1984; Wewer *et al.*, 1987; Aresu *et al.*, 1991; Tashiro *et al.,* 1991), type I collagen (Tchao, 1982; Faassen *et al.*, 1992; Mooradian *et al.*, 1992), type IV collagen (Chelberg *et al.*, 1989; Aznavoorian *et al.*, 1990b), and thrombospondin (Taraboletti *et al.*, 1987). A few of these extracellular matrix proteins have been demonstrated to induce a motility response in tumor cells through integrin receptors (Leavesly *et al.*, 1990; Tashiro *et al.*, 1991). Several fragments of these multidomain proteins have been shown to inhibit the formation of metastases when experimentally co-injected with tumor cells into mice (Humphries *et al.*, 1986; Iwamoto *et al.*, 1987; McCarthy *et al.,* 1988).

The human melanoma cell line, A2058, has been shown to respond in a locomotory fashion to multiple extracellular matrix proteins: laminin, fibronectin, type IV collagen

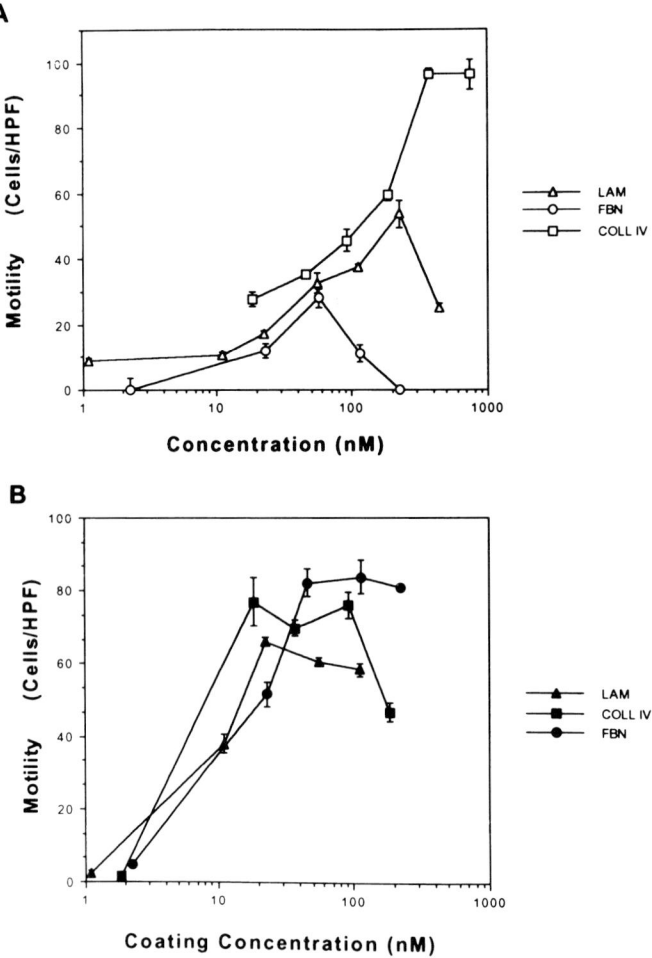

Fig. 8. Concentration curves of A2058 motility responses to extracellular matrix components. A. Chemotactic response. Increasing concentrations of laminin (△), type IV collagen (□), and fibronectin (○) were diluted in medium supplemented with 0.1% bovine serum albumin and added to the lower wells of microchemotaxis modified Boyden chambers. Polycarbonate filters were precoated on both sides with a noninteracting adhesive protein (gelatin for chemotaxis to laminin and type IV collagen; type IV collagen for chemotaxis to fibronectin). B. Haptotactic response. Polycarbonate membranes were precoated with a step gradient of increasing concentrations of laminin (▲), type IV collagen (■) or fibronectin (●). Filters were placed in the modified Boyden chambers with the higher protein side facing the lower compartment. Data are expressed as (mean ± SEM) for triplicate samples of a representative experiment.

and thrombospondin (Taraboletti *et al.*, 1987 Aznavoorian *et al.*, 1990b). These ECM proteins stimulate chemotaxis when they are in solution and haptotaxis when they are insoluble or substratum-bound (Fig. 8). However, chemotactic and haptotactic stimulation by ECM proteins may act through different cell surface receptors and post-receptor signal transduction pathways (Aznavoorian *et al.*, 1990b). When cells are pre-

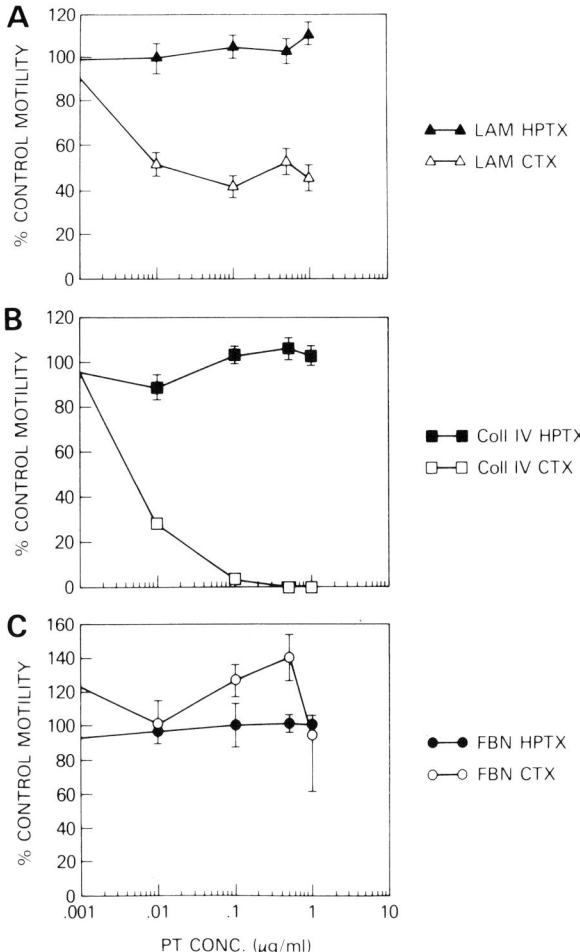

Fig. 9. The effect of pertussis toxin on motility to extracellular matrix components. Cells were preincubated for 2 hr at 25°C in different concentrations of pertussis toxin prior to the start of the assay. Cell chemotactic (open symbols) and haptotactic (closed symbols) responses were assessed for each ECM component in parallel experiments on the same day. A. Laminin-induced motility responses. The haptotactic response was assessed for a filter-coating concentration of 20 nM; chemotaxis was measured in response to 110 nM soluble laminin. B. Type IV collagen-induced motility responses. The haptotactic response was assessed for a filter-coating concentration of 20 nM; chemotaxis was measured in response to 280 nM soluble type IV collagen. C. Fibronectin-induced motility response. The haptotactic response was assessed for a filter-coating concentration of 45 nM; chemotaxis was measured in response to 110 nM soluble fibronectin. Results are expressed as a per cent of control stimulated migration run concurrently in the absence of pertussis toxin.

treated with pertussis toxin, the chemotactic response to laminin is diminished and the response to type IV collagen is abolished (Fig. 9). In contrast, haptotaxis to these same proteins is insensitive to pertussis toxin. A likely explanation of these data is that chemotaxis and haptotaxis to the same ECM protein are mediated by distinct cell surface

Table 2. *Inhibition of thrombospondin-induced motility by site-directed monoclonal antibodies*

Antibody	Concentration (µg/ml)	Chemotaxis (% control ± SEM)	Haptotaxis (% control ± SEM)
C6.7[1]	10	80.8±17.6	10.0±5.0
	1	92.7±12.5	63.0±11.0
	0.1	85.0±18.1	91.5±8.5
A2.5[2]	10	38.0±7.2	90.7±21.9
	1	68.7±8.2	100±21.3
	0.1	59.0±12.2	95.0±70.0

[1]Monoclonal antibody C6.7 recognizes the globular carboxy terminus of thrombospondin.
[2]Monoclonal antibody A2.5 recognizes the amino terminus of thrombospondin.

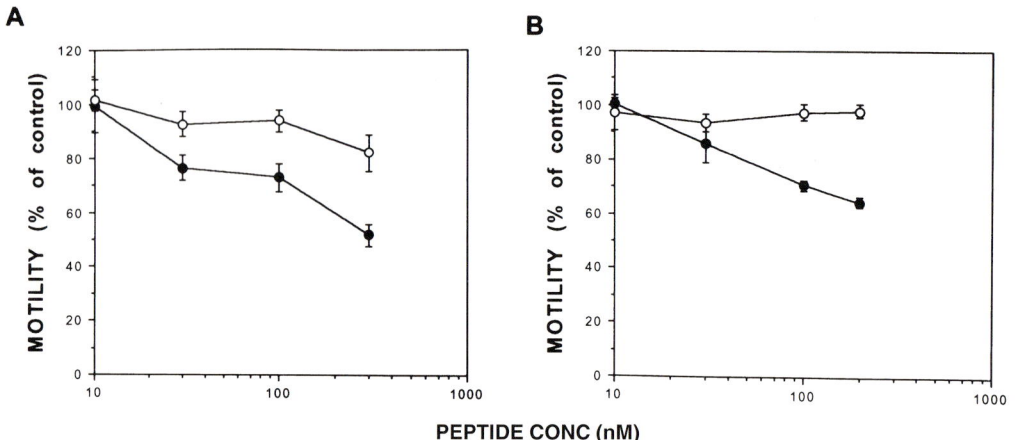

Fig. 10. Effect of GRGDS on fibronectin-stimulated motility responses. Cells were preincubated with varying concentrations of GRGDS (closed circles) or GRGES (open circles) for 1 hr at 25°C prior to the start of the assays. A. The haptotactic responses were assessed to a fibronectin coating concentration of 45 nM. B. Chemotaxis was measured in response to 110 nM soluble fibronectin. Results are the mean of triplicate determinations for a single representative experiment and are expressed as per cent of control stimulated migration in the absence of peptides (mean ± SEM).

receptors, which recognize different domains of these large, multi-domain matrix proteins. This hypothesis is further supported by the work of Taraboletti *et al.* (1987) who showed that the chemotaxis and haptotaxis promoting domains of thrombospondin are on opposite ends of the molecule (Table 2). The carboxy-terminal region of thrombospondin appeared to stimulate haptotaxis and was inhibited by a specific monoclonal antibody (C6.7) as well as by the synthetic peptide GRGDS. The amino-terminal heparin-binding domain of thrombospondin appeared to stimulate chemotaxis and was inhibited by a specific monoclonal antibody (A2.5) as well as by the sulfatides heparin and fucoidan. In contrast, neither chemotaxis nor haptotaxis to fibronectin is inhibited by pertussis toxin (Fig. 8), but both types of motile response are partially inhibited by GRGDS (Fig. 10)

(Aznavoorian *et al.*, 1990b). These data suggest that the fibronectin motility response by these cells may be partially activated through an integrin receptor.

Perhaps, during the initial phases of metastasis, the insoluble matrix proteins might provide tumor cells with a pathway of activation, a very important haptotactic stimulation which would lead the cells along the basement membranes of local blood vessels. Proteolytic enzymes secreted either by the tumor cells or by the host, including type IV (Liotta *et al.*, 1979; Liotta *et al.*, 1982) and interstitial collagenases (Woolley, 1984), cathepsin B (Sloane and Honn, 1984), and plasminogen activator (Mignatti *et al*, 1986), could then result in localized pools of soluble, partially degraded matrix proteins. These soluble pools would then provide an additional chemotactic stimulus to motility.

Growth factors and other host-derived cytokines

Formation of a successful metastatic nidus requires that the tumor cell find a microenvironment capable of supporting cell growth. Certain highly metastatic cell lines have been shown to produce their own necessary autocrine growth factors (Todaro *et al.*, 1980; Anzano *et al.*, 1983; Huff *et al.*, 1986; Rodeck *et al.*, 1987; Halaban *et al.*, 1988; El-Badry *et al.*, 1990; Williams *et al.*, 1992). However, host-secreted, growth-supporting factors may still be advantageous to establishing colonies of metastastic cells. Several growth factors have now been demonstrated to stimulate chemotactic motility in tumor cells. These growth factors appear to be somewhat specific for cellular origin (Table 3). Many of these cytokines also appear to act as mitogens for the same tumor cells that they stimulate to migrate (El-Badry *et al.*, 1990; Liapi *et al.*, 1990; Tilly *et al.*, 1990; Vallés *et al.*, 1990; Schofield *et al.*, 1992).

The insulin-like growth factors (IGF) and insulin stimulate a pertussis toxin insensitive chemotactic response in A2058 cells (Stracke *et al.*, 1988). This response is strongest to IGF-I (Fig. 11) and appears to activate the cells through a type I IGF receptor since a monoclonal antibody, specific for the type I IGF receptor (Jacobs *et al.*, 1986), inhibits both ^{125}I-labelled IGF-I binding and IGF-I-induced motility in A2058 cells (Fig. 12)

Table 3. *Growth factors that affect tumor cell motility*

Factor	Cell types (References)
Bombesin	Small cell lung carcinoma (Ruff *et al.*, 1985)
acidic FGF	Bladder carcinoma (Vallés *et al.*, 1990)
basic FGF	Prostatic carcinoma (Pienta *et al.*, 1991)
	Teratocarcinoma (Schofield *et al.*, 1992).
HGF-SF	Carcinomas (Stoker and Gherardi, 1989; Weidner *et al.*, 1990)
Histamine	Melanoma/Carcinoma (Tilly *et al.*, 1990)
IGF-I	Melanoma (Stracke *et al.*, 1988)
IGF-II	Rhabdomyosarcoma (El-Badry *et al.*, 1990)
IL-6	Ductal breast carcinoma (Tamm *et al.*, 1989)
IL-8	Melanoma (Wang *et al.*, 1990)
NGF	Embryonal carcinoma (Kahan and Kramp, 1987)
PDGF	Teratocarcinoma (Liapi *et al.*, 1990)
TGF-β1	Adenocarcinoma of lung (Mooradian *et al.*, 1992)

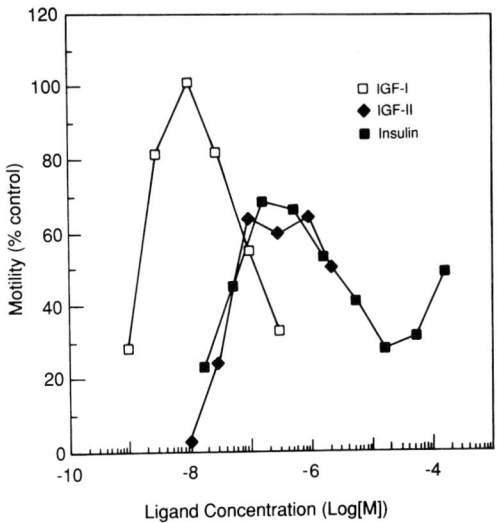

Fig. 11. Dose response curves for the IGF's and insulin in A2058 cells. Increasing concentrations of IGF-I (□), IGF-II (◆), and insulin (■) were tested for motile response by A2058 cells. The autocrine motility factor served as a positive control with results normalized to this value. SEM were less than 10%.

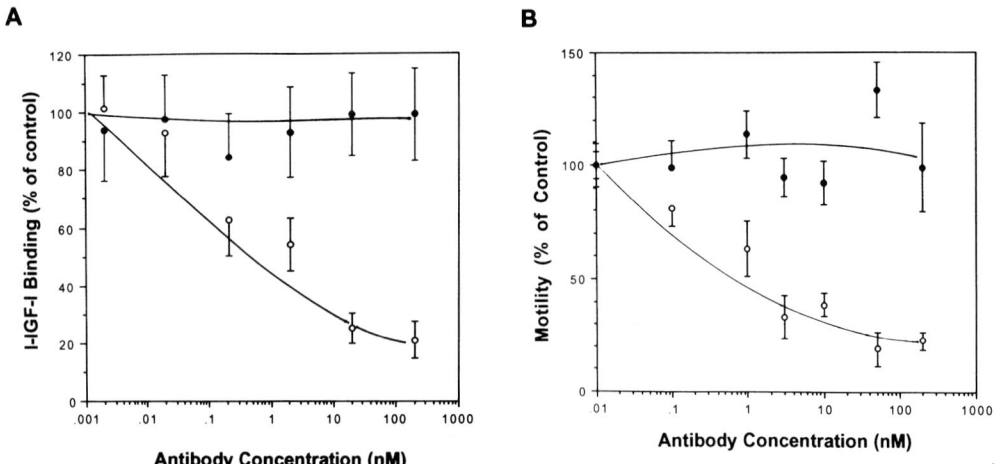

Fig. 12. Effect of monoclonal antibodies on IGF-I binding and motility stimulation in A2058 cells. For each experiment, the effect of a specific anti-type I IGF receptor (open symbols) was compared to a control monoclonal antibody (closed symbols) matched for antibody subclass. A. Effect on ^{125}I-IGF-I binding to melanoma cells. Varying concentrations of antibodies along with ^{125}I-IGF-I were added to cells at 15°C for two hours. Results are shown as a per cent of total binding (mean ± SEM) for a representative experiment. B. Effect on IGF-I-induced motility. Cell motility was stimulated with 10 nM IGF-I in the presence of varying concentrations of antibodies. Results are expressed as a per cent of control migration in the absence of antibodies (mean ± SEM).

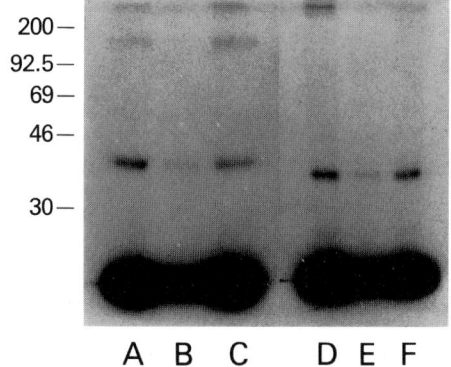

Fig. 13. Cross-linking of ^{125}I-IGF-I to its receptor on melanoma cells. Cells were labeled with ^{125}I-IGF-I, surface proteins were cross-linked with disuccinimidyl suberate, then whole cells were extracted with detergent. The extract was subjected to 10% polyacrylamide gel electrophoresis under reducing conditions (A-C), and non-reducing conditions (D-F). Subsequently the gel was analyzed by autography. Lanes A and F, extract of cells labeled in the presence of 5000-fold molar excess of insulin; lanes B and E, extract of cells labeled in the presence of 100-fold molar excess of unlabeled IGF-I; lanes C and F, labeled cell extract run without competition.

(Stracke *et al.*, 1989). In addition, cross-linking experiments with ^{125}I-labelled IGF-I reveal a typical heterotetrameric receptor with IGF-I bound to the larger (α) subunit (Fig. 13). These experiments also demonstrate that A2058 cells produce an IGF-I specific binding protein. Interestingly, both insulin and IGF-I have been implicated as necessary growth factors for culture of primary human melanoma cells (Rodeck *et al.*, 1987). In similar experiments, IGF-II has been found to stimulate motility through a type II IGF/Mannose-6-phosphate receptor in human rhabdomyosarcoma cells (Minniti *et al.*, 1992). IGF-II stimulates mitogenesis in these same cells through a type I IGF receptor. Thus, these growth factors, acting through "normal" receptor mechanisms, may serve as "homing" factors for tumor cells which have reached the vasculature, directing the tumor cells to extravasate into a secondary site which provides a suitable microenvironment for growth.

Tumor-secreted and tumor-induced factors that stimulate host cell motility

Tumor cells interact with the host in various ways that can influence their metastatic capability. For example, tumor cells can produce chemotactic factors that affect host cell motility or induce the synthesis of chemotactic factors by the host cells themselves.

Several breast carcinoma cell lines have been shown to secrete a protein factor that stimulates fibroblast motility (Gleiber and Schiffmann, 1984) and may influence the fibrosis seen around tumors. The skin metastases of human breast carcinomas have also been demonstrated to synthesize a factor chemotactic for melanocytes (Konomi *et al.*, 1992). Several different tumor cell lines have been shown to secrete monocyte chemotactic and activating factor (MCAP), also known as tumor-derived chemotactic

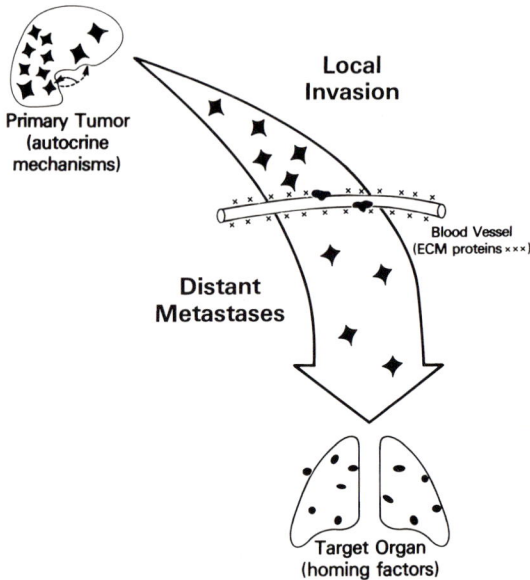

Fig. 14. Tumor Cell Motility. Tumor cells are capable of responding to a variety of stimuli in a motile fashion. These different stimuli may take on greater or lesser significance during different stages of the metastatic cascade. The initial motile stimulus in or near the primary tumor may be an autocrine factor such as autotaxin. As the tumor cell moves into the stroma and invades local blood vessels and lymphatics, extracellular matrix proteins, either insoluble within the matrix or partially degraded in solution, may play a part in directing this motility. Once inside the vascular tree, the "homing" of the tumor cell to an appropriate metastatic site may depend on cytokines such as IGF-1 which stimulate growth as well as a chemotactic response.

factor (Mantovani *et al.*, 1986; Bottazzi *et al.*, 1990; Zachariae *et al.*, 1990). This 10-12 kDa peptide may play a role in stimulating tumor-associated macrophages. Human melanoma cells have been shown to produce IL-8 (Zachariae *et al.*, 1991) and human glioma cells to synthesize an endothelial cell chemotactic factor (Takaki *et al.*, 1992). Production of these chemotactic factors by the tumor cells may influence such diverse processes as angiogenesis and production of essential growth factors by host cells. In addition, fetal and cancer patient fibroblasts have been shown to produce a migration-stimulating factor which is not made by normal adult fibroblasts (Schor *et al.*, 1988). This 70 kDa protein (Grey *et al.*, 1989) is an autocrine motility factor for fibroblasts.

Conclusion

Tumor cells have proved capable of responding in a motile fashion to a wide variety of stimuli (Fig. 14). These different stimuli could take on greater or lesser importance during different stages of tumor cell invasion and progression, influencing both the time course of each step and the ultimate site of the metastatic nidus. In addition, this flexibility of cellular activation indicates a likely mechanism by which tumor cells adapt to the different microenvironments which they encounter during the metastatic cascade.

References

Anzano, M. A., A. B. Roberts, J. M. Smith, M. B. Sporn and J. E. de Larco (1983). Sarcoma growth factor from conditioned medium of virally transformed cells is composed of both type a and type transforming growth factors. *Proc. Natl. Acad.* Sci. *USA* **80**, 6264-6268.

Aresu, O., G. Nicolo, G. Allavena, A. Melchiori, J. Schmidt, J. B. Kopp, E. d'Amore, G. D. Chader and A. Albini (1991). Invasive activity, spreading on and chemotactic response to laminin are properties of high but not low metastatic mouse osteosarcoma cells. *Invasion Metastasis.* **11**, 2-13.

Atnip, K. D., L. M. Carter, G. L. Nicolson and M. K. Dabbous (1987). Chemotactic response of rat mammary adenocarcinoma cell clones to tumor-derived cytokines. *Biochem. Biophys. Res. Comm.* **146**, 996-1002.

Aznavoorian, S., L. A. Liotta and H. Z. Kupchik (1990a). Characteristics of invasive and noninvasive human colorectal adenocarcinoma cells. *J. Natl. Cancer Inst.* **82**, 1485-1492.

Aznavoorian, S., M. L. Stracke, H. C. Krutzsch, E. Schiffmann and L. A. Liotta (1990b). Signal transduction for chemotaxis and haptotaxis by matrix molecules in tumor cells. *J. Cell Biol.* **110**, 1427-1438.

Barsky, S. H., G. P. Siegal, F. Jannotta and L. Liotta (1983). Loss of basement membrane components by invasive tumors but not by their benign counterparts. *Lab Invest.* **49**, 140-147.

Basara, M. L., J. B. McCarthy, D. W. Barnes and L. T. Furcht (1985). Stimulation of haptotaxis and migration of tumor cells by serum spreading factor. *Cancer Res.* **45**, 2487-2494.

Bottazzi, B., F. Colotta, A. Sica, N. Nobili and A. Mantovani (1990). A chemoattractant expressed in human sarcoma cells (tumor-derived chemotactic factor, TDCF) is identical to monocyte chemoattractant protein-1/monocyte chemotactic and activating factor (MCP-1/MCAF). *Int. J. Cancer* **45**, 795-797.

Chelberg, M. K., E. C. Tsilibary, A. R. Hauser and J. B. McCarthy (1989). Type IV collagen-mediated melanoma cell adhesion and migration: involvement of multiple, distinct domains of the collagen molecule. *Cancer Res.* **49**, 4796-4802.

El-Badry, O. M., C. Minniti, E. C. Kohn, P. J. Houghton, W. H. Daughaday and L. J. Helman (1990). Insulin-like growth factor II acts as an autocrine growth and motility factor in human rhabdomyosarcoma tumors. *Cell Growth & Differentiation* **1**, 325-331.

Evans, C. P., D. S. Walsh and E. C. Kohn (1991). An autocrine motility factor secreted by the Dunning R-3327 rat prostatic adenocarcinoma cell subtype AT2.1. *Int. J. Cancer* **49**, 109-113.

Faassen, A. E., J. A. Schrager, D. J. Klein, T. R. Oegema, J. R. Couchman and J. B. McCarthy (1992). A cell surface chondroitin sulfate proteoglycan, immunologically related to CD44, is involved in Type I collagen-mediated melanoma cell motility and invasion. *J. Cell Biol.* **116**, 521-531.

Fidler, I. J. and I. R. Hart (1982). Biologic diversity in metastatic neoplasms-origins and implications. *Science* **217**, 998-1001.

Gleiber, W. E. and E. Schiffmann (1984). Identification of a chemoattractant for fibroblasts produced by breast carcinoma cell lines. *Cancer Res.* **44**, 3398-3402.

Grey, A.-M., A. M. Schor, G. Rushton, I. Ellis and S. L. Schor (1989). Purification of the migration stimulating factor produced by fetal and breast cancer patient fibroblasts. *Proc. Natl. Acad. Sci. USA.* **86**, 2438-2442.

Guirguis, R., I. Margulies, G. Taraboletti and L. Liotta (1987). Cytokine-induced pseudopodial protrusion is coupled to tumour cell migration. *Nature* **329**, 261-263.

Haemmerli, G., B. Arnold and P. Strauli (1982). Cell locomotion, a contributing factor in spread of the V2 rabbit carcinoma. *Int. J. Cancer* **29**, 223-227.

Halaban, R., B. S. Kwon, S. Ghosh, P. D. Bovis and A. Baird (1988). bFGF as an autocrine growth factor for human melanomas. *Oncogene Res.* **3**, 177-186.

Hayashi, H., K. Yoshida, T. Ozaki and K. Ushijima (1970). Chemotactic factor associated with invasion of cancer cells. *Nature* **226**, 174-175.

Huff, K. K., D. Kaufman, K. H. Gabbay, E. M. Spencer, M. E. Lippman and R. B. Dickson (1986). Secretion of an insulin-like growth factor-I-related protein by human breast cancer cells. *Cancer Res.* **46**, 4613-4619.

Humphries, M. J., K. Olden and K. M. Yamada (1986). A synthetic peptide from fibronectin inhibits experimental metastasis of murine melanoma cells. *Science* **233**, 467-470.

Iwamoto, Y., F. A. Robey, J. Graf, M. Sasaki, H. K. Kleinman, Y. Yamada and G. R. Martin

(1987). YIGSR, a synthetic laminin pentapeptide, inhibits experimental metastasis formation. *Science* **238**, 1132-1134.

Jacobs, 5., 5. Cook, M. E. Svoboda and J. J. Van Wyk (1986). Interaction of the monoclonal antibodies aIR-1 and aIR-3 with insulin and somatomedin-C receptors. *Endocrinol.* **118**, 223-226.

Kahan, B. W. and D. C. Kramp (1987). Nerve growth factor stimulation of mouse embryonal cell migration. *Cancer Res.* **47**, 6324-6328.

Kohn, E. C., E. A. Francis, L. A. Liotta and E. Schiffmann (1990). Heterogeneity of the motility response in malignant tumor cells: a biological basis for the diversity and homing of metastatic cells. *Int. J. Cancer* **46**, 287-292.

Konomi, K., S. Imayama, S. Nagae, R. Terasaka, K. Chijiwa and Y. Yashima (1992). Melanocyte chemotactic factor produced by skin metastases of a breast carcinoma. *J. Surg. Oncol.* **50**, 62-66.

Leavesly, D. I., G. D. Ferguson, E. A. Wayner and D. A. Cheresh (1990). Requirement of the integrin β3 receptor for carcinoma cell spreading or migration on vitronectin and fibronectin. *J. Cell Biol.* **117**, 1101-1107.

Liapi, C., F. Raynaud, W. B. Anderson and D. Evain-Brion (1990). High chemotactic response to platelet-derived growth factor of a teratocarcinoma differentiated mesodermal cell line. *In Vitro Cell. Dev. Biol.* **26**, 388-392.

Liotta, L. A., S. Abe, P. Robey and G. Martin (1979). Preferential digestion of basement membrane collagen by an enzyme derived from a metastatic murine tumor. *Proc. Nat'l. Acad. Sci. USA* **76**, 2268-2272.

Liotta, L. A., R. Mandler, G. Murano, D. A. Katz, R. K. Gordon, P. K. Chiang and E. Schiffmann (1986). Tumor cell autocrine motility factor. *Proc. Natl. Acad. Sci. USA* **83**, 3302-3306.

Liotta, L. A., C. N. Rao and S. H. Barsky (1983). Tumor invasion and the extracellular matrix. *Lab. Invest.* **49**, 636-649.

Liotta, L. A. and E. Schiffmann (1988). Tumor motility factors. *Cancer Surveys* **7**, 631-652.

Liotta, L. A., U. P. Thorgeirsson and S. Garbisa (1982). Role of collagenases in tumor cell invasion. *Cancer Metastasis Rev.* **1**, 277-288.

Makabe, T., I. Saiki, J. Murata, Y. Ohdate, Y. Kawase, Y. Taguchi, T. Shimojo, F. Kumizuka, I. Kato and I. Azuma (1990). Modulation of haptotactic migration of metastatic melanoma cells by the interaction between heparin and heparin-binding domain of fibronectin. *J. Biol. Chem.* **265**, 14270-14276.

Mantovani, A., W. J. Ming, C. Balotta, B. Abdeljalil and B. Bottazzi (1986). Origin and regulation of tumor-associated macrophages: the role of tumor-derived chemotactic factor. *Biochim. Biophys. Acta* **865**, 59-67.

McCarthy, J. B. and L. T. Furcht (1984). Laminin and fibronectin promote the haptotactic migration of Bl 6 melanoma cell *in vitro*. *J. Cell Biol.* **98**, 1474-1480.

McCarthy, J. B., S. T. Hagen and L. T. Furcht (1986). Human fibronectin contains distinct adhesion- and motility-promoting domains for metastatic melanoma cells. *J. Cell Biol.* **102**, 179-188.

McCarthy, J. B., S. L. Palm and L. T. Furcht (1983). Migration by haptotaxis of a Schwann cell tumor line to the basement membrane glycoprotein laminin. *J. Cell Biol.* **97**, 772-777.

McCarthy, J. B., A. P. N. Skubitz, S. L. Palm and L. T. Furcht (1988). Metastasis inhibition of different tumor types by purified laminin fragments and a heparin-binding fragment of fibronectin. *J. Natl. Cancer Inst.* **80**, 108-115.

Mensing, H., A. Albini, T. Krieg, B. F. Pontz and P. K. Muller (1984). Enhanced chemotaxis of tumor-derived and virus-transformd cells to fibronectin and fibroblast-conditioned medium. *Int. J. Cancer* **33**, 43-48.

Mignatti, P., E. Robbins and D. B. Rifkin (1986). Tumor invasion through the human amniotic membrane: Requirement for a proteinase cascade. *Cell* **47**, 487-498.

Minniti, C. P., E. C. Kohn, J. H. Grubb, W. S. Sly, Y. Oh, H. L. Muller, R. G. Rosenfeld and L. J. Helman (1992). The insulin-like growth factor-II (IGF-II)/mannose 6-phosphate receptor mediates IGF-II-induced motility in human rhabdomyosarcoma cells. *J. Biol. Chem.* **267**, 9000-9004.

Mooradian, D. L., J. B. McCarthy, K. V. Komanduri and L. T. Furcht (1992). Effects of transforming growth factor-β1 on human pulmonary adenocarcinoma cell adhesion, motility, and invasion *in vitro*. *J. Natl. Cancer Inst.* **84**, 523-527.

Nabi, I. R., H. Watanabe and A. Raz (1990). Identification of B16-F1 melanoma autocrine motility-like receptor. *Cancer Res.* **50**, 409-414.

Ohnishi, T., N. Arita, T. Hayakawa, S. Izumoto, T. Taki and H. Yamamoto (1990). Motility factor produced by malignant glioma cells: Role in tumor invasion. *J. Neurosurg.* **73**, 881-888.

Ozaki, T., K. Yoshida, K. Ushijima and H. Hayashi (1971). Studies on the mechanisms of invasion in cancer. II. In vivo effects of a factor chemotactic for cancer cells. *Int. J. Cancer* **7**, 93-100.

Pienta, K. J., W. B. Isaacs, D. Vindivich and D. S. Coffey (1991). The effects of basic fibroblast growth factor and suramin on cell motility and growth of rat prostate cancer cells. *J. Urol.* **145**, 199-202.

Rodeck, U., M. Herlyn, H. D. Menssen, R. W. Furlanetto and H. Koprowski (1987). Metastatic but not primary melanoma cell lines grow in vitro independently of exogenous growth factors. *Int. J. Cancer* **40**, 687-690.

Ruff, M., E. Schiffmann, V. Terranova and C. B. Pert (1985). Neuropeptides are chemoattractants for human tumor cells and monocytes: A possible mechanism for metastasis. *Clin. Immunol. Immunopath.* **37**, 387-396.

Schirrmacher, V. (1985). Experimental approaches, theoretical concepts, and impacts for treatment strategies. *Cancer Res.* **43**, 1-32.

Schofield, P. N., M. Granerus, A. Lee, T. J. Ektrom and W. Engstrom (1992). Concentration-dependent modulation of basic fibroblast growth factor action on multiplication and locomotion of human teratocarcinoma cells. *FEBS Letters* **298**, 154-158.

Schor, S. L., A. M. Schor, A. M. Grey and G. Rushton (1988). Foetal and cancer patient fibroblasts produce an autocrine migrationstimulating facator not made by normal adult cells. *J. Cell Sci.* **90**, 391-399.

Seiki, M., H. Sato, L. A. Liotta and E. Schiffmann (1991). Comparison of autocrine mechanisms promoting motility in two metastatic cell lines: human melanoma and *ras*-transfected NIH3T3 cells. *Int. J. Cancer* **49**, 717-720.

Siletti, S., H. Watanabe, V. Hogan, I. R. Nabi and A. Raz (1991). Purification of B16-Fl melanoma autocrine motility factor and its receptor. *Cancer Res.* **51**, 3507-3511.

Situ, R., E. C. Lee, J. P. McCoy Jr. and J. Varani (1984). Stimulation of murine tumour cell motility by laminin. *J. Cell Sci.* **70**, 167-176.

Sloane, B. R. and K. V. Honn (1984). Cysteine proteinases and metastasis. *Cancer Metastasis Rev.* **3**, 249-263.

Stoker, M. and E. Gherardi (1989). Scatter factor and other regulators of cell mobility. *Brit. Med. Bull.* **45**, 481-491.

Stracke, M. L., J. D. Engel, L. L. Wilson, M. M. Rechler, L. A. Liotta and E. Schiffmann (1989). The type I insulin-like growth factor receptor is a motility receptor in human melanoma cells. *J. Biol. Chem.* **264**, 21544-21549.

Stracke, M. L., R. Guirguis, L. A. Liotta and E. Schiffmann (1987). Pertussis toxin inhibits stimulated motility independently of the adenylate cyclase pathway in human melanoma cells. *Biochem. Biophys. Res. Comm.* **146**, 339-345.

Stracke, M. L., E. C. Kohn, 5. Aznavoorian, L. L. Wilson, D. Salomon, H. C. Krutzsch, L. A. Liotta and E. Schiffmann (1988). Insulin-like growth factors stimulate chemotaxis in human melanoma cells. *Biochem. Biophys. Res. Comm.* **153**, 1076-1083.

Stracke, M. L., H. C. Krutzsch, E. J. Unsworth, A. Årestad, V. Cioce, E. Schiffmann and L. A. Liotta (1992). Identification, purification, and partial sequence analysis of Autotaxin, a novel motility-stimulating protein. *J. Biol. Chem.* **267**, 2524-2529.

Stracke, M. L., M. Soroush, L. A. Liotta and E. Schiffmann (1993). Cytoskeletal agents inhibit motility and adherence of human tumor cells. *Kidney International.* **43**, 151-157.

Strauli, P. and G. Haemmerli (1984). The role of cancer cell motility in invasion. *Cancer Metastasis Rev.* **3**, 127-141.

Takaki, 5., J.-I. Kuratsu, Y. Mihara, M. Yamada and Y. Ushio (1992). Endothelial cell chemotactic factor derived from human glioma cell lines. *J. Neurosurg.* **76**, 822-829.

Tamm, I., I. Cardinale, J. Krueger, J. S. Murphy, L. T. May and P. B. Sehgal (1989). Interleukin 6 decreases cell-cell association and increases motility of ductal breast carcinoma cells. *J. Exp. Med.* **170**, 1649-1669.

Taraboletti, G., D. D. Roberts and L. A. Liotta (1987). Thrombospondin-induced tumor cell migration: haptotaxis and chemotaxis are mediated by different domains . *J. Cell Biol* . **105**, 2409-2415.

Tashiro, K.-I., G. C. Sephel, D. Greatorex, M. Sasaki, N. Shirashi, G. R. Martin, H. K. Kleinman

and Y. Yamada (1991). The RGD containing site of the mouse laminin A chain is active for cell attachment, spreading, migration and neurite outgrowth. *J. Cell. Physiol.* **146**, 451-459.

Tchao, R. (1982). Novel forms of epithelial cell motility on collagen and on glass surfaces. *Cell Motil.* **4**, 333-341.

Tilly, B. C., L. G. J. Tertoolen, R. Remorie, A. Ladoux, I. Verlaan, S. W. de Laat and W. H. Moolenaar (1990). Histamine as a growth factor and chemoattractant for human carcinoma and melanoma cells: action through Ca^{2+}-mobilizing H1 receptors. *J. Cell Biol.* **110**, 1211-1215.

Todaro, G. J., C. Fryling and J. E. DeLarco (1980). Transforming growth factors produced by certain human tumor cells: polypeptides that interact with epidermal growth factor receptors. *Proc. Natl. Acad. Sci. USA* **77**, 5258-5262.

Turley, E. A., K. Vandeligt and C. Clary (1991). Hyaluronan and a cell-associated hyaluronan binding protein regulate the locomotion of *ras*-transformed cells. *J. Cell Biol.* **112**, 1041-1047.

Valles, A. M., B. Boyer, J. Badet, G. C. Tucker, D. Barritault and J. P. Thiery (1990). Acidic fibroblast growth factor is a modulator of epithelial plasticity in a rat bladder carcinoma cell line. *Proc. Natl. Acad. Sci. USA* **87**, 1124-1128.

Wang, J. M., G. Taraboletti, K. Matsushima, J. Van Damme and A. Montovani (1990). Induction of haptotactic migration of melanoma cells by neutrophil activating protein/interleukin-8. *Biochem. Biophys. Res. Comm.* **169**, 165-170.

Watanabe, H., I. R. Nabi and A. Raz (1991). The relationship between motility factor receptor internalization and the lung colonization of murine melanoma cells. *Cancer Res.* **51**, 2699-2705.

Weidner, K. M., J. Behrens, J. Vandekerckhove and W. Birchmeier (1990). Scatter factor: Molecular characteristics and effect on the invasiveness of epithelial cells. *J. Cell Biol.* **111**, 2097-2108.

Wewer, U. M., G. Taraboletti, M. E. Sobel, R. Albrechtsen and L. A. Liotta (1987). Role of laminin receptor in tumor cell migration. *Cancer Res.* **47**, 5691-5698.

Williams, N. N., T. Gyorfi, D. Iliopoulos, D. Herlyn, D. Greenstein, A. J. Linnenbach, J. M. Daly, P. Jensen, U. Rodeck and M. Herlyn (1992). Growth factor-independence and invasive properties of colorectal carcinoma cells. *Int. J. Cancer* **50**, 274-280.

Woolley, D. E. (1984). Collagenolytic mechanism in tumor cell invasion. *Cancer Metastasis Rev.* **3**, 361-372.

Zachariae, C. O. C., A. O. Anderson, H. L. Thompson, E. Appella, A. Mantovani, J. J. Oppenheim and K. Matsushima (1990). Properties of monocyte chemotactic and activating factor (MCAF). purified from a human fibrosarcoma cell line. *J. Exp. Med.* **171**, 2177-2182.

Zachariae, C. O. C., K. Thestrup-Pedersen and K. Matsushima (1991). Expression and secretion of leukocyte chemotactic cytokines by normal human melanocytes and melanoma cells. *J. Invest. Dermatol.* **97**, 593-599.

Printed in Great Britain © The Society of Experimental Biology 1993   215

IN VITRO REGULATION OF HGF/SF EXPRESSION

T. KAMALATI[1], *B. THIRUNAVUKARASU*[1] and *L. BULUWELA*[2]

[1]Institute of Cancer Research, Haddow laboratories, Sutton, Surrey SM2 5NG, UK
[2]Dept. Biochemistry, Charing Cross and Westminster Medical School, University of London,
St. Dunstan Road, London W6 8RP, UK

Summary

Studies of parameters which affect cellular proliferation, cellular differentiation and cell-cell interactions influencing cell behaviour are of particular interest. They can be used to identify and characterise molecules which, through changes in gene expression, induce or inhibit cell proliferation, differentiation and movement. Such studies are crucial, not only in the context of understanding growth and development, but also in understanding the processes of wound healing and regeneration, tumour invasion and metastasis. Here we present a summary of some cell culture models which we have developed for the study of the above-mentioned phenomena, together with their application to studies of the regulation of HGF/SF expression.

Introduction

The processes of cellular proliferation, differentiation and morphogenesis in multicellular systems such as tissues and organs are highly coordinated and entail inter-cellular communication as well as intra-cellular programming. A number of variables including physical parameters, extracellular matrix components, cell adhesion molecules and membrane junctional complexes between opposing cells can influence cellular proliferation and differentiation. Further, a crucial role is played by soluble factors produced by cells and carried in interstitial fluid or in the circulation. Fractionation of serum and tissue extracts has led to the isolation and characterisation of a diverse group of secreted regulatory proteins known as polypeptide growth factors. Growth factors are one of many factors which can modulate and mediate growth, differentiation and positioning of cells, not only during development, but throughout adult life in the processes of wound healing and tissue regeneration, metastasis and basic cellular homeostasis.

Recently, Hepatocyte Growth Factor, HGF, the most potent mitogen for primary hepatocytes (Nakamura *et al.*, 1984) has been shown to be identical to Scatter Factor, SF, (Gherardi and Stoker, 1991, Weidner *et al.*, 1991) a cytokine independently characterised as a motility factor (Stoker and Perryman 1985, Stoker and Gherardi, 1989, Gherardi *et al.*, 1989). Hence, in acknowledgment of this identity, the factor is now denoted as HGF/SF.

Extensive work has shown HGF/SF to have a repertoire of functions all mediated at physiological concentrations. HGF/SF is mitogenic not only for primary hepatocytes (Nakumura *et al.*, 1984) but for a broad spectrum of epithelial cells (Gherardi and Stoker,

Key words: HGF/SF expression, cell proliferation, cell-cell interaction, model.

1991). In addition HGF/SF can stimulate the motility of a variety of epithelial cells to varying degrees (Stoker, 1989). Further, HGF/SF can induce morphogenesis in MDCK cells (Montesano *et al.*, 1991) as well as invasiveness in a variety of epithelial cells (Weidner *et al.*, 1990). Finally, HGF/SF has been shown to be cytotoxic to a variety of tumour cells (Higashio *et al.*, 1990, Shiota *et al.*, 1992).

Recently HGF/SF has been shown to be the ligand for the transmembrane receptor encoded by the *c-met* proto-oncogene (Naldini *et al.*, 1991, Bottaro, 1991). Although expression of the HGF/SF transcript and that of *c-met* has been demonstrated in a variety of tissues (Gherardi and Stoker, 1991) very little is known about the regulation of expression of this factor and the molecular mechanisms by which it appears to elicit the five distinct functions outlined above, via a single receptor. The aim of this article is to discuss the role of cell-cell interactions in a cell culture model system for studying the regulation of expression of HGF/SF *in vitro*. The implications of such regulatory interactions in understanding the molecular mechanisms that regulate HGF/SF expression *in vivo* will also be discussed.

HGF/SF is a unique cytokine

Although the term growth factor implies an ability to stimulate growth, some growth factors have growth inhibitory effects such as TGF β (Roberts *et al.*, 1985), while others can affect differentiation and cell movement as well as proliferation, depending on the cell type and/or conditions of the assay system. Examples of these are Nerve Growth Factor (NGF), Epidermal Growth Factor (EGF), Platelet Derived Growth Factor (PDGF), Basic Fibroblast Growth Factor (bFGF) and Acidic Fibroblast Growth Factor (aFGF), which can all stimulate mitogenesis, motility and differentiation of a broad spectrum of cell types (Blay and Brown, 1985; Barrandon and Green, 1987; Grotendorst *et al.*, 1981; Klagsbrun, 1990).

Like many other growth factors HGF/SF is a multifunctional cytokine. However, unlike many of the above mentioned factors HGF/SF is a paracrine effector of epithelial cells (Warn and Dowrick, 1989). Specifically, HGF/SF is mitogenic for epithelial cell types only (Gherardi and Stoker, 1991), while most growth factors including the above-mentioned, will stimulate proliferation in multiple cell types. The only other example of a paracrine epithelial cell specific growth factor is Keratinocyte Growth Factor (KGF) (Rubin *et al.*, 1989; Finch *et al.*, 1989). However, unlike HGF/SF, KGF is not known to affect cell movement nor does it possess the functional repertoire of HGF/SF.

HGF/SF exerts its motility effect on epithelial cells by breaking their intercellular connections and increasing their mobility, thus resulting in the 'scattering' of cohesive epithelial cell colonies into isolated cells. Further, HGF/SF changes the cell morphology from a polygonal appearance typical of epithelial cells to a multipolar fibroblastic morphology (Stoker, 1989). Although sensitivity to the motility effects of HGF/SF is restricted to normal epithelial cells (both freshly isolated and as cell lines) the degree of sensitivity to the factor is variable within epithelial cell strains (Stoker *et al.*, 1987). Madin Darby Canine Kidney (MDCK) cells have been found to be particularly sensitive to the motility effects of HGF/SF and begin to scatter as early as 15 minutes after addition of the factor. Most growth factors including those mentioned above stimulate the movement of epithelial cells by chemotaxis and/or chemokinesis, without affecting cellular junctions. In this context the effects of HGF/SF on epithelial cell motility are not only dramatic but unique. In summary, in comparison with other growth factors, HGF/SF is a unique cytokine as it is a paracrine mitogen and motogen, specific to epithelial cells.

Mitogenic effects of HGF/SF on non-hepatic epithelia

Although the original purification of HGF was based on its stimulation of DNA synthesis in mature rat hepatocytes and the name of the factor depicts its biological activity, the biological effects of HGF/SF are not restricted to hepatocytes. Northern blot analysis of rat tissue has revealed that the HGF/SF transcript is expressed in a variety of tissues, including lung, liver, thymus, kidney and brain. (Tashiro *et al.*, 1990). *In-situ* hybridisation experiments have identified the cell types responsible for the expression of this transcript to be of mesenchymal origin. For example, Kupffer and endothelial cells in the liver, endothelial cells in the kidney and alveolar macrophages, endothelial cells and embryonic fibroblasts in the lung are the sources of HGF/SF message in these organs (Noji *et al.*, 1990; Rubin *et al.*, 1991). Further, expression of the *c-met* transcript, the cellular receptor for HGF/SF, in human tissue has been found to be highest in kidney, thyroid, liver and stomach as assessed by northern blot analysis (Gherardi and Stoker, 1991). Together, these findings indicate that HGF/SF may act as a potent paracrine mitogen for a broad spectrum of epithelial cells as well as mature hepatocytes. Indeed, HGF/SF has now been shown to be a potent mitogen for rabbit renal tubules (Igawa *et al.*, 1991), human melanocytes (pigmented cells of the skin) (Matsumoto *et al.*, 1991a) and human keratinocytes (epithelial cells of the skin) in which HGF/SF also has the potential to promote cell migration (Matsumoto *et al.*, 1991b).

Since cultured dermal fibroblasts of the skin express the HGF/SF transcript it is likely that they are the cellular source of HGF/SF in skin (Kamalati, unpublished observation). In this context HGF/SF may be a dermal fibroblast-derived factor which enhances keratinocyte growth through a paracrine mechanism. Similarly, expression of HGF/SF in the mesenchymal components of the liver has demonstrated the paracrine mode of action of this cytokine in liver (Noji *et al.*, 1990; Kinoshita *et al.*, 1989). However, the precise molecular mechanisms involved in the regulation of expression of HGF/SF remain unclear. Human keratinocytes can be cultivated serially and retain the potential to undergo terminal differentiation, reminiscent of the defined programme of terminal differentiation in skin. Hence they are used extensively in studies of the regulatory mechanisms of proliferation and differentiation in human cells. In this context and in view of the relative simplicity of keratinocyte cultivation in comparison with hepatocytes, human keratinocytes offer a useful model with which to identify factors important in regulating HGF/SF expression

A cell culture model for epithelial cell growth and differentiation

We have chosen keratinocytes as a model system for our studies of epithelial cell growth and differentiation since the conditions for their serial propagation in culture are well established (Rheinwald and Green, 1975) and their differentiation is well characterised both morphologically and by changes in gene expression. Further, the terminal differentiation of keratinocytes in culture resembles keratinocyte differentiation in skin. In our studies of keratinocyte cell growth and differentiation, we have chosen SVK14, an SV40 transformed human keratinocyte cell line (Taylor-Papadimitriou *et al.*, 1982). Although transformed, SVK14 cells are responsive to external stimuli such as growth factors (Kamalati *et al.*, 1989a) and mesenchymal cues (Kamalati *et al.*, 1989b) as demonstrated by re-expression of markers of keratinocyte differentiation in response to exposure to IGF I, EGF and mesenchymal cells.

Inhibition of expression of HGF/SF by heterotypic intercellular interactions

Mesodermal induction of epithelial cell differentiation has been well documented; however, epithelial cells also have the capacity to coordinately regulate the expression of mesenchymal activities. In this context, in view of the competence of SVK14 cells to respond to mesenchymal cues, we began to investigate whether SVK14 cells could influence mesenchymal cell behaviour. To evaluate this possibility, mesenchymal specific markers, whose expression can be readily assessed, are required.

MRC 5 cells, a human foetal lung fibroblast strain, are a rich source of HGF/SF in culture (Stoker *et al.*, 1987). Further, MRC 5 cells are one of a variety of mesenchymal cells able to induce the expression of markers of differentiation in SVK14 cells (Kamalati *et al.*, 1989b). In this light, we co-cultured MRC 5 with SVK14 cells, with the aim of examining the capacity of SVK14 to modulate the expression of HGF/SF in MRC 5 cells. Our data shows that HGF/SF activity released into the culture medium by MRC 5 cells is greatly reduced when MRC 5 cells are in co-culture with SVK14 cells (Fig. 1). SVK14 cells do not produce HGF/SF nor degrade or remove it from the culture medium. HGF/SF is therefore stable in the presence of SVK14 cells for at least 72 hrs (Kamalati *et al.*, 1992).

The greatly reduced HGF/SF activity in MRC 5/SVK14 co-cultures does not appear to be due to release of some factor(s) by SVK14 cells or the co-culture, which can effectively remove HGF/SF from the co-culture medium. Further, SVK14 cells do not release a factor which inhibits the release of HGF/SF by MRC 5 cells, since MRC 5 cells can condition SVK14 conditioned medium to precisely the same extent as fresh culture medium. This observation also demonstrates that the absence of HGF/SF activity in the co-culture medium is not due to the inability of MRC 5 cells to produce HGF/SF as a consequence of medium being depleted by SVK14 cells.

We have demonstrated that SVK14 cells will inhibit the expression of HGF/SF by MRC 5 cells only when the co-culture is initiated as a well mixed suspension of both cell types. When the two cell types are seeded independently and on different areas of the dish, co-culture does not lead to an inhibition of HGF/SF expression. This is irrespective of whether the two cell types have no cell-cell contact or whether they are allowed a limited degree of contact. This illustrates that inhibition of expression of HGF/SF requires more than merely having the two cell types present in the same co-culture dish, suggesting a requirement for intimate cell-cell associations (Fig. 2).

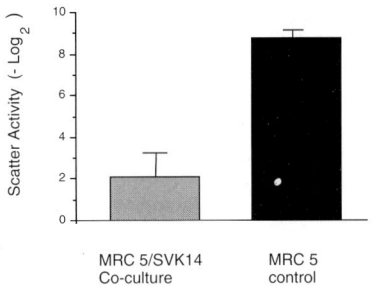

FIG. 1. Inhibition of scatter factor activity released by MRC 5 cells in co-culture with SVK14 cells.
(⬤) Scatter factor activity in MRC 5/SVK14 co-cultures, and (●) MRC 5 controls after 7 days of culture. As the assay is performed in two fold serial dilutions the scatter factor activity is presented as Log_2. The bars represent the standard deviation with n=12.

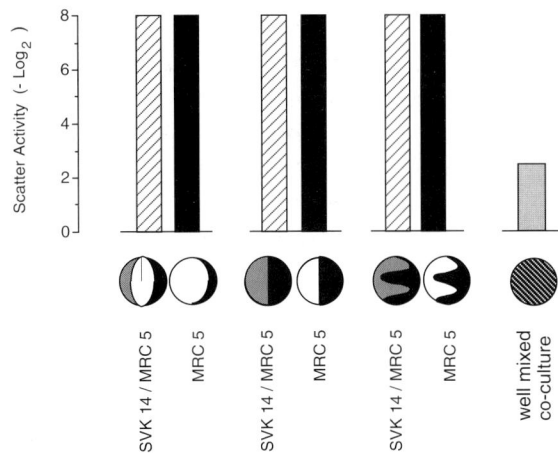

Fig. 2. Extensive cell-cell contact is required for inhibition of scatter activity in MRC 5/SVK14 co-cultures.

Segregated co-cultures of SVK14 (⬤) and MRC 5 (●) cells were initiated by seeding exponentially growing cells independently, on different areas of the culture dish in the patterns shown above, in order to allow different degrees of cell-cell contact between the two cell types. Well mixed co-cultures (◍). The controls were composed of MRC 5 cells only seeded in the same pattern as that of their co-culture partner. (▨) Scatter activity of segregated co-cultures, (▫) scatter activity of well mixed co-cultures and (■) scatter activity of MRC 5 control after 7 days of culture.

The requirement for extensive and intimate cell-cell contact for inhibition of HGF/SF activity in MRC 5/SVK14 co-cultures to occur is further illustrated by the dose-dependent inhibition of HGF/SF activity by SVK14 cells. In a series of co-cultures, when the number of MRC 5 cells is kept constant while that of SVK14 cells is varied, an increasing inhibition of HGF/SF activity is observed with increasing SVK14 cell numbers (Fig. 3). This indicates that either direct cell-cell contact or short-range, short-

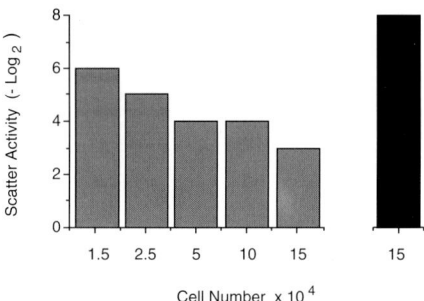

Fig. 3. Dose dependent inhibition of scatter factor activity by
SVK14 cells.
Keeping the MRC 5 cell numbers constant while varying that
of SVK14, 1.5 X 10⁵ MRC 5 cells were co-cultured with
1.5 X 10⁴ - 1.5 X 10⁵ SVK14 cells (▨). The control
cultures were composed of 1.5 X 10⁵ MRC 5 cells only
(■). The cultures were assayed after 7 days.

lived interactions, which require intimate cell-cell associations, are likely to be involved.

Using northern blot analysis we have demonstrated that MRC 5 cells cease to express the HGF/SF transcript when co-cultured with SVK14 cells. Fig. 4 presents a northern blot where total RNA from co-cultures has been probed with a DNA probe to the β subunit of human HGF/SF. It shows that this probe detects a 6 Kb transcript in the RNA prepared from MRC 5 cells. This is in agreement with the size of the HGF/SF transcript detected in human liver (Nakumura *et al.*, 1989) and placenta (Miyazawa *et al.*, 1989). Further, this probe is able to detect HGF/SF transcripts from MRC 5 cells equivalent in number to those present during the initial 3 days of co-culture, that is 8.9, 7.4 and 4.6 μg total RNA respectively.

From this blot it is evident that SVK14 do not produce any HGF/SF transcripts, in agreement with their inability to release HGF/SF as mentioned above. Most importantly, it appears that the HGF/SF transcript is absent in MRC 5/SVK14 co-cultures as early as 24 hrs after culture initiation, and remains absent for the following 48 hrs. The presence of MRC 5-derived RNA in the co-cultures is confirmed by detection of the transcript for vimentin, a cytoskeletal protein expressed by fibroblasts but not SVK14 (Fig. 4). The disappearance of the 6Kb HGF/SF message in MRC 5/SVK14 co-cultures correlates with the greatly reduced HGF/SF activity observed in these co-cultures, indicating that the inhibition of HGF/SF activity in MRC 5 cells involves transcriptional regulation of HGF/SF message (Kamalati *et al.*, 1992).

Cell type specifity and the ability to inhibit MRC 5 HGF/SF expression

Using a variety of epithelial cell strains as co-culture partners for MRC 5 cells, we have observed that the capacity to inhibit HGF/SF expression in MRC 5 cells is not unique to SVK14 cells. Indeed all epithelial cells tested, normal or transformed, appear to be able to inhibit the above activity by more than 90%, when co-cultured with MRC 5 cells (Fig. 5b). However, the two fibroblastic cell lines used as co-culture partners for MRC 5 cells

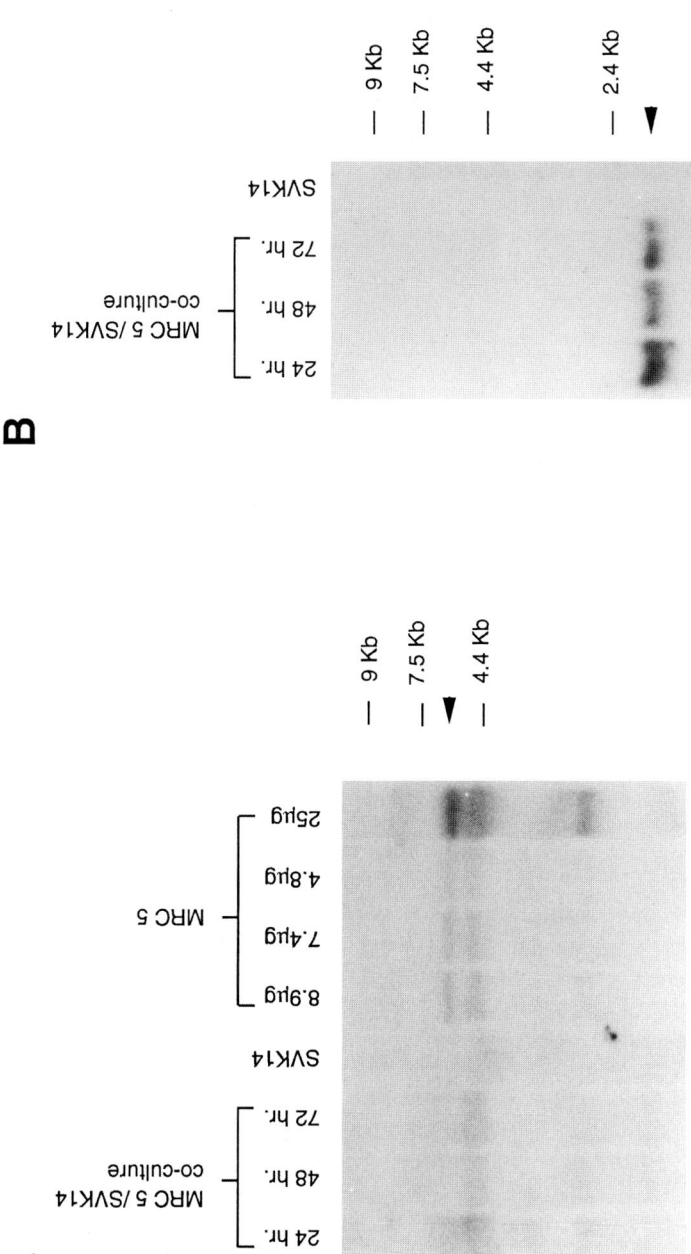

Fig. 4. Expression of SF/HGF in MRC 5/SVK14 co-cultures. Northern blots, **A** -Total RNA extracted from MRC 5/SVK14 co-cultures, SVK14 and MRC 5 cells ($25 \mu g$) was probed with the h-HGF (30 kD) probe. 8.9, 7.4 and 4.8 μg RNA represent the RNA equivalent of MRC 5 cells present in co-cultures on days 1,2 and 3 respectively. The 4.2 Kb band is ribosomal RNA. **B** - The first four tracks of the same blot rehybridised to a 1 Kb cDNA fragment of the human vimentin gene.

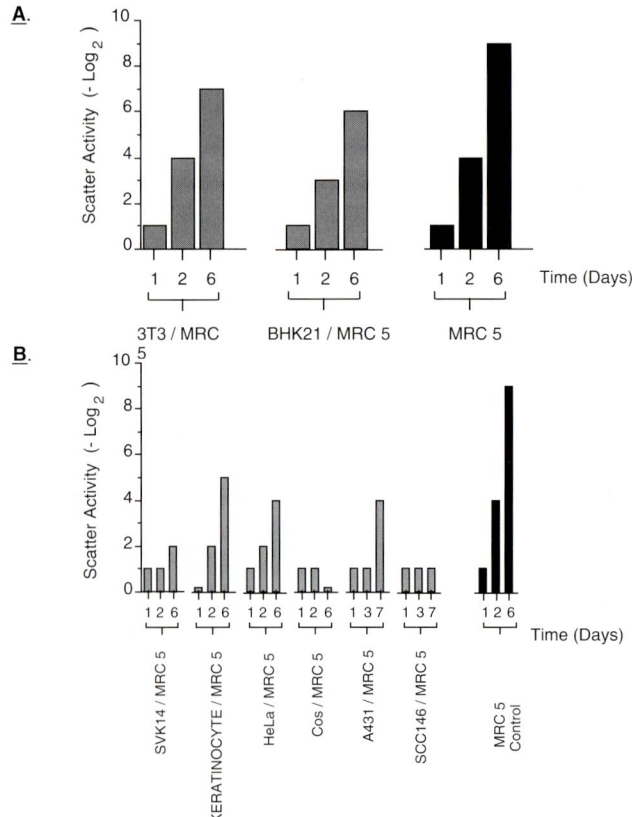

Fig. 5. Scatter factor activity in co-cultures of MRC 5 cells with a variety of mesenchymally derived and epithelial cell strains.
A: Scatter factor activity in co-cultures of MRC 5 with two fibroblast cell lines (▨), in comparison to MRC 5 controls (■) after 7 days in culture.
B: Scatter factor activity in co-cultures of MRC 5 with a variety of epithelial cell lines (▨) in comparison to MRC 5 control cultures (■), after 7 days in culture.
All co-cultures were initiated as in MRC 5/SVK14 co-cultures.

did not exhibit a capacity to inhibit HGF/SF activity in MRC 5 cells to any degree (Fig. 5a) (Kamalati *et al.*, 1992). These results are interesting in that they demonstrate the regulation of expression of a mesenchymal activity by epithelial cells. Further, these observations illustrate that the competence of the epithelial cells to participate in such instructive interaction is not specified by their strain nor state of transformation. Currently very little is known about the regulation of expression of HGF/SF. In this context, our observations are significant in that they relate to a unique culture model for the identification and characterisation of molecules and mechanisms of interactions involved in the regulation of expression of HGF/SF.

The capacity of liver to regenerate in most vertebrate organisms including man, even after surgical removal of two thirds of its mass, has been well documented over the years. However, the detailed nature of the controlling factors that trigger or modulate this phenomenon has only recently begun to be understood. Recently it has been demonstrated that HGF/SF levels in serum rise 3-9 fold within 3-6 hours as a result of hepatic injury. Further, the expression of HGF/SF transcripts in the lung increases within 6 hours of unilateral nephrectomy or hepatic injury (Yanagita *et al.*, 1992) suggesting that a circulating humoral factor is responsible for signalling injury to distal organs. Furthermore, the sera from rats with hepatic injury or partial nephrectomy has been shown to increase the expression of HGF/SF transcripts in the lungs of healthy control rats as well as in MRC 5 cells in culture, reinforcing the concept of a humoral factor able to induce expression of HGF/SF in the injured organ as well as in intact organs distant from the injured site. Hence Matsumoto *et al.*, have named this factor "Injurin" and demonstrated it to be a 10-20 kDa protein, not related to the Tumour Necrosis Factor (TNF), Interleukin (IL), Fibroblast Growth Factor (FGF) or Transforming Growth Factor (TGF) families of proteins (Matsumoto *et al.*, 1992a).

More recently Interleukin-1α (IL-1α) and Interleukin-1β (IL-1β) have been shown to strongly enhance HGF/SF gene expression in human dermal fibroblasts (Matsumoto *et al.*, 1992b) while TGF-β1 and glucocorticoids down-regulate HGF/SF expression in MRC 5 cells (Matsumoto *et al.*, 1992c). However, the suppression of HGF/SF gene expression found in our heterotypic co-cultures of MRC 5 fibroblasts with epithelial cells are unlikely to be solely due to such soluble factors since this inhibition requires close cell - cell contact between heterotypic cells, raising the interesting possibility that a specific cell surface mechanism may regulate HGF/SF in this model system. In view of the above mentioned data, it appears that HGF/SF gene expression may be positively or negatively regulated not only by soluble factors such as Injurin, IL-1α, IL-1β, TGF-β1 and glucocorticoids, but also by direct or close cell associations.

Epithelial-mesenchymal interactions

Some of the best studied cases of proximate interactions are the induction of differentiation of epithelial cells by mesenchymal cells in tissue recombination experiments (Gilbert, 1988). However, very little is known about the molecular species and the precise mechanisms involved in such interactions.

In further studies of epithelial-mesenchymal inteactions, we have demonstrated that SVK14 cells are not only responsive to external stimulation but also have the capacity to in turn influence gene expression in a mesenchymal cell strain. This is particularly interesting since such inductive phenomena are common features of early development where gene control signals requiring direct cell-cell contact are a prerequisite for the embryonic determination of most specialised cells. Further, such regulatory cell-cell interactions remain central to the maintenance of basic tissue homeostasis throughout adult life.

The expression of HGF/SF (Kinoshita *et al.* 1989, Rubin *et al.* 1991, Tashiro *et al.*, 1990, Zarnegar *et al.*, 1990) together with that of its cellular receptor, the c-*met* proto-oncogene, (Bottaro *et al.*, 1991, Naldini *et al.*, 1991) has been demonstrated in a variety of tissues. However, very little is known about the control of regulation of expression of this factor. We have demonstrated that co-culture of SVK14 cells with MRC 5 cells results in an inhibition of expression of HGF/SF by MRC 5 at transcription level and that this inhibition requires intimate cell-cell association between the mesenchymal MRC 5 and epithelial SVK14 cells. The expression of mHGF/SF in the non-parenchymal cells of the liver has been shown to be down-regulated by the presence of parenchymal hepatocytes and up-regulated by their absence (Noji *et al.*, 1990, Kinoshita *et al.*, 1989). In this context the down-regulation of expression of the HGF/SF transcript in MRC 5 cells, observed as a result of co-culture with SVK14 cells, may be analogous to the mode of control of mHGF/SF expression in the liver. In this light, the use of MRC 5/SVK14 co-culture as a model for studies of identification of factors which are important in regulating HGF/SF expression, together with further characterisation of the interactions of SVK14 with MRC 5 cells may provide insight into the regulation of expression of SF/HGF *in vivo* and elucidate general features which govern the control of epithelial-mesenchymal interactions during development and differentiation.

References

Barrandon, Y. and Green, H. (1987) Cell migration is essential for sustained growth of keratinocyte colonies, The role of transforming growth factors and epidermal growth factor. *Cell* **50**, 1131-1137.

Blay, J. and Brown, K.D. (1985) Epidermal growth factor promotes the chemotactic migration of cultured rat intestinal epithelial cells. *J. Cell Physiol.* **124**, 107-112.

Bottaro., D.P., Rubin, J.S., Faletto, D.L., Chan, A. M-L., Kmiecik, T.E., Vande Woude, G.F. and Aaronson, S.A. (1991) Identification of the hepatocyte growth factor receptor as the c-met proto-oncogene product. *Science* **251**, 802-804.

Finch, P.W., Rubin, J.S., Miki, T., Ron, D. and Aaronson, S.A. (1989) Human KGF is FGF-related with properties of a paracrine effector of epithelial cell growth. *Science* **245**, 752-755.

Gherardi, E. and Stoker, M. (1991) Hepatocyte growth factor-scatter factor: Mitogen, Motogen and Met. *Cancer Cells*, **3**, 227-232.

Gherardi, E., Gray, J., Stoker, M., Perryman, M. and Furlong, R. (1989) Purification of scatter factor, a fibroblast-derived basic protein that modulates epithelial interactions and movement. *Proc. Natl. Acad. Sci.* **86**, 5844-5848.

Gilbert, S.F. (1988) Epitheliomesenchymal interactions. In *Developmental biology*, 2nd edition. Sinauer Associates, Inc. 560-562

Grotendorst, G.R., Seppa, H. E. J., Kleinman, H.K. and Martin, G.R. (1981) Attachment of smooth muscle cells to collagen and their migration towards platelet derived growth factor. *Proc. Natl. Acad. Sci.* **78**, 3669-3672.

Higashio, K., Shima, N., Goto, M., Itagaki, Y., Nagao, M., Yasudo, H. and Morinaga, T. (1990) Identity of a tumour cytotoxic factor from human fibroblasts and hepatocyte growth factor. *Biochem. Biophys. Res. Commun.* **170**, 397-404.

Igawa, T., Kanda, S., Kanetake, H., Saitoh, Y., Ichihara, A., Tomita, Y. and Nakamura, T. (1991) Hepatocyte growth factor is a potent mitogen for cultured rabbit renal tubular epithelial cells. *Biochem. Biophys. Res. Commun.* **174**, 831-838.

Kamalati. T., Howard, M. and Brooks, R.F. (1989a) IGF I induces differentiation in a transformed human keratinocyte line. *Development* **106**, 283-293.

Kamalati, T., McIvor, Z., Howard, M., Green, M. and Brooks, R.F. (1989b) Expression of markers of differentiation in a transformed human keratinocyte line induced by co-culture with a fibroblast line. *Exp. Cell Res.* **185**, 453-463.

Kamalati, Thirunavukarasu, B., Wallace, A., Holder, N., Brooks, R.F., Nakamura, T., Stoker, M., Gherardi, E. and Buluwela, L. (1992) Down-regulation of scatter factor in MRC 5 fibroblasts by epithelial derived cells, A model for scatter factor modulation. *J. Cell Sci.* **101**, 323-332.

Kinoshita, T., Tashiro, K. and Nakamura, T. (1989) Marked increase of HGF mRNA in non-parenchymal liver cells of rats treated with hepatotoxins. *Biochem. Biophys. Res. Commun.* **165**, 1229-1234.

Klagsbrun, M. (1990) The affinity of fibroblast growth factors (FGFs) for heparin; FGF heparan sulphate interactions in cells and extracellular matrix. *Curr. Opin. Cell Biol.* **2**, 857-863.

Matsumoto, K., Tajima, H. and Nakamura, T. (1991a) Hepatocyte growth factor is a potent stimulator of human melanocyte DNA synthesis and growth. *Biochem. Biophys. Res. Commun.* **176**, 45-51.

Matsumoto, K., Hashimoto, K., Yoshikawa, K. and Nakamura, T. (1991b) Marked stimulation of growth and motility of human keratinocytes by hepatocyte growth factor. *Exp. Cell Res* . **196**, 114-120.

Matsumoto, K., Tajima, H., Hamanoue, M., Kohno,S., Kinoshita, T. and Nakamura, T. (1992a) Identification and characterisation of "Injurin", an inducer of expression of the hepatocyte growth factor. *Proc. Natl. Acad. Sci.* **89**, 3800-3804.

Matsumoto, K., Okazaki, H., and Nakamura, T. (1992b) Up regulation of hepatocyte growth factor gene expression by Interleukin-1 in human skin fibroblasts. *Biochem. Biophys. Res. Commun.* **188**, 235-243.

Matsumoto, K., Hisao, T., Hiroko, O. and Nakamura, T. (1992c) Negative regulation of hepatocyte growth factor gene expression in human lung fibroblasts and leukemic cells by transforming growth factor - β1 and glucocorticoids. *J. Biol. Chem.* **267**, 24917-24920.

Miyazawa, K., Tsubouchi, H., Naka, D., Takahashi, K., Okigaki, M., Arakaki, N., Nakayama, H., Hirono, S., Sakiyama, O., Takahashi, K., Gohda, E., Daikuhara, Y. and Kitamura, N. (1989) Molecular cloning and sequence analysis of cDNA for human hepatocyte growth factor. *Biochem. Biophys. Res. Commun.* **163**, 967-973.

Montesano, R., Matsumoto, K., Nakamura, T. and Orci, L. (1991) Identification of a fibroblast - derived epithelial morphogen as hepatocyte growth factor. *Cell* **67**, 901-908.

Nakamura, T., Nawa, K. and Ichihara, A. (1984) Partial purification and characterisation of hepatocyte growth factor from serum of hepatectomized rats. *Biochem. Biophys. Res. Commun.* **122**, 1450-1459.

Nakamura, T., Nishizawa, T., Hagiya, M., Seki, T., Shimonishi, M., Sugimura, A., Tashiro, K. and Shimizu, S. (1989) Molecular cloning and expression of human hepatocyte growth factor. *Nature* **342**, 440-443.

Naldini, L., Vigna, E., Narsimhan, R., Gaudino, G., Zarnegar, R., Michalopoulos, G. K. and Comoglio, P.M. (1991) Hepatocyte growth factor (HGF) stimulates the tyrosine kinase activity of the receptor encoded by the proto-oncogene c-Met. *Oncogene*, **6**, 501-504.

Noji, S., Tashiro, K., Koyama, E., Nohno, T., Ohyama, K., Taniguchi, S. and Nakamura, T. (1990) Expression of hepatocyte growth factor gene in endothelial and Kupffer cells of damaged rat livers as revealed by in-situ hybridisation. *Biochem. Biophys. Res. Commun.* **173**, 42-47.

Rheinwald, J.G. and Green, H. (1975) Serial cultivation of strains of human epidermal keratinocytes, the formation of keratinising colonies from single cells. *Cell* **6**, 331-344.

Roberts, A.B., Anzano, M.A., Wakefield, L.M., Roche, N.S., Stern, D.F. and Sporn, M.B. (1985) Type beta transforming growth factor, A bifunctional regulator of cellular growth. *Proc. Natl. Acad. Sci.* **82**, 119-123.

Rubin, J.S., Osada, H., Finch, P.W., Taylor, W.g., Rudikoff, S. and Aaronson, S.A. (1989) Purification and characterisation of a newly identified growth factor specific for epithelial cells. *Proc. Natl. Acad. Sci.* **86**, 802-806.

Rubin, J.S., Chan, M.-L., Bottaro, D.P., Burgess, W.H., Taylor, W.G., Cech, A.C., Hirschfield, D.W., Wong, J., Miki, T., Finch, P.W. and Aaronson, S.A. (1991) A broad-spectrum human lung fibroblast-derived mitogen is a variant of hepatocyte growth factor. *Proc. Natl. Acad. Sci. USA.* **88**, 415-419.

Shiota, G., Rhoads, D.B., Wang, T.C. and Nakamura, T. (1992) Hepatocyte growth factor inhibits growth of hepatocellular carcinoma cells. *Proc. Natl. Acad. Sci.* **89**, 373-377.

Stoker, M. (1989) Effects of scatter factor on motility of epithelial cells and fibroblasts. *J. Cell Physiol.* **139**, 565-569.

Stoker, M. and Gherardi, E. (1989) Scatter factor and regulators of cell mobility. *British Med. Bulletin* **45**, 481-491.

Stoker, M., Gherardi, E., Perryman, M. and Gray, J. (1987) Scatter factor is a fibroblast-derived modulator of epithelial cell mobility. *Nature* **327**, 239-242.

Stoker, M. and Perryman, M. (1985) An epithelial scatter factor released by embryo fibroblasts. *J.Cell Sci.* **77**, 209-223.

Tashiro, K., Hagiya, M., Nishizawa, T., Seki, T., Shimonishi, M., Shimizu, S, and. Nakamura, T.

(1990). Deduced primary structure of rat hepatocyte growth factor and expression of the mRNA in rat tissue. *Proc. Natl. Acad. Sci. USA.* **87**, 3200-3204.

Taylor-Papadimitriou, J., Purkis, P., Lane, E.B., Mckay, I.A. and Chang, S.E. (1982) Effect of SV40 transformation on the cytoskeleton and behavioural properties of human keratinocytes. *Cell Diff.* **11**, 169-180.

Warn, R.M. and Dowrick, P. (1989) Motility factors on the march. *Nature* **340**, 186-187.

Weidner, K.M., Behrens, J., Vanderkerckhove, J. and Birchmeir. (1990) Scatter factor, Molecular charectarisation and effects on the invasiveness of epithelial cells. *J.Cell Biol.* **111**, 2097-2108.

Weidner, K.M., Araki, N., Vanderkerchove, J., Weingart, S., Hartmann, G., Rieder, H., Fonatsch, C.,Tsubouchi, H., Hishida, T., Daikuhara, Y. and Birchmeier, W. (1991). Evidence for the identity of human scatter factor and human hepatocyte growth factor. *Proc. Natl. Acad. Sci.* **88**, 7001-7005.

Yanagita, K., Nagaike, M., Ishibashi, H., Niho, y., Matsumoto, K. and Nakamura, T. (1992) Lung may have an endocrine function producing hepatocyte growth factor in response to injury of distal organs. *Biochem. Biophys. Res. Commun.* **182**, 802-809.

Zarnegar, R., Muga, S., Rahija, R., and Michalopoulos, G.K. (1990) Tissue distribution of hepatopoientin A, A heparin-binding polypeptide growth factor for hepatocytes. *Proc. Natl. Acad. Sci. USA.* **87**, 1252-1256.

Printed in Great Britain © The Society of Experimental Biology 1993   227

SCATTER FACTOR (HEPATOCYTE GROWTH FACTOR) IS A POTENT ANGIOGENESIS FACTOR *IN VIVO*

ELIOT M. ROSEN[1,][*], *DERRICK S. GRANT*[2], *HYNDA K. KLEINMAN*[2],
ITZHAK D. GOLDBERG[2], *MAHDU M. BHARGAVA*[2], *BRIAN J. NICKOLOFF*[3],
JAMES L. KINSELLA[4] *and PETER POLVERINI*[5]

[1]Department of Therapeutic Radiology, Yale University School of Medicine, HRT 132, 333 Cedar Street, New Haven, CT 06510, USA

[2]Laboratory of Developmental Biology, National Institute of Dental Research, NIH, Bldg 30/430, 9000 Rockville Pike, Bethesda, MD 20892, USA

[3]Department of Radiation Oncology, Long Island Jewish Medical Center, New Hyde Park, NY 11042, USA

[4]Department of Pathology, University of Michigan Medical School, Ann Arbor, Michigan 48109, USA

[5]Laboratory of Cardiovascular Science, Gerontology Research Center, NIA 4940, Eastern Avenue, Baltimore, MD 21224, USA

[6]Department of Pathology, Northwestern University 303 E. Chicago Ave., Chicago, Illinois 60611, USA

Summary

Scatter factor (SF), a fibroblast-derived cytokine characterized by its ability to convert non-motile epithelial cells to a motile fibroblast-like phenotype, is identical to hepatocyte growth factor (HGF), a broad-spectrum mitogen. SF is a heterodimeric glycoprotein that is homologous to plasminogen and other blood coagulation proteases but lacks proteolytic activity. Its receptor is the *c-met* proto-oncogene product, a growth factor receptor-like transmembrane tyrosine kinase. This unique cytokine is also synthesized and secreted by vascular smooth muscle cells and acts on endothelial cells to stimulate migration, protease production, invasion, proliferation, and differentiation into capillary-like tubes *in vitro*. SF-containing implants in mouse subcutaneous tissue and rat cornea induce directed ingrowth of new blood vessels from surrounding tissue, with maximal angiogenic responses at doses of 100-200 ng of SF. Immunoreactive SF is expressed at sites of neovascularization within human psoriatic plaques. These findings suggest that SF may play a significant role in the formation and repair of blood vessels under physiologic and pathologic conditions.

Introduction

The following short review outlines important developments in the scatter factor field, followed by a description of some of our own work relating to scatter factor and its activity on vascular endothelium.

*To whom reprint requests should be addressed.

Key words: scatter factor, hepatocyte growth factor, angiogenesis, endothelial cells, cell migration, urokinase, capillaries, cell proliferation.

Scatter factor, hepatocyte growth factor and *met*

Scatter factor (SF) was characterized by Stoker and his colleagues at the University of Cambridge as a fibroblast-derived cytokine that induces spreading, dispersion into individual cells, conversion to a fibroblastic morphology, and motility in normally stationary, cohesive epithelium (Stoker and Perryman, 1985; Stoker *et al.*, 1987; Stoker, 1989; Gherardi et al., 1989). Based on these properties, it was suggested that SF might be involved in embryogenesis and wound healing. During the same time, several other groups of investigators *independently* studied an hepatocyte growth factor (HGF), thought to be a mediator of liver regeneration (Nakamura *et al.*, 1987; Gohda *et al.*, 1988; Zarnegar and Michalopoulos, 1989; Selden and Hodgson, 1989). Human HGF was cloned, and its entire amino acid sequence was deduced from the cDNA sequence (Nakamura *et al.*, 1989; Miyazawa *et al.*, 1989). These studies revealed that, unlike all other known growth factors, HGF is closely related to the blood coagulation family of proteases. Thus, HGF has 38% amino acid sequence identity and similar protein domain structure to the serum proenzyme plasminogen. Subsequent studies revealed that based on amino acid sequence, biologic activities, immunologic cross-reactivity, and activation of the same cell surface receptor (*vide infra*), SF and HGF are one and the same protein (Weidner *et al.*, 1991; Furlong *et al.*, 1991; Naldini *et al.*, 1991: Bhargava *et al.*, 1992).

SF (HGF) is a complex, high molecular weight glycoprotein that is synthesized as a single polypeptide precursor and is proteolytically processed to yield a 728 amino acid heterodimer consisting of a heavy (α) subunit ($M_r \approx 60$ kDa) and a light (β) subunit ($M_r \approx 30$ kDa). The cleavage of an arg_{495}-val_{496} bond generates the two subunits. A similar cleavage of plasminogen results in conversion to plasmin, suggesting that plasminogen activators might also be involved in SF (HGF) processing. The α chain of SF contains four kringle domains, folded regions believed to mediate binding of proteases to cells and to other proteins. The β chain is homologous to the protease domain of plasminogen, but lacks proteolytic activity due to substitutions in two of three critical amino acids at the serine catalytic center (Nakamura et al., 1989).

The *c-met* proto-oncogene protein product was characterized as a transmembrane growth factor receptor-like tyrosine kinase (Park *et al.*, 1987; Gonzatti-Haces *et al.*, 1988). The major *c-met* protein (p190*c-met*) is a heterodimeric glycoprotein consisting of a 50 kDa α chain that is expressed at the cell surface and a 140 kDa β chain that contains the ligand-binding membrane kinase and phosphate acceptor sites. Bottaro and colleagues (Bottaro *et al.*, 1991) demonstrated that the β chain of p190*c-met* is a functional HGF receptor. A low molecular weight (M_r 28 kDa) form of HGF, consisting of only the first two kringle domains of the α chain, binds to the *c-met* protein and inhibits HGF-stimulated mitogenesis of mammary epithelial cells, suggesting that the receptor-binding site may reside within the first two kringles (Chen *et al.*, 1991; Hartmann *et al.*, 1993). Thus, by the end of 1991, three initially independent subjects of investigation - SF, HGF, and *c-met* - converged into a single field (Rosen *et al.*, 1991c).

Scatter factor and the vascular endothelium

We first became interested in SF when we found that blood vessel-wall derived smooth muscle cells secrete a protein with similar characteristics to the fibroblast-derived SF described by Stoker (Rosen *et al.*, 1989, 1990c). Bovine aorta and human iliac artery smooth muscle cells produced SF activity at rates comparable to the highest producer fibroblast lines (128-256 units/10^6 cells/48 hr). Both the smooth muscle and fibroblast SFs bound tightly to immobilized heparin. Soluble heparin inhibited the scatter effect, and this inhibition could be reversed with protamine, suggesting that SFs are basic heparin-binding proteins. Production of SF by vascular smooth muscle cells suggested to us that this factor be important in vascular physiology, and that it might act on a vascular target cell, the endothelium. By analogy, members of the fibroblast growth factor family of heparin-binding glycoproteins stimulate proliferation and motility of vascular endothelial cells *in vitro* and induce angiogenesis, the formation of new blood vessels, *in vivo* (Thomas, 1987).

Regulation of vascular endothelial cell motility

We subsequently found that SF stimulates motility of a variety of large vessel- and microvessel-derived lines of endothelial cells (Rosen *et al.*, 1990a,b,d). SF was exquisitively active on endothelial cells in a variety of assays which measure different aspects of motility. In standard Boyden chamber assays of cell migration, SF stimulated chemotactic and random migration of endothelial cells from about three- to 20-fold, depending upon the cell line. *Maximum* responses were observed at concentrations in the range of 4-10 ng/ml, which corresponds to about 50-110 *picomolar*. To investigate the means by which endothelial responsiveness to SF is regulated, we devised a novel migration assay in which movement of cells *off* microcarrier beads *onto* flat culture surfaces is quantitated (Rosen *et al.*, 1990d). This assay appeared to be relatively *selective* for SF, since a variety of other factors (including basic and acidic FGFs, endothelial cell growth factor, platelet-derived growth factor, epidermal growth factor, and tumor cell autocrine motility factor) had little or no effect on migration.

Utilizing the microcarrier bead migration assay, we found that SF altered the migration of vascular endothelial cells in three phases: (i) a "lag" phase (from T=0 to T=5-6 hr) in which migration was similar to or only slightly faster than control migration; (ii) an "accelerated" phase (from T=6 to T=18 hr) during which the rate of migration was about five times greater than controls; and a final phase (T>18 hr) during which the migration rate returned to control levels. We used the bead assay to investigate the regulation of SF-stimulated migration of bovine brain endothelial cells (BBEC) (Rosen *et al.*, 1991b). SF-stimulated migration was abolished in the presence of cycloheximide, but was unaffected by hydroxyurea, indicating a requirement for protein synthesis but *not* for DNA synthesis. Agents that *activate* the adenylate cyclase signalling pathway by various mechanisms (cholera toxin, forskolin, dibutyryl cyclic AMP, theophylline) all inhibited SF-stimulated migration. Transforming growth factor-β inhibited both basal and SF-stimulated migration of BBEC cells. BBEC migration was also blocked by a variety of protein kinase inhibitors (eg., staurosporine,

K252a, H-7), and by anti-microfilament (cytochalasin B) and anti-microtubule (colcemid, vincristine) agents. Thus, detachment of endothelial cells from carrier beads and reattachment to culture surfaces requires protein phosphorylation and an intact cytoskeletal system.

The tumor-promoting phorbol ester PMA (phorbol myristate acetate), also stimulated migration of BBEC. PMA induces redistribution of protein kinase C (PKC) from the cytosol to the cell membrane and subsequent down-modulation of PKC. Pre-incubation of cells with a high dose of the PMA (150 ng/ml) in order to deplete PKC resulted in loss of responsiveness to PMA with no change in responsiveness to SF. Treatment of BBEC with SF did not induce redistribution or down-modulation of ^{3}H-phorbol ester binding sites or of immunoreactive α-PKC, while treatment with phorbol ester induced redistribution and subsequent down-modulation of both ^{3}H-phorbol ester binding sites and α-PKC. These findings suggest that PKC is involved in phorbol ester-mediated BBEC motility but *not* in SF-mediated motility.

Invasion and protease production

During the early stages of angiogenesis, endothelial cells focally degrade and invade through subendothelial basement membrane of the parent vessel and migrate toward the angiogenic stimulus (Folkman, 1985). Mouse and human SFs markedly stimulated invasion of endothelial cells across a basement membrane-like matrix (Rosen *et al.*, 1991a). Maximal invasion rates of 30-60 times control values were observed at SF concentrations of only 2-10 ng/ml. Treatment of endothelial cells with SF also induced large increases in both secreted and cell associated urokinase-type plasminogen activator (uPA) activity (Rosen *et al.*, 1991a; Grant *et al.*, 1993). This finding may be relevant to angiogenesis, since uPA bound to its specific cell surface receptor is thought to mediate directed extracellular proteolysis and cell migration through tissue (Saksela and Rifkin, 1988).

Proliferation

Although limited capillary formation can occur without endothelial cell proliferation, the continued formation of anastomosing networks of capillaries requires DNA synthesis and cell division. Rubin and colleagues (Rubin *et al.*, 1991), while studying fibroblast-derived mitogens, isolated and cloned a variant of HGF from human lung fibroblasts. They showed that HGF is mitogenic for a variety of normal cell types, including various human umbilical vein endothelium. The observation that HGF is mitogenic for endothelial cells was confirmed by Morimoto *et al.* (1991), who reported that HGF stimulates proliferation of human omental microvessel-derived endothelial cells.

Differentiation into capillary-like structures

Endothelial cells inoculated onto a matrix of reconstituted basement membrane (Matrigel) stop dividing, align in tandem, and organize into capillary-like structures within 12-24 hr (Kubota *et al.*, 1988). This *in vitro* behavior reflects a distinct phase in angiogenesis, in which endothelial cells re-align, develop lumens, and organize into capillaries (Folkman, 1985). We utilized a quantitative digital imaging system (Grant *et al.*, 1989) to show that SF stimulates the formation of capillary-like tubes *in vitro* by human umbilical vein and bovine brain endothelial cells. A two-fold stimulation of tube formation was observed at SF concentrations of 2-20 ng/ml (Rosen *et al.*, 1991a). The ability of SF (HGF) to induce endothelium to organize into vascular structures is reminiscent of two other recently described biologic activities of this remarkable cytokine: (i) SF induces kidney epithelial cells to organize into three-dimensional networks of branching tubules within a collagen matrix (Montesano *et al.*, 1991); and (ii) SF causes mammary epithelial cells to form duct-like structures in suspension culture (Tsarfaty *et al.*, 1992). Immunohistochemical analysis revealed that the predominent *c-met* staining was localized to the lumenal surface of mammary duct epithelium, implicating the SF and the *c-met* receptor in epithelial differentiation leading to duct formation (Tsarfaty *et al.*, 1992).

In vivo angiogenesis

Since SF stimulates vascular endothelial cell motility, invasiveness, proliferation, and differentiation into capillary-like structures *in vitro*, it has all of the properties expected of a direct-acting angiogenesis factor. We used two experimental models to show that SF induces the formation of new blood vessels *in vivo* (Grant *et al.*, 1992, *In press*). In the first model, angiogenesis was assayed as the growth of blood vessels from surrounding subcutaneous tissue into a solid gel of basement membrane (Matrigel) containing SF. SF was mixed with Matrigel in liquid form at 4°C and injected into the subcutaneous tissue of mice. Matrigel rapidly solidifies at body temperature, trapping the factor and allowing prolonged exposure to surrounding tissue during the *in vivo* incubation period. After ten days, the mice were sacrificed for histologic and morphometric analysis of the Matrigel plugs. The SF-containing plugs were bright red and contained superficial blood vessels, a prominent cellular infiltrate of Factor VIII-positive endothelial cells, capillaries, and larger vessels with smooth muscle cells. In contrast, control plugs were pale pink and showed little cellularity or vessel formation. The angiogenic response was dose-dependent, reaching half-maximal and maximal intensity at about 20 and 200 ng SF, respectively.

In the second assay, Hydron pellets containing SF were implanted into the avascular rat cornea and ingrowth of blood vessels from the limbus toward the implant was followed by daily examination using a dissecting microscope (Polverini and Liebovich, 1984). SF induced dose-dependent corneal angiogenesis, with no reponse at 0 (control) or 5 ng; submaximal response at 50 ng; and maximal reponse at 100 ng or higher (Grant *et al.*, *In press*). Purified native mouse SF and recombinant human HGF (a generous gift of Dr. Toshikazu Nakamura, Kyushu University, Fukuoka, Japan) gave similar angiogenic responses. These responses were substantially inhibited or blocked by specific chicken or rabbit antibodies to SF.

Chronic inflammatory diseases are often associated with a significant component of angiogenesis. We performed immunohistochemical staining of skin biopsies from patients with psoriasis, a common inflammatory skin disorder characterized by epidermal hyperplasia and prominent elongation of small blood vessels within the dermal papillae and papillary dermis. Antisera to human SF gave stained spindle-shaped and mononuclear cells surrounding microvessels in psoriatic plaques (Grant *et al.*, 1993). The cells of the microvessel wall (endothelium and pericytes) did *not* stain for SF. Normal skin from psoriasis patients and from normal subjects gave little or no staining for SF. These studies suggest a possible role for SF as a paracrine mediator of pathologic angiogenesis in human inflammatory disease.

Supported in part by research grants from the United States Public Health Service (CA50516) and the American Cancer Society (BE-7). Drs. Bhargava and Goldberg were supported by the Finkelstein Foundation at Long Island Jewish Medical Center. Dr. Rosen is an Established Investigator of the American Heart Association.

References

Bhargava, M., Joseph, A., Knesel, J., Halaban, R., Li, Y., Pang, S., Goldberg, I., Setter, E., Donovan, M.A., Zarnegar, R., Michalopoulos, G.A., Nakamura, T., Faletto, D., and Rosen, E.M. (1992). Scatter factor and hepatocyte growth factor: Activities, properties, and mechanism. *Cell Growth & Different* 3, 11-20.

Bottaro, D.P., Rubin, J.S., Faletto, D.L., Chan, A.M.-L., Kmiecik, T.E., Vande Woude, G.F., and Aaronson, S.A. (1991). Identification of the hepatocyte growth factor receptor as the *c-met* proto-oncogene product. *Science* 251, 802-804.

Chen, A.M.-L., Rubin, J.S., Bottaro, D.P., Hirschfield, D.W., Chedid, M., and Aaronson, S.A. (1991). Identification of a competitive antagonist encoded by an alternative transcript. *Science* 254, 1382-1385.

Dean, M., Park, M., Le Beau, M.M., Robins, T.S., Diaz, M.O., Rowley, J.D., Blair, D.G., and Vande Woude, G.F. (1985). The human *met* oncogene is related to the tyrosine kinase oncogenes. *Nature* 318, 385-388.

Folkman, J. (1985). Tumor angiogenesis. *Adv Cancer Res* 43, 173-203.

Furlong, R.A., Takheara, T., Taylor, W.G., Nakamura, T., and Rubin, J.S. (1991). Comparison of biologic and immunochemical properties indicate that scatter factor and hepatocyte growth factor are indistinguishable. *J Cell Sci* 100, 173-177.

Gherardi, E., Gray, J., Stoker, M., Perryman, M., and Furlong, R. (1989). Purification of scatter factor, a fibroblast-derived basic protein which modulates epithelial interactions and movement. *PNAS USA* 86, 5844-5848.

Gohda, E., Tsubouchi, H., Nakayama, H., Hirono, S., Sakiyama, O., Takahashi, K., Miyazaki, H., Hashimoto, S., and Daikuhara, Y. (1988). Purification and partial characterization of hepatocyte growth factor from plasma of a patient with fulminant hepatitis. *J Clin Invest* 81, 414-419.

Gonzatti-Haces, M., Seth, A., Park, M., Copeland, T., Oroszlan, S., and Vande Woude, G.F. (1988). Characterization of the TPR-MET oncogene p65 and the MET protooncogene p140 protein tyrosine kinases. *PNAS USA* 85, 21-25.

Grant, D.S., Kleinman, H.K., Goldberg, I.D., Bhargava, M.M., Nickoloff, B.J., Kinsella, J.L., Polverini, P.J., and Rosen, E.M. (1993). Scatter factor induces blood vessel formation *in vivo*. *PNAS USA, In press.*

Grant, D.S., Tashiro, K.-I., Segui-Real, B., Yamada, Y., Martin, G.R., and Kleinman, H. (1989). Two different laminin domains mediate the differentiation of human endothelial cells into capillary-like structures *in vitro*. *Cell* 58, 933-943.

Hartmann, G., Naldini, L., Weidner, K.M., Sachs, M., Vigna, E., Comoglio. P.M., and Birchmeier, W. (1993). A functional domain in the heavy chain of scatter factor/hepatocyte growth factor binds and activates the c-Met receptor, and induces cell dissociation but not mitogenesis. *PNAS USA, In press.*

Kubota, Y., Kleinman, H.K., Martin, G.R., and Lawley, T.J. (1988). Role of laminin and basement membrane in the morphological differentiation of human endothelial cells into capillary-like structures. *J Cell Biol* 107, 1589-1598.

Miyazawa, K., Tsubouchi, H., Naka, D., Takahashi, K., Okigaki, M., Arakaki, N., Nakayama, H., Hirono, S., Sakiyama, O., Gohda, E., Daikuhara, Y., and Kitamura, N. (1989). Molecular cloning and sequence analysis of cDNA for human hepatocyte growth factor. *Biochem Biophys Res Commun* **163**, 967-973.

Montesano, R., Matsumoto, K., Nakamura, T., and Orci, L. (1991). Identification of a fibroblast-derived epithelial morphogen as hepatocyte growth factor. *Cell* **67**, 901-908.

Morimoto, A., Okamura, K., Hamanaka, R., Sato, Y., Shima, N., Higashio, K., and Kuwano, M. (1991). Hepatocyte growth factor modulates migration and proliferation of human microvascular endothelial cells in culture. *Biochem Biophys Res Commun* **179**, 1042-1049.

Nakamura, T., Nawa, K., Ichihara, A., Kaise, N., and Nishino, T. (1987). Purification and subunit structure of hepatocyte growth factor from rat platelets. *FEBS Lett* **224**, 311-316.

Nakamura, T., Nishizawa, T., Hagiya, M., Seki, T., Shimonishi, M., Sugimura, A., and Shimizu, S. (1989). Molecular cloning and expression of human hepatocyte growth factor. *Nature* **342**, 440-443.

Naldini, L., Weidner, K.M., Vigna, E., Guadino, G., Bardelli, A., Ponzetto, C., Narsimhan, R.P., Hartmann, G., Zarnegar, R., Michalopoulos, G., Birchmeier, W., and Comoglio, P.M. (1991). Scatter factor and hepatocyte growth factor are indistinguishable ligands for the MET receptor. *EMBO J* **10**, 2867-2878.

Park, M., Dean, M., Kaul, K., Braun, M.J., Gonda, M.A., and Vande Woude, G.F. (1987). Sequence of MET protooncogene cDNA has features characteristic of the tyrosine kinase family of growth factor receptors. *PNAS, USA* **84**, 6379-6383.

Polverini, P.J., and Leibovich, S.J. (1984). Induction of neovascularization *in vivo* by tumor-associated macrophages. *Lab Invest* **51**, 635-642.

Rosen, E.M., Goldberg, I.D., Kacinski, B.M., Buckholz, T., and Vinter, D.W. (1989). Smooth muscle releases an epithelial cell scatter factor which binds to heparin. *In Vitro Cell Dev Biol* **25**, 163-173.

Rosen, E.M., Carley, W., and Goldberg, I.D. (1990a). Scatter factor regulates vascular endothelial cell motility. *Cancer Invest* **8**, 647-650.

Rosen, E.M., Meromsky, L., Setter, E., Vinter, D.W., and Goldberg, I.D. (1990b). Purification and migration-stimulating activities of scatter factor. *Proc Soc Exp Biol Med* **195**, 34-43.

Rosen, E.M., Meromsky, L., Setter, E., Vinter, D.W., and Goldberg, I.D. (1990c). Smooth muscle-derived factor stimulates mobility of human tumor cells. *Invasion Metastasis* **10**, 49-64.

Rosen, E.M., Meromsky, L., Romero, R., Setter, E., and Goldberg, I. (1990d). Human placenta contains an epithelial scatter protein. *Biochem Biophys Res Commun* **168**, 1082-1088.

Rosen, E.M., Meromsky, L., Setter, E., Vinter, D.W., and Goldberg, I.D. (1990e). Quantitation of cytokine-stimulated migration of endothelium and epithelium by a new assay using microcarrier beads. *Exp Cell Res* **186**, 22-31.

Rosen, E.M., Grant, D., Kleinman, H., Jaken, S., Donovan, M.A., Setter, E., Luckett, P.M., and Carley, W. (1991a). Scatter factor stimulates migration of vascular endothelium and capillary-like tube formation. In *Cell Motility Factors*, Goldberg, I.D., Rosen, E.M., eds., Birkhauser-Verlag, Basel, pp 76-88.

Rosen, E.M., Jaken, S., Carley, W., Setter, E., Bhargava, M., and Goldberg, I.D. (1991b). Regulation of motility in bovine brain endothelial cells. *J Cell Physiol* **146**, 325-335.

Rosen, E.M., Knesel, J., and Goldberg, I.D. (1991c). Scatter factor and its relationship to hepatocyte growth factor and *met*. Research Capsule. *Cell Growth & Different* **2**, 603-607.

Rubin, J.S., Chan, A.M.-L., Bottaro, D.P., Burgess, W.H., Taylor, W.G., Cech, A.C., Hirschfield, D.W., Wong, J., Miki, T., Finch, P.W., and Aaronson, S.A. (1991). A broad spectrum human lung fibroblast-derived mitogen is a variant of hepatocyte growth factor. *PNAS USA* **88**, 415-419.

Saksela, O., and Rifkin, D.M. (1988). Cell-associated plasminogen activation: Regulation and physiologic functions. *Ann. Rev. Cell Biol.* **4**, 93-126.

Selden, C., and Hodgson, H. (1989). Further characterization of hepatotropin, a high molecular weight hepatotropic factor in rat serum. *J Hepatol* **9**, 167-176, 1989.

Stoker, M., and Perryman, M. (1985). An epithelial scatter factor released by embryo fibroblasts. *J Cell Sci* **77**, 209-223.

Stoker, M. (1989). Effect of scatter factor on motility of epithelial cells and fibroblasts. *J Cell Physiol* **139**, 565-569.

Stoker, M., Gherardi, E., Perryman, M., and Gray, J. (1987). Scatter factor is a fibroblast-derived modulator of epithelial cell mobility. *Nature* **327**, 238-242,

Thomas, K.A. (1987). Fibroblast growth factors. *FASEB J* **1**, 434-440.

Tsarfaty, I., Resau, J.H., Rulong, S., Keydar, I., Faletto, D., and Vande Woude, G.F. (1992). The *met* proto-oncogene receptor and lumen formation. *Science* **257**, 1258-1261.

Weidner, K.M., Arakaki, N., Vandekereckhove, J., Weingart, S., Hartmann, G., Rieder, H., Fonatsch, C., Tsubouchi, H., Hishida, T., Daikuhara, Y., and Birchmeier, W. (1991). Evidence for the identity of human scatter factor and human hepatocyte growth factor. *PNAS USA* **88**, 7001-7005.

Zarnegar, R., and Michalopoulos, G. (1989). Purification and biologic characterization of human hepatopoietin A, a polypeptide growth factor for hepatocytes. *Cancer Res* **49**, 3314-3320.

MIGRATION STIMULATING FACTOR (MSF): ITS STRUCTURE, MODE OF ACTION AND POSSIBLE FUNCTION IN HEALTH AND DISEASE

SETH L. SCHOR[1,*], *ANNE MARIE GREY*[1], *IAN ELLIS*[1], *ANA M. SCHOR*[2], *BRIAN COLES*[3] *and RUTH MURPHY*[2]

[1]Department of Cell and Structural Biology, Stopford Building, Oxford Road, University of Manchester, Manchester, M13 9PT, UK
[2]CRC Department of Medical Oncology, Christie Hospital, Wilmslow Road, Manchester, M20 9BX, UK
[3]CRC Molecular Toxicology Group, University College and Middlesex Hospital, Windeyer Building, Cleveland St, London W1P 6DB, UK

Summary

We have previously reported that (a) fetal fibroblasts migrate into 3-dimensional collagen matrices to a significantly greater extent that do adult cells, (b) this difference in migratory behaviour results from the secretion by fetal fibroblasts of a "migration stimulating factor" (MSF), and (c) adult fibroblasts retain responsiveness to MSF, this providing the basis of a bioassay for monitoring factor activity. Using a recently modified purification protocol, MSF isolated from fetal fibroblast conditioned medium elutes as a single activity peak in the penultimate Mono Q anion exchange chromatography step. Analysis of this material by SDS-PAGE indicates that it consists of three proteins, one with an apparent molecular mass of 119 kDa and a doublet with molecular masses of approximately 43 and 33 kDa, respectively. Our data suggest that the two proteins comprising the doublet result from the degradation of the larger molecule during the purification procedure. Both the 119 kDa species and lower molecular weight doublet stimulate fibroblast migration (with half maximal activity in the region of 1-10 pg/ml) and contain a structural domain exhibiting significant amino acid sequence homology with the gelatin-binding fragment (GBF) of fibronectin. Bona fide preparations of GBF, obtained by the limited proteolysis of plasma fibronectin, also stimulate the migration of adult fibroblasts in a similar dose-dependent manner to that of MSF. In spite of this similarity, MSF and GBF differ in terms of a number of biological and biochemical parameters, thereby suggesting that MSF is a distinct gene product and not a proteolytic degradation fragment of fibronectin.

MSF stimulates the synthesis of a high molecular weight species of hyaluronic acid (HA). Our current data suggest that the observed effect of MSF on cell migration is actually a secondary consequence of the accumulation of this HA in the collagen matrix. TGF-β is a potent inhibitor of MSF, both in terms of its effects on cell migration and HA synthesis. As MSF is present in wound fluid, we have suggested that the inhibition of

*Author for correspondence.

Key words: cell migration, fibronectin, cytokine, TGF-β, wound healing, cancer progression.

MSF activity by TGF-β may reflect the antagonistic interaction of these two cytokines in the control of the wound healing process.

Our recent data indicate that discrete minority subpopulations of MSF-secreting fibroblasts are also present at specific sites in the healthy adult and that these may undergo a transient and local expansion during wound healing. We have also documented an atypical (more widespread) distribution of MSF-secreting fibroblasts in patients with breast cancer and the presence of detectable levels of MSF activity in the serum of these individuals. On the basis of these and related data, we suggest that the relative proportion of MSF-secreting fibroblasts in cancer patients may be elevated either as a result of the persistence of "fetal-like" cells into adult life and/or their systemic clonal expansion (possibly in response to cytokine signals originating from an emerging pre-neoplastic cell population or environmental insult). Irrespective of their mode of origin, we further postulate that the inappropriate secretion of MSF by these cells may contribute directly to cancer pathogenesis as a consequence of the resultant perturbations in normal epithelial-mesenchymal interactions.

Cell migration: an overview

Cell migration is a prominent feature of embryonic development, entailing both the coordinate movement of cell sheets and the translocation of single, generally neural crest-derived, cells. Tissue cells in the adult tend to lead considerably more sedentary lives, although they may be coaxed into re-expressing the more mobile attributes of their fetal progenitors in processes such as wound healing. Inappropriate cell migration is also a key event in various pathological processes, such as tumour invasion and metastasis, this often being accompanied by the lysis and remodelling of connective tissue matrices.

In view of the involvement of cell motility in such diverse events, it is not surprising that there has been a considerable amount of effort directed towards elucidating the basic intracellular "machinery" driving cell movement (Bray, 1992), as well as the extracellular signals regulating the expression of a motile phenotype. These latter studies have made it abundantly clear that the initiation, directionaleity and cessation of cell motility are regulated by the complex interplay of various soluble (cytokine) and insoluble (matrix) effector molecules (Rosen and Goldberg, 1989; Zetter and Brightman, 1990; Humphries, Mould and Yamada, 1991; Stoker and Gherardi, 1991).

These "cytokine-matrix" interactions are central to the control of various fundamental aspects of cell behaviour and may involve (a) the modulation of cellular response to cytokines by cell adhesion to specific matrix constituents (Gospodarowicz, Greenburg and Birdwell, 1978; Ingber and Folkman, 1989; Sutton et al., 1991), and (b) the binding of cytokines to matrix macromolecules and their subsequent presentation to specific receptors on target cells in this form (Rogelj et al., 1989). On the basis of these observations, it is now evident that the particular effect of a potentially multi-functional cytokine on cell behaviour is regulated by the precise "context" of the extracellular matrix (Nathan and Sporn, 1991).

The effects of cytokines and matrix macromolecules on cell migration have generally been studied in Boyden chamber assays in which cells move through the pores of a polycarbonate membrane separating an upper and lower medium compartment. By using the "checkerboard" method of analysis (which entails the measurement of cell migration in the presence of different concentrations of a potential soluble effector molecule in the upper and lower compartments) it is possible to distinguish between chemotactic and chemokinetic modes of action (Zigmond and Hirsch, 1973); the potential involvement of haptotactic mechanisms in the control of cell migration may similarly be investigated by establishing a concentration gradient of adsorbed (matrix) macromolecules across the membrane (Aznavoorian *et al.*, 1990). Although these assays are very useful for studying particular aspects of cell motility *in vitro*, it is difficult to know precisely to what extent the accrued results may be extrapolated to cell behaviour in the considerably more complex milieu encountered by cells *in vivo*.

Fibroblasts *in vivo* migrate through a 3-dimensional fibrillar meshwork of matrix macromolecules (eg. collagen) containing interspersed "amorphous" matrix constituents, such as hyaluronic acid. In order to model this environment more accurately than can be achieved in Boyden chamber assays, we have developed a migration assay involving the assessment of cell movement through a 3-dimensional meshwork of type I collagen fibres (Schor, 1980). In this assay system, cells are plated onto the surface of the collagen substratum and their migration down into the 3-dimensional fibrillar matrix measured by microscopic observation after a standard four day incubation period. Our intention in establishing this assay was to provide a more physiologically relevant matrix "context" in which to study the effects of cytokines on cell motility. Indeed, we suggested that particular cytokines may exert quite different effects on cell motility in 3-dimensional collagen matrices compared to those observed in Boyden chamber assays and that such apparent discrepancies may ultimately lead to a better understanding of the precise manner in which these cytokines function during key physiological events. Experimental verification of this possibility has come from recent studies in which we demonstrated that TGF-β1 inhibits fibroblast migration into 3-dimensional collagen matrices, as opposed to its reported stimulatory action when assessed in Boyden chamber assays (Postlethwaite *et al.*, 1987).

In our early studies with the collagen matrix assay, we observed that the migration of adult skin fibroblasts into the collagen matrix was dependent upon cell density, this involving a significant down regulation of migration at confluence (Schor *et al.*, 1985a). This study also revealed that fetal fibroblasts displayed a distinct migratory phenotype characterised by the persistence of elevated levels of migration at confluent cell densities. Subsequent work revealed that this behavioural difference between the two cell types resulted from the secretion by fetal fibroblasts of a soluble "migration stimulating factor" (MSF) which was not produced by their adult counterparts (Schor *et al.*, 1988a). Interestingly, confluent adult fibroblasts retain responsiveness to MSF, as evidenced by the significant stimulation of migration when exposed to it. Using this migratory response of adult fibroblasts as a convenient bioassay for monitoring MSF activity, we have developed a biochemical purification protocol capable of providing sufficient quantities of MSF for further investigation.

Data are presented in this review summarising our current state of knowledge regarding the biochemical characterisation of MSF, including its (a) partial amino acid sequence, (b) mode of action, and (c) potential involvement in various pathologic processes, such as wound healing and cancer progression.

Structural homology of MSF with the gelatin-binding fragment of fibronectin and a comparison of their respective biological activities

We have previously described a protocol for the purification of MSF from fetal fibroblast conditioned medium (Grey *et al.*, 1989). Briefly, this involves (a) an initial precipitation of MSF activity at 20% ammonium sulphate, (b) heparin affinity chromatography (with MSF activity eluting at 0.3-0.6 M NaCl), (c) FPLC gel filtration chromatography, and (d) FPLC reverse phase chromatography. The precipitation of all MSF activity at such a low concentration of ammonium sulphate distinguishes it from the majority of other proteins in fibroblast conditioned medium and results in its efficient early purification.

In view of the apparent instability of MSF following exposure to the organic buffers used in the final reverse phase chromatography, we have replaced this procedure with a milder anion exchange (Mono Q) FPLC step. MSF obtained by this scheme has been characterised in terms of a number of biochemical criteria and with respect to its biological activity (Grey *et al.*, manuscript in preparation). These results indicate that all of the applied MSF activity is unbound to Mono Q and recovered in the initial column wash (Figure 1). SDS-PAGE indicates that this material consists of a protein with a molecular mass of approximately 119,000 kDa and a doublet with molecular masses of 33 and 43 kDa, respectively (Figure 2). Our data further suggest that the two lower molecular weight bands are derived from the degradation of the higher molecular mass species, and that at least one of the lower molecular doublet proteins may subsequently

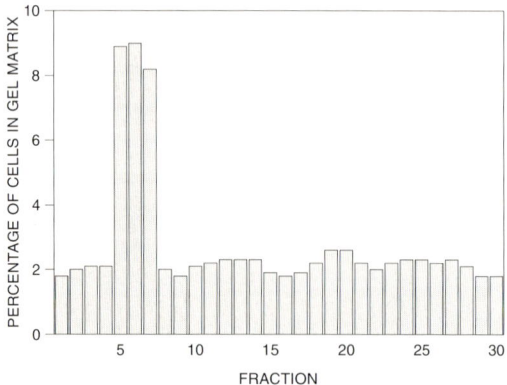

Fig. 1. The elution profile of MSF activity on Mono Q anion exchange chromatography. Partially purified MSF activity following Superose 12 gel filtration chromatography was applied to a Mono Q column and eluted with a linear gradient of 0-1.0 N NaCl between 4-24 ml, and a decreasing linear gradient of 1.0-0 N NaCl between 29-32 ml. One ml fractions were collected and analyzed for MSF activity using the collagen matrix migration assay.

break down into a biologically active molecule with a molecular mass of approximately 25 kDa.

All three protein bands in the Mono Q unbound preparation of MSF are immunoprecipitated with a polyclonal anti-fibronectin antibody, indicating that each one contains epitopes also present in fibronectin. In keeping with this observation, amino acid sequence analysis has indicated that the 33-43 kDa molecular weight doublet exhibits striking N-terminal sequence homology with the gelatin-binding domain of fibronectin, as follows:

```
    -  P  Y  G  H  -  V  T  D  S  G  V  V  Y  G  V  T  M      MSF
       1           5              10             15
    Q  P  P  P  Y  G  H  C  V  T  D  S  G  V  V  Y  S  V  G  M  Q    fibronectin.
    270        275           280           285           290
```

These observations raise the intriguing possibility that MSF may be a proteolytic degradation product of fibronectin. In order to explore this possibility, we have compared MSF with bona fide preparations of the gelatin-binding fragment (GBF) of fibronectin in terms of a number of biological and biochemical criteria. Preparations of GBF were produced by the limited thermolysin degradation of human plasma fibronectin and subsequent separation of the generated peptide fragments by hydroxyapatite chromatography (Zardi *et al.*, 1985). The GBF generated by this protocol had a molecular mass of approximately 43 kDa and amino terminus at ala$_{262}$ (ie. only ten amino acids removed from that of MSF).

In the first instance, the effects of these molecules on fibroblast migration were examined. Our results indicated that MSF *and* GBF stimulated the migration of confluent adult fibroblasts with a similar biphasic dose-response (Figure 3 and Schor *et al.*, 1989; Schor *et al.*, manuscript submitted); in both cases, significant migration stimulating activity was apparent at concentrations of effector molecule as low as *1.0-10.0 pg/ml* and returned to control (unstimulated) levels at concentrations greater than 25 ng/ml. These studies also indicated that native fibronectin and all other of its peptide degradation products were completely devoid of migration stimulating activity at concentrations up to 1 µg/ml. In marked contrast to these findings, previous studies

A

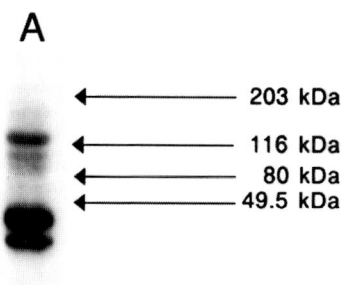

Fig. 2. The composition of MSF eluted from the Mono Q anion exchange column. The peak of MSF biological activity recovered following anion exchange chromatography was iodinated and the constituent proteins visualised by SDS-PAGE and fluorography.

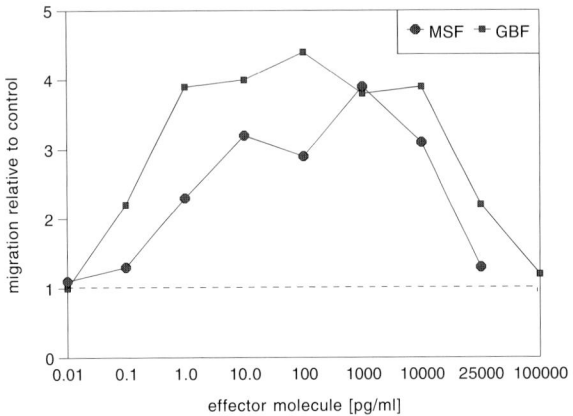

Fig. 3. Dose-response of MSF and GBF on the migration of confluent adult fibroblasts. Confluent adult fibroblasts were exposed to different concentrations of MSF and the cultures incubated for 4 days. The percentages of cells within the collagen matrix were then determined and these data expressed relative to the control (dotted line). Under these confluent conditions, adult fibroblasts displayed a relatively low level of migration in control cultures (2.3±0.6%).

using Boyden chamber assays have indicated that (a) native fibronectin stimulated fibroblast migration, but only when present at the relatively high concentrations of 10-100 µg/ml (Postlethwaite *et al.*, 1981; Aznavoorian *et al.*, 1990), and (b) similarly high concentrations of fibronectin proteolytic fragments containing the central cell-binding domain could replace native fibronectin in this assay, whilst fragments containing the gelatin-binding domain were completely inactive (Seppä *et al.*, 1981). The very potent migration stimulating activity of GBF in the collagen matrix assay was most unexpected in the light of these reports and once again serves to underscore the importance of the substratum in determining cellular responsiveness to potential soluble effector molecules. The extremely low concentration at which GBF stimulates fibroblast migration in the collagen gel assay system implies that it is acting in a "cytokine-like" fashion involving its interaction with an hitherto unrecognised cell surface receptor; our current data suggest that this is indeed that case and relevant supporting information will be presented in a forthcoming publication.

The molecular basis of the observed bell-shaped dose-response of MSF and GBF on fibroblast migration is not understood; we have postulated that it may reflect the involvement of distinct classes of cell surface receptors and/or relative receptor occupancy in determining cellular response. Similar biphasic dose-response curves have been observed with respect to the effects of other cytokines on various aspects of cell behaviour, including migration (Pierce *et al.*, 1989), although the nature of the responsible mechanisms remains equally obscure.

In spite of this striking similarity, MSF and GBF differ from each other in terms of their respective effects upon the migration of subconfluent cells; as indicated in Figure 4, MSF has no effect upon the inherently elevated migration of subconfluent skin fibroblasts,

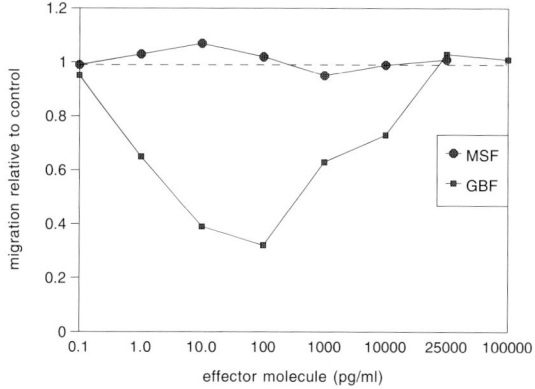

Fig. 4. Dose-response of MSF and GBF on the migration of subconfluent adult fibroblasts. Subconfluent adult fibroblasts were exposed to different concentrations of MSF and GBF and the cultures incubated for 4 days. The percentages of cells within the collagen matrix were then determined and these data expressed relative to the control (dotted line). Under these subconfluent conditions, adult fibroblasts displayed an elevated absolute level of migration in control cultures (25.6±1.6%).

whereas GBF actually *inhibits* cell migration, again at extremely low concentrations (to be discussed in further detail in Grey *et al.,* in preparation). This represents a significant difference in the biological activity of MSF and GBF, and supports the view that they are structurally related, but distinct molecules. The underlying mechanisms responsible for the diametrically opposed effects of GBF on the migration of confluent and subconfluent fibroblasts is currently under investigation and may ultimately be related to the role of fibronectin (fragments) in determining the alternative initiation and cessation of active cell motility which characterizes cell behaviour *in vivo* (Hynes, 1990).

MSF and GBF also differed in terms of their respective recognition by a recently generated rabbit polyclonal anti-MSF antibody. The antibody was raised against the Mono Q unbound preparation of MSF and purified by sequential protein A, fibronectin and GBF affinity chromatography. The partially purified IgG fraction of antibody obtained after protein A affinity chromatography recognised MSF, as well as native fibronectin and GBF in an ELISA (Figure 5); this result is consistent with the shared epitopes present in MSF and GBF. Subsequent purification of the antibody by sequential fibronectin and GBF affinity chromatography resulted in the expected elimination of antibody recognition of these two ligands in the ELISA; such affinity purified antibody still recognised MSF in a specific fashion, indicating that it contains epitopes *not* shared by either native fibronectin or GBF. Related experiments using partially degraded, but unfractionated, preparations of fibronectin confirmed that the fibronectin- and GBF-affinity purified anti-MSF antibody no longer recognised any epitopes present in fibronectin.

Interestingly, the stimulation of fibroblast migration by MSF was completely neutralised by the fibronectin- and GBF-affinity purified anti-MSF antibody at dilutions of up to 1:500, whilst considerably higher concentrations of antibody (1:10

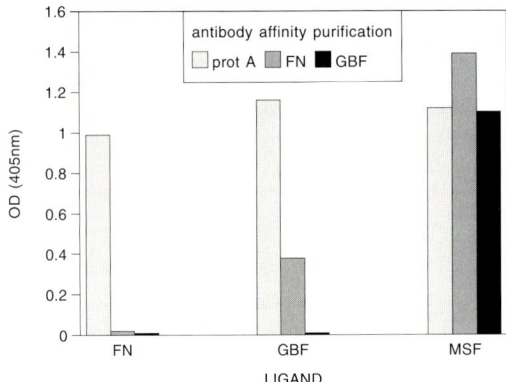

Fig. 5. The recognition of fibronectin, the gelatin-binding fragment of fibronectin and MSF by variously purified anti-MSF polyclonal antibody. Anti-MSF polyclonal antibody was purified by sequential chromatography on protein A (prot A), fibronectin (FN) and gelatin-binding fragment (GBF) affinity columns. The binding of these different antibody preparations to the different ligands was then determined by ELISA using an AP-conjugated secondary antibody. Binding is measured by the OD (405 nm).

dilution) had no effect on the comparable biological activity of GBF (Figure 6). Related data indicated that the activity of both MSF and GBF were completely neutralised by two monoclonal antibodies recognising epitopes in the gelatin binding domain, but were completely unaffected by a monoclonal antibody recognising an epitope in the cell-binding domain.

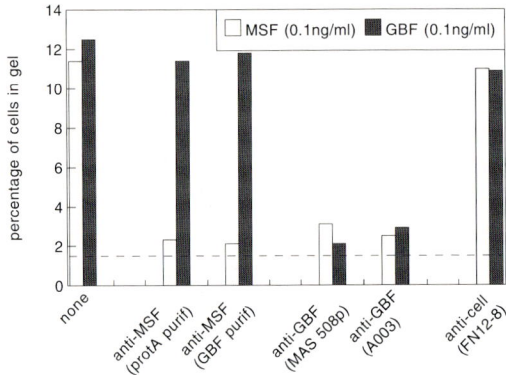

Fig. 6. The effects of various antibodies on the stimulation of fibroblast migration by MSF and GBF. Fibroblasts were co-incubated with various antibodies (indicated on the x-axis) in the presence of either MSF or GBF in the standard migration assay. Anti-MSF polyclonal antibody was used following both the protein A (protA purif) and GBF (GBF purif) affinity chromatography steps at final dilutions 1:10. In addition, two monoclonal antibodies were used which recognize epitopes present in the gelatin-binding domain of fibronectin (MAS 508p and A003 obtained from Sera Lab, Crawley Down and Bioquote Ltd, York, respectively), as was a third which recognized an epitope present in the cell-binding domain (FN12-8 obtained from Pierce and Warriner, Chester); all of these three antibodies were used at a final concentration of 1 μg/ml.

Mode of action of MSF

We have previously reported that MSF stimulates the synthesis of a high molecular weight species of hyaluronic acid (HA) (Ellis *et al.,* 1992). HA is a linear glycosaminoglycan consisting of repeating disaccharide subunits of glucuronic acid and N-acetylglucosamine. It is a major biosynthetic product of fibroblasts and has been shown to promote the migration of a number of cell types both during embryonic development (Toole and Trelstad, 1971; Markwald *et al.,* 1978) and *in vitro* (Turley and Torrance, 1984; Docherty *et al.,* 1989). The involvement of HA in modulating cell migration *in vivo* appears to continue in the adult, with various lines of evidence implicating it in the regulation of cell movement in various pathological processes, such as wound healing and tumour invasion (Knudson *et al.,* 1989).

HA is a polydisperse macromolecule exhibiting significant tissue-dependent variation in molecular mass. The biological activity of HA with respect to the control of a number of physiological and pathological processes appears to be critically dependent upon molecular mass (Feinberg and Beebe, 1983). In this regard, it is important to note that our data indicate that MSF specifically stimulates the synthesis of a high molecular mass (greater than 10^6 kDa) size-class of HA (Ellis *et al.,* 1992).

Our data indicate that the stimulation of fibroblast migration by MSF is in fact a secondary consequence of its primary effect upon HA synthesis. This was first suggested by the observation that the stimulation of both cell migration and HA synthesis by MSF follow parallel biphasic dose-response curves (Figure 7). In a related series of experiments, we noted that the stimulatory effect of MSF on cell migration was completely blocked by co-exposure of fibroblasts to *Streptomyces* hyaluronidase during the four day duration of the assay (Schor *et al.,* 1989). Finally, the addition of exogenous high molecular mass HA to control adult fibroblasts was found to induce the stimulation

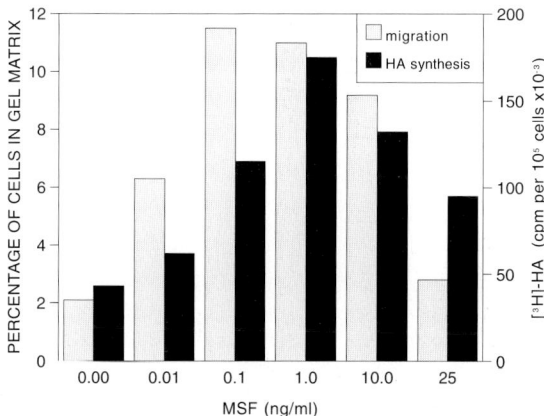

Fig. 7. The dose-response of MSF on the migration of adult fibroblasts and their synthesis of hyaluronic acid (HA). Confluent cultures of adult fibroblasts on collagen gels were exposed to different concentrations of MSF. Fibroblast migration into the collagen matrix and the incorporation of ^{3}H-glucosamine into HA was determined as indicated in Ellis *et al* (1992).

of their migration in a biphasic dose-dependent fashion (Figure 8) similar to that produced by MSF.

The stimulation of HA synthesis by MSF and the apparent subsequent effect of this HA on cell migration should caution us that the ascription of a biological function in the naming of a cytokine (eg. *migration* stimulating factor) generally reflects its activity in the particular bioassay first used in its identification rather than being a necessarily accurate description of its principal physiological function *in vivo*. In this regard, many well-characterized "growth factors" (such as PDGF and EGF) also affect a variety of aspects of cell behaviour in addition to proliferation, these including cell migration (Rosen and Goldberg, 1989). With this multi-functionality of cytokine action in mind, it should be noted that the principal biological activity of MSF *in vivo* may not relate to cell motility *per se*, but rather some other aspect of cell function, such as HA production.

In addition to modulation by the extracellular matrix, it is now clear that the biological activity of a particular cytokine is influenced by the presence of other cytokines in the microenvironment. In view of this interdependence of cytokine function, we have been particularly interested in ascertaining the interaction of MSF with other potentially relevant cytokines. In the first instance we have reported that the stimulatory effects of MSF on both cell migration and HA biosynthesis are inhibited by TGF-β1 (Figure 9, and Ellis *et al.*, 1992). Such an apparently antagonistic effect of these two cytokines may reflect their "balancing" role in the control of cell behaviour during various physiological and pathological events, such as wound healing. In this regard, MSF activity has been observed in 16/17 (94.1%) wound fluid samples (Picardo *et al.*, 1992). The directed migration of fibroblasts into the wound site and the transient increase of HA in granulation tissue during the wound healing response are both

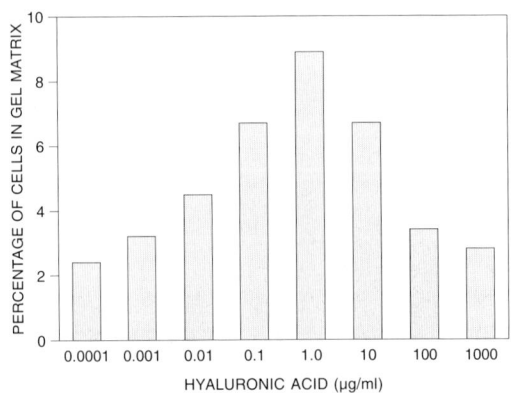

Fig. 8. The dose-response of exogenous high molecular weight HA on the migration of adult fibroblasts. Confluent adult fibroblasts were exposed to different concentrations of high molecular weight ($>10^6$ kDa) hyaluronic acid and their migration into the collagen matrix measured in the standard bioassay. Control cells incubated in the absence of exogenous MSF (dotted line) displayed a relatively low level of migration (2.1±0.2%).

consistent with the involvement of MSF. Although the source of MSF in wound fluid is not known, its absence from matched serum samples collected from the same patients suggests that it is not released from degranulating platelets nor derived from a plasma transudate.

Heterogeneity in fibroblast phenotype and the potential involvement of MSF-producing fibroblasts in cancer pathogenesis

Our recent data indicate that subpopulations of MSF-secreting fibroblasts are in fact present in the adult and that these display both inter- and intra-site heterogeneity with respect to their tissue distribution. For example, fibroblasts obtained from 15/20 (75%) of oral mucosal biopsies produced detectable amounts of MSF compared to only 2/20 (10%) of paired forearm skin fibroblasts obtained from the same individuals (Picardo *et al.,* in preparation). Interestingly, wound healing in the oral mucosa is clinically distinguished from dermal healing in terms of both its rapidity and lack of scar formation. It is possible that the presence of these MSF-producing fibroblasts contribute to this regenerative and characteristically "fetal-like" mode of wound healing.

Related studies have revealed the existence of intra-site heterogeneity in the oral mucosa with respect to the tissue distribution of MSF-producing fibroblasts. This involved the separation of gingival lamina propria (connective tissue) from its overlying epithelium by exposure to trypsin and the subsequent microdissection of the lamina propria to allow the selective culture of fibroblasts derived from the tips of the papillae and deeper reticular tissue. Only fibroblasts derived from the papillae produced MSF (Irwin *et al.,* manuscript in preparation). Prolonged subculture of papillary fibroblasts resulted in their cessation in MSF production and their adoption of a reticular fibroblast

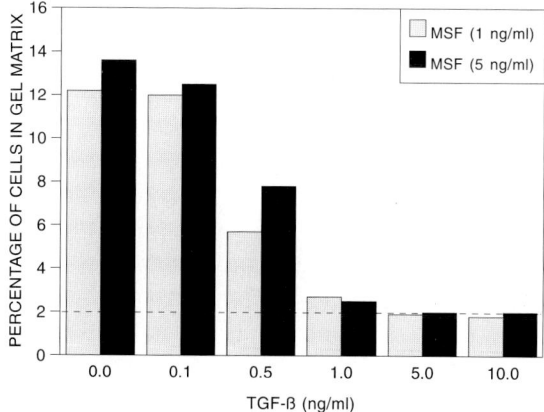

Fig. 9. The effects of TGF-β on the stimulation of fibroblast migration by MSF. Confluent adult fibroblasts were incubated with 0, 1.0 and 5.0 ng/ml MSF in the presence of various concentrations of TGF-β1 (British BioTechnology, Cowley) and the percentage of cells within the collagen matrix determined 4 days later. Control cells incubated in the absence of both MSF and TGF-β (dotted line) displayed an low level of migration (2.0±0.2%).

phenotype. Staining of gingival tissue with affinity-purified anti-MSF antibody clearly demonstrated the preferential localisation of MSF in the papillae; interestingly, this particular pattern of MSF distribution is identical to that previously described for tenascin (Sloan, Schor and Lopes, 1990), another "fetal-associated" fibroblast product (Chiquet-Ehrismann et al., 1986).

Much of our previous work has been concerned with documenting the presence of "fetal-like" (MSF-producing) fibroblasts in breast cancer patients. These studies have demonstrated that (a) tumour-derived fibroblasts obtained from approximately 50% of sporadic breast cancer patients expressed a fetal-like migratory phenotype (Durning et al., 1984; Schor et al., 1985b), (b) paired skin fibroblasts obtained from the same individuals also expressed a fetal-like migratory phenotype, thereby indicating the systemic nature of this stromal cell abnormality (Durning et al., 1984), and (c) skin fibroblasts obtained from approximately 90% of patients with familial breast cancer behaved in a similar fetal-like fashion, as did greater than 50% of their unaffected first-degree relatives (Schor et al., 1986; Haggie et al., 1987). This latter finding is of particular significance as it indicates that the systemic presence of fetal-like fibroblasts may precede the development of overt malignant disease in this population with a clearly documented elevated risk of developing breast cancer (Ottman et al., 1983).

Subsequent work indicated that the fetal-like fibroblasts obtained from breast cancer patients also produce MSF (Schor et al., 1988b). This MSF is indistinguishable from that produced by fetal fibroblasts with respect to all of the biological and biochemical parameters we have investigated to date. We have recently observed that detectable levels of MSF are present in the serum of sporadic breast cancer patients (Picardo et al., 1991). Serum was collected from two groups of patients; the first (untreated) group consisted of newly diagnosed patients (n=12) from whom serum was collected both 24 hours prior to surgical resection of the primary tumour and 4 days post-operatively, whilst the second (treated) group consisted of patients (n=14) at various times after tumour resection who had received adjuvant therapy. Serum samples were also collected from age-matched healthy controls with no family history of breast cancer (n=20). Serum samples were fractionated according to our protocol for MSF and then assessed for migration stimulating activity in our standard collagen gel assay.

Our data indicate that MSF activity was present in 10/12 (83.3%) of serum samples obtained from untreated patients prior to surgery and 9/12 (75%) of these same individuals four days post-operatively (Figure 10). Corresponding data obtained from the treated group indicated that detectable MSF activity was present in 13/14 (93%) serum samples. In marked contrast to this relatively high incidence of MSF in the patient sera, we only detected MSF activity in 2/20 (10%) of the control sera. Biochemical characterisation of the serum-derived MSF indicated that it was indistinguishable from its fibroblast-produced counterpart.

The presence of MSF in the serum of the post-operative patient group, these individuals having been free of detectable residual disease for periods of time up to 13 years, clearly distinguishes MSF from previously described onco-fetal proteins which are produced by the tumour and consequently function as markers of tumour burden. Taken together with our previous results, the observed presence of MSF in the serum of these patients may reflect the systemic and persistent presence of a population of fetal-like (MSF-producing) fibroblasts.

The detection of MSF-secreting fibroblasts in a significant proportion of breast cancer patients is consistent with published literature documenting the presence of aberrant skin and tumour-derived fibroblasts in patients with a variety of both sporadic and hereditary cancers (reviewed in Schor *et al.,* 1987); such cells have been reported to express a number of phenotypic characteristics commonly associated with transformation, such as colony formation in semi-solid medium and reduced serum requirement for growth. In the context of the present discussion, it should be noted that many of these purportedly transformation-specific characteristics are also commonly displayed by fetal cells (Nakano and Ts'O, 1981). The "fetal-like" nature of tumour-associated fibroblasts has been further indicated by Bartal *et al* (1986) who identified a subpopulation of fibroblasts in the stroma associated with several types of tumours which stain positively with an antibody recognising fetal, but not adult, cells. Similarly, Chambon and colleagues have recently reported that a proportion of fibroblasts in the stroma of breast cancers resemble fetal cells in terms of their production of stromelysin-3, a novel protease not apparently produced by normal adult fibroblasts (Basset *et al.,* 1990).

We have suggested that these "fetal-like" fibroblasts in the cancer patients contribute directly to tumour progression by virtue of their perturbation of normal epithelial-mesenchymal interactions (Schor *et al.,* 1987). Independent experimental support of this possibility has been provided by Sakakura (1983); she reported that implantation of fetal (but not adult) fibroblasts into the adult rat mammary gland induced the hyperplastic growth of the epithelial elements and rendered these exquisitely sensitive to overt transformation by carcinogenic agents. Fetal-like fibroblasts may facilitate tumour progression by creating a milieu which promotes the clonal expansion and invasive

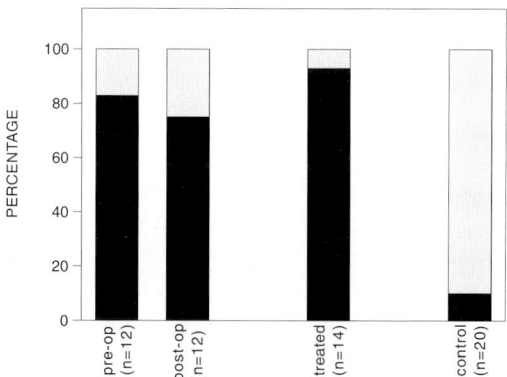

Fig. 10. The presence of detectable levels of MSF in the serum of both treated and untreated breast cancer patients. The presence of MSF activity in the serum of breast cancer patients and age-matched controls was determined using the collagen matrix bioassay. The percentage of each patient group containing detectable MSF activity is indicated by the black bar and the percentage devoid of activity by the stacked grey bar. Details regarding patients and experimental procedures may be found in Picardo *et al* (1991).

behaviour of the emerging neoplastic cell population. The MSF produced by these fibroblasts may contribute to these processes, perhaps as a consequence of its effect on HA synthesis. Various studies have noted elevated levels of HA associated with the stroma of different types of tumours and demonstrated that this is correlated with more aggressive invasive behaviour (Toole, Biswas and Gross, 1979). HA has also been reported to modulate a number of other processes of potential relevance to tumour progression, including the proliferation of mammary epithelial cells (Elstad and Hosick, 1987) and angiogenesis (Feinberg and Beebe, 1983).

The apparent involvement of MSF in both cancer pathogenesis and the wound healing response may be relevant to the observation of Dvorak (1986) that the tumour-host interface resembles a "wound that does not heal". In this context, the same effects of MSF which contribute positively to wound healing when expressed *locally* and in a *transient* manner may facilitate cancer progression when they are *systemic* and/or *prolonged* in nature. This postulated direct involvement of fetal-like (MSF-producing) fibroblasts in cancer progression is fundamentally different from previous "non-involvement" interpretations; according to these models, the expression of transformation-associated phenotypic characteristics by cancer patient fibroblasts was regarded merely as a marker of a genetic lesion which only contributed to cancer development when expressed by the relevant target cell population.

We have proposed a "clonal modulation" model to account for the origin of fetal-like fibroblasts in cancer patients (Schor and Schor, 1987). This model is based on our observations relating to the existence of heterogeneity in fibroblast phenotype and suggests that there is an inappropriate and long-term clonal expansion of MSF-producing fibroblasts in these individuals in response to as yet unidentified internal or environmental stimuli. According to this epigenetic model, the MSF-secreting fibroblasts detected in cancer patients are not considered to be intrinsically aberrant, but rather an expanded subpopulation of cells also present in the healthy adult. It is also possible that the MSF-secreting fibroblasts of breast cancer patients result from the persistence of such "fetal-like" cells into adult life.

Conclusions and prospectives for future studies

Our present data indicate that MSF is a novel gene product containing a region of amino acid sequence homology with the gelatin-binding domain of fibronectin. Ongoing studies are concerned with (a) the further molecular characterization of MSF, including determination of its complete amino acid sequence, (b) characterising the cell surface receptor for MSF, and (c) quantitating the serum levels of MSF in breast cancer patients and correlating these with various clinically relevant parameters relating to risk. This information may provide further insight into the involvement of MSF in the complex process of cancer progression and ultimately lead to the development of novel therapeutic modalities targeted at normalizing epithelial-mesenchymal interactions.

The observation that picomolar concentrations of the gelatin-binding fragment of fibronectin affects cell migration migration in the collagen matrix assay was totally unexpected and opens up an entirely new direction of study concerned with identification of the active amino acid motif in this domain and characterization of its putative cell surface receptor.

This work was supported by grants from the Cancer Research Campaign and Christie Hospital Endowment Fund.

References

Aznavoorian, S., Stracke, M.L., Krutzsch, H., Schiffmann, E. and Liotta, L.A. (1990) Signal transduction for chemotaxis and haptotaxis by matrix molecules in tumor cells. *J. Cell Biol.* **110**, 1427-1438

Bartal, A.H., Lichtig, C., Cardo, C.C., Feit, C., Robinson, E. and Hirshaut, Y. (1986) Monoclonal antibody defining fibroblasts appearing in fetal and neoplastic tissue. *J. natl. Cancer Inst.* **76**, 415-419

Basset, P., Bellocq, J.P., Wolf, C., Stoll, I., Hutin, P., Limacher, J.M., Podhajcer, O.L., Chenard, M.P., Rio, M.C. and Chambon, P. (1990) A novel metalloproteinase gene specifically expressed in stromal cells of breast carcinomas. *Nature* **348**, 699-704

Bray, D. (1992) Cell Movements, Garland Publishing, Inc, (New York)

Chiquet-Ehrismann, R., Mackie, E.J., Pearson, C.A. and Sakaura, T. (1986) Tenascin: an extracellular matrix protein involved in tissue interactions during fetal development and oncogenesis. *Cell* **47**, 131-139

Docherty, R., Forrester, J.V., Lackie, J.M. and Gregory, D.W. (1989) Glycosaminoglycans facilitate the movement of fibroblasts through 3D collagen matrices. *J. Cell Sci.* **92**, 263-267

Durning, P., Schor, S.L. and Sellwood, R.A.S. (1984) Fibroblasts from patients with breast cancer show abnormal migratory behaviour *in vitro*. *Lancet* **ii**, 890-892

Dvorak, H.F. (1986) Tumors: wounds that do not heal. *New Eng. J. Med* **315**, 1650-1659

Ellis, I., Grey, A.M., Schor, A.M. and Schor, S.L. (1992) Antagonistic effects of transforming growth factor beta and MSF on fibroblast migration and hyaluronic acid synthesis: possible implications for dermal wound healing. *J. Cell Sci.* **102**, 447-456

Elstad, C.A. and Hosick, H.L. (1987) Contribution of the extracellular matrix to the growth properties of cells from a preneoplastic outgrowth: possible role of hyaluronic acid. *Exp. Cell Res.* **55**, 313-321

Feinberg, R.N. and Beebe, D.C. (1983) Hyaluronate in vasculargenesis. *Science* **220**, 1177-1179

Gospodarowicz, D., Greenburg, G. and Birdwell, C.R. (1978) Determination of cellular shape by the extracellular matrix and its correlation with the control of cellular growth. *Cancer Res.* **38**, 4155-4171

Grey, A.M., Schor, A.M., Rushton, G., Ellis, I. and Schor, S.L. (1989) Purification of the migration stimulating factor produced by fetal and breast cancer patient fibroblasts. *Proc. Nat. Acad. Sci. (USA)* **86**, 2438-2442

Haggie, J., Schor, S.L., Howell, A., Birch, J.M. and Sellwood, R.A.S. (1987) Fibroblasts from relatives of hereditary breast cancer display foetal-like behaviour *in vitro*. *Lancet* **i**, 1455-1457

Humphries, M.J., Mould, A.P. and Yamada, K.M. (1991) Matrix receptors in cell migration. In: Receptors for Extracellular Matrix, pp. 195-253 (Academic Press: San Diego)

Hynes, R. (1990) Fibronectins. Springer-Verlag (New York)

Ingber, D.E. and Folkman, J. (1989) Mechanochemical switching between growth and differentiation during fibroblast growth factor-stimulated angiogenesis in vitro: role of extracellular matrix. *J. Cell Biol.* **109**, 317-330

Knudson, W., Biswas, C., Li, X.-Q., Nemec, R.E. and Toole, B.P. (1989) The role and regulation of tumor-associated hyaluronan. In: *The Biology of Hyaluronan*, Ciba Foundation Symposium 143, pp. 150-159 (John Wiley and Sons: Chichester)

Markwald, R.R., Fitzharris, T.P., Bank, H. and Bernanke, D.H. (1978) Structural analysis on the matrical organization of glycosaminoglycans in developing endocardial cushions. *Devl. Biol.* **62**, 292-316

Nakano, S. and Ts'O, P.O. (1981) Cellular differentiation and neoplasia: characterization of subpopulations of cells that have neoplasia related growth properties in Syrian hamster embryo cell cultures. *Proc. natl. Acad. Sci. (USA)* **78**, 4995-4999

Nathan, C. and Sporn, M. (1991) Cytokines in context. *J Cell Biol* **113**, 981-986

Ottman, R., King, M.C., Pike, M.C. and Henderson, B.E. (1983) Practical guide for estimating risk for familial breast cancer. *Lancet* **ii**, 556-558

Picardo, M., Schor, S.L., Grey, A.M., Howell, A., Laidlow, I., Redford, J. and Schor, A.M. (1991) Migration stimulating activity in serum of breast cancer patients. *Lancet* **337**, 130-133

Picardo, M., Grey, A.M., McGurk, M. Ellis, I. and Schor, S.L. (1992) Identification of migration stimulating factor in wound fluid. *Exp Mol. Path.* **57**, 8-21

Pierce, C.F., Mustoe, T.A., Lingelbach, J., Masakowski, V.R., Griffin, G.L., Senior, R.M. and Deuel, T.F. (1989) Platelet derived growth factor and transforming growth factor beta enhance tissue repair activities by unique mechanisms. *J. Cell Biol.* **109**, 429-499

Postlethwaite, A.H., Keski-Oja, J., Balian, G. and Kang, A.H. (1981) Induction of fibroblast chemotaxis by fibronectin. Localization of the chemotactic region to a 140,000 molecular weight non-gelatin-binding fragment. *J. Exp. Med.* **153**, 494-499

Postlethwaite, A.H., Keski-Oja, J., Moses, H.L. and Kang, A. (1987) Stimulation of chemotactic migration of human fibroblasts by transforming growth factor β. *J. Exp. Med.* **165**, 251-256

Rogelj, S., Klagsbrun, M., Atzmon, R., Kurokawa, M., Haimovitz, A., Fuks, Z. and Vlodavsky, I. (1989) Basic fibroblast growth factor is an extracellular matrix component required for promoting the proliferation of vascular endothelial cells and the differentiation of PC12 cells. *J. Cell Biol.* **109**, 823-831

Rosen, E.M. and Goldberg, I.D. (1989) Protein factors which regulate cell motility. *In Vitro Cellul. Devel. Biol.* **25**, 1079-1087

Sakakura, T. (1983) Epithelial-mesenchymal interactions in mammary gland development and its perturbation in tumorigenesis. In: *Understanding Breast Cancer* (eds. MA Rich, JC Hager and P Furmanski) pp 261-284 (Marcel Dekker, Inc: New York)

Schor, S.L. (1980) Cell proliferation and migration within three-dimensional collagen gels. *J. Cell Sci.* **41**, 159-175

Schor, S.L. and Schor, A.M. (1987) Foetal-to-adult transitions in fibroblast phenotype: their possible relevance to the pathogenesis of cancer. *J Cell Sci* **Suppl. 8**, 165-180

Schor, S.L., Schor, A.M., Rushton, G. and Smith, L. (1985a) Adult, foetal and transformed fibroblasts display different migratory phenotypes on collagen gels. *J. Cell Sci.* **73**, 221-234

Schor, S.L., Schor, A.M., Durning, P. and Rushton, G. (1985b) Skin fibroblasts obtained from cancer patients display foetal-like migratory behaviour on collagen gels. **J. Cell Sci. 73**, 235-244

Schor, S.L., Haggie, J., Durning, P., Howell, A., Sellwood, R.A.S. and Crowther, D. (1986) The occurrence of a foetal fibroblast phenotype in breast cancer. *Int. J. Cancer* **37**, 831-836

Schor, S.L., Schor, A.M., Howell, A. and Crowther, D. (1987) Hypothesis: persistent expression of fetal phenotypic characteristics by fibroblasts is associated with an increased susceptibility to neoplastic disease. *Exp. Cell Biol.* **55**, 11-17

Schor, S.L., Schor, A.M., Grey, A.M. and Rushton, G. (1988a) Foetal and cancer patients fibroblasts produce an autocrine migration stimulating factor not made by normal adult cells. *J. Cell Sci.* **90**, 391-399

Schor, S.L., Schor, A.M. and Rushton (1988b) Fibroblasts from cancer patients display a mixture of both foetal and adult-like phenotypic characteristics. *J. Cell Sci.* **90**, 401-407

Schor, S.L., Schor, A.M., Grey, A.M., Chen, J., Rushton, G. and Ellis, I. (1989) Mechanism of action of the migration stimulating factor produced by fetal and cancer patient fibroblasts: effect on hyaluronic acid synthesis. *In Vitro* **25**, 737-746

Seppä, H.E., Yamada, K.M., Seppä, S.T., Silver, M.H., Kleinman, H.K. and Schiffman, E. (1981) The cell binding fragment of fibronectin is chemotactic for fibroblasts. *Cell. Biol. Int. Rep.* **5**, 813-819

Sloan, P., Schor, S.L. and Lopes, V. (1990) Immunohistochemical study of the heterogeneity of tenascin distribution within the oral mucosa of the mouse. *Archs. oral Biol.* **35**, 67-70

Stoker, M. and Gherardi, E. (1991) Regulation of cell movement: the motogenic cytokines. *Biochim. Biophys. Acta.* **1072**, 81-102

Sutton, A. M., Canfield, A. E., Schor, S. L., Grant, M. E. and Schor, A. M. (1991). The response of endothelial cells to TGF-β1 is dependent upon cell shape, proliferative state and the nature of the substratum. *J. Cell Sci.* **99**, 777-787

Toole, B.P., Biswas, C. and Gross, J. (1979) Hyaluronate and invasiveness of the rabbit V2 carcinoma. *Proc. natl. Acad. Sci. (USA)* **76**, 6299-6303

Toole, B.P. and Trelstad, R.L. (1971) Hyaluronate production and removal during corneal development in the chick. *Devl. Biol.* **26**, 28-35

Turley, E.A. and Torrance, J. (1984) Localization of hyaluronate and hyaluronate-binding protein on motile and non-motile fibroblasts. *Expl. Cell Res.* **161**, 17-28

Zardi, L., Carnemolla, B., Balza, E., Borsi, L., Castellani, P., Rocco, M. and Siri, A. (1985) Elution of fibronectin proteolytic fragments from a hydroxyapatite column. *Eur. J. Biochem* **146**, 571-579

Zetter, B.R. and Brightman, S.E. (1990) Cell motility and the extracellular matrix. *Curr. Opin. Cell Biol.* **2**, 850-856

Zigmond, S. and Hirsch, J.G. (1973) Leukocyte locomotion and chemotaxis. *J. Exp. Med.* **137**, 387-410

Printed in Great Britain © The Society of Experimental Biology 1993   253

DYNAMIC ASPECTS OF MICROFILAMENT-MEMBRANE ATTACHMENTS

B.M. JOCKUSCH*, CH. WIEGAND, C.J. TEMM-GROVE and G. NIKOLAI

Cell Biology Group, University of Bielefeld, W-4800 Bielefeld 1, Germany

Summary

Microfilament-membrane attachment sites are complex structures that are essential for tissue differentiation in animals. In this article, we focus on the assembly and dynamics of such contact sites as seen in two cell types differentiating in cultures of the embryonic chicken heart, cardiocytes and fibroblasts. Concentrating on the cytoplasmic domain, we refer to previous biochemical, light, and electron microscopic studies on the structure and dynamics of these regions and supplement them with our own recent data. Although many details are still to be elucidated, we would like to propose the following model. Actin, α-actinin and vinculin are the major structural components of all microfilament-membrane contacts. Various subtypes of junctions are characterised by additional structural components or by specific isoforms. Temporal regulation of contact sites is linked to assembly and disassembly of microfilaments and might be controlled by special regulatory proteins. Finally, the cytoplasmic domains of junctional complexes may serve as structural matrices for the positioning of proteins involved in signal transduction pathways.

Introduction

Tissue formation in the vertebrate organism involves cellular differentiation, cell-cell recognition and the establishment of well defined cell-cell and cell-matrix contacts. During the last twenty years, some progress in understanding these events on a molecular level has been made. In general, it was found that the interaction of actin-based microfilaments with the plasmalemma is an essential step. Light microscopic analysis of living cells in tissue culture that are amenable to manipulation has given insight into the dynamics of microfilament-membrane contacts, and labeling with specific antibodies has revealed a catalogue of proteins that are common to all these junctions and some that are specific for certain types. Yet, neither all the molecular components nor the exact spatial and temporal mechanisms of their assembly into such junctional complexes are known today.

In this article, we report on microfilament-plasma membrane contact sites assembled in cardiocytes and fibroblasts, emphasizing compositional differences in morphologically similar junctions. These differences in protein composition at the cytoplasmic face of the attachment site can be correlated with differential functions.

*Author for correspondence.

Key words: junctional complexes, adherens junctions, costameres, cardiocytes, fibroblasts.

Development and morphology of microfilament-membrane attachment sites in cultured cells

When seeded from suspension or migrating out of an explant, tissue-forming vertebrate cells attach to the culture dish, spread, and develop several types of contacts with their environment. Those contacts that apparently involve a chain of molecular interactions between extracellular molecules, integral membrane proteins and a compact bundle of actin filaments have been classified as "adherens junctions". This term implies the involvement of these structures in cell attachment (for reviews, see Geiger et al., 1987; Geiger and Ginsberg, 1991). Depending on the nature of the extracellular partner, adherens junctions as well-defined, structural entities are divided into those that link cells with the substratum, i.e. the extracellular matrix (cell-substratum contacts) and those that are established between cells (cell-cell contacts). The latter are only developed in certain cell types, such as epithelial and cardiac cells.

Both types of adherens junctions are subject to cell- and developmental stage-specific modifications. This can easily be observed in co-cultures of embryonic cardiac and fibroblastic cells, as obtained from embryonic chicken heart. Numerous studies on the development and morphology of junctional complexes in these cell types have been published, whose main results are briefly summarised in the following.

Cardiac cells, as obtained from trypsin-digested, 7-10 day old chicken embryonic heart, attach to the substratum, spread and resorb most of the preexisting myofibrils. During their period of redifferentiation (which may last several days), these cells do not divide, but express stress fibre-like microfilament bundles and also new myofibrils that gradually increase in length and thickness from smaller precursors (cf. Sanger, 1977; Sanger et al., 1984; Lin et al., 1989; Schultheiss et al., 1990). The structural relationship between the stress fibre-like, nonstriated bundles and the striated, sarcomeric myofibrils is still a matter of debate. The main issue is the question of whether the stress fibre-like structures containing nonmuscle isoforms of microfilament proteins can serve as scaffolds for nascent myofibrils, or whether those are built de novo from freshly synthesized muscle-specific isoforms. There is evidence that the former is indeed the case for at least a subset of differentiating myofibrils (for the controversal discussion, see Dlugosz et al., 1984; Antin et al., 1986; Sanger et al., 1986, 1989; Handel et al., 1991; Hu et al., 1992). It is, however, generally agreed that the cardiocytes, once mature, display numerous prominent, spontaneously contracting myofibrils and no or very few stress fibre-like structures. With respect to microfilament-membrane attachment sites, one can easily recognise by light and electron microscopy the areas of anchorage between the terminal portions of myofibrils and the ventral sarcolemma, which are co-localised with the contact sites between the sarcolemma and the substratum (cardiac cell-substratum-junctions, CSJ) and those where myofibrils attach to cell-cell-contact sites (intercalated discs, ICD). The development and protein content of these structures has recently been analysed extensively by Hu et al. (1992).

In addition, however, there is evidence that at least some of the I-Z-I bands within the myofibrils also make contact with the sarcolemma. In sections of heart tissue, such junctions between the subsarcolemmal myofibrils and the cell membrane are regular, periodical elements aligned in a two-dimensional lattice which links all peripheral Z-discs to the plasma membrane. These connections have been termed costameres (Pardo *et al.*, 1983a, 1983b; Koteliansky and Gneushev, 1983). In cultured cardiocytes, costamere-like structures (Z-disc junctions, ZDJ) are also expressed, but only within some regions of a given subsarcolemmal myofibril. This may be due either to differences in geometry between cardiocytes in situ and in culture (the latter having their myofibrils much less exactly aligned in parallel with the plasma membrane), or to a less mature state of differentiation aquired in culture. CSJ, ICD and ZDJ comprise the most obvious microfilament-membrane contacts seen in cultured cardiocytes.

Primary chicken heart fibroblasts, obtained as a byproduct in the same cultures, have been the object of numerous studies of microfilament-membrane attachment sites initiated by M. Abercrombie and coworkers, mainly in their attempt to elucidate the role of such structures in cell locomotion (for recent review, see Heath and Holifield, 1991). After several days in culture, however, most of these fibroblasts are non-motile, well spread and demonstrate prominent stress fibres that terminate in spear-shaped cell-substratum junctions (fibroblastic CSJ; cf. Heath and Dunn, 1978). These are morphologically similar to the cardiac CSJ, and are the most obvious contact sites in fibroblasts. Fig. 1 gives a schematic view on the four different junctional complexes expressed in these two cell types.

Structural proteins of the cytoplasmic domains

There are protein constituents common to all four junctional complexes described above. Among these are actin filaments and the already "classical" microfilament proteins α-actinin and vinculin. Fig. 2 gives an example of the co-localisation of α-actinin and vinculin in the cardiac junctions. In analogy to biochemical data obtained with isolated proteins, it is inferred that at the junctional site α-actinin crosslinks the actin filaments, thus stabilising them in a position suitable for interaction with the cytoplasmic face of the membrane. In addition, α-actinin binds to integral membrane proteins at CSJ (integrins, Otey *et al.*, 1990) and intercellular adhesion molecules at cell-cell contact sites (ICAM-1, Carpén *et al.*, 1992), as well as to vinculin (Belkin and Koteliansky, 1987; Pavalko and Burridge, 1991). Moreover, our own studies have demonstrated a direct interaction between actin filaments and vinculin (Jockusch and Isenberg, 1981a, 1981b; Isenberg *et al.*, 1982; Westmeyer *et al.*, 1990) that is located in the rod-like C-terminal portion of the molecule (Menkel *et al.*, manuscript in preparation). There is also evidence that α-actinin and vinculin can interact directly with phospholipids (reviewed in Isenberg, 1991). Thus, all junctional complexes considered here have common structural elements that are interconnected. Multiple interactions of moderate or even weak affinities between actin, α-actinin and vinculin may be advantageous for the assembly of the cytoplasmic domains of junctions with high precision, reasonable stability, and in relation to regulatory mechanisms.

However, upon looking closer, one finds that even these three common structural components are not identical in all these junctions. For example, even the morphologically very similar CSJ in mature cardiocytes and fibroblasts (Fig. 1) differ

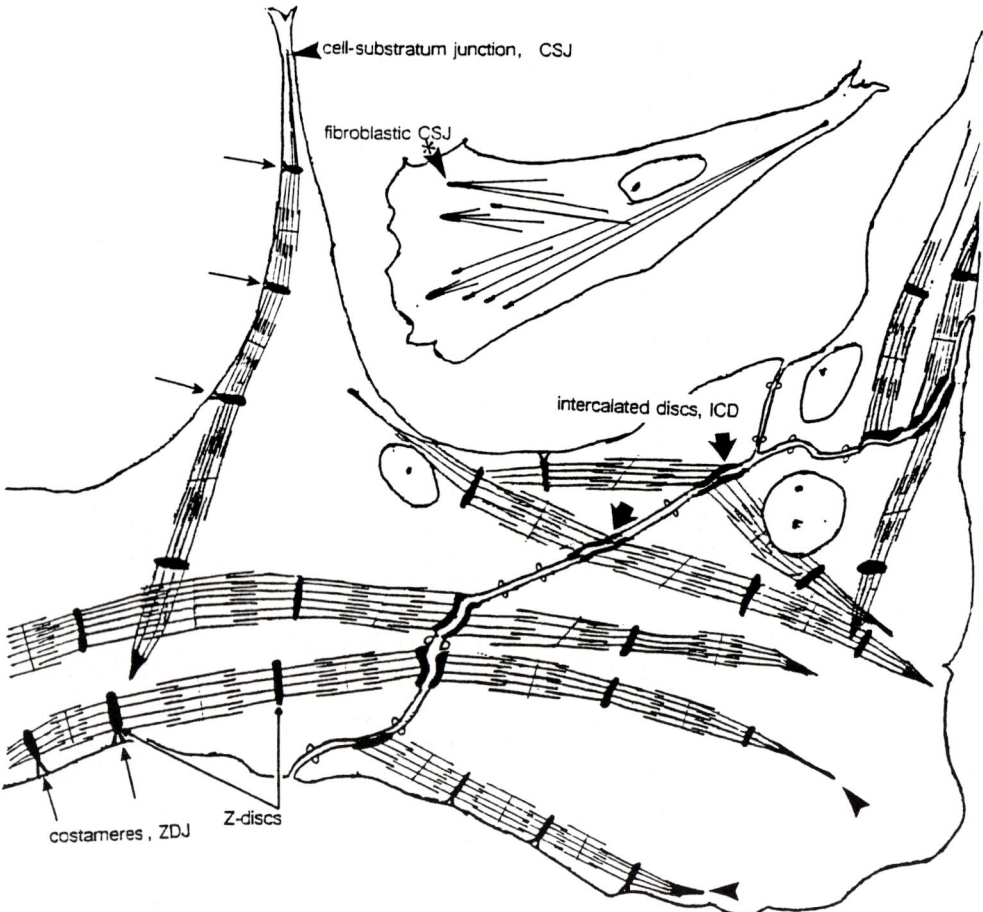

Fig. 1. Schematic drawing of the different types of microfilament-membrane attachment sites, as seen in cardiocytes and fibroblasts grown in culture from an explant of chicken embryonic heart. The cardiocytes express three types of contacts: cell-substratum junctions (CSJ) that serve as anchoring points for myofibrils, cell-cell contacts (intercalated discs, ICD) and, occasionally, junctional complexes between Z-lines and the cardiolemma (costameres, ZDJ). The fibroblasts develop CSJ that connect stress fibres with the ventral membrane. However, in a locomoting fibroblast, as depicted here, not all fibroblastic CSJ are equal with respect to their molecular composition (see text).

slightly: The heart cells use exclusively the muscle-specific actin and α-actinin isoforms for anchoring their myofibrils, as can be demonstrated with specific antibodies (Hu *et al.*, 1992; Nikolai *et al.*, 1992). Remarkably, this is not so for vinculin: There is only one isoform found in all microfilament-based junctional complexes and cell types so far.

There are also protein components that are present either in cell-substratum or cell-cell contacts. Talin, which shows multiple interactions with actin, vinculin and members of the integrin transmembrane receptor family is exclusively found at CSJ, and so is paxillin, another vinculin-binding protein (Burridge and Connell, 1983; Turner *et al.*, 1990). Corresponding vinculin partners at ICD (or other cell-cell contact sites) and ZDJ,

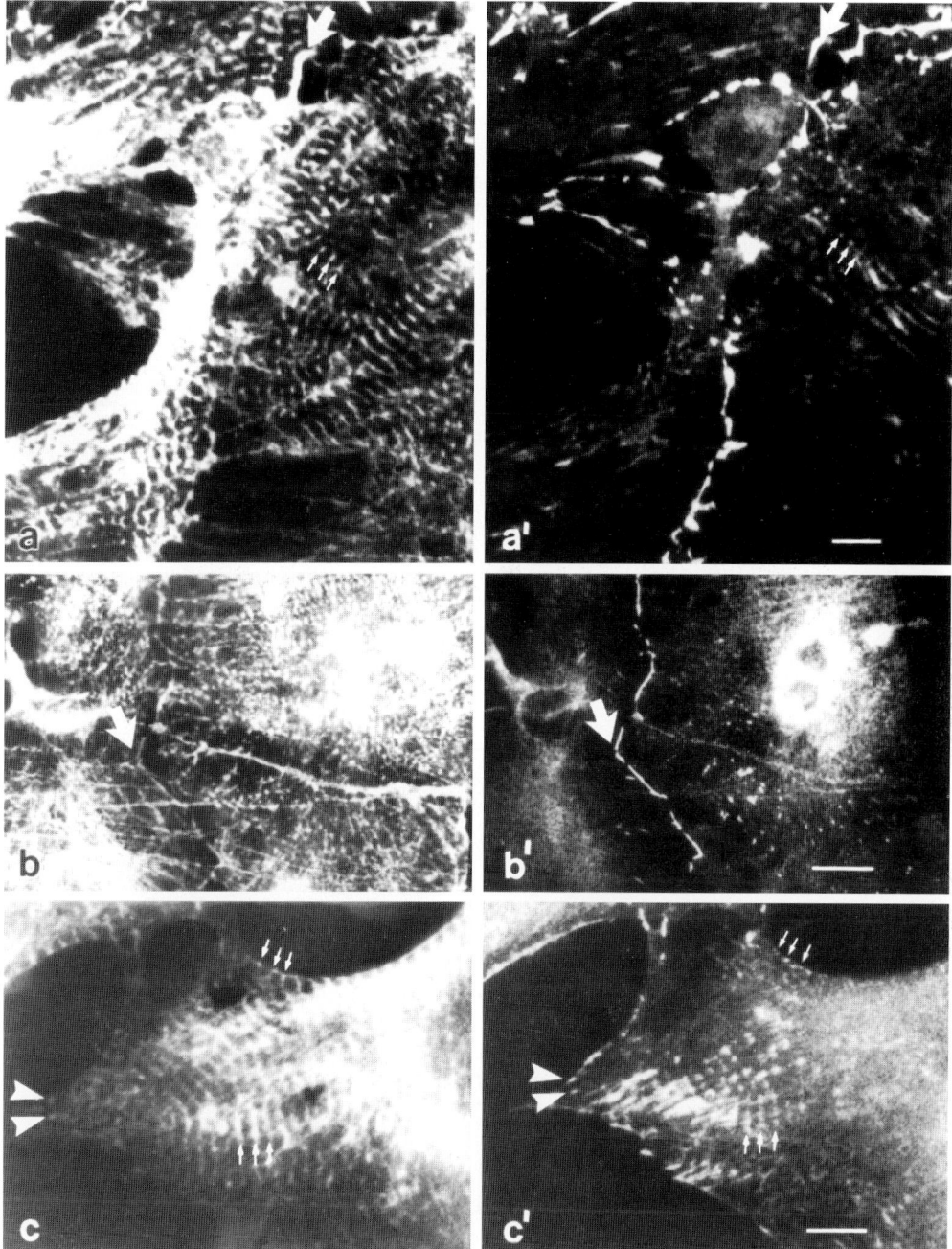

Fig. 2. Chicken embryonic cardiocytes double labelled for α-actinin (a-c) and vinculin (a'-c'). CSJ (a, a' and c, c', arrowheads) and ICD (a, a' and b, b', large arrows) contain both proteins, and vinculin and α-actinin positive structures are seen occasionally within myofibrils, indicating the existence of costameres (a, a' and c, c', small arrows). Note, however, that most of the Z-lines lack vinculin staining. Bars: 10 μm.

Corresponding vinculin partners at ICD (or other cell-cell contact sites) and ZDJ, however, have not been identified.

In addition to blueprints specifying the type of junction and the organisation of their structural constituents, there must be factors regulating their dynamics. These might involve modifications of the bulk components that would change the interaction between the different components, possibly by strengthening or weakening the various affinities. There are some data indicating that such mechanisms exist. Vinculin, for example, can be subject to covalent addition of lipids (Burn and Burger, 1987), fatty acylation (Kellie and Wigglesworth, 1987) and phosphorylation by a kinase that is also localised at CSJ (Shriver and Rohrschneider, 1981). All three modifications might be envisioned to alter vinculin conformation and affinity for partner molecules.

Regulatory proteins involved in junctional dynamics

There is also increasing evidence that the junctional complexes harbour microfilament-associated components that have primarily a regulatory function. One of these proteins is zyxin. Like α-actinin and vinculin, this protein was originally isolated from chicken gizzard and detected in CSJ of stationary fibroblasts by immunocytochemistry (Beckerle, 1986). More recently, this 82 kD protein has also been described as a component of the apical cell-cell contacts in pigmented retinal epithelial cells, with the help of a polyclonal antibody (Crawford and Beckerle, 1991). Biochemical analyses of the purified protein have demonstrated its interaction with α-actinin (Crawford et al., 1992). We have characterised a monoclonal antibody, produced against fractionated gizzard proteins, as zyxin-specific by the following criteria: The antigen was indistinguishable from zyxin in its gel chromatographic behaviour, location on 2D gels, proteolytic fragmentation patterns, and localisation in fibroblastic cells. Fig. 3 shows the co-localisation of vinculin and zyxin at CSJ in cardiac fibroblasts. It is obvious that each vinculin containing contact site is also positive for zyxin. However, as there is much less zyxin found in cell homogenates than vinculin, and because zyxin contains multiple phosphorylation sites, it has been suggested that zyxin is involved in the regulation and dynamics of junctional complexes, rather than being a major structural component (Crawford and Beckerle, 1991; Crawford et al., 1992). This hypothesis is strengthened by our observation that zyxin is only present in a subset of contact sites. Fig. 4 demonstrates that the antibody described above does not detect zyxin in cardiocyte CSJ, and also not in ICD or ZDJ (not shown). Thus, zyxin is either not expressed in cardiocytes, or not directed towards contact sites in these cells.

A more complicated situation holds for another putative regulatory protein, tensin. This is a high molecular weight (200 kD) component which is thought to be involved in junctional dynamics, because it contains a domain common to many signal transduction proteins (Davis et al., 1991). Polyclonal antibodies against several proteolytic fragments of tensin were found to label tensin in fibroblastic CSJ (Wilkins et al., 1986; Davis et al., 1991), but also to decorate cardiocyte CSJ, ICD and all Z-lines, including ZDJ (Wilkins et al., 1986). Recently, we obtained a monoclonal antibody against chicken gizzard insertin, a protein almost identical with a 350 amino acid internal region of the 1733 residue tensin

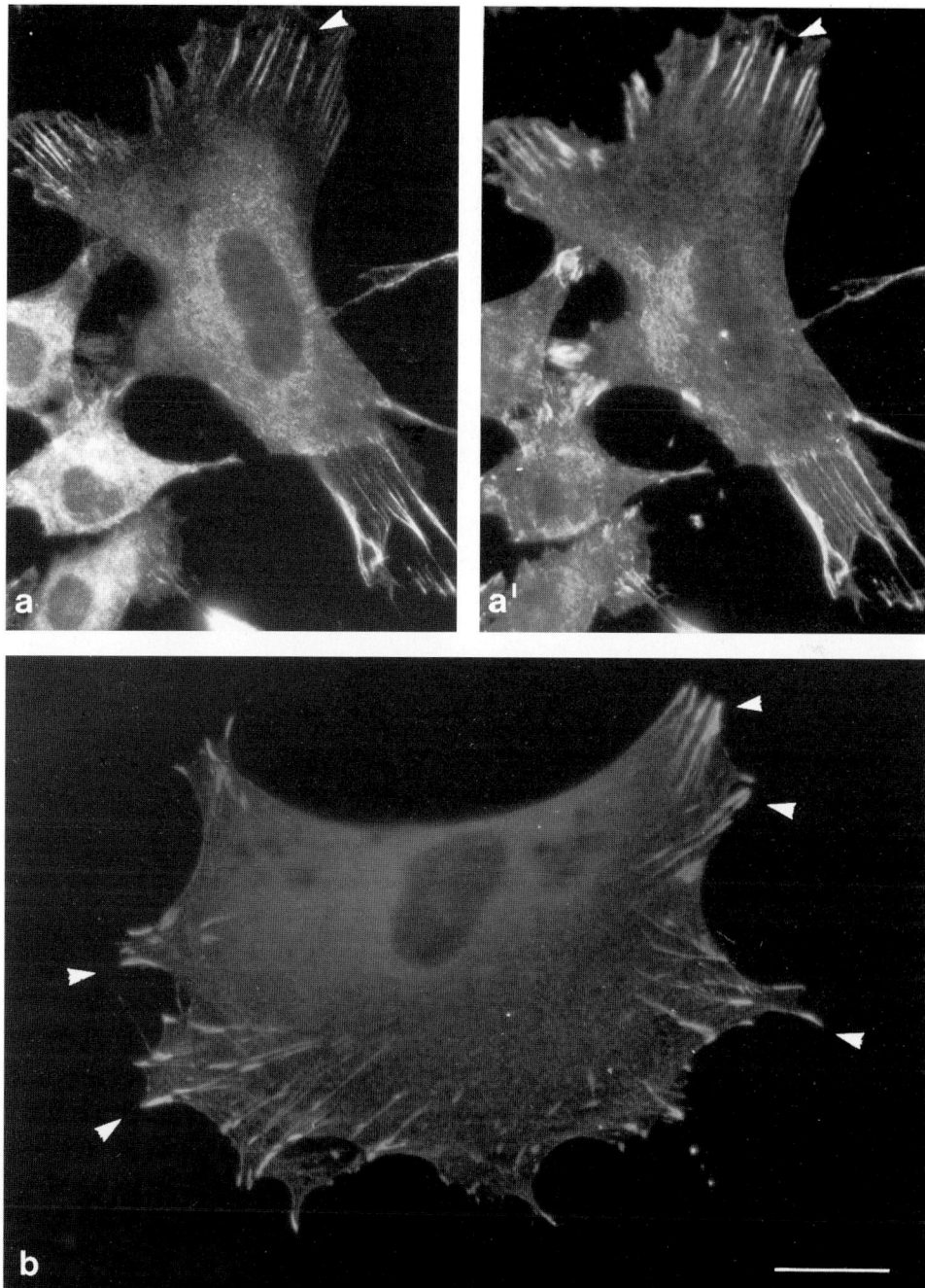

Fig. 3. Localization of zyxin in chicken embryonic fibroblasts. (a. a') Double label with antibodies to zyxin (a) and vinculin (a'). Note that in this stationary fibroblast there is an exact correspondence between vinculin- and zyxin-positive CSJ (arrowheads). (b) A well spread cell showing zyxin localized in many focal adhesions. Bar: 10 μm.

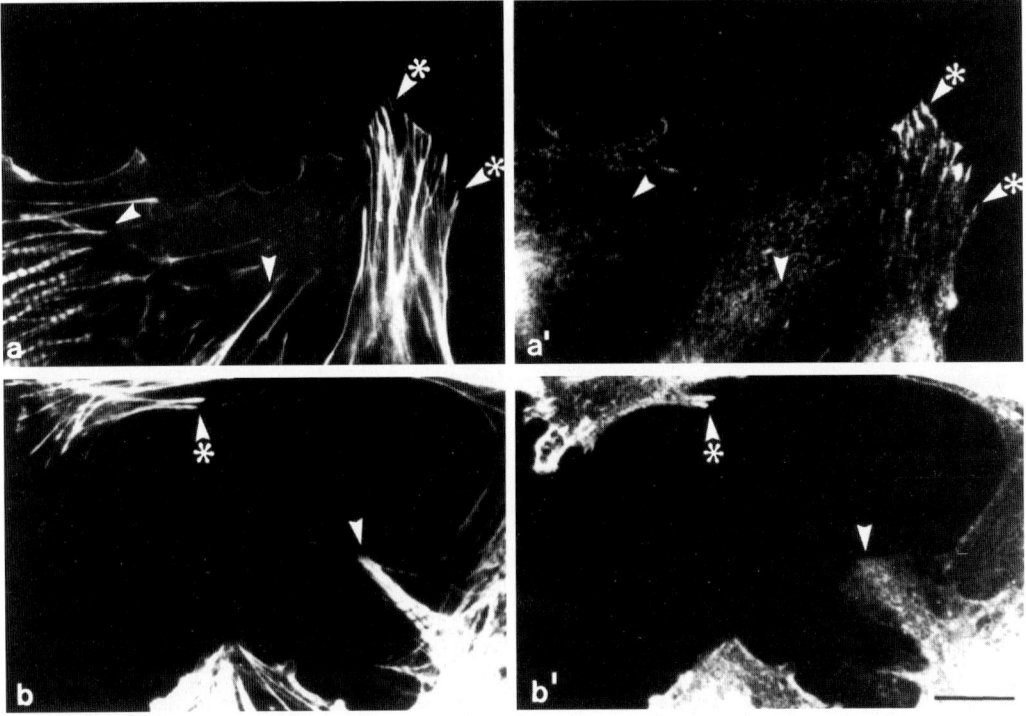

Fig. 4. Cardiocytes and fibroblasts double labelled for actin with FITC-phalloidin (a, b) and zyxin (rhodamine, a' b'). Note that the cardiocyte CSJ (arrowheads) are not decorated by antizyxin, while the corresponding structures in fibroblasts (arrowheads/asterisks) contain this protein. Bar: 10 μm.

polypeptide chain (Weigt *et al.*, 1992). Insertin had been shown to bind strongly to those ends of actin filaments that are apposed to the plasma membrane (Ruhnau *et al.*, 1989). At present, the relationship between tensin and insertin is not clear. One possibility, of course, is that insertin is a proteolytic breakdown product of tensin, harbouring an actin-binding domain. However, the immunofluorescence patterns obtained with our monoclonal anti-insertin hint at an alternative explanation. In contrast to the polyclonal antibodies against several genuine tensin fragments mentioned above, it does not decorate any junctional complexes in cardiocytes, but selectively outlines fibroblastic CSJ (Fig. 5). Thus, anti-insertin shows the same preference as anti-zyxin with respect to the various contact sites, but there are distinct differences: The insertin-labelled "spear tips" at the fibroblastic CSJ appear finer and longer than the zyxin-labelled structures (cf. fibroblastic CSJ in Fig. 4 a', b' with Fig. 5 a', b'). The selectiveness of anti-insertin for fibroblastic junctions, as compared to the nonselective labelling found with polyclonal antibodies against tensin fragments may suggest that insertin is a discrete splicing product expressed in fibroblasts and smooth muscle, rather than a proteolytic breakdown product. The finding that the N-terminus of isolated insertin is blocked (A. Wegner, personal communication) supports this model.

The existence of components specific for fibroblast cell-substratum adhesion sites, like

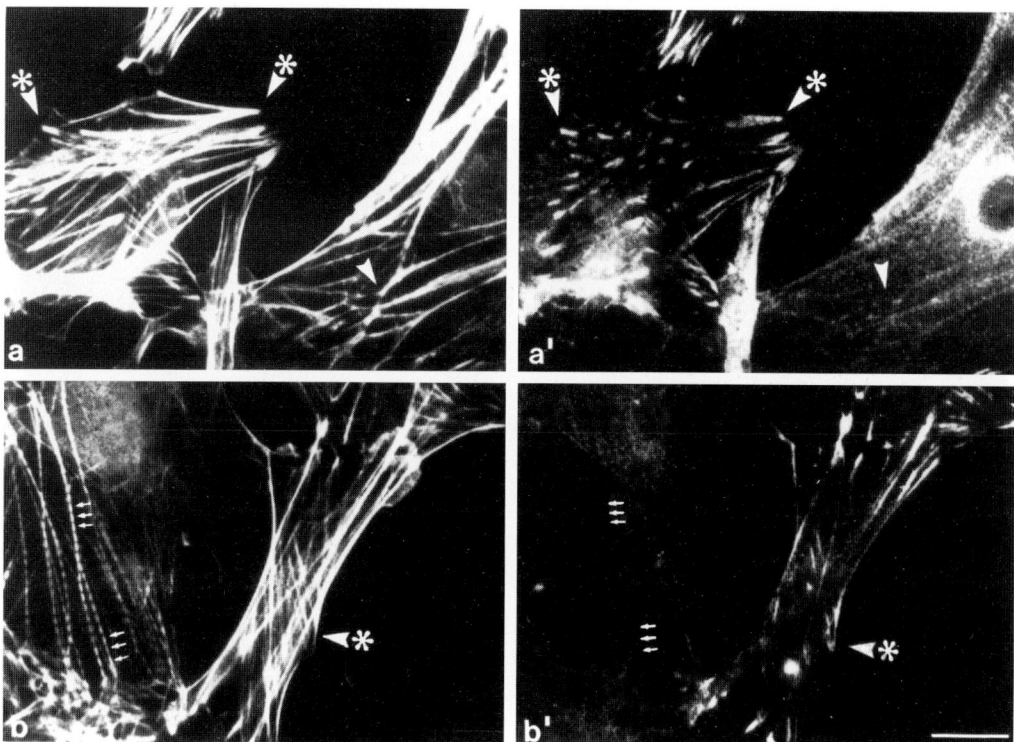

Fig. 5. Cardiocytes and fibroblasts double labelled for actin (FITC-phalloidin, a, b) and insertin (rhodamine, a′, b′). Similar to zyxin (cf. Fig. 4), this tensin-related protein (see text) is a marker for fibroblast CSJ (arrowheads/asterisks), and is not associated with contact sites or myofibrils (b, b′, small arrows) in cardiocytes. Bar: 10 μm.

zyxin and insertin, may be related to differences in dynamics between this type of contact in fibroblasts as opposed to cardiocytes. The latter assemble all contact sites considered here at the postmitotic stage, as permanent structures. In contrast, CSJ in fibroblasts, together with their stress fibres, are subject to rapid disassembly/assembly processes, as is for example seen during the cell cycle. With entry into mitosis, CSJ and stress fibres disappear and are reformed in late telophase/early G_1 phase. It seems conceivable that special regulatory proteins mediate between external signals and the microfilament system, and zyxin and insertin are good candidates for such a function.

With respect to dynamics, there are different subsets of CSJ to be considered in fibroblasts. In the cardiac cultures discussed so far, the fibroblastic cells are mostly stationary and are, through most of their mitotic cycle, firmly anchored to the substratum in prominent adhesion plaques at the terminal portions of very thick stress fibres (cf. Figs. 4a, 5a, b). In contrast, locomoting fibroblasts that are seen early after taking embryonic heart or skin into culture, as well as spreading or migrating fibroblasts observed in fibroblast cultures within a few hours after plating, show much of their actin filaments organised in smaller microfilament bundles that extend with their distal portions into the leading lamella and also in "actin ribs" that radially traverse the cortical cytoplasm to the cell margin. Their contacts with the inner face of the ventral plasma membrane coincide with small, punctate adhesion plaques at the underside of the cell. Such CSJ are short-lived and disassemble rapidly, while new actin subunits are added within seconds at the tip of the associated short microfilament bundles (Wang, 1985). Obviously, such a rapid assembly of actin filaments at the junctional complex would need a high local concentration of actin for polymerisation. Recently, we reported on the association of the G-actin-binding protein profilin with the terminal portions of the microfilament bundles in lamellipodia, while stress fibres and CSJ in less dynamic regions of the same cells contained very little profilin (Buß et al., 1992). This finding suggests that local differences in profilin (and profilin-actin complexes) may modulate life-span and dynamics of CSJ in different regions of the same cell, by the controlled release of polymerisation competent actin.

The microfilament system as a cytoplasmic matrix

In addition to proteins that serve as structural components or regulate the dynamics of junctional complexes, one might expect the association of proteins with the microfilament system and its junctional complexes for quite different reasons. The abundance of microfilaments, the diversity of their organisational patterns that are rapidly interconvertible in many cell types, and the high precision of their structural organisation are parameters that not only need proteins to assemble and regulate them, but there may also be proteins that take advantage of them as a scaffold. The microfilament system may be a convenient structural basis for the compartmentalisation of certain proteins. This concept has not yet been explored in detail, but evidence for its validity is accumulating. An example for this may be the observed localisation of several glycolytic enzymes with myofibrils and stress fibres (Dölken et al., 1975; Clarke et al., 1983; Buß et al., 1988; Minaschek et al., 1992). High local concentrations of enzyme-substrate complexes should be the result of this association, leading to a dramatic increase in the efficiency of the glycolytic cycle.

Similarly, the microfilament system may be used to concentrate specific proteins of signal transduction pathways that do not directly or not exclusively control microfilament organisation. We have recently reported on the microfilament association of a protein that might be a representative of this class (Reinhard et al., 1992). This "Vasodilator stimulated phosphoprotein" (VASP) had originally been identified in human platelets, as a substrate for cAMP- and cGMP-kinase. VASP phosphorylation in a single serine residue is seen as a result of raising cAMP or cGMP levels by various physiological agents in intact platelets. This event correlates precisely with the inhibition of platelet activation which includes dramatic changes in platelet morphology, microfilament organisation and adhesive properties. Thus, VASP may be a key factor in physiological control of thrombosis (Walter et al., 1992). Purified VASP was found to sediment with actin filaments, suggesting that VASP is another actin-binding protein. Immunofluorescence demonstrated that VASP is not only localised at the distal portions of platelets spread on glass, but also in CSJ junctions in fibroblasts (Reinhard et al., 1992) and in cardiocytes. In addition, it is located in cell-cell contacts like ICD, at all Z-lines of

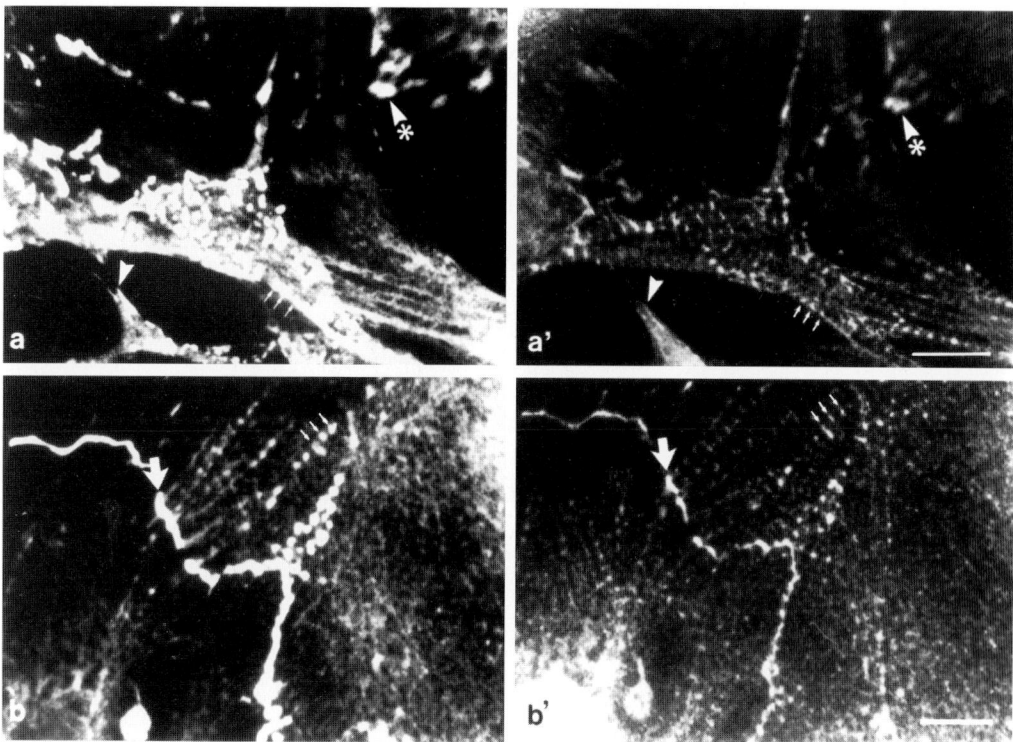

Fig. 6. Cardiocytes and fibroblasts double labelled for vinculin (a, b) and the vasodilator stimulated phosphoprotein VASP (a', b'). Both proteins are localised in cardiocyte and fibroblast CSJ (a, a', arrowheads and arrowheads/asterisks, respectively), in ICD (b, b', large arrows) and in costameric structures (ZDJ; a, a' and b, b', small arrows). Note that VASP is seen in all myofibrillar Z-lines similar to α-actinin (cf. Fig. 6 a', b' with Fig. 2 a-c). Bars: 10

myofibrils (Fig. 6), and periodically distributed in stress fibres (Reinhard *et al.*, 1992; see also cover of *EMBO J.* **11**/6). In fact, at the light microscopic level, the distribution of this protein is identical to that of α-actinin in all these locations. However, cAMP kinase mediated VASP phosphorylation in fibroblasts does not markedly alter the microfilament pattern. Thus, VASP is probably not directly involved in the regulation of microfilament organisation, but may be positioned at these places to effectively localise with various other signal transduction components. For example, pp60[src], protein kinase C and several kinases of the phospholipid metabolism have all been shown to interact with the microfilament system and the cytoplasmic domain of contact sites (Rohrschneider, 1980; Jaken *et al.*, 1989; Grondin *et al.*, 1991), and VASP might link the activity of these enzymes to that of cyclic nucleotide activated kinases at these sites.

Conclusion

In summary, we have seen that membrane-microfilament contact sites are discrete structures essential for cell adhesion, tissue formation and motility. The numerous protein components considered here are all located at the cytoplasmic face of these junctions, and various transmembrane molecules as well as extracellular components of CSJ and ICD add to the junctional diversity. Since Abercrombie's observations on fibroblast adhesion plaques (Abercrombie and Dunn, 1975), we have made progress in identifying the protein factors involved, but have yet to arrive at a complete identification of junctional components. However, a few rules regarding the architecture and dynamics of these structures have begun to emerge: First, there are bulk proteins that serve as building blocks in all these complexes, with possible cell specific variations based on protein isoforms. These major structural components show multiple interactions with each other. Second, there are additional components probably regulating junctional dynamics, and the intracellular members of this class themselves must be under the control of signal transduction factors. Third, the microfilament system and its contact sites with the plasma membrane in particular may be exploited as a means for intracellular compartmentalisation.

We thank the German Research Council (DFG) for supporting our own work reported in this article, Dr. U. Walter and M. Reinhard (Würzburg) for antibody to VASP, Dr. A. Wegner (Bochum) for insertin, and Mrs. R. Klocke for expert typing.

References

Abercrombie, M. and Dunn, G.A. (1975). Adhesions of fibroblasts to substratum during contact inhibition observed by interference reflexion microscopy. *Exptl. Cell Res.* **92**, 57-62.

Antin, P., Tokunaka, S., Nachmias, V. and Holtzer, H. (1986). Role of stress fiber-like structures in assembling nascent myofibrils in myosheets recovering from exposure to ethyl methanesulfonate. *J. Cell Biol.* **102**, 1464-1479.

Beckerle, M.C. (1986). Identification of a new protein localized at sites of cell-substrate adhesion. *J. Cell Biol.* **103**, 1679-1687.

Belkin, A.M. and Koteliansky, V.E. (1987). Interaction of iodinated vinculin, metavinculin and α-actinin with cytoskeletal proteins. *FEBS Lett.* **220**, 291-294.

Burn, P. and Burger, M.M. (1987). The cytoskeletal protein vinculin contains transformation-sensitive, covalently bound lipid. *Science* **235**, 476-479.

Burridge, K. and Connell, L. (1983). A new protein of adhesion plaques and ruffling membranes. *J. Cell Biol.* **97**, 359-367.

Buß, F., Hinssen, H. and Jockusch, B.M. (1988). Immunological and biochemical studies on the relationship between two actin-binding proteins, phosphofructokinase and gelsolin. *Eur. J. Biochem.* **175**, 251-257.

Buß, F., Temm-Grove, C., Henning, S. and Jockusch, B.M. (1992). Distribution of profilin in fibroblasts correlates with the presence of highly dynamic actin filaments. *Cell Motil. Cytoskel.* **22**, 51-61.

Carpén, O., Pallai, P., Staunton, D.E. and Springer, T.A. (1992). Association of intercellular adhesion molecule-1 (ICAM-1) with actin-containing cytoskeleton and α-actinin. *J. Cell Biol.* **118**, 1223-1234.

Clarke, F., Stephan, P., Morton, D. and Weidemann, J. (1983). The role of actin and associated structural proteins in the organization of glycolytic enzymes. In: Actin. Structure and Function in Muscle and Non-muscle Cells (eds. C.G. dos Remedios, J.A. Barden), pp. 249-257, Academic Press, Sydney.

Crawford, A.W. and Beckerle, M.C. (1991). Purification and characterization of zyxin, an 82,000 dalton component of adherens junctions. *J. Biol. Chem.* **266**, 5847-5853.

Crawford, A.W., Michelsen, J.W. and Beckerle, M.C. (1992). An interaction between zyxin and α-actinin. *J. Cell Biol.* **116**, 1381-1393.

Davis, S., Lu, M.L., Lo, S.H., Lin, S., Butler, J.A., Druker, B.J., Roberts, T.M., An, Q. and Chen, L.B. (1991). Presence of an SH2 domain in the actin-binding protein tensin. *Science* **252**, 712-715.

Dlugosz, A., Antin, P., Nachmias, V. and Holtzer, H. (1984). The relationship between stress fiber-like structures and nascent myofibrils in cultured cardiac myocytes. *J. Cell Biol.* **99**, 2268-2278.

Dölken, G., Leisner, E. and Pette, D. (1975). Immunofluorescent localization of glycogenolytic and glycolytic enzyme proteins and of malate dehydrogenase isozymes in cross-striated skeletal muscle and heart of rabbit. *Histochemistry* **43**, 113-121.

Geiger, B. and Ginsberg, D. (1991). The cytoplasmic domain of adherens-type junctions. *Cell Motil. Cytoskel.* **20**, 1-6.

Geiger, B., Volk, T., Volberg, T. and Bendori, R. (1987). Molecular interactions in adherens type contacts. *J. Cell Sci. Suppl.* **8**, 251-272.

Grondin, P., Plantavid, M., Sultan, C., Breton, M., Mavco, G. and Chap, H. (1991). Interaction of pp60$^{c\text{-src}}$, phospholipase C inositol-lipid, and diacylglycerol kinases with the cytoskeletons of thrombin-stimulated platelets. *J. Biol. Chem.* **266**, 15705-15709.

Handel, S.E., Greaser, M.L., Schultz, E., Wang, S.M., Bulinski, J.C., Lin, J.J.C. and Lessarde, J.L. (1991). Chicken cardiac myofibrillogenesis studied with antibodies specific for titin and the muscle and nonmuscle isoforms of actin and tropomyosin. *Cell Tissue Res.* **263**, 419-430.

Heath, J.P. and Dunn, G.A. (1978). Cell to substratum contacts of chick fibroblasts and their relation to the microfilament system. A correlated interference-reflexion and high-voltage electron-microscope study. *J. Cell Sci.* **29**, 197-212.

Heath, J.P. and Holifield, B.F. (1991). Cell locomotion: New research tests old ideas on membrane and cytoskeletal flow. *Cell Motil. Cytoskel.* **18**, 245-257.

Hu, M.H., DiLullo, C., Schultheiss, T., Holtzer, S., Murray, J.M., Choi, J., Fischman, D.A. and Holtzer, H. (1992). The vinculin/sarcomeric-α actinin/α actin nexus in cultured cardiac myocytes. *J. Cell Biol.* **117**, 1007-1022.

Isenberg, G. (1991). Actin binding proteins-lipid interactions. *J. Muscle Res. Cell Motil.* **12**, 136-144.

Isenberg, G., Leonard, K. and Jockusch, B.M. (1982). Structural aspects of vinculin-actin interactions. *J. Mol. Biol.* **158**, 231-249.

Jaken, S., Leach, K. and Klauck, T. (1989). Association of type 3 protein kinase C with focal contacts in rat embryo fibroblasts. *J. Cell Biol.* **109**, 697-704.

Jockusch, B.M. and Isenberg, G. (1981a). Interaction of α-actinin and vinculin with actin: Opposite effects on filament network formation. *Proc. Natl. Acad. Sci. USA* **78**, 3005-3009.

Jockusch, B.M. and Isenberg, G. (1981b). Vinculin and α-actinin: Interaction with actin and effect on microfilament network formation. *Cold Spring Harbor Symp.* **46**, 613-623.

Kellie, S. and Wigglesworth, N.M. (1987). The cytoskeletal protein vinculin is acylated by myristic acid. *FEBS Lett.* **213**, 428-432.

Koteliansky, V.E. and Gneushev, G.N. (1983). Vinculin localization in cardiac muscle. *FEBS Lett.* **159**, 158-160.

Lin, Z., Holtzer, S., Schultheiss, T., Murray, J., Masaki, T., Fischman, D.A. and Holtzer, H. (1989). Polygons and adhesion plaques and the disassembly and assembly of myofibrils in cardiac myocytes. *J. Cell Biol.* **108**, 2355-2367.

Minascheck, G., Gröschel-Stewart, U., Blum, S. and Bereiter-Hahn, J. (1992). Microcompartmentation of glycolytic enzymes in cultured cells. *Eur. J. Cell Biol.* **58**, 418-428.

Nikolai, G., Temm-Grove, C., Wiegand, Ch., Citi, S., Reinhard, M., Walter, U. and Jockusch, B.M. (1992). Microfilament-membrane attachment sites in chicken embryonic cardiocytes. *J. Muscle Res. Cell Motil.*, in press.

Otey, C.A., Pavalko, F.M. and Burridge, K. (1990). An interaction between α-actinin and the β1 integrin subunit in vitro. *J. Cell Biol.* **111**, 721-729.

Pardo, J.V., D'Angelo Siciliano, J. and Craig, S.W. (1983a). Vinculin is a component of an extensive network of myofibril-sarcolemma attachment regions in cardiac muscle fibers. *J. Cell Biol.* **97**, 1081-1088.

Pardo, J.V., D'Angelo Siciliano, J. and Craig, S.W. (1983b). A vinculin-containing cortical lattice in skeletal muscle: Transverse lattice elements ("costameres") mark sites of attachment between myofibrils and sarcolemma. *Proc. Natl. Acad. Sci. USA* **80**, 1008-1012.

Pavalko, F.M. and Burridge, K. (1991). Disruption of the actin cytoskeleton after microinjection of proteolytic fragments of α-actinin. *J. Cell Biol.* **114**, 481-491.

Reinhard, M., Halbrügge, M., Scheer, U., Wiegand, Ch., Jockusch, B.M. and Walter, U. (1992).

The 46/50 kDa phosphoprotein VASP purified from human platelets is a novel protein associated with actin filaments and focal contacts. *EMBO J.* **11**, 2063-2070.

Rohrschneider, L.R. (1980). Adhesion plaques of Rous sarcoma virus-transformed cells contain the src gene product. *Proc. Natl. Acad. Sci. USA* **77**, 3514-3518.

Ruhnau, K., Gaertner, A. and Wegner, A. (1989). Kinetic evidence for insertion of actin monomers between the barbed ends of actin filaments and barbed end-bound insertin, a protein purified from smooth muscle. *J. Mol. Biol.* **210**, 141-148.

Sanger, J., Mittal, B., Meyer, T. and Sanger, J. (1989). Use of fluorescent probes to study myofibrillogenesis. In *Cellular and Molecular Biology of Muscle Development* (eds. L. Kedes and F. Stockdale), pp. 221-235.

Sanger, J.M., Mittal, B., Pochapin, M.B. and Sanger, J.W. (1986). Myofibrillogenesis in living cells microinjected with fluorescently labeled α-actinin. *J. Cell Biol.* **102**, 2053-2066.

Sanger, J.W. (1977). Mitosis in beating cardiac myoblasts treated with cytochalasin B. *J. Exp. Zool.* **201**, 403-409.

Sanger, J.W., Mittal, B. and Sanger, J.M. (1984). Formation of myofibrils in spreading chick cardiac myocytes. *Cell Motil. Cytoskel.* **4**, 405-416.

Schultheiss, T., Lin, Z., Lu, M.H., Murray, J., Fischman, D.A., Weber, K., Masaki, T., Imamura, M. and Holtzer, H. (1990). Differential distribution of subsets of myofibrillar proteins in cardiac nonstriated and striated myofibrils. *J. Cell Biol.* **110**, 1159-1172.

Shriver, K. and Rohrschneider, L. (1981). Organization of pp60[src] and selected cytoskeletal proteins within adhesion plaques and junctions of Rous Sarcoma Virus-transformed rat cells. *J. Cell Biol.* **89**, 525-535.

Turner, C.E., Glenney Jr., J.R. and Burridge, K. (1990). Paxillin: A new vinculin-binding protein present in focal adhesions. *J. Cell Biol.* **111**, 1059-1068.

Walter, U., Eigenthaler, M., Geiger, J. and Reinhard, M. (1992). Role of cyclic nucleotide-dependent protein kinases and their common substrate VASP in the regulation of human platelets. In *Mechanisms of Platelet Activation and Control* (eds. K.S. Anthi, S.P. Watson and V.V. Kakkar), Plenum Press, In press.

Wang, Y.L. (1985). Exchange of actin subunits at the leading edge of living fibroblasts: Possible role of treadmilling. *J. Cell Biol.* **101**, 597-602.

Weigt, Ch., Gaertner, A., Wegner, A., Korte, H. and Meyer, H.E. (1992). Occurrence of an actin-binding domain in tensin. *J. Mol. Biol.* **227**, In press.

Westmeyer, A., Ruhnau, K., Wegner, A. and Jockusch, B.M. (1990). Antibody mapping of functional domains in vinculin. *EMBO J.* **9**, 2071-2078.

Wilkins, J.A., Risinger, M.A. and Lin, S. (1986). Studies on proteins that co-purify with smooth muscle vinculin: Identification of immunologically related species in focal adhesions of nonmuscle and Z-lines of muscle cells. *J. Cell Biol.* **103**, 1483-1494.

Printed in Great Britain © The Society of Experimental Biology 1993   267

THE INTERACTION OF THE TYROSINE KINASE pp60^{src} WITH MEMBRANE AND CYTOSKELETAL COMPONENTS

S. KELLIE[1]*, A.R. HORVATH[2], G. FELICE[4], R. ANAND[5], C. MURPHY[3] and J. WESTWICK[3]

[1]Yamanouchi Research Institute, Littlemore Hospital, Oxford, OX4 4XN, UK

[2]Dept. of Clinical Chemistry, University School of Medicine, PO Box 40, H 4012, Debrecen, Hungary

[3]School of Pharmacy and Pharmacology, University of Bath, Claverton Down, Bath BA2 7AY UK

[4]Surgical Dept., St. Luke's Hospital, Guardamangia, Malta

[5]Dept. of Biochemical Pharmacology, William Harvey Research Institute, St. Bartholomew's Medical College, London EC1M 6BQ UK

Summary

To study the mechanism of oncogenic transformation we have investigated the association of pp60^{v-src} with the cytoskeleton using a variety of mutants. Transformation is associated with the interaction of pp60^{v-src} with the cytoskeleton, specifically in adhesion plaques. Biochemical analysis has shown a correlation between tyrosine-specific phosphorylation of the fibronectin receptor and the loss of surface-bound fibronectin, but no such correlation with phosphorylation of vinculin or talin. The role of the nontransforming protooncogene product pp60^{c-src} was studied in platelets, which express large amounts of this protein. In resting platelets pp60^{c-src} was soluble in detergent-containing buffers, however it became associated with the cytoskeleton, but not the membrane skeleton, after platelet activation. This association was inhibited by EDTA or RGDS peptides and was therefore dependent on occupancy of a platelet integrin, gpIIb/IIIa. The role of pp60^{c-src}-associated tyrosine phosphorylation in platelet activation was also investigated. Inhibitors of tyrosine phosphorylation inhibited integrin-dependent platelet aggregation and second messenger production, indicating a close linkage between matrix receptor occupancy, pp60^{c-src} association with the cytoskeleton and platelet function.

Introduction

In recent years there has been intense investigation into the role of tyrosine phosphorylation in cell function (Cooper and Hunter, 1983). It is now well established that many growth factor receptors are tyrosine kinases, however the signal transduction pathways and the mechanism by which oncogenic tyrosine kinases induce cell growth and neoplastic changes, and nononcogenic kinases regulate cell function in most cases remain elusive (Jove and Hanafusa, 1987; Yarden and Ullrich, 1988). Until recently the *src* gene product of Rous sarcoma virus (RSV) was probably the best characterised tyrosine kinase, and in this paper we will review the role of the oncogenic *src* gene

*Author for correspondence.

Key words: tyrosine kinase, cytoskeleton, *src*.

product pp60$^{v\text{-}src}$ in the disruption of the cytoskeleton and matrix components which characterise oncogenic transformation. We will also review the function of its normal cellular counterpart, pp60$^{c\text{-}src}$, in the regulation of a normal cell type, the blood platelet.

Tumour formation *in vivo* and transformation of cells *in vitro* by RSV is due to the expression of the viral *src* gene, which codes for a 60kD membrane-associated phosphoprotein whose only known function is that of a tyrosine kinase (pp60$^{v\text{-}src}$). Cells transformed by RSV are characterised by anchorage independent growth, loss of growth regulation and changes in metabolite transport and utilization. In addition, transformed cells have a rounded cell shape and exhibit cytoskeletal changes including a reduction in microfilament bundles, a redistribution of alpha-actinin, vinculin and talin.

The investigation of targets for pp60^{src} within whole cells has proved complex. However the use of mutants, combined with educated guesses as to what the likely candidates for pp60$^{v\text{-}src}$ targets within cells might be, has led to the conclusions that cytoskeletal interactions are important in tyrosine kinase function. Inducible transforming mutants of pp60$^{v\text{-}src}$ have demonstrated that changes in the cytoskeleton occur early in the transformation process, and disruption of polymeric actin can be seen within 15 minutes of activation of pp60$^{v\text{-}src}$ (Boschek *et al.*, 1983; Felice *et al.*, 1990). pp60$^{v\text{-}src}$ associates with the cytoskeleton and its localisation in cell-cell and cell-matrix contacts suggested that phosphorylation of cytoskeletal target proteins such as vinculin and talin on tyrosine residues may result in the disassembly of adhesion plaques (Rorhschneider, 1980). However, tyrosine phosphorylation of these and other cytoskeletal substrates so far investigated appears to be irrelevant to the induction or maintenance of the transformed phenotype (Rohrschneider and Rosok, 1983; Antler *et al.*, 1985; Kellie *et al.*, 1986a,b; Stoker *et al.*, 1986; DeClue and Martin, 1987). Another component of the adhesion plaque, the transmembrane receptor for fibronectin, has been shown to be disrupted by RSV transformation. In CEF the fibronectin receptor bands 2 and 3 are phosphorylated on tyrosine in cells transformed by pp60$^{v\text{-}src}$ and other oncogenic tyrosine kinases (Hirst *et al.*, 1986). The tyrosine phosphorylation site of CSAT band 3 (integrin β_1 subunit) is in the cytoplasmic domain, in a region apparently involved in talin binding (Tapley *et al.*, 1989). Furthermore, *in vitro* studies have demonstrated that tyrosine phosphorylated integrins from RSV-transformed cells have a reduced capacity to interact with either fibronectin or talin and it has been suggested that this phosphorylation is responsible for the reduction in cell-matrix interactions and the cytoskeletal changes associated with transformation (Tapley *et al.*, 1989). We have used a series of mutants of RSV which confer differing phenotypes on the cells to investigate (a) the relevance of pp60$^{v\text{-}src}$ association with the cytoskeleton for transformation and (b) the role of tyrosine phosphorylation of membrane-cytoskeletal proteins in the transformed phenotype.

The role of the cellular homologue of pp60$^{v\text{-}src}$, pp60$^{c\text{-}src}$ remains to be elucidated. High levels of this cellular kinase are found in platelets, giving rise to suggestions that it plays a role in platelet activation, and some evidence for this has now been reported since there is a rapid induction of tyrosine phosphorylation when platelets are activated, although it is not clear whether pp60$^{c\text{-}src}$ is responsible for this and not some other kinase such as *c-fyn*. Another discrepancy is that although pp60$^{v\text{-}src}$ is associated with the cytoskeleton, pp60$^{c\text{-}src}$ (at least in overexpressing fibroblasts) does not, although they both contain identical SH2 domains thought to be important in this cytoskeletal interaction. One explanation may be that overexpressing fibroblasts are not a suitable model for protein localisation, as the protein of interest may be driven into the wrong subcellular compartment due to the lack of appropriate translocation machinery, therefore in conjunction with studies on the activation of tyrosine kinase activity we have investigated the localisation of pp60$^{c\text{-}src}$ in platelets where this protein occurs naturally at high levels. We will review the results in the context of platelet function and responses to agonist-induced activation.

Results

A. Localisation and phosphorylating activities of pp60$^{v\text{-}src}$ in fibroblasts

1. The use of conditional mutants to study pp60$^{v\text{-}src}$ localisation and activity

We have used a temperature-sensitive mutant of pp60$^{v\text{-}src}$, *ts*LA29, to study its localisation, activation, and oncogenic activity. Permanent Rat-1 cell lines transformed by *ts*LA29 were used in these studies (Felice *et al.,* 1990; Welham and Wyke, 1988). These are phenotypically normal at the restrictive temperature of 39°C (Fig. 1A) but are phenotypically transformed at the permissive temperature of 35°C (Fig. 1B). When these cells were grown at the restrictive temperature the pp60$^{v\text{-}src}$ was kinase inactive (Fig. 2, lane 8). If these cells were then switched to the permissive temperature there was a progressive increase in the kinase activity, detectable within 1-2 hours of temperature switch (Fig. 2). The localisation of pp60$^{v\text{-}src}$ in the inactive and active states was investigated in these cells by fractionation into detergent-insoluble (cytoskeleton) or detergent-soluble phases followed by immune precipitation and pp60$^{v\text{-}src}$ kinase assays. These showed that at the restrictive temperature almost all the pp60$^{v\text{-}src}$ was detergent soluble (Fig. 3). Concurrent with induction of kinase activity was the initiation of association with the cytoskeleton (Fig. 3), thus interaction of pp60$^{v\text{-}src}$ with the cytoskeleton is also an early event in oncogenic transformation. The localisation of pp60$^{v\text{-}src}$ was further investigated by immunofluorescence. Staining of cells at 35°C showed that LA29 pp60$^{v\text{-}src}$ was concentrated in aggregates at the plane of the ventral membrane which have a structure similar to those which have been previously described as podosomes or rosettes in other cell types (Fig. 1F). Extraction of these cells revealed that both the pp60$^{v\text{-}src}$ and actin in these podosomes were detergent insoluble and thus associated with cytoskeletal structures (Felice *et al.,* 1990).

At 39°C, however, LA29pp60$^{v\text{-}src}$ was mainly found in a cytosolic localisation (Fig. 1E) which was lost after detergent extraction (data not shown). These experiments indicate that at 39°C LA29pp60$^{v\text{-}src}$ has a decreased ability to associate with the cytoskeleton, concurrent with a decrease in its kinase and transforming activity.

The localisation of phosphotyrosine-containing proteins was also investigated by immunofluorescent staining. At 35°C phosphotyrosine-containing proteins localised to adhesion plaque-like or podosome structures, many of which coincided with F-actin (Fig. 1D), however there was in addition strong staining within the nucleus of the cells. At the restrictive temperature phosphotyrosine staining was faint and where detectable concentrated in adhesion plaques where microfilament bundles terminate (Fig. 1C). Thus there was a reduction in tyrosine phosphorylation of adhesion plaque components which correlated with a reduction in the kinase activity of pp60$^{v\text{-}src}$.

2. The use of nonconditional mutants to examine pp60$^{v\text{-}src}$ substrates

We have used a nonconditional mutant of RSV to investigate the relationship

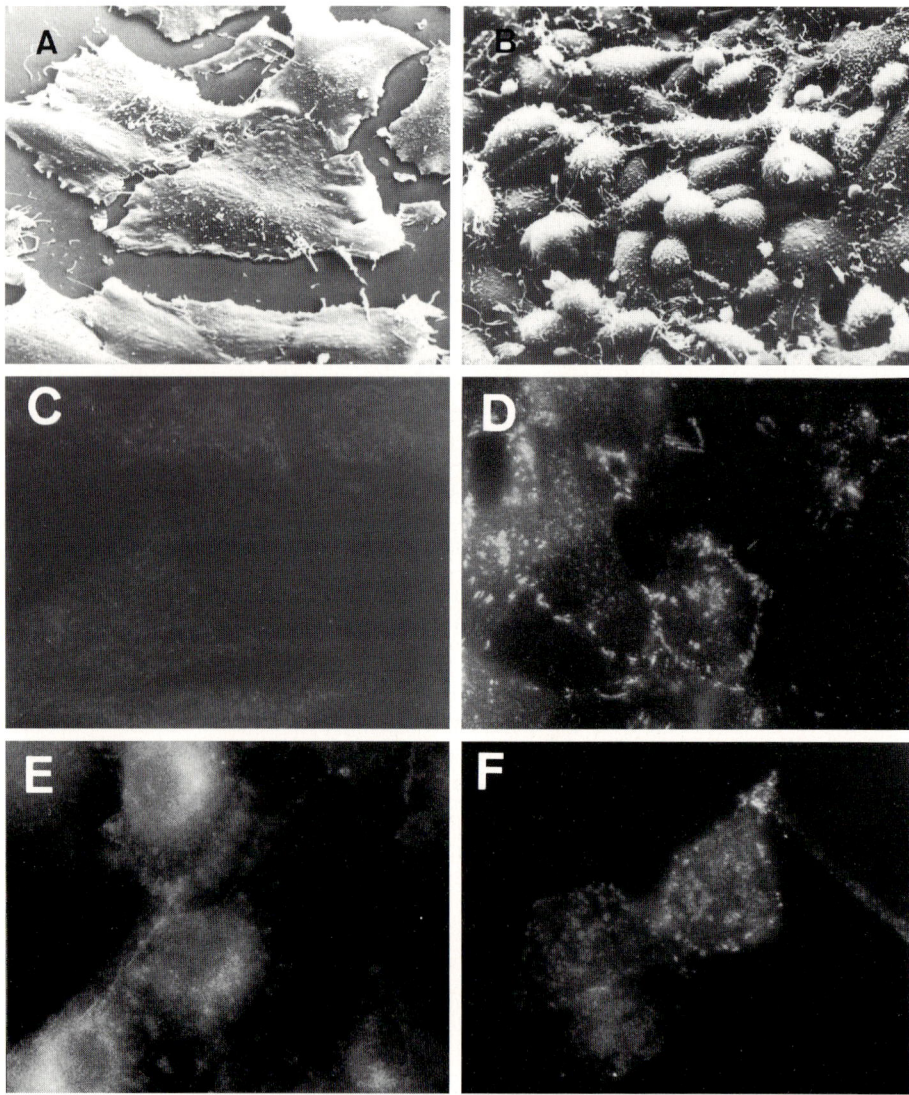

Fig. 1. Morphology and immunofluorescent staining of pp60$^{v\text{-}src}$ and phosphotyrosine in Rat-1 fibroblasts transformed by tsLA29 grown at 35°C (B,D,F) or 39°C (A,C,E). A,B: Scanning electron micrographs showing cell morphology. C,D: Immunofluorescence of phosphotyrosine showing light staining in the untransformed cells but strong staining of adhesion structures in the transformed cells. E,F: Immunofluorescence of pp60$^{v\text{-}src}$ showing cytosolic staining in the untransformed cells and adhesion plaque staining in the transformed cells.

between pp60^v-src^, integrins and phosphotyrosine in normal and RSV-transformed fibroblasts. rASV2234.3 is a recovered avian sarcoma virus which expresses a pp60^v-src^ with fully active kinase activity and is transforming *in vivo* (Enrietto *et al.*, 1983). However cells transformed by this virus are morphologically similar to untransformed cells in that they are flat, contain stress fibres and vinculin-containing adhesion plaques (Kellie *et al.*, 1986b), although the cells will overgrow in culture. Thus although these

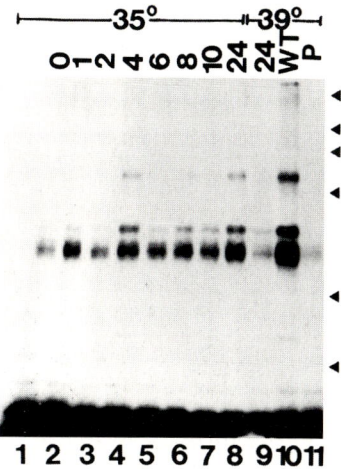

Fig. 2. Time course of pp60^v-src^-associated kinase activity in LA29 cells. Cells were grown at 39°C for 48 hours then switched to 35°C. At the indicated times pp60^v-src^ was immune precipitated and in vitro kinase assay performed (Felice *et al.*, 1990). WT: wild-type RSV-transformed cells. P: Pre-immune serum control. Numbers at the top are hours after temperature switch from 39°C to 35°C.

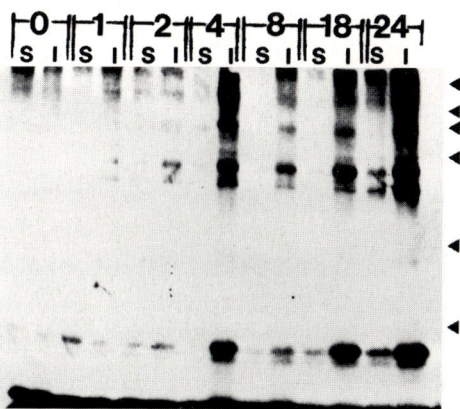

Fig. 3. Time course of association of kinase active pp60^v-src^ with the cytoskeleton. Cells were grown as described for Fig. 2 and at the indicated times fractionated into cytoskeletal and detergent-soluble phases (Felice *et al.*, 1990). PP60^v-src^ was immune precipitated from each fraction and kinase assay performed. S: detergent-soluble phase. I: detergent-insoluble (cytoskeleton) phase.

cells are neoplastically transformed they morphologically resemble untransformed cells, making them an ideal tool to investigate the role of tyrosine phosphorylation of specific proteins in the changes in cell morphology characteristic of transformation. Several adhesion plaque proteins have been identified as substrates of pp60$^{v\text{-}src}$. These include talin, vinculin and the fibronectin receptor, all of which physically associate with each other in vitro and possibly within cells. Previous studies have shown no correlation between tyrosine-specific phosphorylation of vinculin and the loss of adhesion plaques after transformation (Rohrschneider and Rosok, 1985; Kellie *et al.*, 1986b), however similar studies on the fibronectin receptor have not been extensively performed. Before studying phosphorylation, however, we studied the physical relationships between the fibronectin receptor, pp60$^{v\text{-}src}$ and phosphotyrosyl-containing proteins in this RSV mutant.

(a) Localisation of pp60$^{v\text{-src}}$*, phosphotyrosyl proteins and the fibronectin receptor in transformed cells.* In untransformed chick embryo fibroblasts (CEF) the fibronectin receptor showed a typical needle eye type distribution corresponding to well-defined

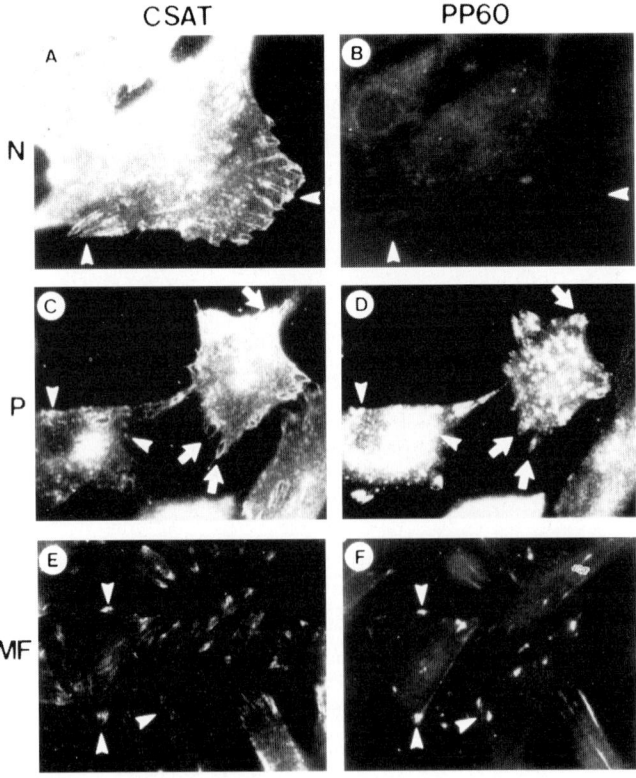

Fig. 4. Colocalisation of integrin and pp60$^{v\text{-}src}$ in normal and RSV transformed chick embryo fibroblasts. Cells were double-labelled with anti-fibronectin receptor monoclonal antibody (left) and a rabbit antiserum against pp60$^{v\text{-}src}$ (right). A,B: normal chick embryo fibroblasts. C,D: CEF transformed by Prague C strain RSV. E,F: CEF transformed by the fusiform mutant rASV2234.3. Bar = 10μm. Figure courtesy of *Oncogene*.

adhesion plaques (Fig. 4A, arrowheads). A similar pattern was observed in the elongated, fusiform mutant RSV-transformed CEF (Fig. 4E, arrowheads), while in wild-type Prague C (PrC) virus transformed rounded CEF integrins formed point contacts (Fig. 4C, arrowheads) or circle-like adhesions at the periphery of the cell (Fig. 4C, arrows). Fig. 4C-F illustrate that pp60^{v-src} codistributed with these adhesion structures in both RSV-transformed cell types. The localisation of tyrosine-phosphorylated proteins was compared using a monoclonal anti-phosphotyrosine antibody (Glenney *et al.*, 1988) (Fig. 5). Phosphotyrosine containing proteins in PrC-transformed CEF showed a discrete distribution corresponding to podosomes or point contacts (Fig. 5B). In rASV2234.3-transformed CEF they correlated with remaining adhesion plaques, or were aligned with portions of stress fibres (Fig. 5C).

(b) Phosphorylation of integrins in RSV-transformed cells. Metabolic and surface labelling revealed that neither the synthesis nor the cell surface expression of integrins were influenced by transformation by RSV (data not shown). Therefore we investigated the phosphorylation of integrin in PrC-transformed CEF which lose fibronectin or in rASV2234.3-transformed CEF which retain a fibronectin matrix. Integrins were immunoprecipitated from cell lysates of untransformed and RSV-transformed CEF

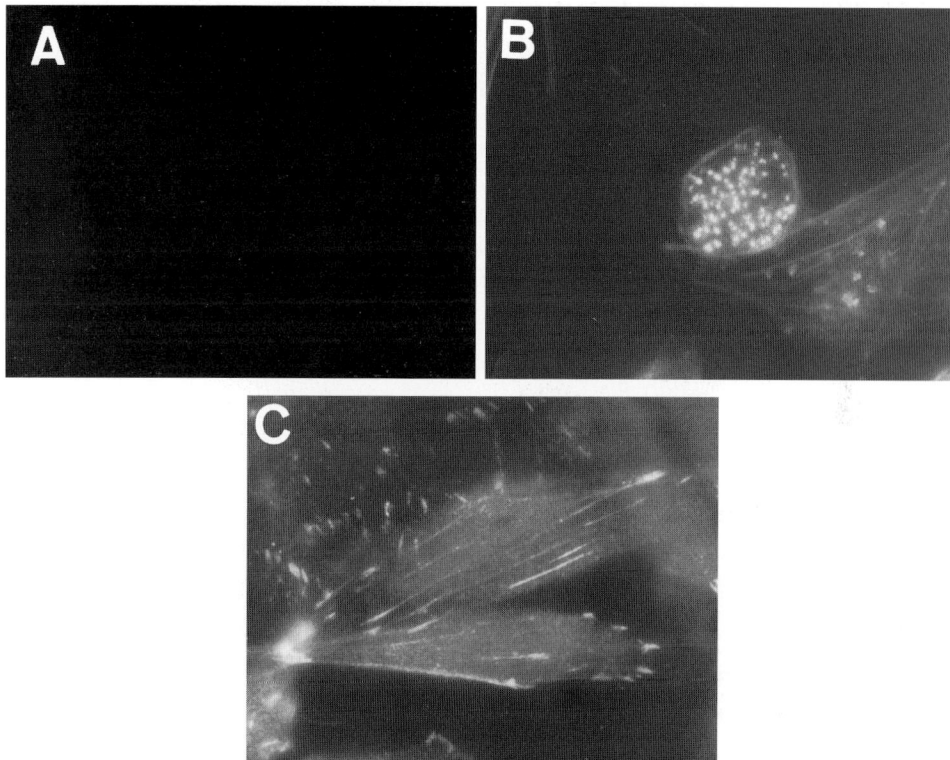

Fig. 5. Localisation of tyrosine-phosphorylated proteins in normal and RSV-transformed chick embryo fibroblasts. Cells were labelled with PY20 monoclonal antibody against phosphotyrosine. (A) normal, (B) Prague C RSV-transformed, or (C) rASV2234.3-transformed CEF.

labelled with [^{32}P] orthophosphate in the absence or presence of sodium orthovanadate. The phosphorylation of bands 2 and 3 of integrin in PrC-transformed CEF were significantly increased compared to that of untransformed or rASV2234.3-transformed CEF (Fig. 6, lanes 2,5). The phosphorylation of band 3 from rASV2234.3-transformed cells was only slightly elevated compared with untransformed cells, and in some experiments there was no observable difference (Fig. 6, lanes 1,3,4,6). To examine which amino-acid residues were phosphorylated [^{32}P]-labelled integrin band 1 or 3 were subjected to partial acid hydrolysis followed by thin layer electrophoresis to separate the phosphoamino acids. Band 1 contained exclusively phosphoserine which was unaffected by transformation. In contrast, a significant increase in phosphotyrosine content of band 3 could be observed in wild-type transformed cells compared to normal and rASV2234.3-transformed counterparts (Horvath *et al.,* 1990).

B. Localisation of pp60$^{c\text{-}src}$ and tyrosine phosphorylation in platelets

Localisation of pp60$^{c\text{-}src}$ in platelets has led to controversial results, with it being variously described in the cytosol, dense bodies and plasma membrane/dense canalicular system. We fractionated resting platelets into cytosol, mixed membrane, and highly purified surface membrane by density centrifugation and immunoblotted these fractions for pp60. Fig. 7 shows that almost all (>90%) of the platelet pp60$^{c\text{-}src}$

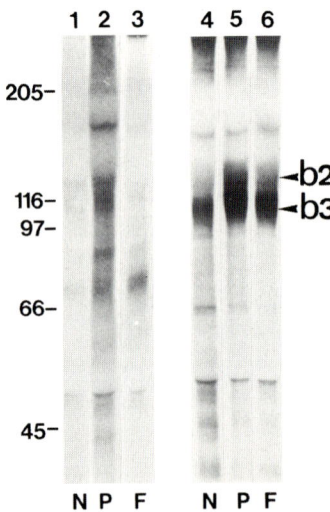

Fig. 6. Identification of tyrosine-phosphorylated integrin from untransformed and RSV-transformed chick embryo fibroblasts by double immunoprecipitation. Autoradiograms of integrin immunoprecipitates from cells labelled with [^{32}P]orthophosphate in the presence of sodium orthovanadate. Phosphorylated material was immunoprecipitated with PY20 anti-phosphotyrosine, eluted with phenylphosphate and the eluate re-immunoprecipitated with CSAT (1-3). Proteins from the lysates were also immunoprecipitated with anti-fibronectin receptor (4-6). N : normal; P : Prague C RSV-transformed CEF; F : rASV2234.3-transformed CEF. Figure courtesy of *Oncogene.*

fractionated with the surface membrane fraction with 10% associated with granule membranes.

Although pp60$^{v\text{-}src}$ binds avidly to the cytoskeleton, in fibroblasts pp60$^{c\text{-}src}$ does not. We decided to investigate the localisation of pp60$^{c\text{-}src}$ in resting and activated platelets. To perform these studies, unactivated and activated platelets were fractionated into cytosol, cytoskeleton and membrane skeleton (Horvath *et al.*, 1992). In resting platelets

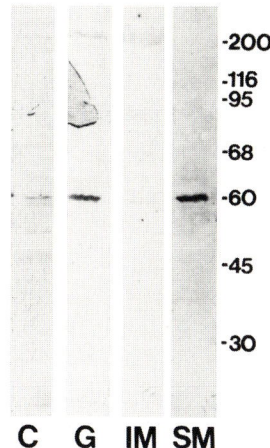

Fig. 7. Localisation of pp60$^{c\text{-}src}$ in platelet membrane fractions. Human blood platelets were fractionated into cyosol (C), granule membranes (G), intracellular membranes, mainly endoplasmic reticulum and golgi (IM), and surface plasma membrane (SM) as described in Elmore *et al.* (1990). These fractions were then immunoblotted for pp60$^{c\text{-}src.}$

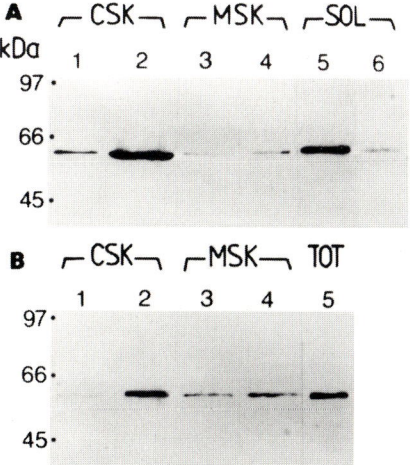

Fig. 8. Subcellular distribution of pp60$^{c\text{-}src}$ in resting and activated platelets. Platelets were fractionated into soluble phase (sol), cytoskeletons (CSK) or membrane skeletons (MSK) as described by Horvath *et al.* (1992). Equal numbers of cell equivalents (A) or cell protein (B) were immunoblotted for pp60$^{c\text{-}src}$. Lanes 1,3,5: Extracts from resting platelets. Lanes 2,4,6: extracts from platelets activated by 1 U/ml thrombin. Figure courtesy of *EMBO J.*

most pp60$^{c\text{-}src}$ was detergent-soluble, with <4% of the total cellular pp60$^{c\text{-}src}$ associated with the cytoskeleton. After stimulation with 1U/ml thrombin there was a significant increase in the amount of pp60$^{c\text{-}src}$ retained in the cytoskeletal fraction (Fig. 8). In contrast there was no increase in the amount of pp60$^{c\text{-}src}$ associated with the membrane skeleton after activation (Fig. 8). Treatment of the platelet cytoskeleton with DNase 1 reduced the amount of pp60$^{c\text{-}src}$ present, indicating that the association of pp60$^{c\text{-}src}$ with the cytoskeleton was dependent on polymeric actin (Horvath *et al.*, 1992). Time course and dose response experiments indicated that pp60$^{c\text{-}src}$-cytoskeleton association did not correlate with pseudopod formation or secretion, but did occur in parallel with aggregation. This was further investigated using reagents which inhibited platelet aggregation without inhibiting other responses. Neither cytochalasin B, which inhibits pseudopod formation but not aggregation or secretion, nor PMA which inhibits degranulation but not aggregation or shape change, had any effect on thrombin-induced cytoskeleton binding of pp60$^{c\text{-}src}$. However, activation of platelets without stirring, EDTA or RDGS peptides all inhibited the pp60$^{c\text{-}src}$-cytoskeleton association induced by thrombin (Fig. 9). EDTA inhibits assembly of the platelet integrin gpIIb/IIIa, and RGDS peptides inhibit the interaction of gpIIb/IIIa with matrix proteins, in particular fibrinogen. Thus there is a strong correlation between the association of pp60$^{c\text{-}src}$ with the platelet cytoskeleton and occupancy of the integrin gpIIb/IIIa.

Several groups have shown that tyrosine phosphorylation accompanies platelet activation, however the functional relevance of this has yet to be established. Tyrosine kinase inhibitors such as genestein have been useful in dissecting the relationships between tyrosine phosphorylation and growth factor receptor occupancy. We investigated the effect of genestein on platelet function. Genestein induced a dose-dependent inhibition of tyrosine phosphorylation in response to platelet activating factor (Fig. 10, inset). At these same concentrations genestein inhibited platelet aggregation in

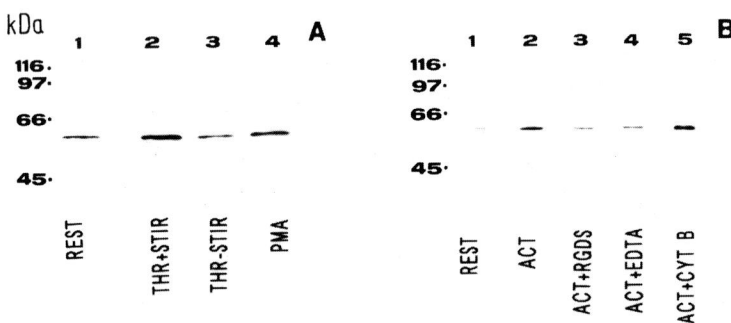

Fig. 9. Association of pp60$^{c\text{-}src}$ with the cytoskeleton during different phases of cell activation. (A) Cytoskeletons from resting platelets (1) and from cells activated by 1U/ml thrombin with (2) or without (3) stirring or by 5 nM phorbol myristate acetate (4). (B) Cytoskeletons from resting cells (1) or from platelets activated by 1 U/ml thrombin (2) in the presence of 200 µg/ml RGDS peptide (3), or 10 mM EDTA (4). or 10 µg/ml cytochalasin B (5). Figure courtesy of *EMBO J.*

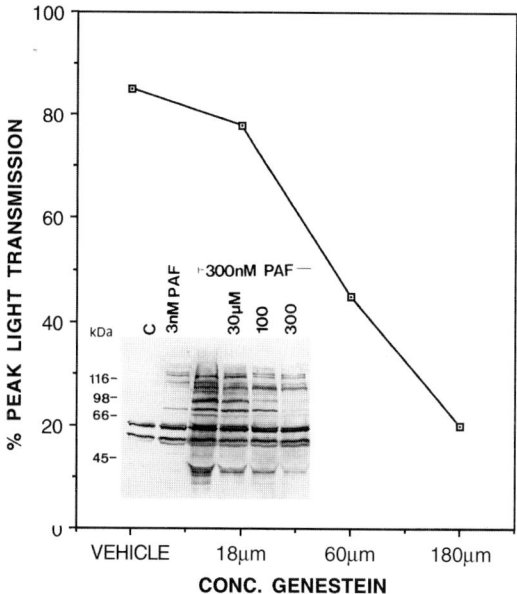

Fig. 10. Dose-response of the effect of genestein on platelet aggregation or tyrosine phosphorylation (inset) induced by 3 nM platelet activating factor. Aggregation was measured by light transmitance in an aggragometer and tyrosine phosphorylation was assayed by immunoblotting platelet lysates with anti-phosphotyrosine PY20.

response to PAF (Fig. 10). Further studies have shown that second messenger production is also inhibited (data not shown).

Discussion

(a) Association of pp60^{v-src} and phosphotyrosyl proteins with the cytoskeleton of transformed cells

Immunofluorescence, electron microscopy, western blotting and kinase assays of detergent-insoluble fractions of LA29Rat-1 cells revealed that gross changes in actin, alpha-actinin and vinculin correlated with the association of pp60^{v-src} with cytoskeletal complexes (Felice *et al.,* 1990). With other RSV mutants there is a strong correlation between the association of pp60^{v-src} with the cytoskeleton and morphological transformation (Hamaguchi and Hanafusa, 1987).

We have demonstrated that in LA29Rat-1 cells pp60^{v-src} localisation to the cytoskeleton was reduced at the restrictive temperature. The temperature-sensitivity of LA29 pp60^{v-src} is due to a single point mutation at amino acid 507 (Welham and Wyke, 1988). This must affect at least two domains of the protein: the enzyme site and the region(s) responsible for interaction with the plasma membrane and/or cytoskeletal structures. Cytoskeletal association and membrane interaction differ since myristoylation is necessary (but not sufficient) for membrane association but is unnecessary for accumulation in adhesion plaques (Cross *et al.,* 1984; Kreuger *et al.,* 1983; Buss *et al.,*

1984). The reorganisation of actin in these RSV-transformed Rat-1 cells also correlates well with the activation of a membrane-bound phosphatidylinositol kinase (Tones *et al.,* 1987). This raises the possibility that some component of the phosphatidylinositol cycle might be important in regulating actin organisation (Lassing and Lindberg, 1985; Meyer *et al.,* 1985; Niggli *et al.,* 1986).

(b) Phosphorylation of integrins in RSV-transformed cells

One explanation for the rounded phenotype is that transformed cells cannot interact with fibronectin due to transformation-dependent changes in adhesion plaque proteins, resulting in a loss of cytoskeletal regulation. We have therefore explored the physiological significance of tyrosine phosphorylation of integrins and its correlation with the altered phenotype in chick embryo fibroblasts transformed by wild type RSV or a variant which induces a flatter, fusiform morphology. Since many proteins in RSV-transformed cells appear to be phosphorylated fortuitously, it is important to identify those proteins whose phosphorylation is a requirement for the transformation process. Many phosphotyrosyl proteins are concentrated in adhesion plaques, however, there is no general correlation between the presence of tyrosine phosphorylated proteins in adhesion structures and the rounded morphology. Recently several proteins have been reported whose phosphorylation on tyrosine correlates with RSV-induced morphological changes (Glenney and Zokas, 1989; Kozma *et al.,* 1990). We have described the correlation of tyrosine-specific phosphorylation of the β_1 chain of integrin in these cells with the loss of fibronectin from the cell surface.

The biosynthesis and surface localisation of fibronectin and integrin in rASV2234.3-transformed CEF were similar to that found in untransformed CEF, indicating that integrin-fibronectin interaction was not altered in these cells. The chicken integrin receptor complex contains several different heterodimers consisting of a common β_1 subunit combined with different α-subunits. Band 1 contains the α_5, band 2 contains both α_3 and α_5, while band 3 corresponds to the highly conserved β_1 subunit. The $\alpha_5\beta_1$ subclass of heterodimers has a high affinity for fibronectin while the $\alpha_3\beta_1$ complex mediates lower affinity adhesion to fibronectin, laminin and collagen (Hynes *et al.,* 1989). The synthesis of the $\alpha_5\beta_1$ complex is markedly reduced in RSV-transformed mammalian cells (Plantefaber and Hynes, 1989), however it is unlikely that reduced synthesis of the high affinity fibronectin receptor subclass in PrC-CEF accounts for the loss of surface fibronectin since we observed no significant reduction in iodinated band 1 (α_5) or band 2 ($\alpha_5 + \alpha_3$) in either PrC- or rASV2234.3-transformed CEF. Others have also found similar levels of all integrin subunits in RSV-transformed CEF (Hirst *et al.,* 1986; Chen *et al.,* 1986). It is therefore likely that the differences observed in cell-matrix interaction between PrC-transformed and rASV2234.3-transformed CEF were due to an altered regulation of receptor function by phosphorylation rather than decreased expression.

Immunofluorescent colocalisation and *in vitro* binding studies suggest that there is a physical association between integral membrane proteins and extracellular matrix or cytoskeletal molecules, however the biochemical basis for these interactions and their regulation are at present unknown. For this reason the functional consequences of integrin phosphorylation on tyrosine residues need further investigation. Integrins isolated from RSV-CEF in the absence of orthovanadate were found to be phosphorylated primarily on serine residues and these have been shown to have a normal capacity to interact with both extra-and intracellular ligands *in vitro*. In the presence of vanadate integrins are phosphorylated mainly on tyrosine and have a reduced capacity to interact with talin and fibronectin (Tapley *et al.*, 1989). Recently it has been reported that there is a transient phosphorylation of CD18 (the leucocyte function antigen-1 β chain) mediated by protein kinase C in neutrophils and monocytes exposed to a chemotactic stimulus (Chatila *et al.*, 1989), suggesting a regulatory role for integrin phosphorylation related to cytoskeletal changes or signal transduction in these cells. Fibronectin binding to integrin can directly lead to tyrosine phosphorylation of a 120 kDa protein, and recently this protein has been identified as a focal adhesion-associated tyrosine kinase whose tyrosine phosphorylation can be modified by pp60^{v-src} (Guan *et al.*, 1991; Guan and Shalloway, 1992), thus adhesion via integrins is intimately involved in some signalling events. The correlation between the morphological changes and tyrosine-phosphorylation of integrins and associated molecules suggests that this posttranslational modification of integrin molecules is a physiologically important event in the induction of the transformed phenotype.

(c) pp60^{c-src} *and tyrosine phosphorylation in platelets*

Whilst the role of several enzymes such as protein kinase C and phospholipase C has been well documented as being important in platelet activation, less is known about the role of tyrosine phosphorylation. Several groups have reported that increases in tyrosine phosphorylated proteins occur in temporal waves when platelets are stimulated and we have also found this (Murphy *et al.*, submitted for publication; Golden and Brugge, 1989; Ferrell and Martin, 1988). Recently tyrosine phosphorylation has been linked to the activation of phospholipase D, a potential source for diacylglycerol generation which might allow activation of protein kinase C (Uings *et al.*, 1992). We investigated the importance of tyrosine phosphorylation in platelet activation using the tyrosine kinase inhibitor genestein. Incubation of platelets with genestein at concentrations which inhibited intracellular tyrosine phosphorylation inhibited agonist-induced platelet aggregation. Indeed, further studies have shown that inhibition of tyrosine kinase activity inhibited the generation of several second messengers including increases in intracellular calcium, phosphoinositide production and prostaglandin synthesis (data not shown). Therefore we conclude that tyrosine kinase activity is essential for platelet acivation. Recently it has been shown that inhibition of tyrosine phosphatase activity is sufficient for platelet activation, therefore tyrosine phosphorylation needs to be tightly regulated (Pumiglia *et al.*, 1992).

Since pp60^{c-src} is the most abundant tyrosine kinase in platelets, and activation results in rapid induction of tyrosine phosphorylation, this enzyme has been implicated in platelet signal transduction. There is strong evidence that pp60^{v-src} associates with the cytoskeleton by its SH2 domain, and that this is required for full transformation as discussed previously. The SH2 domain is conserved in pp60^{c-src}, however several groups have reported that this cellular homologue does not associate with the cytoskelton in fibroblasts. Platelet activation also leads to a rapid change in cytoskeletal architecture, including the rearrangement of filamentous actin and associated proteins such as vinculin and talin, therefore we investigated whether pp60$^{c\text{-}src}$ interacted with the cytoskeleton of platelets in a manner similar to that found in fibroblasts. We found that in resting platelets pp60$^{c\text{-}src}$ was mainly detergent-soluble, a situation analagous to that of pp60$^{c\text{-}src}$ in fibroblasts. However after activation there was a translocation of pp60$^{c\text{-}src}$ to the detergent-insoluble cytoskeleton, and this process appeared to be dependent on the occupancy of the platelet integrin gpIIb/IIIa (Horvath *et al.,* 1992). Since this translocation occurs concurrently with aggregation it is possible that the interaction of pp60$^{c\text{-}src}$ with the cytoskeleton results in phosphorylation of cytoskeletal or membrane components such as integrin molecules. In support of this a close association between tyrosine phosphorylation and gpIIb/IIIa occupancy has been found (Ferrell and Martin, 1989; Golden *et al.,* 1990), and others have shown that the IIIa chain of gpIIb/IIIa can be a substrate for platelet pp60$^{c\text{-}src}$ (Elmore *et al.,* 1990; Findik *et al.,* 1990). Thus tyrosine phosphorylation of cytoskeletal-associated components may be important for normal platelet function.

References

Antler, A.M., Greenberg, M.E., Edelman, G.M. and Hanafusa, H. (1985). Increased phosphorylation of tyrosine in vinculin does not occur upon transformation by some avian sarcoma viruses. *Mol. Cell. Biol.,* **5**, 263-267.

Boschek, C.B., Jockusch, B.M., Friis, R.R., Back,R., Gandemann, E. and Bauer, H. (1981). Early changes in the distribution and organisation of micro-filament proteins during cell transformation. *Cell* **24**, 175-185.

Buss, J.E., Kamps, M.P. and Sefton, B.M. (1984) Myristic acid is attached to the transforming protein of Rous sarcoma virus during or immediately after synthesis and is present in both soluble and membrane-bound forms. *Mol. Cell. Biol.* **4**, 454-467.

Chatila, T. A., Geha, R.S. and Arnaout, M. A. (1989). Constitutive and stimulus-induced phosphorylation of CD11/CD18 leukocyte adhesion molecules. *J. Cell Biol.* **109**, 3435-3444.

Chen, W-T., Wang, J., Hasegawa, T., Yamada, S. S. and Yamada, K. M. (1986). *J. Cell Biol.* **103**, 1649-1661.

Cooper, J.A. and Hunter, T. (1983). Regulation of cell growth and transformation by the tyrosine-specific protein kinases: the search for important cellular substrate proteins. *Curr. Top. Microbiol. Immunol.* **107**, 125-162.

Cross, F.R., Garber, E.A., Pellman, D. and Hanafusa, H. (1984). A short sequence in the pp60^{src} N terminus is required for myristylation and membrane association and for cell transformation. *Mol. Cell Biol.* **4**, 1834-1842.

Declue, J. E., and Martin, G. S. (1987). Phosphorylation of talin at tyrosine in Rous sarcoma virus transformed cells. *Mol. Cell. Biol.* **7**, 371-378.

Elmore, M.A., Anand, R., Horvath, A.R. and Kellie, S. (1990). Tyrosine-specific phosphorylation of gpIIIa in platelet membranes. *FEBS Letts.* **269**, 283-287.

Enrietto, P., Payne, L.N. and Wyke, J.A. (1983) Analysis of the pathogenicity of transformation defective mutants of avian sarcoma virus: characterisation of recovered viruses which encode novel *src* specific proteins.*Virology,* **127**, 397-411.

Felice, G.R., Eason, P., Nermut, M.V. and Kellie, S. (1990). pp60$^{v\text{-}src}$ association with the cytoskeleton induces actin reorganization without affecting polymerisation status. *Eur. J. Cell Biol.* **52**, 47-59.

Ferrell, J.E. and Martin, G.S. (1989). Tyrosine-specific protein phosphorylation is regulated by glycoprotein IIb-IIIa in platelets. *Proc. Natl. Acad. Sci. USA* **86**, 2234-2238.

Findik, D., Reuter, C. and Presek, P. (1990). Platelet membrane glycoproteins IIb and IIIa are substrates of purified pp60$^{c\text{-}src}$ protein tyrosine kinase. *FEBS Letts.* **262**, 1-4.

Glenney, J.R., Zokas, L. and Kamps, M.P. (1988). Monoclonal antibodies to phosphotyrosine. *J. Immunol. Meth.* **109**, 277-288.

Glenney, J.R. and Zokas, L. (1989). Novel tyrosine kinase substrates from Rous sarcoma virus-transformed cells are present in the membrane skeleton. *J. Cell Biol* **108**, 2401-2408.

Golden, A. and Brugge, J.S. (1989). Thrombin treatment induces rapid changes in tyrosine phosphorylation in platelets. *Proc. Natl. Acad. Sci. USA* **86**, 901-905.

Golden, A., Brugge, J.S. and Shattil, S.J. (1990). Role of platelet glycoprotein IIb-IIIa in agonist-induced tyrosine phosphorylation of platelet proteins. *J. Cell Biol.* **111**, 3117-3127.

Guan, J-L., Trevithick, J.E. and Hynes, R.O. (1991). fibronectin/integrin interaction induces tyrosine phosphorylation of a 120-kDa protein. *Cell Reg.* **2**, 951-964.

Guan, J-L. and Shalloway, D. (1992) Regulation of focal adhesion-associated protein tyrosine kinase by both cellular adhesion and oncogenic transformation. *Nature* **358**, 690-692.

Hamaguchi, M. and Hanafusa, H. (1987). Association of pp60^{vsrc} with triton X-100-resistant cellular structure correlates with morphological transformation. *Proc. Natl. Acad. Sci. USA* **84**, 2312-2316.

Hirst, R., Horwitz, A., Buck, C. and Rohrschneider, L. (1986). Phosphorylation of the fibronectin receptor complex in cells transformed by oncogenes that encode tyrosine kinases. *Proc. Natl. Acad. Sci. USA* **83**, 6470-6474.

Horvath, A. R., Elmore, M. A. and Kellie, S. (1990). Differential tyrosine-specific phosphorylation of integrin in Rous sarcoma virus transformed cells with differing transformed phenotype. *Oncogene* **5**, 1349-1357.

Horvath, A.R., Muszbek, L. and Kellie, S. (1992). Translocation of pp60^{c-src} to the cytoskeleton during platelet aggrgation. *EMBO J.* **11**, 855-861.

Hynes, R.O., Marcantonio, E.E., Stepp, M.A., Urry, L.A. and Yee, G.H. (1989) Integrin heterodimer and receptor complexity in avian and mammalian cells. *J. Cell Biol.*, **109**, 409-420.

Jove, J.A. and Hanafusa, H. (1987). Cell transformation by the viral *src* gene. *Ann. Rev. Cell Biol.* **3**, 31-56.

Kellie, S., Patel, B., Wigglesworth, N.M., Critchley, D. R. and Wyke, J. A. (1986a). The use of Rous sarcoma virus transformation mutants with differing tyrosine kinase activities to study the relationships between vinculin phosphorylation, pp60^{v-src} location and adhesion plaque integrity. *Exp. Cell Res.* **165**, 216-228.

Kellie, S., Patel, B., Wigglesworth, N.M., Mitchell, A., Critchley, D. R. and Wyke, J. A. (1986b). Comparison of the relative importance of tyrosine-specific vinculin phosphorylation and the loss of surface-associated fibronectin in the morphology of cells transformed by Rous sarcoma virus. *J. Cell Sci.* **82**, 129-142.

Kozma, L. M., Reynolds, A. B. and Weber, M. J. (1990) Glycoprotein tyrosine phosphorylation in Rous sarcoma virus-transformed cell lines. *Mol. Cell Biol.* **10**, 837-841.

Kreuger, J.G., Garber, E.A., Chin, S.S., Hanafusa, H. and Goldberg, A. (1983). Size-variant pp60^{src} proteins of recovered avian sarcoma viruses interact with adhesion plaques as peripheral membrane proteins: effect on cell transformation. *Mol. Cell. Biol.* **4**, 454-467.

Lassing, I. and Lindberg, U. (1985). Specific interaction between phosphatidylinositol 4,5-bisphosphate and profilactin. *Nature* **314**, 472-474.

Meyer, R.K., Schindler, H. and Burger, M.M. (1985). alpha-actinin interacts specifically with model membranes containing glycerides and fatty acids. *Proc. Natl. Acad. Sci. USA* **79**, 4280-4284.

Niggli, V., Dimitrov, V.P., Brunner, J. and Burger, M.M. (1986). Interaction of the cytoskeletal component vinculin with bilayer structures analysed by a photoactivatable phospholipid. *J. Biol. Chem.* **261**, 6912-6918.

Plantefaber, L.C. and Hynes, R.O. (1989). Changes in integrin receptors on oncogenically transformed ccll lines. *Cell* **56**, 281-290.

Pumiglia, K.M., Lau, L-F., Huang, C-K., Burroughs, S. and Feinstein, M.B. (1992). Activation of signal transduction in platelets by the tyrosine phosphatase inhibitor pervanadate (vanadyl hydroperoxide). *Biochem. J.* **286**, 441-449.

Rohrschneider, L.R. (1980). Adhesion plaques of Rous sarcoma virus-transformed cells contain the *src* gene product. *Proc. Natl. Acad. Sci. USA* **77**, 3514-3518.

Rohrschneider, L. and Rosok, M.J. (1983). Transformation parameters and pp60^{v-src} localisation in cells infected with partial transformation mutants of Rous sarcoma virus. *Mol. Cell. Biol.* **3**, 731-746.

Stoker, A., Kellie, S. and Wyke, J.A. (1986). Intracellular localisation and processing of pp60^{v-src}

proteins expressed by two distinct temperature-sensitive mutantsof Rous sarcoma virus. *J.Virol.* **58**, 876-883.

Tapley, P., Horwitz, A., Buck, C., Duggan, K. and Rohrschneider, L. (1989). Integrins isolated from Rous sarcoma virus-transformed chicken embryo fibroblasts. *Oncogene* **4**, 325-333.

Tones, M., Kellie, S. and Hawthorne, J.N. (1987). Elevated phosphatidylinositol kinase activity in Rous sarcoma virus-transformed cells. Lack of evidence for enzyme translocation. *Biochim. Biophys. Acta* **931**, 165-169.

Uings, I.J., Thompson, N.T., Randall, R.W., Spacey, R.W., Bonser, R.W., Hudson, A.T. and Garland, L.G. (1992). Tyrosine phosphorylation is involved in receptor coupling to phospholipase D but not phospholipase C in the human neutrophil. *Biochem. J.* **281**, 597-600.

Welham, M.J. and Wyke, J.A. (1988). A single point mutation has pleiotropic effects on pp60[v-src] function. *J. Virol.* **62**, 1898-1906.

Yarden and Ullrich (1988). Growth factor receptor tyrosine kinases. *Ann. Rev Biochem.* **57**, 443-478.

Printed in Great Britain © The Society of Experimental Biology 1993   283

ROLE OF CYCLIC GMP IN SIGNAL TRANSDUCTION TO CYTOSKELETAL MYOSIN

GANG LIU and PETER C. NEWELL

Department of Biochemistry, University of Oxford, South Parks Road, Oxford, OX1 3QU, UK

Summary

Evidence is presented for cyclic GMP having a role as a secondary messenger connecting the cell surface cyclic AMP receptors and cytoskeletal myosin II involved in chemotaxis of amoebae of *Dictyostelium*.

Studies were conducted using mutants whose primary defect is in the structural gene for the cyclic GMP-specific phosphodiesterase (streamer F mutants). These mutants show abnormally prolonged accumulation of cyclic GMP in response to stimulation with the chemoattractant cyclic AMP. Investigation of signal transduction in these mutants indicated that, while events associated with production and relay of cyclic AMP signals were normal, certain events associated with movement were (like the cyclic GMP response) abnormally prolonged and these included myosin II association with the cytoskeleton and inhibition of myosin heavy and light chain phosphorylation. These events can be correlated with the amoebae becoming elongated and transiently decreasing their locomotive speed after chemotactic stimulation. Other mutants studied in which the accumulation of cyclic GMP was reduced or absent produced correspondingly reduced or absent myosin responses.

We propose a model in which cyclic GMP (transiently accumulated intracellularly in response to stimulation with extracellular cyclic AMP) induces accumulation of myosin II on the cytoskeleton by inhibiting phosphorylation of the myosin heavy chain. As a consequence, bending of the myosin tail and its dissociation from the cytoskeleton are inhibited.

Introduction

Stimulation of *Dictyostelium discoideum* amoebae with the chemoattractant cyclic AMP results in movement of the cells to the signal source. The first major observable event is the formation of a pseudopodium directed towards the incoming signal (Gerisch *et al.,* 1975). This event is correlated with changes in the actin content of the Triton-insoluble cytoskeleton. Normally the cytoskeleton contains about one third of the cell's actin but this doubles within five seconds after stimulation and then rapidly decreases again. (McRobbie and Newell, 1983, 1984a,b, 1985; Newell, 1986, Condeelis *et al.* 1988). Later (at about 40-60 sec) the amoebae begin to elongate and move in the direction of the stimulus. The period of elongation is correlated in its timing with the association of another macromolecule, myosin II, with the cytoskeleton (Liu and Newell, 1988).

A number of investigations have been carried out in several laboratories to

Key words: cyclic GMP, myosin, chemotaxis, *Dictyostelium*.

understand the role of myosin II in chemotaxis and the mechanism of its regulation. Yumura and Fukui (1985) reported that *Dictyostelium* myosin exists as thick filaments *in vivo* and these thick filaments translocate to the cortex at the rear of the cell in response to stimulation with cyclic AMP. Myosin has been shown to be non-essential for chemotaxis by the isolation of myosin heavy chain (MHC) null mutants (produced by gene disruption) (De Lozanne and Spudich 1987; Wessels *et al.* 1988) and by anti-sense RNA (Knecht and Loomis, 1987, 1988). However, these mutants were less efficient at chemotactic locomotion, and their more rounded shape suggests that the role of myosin II is to induce polarity for efficient chemotaxis rather than being the chemotactic motor.

Evidence for cyclic GMP being an important part of the chemotactic signal transduction chain comes from its transient formation in response to various chemoattractants in several species (Mato et al., 1977; Wurster *et al.*, 1977; Van Haastert and Konijn, 1982) and its formation in large amounts in the "streamer F" mutants (Ross and Newell, 1979, 1981). In the wild-type strain cyclic GMP is transiently formed in response to a cyclic AMP signal with a peak at 10 sec. In streamer F mutants the cyclic GMP is not destroyed so rapidly but persists for approximately five-fold longer (Figure 1A). The persistence of the cyclic GMP results from a defect in the structural gene for the cyclic GMP-specific phosphodiesterase, and several independent isolates in the same complementation group were found to have the same phenotype (Ross and Newell, 1981; Van Haastert *et al.*, 1982; Coukell and Cameron, 1986). The most obvious effect of this defective gene on the visible phenotype is that the amoebae remain in the elongated state during chemotaxis for approximately 5-fold longer than the parental strain XP55.

The evidence for the molecular link between cyclic GMP accumulation and the observed changes in cell shape and myosin II attachment to the cytoskeleton is described below.

Changes in myosin and actin in the streamer F mutants

The correlation of the prolonged cyclic GMP response with the prolonged period of elongation of the streamer F cells suggested that this nucleotide might affect some event connected with the (Triton-insoluble) cytoskeleton. However, actin, the first molecule to be investigated, was found by McRobbie and Newell (1984b) to be unaffected. It was found that the rapid accumulation of F-actin in the cytoskeleton that peaks at 5 seconds after a cyclic AMP stimulus was observed in the streamer F mutants and the parental strain in identical fashion. In contrast, when myosin was studied in the streamer mutants by Liu and Newell (1988) it was found that its association with the cytoskeleton was dramatically different. After an initial small drop (seen in both mutants and parental strain) the myosin associated with the cytoskeleton rapidly increased and in the parental and wild type strains formed a peak at about 25 sec. In the streamer mutants, however, this peak was persistent and only slowly declined to basal values (Figure 1B). Such changes correlate well with the changes in cyclic GMP formation.

It is also of interest to note that during the period of myosin association with the cytoskeleton the speed of the amoebae dramatically decreases (Figure 2), an effect that is also prolonged in the streamer F mutants (Segall, 1992).

The myosin response in cyclic GMP-deficient mutants

In order to test the correlation between cyclic GMP and myosin changes seen in the streamer F mutants, other mutants were studied in which the cyclic GMP reponse was abnormal. One such class is the *ras* mutants. *D. discoideum* has been found to possess two *ras* genes (*Ddras* and *DdrasG*) that code for polypeptides highly homologous to

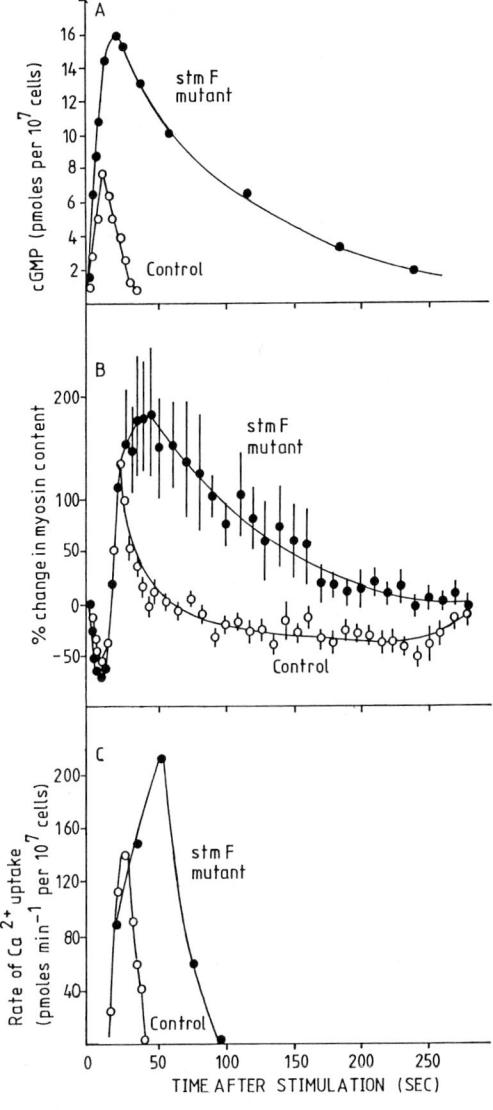

Fig. 1. Effect of cyclic AMP stimulation of XP55 and a streamer F mutant on (A) formation of cyclic GMP, (B) association of myosin II with the triton X-100 insoluble cytoskeleton; and (C) the rate of calcium uptake from the medium. From Newell and Liu (1992). The data for cyclic GMP are the combined data taken from Ross and Newell, (1981), Van Haastert *et al.* (1983) and Segall (1992). The myosin II figure is from Liu and Newell (1988), and the calcium data from Menz, *et al.* (1991).

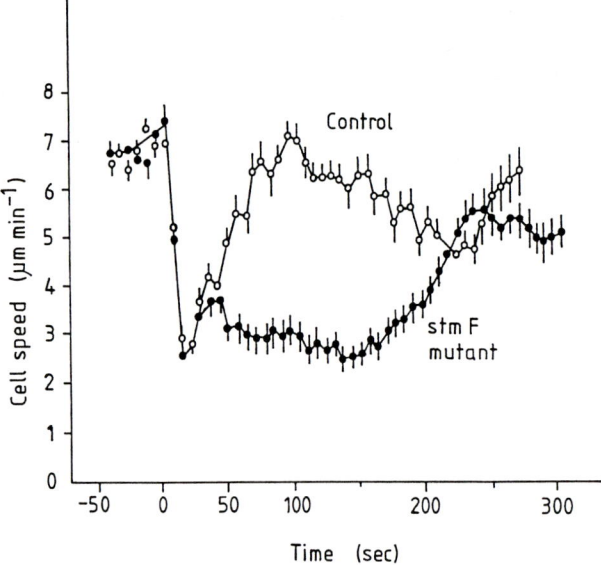

Fig. 2. Cell speed in response to cyclic AMP stimulation for strain XP55 and a streamer F mutant. Cells were incubated in a flow cell and 1 μM cyclic AMP introduced at time 0. Data are the mean and S.E.M (for XP55) of 119 cells from 14 flows on 3 different days and (for NP368) 140 cells from 15 flows on 5 different days. From Newell and Liu (1992). The data were redrawn from Segall (1992).

mammalian *ras* proteins (Reymond *et al.* 1984; Pawson *et al.* 1985; Weeks and Pawson, 1987; Robbins *et al.* 1989). Ddras has been cloned and re-introduced into *D. discoideum* and shown to be expressed (Reymond *et al.* 1985). Extra copies of the wild-type Ddras gene (*ras*-Gly12) or of a gene carrying a missense mutation in codon 12 (*ras*-Thr12) have been introduced into the amoebae by transformation (Reymond *et al.* 1986). The total *ras* protein levels are about fourfold higher in the high-copy *ras*-Gly12 and *ras*-Thr12 transformants than in control cells. Despite such abnormally high levels of *ras* protein, a study of the phenotypes of the cells transformed with multiple copies of the wild type *ras*-Gly12 gene revealed no measurable difference from the untransformed control. In contrast, amoebae transformed with multiple copies of the mutant *ras*-Thr12 gene showed reduced chemotactic sensitivity (and aberrant later development with the formation of multiple organising tips). When various components of the signal relay system connected with adenylate cyclase were measured, no aberration could be detected in the *ras*-Thr12-transformed cells that could explain the mutant phenotype. However, the formation of cyclic GMP in response to pulses of cyclic AMP was found consistently to be approximately 50-60% of the value in the *ras*-Gly12 cells (Reymond *et al.* 1986, 1989; Van Haastert *et al.* 1987) (Figure 3A,B). When the cytoskeletal actin response was measured in these cells, cyclic AMP stimulation of *ras*-Thr12 induced a normal response that was similar to the *ras*-Gly12 strain (Figure 3C,D). In contrast, the cytoskeletal MHC response in the *ras*-Thr12 cells showed only a weak MHC response (about half of the peak of ras-Gly12 in terms of maximum response) as might have been predicted from its weak cyclic GMP response (Figure 3E,F).

A mutant has also been found (KI-10) that lacks the cyclic GMP response completely and is unable to move chemotactically to cyclic AMP. While its cyclic AMP response and formation of Ins(1,4,5)P_3 are normal, this mutant fails to accumulate any significant amount of cyclic GMP in response to stimulation with 100 nM cyclic AMP (H. Kuwayama and S. Ishida, personal communication). While the mutant's actin response is similar to its parental strain, its MHC response is completely abolished (G. Liu, H.

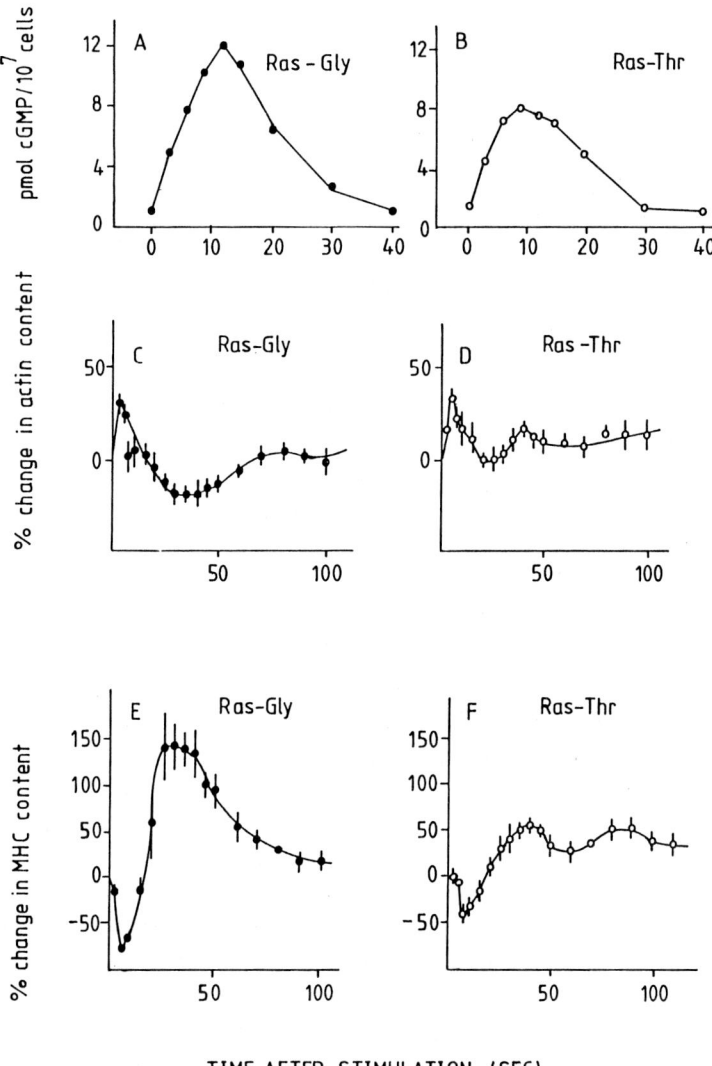

TIME AFTER STIMULATION (SEC)

Fig. 3. Comparison of responses to stimulation of amoebae with cyclic AMP in mutants *ras*-Gly$_{12}$ (left column) *ras*-Thr$_{12}$ (right column). The responses shown are: formation of cyclic GMP (A,B); change in actin content (C,D), and change in myosin heavy chain (MHC) content of the cytoskeleton. The data in panels A and B are redrawn from Van Haastert *et al.* (1987). Panels C,D,E and F show unpublished data of Liu and Newell.

Kuwayama and P.C. Newell, manuscript in preparation) further strengthening the notion that cyclic GMP regulates myosin in *Dictyostelium* cells.

The role of myosin heavy chain phosphorylation

Phosphorylation *in vitro* of *Dictyostelium* myosin heavy chain was first reported by Rahmsdorf *et al.* (1978) and later further studied by Malchow *et al.* (1981). Results of *in vivo* studies by Berlot *et al.* (1985, 1987) confirmed the existence of a transient phosphorylation of MHC (primarily on threonine residues) peaking at 30-40 sec after cyclic AMP stimulation. Based on analysis of the kinetics of myosin kinases (Berlot *et al.* 1987; Griffith *et al.* 1987), the authors suggested that the phosphorylation of MHC is substrate-regulated.

Kuczmarski and Spudich (1980) reported that dephosphorylation of *D. discoideum* MHC *in vitro* favoured the formation of myosin thick filaments and increased the activity of actin-activated myosin ATPase, and may, therefore, regulate cell contractile events. Several phosphorylation sites have been found in the tail region of MHC (Peltz *et al.* 1981; Pagh *et al.* 1984; Vaillancourt *et al.* 1988; Kuczmarski *et al.* 1988) and data from detailed investigations of *in vitro* phosphorylation of MHC all suggested that phosphorylation of MHC inhibits the formation of myosin thick filaments and imparts a bend in the tail region of the molecule (Kuczmarski *et al.* 1987; Cote and McCrea, 1987; Pasternak *et al.* 1989). In *Acanthamoebae*, *in vitro* phosphorylation of MHC has been shown to inhibit myosin thick filament formation (Collins *et al.* 1982). It is in connection with such phosphorylation of myosin *in vivo* that cyclic GMP may be involved in *Dictyostelium*.

Recent studies by Liu and Newell (1991) have revealed that phosphorylation of the myosin heavy chain is abnormal in the streamer F mutants in being considerably delayed compared to the parental strain, with a peak at about 80 sec rather than 30 sec (Figure 4). The timing of this phosphorylation correlates, not with the association of the myosin with the cytoskeleton, but rather with its dissociation. A bending of the molecule due to phosphorylation would greatly reduce its ability to associate or remain associated in the form of thick filaments. On this basis it was suggested by Liu and Newell that phosphorylation of the myosin heavy chain removes it from the cytoskeleton. In support of this it was found that little of the myosin that was on the cytoskeleton was phosphorylated compared to that in the soluble cell fraction. Recently it has also been found that in mutant KI-10 (which lacks cyclic AMP-induced cyclic GMP formation and cytoskeletal myosin accumulation) cyclic AMP also fails to stimulate any observable myosin heavy chain phosphorylation (G. Liu, H. Kuwayama and P.C. Newell, manuscript in preparation).

A model was proposed (Figure 5) in which (in the parental strain) the enzyme that phosphorylates the myosin (myosin heavy chain kinase) is transiently inhibited by the formation of peak of cyclic GMP. In the unstimulated cell the addition of myosin to the cytoskeleton is in equilibrium with its removal. Transient inhibition of its removal would have the effect of allowing more myosin to be added than removed during this time. After the cyclic GMP is hydrolysed the original state would be rapidly restored as

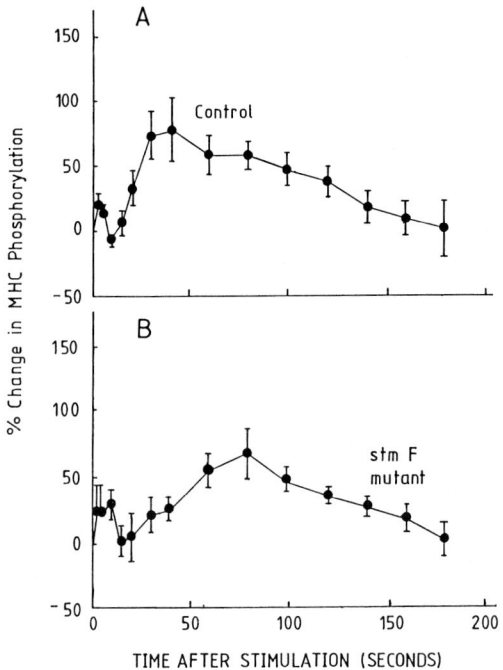

Fig. 4. Changes in myosin heavy chain phosphorylation after cyclic AMP stimulation in parental strain XP55 and a streamer F mutant. Results are expressed as percentage change over prestimulus value. Error bars represent S.E.M. (A) XP55, and (B) a streamer F mutant. From Liu and Newell, 1991.

phosphorylation restarted and removed the excess myosin from the cytoskeleton. In the streamer mutants, the extended period during which cyclic GMP is present would delay this renewed phosphorylation of the myosin and would extend the period during which myosin was associated with the cytoskeleton, as observed.

The link between cyclic GMP and myosin heavy chain phosphorylation

The mechanism of the proposed regulation of myosin phosphorylation by cyclic GMP in *Dictyostelium* may be direct or indirect via (for example) metal ion movements (Liu and Newell, 1991). We discuss the implications of the two types of models below.

(a) Direct models

We postulate that in such models the cyclic GMP acts directly on myosin heavy chain kinase and inhibits its activity (Figure 5). (The effect could, in principle, be by stimulation of a specific phosphatase activity, but from experience of other regulatory systems this seems less likely and no such specific phosphatase has yet been reported). Myosin heavy chain kinases from the insoluble cell fraction and from the cytoplasmic fraction have been detected, but while they can undergo autophosphorylation, those studied by Cote and Bukiejko (1987) and Ravid and Spudich (1989) seemed not to be affected by any of the second messengers tried, including cyclic GMP.

(b) Indirect models

The impetus for models in which cyclic GMP acts indirectly via metal ions such as calcium come from the report of Menz *et al.* (1991) that in the streamer F mutants the transient uptake of calcium ions normally seen in response to stimulation by cyclic AMP is prolonged (Figure 1C). As the primary defect in the streamer F mutants is in the structural gene for cyclic GMP phosphodiesterase the effect on calcium uptake must in some way be related to the level of cyclic GMP but the mechanism for such a link is unknown. The other component of such a model is the presence of a MHC kinase that is inhibited by calcium ions (Figure 5). Such an MHC kinase (inhibited by Ca^{2+}/calmodulin) has been reported by Maruta *et al.* (1983) but it has not been shown that this is the enzyme that phosphorylates myosin on the cytoskeleton.

Until further clarification of the properties of the functional MHC kinase has been achieved, it will be difficult to distinguish between the two types of models. In this connection it is of interest to note that a gene for myosin heavy chain kinase has recently been cloned (Ravid and Spudich, 1992). Surprisingly, this gene was found to contain the structural motif of protein kinase C (PKC) although it was distinct from all known PKCs. The purified enzyme however was not regulated by calcium, phospholipid and

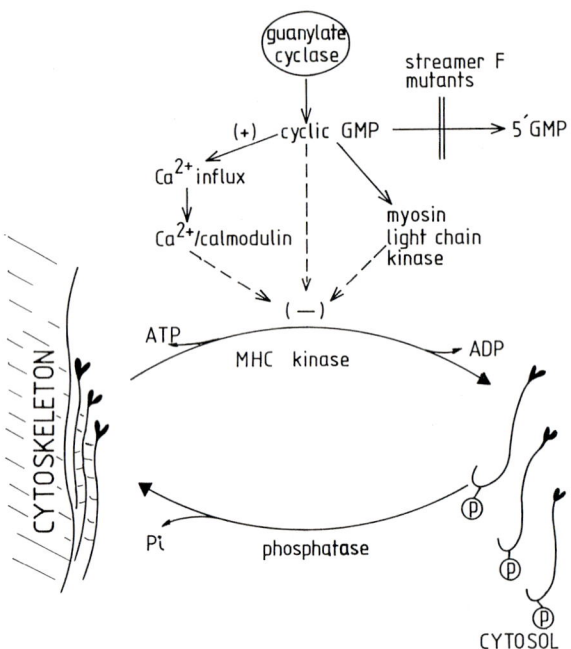

Fig. 5. Model of cyclic GMP regulation of the phosphorylation/dephosphorylation cycle of MHC. The phosphorylated MHC molecules are presented as bent monomers while the dephosphorylated MHC are in the form of parallel dimers involved in the formation of thick filaments. Cyclic GMP is shown inhibiting MHC phosphorylation via (a) a Ca^{2+}/calmodulin sensitive MHC kinase or (b) via a direct effect on the kinase, or (c) via effects on myosin light chain kinase. It is postulated that inhibition of myosin heavy chain phosphorylation induces a shift towards the accumulation of myosin on the cytoskeleton.

diacylglycerol as is normal for most mammalian enzymes (although it is possible that some factor allowing such regulation was lost during the enzyme purification).

Based on a recent study using immunofluorescence and ^{32}P-labelling of cytoskeletons, Yumura and Kitanishi-Yumura (1992) propose that during localised actin-myosin based contraction in the cytoskeleton, the MHC molecules move towards actin foci on the cell membrane and are phosphorylated by a specific MHC kinase. They also propose that this induces disassembly of filaments which are released into the endoplasm from the membrane-cytoskeleton region.

The role of myosin light chain phosphorylation

Myosin light chain phosphorylation is thought to be involved in various forms of motility of non-muscle cells (Collins *et al.*, 1982) and *Dictyostelium* has been found to possess a phosphorylatable 18 kDa "regulatory" light chain in addition to a 16 kDa

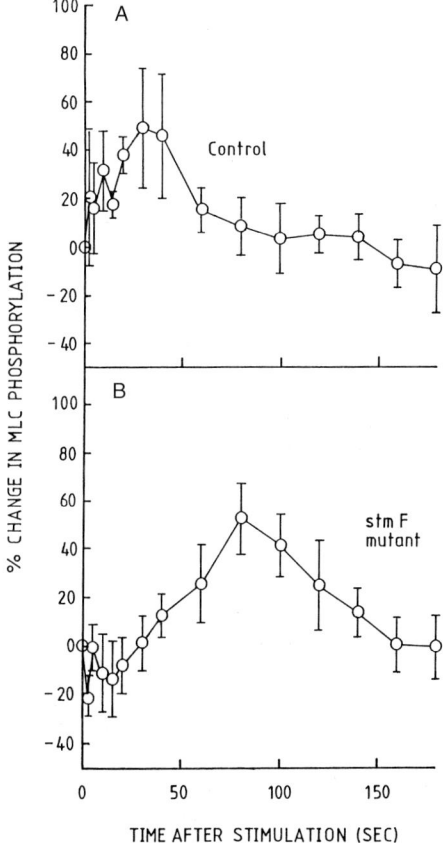

Fig. 6. Changes in myosin light chain phosphorylation after cyclic AMP stimulation in parental strain XP55 and a streamer F mutant. Results are expressed as percentage change over prestimulus value. Error bars represent S.E.M. (A) XP55, and (B) a streamer F mutant (from Liu and Newell, in preparation).

"essential" non-phosphorylated myosin light chain (Kuczmarski and Spudich, 1980). Berlot *et al.* (1985, 1987) observed transient increases in the level of myosin light chain phosphorylation after cyclic AMP stimulation which it was deduced resulted from activation of a myosin light chain kinase. In most systems studied, phosphorylation of myosin light chain is mediated by calcium/calmodulin activation of myosin light chain kinase (MLCK). However, recent studies of Tan and Spudich (1990a,b, 1991) who purified the enzyme in *Dictyostelium* to near homogeneity and obtained a full-length cDNA clone, have revealed that, although the enzyme does have some sequence homology to the calcium/calmodulin binding domain of other MLCK enzymes, this region does not satisfy the criteria of forming the basic amphiphilic alpha-helical structure that is essential for calcium/calmodulin binding. However, the enzyme does contain concensus cyclic GMP-dependent and cyclic AMP protein kinase phosphorylation sites defined by the sequences Ser/Thr-X-B and B-B-X-Ser/Thr (respectively), (where X is any residue and B is a basic amino acid). A possible mechanism of regulation of myosin light chain phosphorylation is therefore by cyclic GMP or cyclic AMP-dependent protein kinases.

It is of interest, therefore, to know whether the pattern of myosin light chain phosphorylation in *Dictyostelium* was altered in streamer F mutants. The results of recent studies of Liu and Newell (manuscript in preparation) are shown in Figure 6. For these experiments the light chain associated with the heavy chain (rather than the total pool of light chain) was examined after phosphorylation *in vivo* in response to stimulation of the amoebae with cyclic AMP. The results show that in the parental strain the light chain is phosphorylated at about 30 sec (in agreement with previous studies of Berlot *et al.*, 1987). In contrast, in the streamer F mutant this phosphorylation was delayed and was seen as a broad peak around 80 sec after stimulation.

Clearly cyclic GMP is directly or indirectly implicated in MLCK regulation, but the detailed significance of these findings is as yet hard to determine. It is possible that cyclic GMP inhibits the myosin heavy chain via regulation of the light chain kinase (as shown as a possibility in Figure 5), although an effect on the MLCK via the state or position of the whole myosin molecule seems equally plausible.

The link between the cell surface cyclic AMP receptors and formation of cyclic GMP

Despite a number of studies, the connection between the cell surface cyclic AMP receptors and guanylate cyclase is far from being understood. The evidence that this connection might involve the inositol pathway comes from work showing that, when amoebae (made permeable with saponin) were treated with 5 μM Ins(1,4,5)P_3 or 60 μM Ca^{2+}, they responded by forming a peak of cyclic GMP that closely resembled that produced by cyclic AMP with non-permeabilized cells in both its timing and duration (Europe-Finner and Newell, 1985; Small et al., 1986). However, guanylate cyclase has been reported to be inhibited rather than activated by Ca^{2+} *in vitro* (Padh and Brenner, 1984; Janssens and Jong, 1988). While this discepancy could conceivably be due to changes in a regulatory factor during the enzyme extraction procedure or possibly be explained by another unknown step between Ca^{2+} and guanylate cyclase, such explanations are rendered less likely by the recent finding that, with electro-permeabilized (rather than saponin-permeabilized) amoebae, calcium was found to inhibit the basal level of cyclic GMP *in vivo* at similar concentrations to its effects on guanylate cyclase *in vitro* (Van Duijn and Van Haastert, 1992). Recent data of Bominaar *et al.* (1991), moreover, have shown that a mutant that was defective in chemotaxis (due to a mutation in the *fgd*C gene) was fully able to respond to cyclic AMP stimulation by accumulating cyclic GMP but failed to show the normal accumulation of Ins(1,4,5)P_3 (instead it showed a drop in the concentration of this signalling intermediate). These data strongly suggest that guanylate cyclase must be capable of being stimulated by a connection to the cell surface receptors other that via the inositol phosphate pathway but details of this connection are as yet obscure.

References

Berlot, C. H., Spudich, J. A., and Devreotes, P. N. (1985). Chemoattractant-elicited increases in myosin phosphorylation in *Dictyostelium*. *Cell*, **43**, 307-314.

Berlot, C. H., Devreotes, P. N., and Spudich, J. A. (1987). Chemoattractant elicited increases in *Dictyostelium* myosin phosphorylation are due to changes in myosin localization and increases in kinase activity. *J.Biol.Chem.* **262**, 3918-3926.

Bominaar, A. A., Kesbeke, F., Snaarjagalska, B. E., Peters, D. J. M., Schaap, P., and Vanhaastert, P. J. M. (1991). Aberrant chemotaxis and differentiation in *Dictyostelium* Mutant *fgdC* with a defective regulation of receptor-stimulated phosphoinositidase-C. *J. Cell Sci.* **100**, 825-831.

Collins, J.H., Kuznicki, J., Bowers, B. and Korn, E.D. (1982) Comparison of the actin binding and filament formation properties of phosphorylated and dephosphorylated *Acanthamoeba* myosin II. *Biochemistry*, **21**, 6910-6915.

Condeelis, J., Hall, A., Bresnick, A., Warren, V., Hock, R., Bennet, H., and Ogihara, S. (1988). Actin polymerization and pseudopod extension during amoeboid chemotaxis. *Cell Motility and Cytoskeleton*, **10**, 77-90.

Cote, G. P., and Bukiejko, U. (1987). Purification and characterization of a myosin heavy chain kinase from *Dictyostelium discoideum*. *J.Biol.Chem.* **262**, 1065-1072.

Cote, G. P., and McCrea, S. M. (1987). Selective removal of the carboxyl-terminal tail end of the *Dictyostelium* myosin 2 heavy chain by chymotrypsin. *J. Biol.Chem.* **262**, 13033-13038.

Coukell, M. B., and Cameron, A. M. (1986). Characterization of revertants of stmf mutants of *Dictyostelium discoideum*: evidence that stmF is the structural gene of the cGMP-specific phosphodiesterase. *Devel.Genet.* **6**, 163-177.

De Lozanne, A., and Spudich, J. A. (1987). Disruption of the *Dictyostelium* myosin heavy chain gene by homologous recombination. *Science*, **236**, 1086-1091.

Europe-Finner, G. N., and Newell, P. C. (1985). Inositol(1,4,5)trisphosphate induces cyclic GMP formation in *Dictyostelium discoideum*. *Biochem. Biophys. Res. Commun.* **130**, 1115-1122.

Gerisch, G., Malchow, D., Huesgen, A., Nanjundiah, V., Roos, W. and Wick, U. (1975). Cyclic AMP reception and cell recognition in *Dictyostelium discoideum*. In D. McMahon and C. F. Fox (Eds), *Developmental Biology, ICN-UCLA Symposia on Molecular and Cellular Biology*. W.A. Benjamin Inc.

Griffith, L. M., Downs, S. M., and Spudich, J. A. (1987). Myosin light chain kinase and myosin light chain phosphatase from *Dictyostelium* : effects of reversible phosphorylation on myosin structure and function . *J.Cell Biol.* **104**, 1309-1323.

Janssens, P. M. W., and De Jong, C. C. C. (1988). A magnesium-dependent guanylate cyclase in cell free preparations of *Dictyostelium discoideum*. *Biochem.Biophys. Res.Commun.*, **150**, 405-411.

Knecht, D. A., and Loomis, W. F. (1987). Antisense RNA inactivation of myosin heavy chain gene expression in *Dictyostelium discoideum*. *Science*, **236**, 1081-1085.

Knecht, D., and Loomis, W. F. (1988). Developmental consequences of the lack of myosin heavy chain in *Dictyostelium discoideum*. *Dev.Biol.*, **128**, 178-184.

Kuczmarski, E. R., Routsolias, L., and Parysek, L. M. (1988). Proteolytic fragmentation of *Dictyostelium* myosin and localization of the *in vivo* heavy chain phosphorylation site. *Cell.Motil.Cytoskel.*, **10**, 471-481.

Kuczmarski, E. R., and Spudich, J. A. (1980). Regulation of myosin self-assembly phosphorylation of *Dictyostelium* heavy chain inhibits formation of thick filaments. *Proc.Natl.Acad.Sci. USA*, **77**, 7292-7296.

Kuczmarski, E. R., Tafuri, S. R., and Parysek, L. M. (1987). Effect of heavy chain phosphorylation on the polymerization and structure of *Dictyostelium* myosin filaments. *J.Cell Biol.*, **105**, 2989-2997.

Liu, G., and Newell, P. C. (1988). Evidence that cyclic GMP regulates myosin interaction with the cytoskeleton during chemotaxis of *Dictyostelium*. *J.Cell Sci.*, **90**, 123-129.

Liu, G., and Newell, P. C. (1991). Evidence that cyclic GMP may regulate the association of Myosin-II heavy chain with the cytoskeleton by inhibiting its phosphorylation. *J Cell Sci.*, **98**, 483-490.

Malchow, D., Bohme, R., and Rahmsdorf, H. J. (1981). Regulation of phosphorylation of myosin heavy chair during the chemotactic response of *Dictyostelium* cells. *Eur.J.Biochem.*, **117**, 213-218.

Maruta, H., Baltes, W., Dieter, P., Marme, D., and Gerisch, G. (1983). Myosin heavy chain kinase inactivated by Ca^{2+}/calmodulin from aggregating cells of *Dictyostelium discoideum*. *EMBO J.*, **2**, 535-542.

Mato, J. M., Krens, F. A., Van Haastert, P. J. M., and Konijn, T. M. (1977). 3':5'-Cyclic AMP-dependent 3':5'-cyclic GMP accumulation in *Dictyostelium discoideum*. *Proc. Natl. Acad. Sci. USA*, **74**, 2348-2351.

McRobbie, S. J., and Newell, P. C. (1983). Changes in actin associated with cytoskeleton following chemotactic stimulation of *Dictyostelium discoideum*. *Biochem. Biophys. Res. Commun.*, **115**, 351-359.

McRobbie, S.J. and Newell, P. C. (1984a). Chemoattractant-mediated changes in cytoskeletal actin of cellular slime moulds. *J.Cell Sci.*, **68**, 139-151.

McRobbie, S. J., and Newell, P. C. (1984b). A new model for chemotactic signal transduction in *Dictyostelium discoideum*. *Biochem. Biophys. Res. Commun.*, **123**, 1076-1083.

McRobbie, S. J., and Newell, P. C. (1985). Effect of cytochalasin B on cell movements and chemoattractant elicited actin changes in *Dictyostelium*. *Exp. Cell Res.*, **160**, 275-286.

Menz, S., Bumann, J., Jaworski, E., and Malchow, D. (1991). Mutant analysis suggests that cyclic GMP mediates the cyclic AMP-induced Ca^{2+} uptake in *Dictyostelium*. *J. Cell Sci.*, **99**, 187-191.

Newell, P. C. (1986). The role of actin polymerization in amoebal chemotaxis. *BioEssays*, **5**, 208-211.

Newell, P.C. and Liu, G. (1992). Streamer F mutants and chemotaxis of *Dictyostelium*. *BioEssays*, **14**, 473-479.

Padh, H., and Brenner, M. (1984). Studies of the guanylate cyclase of the social amoeba *Dictyostelium discoideum*. *Arch. Biochem. Biophys.*, **229**, 73-80.

Pagh, K., Maruta, H., Claviez, M., and Gerisch, G. (1984). Localization of two phosphorylation sites adjacent to a region important for polymerization on the tail of *Dictyostelium* myosin. *EMBO J.*, **3**, 3271-3278.

Pasternak, C., Flicker, P. F., Ravid, S., and Spudich, J. A. (1989). Intermolecular versus intramolecular interactions of *Dictyostelium* myosin: Possible regulation by heavy chain phosphorylation. *J.Cell Biol.*, **109**, 203-210.

Pawson, T., Amiel, T., Hinze, E., Auersperg, N., Neave, N., Sobolewski, A., and Weeks, G. (1985). Regulation of a ras-related protein during development of *Dictyostelium discoideum*. *Mol. Cell. Biol.*, **5**, 33-39.

Peltz, G., Kuczmarski, E. R., and Spudich, J. A. (1981). *Dictyostelium* myosin: Characterization of chymotryptic fragments an localization of the heavy-chain phosphorylation site. *J. Cell Biol.*, **89**, 104-108.

Rahmsdorf, H. J., Malchow, D., and Gerisch, G. (1978). Cyclic AMP-induced phosphorylation in *Dictyostelium discoideum* of polypeptide comigrating with myosin heavy chains. *FEBS lett.*, **88**, 322-326.

Ravid, S., and Spudich, J. A. (1989). Myosin heavy chain kinase from developed *Dictyostelium* cells purification and characterization. *J. Biol. Chem.*, **264**, 15144-15150.

Ravid, S., and Spudich, J. A. (1992). Membrane-bound *Dictyostelium* myosin heavy chain kinase: A developmentally regulated substrate-specific member of the protein kinase C family. *Proc. Natl. Acad. Sci. USA*, **89**, 5877-5881.

Reymond, C. D., Gomer, R. H., Mehdy, M. C., and Firtel, R. A. (1984). Developmental regulation of a Dictyostelium gene encoding a protein homologous to mammalian *ras* protein. *Cell*, **39**, 141-148.

Reymond, C. D., Nellen, W., and Firtel, R. A. (1985). Regulated expression of *ras* gene constructs in Dictyostelium transformants. *Proc. Natl. Acad. Sci. USA*, **82**, 7005-7009.

Reymond, C. D., Gomer, R.H., Nellen, W., Theibert, A., Devreotes, P. and Firtel, R.A. (1986). Phenotypic changes induced by a mutated *ras* gene during the development of *Dictyostelium* transformants. *Nature*, **323**, 340-343.

Reymond, C. D., Luderus, M. E. E., Europe-Finner, G. N., Thompson, N. A., Burki, E., Van Driel, R., and Newell, P. C. (1989). Analysis of the ras gene function in *Dictyostelium discoideum*. In L. Bosch, B. Kraal, and A. Parmeggiani (Eds.), *The Guanine-Nucleotide Binding Proteins* (pp. 265-272). Plenum Publ. Corp.

Robbins, S. M., Williams, J. G., Jermyn, K. A., Spiegelman, G. B., and Weeks, G. (1989). Growing and developing *Dictyostelium* cells express different ras genes. *Proc. Natl. Acad. Sci. USA*, **86**, 938-942.

Ross, F. M., and Newell, P. C. (1979). Genetics of aggregation pattern mutations in the cellular slime mould *Dictyostelium discoideum*. *J. Gen. Microbiol.*, **115**, 289-300.

Ross, F. M., and Newell, P. C. (1981). Streamers: Chemotactic mutants of *Dictyostelium discoideum* with altered cyclic GMP metabolism. *J. Gen. Microbiol.*, **127**, 339-350.

Segall, J. E. (1992). Behavioral responses of streamer F mutants of *Dictyostelium discoideum* - Effects of cyclic GMP on cell motility. *J. Cell Sci.*, **101**, 589-597.

Small, N. V., Europe-Finner, G. N., and Newell, P. C. (1986). Calcium induces cyclic GMP formation in *Dictyostelium*. *FEBS Lett.*, **203**, 11-14.

Tan, J. L., and Spudich, J. A. (1990a). Developmentally Regulated Protein-Tyrosine Kinase Genes in *Dictyostelium discoideum*. *Mol. Cell Biol.*, **10**, 3578-3583.

Tan, J. L., and Spudich, J. A. (1990b). *Dictyostelium* myosin light chain kinase - purification and characterization. *J. Biol. Chem.*, **265**, 13818-13824.

Tan, J. L., and Spudich, J. A. (1991). Characterization and bacterial expression of the *Dictyostelium* myosin light chain kinase cDNA - Identification of an autoinhibitory domain. *J. Biol. Chem.* **266**, 16044-16049.

Vaillancourt, J. P., Lyons, C., and Cote, G. P. (1988). Identification of two phosphorylated threonines in the tail region of *Dictyostelium* myosin II. *J. Biol. Chem.*, **263**, 10082-10087.

Van Duijn, B. and Van Haastert, P.J.M. (1992). Independent control of locomotion and orientation during *Dictyostelium discoideum* chemotaxis. *J. Cell Sci.*, **102**, 763-768.

Van Haastert, P. J. M., Kesbeke, F., Reymond, C.D., Firtel, R.A., Luderus, E. and Van Driel, R. (1987). Aberrant transmembrane signal transduction in *Dictyostelium* cells expressing a mutated *ras* gene. *Proc. Natl. Acad. Sci.,USA*, **84**, 4905-4909.

Van Haastert, P. J. M., and Konijn, T. M. (1982). Signal transduction in the cellular slime molds. *Mol. Cell. Endocrinol.*, **26**, 1-17.

Van Haastert, P. J. M., Van Lookeren Campagne, M. M., and Kesbeke, F. (1983) Multiple degradation pathways of chemoattractant mediated cyclic GMP accumulation in *Dictyostelium*. *Biochim. et Biophys. Acta*, **756**, 67-71.

Van Haastert, P. J. M., Van Lookeren Campagne, M. M., and Ross, F. M. (1982). Altered cGMP phosphodiesterase activity in chemotactic mutants of *Dictyostelium discoideum*. *FEBS Lett.*, **147**, 149-152.

Weeks, G., and Pawson, T. (1987). The synthesis and degradation of ras-related gene products during growth and differentiation in *Dictyostelium discoideum*. *Differentiation*, **33**, 207-213.

Wessels, D., Soll, D. R., Knecht, D., Loomis, W. F., De Lozanne, A., and Spudich, J. (1988). Cell motility and chemotaxis in *Dictyostelium* amebae lacking myosin heavy chain. *Dev. Biol.*, **128**, 164-177.

Wurster, B., Schubiger, K., Wick, U., and Gerisch, G. (1977). Cyclic GMP in *Dictyostelium discoideum*; Oscillations and pulses in response to folic acid and cyclic AMP signals. *FEBS Lett.*, **76**, 141-144.

Yumura, S., and Fukui, Y. (1985). Reversible cAMP-dependent change in distribution of myosin thick filaments in *Dictyostelium*. *Nature*, **314**, 194-196.

Yumura, S., and Kitanishi-Yumura, T. (1992). Release of Myosin II from the membrane-cytoskeleton of *Dictyostelium discoideum* mediated by heavy chain phosphorylation at the foci within the cortical actin network. *J. Cell Biol.*, **117**, 1231-1239.

Printed in Great Britain © The Society of Experimental Biology 1993   297

ACTIN-ASSOCIATED PROTEINS IN MOTILITY AND CHEMOTAXIS OF *DICTYOSTELIUM* CELLS

G. GERISCH, R. ALBRECHT, E. DE HOSTOS, E. WALLRAFF, C. HEIZER, M. KREITMEIER, *and* A. MÜLLER-TAUBENBERGER

Max-Planck-Institut für Biochemie, D-82143 Martinsried, Germany

Summary

The amoeboid cells of *Dictyostelium discoideum* are amenable to a combined biochemical, genetic, and cell biological approach that can be focussed to the study of molecular interactions underlying the chemotactic responses of eukaryotic cells. In these responses the actin-based motility system is involved. This system is characterised in *Dictyostelium* cells by a large number and variety of regulatory proteins. Most of these proteins belong to families that are likewise represented in the cytoskeletons of higher eukaryotes including man. Elimination of some of these actin-binding proteins by chemical mutagenesis or gene disruption is being used to simplify the system by separating essential proteins from non-essential ones. These studies are complemented by the selection and analysis of mutants with altered motility or chemotaxis. Quantitative motion analysis of mutants is employed to establish a link between defects on the molecular level and alterations in cell behaviour. *Dictyostelium* cells respond to local stimulation by extending a newly formed leading edge towards a chemoattractant within less than a minute, thereby changing their polarity. The leading edge is formed by the recruitment of soluble proteins from the cytoplasm and their coassembly with actin into a complicated framework of microfilaments. Patterns of assembly are shown in this report for two proteins, the talin-like filopodin and coronin. Elucidation of the control mechanisms of this ordered assembly will provide the key for understanding the molecular processes responsible for a chemotactic response.

Introduction

The chemotactic responses in aggregating cells of *Dictyostelium* and in neutrophilic granulocytes resemble each other with respect to cell behaviour, signal transduction, and changes in the actin skeleton that are thought to underlie cell orientation in a gradient of attractant (Devreotes and Zigmond, 1988). In *Dictyostelium discoideum* the chemotactic orientation towards cyclic AMP has been analysed by combining molecular genetics, biochemistry of signal transducing and actin-associated proteins, and quantification of cell behaviour in wild-type and mutants. Cells of this microorganism have two major advantages for these investigations. First, they can be easily propagated in mass culture to obtain sufficient amounts of proteins in order to assay their function in vitro. Second,

Key words: chemotaxis, actin-binding proteins, cytoskeleton, cell motility, *Dictyostelium*, supramolecular structures.

since *D. discoideum* cells are haploid, phenotypic changes are immediately apparent in mutants that are produced by gene disruption or gene replacement.

Chemotactic signals are recognized on the cell surface by cAMP receptors which are members of a family of rhodopsin-related proteins that have seven putative trans-membrane regions (Klein *et al.*, 1988). The activated receptors interact at the inside of the

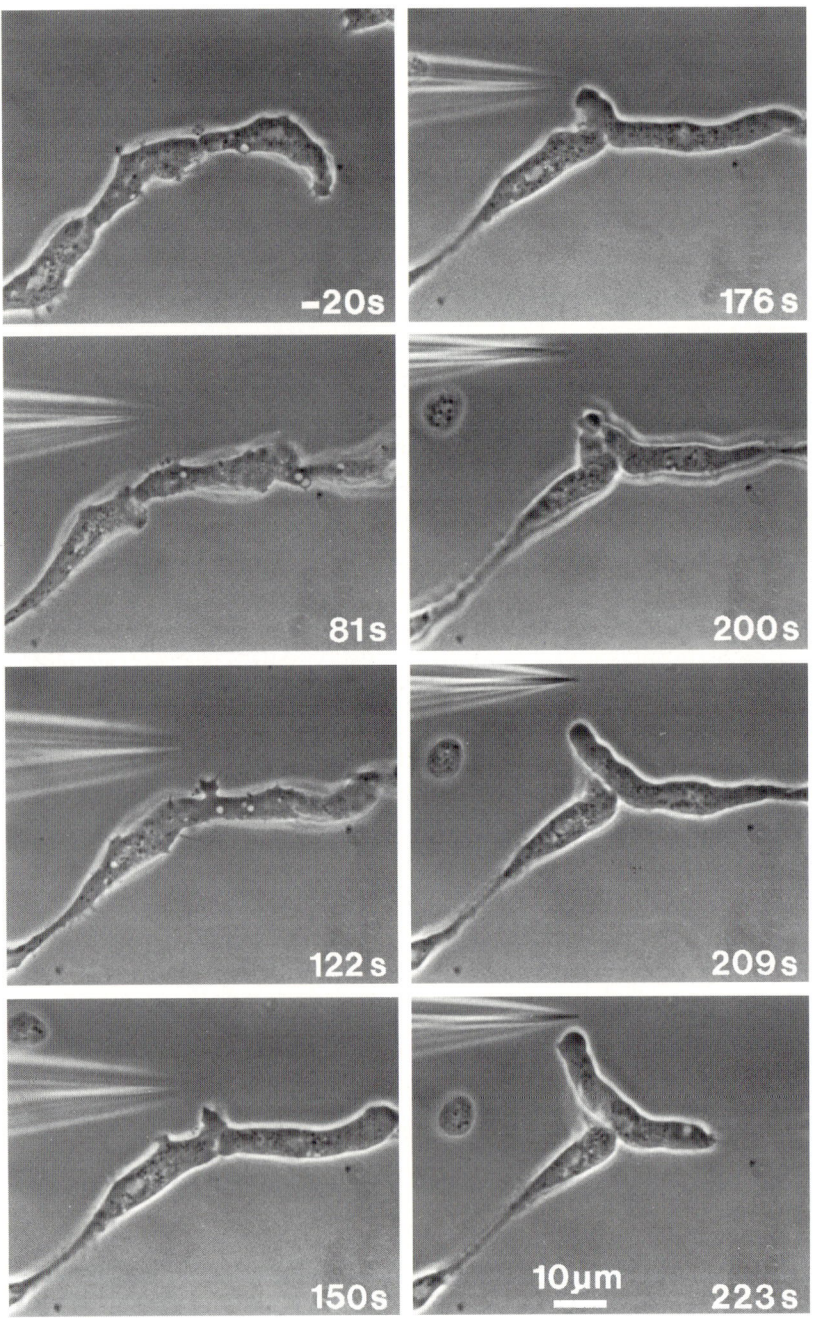

plasma membrane with heterotrimeric G-proteins (Pupillo *et al.*, 1989; Firtel, 1991; Kumagai *et al.*, 1991). Evidence has been accumulated that signals are intracellularly propagated through PIP_2, Ca^{2+}, and cAMP-elicited activation of guanylate cyclase (Ross and Newell, 1981; Newell *et al.*, 1990; Segall, 1992; Liu and Newell, 1991). Transient increases in cGMP and the activity of a protein kinase C-related myosin heavy-chain kinase (Ravid and Spudich, 1989; 1992) are involved in linking the regulation of myosin activities to the reception of extracellular cAMP signals.

When a cell is exposed to a gradient of cAMP, the cytoskeleton is reorganised such that either the leading edge of the cell is turned towards the cAMP source, or new extensions are protruded towards the source whereas the previous front of the cell is retracted (Fig. 1). Upon strong local stimulation with cAMP delivered from a micropipette, protrusions, indicating reorientation, are already seen within 5 seconds (Gerisch *et al.*, 1975). The possibility of inducing any part of the cell surface to take over the function of a front shows that the activities of a leading edge and those of a contracting rear end can be elicited anywhere on the cell surface (Segall and Gerisch, 1989). This observation indicates that the cells have no stable polarity.

The chemotactic response to defined stimuli can be quantified by placing cells in a chamber in which stable linear gradients can be established between two microdialysis tubes that provide a constant source and a sink of attractant (Fig. 2). The response of a large number of cells to these gradients can be simultaneously monitored by an image processing system permitting computer-based analysis of the tracks of moving cells. In an optimal stationary gradient, the cells show detectable orientation to concentration differences of cAMP that do not exceed 2 percent of the average concentration of this attractant between the front and rear end of a cell, this means over a distance of 10 to 20 µm (Fisher *et al.*, 1989).

We will show in this report that establishing a new front means that not only actin is redistributed within the cell but also a set of actin-binding proteins. These proteins are recruited from the cytoplasmic pool of soluble proteins and are assembled in an ordered way, thus forming the lamellipods and filopods that make up the leading edge.

Regulation of the microfilament system by covalent and non-covalent modification of its components

The reorientation of ameoboid cells during chemotaxis requires regulatory mechanisms that are subject to fast alterations when the polarity of a cell is changed. In this report we will concentrate on the regulation of activities in the actin-myosin system, the major motor

---

Fig. 1. Reversal of cell polarity during the chemotactic response. *D. discoideum* cells that had started to aggregate into streams were stimulated by a micropipette filled with cAMP. Time in seconds before and after insertion of the micropipette is indicated. The 176 s and 200 s frames show that end-to-end adhesion of the elongated cells is disrupted in response to a lateral gradient, and that the cell on the right-hand side forms a front at its previous rear end. Cells were incubated on a glass surface in 17 mM Na/K-phosphate buffer, pH 6.0, at 23°C as described for leukocytes and *D. discoideum* cells (Claviez *et al.*, 1986; Gerisch and Keller, 1981).

system in the amoeboid cells of *D. discoideum*. This system can be regulated in different ways. (1) By covalent modification, specifically phosphorylation, of the two major proteins, myosin and actin; (2) by the binding of other proteins to one of these proteins. About 20 actin-binding proteins purified from *Dictyostelium* cells have already been characterised (Luna and Condeelis, 1990; Schleicher and Noegel, 1992). The activities of some of these proteins are known to be controlled by non-covalent binding of ligands. Several proteins are activated or inactivated by Ca^{2+}, e.g. the F-actin fragmenting protein severin, the F-actin crosslinker α-actinin (for review see Gerisch *et al.*, 1991), and a recently identified 100 kDa capping protein (Hofmann *et al.*, 1992). A surprisingly large number of actin-binding proteins are regulated by PIP_2, which may link the actin system to signal transduction pathways (Kwiatkowski *et al.*, 1989; Goldschmidt-Clermont *et al.*, 1990; Haus *et al.*, 1991; Eichinger and Schleicher, 1992; Hofmann *et al.*, 1992).

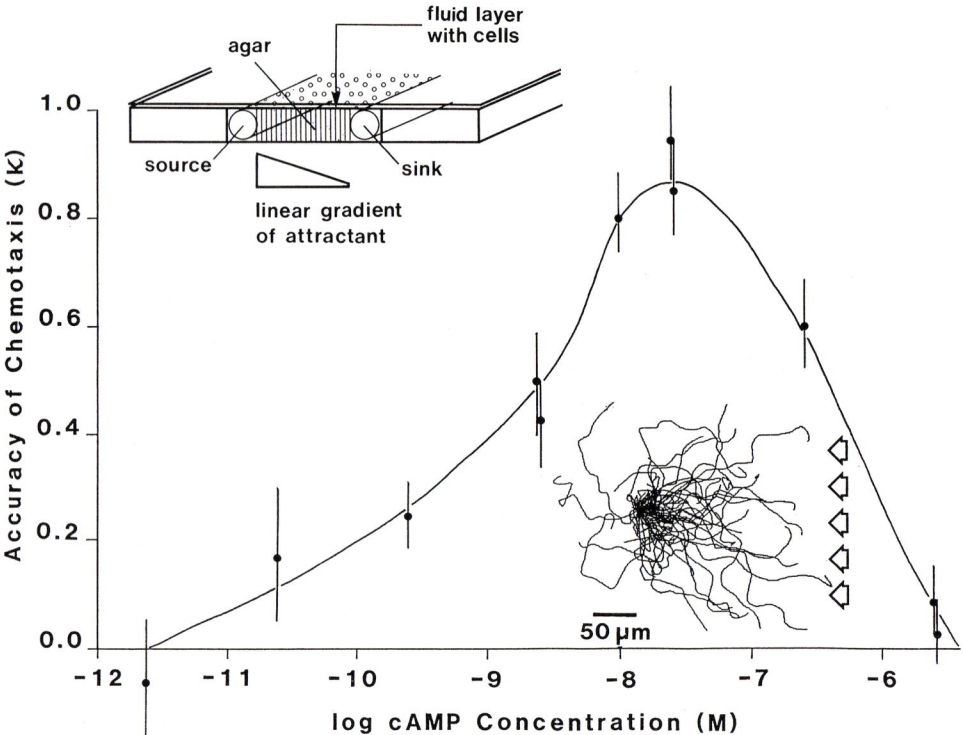

Fig. 2. Quantification of the chemotactic response of *D. discoideum* cells in a stationary gradient of cAMP. The accuracy of chemotactic orientation is plotted as a function of the mean cAMP concentration at the location of each cell (from Fisher *et al.*, 1989). The cells recorded were located in the middle of a source and a sink of the attractant, supplied by two microdialysis tubes that were separated by an agar bridge of 2 mm (insert on top). The steepness of the gradient was limited by that distance and dependent on the concentrations at the source and the sink (the latter was zero in the experiment shown). The insert below the curve shows tracks of cells in the optimum gradient of 25 nM/mm with midpoint 25 nM. The origins of the trails are all plotted to the same point. The arrows indicate direction of the gradient from high to low concentration.

An actin-binding protein with multiple regulatory inputs is hisactophilin (Scheel *et al.*, 1989). This protein exists in a soluble cytoplasmic form and also in a state attached to the inner side of the plasma membrane where it might link the actin skeleton to receptors sensing attractants (Fig. 3A). Hisactophilin is extremely rich in histidine residues which are located along flexible loops surrounding a rigid kernel that is made up of barrels of β-sheets (Habazettl *et al.*, 1992) (Fig. 3B). In accord with the pH of histidine, hisactophilin switches around pH 7.0 from an actin-binding state at lower pH to a non-binding state at higher pH (Scheel *et al.*, 1989). Since chemoattractants cause a change in the intracellular pH (Van Duijn and Inouye, 1991), the switch in the activity of hisactophilin may provide one of the links between chemotactic stimuli and changes in the actin system. When it is purified from *D. discoideum* cells, hisactophilin is found to be modified by phosphorylation and fatty-acid acylation (A. Müller-Taubenberger, unpublished results). Work is in progress to determine how these covalent modifications of hisactophilin regulate its activities and its attachment to the plasma membrane.

Actin and myosin phosphorylation

Dictyostelium actin can be phosphorylated in vivo at tyrosine and serine residues (Schweiger *et al.*, 1992). Regulation of the serine phosphorylation has not yet been investigated. The tyrosine phosphorylation of actin, which can be monitored by a phosphotyrosine-specific antibody, has been shown to change in response to growth conditions: the degree of tyrosine phosphorylation increases in form of a pulse when starving cells are retransferred to nutrient medium. It has not yet been established that this phosphorylation of actin at tyrosine residues plays a role in the control of cell movement. But the reversible rounding up of cells and the disappearance of actin filaments under conditions of reduced phosphotyrosine phosphatase activity, suggests that tyrosine phosphorylation of some important protein in the actin skeleton is involved in the control of cell motility (Schweiger *et al.*, 1992; Howard *et al.*, 1993).

A role of myosin II in motility and chemotaxis of *Dictyostelium* cells is well established (Wessels *et al.*, 1988; Spudich, 1989; Fukui *et al.*, 1990). In addition to the conventional double-headed myosin II, *D. discoideum* cells contain at least five members of the myosin I family that are membrane-associated proteins (Endow and Titus, 1992). Their specific motor functions and mode of regulation are difficult to investigate because the myosin I isoforms can apparently replace each other, at least partially (Wessels *et al.*, 1991).

Myosin II is phosphorylated at both threonine and serine residues of its heavy chains and at serine residues of its regulatory light chains (Berlot *et al.*, 1987; for threonine phosphorylation at heavy chains see Vaillancourt *et al.*, 1988; Lück-Vielmetter *et al.*, 1990). The changes in myosin II phosphorylation that occur in response to stimulation of the cells by cAMP are covered in this volume by Peter C. Newell. We concentrate here on studies based on the replacement of the three phosphorylatable threonine residues in myosin II heavy chains by alanine residues (Fig. 4). In order to evaluate the role of threonine phosphorylation, the phenotype of a "Triple Ala" mutant has to be compared not only with the wild-type phenotype but also with the phenotype of myosin null mutants. Cells without any myosin II heavy chains move more slowly, their chemotactic responsiveness is reduced, and development is arrested at the end of aggregation (Soll *et al.*, 1990; Wessels *et al.*, 1988; Peters *et al.*, 1988). The most striking effect of eliminating myosin II heavy chains is observed in suspension culture, where an impairment of cytokinesis leads to the

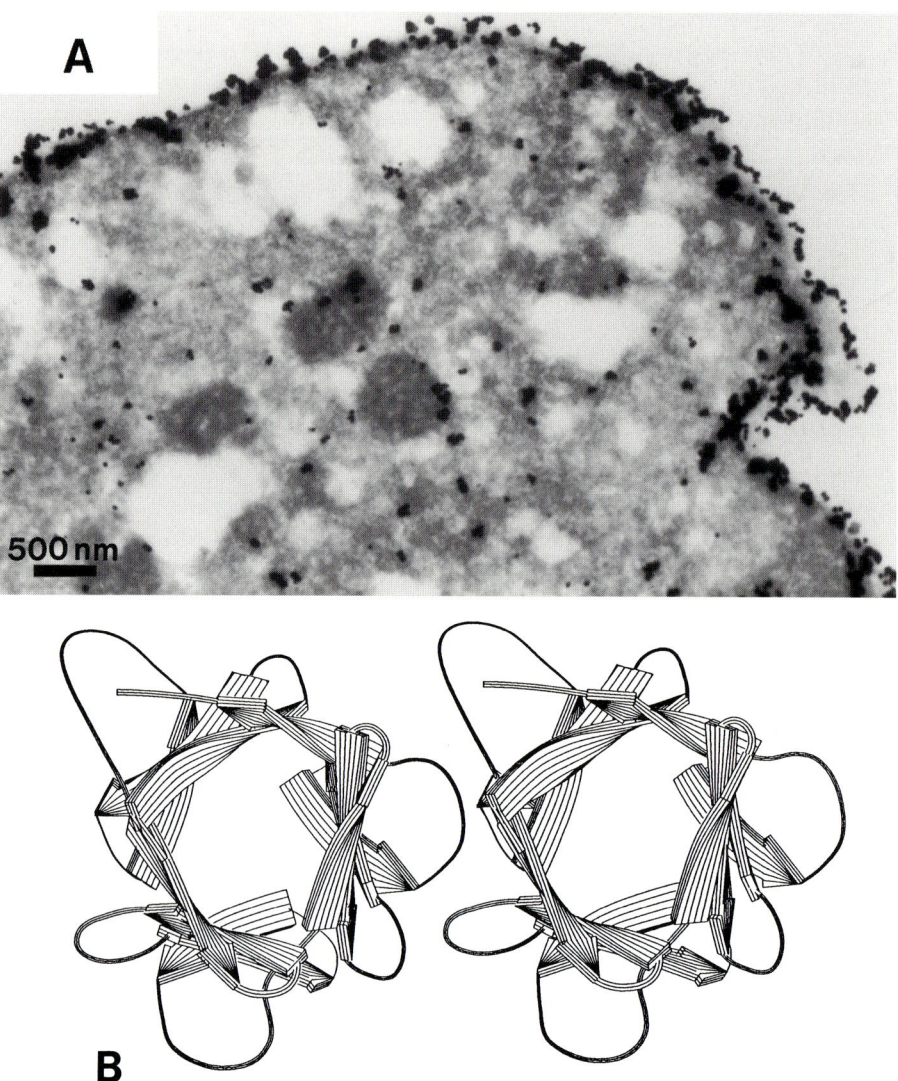

Fig. 3. Intracellular localization and structure of hisactophilin, a pH-regulated actin-binding protein of *D. discoideum*. A, Cells labelled with hisactophilin-specific mAb 11 followed by silver enhanced ultra-small gold immunolabelling as described by Humbel and Biegelmann (1992). The electron micrograph shows that hisactophilin is associated with the plasma membrane. In addition, scattered label is found in the cytoplasm but not at the membranes of intracellular organelles (preparation and photograph Dr. Bruno Humbel). B, Structure of hisactophilin as determined by NMR. The stereoview of a ribbon drawing is reproduced from Habazettl *et al.* (1992). Hisactophilin, a 13.5 kDa protein, contains 31 histidine residues which are concentrated in the flexible loops together with glycine residues. The loops extend from a backbone that is formed by anti-parallel β-sheets.

accumulation of large, multinucleate cytoplasmic masses (Knecht and Loomis, 1987; De Lozanne and Spudich, 1987). Transformation with DNA that encodes the complete coding region for "Triple Ala" heavy chains supplements myosin null cells with myosin II that cannot be phosphorylated at their threonine residues (Lück-Vielmetter, 1992).

Previous work has shown that heavy-chain phosphorylation reduces the actin-activated Mg-ATPase activity of myosin II (Truong *et al.*, 1992) and causes disassembly of myosin filaments (Kuczmarski and Spudich, 1980). The myosin molecules arrested in the monomeric state are characterised by a bend in their tail region (Pasternak *et al.*, 1989). From these findings it can be predicted that the majority of "Triple Ala" myosin is assembled into filaments (Fig. 4). In accord with this prediction, cells containing the mutated myosin show a strong enrichment of myosin II in the cell cortex where F-actin forms a dense network (Fig. 5A). Wild-type myosin II is less strongly enriched in the cortex, a considerable portion being distributed throughout the entire cytoplasm, most of it presumably in the form of monomers.

It should be pointed out that the differences in localisation of normal and "Triple Ala" myosin II are only obvious in cells freely moving in a fluid layer on a support such as glass. Under these conditions, myosin II is found everywhere in the cytoplasm of wild-

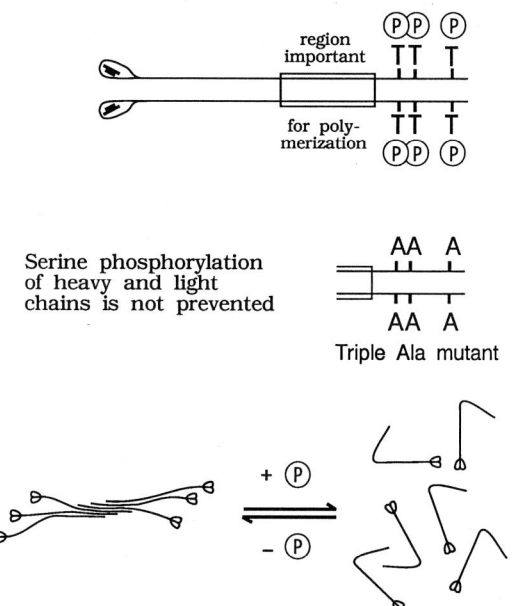

Fig. 4. Scheme of threonine phosphorylation and its implications on myosin II structure and function in *D. discoideum*. <u>Top</u>: Three phosphorylatable threonine residues have been identified in the distal region of the myosin tail (Lück-Vielmetter *et al.*, 1990). This cluster of phosphorylation sites is closely connected to a region important for polymerisation of the myosin (Pagh *et al.*, 1984). <u>Middle</u>: To specifically evaluate the role of phosphorylation at these sites, the three threonine residues have been replaced by alanine residues: "Triple Ala mutation" (Lück-Vielmetter, 1992). <u>Bottom</u>: As indicated by EM studies, the tails of myosin II assume a bent shape upon phosphorylation, which prevents the myosin to polymerise into filaments (Pasternak *et al.*, 1989).

type cells, including the leading edges and pseudopods, with a moderate enrichment in the cell cortex. If cells are squeezed by sandwiching them between a glass and agar surface, the myosin strongly accumulates within a few minutes at the periphery of the cells (Fig. 6). Often the myosin is concentrated at one end of the cells as previously observed under these "agar overlay" conditions (Nachmias *et al.*, 1989). This change in myosin II distribution appears to be a response of the cells to mechanical stress. Under agar overlay conditions, the distribution of myosin II in wild-type cells is no longer clearly distinguishable from that of "Triple Ala" myosin II in the mutants.

The change in myosin II distribution observed in the "Triple Ala" mutants is accompanied by alterations in cell behavior. Motility and precision of the chemotactic response is almost as strongly impaired as in myosin II null cells (Fig. 7). Nevertheless, the "Triple Ala" heavy chains enable the cells to divide in suspension cultures and to complete development (Lück-Vielmetter, 1992). These results indicate that threonine phosphorylation of the myosin II heavy chains is primarily required for optimising motility and orientation.

Attempts to identify proteins important for motility and chemotaxis by genetic methods

Selection for behavioral mutants

One strategy to identify proteins that are important for motility and chemotaxis centers

---

Fig. 5. Localisation of myosin II and coronin. A, normal and Triple Ala myosin II in confocal sections of interphase cells. Left panel: An untransformed cell of the *D. discoideum* strain AX2. Middle panel: A cell of strain HG1555 obtained by transforming the myosin II heavy-chain null strain HS2205 with a vector encoding Triple Ala heavy chains. Right panel: A cell of strain HG1554 obtained by transforming HS2205 with a vector encoding wild-type heavy chains. Growth-phase cells moving in a fluid layer on a glass surface were fixed with picric acid / formaldehyde (Humbel and Biegelmann, 1992), and indirectly labelled with heavy-chain specific mAb 396 (Pagh and Gerisch, 1986) and TRITC-conjugated goat anti-mouse IgG. mAb 396 recognizes myosin II in its monomeric as well as polymerised state. Using a Zeiss LSM 10 fluorescence laser-scanning microscope, four confocal sections of each cell were taken at intervals of 1.2 μm. Data are given in a combined code: fluorescence intensities are quantified as peaks in the z-axis of a landscape, and are colour-labelled from dark blue (below threshold) to green, orange, and light yellow. The left and right panels show that wild-type myosin II is distributed throughout the cell with a moderate enrichment in the cortical region. The middle panel shows that the central portion of a cell is almost depleted of Triple Ala myosin, which is strongly enriched in the cortical region. B, localisation of coronin and myosin II in a dividing cell of wild-type AX2. The cell is shown in phase contrast, with blue DAPI staining showing the daughter nuclei, with green myosin II label, and with red coronin label. C, an early (top) and late (bottom) metaphase cell of AX2 showing together myosin II in green and coronin in red. B and C illustrate accumulation in the course of cytokinesis of myosin II in the cleavage furrow and of coronin in the distal portions of the incipient daughter cells. To promote accumulation of myosin II in the cleavage furrow, the dividing cells were subjected to an agar overlay (Fukui *et al.*, 1987), fixed with picric acid / formaldehyde, incubated with rabbit anti-myosin II serum and FITC-conjugated goat anti-rabbit IgG, and subsequently with anti-coronin mAb 176-3-6 (de Hostos *et al.*, 1993) and TRITC-conjugated goat anti-mouse IgG. The bars indicate 10 μm.

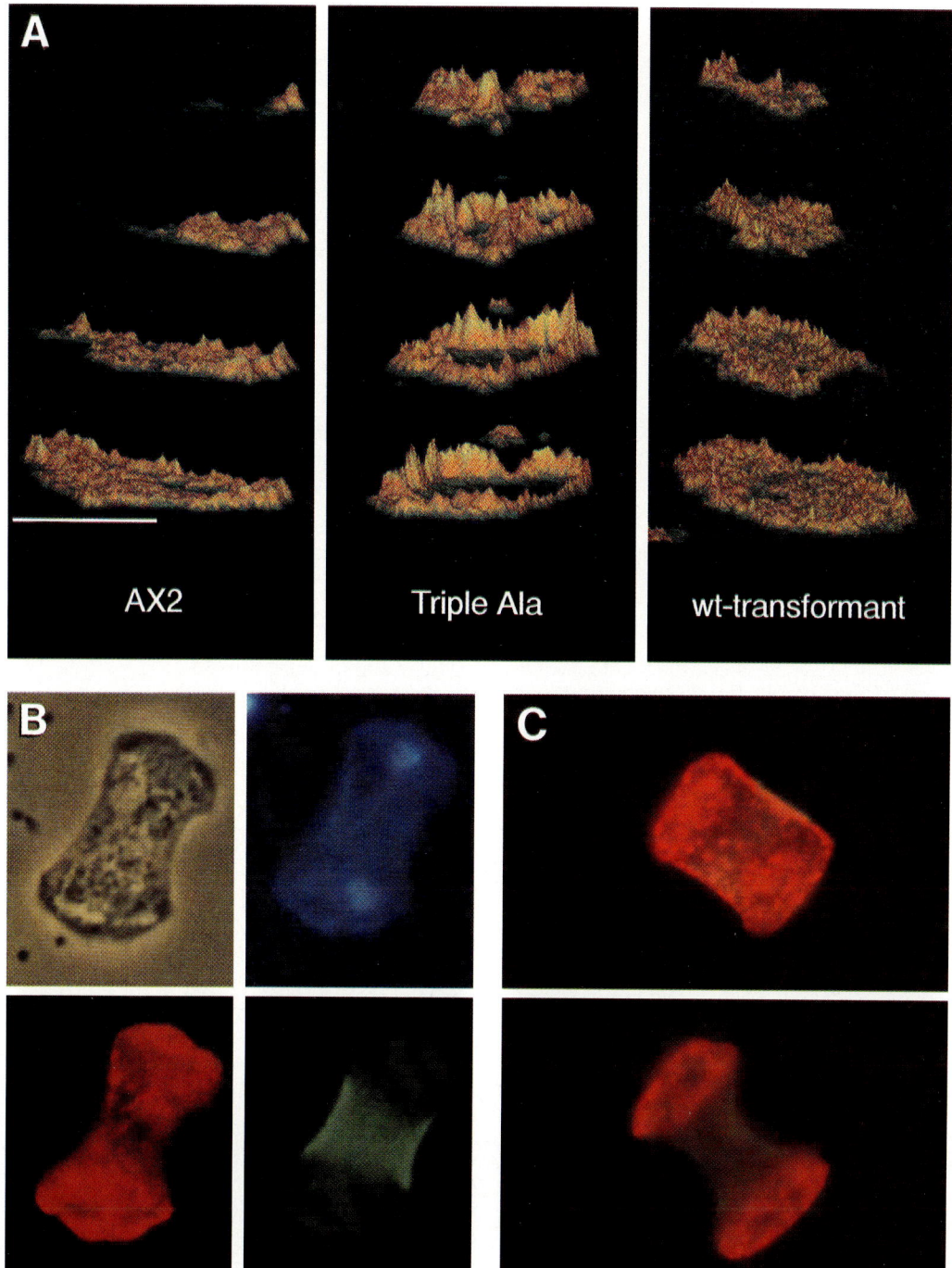

AX2   Triple Ala   wt-transformant

on a selection technique for mutant cells that are less motile or less precisely oriented in an attractant gradient than wild-type cells. A chamber that has successfully been used for selection of mutants defective in responses to the chemoattractants cAMP and folate is shown in Fig. 8.

The problem in using this strategy resides in the difficulty to identify the altered protein responsible for the behavioral defect of a particular mutant. If a hypothetical scheme of

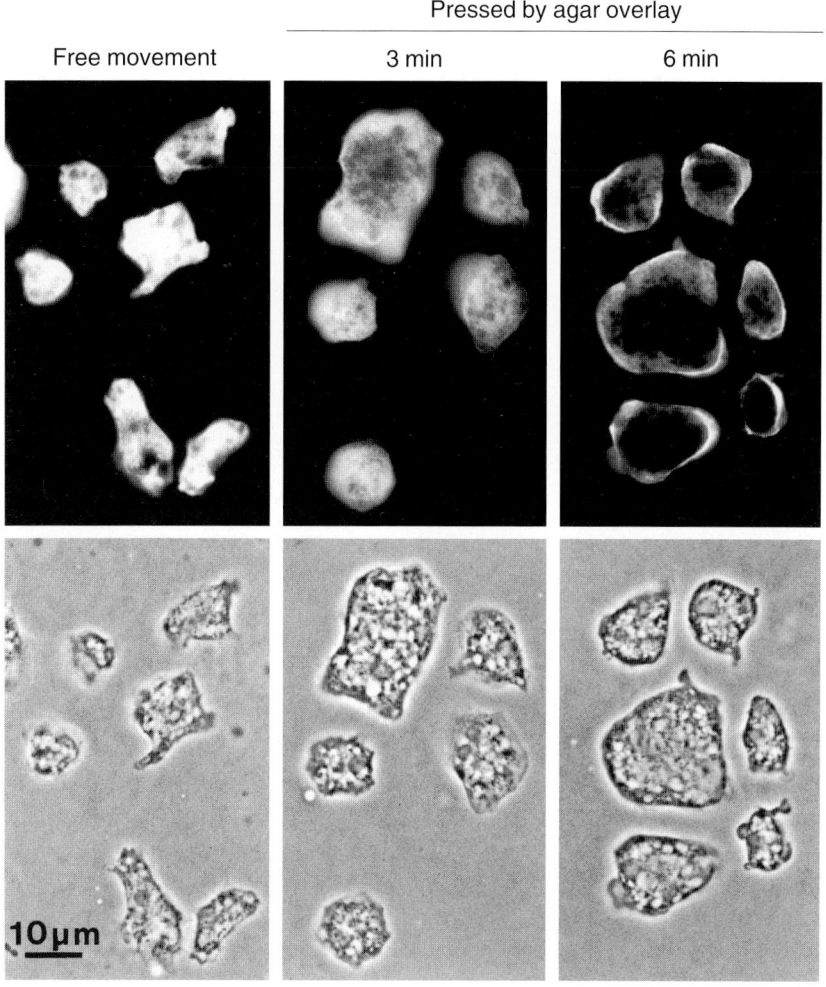

Fig. 6. Redistribution of myosin II in cells subjected to an agar overlay (Fukui *et al.*, 1987). Growth-phase cells are shown in phase-contrast optics (bottom) and labelled with mAb 396 and TRITC-conjugated goat anti-mouse IgG as in Fig. 5 A. Cells freely moving in a high fluid layer on a glass surface show myosin II to be present throughout the cell including the leading edges (left panel). An aliquot of these cells was placed between agar and coverslip and squeezed by removing excess fluid. This squeezing procedure took less than 3 minutes; immediately thereafter the myosin had maintained within the squeezed cells its initial distribution. After 3 more minutes almost all of the myosin was redistributed to the cell periphery, sometimes forming a cap at one side of the cell.

signal transduction chains and target proteins has been set up, one can test for functional defects of the putative components. Thus, one of the behavioural mutants selected has been shown to be defective in the folate receptor system (Segall *et al.*, 1988). In a number of other mutants the defect has been assigned by genetic recombination to the *frigid* A complementation group, which has been shown before to comprise mutations in the gene encoding Gα2 (Kumagai *et al.*, 1991; Firtel, 1991), supporting the hypothesis that this receptor-linked GTP-binding protein is one of the critical elements in chemotactic signal transmission.

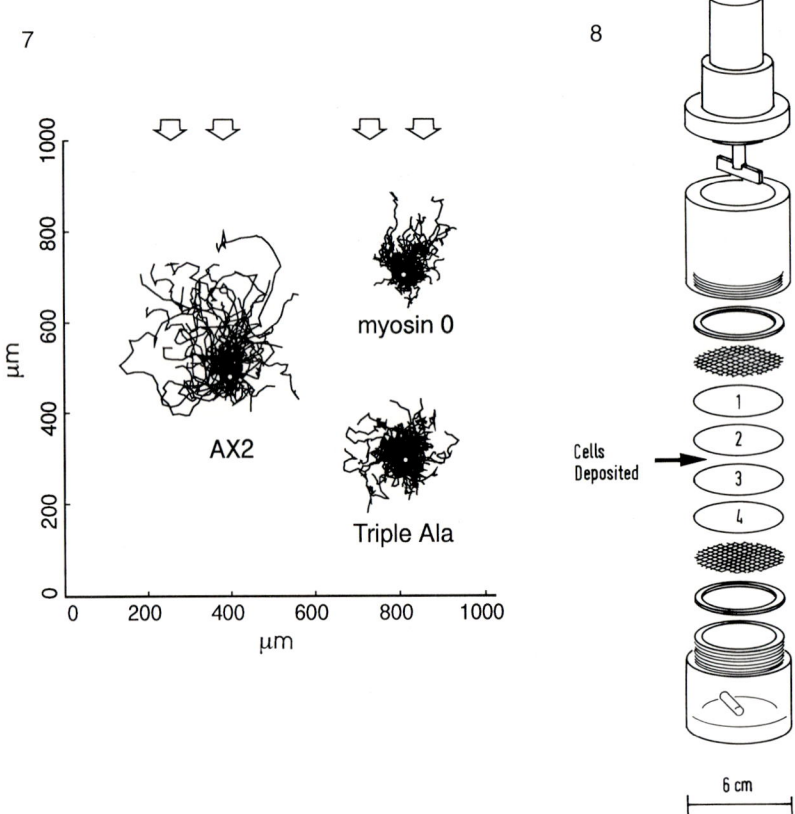

Fig. 7. Cell tracks of wild-type (AX2), of HS2205 lacking myosin II heavy chains (myosin O), and of HG1555, expressing heavy chains that are non-phosphorylatable at threonine residues (Triple Ala) (data from Lück-Vielmetter, 1992). Response of cells to a cAMP gradient as in the insets of Fig. 2 is shown. The arrows indicate direction of the gradient from high to low concentration, the white circles the origins of the trails. The data indicate that threonine phosphorylation is important for maintaining normal motility and chemotactic orientation.

Fig. 8. Chamber used for selection of chemotaxis and motility mutants. The chamber is composed of two compartments separated by a layer of four nitrocellulose filters with 8 μm pore size that allow motile *D. discoideum* cells to pass through. Mutagenized cells are placed between filters 2 and 3. The upper compartment is filled with attractant and the lower with buffer, so that responsive cells move out to the top and mutants become enriched at the second and third filter. From Segall *et al.* (1987).

Screening for mutants defective in specific actin-binding proteins

An alternative strategy to trace alterations in cell behaviour to defects in specific proteins starts from proteins with a known *in-vitro* function. In this approach either mutants obtained by conventional mutagenesis with nitrosoguanidine or transformants carrying vectors designed for homologous recombination have been employed (for review see Egelhoff *et al.*, 1991; Gerisch *et al.*, 1991; Noegel and Schleicher, 1991). Targeted gene replacement occurs after transformation with appropriate vectors undergoing homologous recombination with the endogenous gene (Manstein *et al.*, 1989). Both by conventional mutagenesis and gene disruption three actin-binding proteins have been eliminated in *D. discoideum* in addition to myosin II and, more recently, coronin (de Hostos *et al.*, 1993): the two F-actin crosslinking proteins α-actinin and 120 kDa gelation factor (ABP-120), and the actin-filament fragmenting protein severin. These three proteins exert strong effects on actin-filament length or cross-linkage when fractions of cell lysates are assayed by viscometry (Brown *et al.*, 1982; Condeelis *et al.*, 1984). Quite in contrast to these in-vitro results, mutants lacking one of these proteins show little impairment of motility or chemotaxis (Wallraff *et al.*, 1986; Schleicher *et al.*, 1988; André *et al.*, 1989; Brink *et al.*, 1990).

We do not argue that the three proteins eliminated lack any effect on motility and chemotaxis under the varying conditions a soil organism is exposed to in nature. In fact, the influence on cell behaviour may be stronger under conditions different from those used by us (Cox *et al.*, 1992), and after the elimination of two actin-crosslinking proteins substantial impairment of morphogenesis has been observed (Witke *et al.*, 1992). We wish to stress the point, however, that none of these proteins exerting strong in-vitro activities is essential for the cells to move and to orient in gradients of attractant.

In an attempt to define the minimal complement of proteins necessary to make the machinery of chemotaxis to work, mutants lacking all three proteins have been produced. Such triple mutants are still capable of aggregating, and some of them even form fruiting bodies under appropriate conditions (E. Wallraff, unpublished results). As shown in Fig. 9, quantifying the chemotactic response in HG1397 has revealed that cells of this triple mutant still show chemotactic responses, similar to the mutants defective in only one of the proteins. Subsequently, we have searched for novel actin-binding proteins that are still present in the triple mutant. This mutant proved to be an excellent source for such proteins because their detection was not complicated by presence of the dominating actin-binding proteins severin, 120 kDa protein, and α-actinin that are already known.

Identification of relevant proteins by affinity methods

Proteins bound to an active column

As a straightforward method to search for new actin-binding proteins of *D. discoideum*, a cytosolic fraction of the triple mutant HG1397 was applied to an F-actin affinity column (Drubin *et al.*, 1988; Miller and Alberts, 1989). The bound proteins were salt-eluted from the column and injected into mice for the production of monoclonal antibodies. The hybridoma culture supernatants were screened by fluorescence microscopy with whole-mount preparations of fixed cells in order to identify target proteins that are localised to certain structures in the actin skeleton. Among the antibodies of interest, some labelled the leading edges of the cells including filopods, and turned out to be directed against the 30 kDa actin-bundling protein characterised by Marcus Fechheimer and his colleagues (Fechheimer *et al.*, 1991; Furukawa *et al.*, 1992).

One antibody, mAb 477, showed an intriguing pattern of labelling. In addition to a uniform labelling of the entire cytoplasm, a distinct enrichment of label was observed at the minute tips of filopods (Fig. 10A). Sometimes accumulation of the antibody label was observed at the base or along the length of the filopods. In blots of total cellular proteins separated by SDS-gel electrophoresis, this antibody specifically recognized a 220 kDa protein.

The 220 kDa protein recognized by mAb 477 has been designated by us as "filopodin". Screening an expression library of *D. discoideum* cDNA with the antibody revealed partial clones. The sequence of 670 amino acids indicates that these clones encode the C-terminal portion of a protein showing along its entire length an overall identity of 22 per cent with

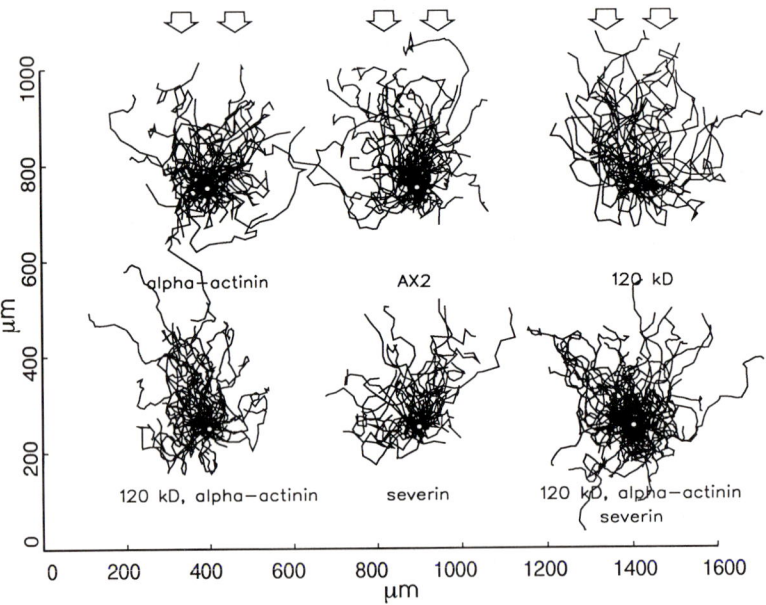

Fig. 9. Tracks of wild-type AX2 cells and of mutant cells defective in actin-binding proteins. Data are plotted as in Fig. 7. Actin-binding proteins that are defective in the mutants are indicated below the tracks. 120 kD stands for lack of the 120 kDa actin-crosslinking protein (ABP120). The mutant strains were: HG1130 (α-actinin), HG1264 (120 kD); double mutant GA1130 (120 kD, α-actinin). HG1132 (severin); and triple mutant HG1397 (120 kD, α-actinin, severin). Generation of the double mutant was described by Witke *et al.* (1992). The triple mutant was obtained by combining mutations introduced by nitrosoguanidine and by gene disruption (E. Wallraff and W. Witke, unpublished). It should be emphasized that the triple mutant proved to be strongly sensitive to slight changes in experimental conditions. Therefore, the results were less reproducible than in the other strains. The example shows an experiment where the cells were highly motile and precisely orientated.

mouse fibroblast talin (Rees *et al.*, 1990). This result is of interest since in fibroblasts, talin is a major component of the complex of proteins linking actin cables to integrins in the adhesion plaques (adherence-type junctions) where the cells are closely attached to each other or to an extracellular matrix (Burridge *et al.*, 1988; Volberg *et al.*, 1992). This, however, is not the only site where talin is found in vertebrate cells, suggesting multiple functions of the protein. In fibroblasts, talin is also accumulated in patches at the leading edges (Hock *et al.*, 1989; DePasquale and Izzard, 1991), similar to the filopodin in *Dictyostelium*. Platelets are extremely rich in talin, which is redistributed upon activation from the cytoplasm towards the plasma membrane (Beckerle *et al.*, 1989).

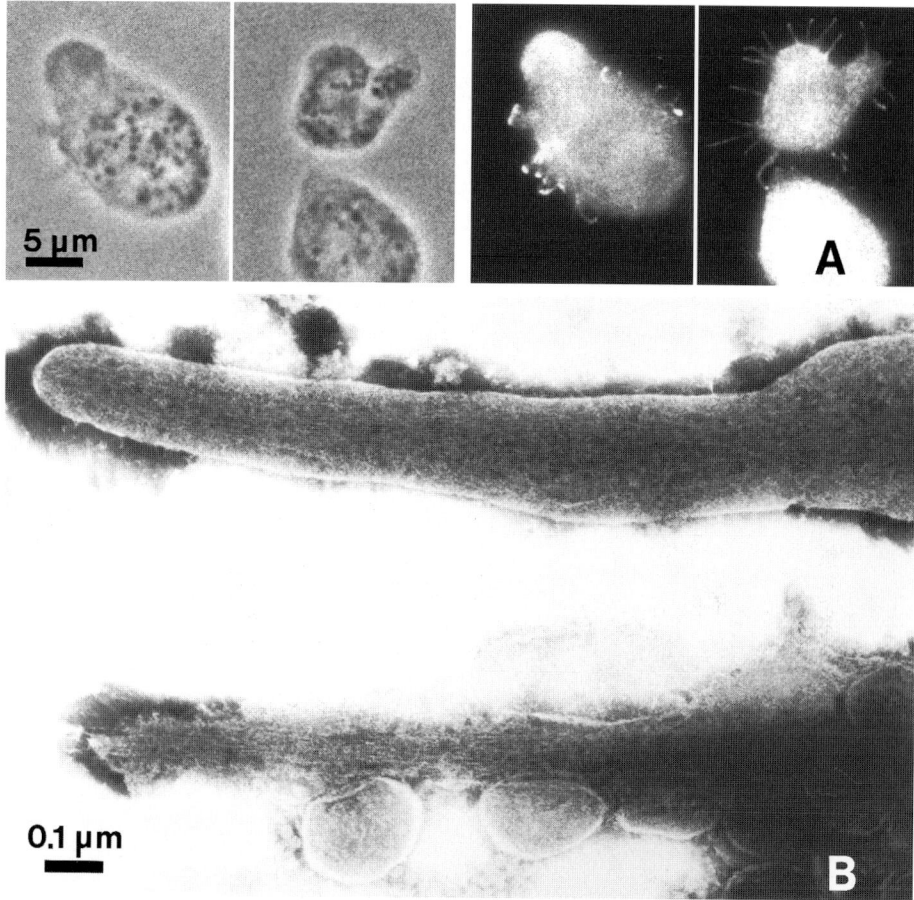

Fig. 10. Localisation of the talin-like protein, filopodin, at the end of actin bundles in filopods. A, Uniform distribution of filopodin in the cytoplasm, and strong enrichment of filopodin at the tip of filopods. Growth-phase cells of *D. discoideum* AX2 cells fixed with methanol at −20°C were indirectly labelled with filopodin-specific mAb 477 and TRITC-conjugated goat anti-mouse IgG. Left panel: phase-contrast image; right panel: fluorescence label showing filopodin. B, Filopods of a *D. discoideum* cell partially extracted with Triton X-100 and fixed with glutaraldehyde as described by Claviez *et al.* (1986). Bundles of actin filaments are seen along the entire length of the filopods.

Purified platelet talin nucleates actin polymerisation and anchors the actin filaments to lipid bilayers (Kaufmann *et al.*, 1992). The filopods in *Dictyostelium* get their shape from a parallel array of actin filaments which join at their ends to the curved plasma membrane at the tip of the filopod (Fig. 10B). Localisation of the talin-like filopodin in *Dictyostelium* at the tip and base of filopods is consistent with a nucleating and membrane-anchoring function, as it has been found for platelet talin.

Proteins associated with an actin-myosin complex reconstituted in vitro

Actin can be precipitated together with myosin by increasing the ionic strength in a cytosolic fraction from *Dictyostelium* (Fechheimer and Furukawa, 1991). Three major proteins have been found to be associated with such a complex. One is a 30 kDa actin-bundling protein first purified from a similar complex by Fechheimer and Taylor (1984). This is the same protein as found by F-actin affinity chromatography (see previous section). The other two proteins are coactosin, a 17 kDa protein that shows sequence similarities to the cofilin family of vertebrate actin-binding proteins (de Hostos *et al.*, submitted), and coronin, a 55 kDa protein (de Hostos *et al.*, 1991). The name coronin has been given because this protein is accumulated in crown-shaped surface extensions of growth-phase cells. Coronin is also strongly enriched at the leading edges of growth-phase cells and at the fronts of the typically elongated aggregating cells. Coronin consists of two domains: an N-terminal one with sequence similarities to the β-subunits of heterotrimeric G-proteins, and a C-terminal domain with predicted α-helical structure.

The localisation of coronin in fixed cells indicates that it is redistributed within the highly motile *Dictyostelium* cells at the same rate as surface extensions are formed and retracted during chemotaxis. The coronin appears to be recruited from a soluble pool of proteins in the cytoplasm in order to be assembled at the network of actin filaments in a newly established front. This localisation at sites where active shape changes are occuring, suggests a role for coronin in the control of cell shape and motility. This conjecture has been corroborated by eliminating coronin through gene replacement (de Hostos et al., 1993). In the coronin null mutants, cell movement is slowed down. Moreover, cytokinesis is impaired, as indicated by the appearance of a high percentage of multinucleate cells. The phenotype of coronin null mutants is similar to that of myosin II null mutants, but in the coronin null mutants cytokinesis is less strongly impaired. More importantly, attachment to a substratum does not help the mutant cells to overcome their defect in coronin. Whereas myosin II null cells divide by traction-mediated pseudofission when spread on a surface (Egelhoff and Spudich, 1991), cytokinesis in coronin null cells is not supported under these conditions. This finding has prompted us to compare the localisation of myosin II and coronin in dividing wild-type cells. Myosin II is assembled in the cleavage furrow, as shown previously by Fukui *et al.* (1989). Coronin, on the contrary, is accumulated at the distal portions of a dividing cell, which are depleted from myosin II (Fig. 5 B, C). Based on these observations it has been hypothesised that myosin II and coronin are separately involved in cytokinesis, the myosin II participating in the constriction that separates the daughter cells, the coronin being involved in drawing the incipient daughter cells apart from each other (de Hostos *et al.*, 1993).

Organelle assembly model of chemotactic orientation

In this overview the leading edge of a cell is discussed as an organelle that is generated by the ordered assembly of a series of proteins. Actin filaments build the groundwork for the coassembly of associated proteins which occupy specific positions within the organelle. This has been exemplified by the localisation of coronin and the talin-like filopodin. These two proteins, and also the actin-bundling protein p30, are stored in the cell in soluble form; for assembly into the organelle they have to be recruited from the cytoplasm.

The most interesting feature of the leading edge is its transient nature: this organelle is generated and retracted on the second or minute time scale. Formation of a new leading edge during chemotactic reorientation is thus dependent on the dynamics of protein assembly and disassembly. The pertinent question here is how this complicated organelle is so rapidly formed in response to external signals: which regulatory mechanisms are involved, which modifications of the participating proteins are necessary, and which interactions determine the position of a protein in the final complex.

Work in the near future will hopefully convert the static image obtained by immunofluorescence labelling of fixed cells into a motion picture reflecting the dynamics of molecular interactions during cell movement and chemotactic orientation. The technique to be applied is fluorescence imaging of labelled proteins with high-resolution intensifier cameras. One critical step in applying this technique is to introduce proteins conjugated with fluorescent dyes into the cells. *Dictyostelium* cells are extremely sensitive to microinjection, but alternative methods of introducing proteins into transiently permeabilised cells have been developed (Furukawa *et al.*, 1992; Schlatterer *et al.*, 1992). Our own attempts to improve microinjection led to the discovery that mutants lacking the Ca^{2+}-activated actin-filament fragmenting protein, severin, can be conveniently microinjected (G. Gerisch and C. Heizer, unpublished). Work is in progress to employ severin mutants to analyse the sequence and pattern of assembly of specific proteins during the formation of a leading edge in a chemotactically stimulated cell.

This overview summarises work in which Drs. Bruno Humbel, Michael Schleicher, Dorothea Lück-Vielmetter, and Jeff Segall were involved. The triple mutant defective in actin-binding proteins was constructed in cooperation with Drs. Angelika Noegel and Walter Witke. Construction of the myosin Triple Ala mutant was part of a cooperation with Drs. James A. Spudich and Thomas Egelhoff, Stanford University. We thank Birgit Bradtke, Maria Ecke, and Bettina Mühlbauer for technical assistance, Maria Zwermann and Barbara Wall for organizing the manuscript, and Dr. Gerard Marriott for critically reading the manuscript.

Experimental work of the authors was supported by the Deutsche Forschungsgemeinschaft (SFB 266), the Humboldt Stiftung, and the Fonds der Chemischen Industrie.

References

André, E., Brink, M., Gerisch, G., Isenberg, G., Noegel, A., Schleicher, M., Segall, J. E. and Wallraff, E. (1989). A *Dictyostelium* mutant deficient in severin, an F-actin fragmenting protein, shows normal motility and chemotaxis. *J. Cell Biol.* **108**, 985-995.

Beckerle, M. C., Miller, D. E., Bertagnolli, M. E. and Locke, S. J. (1989). Activation-dependent redistribution of the adhesion plaque protein, talin, in intact human platelets. *J. Cell Biol.* **109**, 3333-3346.

Berlot, C. H., Devreotes, P. N., and Spudich, J. A. (1987). Chemoattractant-elicited increases in *Dictyostelium* myosin phosphorylation are due to changes in myosin localization and increases in kinase activity. *J. Biol. Chem.* **262**, 3918-3926.

Brink, M., Gerisch, G., Isenberg, G., Noegel, A. A., Segall, J. E., Wallraff, E. and Schleicher, M. (1990). A *Dictyostelium* mutant lacking an F-actin cross-linking protein, the 120-kD gelation factor. *J. Cell Biol.* **111**, 1477-1489.

Brown, S. S., Yamamoto, K. and Spudich, J. A. (1982). A 40,000-dalton protein from *Dictyostelium discoideum* affects assembly properties of actin in a Ca^{2+}-dependent manner. *J. Cell Biol.* **93**, 205-210.

Burridge, K., Fath, K., Kelly, T., Nuckolls, G. and Turner, C. (1988). Focal adhesions: Transmembrane junctions between the extracellular matrix and the cytoskeleton. *Ann. Rev. Cell Biol.* **4**, 487-525.

Claviez, M., Brink, M. and Gerisch, G. (1986). Cytoskeletons from a mutant of *Dictyostelium discoideum* with flattened cells. *J. Cell Sci.* **86**, 69-82.

Condeelis, J., Vahey, M., Carboni, J. M., DeMey, J. and Ogihara, S. (1984). Properties of the 120,000- and 95,000-dalton actin-binding proteins from *Dictyostelium discoideum* and their possible functions in assembling the cytoplasmic matrix. *J. Cell Biol.* **99**, 119s-126s.

Cox, D., Condeelis, J., Wessels, D., Soll, D., Kern, H. and Knecht, D. A. (1992). Targeted disruption of the ABP-120 gene leads to cells with altered motility. *J. Cell Biol.* **116**, 943-955.

De Hostos, E. L., Bradtke, B., Lottspeich, F., Guggenheim, R. and Gerisch, G. (1991). Coronin, an actin binding protein of *Dictyostelium discoideum* localized to cell surface projections, has sequence similarities to G protein ß subunits. *EMBO J.* **10**, 4097-4104.

De Hostos, E. L., Rehfueß, C., Bradtke, B., Waddell, D. R., Albrecht, R., Murphy, J. and Gerisch, G. (1993). *Dictyostelium* mutants lacking the cytoskeletal protein coronin are defective in cytokinesis and cell motility. *J. Cell Biol.* **120**, 163-173.

De Lozanne, A. and Spudich, J. A. (1987). Disruption of the *Dictyostelium* myosin heavy chain gene by homologous recombination. *Science* **236**, 1086-1091.

DePasquale, J. A. and Izzard, C. S. (1991). Accumulation of talin in nodes at the edge of the lamellipodium and separate incorporation into adhesion plaques at focal contacts in fibroblasts. *J. Cell Biol.* **113**, 1351-1359.

Devreotes, P. N. and Zigmond, S. H. (1988). Chemotaxis in eukaryotic cells: a focus on leukocytes and *Dictyostelium. Ann. Rev. Cell Biol.* **4**, 649-686.

Drubin, D. G., Miller, K. G. and Botstein, D. (1988). Yeast actin-binding proteins: evidence for a role in morphogenesis. *J. Cell Biol.* **107**, 2551-2561.

Egelhoff, T. T. and Spudich, J. A. (1991). Molecular genetics of cell migration: *Dictyostelium* as a model system. *Trends in Genetics* **7**, 161-166.

Egelhoff, T. T., Titus, M. A., Manstein, D. J., Ruppel, K. M. and Spudich, J. A. (1991). Molecular genetic tools for study of the cytoskeleton in *Dictyostelium. Methods in Enzymology* **196**, 319-334.

Eichinger, L. and Schleicher, M. (1992). Characterization of actin- and lipid-binding domains in severin, a Ca^{2+}-dependent F-actin fragmenting protein. *Biochem.* **31**, 4779-4787.

Endow, S. A. and Titus, M. A. (1992). Genetic approaches to molecular motors. *Ann. Rev. Cell Biol.* **8**, 29-66.

Fechheimer, M. and Taylor, D. L. (1984). Isolation and characterization of a 30,000-dalton Calcium-sensitive actin cross-linking protein from *Dictyostelium discoideum. J. Biol. Chem.* **259**, 4514-4520.

Fechheimer, M. and Furukawa, R. (1991). Preparation of 30,000-Da actin cross-linking protein from *Dictyostelium discoideum. Meth. Enzym.* **196**, 84-91.

Fechheimer, M., Murdock, D., Carney, M. and Glover, C. V. C. (1991). Isolation and sequencing of cDNA clones encoding the *Dictyostelium discoideum* 30,000-dalton actin-bundling protein. *J. Biol. Chem.* **266**, 2883-2889.

Firtel, R. A. (1991). Signal transduction pathways controlling multicellular development in *Dictyostelium. Trends in Genetics* **7**, 381-388.

Fisher, P. R., Merkl, R., and Gerisch, G. (1989) Quantitative analysis of cell motility and chemotaxis in *Dictyostelium discoideum* by using an image processing system and a novel chemotaxis chamber providing stationary chemical gradients. *J. Cell Biol.* **108**, 973-984.

Fukui, Y., Yumura, S. and Yumura, T. K. (1987). Agar-overlay immunofluorescence: High-resolution studies of cytoskeletal components and their changes during chemotaxis. *Methods Cell Biol.* **28**, 347-356.

Fukui, Y., Lynch, T. J., Brzeska, H. and Korn, E. D. (1989). Myosin I is located at the leading edges of locomoting *Dictyostelium* amoebae. *Nature* **341**, 328-331.

Fukui, Y., De Lozanne, A. and Spudich, J. A. (1990). Structure and function of the cytoskeleton of a *Dictyostelium* myosin-defective mutant. *J. Cell Biol.* **110**, 367-378.

Furukawa, R., Butz, S., Fleischmann, E. and Fechheimer, M. (1992). The *Dictyostelium discoideum* 30,000 dalton protein contributes to phagocytosis. *Protoplasma* **169**, 18-27.

Gerisch, G. and Keller, H. U. (1981). Chemotactic reorientation of granulocytes stimulated with micropipettes containing fMet-Leu-Phe. *J. Cell Sci.* **52**, 1-10.

Gerisch, G., Malchow, D. Huesgen, A., Nanjundiah, V., Roos, W., Wick, U. and Hülser, D. (1975). Cyclic-AMP reception and cell recognition in *Dictyostelium discoideum*. In *Developmental Biology, ICN-UCLA Symposia on Molecular & Cellular Biology* **2**, 76-88.

Gerisch, G., Noegel, A. A. and Schleicher, M. (1991). Genetic alteration of proteins in actin-based motility systems. *Annu. Rev. Physiol.* **53**, 607-628.

Goldschmidt-Clermont, P. J., Machesky, L. M., Baldassare, J. J. and Pollard, T. D. (1990). The actin-binding protein profilin binds to PIP$_2$ and inhibits its hydrolysis by phospholipase-C. *Science* **247**, 1575-1578.

Habazettl, J., Gondol, D., Wiltscheck, R., Otlewski, J., Schleicher, M. and Holak, T. A. (1992). Structure of hisactophilin is similar to interleukin-1β and fibroblast growth factor. *Nature* **359**, 855-858.

Haus, U., Hartmann, H., Trommler, P., Noegel, A. A. and Schleicher, M. (1991). F-actin capping by cap32/34 requires heterodimeric conformation and can be inhibited with PIP$_2$. *Biochem. Biophys. Res. Commun.* **181**, 833-839.

Hock, R. S., Sanger, J. M. and Sanger, J. W. (1989). Talin dynamics in living microinjected nonmuscle cells. *Cell Motility and the Cytoskeleton* **14**, 271-287.

Hofmann, A., Eichinger, L., André, E., Rieger, D., and Schleicher, M. (1992). Cap 100, a novel phosphatidylinositol 4,5-bisphosphate-regulated protein that caps actin filaments but does not nucleate actin assembly. *Cell Motility and the Cytoskeleton* **23**, 133-144.

Howard, P. K., Sefton, B. M. and Firtel, R. A. (1993). Tyrosine phosphorylation of actin in *Dictyostelium* associated with cell-shape changes. *Science* **259**, 241-244.

Humbel, B. M. and Biegelmann, E. (1992). A preparation protocol for postembedding immunoelectron microscopy of *Dictyostelium discoideum* cells with monoclonal antibodies. *Scan. Microscopy* **6**, 817-825.

Kaufmann, S., Käs, J., Goldmann, W. H., Sackmann, E. and Isenberg, G. (1992). Talin anchors and nucleates actin filaments at lipid membranes: a direct demonstration. *FEBS Lett.* **314**, 203-205.

Klein, P. S., Sun, T. J., Saxe III, C. L., Kimmel, A. R., Johnson, R. L., and Devreotes, P. N. (1988). A chemoattractant receptor controls development in *Dictyostelium discoideum*. *Science* **241**, 1467-1472.

Knecht, D. A. and Loomis, W. F. (1987). Antisense RNA inactivation of myosin heavy chain gene expression in *Dictyostelium discoideum*. *Science* **236**, 1081-1086.

Kuczmarski, E. R. and Spudich, J. A. (1980). Regulation of myosin self-assembly: Phosphorylation of *Dictyostelium* heavy chain inhibits formation of thick filaments. *Proc. Natl. Acad. Sci. USA* **77**, 7292-7296.

Kumagai, A., Hadwiger, J. A., Pupillo, M. and Firtel, R. A. (1991). Molecular genetic analysis of two G$_\alpha$ protein subunits in *Dictyostelium*. *J. Biol. Chem.* **266**, 1220-1228.

Kwiatkowski, D. J., Janmey, P. A. and Yin, H. L. (1989). Identification of critical functional and regulatory domains in gelsolin. *J. Cell Biol.* **108**, 1717-1726.

Liu, G. and Newell, P. C. (1991). Evidence that cyclic GMP may regulate the association of myosin II heavy chain with the cytoskeleton by inhibiting its phosphorylation. *J. Cell Sci.* **98**, 483-490.

Lück-Vielmetter, D., Schleicher, M., Grabatin, B., Wippler, J. and Gerisch, G. (1990). Replacement of threonine residues by serine and alanine in a phosphorylatable heavy chain fragment of *Dictyostelium* myosin II. *FEBS Lett.* **269**, 239-243.

Lück-Vielmetter, D. (1992). Molekulargenetische Untersuchungen zur Rolle von Myosin II bei der Chemotaxis. Doctoral thesis, Ludwig-Maximilians-Universität, München.

Luna, E. J. and Condeelis, J. S. (1990). Actin-associated proteins in *Dictyostelium discoideum. Dev. Gen.* **11**, 328-332.

Manstein, D. J., Titus, M. A., De Lozanne, A. and Spudich, J. A. (1989). Gene replacement in *Dictyostelium*: generation of myosin null mutants. *EMBO J.* **8**, 923-932.

Miller, K. G. and Alberts, B. M. (1989). F-actin affinity chromatography: technique for isolating previously unidentified actin-binding proteins. *Proc. Natl. Acad. Sci. USA* **86**, 4808-4812.

Nachmias, V. T., Fukui, Y. and Spudich, J. A. (1989). Chemoattractant-elicited translocation of myosin in motile *Dictyostelium. Cell Mot. and the Cytoskeleton* **13**, 158-169.

Newell, P. C., Europe-Finner, G. N., Liu, G. , Gammon, B. and Wood, C. A. (1990). Signal transduction for chemotaxis in *Dictyostelium* amoebae. *Cell Biology.* **1**, 105-113.

Noegel, A. A. and Schleicher, M. (1991). Phenotypes of cells with cytoskeletal mutations. *Curr. Op. Cell Biol.* **3**, 18-26.

Pagh, K., Maruta, H., Claviez, M. and Gerisch, G. (1984). Localization of two phosphorylation sites adjacent to a region important for polymerization on the tail of *Dictyostelium* myosin. *EMBO J.* **3**, 3271-3278.

Pagh, K. and Gerisch, G. (1986). Monoclonal antibodies binding to the tail of *Dictyostelium discoideum* myosin: their effects on antiparallel and parallel assembly and actin-activated ATPase activity. *J. Cell Biol.* **103**, 1527-1538.

Pasternak, C., Flicker, P. F., Ravid, S. and Spudich, J. A. (1989). Intermolecular versus intramolecular interactions of *Dictyostelium* myosin: possible regulation by heavy chain phosphorylation. *J. Cell Biol.* **109**, 203-210.

Peters, D. J. M., Knecht, D. A., Loomis, W. F., De Lozanne, A., Spudich, J. and Van Haastert, P. J. M. (1988). Signal transduction, chemotaxis, and cell aggregation in *Dictyostelium discoideum* cells without myosin heavy chain. *Dev. Biol.* **128**, 158-163.

Pupillo, M., Kumagai, A., Pitt, G. S., Firtel, R. A., and Devreotes, P. N. (1989). Multiple α subunits of guanine nucleotide-binding proteins in *Dictyostelium. Proc. Natl. Acad. Sci. USA.* **86**, 4892-4896.

Ravid, S. and Spudich, J. A. (1989). Myosin heavy chain kinase from developed *Dictyostelium* cells. *J. Biol. Chem.* **264**, 15144-15150.

Ravid, S. and Spudich, J. A. (1992). Membrane-bound *Dictyostelium* myosin heavy chain kinase: A developmentally regulated substrate-specific member of the protein kinase C family. *Proc. Natl. Acad. Sci.* **89**, 5877-5881.

Rees, D. J. G., Ades, S. E., Singer, S. J. and Hynes, R. O. (1990). Sequence and domain structure of talin. *Nature* **347**, 685-689.

Ross, F. M. and Newell, P. C. (1981). Streamers: Chemotactic mutants of *Dictyostelium discoideum* with altered cyclic GMP metabolism. *J. General Microbiol.* **127**, 339-350.

Scheel, J., Ziegelbauer, K., Kupke, T., Humbel, B. M., Noegel, A. A., Gerisch, G. and Schleicher, M. (1989). Hisactophilin, a histidine-rich actin-binding protein from *Dictyostelium discoideum. J. Biol. Chem.* **264**, 2832-2839.

Schlatterer, C., Knoll, G. and Malchow, D. (1992). Intracellular calcium during chemotaxis of *Dictyostelium discoideum:* a new fura-2 derivative avoids sequestration of the indicator and allows long-term calcium measurements. *Europ. J. Cell Biol.* **58**, 172-181.

Schleicher, M. and Noegel, A. A. (1992). Dynamics of the *Dictyostelium* cytoskeleton during chemotaxis. *The New Biologist* **4**, 461-472.

Schleicher, M., Wallraff, E., Gerisch, G. and Isenberg, G. (1988). Construction and analysis of *Dictyostelium* mutants with defects in actin-binding proteins. *Protoplasma* **Suppl. 2**, 22-26.

Schweiger, A., Mihalache, O., Ecke, M., and Gerisch, G. (1992). Stage-specific tyrosine phosphorylation of actin in *Dictyostelium discoideum* cells. *J. Cell Sci.* **102**, 601-609.

Segall, J. E. (1992) Behavioral responses of streamer F mutants of *Dictyostelium discoideum*: effects of cyclic GMP on cell motility. *J. Cell Sci.* **101**, 589-597.

Segall, J. E. and Gerisch, G. (1989). Genetic approaches to cytoskeleton function and the control of cell motility. *Curr. Op. Cell Biol.* **1**, 44-50.

Segall, J. E., Fisher, P. R. and Gerisch, G. (1987). Selection of chemotaxis mutants of *Dictyostelium discoideum. J. Cell Biol.* **104**, 151-161.

Segall, J. E., Bominaar, A. A., Wallraff, E. and De Wit, R. J. W. (1988). Analysis of a *Dictyostelium* chemotaxis mutant with altered chemoattractant binding. *J. Cell Sci.* **91**, 479-489.

Soll, D. R., Wessels, D., Murray, J., Vawter, H., Voss, E. and Bublitz, A. (1990). Intracellular vesicle movement, cAMP and myosin II in *Dictyostelium. Dev. Gen.* **11**, 341-353.

Spudich, J. A. (1989). In pursuit of myosin function. *Cell Regulation* **1**, 1-11.

Truong, T., Medley, Q. G. and Côté, G. P. (1992). Actin-activated Mg-ATPase activity of *Dictyostelium* myosin II. *J. Biol. Chem.* **267**, 9767-9772.

Vaillancourt, J. P., Lyons, C. and Côté, G. P. (1988). Identification of two phosphorylated threonines in the tail region of *Dictyostelium* myosin II. *J. Biol. Chem.* **263**, 10082-10087.

Van Duijn, B. and Inouye, K. (1991). Regulation of movement speed by intracellular pH during *Dictyostelium discoideum* chemotaxis. *Proc. Natl. Acad. Sci. USA* **88**, 4951-4955.

Volberg, T., Zick, Y., Dror, R., Sabanay, I., Gilon, C., Levitzki, A. and Geiger, B. (1992). The effect of tyrosine-specific protein phosphorylation on the assembly of adherens-type junctions. *EMBO J.* **11**, 1733-1742.

Wallraff, E., Schleicher, M., Modersitzki, M., Rieger, D., Isenberg, G. and Gerisch, G. (1986). Selection of *Dictyostelium* mutants defective in cytoskeletal proteins: use of an antibody that binds to the ends of α-actinin rods. *EMBO J.* **5**, 61-67.

Wessels, D., Soll, D. R., Knecht, D., Loomis, W. F., De Lozanne, A. and Spudich, J. (1988). Cell motility and chemotaxis in *Dictyostelium* amebae lacking myosin heavy chain. *Dev. Biol.* **128**, 164-177.

Wessels, D., Murray, J., Jung, G., Hammer III, J. A. and Soll D. R. (1991). Myosin IB null mutants of *Dictyostelium* exhibit abnormalities in motility. *Cell Mot. and the Cytoskeleton.* **20**, 301-315.

Witke, W., Schleicher, M. and Noegel, A. A. (1992). Redundancy in the microfilament system: abnormal development of *Dictyostelium* cells lacking two F-actin cross-linking proteins. *Cell* **68**, 53-62.

Printed in Great Britain © The Society of Experimental Biology 1993   317

CELL SIGNALLING AND MOTILE ACTIVITY

ALAN W. HELDMAN and PASCAL J. GOLDSCHMIDT-CLERMONT

Cardiology Division, Department of Medicine and Department of Cell Biology and Anatomy,
Johns Hopkins University School of Medicine, Baltimore, MD 21287, USA

Summary

The actin cytoskeleton is a remarkably dynamic structure in non-muscle cells which responds to cell stimulation by a variety of cytokines. A paradigm for such cytokine-induced reorganization of the actin superstructure is the actin response to growth factor of the EGF/PDGF family. This paper reviews the role of the polyphosphoinositide-metabolism, small GTP-binding proteins and profilin in the reorganization of the actin cytoskeleton in cells activated by growth factors.

Actin and Ras, two nucleotide triphosphatases regulated by cytokines

A myriad of cytokines and growth factors are known to affect the biology of responsive cells. For many cells, responses to these agonists include profound reorganization of the actin cytoskeleton, with associated changes in a number of processes of cell motility (Cooper, 1991). Motile responses to growth factor stimulation of responsive cells typically occur within minutes, and include the intense membrane activity known as ruffling (which corresponds to fluid phase endocytosis), receptor-mediated endocytosis, the intracellular migration of organelles, shape changes, and directional locomotion. DNA replication, necessary for the proliferative response of the cell, occurs only after a delay of several hours (Cantley *et al.*, 1991). We are characterizing the signalling pathways and actin-regulating proteins for these motile responses to agonist stimulation.

Actin and Ras both belong to the nucleotide triphosphatase super-family (Goldschmidt-Clermont *et al.*, 1992a). As such they share some key characteristics (Figure 1). Both bind to their respective nucleotides (adenine nucleotide for actin and guanine nucleotide for Ras) with high affinity, at a K_d well below the intracellular concentration of these nucleotides. Therefore, these nucleotide triphosphatases are not regulated by the concentrations of their substrates in physiological conditions. The hydrolysis of the nucleotide is irreversible, and controls the direction of cycling through the free energy levels resulting from the hydrolysis of nucleotides (Figure 1).

Two steps of these cycles are rate limiting. First, the hydrolysis of the nucleotide is usually slow, and subject to regulation. In the Ras system, GTPase activating proteins (GAPs) markedly increase the GTPase activity of the Ras-GTP complex. In the case of actin, binding to other actin subunits within polymers increases the hydrolysis of the ATP nucleotide by 7,000-fold over monomeric actin. Second, the dissociation rate of the nucleotide from the triphosphatase is slow, and is also subject to regulation. Recently proteins have been identified whose role is to increase the rate of nucleotide exchange. In

Key words: cell signalling, cytokine, growth factor, actin, Ras, profilin, receptor tyrosine kinase.

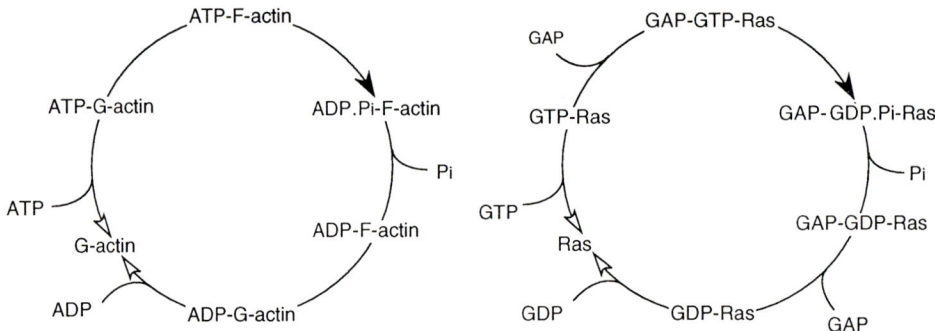

Figure 1. The nucleotide triphosphatase cycle of actin and Ras. The solid arrows indicate the irreversible step for each cycle. The open arrows correspond to the reactions which are accelerated by the exchanger proteins. G- and F- represent the monomeric and filamentous conformation of actin, respectively. GAP corresponds to the GTPase activating protein.

the Ras system, proteins with related sequences like GTP releasing factor, Son of Sevenless, and CDC25 correspond to proteins which all accelerate the dissociation rate of the bound-nucleotide from Ras (Downward, 1992). In the actin system, the small actin binding protein profilin increases nucleotide dissociation from actin (Mockrin and Korn, 1980).

Small GTP binding proteins as a link between receptors and the actin cytoskeleton

The membrane receptors which recognize growth factors like epidermal growth factor and platelet derived growth factor are tyrosine kinases, activated by agonist binding (Cantley *et al.*, 1991). Activated receptors regulate the GAPs and nucleotide exchange proteins responsible for altering the interaction of Ras with its bound nucleotide (Figure 2; Cantley *et al.*, 1991; Downward, 1992). Nucleotide regulating proteins can affect concurrently more than one small GTP binding protein systems, either through multiple domains in their sequence which are able to affect GTP binding proteins, or through complexing with other proteins which also affect the interaction of small GTP binding proteins with their respective nucleotide (Downward, 1992). This dispersion of signal to multiple systems might be important to provide coherent response of a variety of pathways to agonist activation.

Our interest in describing together the actin system and the Ras system rests not simply in insights into biological patterns of organization, but also, in the direct control which small GTP binding proteins have been shown to exert upon specific actin structures. Rho controls the stability of focal adhesions and actin stress fibers, whereas Rac induces membrane ruffling (Ridley and Hall, 1992; Ridley *et al.*, 1992). GTP binding proteins in general have seemed to serve as biological timers for the stabilization of a variety of complex structures. These proteins are therefore poised to regulate complex actin structures inside cells. We have performed experiments to analyze the mechanism(s) by which GTP-binding proteins control the actin cytoskeleton.

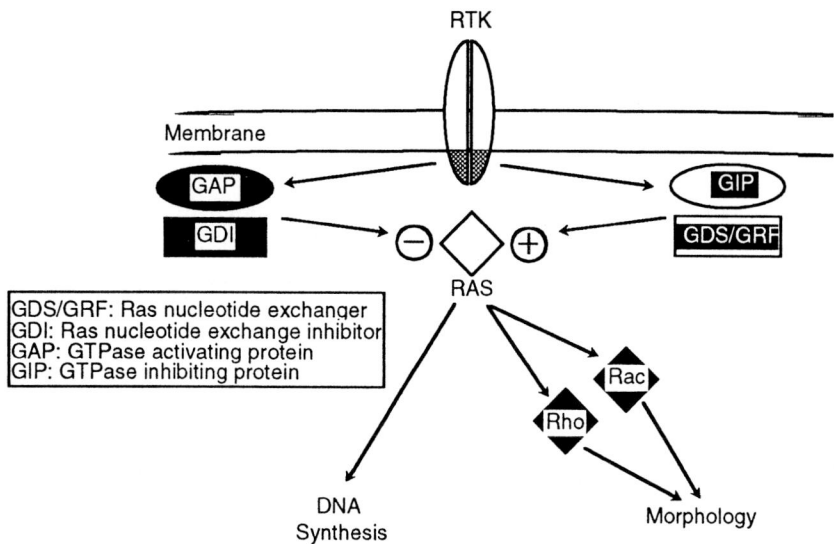

Figure 2. Cell activation and regulation of small GTP-binding proteins. Fine tuning of Ras activity by receptor tyrosine kinase (RTK) results from activation of proteins controlling the exchange of Ras nucleotide (GDS/GRF and GDI), and proteins controlling Ras nucleotide hydrolysis. The effector pathway of Ras may in turn include other small GTP-binding proteins like Rho and Rac, which exert a regulatory function on the actin cytoskeleton, and additional molecules mediating Ras control of DNA synthesis.

We used NIH 3T3 cell lines transformed with oncogenic Ha-Ras to study the control of the actin cytoskeleton by the receptor tyrosine kinase and Ras signalling pathways. NIH 3T3 cells stably transformed by Ha-Ras display markedly increased motile activity. The motile activity is particularly evident in terms of membrane ruffling and, to a lesser degree, locomotion. The shape of the transfected cells is more retracted, and adhesion to the substrate is weaker.

NIH 3T3 cells cultured in medium containing 0.5% of fetal calf serum develop a very complex network of actin filaments. The distribution of the filaments is typically dense beneath the cell membrane (cortical filaments), with another network connecting the cortical filaments to the peri-nuclear region. The filaments observed in the non-transformed cells are relatively thick and contain several actin binding proteins with actin, forming structures known as stress fibers. In contrast, in Ha-Ras transformed cells, actin filaments are practically absent from the center of the cell, and are highly concentrated underneath the cell membrane. It is quite interesting to observe that profilin, the exchanger protein for actin-associated nucleotides, is also localized in the sub-membranous area (Figure 3; ßub *et al.*, 1992).

Profilin, a nucleotide exchanger protein for actin

Profilin accelerates the nucleotide exchange by increasing the off-rate of the nucleotide by 2 to 3 orders of magnitude (Mockrin and Korn, 1980; Goldschmidt-Clermont *et al.*,

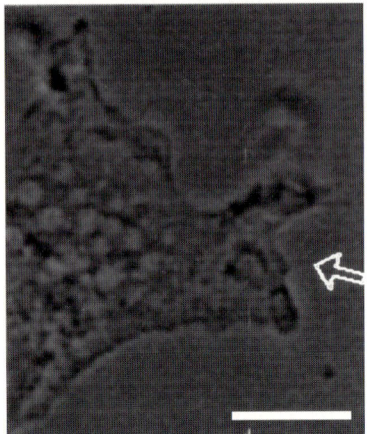 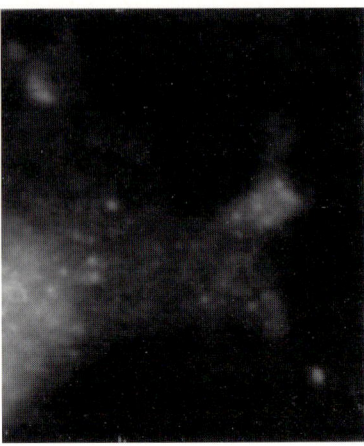

Figure 3. Localization of profilin in overexpressing CHO cells. Matching phase contrast (left panel) and fluorescent (right panel) micrographs of a cell from CHO clonal cell line 1-V (Finkel *et al.*, 1993). The arrow represents a typical area of membrane ruffling and the bar is 10 µm. The fluorescent image was obtained by three-dimensional confocal microscopy and represents the linear projection of 17 frames corresponding to individual parallel planes across the cells, with a distance of 1 µm between planes. Fluorescent patches were particularly concentrated within the perinuclear region and subcortical area at the level of intense membrane ruffling. Permeabilized cells were stained using an affinity purified anti-human platelet profilin polyclonal rabbit first antibody and a FITC-labeled goat anti-rabbit secondary antibody. Control micrographs (with pre-immune serum as first antibody) indicated that the staining of the perinuclear regions or subcortical area is specific.

1992b). Profilin is the only actin binding protein known to increase nucleotide exchange. However, profilin has to compete with other monomeric actin binding proteins for interaction with actin. One such binding protein is thymosin β_4, an inhibitor of actin nucleotide exchange, which is more abundant than profilin in most cells studied. Because of the kinetics of interaction between profilin, actin and thymosin β_4, profilin is able to overcome the inhibitory effect of thymosin β_4 on the actin nucleotide exchange (Goldschmidt-Clermont *et al.*, 1992b). Both proteins bind to actin monomers in a transient manner. It is the transient character of the interaction of actin with other actin binding proteins that allow profilin efficiently to promote the exchange of the nucleotide, in spite of the presence of other actin binding proteins. This effect of profilin is expected to be particularly important in areas of the cell where actin filament turnover is very rapid. Both the density of filaments and the rapid turnover of the monomers inside filaments present underneath the cell membrane are likely to produce large amount of actin monomers bound to ADP. These ADP actin monomers are less appropriate for rapid polymerization and formation of stable filaments (Pollard, 1986; Carlier, 1989). Therefore, the presence of profilin in these areas of rapid cycling of actin subunits through the triphosphatase energy levels is consistent with profilin's role as the exchange protein of the actin bound nucleotide (Goldschmidt-Clermont *et al.*, 1992b). In some conditions, the concentration of profilin could become limiting for actin polymerization and therefore, we hypothesize that the regulation of profilin concentration, possibly through the metabolism of inositol phospholipids (Goldschmidt-Clermont *et al.*, 1990), could be used by receptor tyrosine kinases and by Ras to regulate actin (Goldschmidt-Clermont and Janmey, 1991; Goldschmidt-Clermont *et al.*, 1992a).

Profilin stabilizes actin filaments *in vivo*

To test this hypothesis *in vivo*, we overexpressed human profilin in Chinese hamster ovary (CHO) cells, in a stable fashion, using a vector with the β-actin promotor and the neomycin resistance gene for selection (Finkel *et al.*, 1993). Profilin overexpression had no detectable effect on the growth rate of transfected cells. The concentration of total and filamentous actin in transfected cells was not detectably affected by increasing profilin concentration up to 4-fold. Photolytic activation of caged resorufin actin was used to quantitate the turnover rate of actin filaments in transfected cells (these experiments were performed by Julie Theriot at UCSF). Clonal lines overexpressing profilin do contain actin filaments which are overall more stable. This finding is consistent with profilin's role as a nucleotide exchanger for actin because the stability of actin filaments may be exquisitely sensitive to the addition of actin monomers bound to ADP instead of ATP (Carlier, 1989). Although wild-type CHO cells contain enough profilin to develop the motile activities required for their survival, the ability of increased profilin concentrations to stabilize even further the actin filaments contained in overexpressing cells may be an indication that the ADP-monomer/ATP-monomer ratio is the key to filament stability.

To test the possibility that profilin's role in actin regulation *in vivo* is at the level of the nucleotide exchange, we treated CHO clonal lines containing various concentrations of profilin with the actin inhibitor cytochalasin D (cyto D; Schliwa, 1982; Cooper, 1987). Cyto D has several effects on actin filaments: it caps the fast growing end of filaments, thereby limiting the turnover rate of actin subunits. In addition, cyto D substantially increases the hydrolytic activity of actin monomers, by inducing the formation of actin dimers and trimers with enhanced ATPase activity (Cooper, 1987). In CHO cell lines, cyto D (2 μM) induces a small (nearly 30%) drop in F-actin concentration. After washing out the cyto D, the concentration of F-actin returns within 120 minutes to pretreatment levels. Profilin overexpression decreased the sensitivity of actin filaments to cyto D and accelerated the recovery of F-actin after cyto D washout (Finkel *et al.*, 1993). This finding is consistent with profilin overcoming cyto D's effect on formation of ADP actin monomers. Clearly, these results could be explained otherwise, but we have ruled out the possibility that cyto D would have differential toxicity for cell lines expressing various concentrations of profilin, as all the clonal lines grew at the same rate after cyto D treatment.

Mechanisms of regulation of actin and Ras by growth factor receptors

In both the Ras signalling pathway and the actin network, receptor tyrosine kinase controls the activity of the nucleotide triphosphatases by regulating the interaction between the nucleotide triphosphatases and their respective nucleotide exchangers and triphosphatase activating proteins (Downward, 1992; Goldschmidt-Clermont *et al.*,

1992a). The next question to be resolved is how growth factors control these exchanger proteins.

Receptors with tyrosine kinase activity, upon interaction with their growth hormone agonists, induce the aggregation underneath the cell membrane of a few key effector proteins (Cantley *et al.*, 1991). These proteins, phosphorylated on specific tyrosine residues by the receptor kinase, and complexed with the receptor as a result of SH$_2$-domain interaction with phosphotyrosine residue(s), represent major elements of specific signalling pathways. Formation of this functional unit at the plasma membrane switches on a cascade of molecular events resulting in a complex cell response. Although receptor tyrosine kinases have the ability profoundly to alter the biology of responsive cells, the amplitude of the tyrosine kinase reaction itself is relatively limited. Therefore, it does not seem that direct phosphorylation on tyrosine of the very abundant actin and actin binding proteins would represent a suitable general way to regulate the actin network. Instead, there is a need for an amplification mechanism to allow receptor tyrosine kinase to induce actin cytoskeletal reorganization (Howard *et al.*, 1993). One of the major substrate proteins for receptor tyrosine kinase is the gamma isoform of phospholipase C (PLC-γ; Cantley *et al.*, 1991).

Polyphosphoinositides and triphosphate nucleotide exchanger proteins

Profilin binds to phosphatidylinositol 4,5 bisphosphate (PIP2), and interaction between this inositol phospholipid and profilin prevents the hydrolysis of PIP2 by the unphosphorylated form of PLC-γ (Goldschmidt-Clermont *et al.*, 1990; Goldschmidt-Clermont and Janmey, 1991). Phosphorylation on tyrosine of PLC-γ allows the lipase to overcome profilin inhibition and to hydrolyze PIP2. The interaction between profilin and PIP2 has been highly conserved throughout evolution; Staiger *et al.* have shown that profilin from maize pollen binds to PIP2 and inhibits PLC from the plasma membrane of bean leaf (Abstract, Cell Behavior: Adhesion and Motility Meeting, 3rd Abercrombie Symposium, Bath, 1992). While profilin seems to be able to regulate the activity of PLC-γ and to make it dependent upon tyrosine phosphorylation by receptor tyrosine kinase, in turn hydrolysis of PIP2 by activated PLC-γ may regulate the interaction of profilin with actin (Goldschmidt-Clermont and Janmey, 1991). Lassing and Lindberg reported in 1985 (Lassing and Lindberg, 1985) that binding of PIP2 to profilin inhibits interaction between profilin and actin. Therefore, PIP2 turnover may link receptor tyrosine kinase activation and actin network reorganization, by modulating the availability of profilin, as the concentration and distribution of PIP2 is altered in response to receptor tyrosine kinase activation.

In addition, Shariff and Luna (1992) have recently shown that diacylglycerol, a byproduct of PIP2 hydrolysis by PLC, can directly enhance the formation of actin nuclei at the membrane level by activating a yet unidentified nucleating protein factor. Small GTP binding proteins like Ras also have the ability to regulate inositol phospholipid metabolism (Cantley *et al.*, 1991). It is possible that regulation of the actin network by small GTP binding proteins requires specific modulation of the local inositol phospholipid concentration at the membrane level (Figure 4).

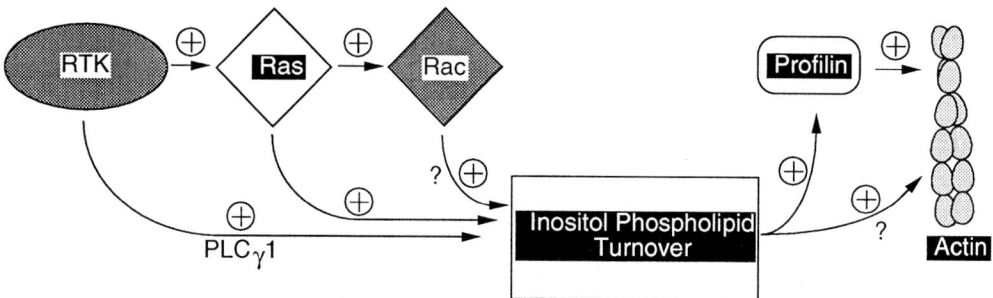

Figure 4. Model for the regulation of cellular actin by receptor tyrosine kinase and Ras. Activation of receptor tyrosine kinase (RTK) results in tyrosine phosphorylation of phospholipase Cγ1 by the cytoplasmic kinase of the receptor. Tyrosine phosphorylated phospholipase Cγ1 hydrolyzes PIP_2, which otherwise sequesters profilin at the membrane level, away from actin. Once off the membrane, profilin is able to accelerate the nucleotide exchange on actin, providing actin subunits charged with ATP which polymerize readily. In addition, RTK switches on Ras and further small GTP binding proteins like Rac, through the activation of GTP nucleotide exchange proteins. The activated small GTP-binding proteins in turn, regulate the stability of specific actin structures, possibly by concentrating profilin in certain areas of the cell. The control of profilin distribution by small GTP-binding proteins may be mediated by the rate of inositol phospholipid turnover (Goldschmidt-Clermont and Janmey, 1991).

Integrated studies of nucleotide triphosphatases like Ras and actin may reveal patterns of organization which will allow us to improve markedly our understanding of the mechanisms linking signalling-induced cell proliferation and the reorganization of the cytoskeleton. These patterns would probably be useful as well for the study of a variety of additional nucleotide triphosphatases like, for example, tubulin and HSP 70.

This research is supported in part by a grant from Syntex, by a grant from the Henry Ciccarone Center for the Prevention of Heart Disease and by the American heart Association (G-I-A, Maryland Affiliate, Inc.). PJG-C was selected as a Syntex Scholar in 1992.

References

Cantley, L. C., Auger, K. R., Carpenter, C., Duckworth, B., Graziani, A., Kapeller, R., and Soltoff, S. (1991). Oncogenes and signal transduction. *Cell* **64**, 281-302.

Carlier, M.-F. (1989). Role of nucleotide hydrolysis in the dynamics of actin filaments and microtubules. *Int. Rev. Cytol.* **115**, 139-170.

Cooper, J. A. (1987). Effects of cytochalasin and phalloidin on actin. *J. Cell. Biol.* **105**, 1473-1478.

Cooper, J. A. (1991). The role of actin polymerization in cell motility. *Ann. Rev. Physiol.* **53**, 585-605.

Downward, J. (1992). Exchange rate mechanisms. *Nature* **358**, 282-283.

Finkel T., Theriot, J. A., Dise, K. R., Tomaselli, G. F., and Goldschmidt-Clermont, P. J. (1993). Dynamic actin structures stabilized by profilin. Unpublished.

Goldschmidt-Clermont, P.J., Machesky, L.M., Baldasarre, J.J., and Pollard, T.D. (1990). The actin-binding protein profilin binds to PIP2 and inhibits its hydrolysis by phospholipase-C. *Science* **247**, 1575-1578.

Goldschmidt-Clermont, P. J. and Janmey, P. A. (1991). Profilin, a weak CAP for actin and RAS. *Cell* **66**, 419-421.

Goldschmidt-Clermont, P. J., Mendelsohn, M. E., and Gibbs, J. B. (1992a). Rac and Rho in control. *Curr. Biol.* **2**, 669-671.

Goldschmidt-Clermont, P.J., Furman, M.I., Wachstock, D., Safer, D., Nachmias, V.T., and Pollard, T.D. (1992b). The control of actin nucleotide exchange by thymosin B4 and profilin. A potential regulatory mechanism for actin polymerization in cells. *Mol. Biol. Cell* **3**, 1015-1024.

Howard, P. K., Sefton, B. M., and Firtel, R. A. (1993). Tyrosine phosphorylation of actin in Dictyostelium associated with cell-shape changes. *Science* **259**, 241-244.

Lassing, I., and Lindberg, U. (1985). Specific interaction between phosphatidylinositol 4,5 bisphosphate and profilactin. *Nature* **318**, 472-474.

Mockrin, S.C., and Korn, E.D. (1980). Acanthamoeba profilin interacts with G-actin to increase the rate of exchange of actin-bound adenosine 5′-triphosphate. *Biochemistry* **19**, 5359-5362.

Pollard, T.D. (1986). Rate constants for the reactions of ATP- and ADP-actin with the ends of actin filaments. *J. Cell. Biol.* **103**, 2747-2754.

Ridley, A.J. and Hall, A. (1992). The small GTP-binding protein rho regulates the assembly of focal adhesions and actin stress fibers in response to growth factors. *Cell* **70**, 389-399.

Ridley, A.J., Paterson, H.F., Johnston, C.L., Diekman, D., and Hall, A. (1992). The small GTP-binding protein rac regulates growth factor-induced membrane ruffling. *Cell* **70**, 401-410.

Schliwa, M.J. (1982). Action of cytochalasin D on cytoskeletal networks. *J. Cell. Biol.* **92**, 79-91.

Shariff A. and Luna, E. J. (1992). Diacylglycerol-stimulated formation of actin nucleation sites at plasma membranes. *Science* **256**, 245-248.

ßub, F., Temm-Grove, C., Henning, S., and Jockusch, B. M. (1992). Distribution of profilin in fibroblasts correlates with highly dynamic actin filaments. *Cell Motil. Cytoskel.* **22**, 51-61.

Printed in Great Britain © The Society of Experimental Biology 1993   325

CYTOSKELETAL CHANGES ASSOCIATED WITH CELL MOTILITY

RICHARD WARN, DENISE BROWN, PAUL DOWRICK, ALAN PRESCOTT
and ALBA WARN

School of Biological Sciences, University of East Anglia, Norwich NR4 7TJ, UK

Summary

This chapter reviews various aspects of changes in cytoskeletal organization which occur upon activating a highly motile cell phenotype. The first of these relates to the rapid formation of F-actin-rich ruffles on the apical cell surfaces following the addition of one of several motility factors. In a number of aspects these ruffles resemble leading edge lamellipodia. The ruffles form upon ligand binding and, at least for one cytokine, the receptors become associated with the ruffles. For several cytokines the ruffles are circular and in all cases they are associated with much increased pinocytosis. The possible significance is considered of these very early markers of a motile cell phenotype.

A second topic covered is the role of microtubules (MTs) in maintaining cell polarity in some cell types but not others. The micro-injection of biotin-tubulin into cells has provided a valuable marker of MT turnover. Using this method it has been found that the microtubule network in secondary chick heart fibroblasts (2° CHFs), where MTs are required to maintain a polarized motility, does not turn over significantly more slowly than in 1° CHFs which do not require an intact MT network for locomotion. In both cases the MT network turns over very quickly and no sub-population of longer-lived MTs has been found. In contrast, in motile epithelial (PtK$_2$) cells, a sub-population of longer-lived microtubules has been identified and these appear to maintain the long cell processes.

Introduction

For nearly all types, cell motility is a feature of development rather than maturity. During embryogenesis there is a phase when all cell types undergo extensive migrations either as sheets or as single cells (Kolega, 1986). Examples of such movements include gastrulation and neural crest migration. As tissues differentiate cell movements cease and nearly all cell types become firmly anchored, either directly to their neighbours or via the extracellular matrix. Only in the form of tumour metastases can most cell types become motile again (Fidler and Hart, 1982; Nicolson, 1988).

Because of the fact that there is, for most cell types, a definite transition from motile to static, and sometimes also the reverse, two important questions can be asked. These are:- what activates cells to become motile, and by what mechanisms does this occur? There is increasing evidence that a variety of proteins - motility factors - can induce or considerably enhance the motility of a wide range of mammalian cell types (for reviews see Rosen and Goldberg, 1989; Stoker and Gherardi, 1991). This review concentrates on the second question posed and considers various aspects of the cellular response to

Key words: cytoskeleton, motility factors, macropinocytosis, microtubule dynamics.

mammalian motility factors, paying particular attention to the way in which the cytoskeleton is activated to generate motility.

Rapid activation of the cytoskeleton in response to motility factors

Most mammalian motility factors are also growth factors, a most intriguing duality of function which implies the possibility of common or related signalling transduction pathways. A number of well characterized growth/motility factors initiate a response by binding to receptors and thereby activating a tyrosine kinase domain located in the cytoplasmic portion of the receptor (Cross and Dexter, 1991). From a variety of viewpoints two receptor systems - that for PDGF and EGF - have been extensively studied. In both cases ligand stimulated receptor autophosphorylation is an initial step in the response cascade (for reviews on PDGF, see Heldin and Westermark, 1989; EGF, Schlessinger, 1986). Once the ligand has bound to the receptor a number of enzymes and other molecules rapidly become associated with the activated receptor. Thus for PDGF, within minutes of ligand binding, a proportion of the receptors bind the following molecules: PI-3 kinase, PLCγ, ras, GAP, src, raf-1, yes and fyn (see Kaplan et al., 1990, Cantley et al., 1991). Most of these proteins are phosphorylated on tyrosines after PDGF binding and at least for PLCγ this phosphorylation causes an increase in its catalytic activity (Nishibe et al., 1990). The evidence suggests that binding of PDGF to its receptor rapidly leads to the formation of a complex of associated attached molecules which act together to initiate the signalling cascade. These complexes have been named "signalling complexes" or "signalsomes" (Kaplan et al., 1990, Ulllrich and Schlessinger, 1990).

Once a growth factor ligand is bound to its receptor it is rapidly internalized. EGF treatment of A431 cells (a model system which carries large numbers of EGF receptors on its surfaces; Fabricant et al., 1977) leads to a clustering of EGF receptors on the surfaces of plasma membranes prior to internalization (Wiegant et al., 1986, Van Belzen et al., 1988). Various electron microscopy methods have demonstrated that EGF receptors are bound to a detergent-insoluble cytoskeleton fraction (Roy et al., 1989, Van Bergen en Henegouwen et al., 1989). Visualization of the distribution of receptors after EGF binding using immunofluorescence and confocal microscopy has demonstrated that a significant proportion of the receptors are associated with membrane ruffles and microvilli soon after ligand binding (Rijken et al., 1991). These structures are very rich in F-actin microfilaments forming a cortical meshwork, but do not contain either microtubules or intermediate filaments. The cytoskeleton-associated receptors show both a functional EGF binding domain and also EGF-induced kinase activity (Landreth et al., 1985, Roy et al., 1989, Van Bergen en Henegouwen et al., 1989) thus demonstrating that a physical association of the EGF receptors to the F-actin cytoskeleton may be part of the signal transduction process, e.g. having a role in signalsome formation or alternatively, involvement in the internalization of the bound receptor. The receptors for PDGF (Zippel et al., 1989) and NGF (Vale and Shooter, 1983) are also associated with the cytoskeleton, demonstrating that this binding of receptors to F-actin is a general response.

The formation of surface ruffles and microvilli is the earliest observed structural change of responsive cells to growth factors. In A431 cells spectacular bursts of ruffling activity occur within 5 minutes of treatment with EGF (Chinkers *et al.*, 1979). Large lamellipodia, up to several microns long, are formed frequently around the edges of the cells. The ruffling lasts only for about 10 minutes. By the end of this time the edges of the cells begin to pull inwards, leaving retraction fibres, and the cell morphology gradually changes into a highly rounded phenotype with cells piled on top of each other (Chinkers *et al.*, 1981; Watts and Marsh, 1992; Hewlett, Prescott and Watts - unpublished observations). Such a phenotype markedly resembles that of many types of transformed cell. The significance of the membrane ruffling in response to EGF is unknown but it is associated with a marked increase in micro-pinocytosis mainly at the cell edges (Haigler *et al.*, 1979). Sometimes large ruffles were seen in TEM sections curling over and seemingly fusing with the underlying cell membrane to form vesicles.

Treatment of responsive cells with PDGF is also followed by ruffling but with this cytokine two types of ruffling have been observed - linear and circular. Linear ruffles form after ligand binding to transfected PAE cells expressing either the A- or the B-type PDGF receptors but circular ruffles form only in cells carrying functional B-type receptors showing tyrosine kinase activity (Westermark *et al.*, 1990, Eriksson *et al.*, 1992). Circular ruffles have been found to be initiated within 1 min. after the addition of PDGF to serum starved glial cells (Mellström *et al.*, 1983). The ruffles stain strongly for F-actin and are present only on the dorsal surfaces. TEM study has shown that the ruffles contain bundles of F-actin forming part of an elongated F-actin meshwork. Their dimensions vary in size and some are so large that they encompass the whole cell. Circular arrangements of plasma membrane folding either form directly on the apical surfaces or alternatively ruffling membrane moves inwards from the lateral edges to participate in forming membrane circles. Complete actin rings (representing ruffles in fixed preparations) could be seen within $3\frac{1}{2}$ minutes after ligand addition to the culture and continued to appear during the first half hour after PDGF treatment. Frequently they persisted for only five minutes but could remain for as long as 20 min. The circular ruffles markedly resemble the structures which form after treatment of BSC-1 cells with TPA, a potent stimulator of protein kinase C (Schliwa *et al.*, 1984). Curiously though, the formation of circular ruffles after PDGF treatment is entirely blocked as a result of the simultaneous addition of TPA. Even so, it still seems likely that protein kinase C is involved in the mechanism of circular ruffle formation.

Hepatocyte growth factor/scatter factor (HGF/SF) is a third cytokine where ruffle formation occurs soon after addition of the cytokine to MDCK cells, a cell type where scattering rather than increased cell division occurs (Dowrick and Warn, 1991, Dowrick *et al.*, in preparation). Scattering is a breakup of colonies leading to the formation of highly motile individual cells (Stoker and Perryman, 1985). Both linear edge ruffles and circular ruffles are seen. In MDCK cells actin rings begin to appear one minute after HGF/SF application frequently appearing as semi-circles which arch inwards from the edges and are fully formed within a further one to three minutes (Fig. 1a-c). The maximum diameters reached by the circular ruffles are 25-30 μm and on occasion they encompass much of the cell (Fig. 1c). Within 1-2 minutes of their formation the circles become closed, and only apical surface protrusions packed with F-actin remain visible

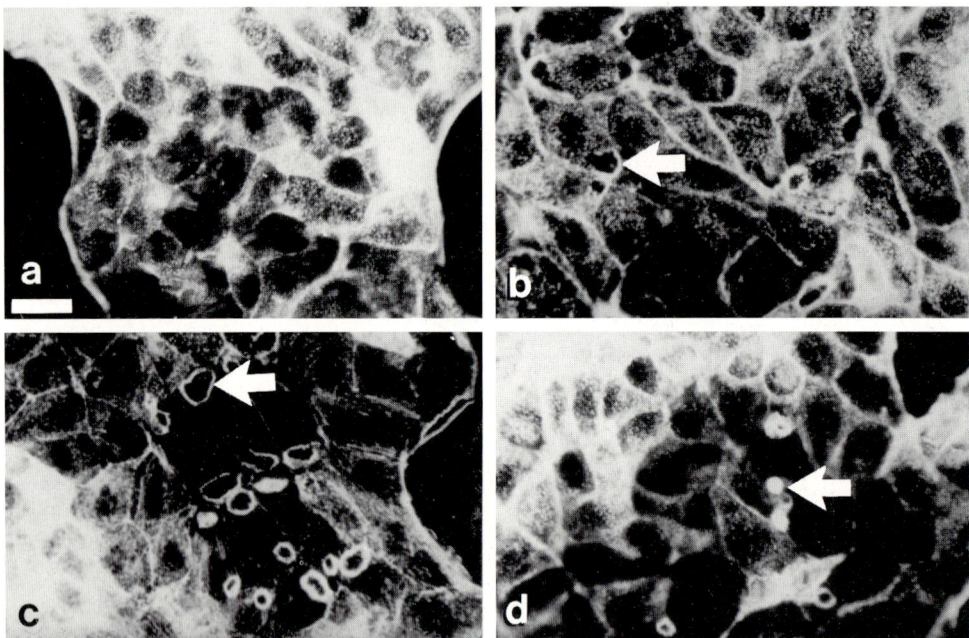

Fig. 1. MDCK cells treated with HGF/SF for the stated times at 37°C, fixed, and stained with rhodamine-phalloidin for F-actin. (a) 1 min incubation (b) 2 min. Arrow, forming circular ruffle. (c) 4 min. Arrow, fully formed ruffle. (d) 6 min. Arrows, closing ruffles. Scale bar = 20 μm.

(Fig. 1d). After about 10 minutes exposure the apical surfaces of the cells return to a state similar to controls. As might be expected, prior treatment with cytochalasin prevents ring formation but nocodazole and taxol have no inhibitory effects on ring formation at concentrations where the microtubule network is largely disassembled (nocodazole) or grossly altered (taxol). Thus F-actin microfilaments have a major role in circular ruffle formation whereas microtubules have none.

Formation of circular ruffles after HGF/SF treatment has been found to be linked to pinocytosis. If MDCK cells are bathed in medium containing FITC-dextran and then factor added, the ruffles are found to correspond to sites of pinocytosis (Fig. 2). Cells which were fixed and observed with epifluorescence optics (Fig. 2a and b) showed the presence of a number of small pinocytotic vesicles associated with the margins of the ruffles whereas once the ruffles had closed, single large vesicles were present (Fig. 2c and d). Soon after the single large vesicles had formed, fingerlike channels became apparent (Fig. 2e and f) suggesting rapid association with the endosome system. Treatment with suramin (a polyanonic detergent that acts to prevent binding of various growth factors to their receptors) blocks scattering by HGF/SF (Adams *et al.*, 1991) and also completely inhibits circular ruffle formation. Thus it seems likely that ruffle formation is an early consequence of scatter factor binding to its receptor. Circular ruffle formation is also inhibited by treatment with 3 mM amiloride (a blocker of Na^+/H^+ exchange) and 250 μM

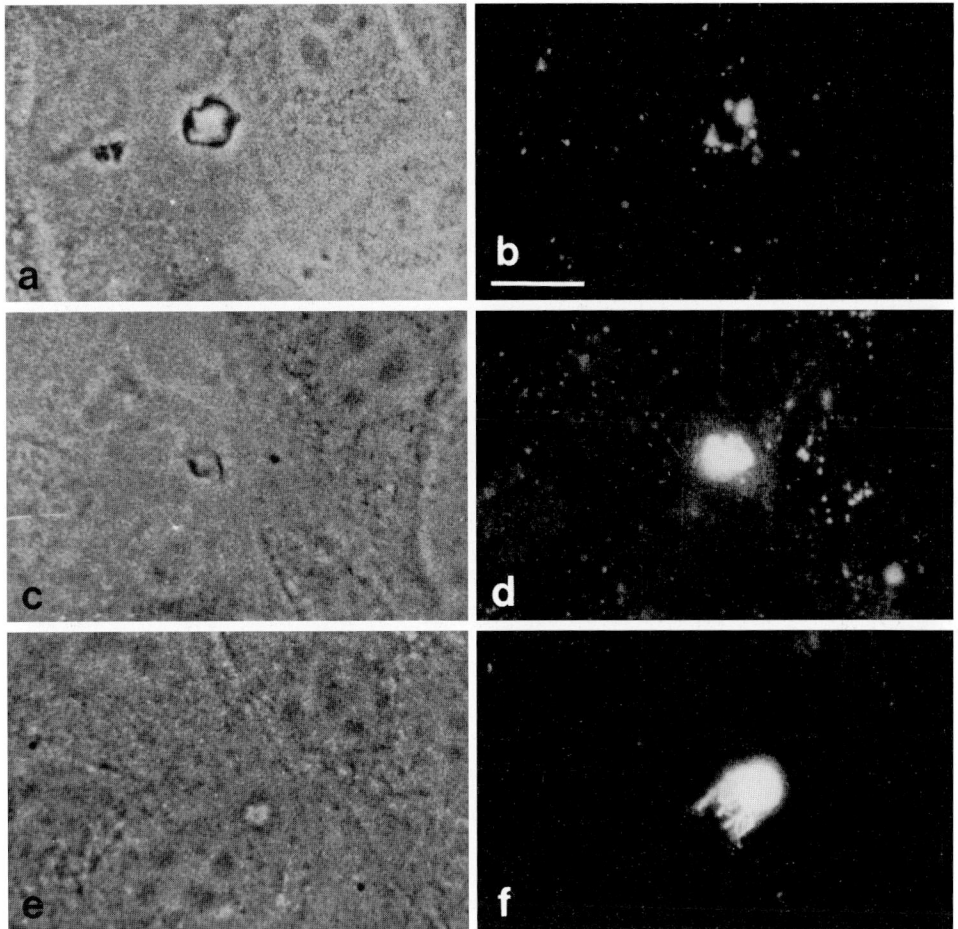

Fig. 2. MDCK cells incubated with HGF/SF in medium containing FITC/dextran for the stated times and then fixed. a, c, e phase contrast. b, d, f FITC channel for dextran fluorescence. Scale bar = 10 μm.

SITS (a blocker of Cl^-/HCO_3^- and $Na^+/Cl^-/HCO_3^-/H^+$ exchange). These results suggest that ion flux is an important step in ruffle formation.

Similar findings have been made with EGF. Both amiloride and the removal of extracellular Na^+ block the stimulation of pinocytosis by EGF (West *et al.*, 1989). However, no effect was found on the internalization of EGF which occurs via coated pits (Hopkins *et al.*, 1985). Furthermore, Defize *et al.* (1989) have found that a monoclonal antibody which specifically blocks EGF binding to its low affinity receptors inhibits the pinocytosis response to EGF binding. This antibody had no effects on EGF binding to the high affinity receptors. Furthermore, a number of significant early cellular changes which occur in response to EGF, including cell rounding, were unaffected. The authors concluded that the enhanced pinocytosis which occurs after EGF application is not essential for the initiation of the signal transduction cascade. However, the significance of the ruffling and macropinocytosis in response to HGF/SF or EGF is puzzling. It would be surprising if they did not reflect some widespread biological function. If the circular ruffles formed as a consequence of PDGF binding are also structures not directly involved in the signal transduction cascade, then the fact that they occur only on cells carrying the B receptors suggests at least a specificity of response. Whether or not the circular ruffles are directly involved in the motility response to PDGF remains to be determined. *Salmonella* has recently been found to enter gut epithelial cells via an EGF receptor dependent macropinocytosis event (Galan *et al.*, 1992). That entry occurs *in vivo* suggests at least a medical significance for the pinocytosis which is induced following EGF binding.

Cytoskeletal protein targets of receptor tyrosine kinase activity

The mechanism by which ruffles and microvilli form as a very early response to growth/motility factors may well prove to be quite similar to the manner in which lamellipodia and microvilli form at the leading edges of motile cells (see Small, 1993; Heath, 1993; this volume). Because EGF, PDGF, and HGF/SF all initiate a signal transduction cascade via the activation of the receptor tyrosine kinase (Cantley *et al.*, 1991) the question must be raised as to how the actin microfilament cytoskeleton is activated and reorganized via receptor binding. Current evidence has suggested several models of a protrusile F-actin meshwork as the driving force for lamellipodium extension (Small, 1993; this volume). Changes in actin dynamics are dependent upon changes in actin binding proteins (ABPs) associated with the microfilaments of the leading edge. Thus the question posed above becomes what ABPs are activated as the result of ligand binding to the receptor and how? This chapter will not consider the recent elegant work of Goldschmidt-Clermont, Pollard, and colleagues relating to profilin release as a consequence of EGF binding to its receptor (see Heldman and Goldschmidt-Clermont, 1993; this volume), rather it will ask the question what potential cortical ABP targets exist for the receptor tyrosine kinases?

Ezrin

At present there are several ABPs which are known to be phosphorylated on tyrosine very shortly after the addition of EGF to responsive cells. The most intriguing of these is ezrin, an 81,000 M_r protein species first isolated from the cores of intestinal microvilli (Bretscher, 1983, 1986). The protein is phosphorylated on tyrosine in response to EGF and is homologous to p81, a substrate of various tyrosine kinases (Gould *et al.*, 1986). In A-431 cells, treatment with EGF leads to the appearance of ezrin on the surfaces of ruffles and microvilli which form soon after application of the ligand (Bretscher, 1989). The time course for ezrin phosphorylation closely correlates with the appearance of the ruffles, first appearing at 30 seconds, reaching a maximum at 2-5 minutes and afterwards declining. Analysis of the deduced sequence has shown that it belongs to a group of proteins known to be involved in the linkage of cortical F-actin microfilaments to the plasma membrane (Gould *et al.*, 1989); among these related proteins are band 4.1 of erythrocytes (Gould *et al.*, 1989) and talin, a component of focal contacts (Rees *et al.*, 1990).

It is not yet known what role ezrin may have in ruffle production. It does not bind to actin at all strongly (Bretscher, 1983). However, the sequence and putative structural relationships with proteins known to be involved in F-actin/membrane linkage and its association with several ABPs upon fractionation all suggest that it is somehow involved with the cortical cytoskeleton re-organization which follows EGF application. Although at present there is no direct evidence that phosphorylation on tyrosine induces ezrin to become involved in the cytoskeletal changes occurring, the circumstantial case for such a role is strong.

Spectrin is also phosphorylated after EGF binding and re-distributes into the ruffles (Bretscher, 1989). However, spectrin is not phosphorylated on tyrosine after EGF treatment, only on serine (ezrin is phosphorylated on both), which suggests an indirect activation via some kinase other than that of the EGF receptor. It is unlikely that ezrin and spectrin interact, judging from sequence considerations. In other cell types they don't even form part of the same cytoskeletal structure e.g. in intestinal cells, ezrin is present in the microvillar cores whereas spectrin forms part of the underlying terminal web (Bretscher, 1983).

Myosin may also be associated with the cytoskeletal changes associated with ruffling. The use of antibodies against the 200 kD myosin heavy chain demonstrated that myosin also re-distributes after EGF treatment, but not into the ruffles (Bretscher, 1989). In untreated cells a rather uniform cortical myosin distribution was seen, but 30 sec after EGF application myosin began to be rearranged into a striated pattern. Myosin heavy chain was not found to be phosphorylated in response to EGF. However, adding PDGF to 3T3 cells induced phosphorylation of the 20 kD myosin light chains within 5 min (Bockus and Stiles, 1984), suggesting that these are also likely targets of EGF.

Annexin II

Annexin II is a second protein which is rapidly phosphorylated on tyrosine in A431 cells after EGF treatment (Hunter and Cooper, 1981). It is a heterotetramer formed of two 36 kD and two 11 kD subunits, also called calpactin I amongst a number of other terms (Crumpton and Dedman, 1990). It has been found to be the same molecule as p36, which is a major substrate of phosphorylation by *src* tyrosine kinase (Gerke and Weber, 1984, 1985; Glenney, 1986). Annexin II has a submembranous distribution in a variety of cell types. In intestinal cells it forms part of the terminal web cytoskeleton (Gerke and Weber, 1984). The protein is a member of a family of Ca^{2+} binding proteins (Crompton *et al.*, 1988) which binds phospholipids and also actin (Glenney, 1985; Glenney *et al.*, 1987). However, in spite of a wealth of data suggesting that it interacts with both the plasma membrane and the F-actin cortical meshwork, a more precise function has yet to be assigned. Like ezrin it is a plausible candidate as a major transducing protein in the very early membrane/cytoskeleton rearrangements which follow the application of a number of motility/growth factors.

Longer term changes in cytoskeleton organization

Striking patterns of changes in stress fibre organization following motility/growth factor treatment have been described for several cytokines and suggested to be casually related to the enhanced motility. These changes have recently been reviewed (Dowrick and Warn, 1991) and will not be considered further here.

However, some new data relating to the enigmatic role of microtubules (MTs) in cell motility has recently emerged. That MTs have a role to play in the onset of cell motility is suggested by the observation that cells at the migrating edge of a wounded monolayer of 3T3 fibroblasts create an asymmetric microtubule array, with virtually all of the putatively stable microtubules orientated towards the free cell edges (Gundersen and Bulinski, 1988). More than 80% of the cells at the wound edge had generated this polarized array of MTs within 2 hours and maintained it for at least a further 12 hours. Formation of these polar arrays of MTs appears to precede the onset of cell motility.

MT disrupting drugs, including nocodazole, colchicine and colcemid, inhibit the polarization of spreading fibroblasts (Goldman and Follet, 1969; Goldman, 1971; Ivanova et al., 1976; Tomasek and Hay, 1984) and cause a loss of polarization in previously spread cells. (Vasiliev et al., 1970; Goldman, 1971; Gail and Boone, 1971). Many polarized epithelial cells respond to MT disrupting drugs in a similar manner (Domnina et al., 1985; Karavanova et al., 1985). However, a number of authors have concluded that the morphological shape changes and spread of epithelial cells are not MT dependent (Di Pasquale, 1975; Downie, 1975; Domnina et al., 1977; Chernoff and Overton, 1979; Middleton, 1982; Euteneuer and Schliwa, 1984; Middleton et al., 1988).

Furthermore, Middleton et al. (1989) found that the polarization of primary chick heart fibroblasts (1° CHFs) was not affected by colcemid or nocodazole whereas the polarization of secondary chick heart fibroblasts (2° CHFs) was. Primary chick heart fibroblasts, obtained by allowing cells to migrate out of embryo explants, are characteristically small and highly polarized, typically fan shaped. In contrast 2° CHFs, derived from 1° cultures after passaging or as the result of changes occurring during 1° culture over 48 hours, are more often polygonal in shape and are slower moving, sometimes bi- or multi-nucleate.

The reason for the difference in response to the drugs is unclear but has been related to surface area. The mean spread area of 1° CHFs is approximately 400 μm^2 and for 2° cells 1300 μm^2 (Middleton et al., 1988). In line with this suggestion is the observation that when 1° cells are maintained in culture they gradually increase their surface area and at the same time become increasingly more sensitive to the effects of MT disassembling drugs on their movements. One possible explanation is that in 1° CHFs the F-actin microfilament network has sufficient rigidity to maintain a polarized shape but in 2° cells MTs are also required (Gelfand et al., 1985).

From this it might be predicted that the microtubules would be significantly longer lived and more stable in the 1° cells, providing a supporting skeletal function. However, analysis of the turnover using injected biotinylated tubulin as a marker has failed to show any significant differences in stability between the MTs of 1° and 2° CHFs, given the sixfold differences in size, although in the larger 2° cells, stable MTs and isolated pieces persist for longer on average (Fig. 3) (Brown and Warn, 1993). No sub-population of slow turning over MTs was detected. For both 1° and 2° cells turnover of the whole network is very fast as compared with other cell types where many stable microtubules

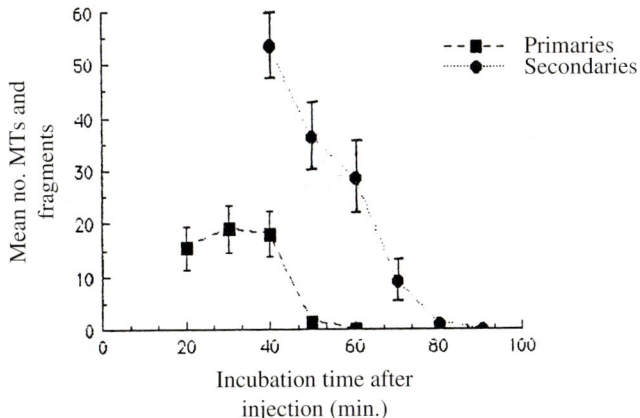

Fig. 3. Graph showing the comparative decrease in the mean number of stable MTs and fragments with time for 1° and 2° CHFs.

are still present after two hours (Schulze and Kirschner, 1987), and some last for 6 hours and longer (Kreis, 1987). In both 1° and 2° CHFs there is no real relationship between microtubule longevity and cell size, and a structural role can be discounted.

What then is the role of microtubules in maintaining the polarity of 2° CHFs but not 1°s? One hypothesis, originally proposed by Kupfer *et al.* (1982), is that the MTs transport new membrane and cortical components from the Golgi apparatus to the leading edge. However, this hypothesis does not explain normal cell motility in drug treated 1° CHFs and other cell types unless it is assumed that an actin based (or other) transport system is sufficient on its own for the supply of lamellar constituents up to a certain size. Recently Kuznetsov *et al.* (1992) have provided evidence for just such an actin based transport system for vesicles. Furthermore the vesicles appeared to jump from microtubules to possible actin networks. Thus shuttling from one part of the cytoskeleton to another may well occur. Cellular redundancy of transport systems, up to a certain size, is a reasonable but by no means proven idea. Based on the observations that cells can attach, spread and form both focal contacts and stress fibres in the absence of MTs, Goldman (1971) and Lloyd *et al.* (1977) have suggested that MTs are not essential for these processes to occur However, the absence of MTs from spreading cells causes a loss of orientation of stress fibre bundles and a loss of persistance at the leading edge, the cells taking 5-10 times longer to spread than controls (Vasiliev and Gelfand, 1981) suggesting that the presence of MTs speeds up the spreading of the cells.

A different situation has been found to occur in another cell type. Polarized motile fibroblasts are characterized by trailing processes rich in MTs (Chen, 1981). These microtubules frequently form a sub-population rich in modified α-tubulin in rapidly moving cells relatively lacking in F-actin stress fibres (Prescott *et al.*, 1989a and b). A similar situation occurs in kidney epithelial cells treated with HGF/SF, where highly motile single cells are released bearing long processes (Dowrick *et al.*, 1991). A sub-population of MTs concentrated in the processes have been found to be both drug stable

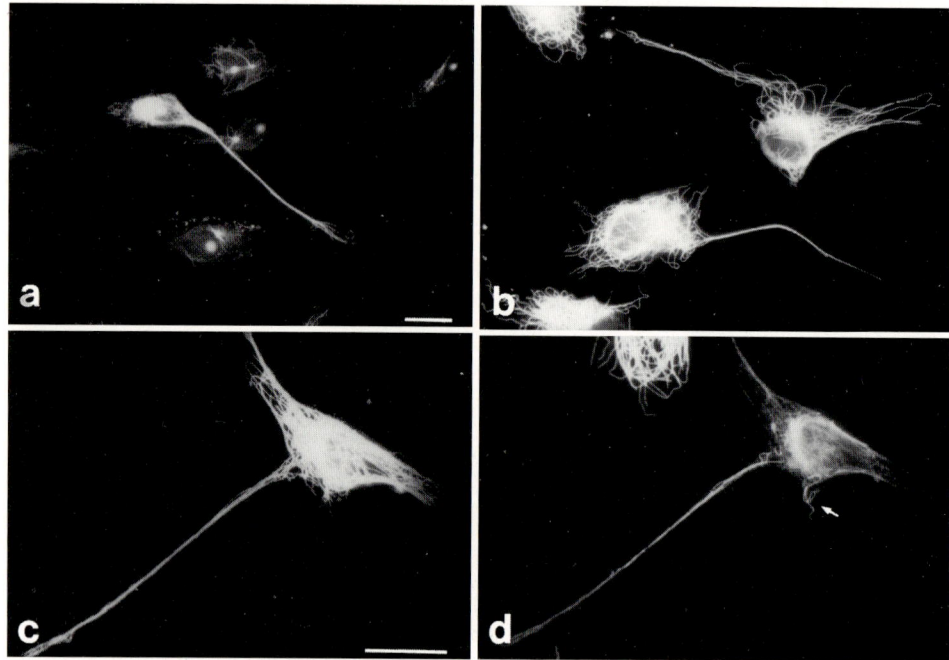

Fig. 4. (a) PtK$_2$ cells treated with HGF/SF for 16 hr and then treated with 5 μg/ml colchicine for 3 hr. (b) PtK$_2$ cells treated as (a) with HGF/SF and then incubated at 4°C for 2 hr. (a) and (b) anti-tubulin antibodies, (c) and (d) PtK$_2$ cells scattered overnight, injected with biotin-tubulin and incubated for 90 min. (c) anti-biotin immunostaining to show turned over MTs (Prescott *et al.*, 1992) (d) counterstaining of (c) with anti-tubulin antibodies to show stable MTs. Arrow in (d) marks short stable MT piece. Scale bar = 25 μm.

and also cold stable (Fig. 4a and b) (Prescott *et al.*, 1992). In addition, micro-injection of biotinylated tubulin has shown that the microtubules turn over more slowly in the processes than elsewhere, persisting for up to 3 hours (Fig. 4c and d). These microtubules are not obligatory structural features of the processes because short processes lacking MTs were found after 3 hour incubation times with nocodazole. However, it is suspected that these processes were in the course of disappearing because the cells gradually became completely rounded. As with 2° CHFs there are several possible explanations as to what roles the MTs may be playing in the processes. Microtubules may be required for transporting membrane/cortical constituents, mitochondria, or other materials down the rather long processes. Alternatively or additionally they may have a structural function to maintain process integrity. However, in either case the size and length of the process may be the deciding factor in the requirement of microtubules to maintain it. Thus, in conclusion, an intact microtubule network would seem to be required only for the polarized motility of larger cells or cells with long processes, and is not indispensable for the motility of a variety of cell types.

We thank the MRC for a project grant to support A.R., the Big C Cancer Charity for financial support for PD, the Wellcome Foundation and SERC for a CASE studentship for DB, and Mrs. J. Gorton for putting the manuscript on disc. The kind permission of the Company of Biologists to reproduce Fig. 4 is acknowledged.

References

Adams, J.F., Furlong, R.A. and Watt, F.M. (1991). Production of scatter factor by NDK, a strain of epithelial cells, and inhibition of scatter factor activity by suramin. *J. Cell Sci.* **98**, 385-394.

Bockus, B.J. and Stiles, C.D. (1984). Regulation of cytoskeletal architecture by platelet derived growth factor, insulin, and epidermal growth factor. *Exp. Cell Res.* **153**, 186-197.

Bretscher, A. (1983). Purification of an 80,000-dalton protein that is a component of the isolated microvillus cytoskeleton, and its localization in non-muscle cells. *J. Cell Biol.* **97**, 425-432.

Bretscher, A. (1986). Purification of the intestinal microvillus cytoskeletal proteins villin, fimbrin and ezrin. *Methods Enzymol.* **134**, 24-37.

Bretscher, A. (1989). Rapid phosphorylation and reorganization of ezrin and spectrin accompany morphological changes induced in A-431 cells by epidermal growth factor. *J. Cell Biol.* **108**, 921-930.

Brown, D.A. and Warn, R.M. (1993) Primary and secondary chick heart fibroblasts: fast and slow moving cells show no significant difference in microtubule dynamics. *Cell Motil. and Cytoskeleton* **24**, 233-244.

Cantley, L.C., Auger, K.R., Carpenter, C., Duckworth, B., Graziani, A., Kapeller, R. and Soltoff, S. (1991). Oncogenes and signal transduction. *Cell* **64**, 281-302.

Chen, W.T. (1981). Mechanism of retraction of the trailing edge during fibroblast movement. *J. Cell Biol.* **90**, 187-200.

Chernoff, E.A.G. and Overton, J. (1979). Organisation of the migrating chick epiblast edge: Attachment sites, cytoskeleton and early developmental changes. *Devl. Biol.* **72**, 291-307.

Chinkers, M., McKanna, J.A. and Cohen, S. (1979). Rapid induction of morphological changes in human carcinoma cells A-431 by epidermal growth factor. *J. Cell Biol.* **83**, 260-265.

Chinkers, M., McKanna, J.A. and Cohen, S. (1981). Rapid rounding of human epidermoid carcinoma cells A-431 induced by epidermal growth factor. *J. Cell Biol.* **88**, 422-429.

Crompton, M.R., Owens, R.J., Totty, N.F., Moss, S.E., Waterfield, M.D. and Crumpton, M.J. (1988) Primary structure of the human membrane-associated Ca^{2+} binding protein p68, a novel member of a protein family. *EMBO J.* **7**, 21-27.

Cross, M. and Dexter, T.M. (1991). Growth factors in development, transformation and tumorigenesis. *Cell* **64**, 271-280.

Crumpton, M.J. and Dedman, J.R. (1990). Protein terminology tangle. *Nature* **345**, 212.

Defize, L.H.K., Boonstra, J., Meisenhelder, J., Kruijer, W., Tertoolen, L.G.J., Tilly, B.C., Hunter, T., van Bergen, P., Moolenaar, W.H. and de Laat, W. (1989). Signal transduction by epidermal growth factor occurs through the subclass of high affinity receptors. *J. Cell Biol.* **109**, 2495-2507.

Di Pasquale, A. (1975). Locomotion of epithelial cells in culture. *Exp. Cell Res.* **95**, 425-439.

Domnina, L.V., Pletjushkina, O.Y., Vasiliev, J.M. and Gelfand, I.M. (1977). Effect of antitubulins on the redistribution of cross-linked receptors on the surface of fibroblasts and epithelial cells. *Proc. Natl. Acad. Sci. USA* **74**, 2865-2868.

Domnina, L.V., Rovensky, J.A., Vasiliev, J.M. and Gelfand, I.M. (1985). Effect of microtubule destroying drugs on the spreading and shape of cultured epithelial cells. *J. Cell Sci.* **74**, 267-282.

Downie, J.R. (1975). The role of microtubules in chick blastoderm expansion - a quantitative study using colchicine. *J. Embryol. exp. Morph.* **34**, 265-277.

Dowrick, P.G. and Warn, R.M. (1991). The cellular responses to factors which induce motility in mammalian cells. In *Cell Motility Factors*. (ed. I.D. Goldberg) pp 89-108. Basel. Birkhäuser Verlag.

Dowrick, P.G., Prescott, A.R. and Warn, R.M. (1991). Scatter factor effects major changes in the cytoskeletal organization of epithelial cells. *Cytokine* **3**, 299-310.

Eriksson, A., Siegbahn, A., Westermark, B., Heldin, C-H and Claesson-Welsh, L. (1992). PDGF α- and β-receptors activate unique and common signal transduction pathways. *EMBO J.* **11**, 543-550.

Euteneuer V. and Schliwa, M. (1984). Persistent, directional motility of cells and cytoplasmic fragments in the absence of microtubules. *Nature* **310**, 58-61.

Fabricant, R.N., De Larco, J.E. and Todaro, G.J. (1977). Nerve growth factor receptors on human melanoma cells in culture. *Proc Natl. Acad. Sci. USA* **74**, 565-569.

Fidler, I.J. and Hart, I.R. (1982). Biological diversity in metastatic neoplasms. Origins and Implications. *Science* **217**, 998-1003.

Gail, M.H. and Boone, C.W. (1971). Effect of colcemid on fibroblast motility. *Exp. Cell Res.* **65**, 221-227.

Galán, J.E., Pace, J. and Hayman, M.J. (1992). Involvement of the epidermal growth factor receptor in the invasion of cultured mammalian cells by *Salmonella typhimurium*. *Nature* **357**, 588-589.

Gelfand, V.I., Glushankova, N.A., Ivanova, O., Mittelman, L.A., Pletyushkina, O., Vasiliev, J.M. and Gelfand, I.M. (1985). Polarization of cytoplasmic fragments microsurgically detached from mouse fibroblasts. *Cell Biol. Int. Rep.* **9**, 883-892.

Gerke, V. and Weber, K. (1984). Identity of p36K phosphorylated upon Rous sarcoma virus transformation with a protein purified from brush borders: calcium dependent binding to non-erythroid spectrin and F-actin. *EMBO J.* **3**, 227-233.

Gerke, V. and Weber, K. (1985). Calcium-dependent conformational changes in the 36-kDa subunit of intestinal protein I related to the cellular 36-kDa target of Rous sarcoma virus tyrosine kinase. *J. Biol. Chem.* **260**, 1688-1695.

Glenney, J.R. (1985). Phosphorylation of p36 *in vitro* with pp60^{src} regulation by Ca^{2+} and phospholipid. *FEBS Lett.* **192**, 79-82.

Glenney, J.R. (1986). Two related but different forms of the 36,000 Mr tyrosine kinase substrate (calpactins) which interact with phospholipid and actin in a Ca^{2+}-dependent manner. *Proc. Natl. Acad. Sci. USA.* **83**, 4258-4262.

Glenney, J.R., Tack, B. and Powell, M.A. (1987). Calpactins: Two distinct Ca^{2+} regulated phospholipid and actin-binding proteins isolated from lung and placenta. *J. Cell Biol.* **104**, 503-511.

Goldman, R.D. (1971). The role of three cytoplasmic fibres in BHK-21 cell motility. 1. Microtubules and the effect of colchichine. *J. Cell Biol.* **51**, 752-762.

Goldman, R.D. and Follet, E.A.C. (1969). The structure of the major cell processes of isolated BHK-21 fibroblasts. *Exp. Cell Res.* **57**, 263-276.

Gould, K.L., Cooper, J.A., Bretscher, A. and Hunter, T. (1986). The protein-tyrosine kinase substrate p81 is homologous to a chicken microvillar core protein. *J. Cell Biol.* **102**, 660-669.

Gould, K.L., Bretscher, A., Esch, F.S. and Hunter, T. (1989). cDNA cloning and sequencing of the protein-tyrosine kinase substrate, ezrin, reveals homology to band 4.1. *EMBO J.* **8**, 4133-4142.

Gundersen, G.G. and Bulinski, J.C. (1988). Selective stabilization of microtubules orientated toward the direction of cell migration. *Proc. Natl. Acad. Sci. USA* **85**, 5946-5950.

Haigler, H.T., McKanna, J.A. and Cohen, S. (1979). Rapid stimulation of pinocytosis in human carcinoma cells A-431 by epidermal growth factor. *J. Cell Biol.* **83**, 82-90.

Heldin, C.H. and Westermark, B, (1989). Platelet derived growth factor: three isoforms and two receptor types. *T.I.G.* **5**, 108-111.

Hopkins, C.R., Miller, K. and Beardmore, J.M. (1985). Receptor-mediated endocytosis of transferrin and epidermal growth factor receptors: a comparison of constitutive and ligand-induced uptake. *J. Cell Sci. Suppl. 3*, 173-186.

Hunter, T. and Cooper, J.A. (1981). Epidermal growth factor induces rapid tyrosine phosphorylation of proteins in A431 human tumour cells. *Cell* **24**, 741-752.

Ivanova, O.Y., Margolis, L.B., Vasiliev, J.M. and Gelfand, I.M. (1976). Effect of colcemid on the spreading of fibroblasts in culture. *Exp. Cell Res.* **101**, 207-219.

Kaplan, D.R., Morrison, D.K., Wong, G., McCormick, F. and Williams, L.T. (1990). PDGF β-receptor stimulates tyrosine phosphorylation of GAP and association of GAP with a signalling complex. *Cell* **61**, 125-133.

Karavanova, I.D., Vasiliev, J.M. and Troyanovsky, S.M. (1985). The role of microtubules and intermediate filaments in the maintenance of epithelial cell shape. *Tsitologiya* **27**, 693-697.

Kolega, J. (1986). The cellular basis of epithelial morphogenesis. In "*Developental Biology*" Vol. 2 (ed. L.W. Browder) pp 103-142, New York, Plenum.

Kreis, T.E. (1987). Microtubules containing detyrosinated tubulin are less dynamic. *EMBO J.* **6**, 2597-2606.

Kupfer, A., Louvard, D. and Singer, S.J. (1982). Polarisation of the Golgi apparatus and the microtubule-organizing centre in cultured fibroblasts at the edge of an experimental wound. *Proc. Natl. Acad. Sci. USA* **79**, 2603-2607.

Kuznetsov, S.A., Langford, G.M. and Weiss, D.G. (1992). Actin-dependent organelle movement in squid axoplasm. *Nature* **356**, 722-725.

Landreth, G.E., Williams, L.K. and Rieser, G.D. (1985). Association of the epidermal growth factor receptor kinase with the detergent-insoluble cytoskeleton of A431 cells. *J. Cell Biol.* **101**, 1341-1350.

Lloyd, C.W., Smith, C.G., Woods, A. and Rees, D.A. (1977). Mechanisms of cellular adhesion II. The interplay between adhesion, the cytoskeleton and morphology in substrate-attached cells. *Exp. Cell Res.* **110**, 427-437.

Mellström, K., Hoglund, A., Nister, M., Heldin, C-H., Westermark, B. and Lindberg, U. (1983). The effect of platelet derived growth factor on morphology and motility of human glial cells. *J. Muscle Res. Cell Motil.* **4**, 589-609.

Middleton, C.A (1982). Cell contacts and the locomotion of tissue cells. In *"Cell Behaviour"* (ed. A.S.G. Curtis, R. Bellairs and G.A. Dunn). pp 159-182. Cambridge C.U.P.

Middleton, C.A., Brown, A.F., Brown, R.M. and Roberts D.J.H. (1988). The shape of cultured epithelial cells does not depend on the integrity of their microtubules. *J. Cell Sci.* **91**, 337-345.

Middleton, C.A., Brown, A.F., Brown, R.M., Karavanova, I.D., Roberts, D.J.H. and Vasiliev, J.M. (1989). The polarisation of fibroblasts in early primary cultures is independent of microtubule integrity. *J. Cell Sci.* **94**, 25-32.

Nicolson, G.L. (1988). Organ specificity of tumor-metastasis. Role of preferential adhesion, invasion, and growth of malignant cells at specific secondary sites. *Cancer Metastasis Rev.* **7**, 143-188.

Nishibe, S., Wahl, M.I., Hernandez-Sotomayor, S.M.T., Tonks, N.K., Rhee, S.G. and Carpenter, G. (1990). Increase of the catalytic activity of phospholipase c-γl by tyrosine phosphorylation. *Science* **250**, 1253-1256.

Prescott, A.R., Vestberg, M. and Warn, R.M. (1989a). Microtubules rich in modified alpha-tubulin characterize the tail processes of motile fibroblasts. *J. Cell Sci.* **94**, 227-236.

Prescott, A.R., Magrath, R. and Warn, R.M. (1989b). *Ras*-transformed cells contain microtubules rich in modified α-tubulin in their trailing edges. In *"ras Oncogenes"* NATO ASI Handbooks pp 243-253. New York. Plenum Press.

Prescott, A.R., Dowrick, P.G. and Warn, R.M. (1992). Stable and slow turning over microtubules characterize the processes of motile epithelial cells treated with scatter factor. *J. Cell Sci.* **102**, 103-112.

Rees, D.J.G., Ades, S.E., Singer, S.J. and Hynes, R.O. (1990). Sequence and domain structure of talin. *Nature* **347**, 685-689.

Rijken, P.J., Hage, W.J., Van Bergen, P., Verkleij, M.J. and Boonstra, J. (1991). Epidermal growth factor induces rapid reorganization of the actin microfilament system in human A431 cells. *J. Cell Sci.* **100**, 491-499.

Rosen, E.M. and Goldberg, I.D. (1989). Protein factors which regulate cell motility. *In Vitro Cell Dev. Biol.* **25**, 1079-1087.

Roy, L.M., Gittinger, C.K. and Landreth, G.E. (1989). Characterization of the epidermal growth factor receptor associated with cytoskeletons of A431 cells. *J. Cell Physiol.* **140**, 295-304.

Schlessinger, J. (1986). Allosteric regulation of the epidermal growth factor receptor kinase. *J. Cell Biol.* **103**, 2067-2072.

Schliwa, M., Nakamura, T., Porter, K. and Euteneuer, V. (1984). A tumor promoter induces rapid and co-ordinated reorganization of actin and vinculin in cultured cells. *J. Cell Biol.* **99**, 1045-1049.

Schulze, E. and Kirschner, M. (1987). Dynamic and stable populations of microtubules in cells. *J. Cell Biol.* **104**, 227-288.

Stoker, M. and Gherardi, E. (1991). Regulation of cell movement: the motogenic cytokines. *Biochim. Biophys. Acta (Cancer Reviews)* **1072**, 81-102.

Stoker, M. and Perryman, M. (1985). An epithelial scatter factor released by embryo fibroblasts. *J. Cell Sci.* **77**, 209-223.

Tomasek, J.J. and Hay, E.D. (1984). Analysis of the role of microfilaments and microtubules in acquisition of bipolarity and elongation of fibroblasts in hydrated collagen gels. *J. Cell Biol.* **99**, 536-549.

Ullrich, A. and Schlessinger, J. (1990). Signal transduction by receptors and tyrosine kinase activity. *Cell* **61**, 203-212.

Vale, R.D. and Shooter, E.M. (1983). Conversion of nerve growth factor-receptor complexes to a slowly dissociating, Triton X-100 insoluble state by anti nerve growth factor antibodies. *Biochemistry* **22**, 5022-5028.

Van Belzen, N., Rijken, P.J., Hage, W.J., De Laat, S.W., Verkleij, A.J. and Boonstra, J. (1988).

Direct visualization and quantitative analysis of epidermal growth factor-induced receptor clustering. *J. Cell Physiol.* **134**, 413-420.

Van Bergen en Henegouwen, P.M.P., Defize, L.H.K., De Kroon, J., Van Damme, H., Verkleij, A.J. and Boonstra, J. (1989). Ligand induced association of epidermal growth factor receptor to the cytoskeleton of A431 cells. *J. Cell Biochem.* **39**, 455-465.

Vasiliev, J.M., Gelfand, I.M., Domnina, L.V., Ivanova, O.Y., Komm, S.G. and Olsherskaya, L.V. (1970). The effect of colcemid on the locomotory behaviour of fibroblasts. *J. Embryol. exp. Morph.* **24**, 625-640.

Vasiliev, J.M. and Gelfand, I.M. (1981). Neoplastic and normal cells in culture. Cambridge, C.U.P.

Watts, C. and Marsh, M. (1992). Endocytosis: what goes in and how? *J. Cell Sci.* **103**, 1-8.

West, M., Bretscher, M. and Watts, C. (1989). Distinct endocytotic pathways in epidermal growth factor stimulated human carcinoma A431 cells. *J. Cell Biol.* **109**, 2731-2739.

Westermark, B., Siegbahn, A., Heldin, C-H. and Claesson-Welsh, L. (1990). B-type receptor for platelet-derived growth factor mediates a chemotactic response by means of ligand-induced activation of the receptor protein-tyrosine kinase. *Proc. Natl. Acad. Sci. USA* **87**, 128-132.

Wiegant, F.A.C., Blok, F.J., Defize, L.H.F., Linnemans, W.A.M., Verkleij, A.J. and Boonstra, J. (1986). Epidermal growth factor receptors associated to cytoskeletal elements of epidermoid carcinoma (A431) cells. *J. Cell Biol.* **103**, 87-94.

Zippel, R., Morello, L., Brambilla, R., Comoglio, P.M., Alberghina, L. and Sturani, E. (1989). Inhibition of phosphotyrosine phosphatases reveals candidate substrates of the PDGF receptor kinase. *Eur. J. Cell Biol.* **50**, 428-434.

Printed in Great Britain © The Society of Experimental Biology 1993   339

MECHANISMS OF ACTIN FILAMENT TURNOVER IN ANIMAL CELLS

PETER SHETERLINE

Department of Human Anatomy and Cell Biology, University of Liverpool, Liverpool L69 3BX, UK

Summary

About half of the total actin in the cytoplasm of cultured animal cells is polymerised into filaments at any time. The filaments are further ordered into 3-dimensional patterns by their interaction with a number of actin-binding proteins (ABPs) to form the functional actin cytoskeleton. Three consistent patterns of organisation can be discerned both by their complement of ABPs, as determined by co-localisation, and by the characteristic arrangement of actin filaments: isotropic arrays, parallel bundles and anti-parallel bundles. These three patterns of organisation appear to have discrete functional properties which, together, give rise to the motile behavior of the cells. The proportions and locations of these actin filament organisations reflect, or give rise to, the particular properties of the cells in which they are found.

In locomoting cells in particular, the extent and precise cellular location of these three classes of organisation are constantly changing during the process of locomotion. This constant adaptation of the actin cytoskeleton appears, from a number of approaches, to result from a continuous cycle of assembly and disassembly of filaments leading to continuous adaptation of the actin cytoskeleton. In some cells at least, net assembly occurs predominantly at the extreme leading edge of the lamellipodium, and it must be presumed that filaments assembled in the isotropic arrays here give rise to the other levels of architecture found in the cell. However, overlying this appears to be a continuous cycle of assembly and disassembly of filaments within all parts of the actin cytoskeleton.

The underlying imperative for turnover of actin filaments derives from the ATPase associated with polymerisation. The implications of this assembly ATPase for both turnover and for the progressive evolution of actin filament architectures is discussed.

The actin cytoskeleton of animal cells

Some 50% of the total actin in animal cells is polymerised into filaments that are organised into complex 3-dimensional architectures by their association with different sets of actin filament binding proteins (Amos and Amos, 1991). At least three types of architecture can be discriminated, both from the organisation of the actin filaments themselves and from the subset of actin-binding proteins associated with them.

The most prominant structures in cultured cells are the stress fibres; loose bundles containing in cross-section up to several hundred actin filaments that are organised predominantly parallel to the major axis of locomotion in motile cells or more randomly

Key words: actin, cytoskeleton, actin-ATPase.

in stationary or poorly motile cells. The characteristics of actin filament organisation in stress fibres mirror those of myofibrils from muscle cells, albeit with a less tightly proscribed symmetry. In stress fibres, actin filaments are organised into oppositely polarised sets that appear to interdigitate with one another (Sanger and Sanger, 1980). The spatial relationships between the sets of actin filaments are maintained by a periodic distribution of a number of actin-binding proteins along the stress fibre that are assumed to both spatially organise the filaments and confer contractile functionality on the structure (De Lanerolle *et al.*, 1981; Herman and Pollard, 1981; Zigmond *et al.*, 1979). The periodicity can be visualised by fluorescence immunocytochemistry as alternate regions, 0.4-0.7 μm that contain respectively α-actinin alternating with regions containing myosin II, myosin light-chain kinase and tropomyosin (Sanger and Sanger, 1980; Goldman *et al.*, 1979). The stress fibres have been shown to contract in the presence of ATP and calcium (Isenberg *et al.*, 1976; Kreis and Birchmeier, 1980) and thus appear both structurally and functionally homologous to muscle myofibrils. Stress fibres probably contain the major proportion of polymerised actin in cells growing on a flat substratum.

At the leading edge of some cells, growth cones of neurons and some fibroblasts in particular, parallel bundles of actin filaments are formed in which the filaments are both tightly packed and uniformly polarised with their barbed ends towards the plasma membrane (Mooseker and Tilney, 1975; Small *et al.*, 1978). In both regards the organisation of actin filaments in these bundles differs from that in stress fibres. These filament bundles form the core of filopodia (microspikes) and can elongate rapidly in motile cells and often protrude beyond the leading edge. Fimbrin co-localises with these bundles and may be the major bundling protein for these structures (Bretscher and Weber, 1980). It is often observed that microspikes appear to originate from the distal end of stress fibres at the boundary between the lamellipodium and the organelle-containing cytoplasm of the rest of the cell. The distal end of the stress fibres is also usually terminated at a focal adhesion to the substratum at a site where talin and vinculin are preferentially located (Geiger *et al.*, 1980).

The third pattern of organisation of actin filaments is observed over the entire cytoplasmic surface of the plasma membrane as a feltwork of short (0.1-0.5 μm) actin filaments that extend for some 0.1-0.2 μm into the cytoplasm and where the barbed ends of filaments are attached to components of the plasma membrane (see Sheterline and Rickard, 1989). Associated with this cortical web of filaments are α-actinin and spectrin (or its homologue fodrin) (Glenney and Glenney, 1983). The combination of actin-binding proteins and the location of the cortical web suggest that it corresponds to the simpler, but well defined structure of the cortical actin cytoskeleton of mammalian erythrocytes (Bennett, 1985). The functions of this cortical structure are not well understood, but it may interact directly with membrane solute transporters (Bennett, 1985) and with the signal transduction apparatus at the plasma membrane (Goldschmidt-Clermont *et al.*, 1991). In erythrocytes, and in other cell types, it may also contribute to the mechanical properties of the cell surface (Bray *et al.*, 1986). From phenomenological observations, it appears likely that the cortical web is the 'unactivated' precursor of the leading lamellipodium and the 'pseudopodia' involved in phagocytosis. From this perspective, the leading lamellae or phagocytic pseudopodia are merely temporarily extended versions of the cortical web. The major line of evidence in favour of this view is the general observation that any region of the cell surface can, in response to polarising or chemotactic stimuli, rapidly expand to form lamellipodia or pseudopods and just as rapidly revert to cortical web. More recently, data have emerged that link the function of actin or of actin-binding proteins in the cortex with the biochemical events of phosphoinositide metabolism suggesting bi-directional exchange of information between the cell surface receptor system and the actin cytoskeleton (Lassing and Lindberg, 1988; Goldschmidt-Clermont *et al.*, 1991; Yin *et al.*, 1988).

The leading lamellipodium is a highly motile organelle-free region of the cytoplasm some 1-3 µm thick that extends some 5-15 µm beyond the organelle-containing cytoplasm of the rest of the cell; the pattern of organisation of actin filaments within the lamellipodium does not readily suggest function. The lamellipodium contains a meshwork of actin filaments that appear to be in excess of 5 µm long (Small, this volume) and that appear to be orientated in an orthogonal array relative to the plasma membrane at the leading edge. These filaments are polarised such that their barbed ends are orientated towards the plasma membrane (Small, this volume). The lamellipodium can rapidly extend or retreat, but in locomoting cells, maintains an approximately constant average width that appears to be independent of the rate of locomotion (1-10 µm min^{-1}). Microspikes seem to form within the lamellipodium by lateral association of lamellipodial actin filaments (Lindberg *et al.*, 1981).

The mechanism of remodelling of actin filament architectures

Many observations of motile cells in culture make it clear that the enormous molecular architectures of the actin cytoskeleton are continuously remodelled. Over a period of minutes, the number, length, thickness and orientation of stress fibres changes; simultaneously, the width, shape and regions of motile activity in the leading lamellipodium change as the overall shape of the cell changes with locomotion. In some cases, as with locomoting cells, the architectures continuously reiterate themselves to drive or accommodate cell locomotion such that the approximate proportions of particular types of actin filament architecture seem to remain constant; as does the average proportion of total actin polymerised. In others, for example, activation of platelets by thrombin, the transient collapse of the interphase actin cytoskeleton and the appearance of the cleavage furrow or the activation of static cells into their motile phenotypes, cells undergo a complete reorganisation of the actin cytoskeleton that leads to changes in the overall structure and behaviour of the cells and may be accompanied by a change in the proportion of total actin assembled into filaments (Fox and Phillips, 1983; Howard and Meyer, 1984). The mechanisms underlying these two 'types' of phenomena need not themselves be substantially different; in one case a region of the actin cytoskeleton may undergo some major structural reorganisation, whilst in the other, the entire cytoskeleton may be remodelled.

The underlying mechanism of reorganisation of the actin cytoskeleton could, in principle, reflect both relocation of existing filaments or disassembly of filaments in redundant parts of the cytoskeleton and assembly of new filaments at growing points. The major source of evidence to discriminate a mechanism has come from experiments where the fate of fluorescently-tagged actin monomers injected into living cells has been followed by light microscopy (see Wang, 1985). These data indicate clearly that the major mechanism of remodelling involves assembly and disassembly. Referring back to the observation that in many cases the proportion of total actin assembled may not change very much throughout these remodelling events, the data imply that net assembly of new filaments must be accompanied by an equivalent disassembly of existing filaments; thus, individual actin molecules shuttle many times through a cycle of assembly and disassembly during their lifetime in a cell. An important corollary of such an underlying mechanism is that there must be a steady-state monomer pool to act as a source of subunits for filament assembly. It is now clear that there is a ubiquitous group of monomer-binding proteins that include the profilins, the thymosins and perhaps the ADF family (Safer, 1992) that serve to maintain approximately 50% of the total actin as a monomer pool by competing with assembly for actin monomers and that may also regulate, in some cases via the phosphoinositide signal transduction pathway, the availability of monomers for assembly (Goldschmidt-Clermont *et al.*, 1991). Such a large monomer pool can provide an adequate reservoir of freely diffusible monomers for new assembly anywhere in the cell that allows the spatial uncoupling (over short time periods) of assembly from disassembly.

There have been two related approaches to the study of the turnover of actin filaments both involving the microinjection of fluorescently-labelled (or biotin-labelled for EM) actin monomers into living motile and non-motile cells. The first and most widely used until recently, involved the injection of (some 5-10% of the existing total) fluorophore-labelled actin into cells and then recording changes in the localisation of labelled actin with time by light microscopy. The key observation of all such experiments has been that the injected actin monomers rapidly co-locate with all elements of the actin cytoskeleton (Wang, 1985). In general, these experiments show a time dependency for the extent of localisation, but no spatial dependency. From such experiments it has been concluded that monomers in the cytoplasm (the injected ones) can become assembled into the actin cytoskeleton, and by using the arguments made above, that monomers must therefore cycle between the assembled and disassembled pools. The time taken for the actin cytoskeleton to incorporate the maximal (steady state?) amount of fluorescence, usually a small number of minutes, suggested that the lifetime of an actin molecule in the assembled pool was of the same order of time (Handel *et al.*, 1990). A more refined extension of this approach follows the kinetics of the recovery of fluorescence (FRAP) in chosen areas of the cytoskeleton bleached by laser light (Wang, 1985). The recovery was assumed to be due to the replacement of bleached molecules with fluorescent molecules from the monomer pool. The recovery rates of such experiments also suggested half-lifetimes for actin monomers within the polymer pool of a few minutes and provided further information on the geographical fate of the bleached spot or region in the context of the whole cell (Wang, 1985).

Some worrying questions remain with regard to this type of experiment: firstly, there is no direct evidence that the fluorescent monomers become assembled into filaments, it can only be observed that they are preferentially located with the actin cytoskeleton. Since many of the actin-binding proteins responsible for the organisation of actin filaments into the cytoskeleton have two binding sites for actin, one of which may bind to a filament in the cytoskeleton and the other that may be free to bind monomer (albeit at lower affinity than polymer), only some of the co-localising fluorescent actin (or the recovery actin in the FRAP experiments) may be actually part of the filament architecture. The general lack of data showing preferential incorporation into spatially-defined regions of the cytoskeleton using this approach, is at variance with the expectation from morphological and other studies that there are preferential sites of assembly (and therefore of disassembly) in cells. Some of these problems appear to have been overcome by observing injected cells that have been fixed, permeabilised and washed rather than observing whole fixed or live cells (Symons and Mitchison, 1991). Under such conditions, comparison between the location of fluorescence due to injected monomer with that of phalloidin attached to a complementary fluorophore showing the entire actin cytoskeleton does reveal localised regions of assembly.

More recently, monomers derivatised with caged-fluorophores that can be photo-activated have been injected into cells (Theriot and Mitchison, 1991; Theriot *et al.*, 1992). This approach has the distinct advantage that the only monomers observed by microscopy are the photo-activated molecules in whatever region of the cell is chosen and so if a proportion of the photoactivated actin is bound weakly by whatever mechanism and rapidly exchangeable, then it will be rapidly diluted into the somewhat larger non-fluorescent pool of the cell and will not confuse interpretation of the fluorescent image. This approach has provided useful information on the sites of polymerisation and depolymerisation.

Turnover of actin in the leading lamellipodium

Use of the ratiometric approach outlined above shows that within 30 s of injection, fluorescent monomer is restricted to the extreme distal edge of the leading lamellipodium in fibroblasts (Symons and Mitchison, 1991). At longer times after injection, the width of the fluorescent region increased whilst the proximal margin of the band of incorporated fluorescent monomer moved in a centripetal direction towards the base of the lamellipodium (Symons and Mitchison, 1991). This data implies that assembly of new filaments occurs at the distal edge of the lamellipodium and that once assembled, filaments move towards the base at a rate of about 4-5 μm min^{-1}. It was further observed that the total amount of actin polymer decreased from a maximal amount at the distal assembly margin of the lamellipodium to about 20% or less at the proximal edge, suggesting that filaments assembled *de novo* at the distal margin and then disassemble as they move through the lamellipodium (Symons and Mitchison, 1991). These observations have been supported and extended by the use of caged fluorescent actin. In keratinocytes which can locomote rapidly (up to 20 or more μm min^{-1}), a photoactivated band of injected actin incorporated at the distal edge arrives at the proximal edge at the same rate as the lamellipodium moves forward over the substratum (Theriot and Mitchison, 1991). The photoactivated band therefore remains stationary relative to the substratum as new actin filaments are assembled distal to it and as the interface between the lamellipodium and the rest of the cytoplasm advances. The intensity of fluorescence in the band also decreases approximately exponentially, suggesting that the band of filaments visualised by photo-activation depolymerises continuously after assembly as it is overtaken by the proximal margin. The half-lifetime of an assembled actin molecule in the lamellipodium is estimated at about 25 s. In contrast to the fibroblast, the density of actin filaments across the lamellipodium in keratinocytes remains roughly constant (Theriot and Mitchison, 1991) suggesting the surprising inference that depolymerisation of the newly assembled filaments is precisely balanced by assembly of other filaments across the lamellipodium (to maintain the constant density observed) or that the exchange is due to treadmilling.

The use of caged actin has also been useful for studying the propulsion of *Listeria monocytogenes* through the cytoplasm of cells that occurs by the induction of assembly by the bacterium of a tail of cross-linked actin filaments. The tail itself consists of a cylinder of actin filament 'gel' that diminishes in actin filament mass exponentially away from the trailing edge of the bacterium. A band of filaments photo-activated near the bacterium remains stationary relative to the cytoplasm but continuously and homogeneously disassembles with an approximately exponential time course, suggesting that the bacterium propels itself forwards, at rates of up to 12 μm min^{-1}, by inducing assembly of actin filaments behind it. The half-life of an actin molecule in the tail is estimated by decay of fluorescence at about 30 s (Theriot *et al.*, 1992).

There are some key conclusions that can be drawn from these and other experiments (Forscher and Smith, 1989): firstly, actin filament assembly occurs predominantly or exclusively (fibroblast and Listeria tail) at the leading edge of the tail/lamellipodium. If the actin filament meshwork into which these filaments are incorporated remains stationary relative to substratum or cytoplasm, then the leading edge moves forwards at the rate of accumulation of actin filament meshwork; if the leading edge remains stationary, then the meshwork moves backwards at the same rate. The attachment or not of the meshwork can be considered as the cellular equivalent of a clutch. In the cases where assembly only occurs at the leading margin, then filaments after being rapidly assembled, disassemble (or exchange) with a half-lifetime of about 20-30 s. The longest surviving actin filament subunits reach the base of the lamellipodium in 1-2 min in any case. These data suggest that forward movement of the leading edge, either by osmotic pumping mechanisms (for a membrane situation) or by random thermal movements (membrane or bacterial propulsion) are consolidated by polymerisation. These data do not imply that the force is created by assembling actin except insofar as the net propensity of actin to polymerise in one direction (because it is linear) consolidates forward movement (however caused) in the polymerisation direction by a mechanism equivalent to a ratchet.

These observations raise some basic questions: firstly, what initiates or sustains assembly at the leading edge of the lamellipodium (bacterial tail) but nowhere else? Secondly, why do filaments stop growing and then switch to disassembly (or exchange) for the rest of their existence in the lamellipodium (bacterial tail)?

In principle, the assembly nuclei could be a *de novo* nucleation site or a free barbed end of an existing or severed filament. Data from a number of sources suggests that the assembly sites are either free barbed ends or mimic free barbed ends (actin-binding proteins alone or an actin-binding protein-stabilised barbed-end nucleus). The evidence includes many observations that protrusion of lamellipodia is highly sensitive to cytochalasins, that the incorporation of monomers into the margin of the lamellipodium of permeabilised cells can be blocked by CapZ, a barbed-end capping protein and that the critical concentration for assembly in the same system is about 0.15 μm, close to the critical concentration for the barbed- but not pointed-ends (Symons and Mitchison, 1991). These data are also consistent with morphological evidence that filaments in the lamellipodium are orientated with their barbed-ends towards the plasma membrane (Small *et al.*, 1978) and that the barbed end has a lower critical concentration (i.e. is therefore the preferred assembly end) *in vitro* (Pollard, 1986). Restriction of assembly to the leading edge could be the result of either spatial or temporal constraints. Spatial models could include the requirement for proteins associated with the plasma membrane (or bacterial tail), localised availability of monomers or ionic or second messenger conditions prevailing close to the leading edge (not so easy to contemplate for Listeria locomotion). Temporal models might consider the role of the intrinsic ATPase 'clock' implicit in actin filament assembly. The observation that assembly is restricted to the leading edge, even in permeabilised cells, suggests that the loss of assembly competence behind the leading edge is the result of some change (chemical or binding of actin-binding protein) in the filaments themselves and not merely a function of ionic conditions that inhibit assembly *per se*.

The inference that filaments assemble and become part of the actin meshwork of the lamellipodium, and thereby consolidate forward movement of the membrane implies that the nuclei must be associated with the meshwork and not with the membrane, since assembly at these nuclei moves the membrane (or bacterium) away from the nuclei. However it has recently been reported that a protein isolated from smooth muscle and named insertin operationally functions as a 'leaking cap' on ends of pre-existing filaments (Gaertner and Wegner, 1991). Insertin remains bound yet still allows monomers to assemble onto the barbed end to which it is bound. Whether such a mechanism would be sensitive to inhibition by CapZ or cytochalasins is not yet known.

The data from all three systems under discussion suggest that an individual actin filament is rapidly assembled at the leading edge for a finite time or to a finite length and then acquires a particular probability for disassembly (which could also be fast). The mechanism of the apparent switch of filaments in the lamellipodium from rapid assembly to disassembly (or exchange) is not known. The time taken for assembly can be estimated in fibroblasts to occur within 15 s from the data showing that assembly is restricted to a margin <1 μm whilst the movement of filaments away from the margin is at a rate of 4-5 μm min[-1] (see Symons and Mitchison, 1991) Put another way, filaments are only present in the polymerising zone for this time. The restriction of assembly nuclei to the distal margin in permeabilised fibroblasts argues against unavailability of monomer as a sufficient mechanism but does not rule out capping of such filaments by an actin-binding protein. An actin-binding protein model, however, simply puts the question one stage back to what discriminates filaments at the leading edge that are not functionally capped from those a little further back in the lamellipodium that are?

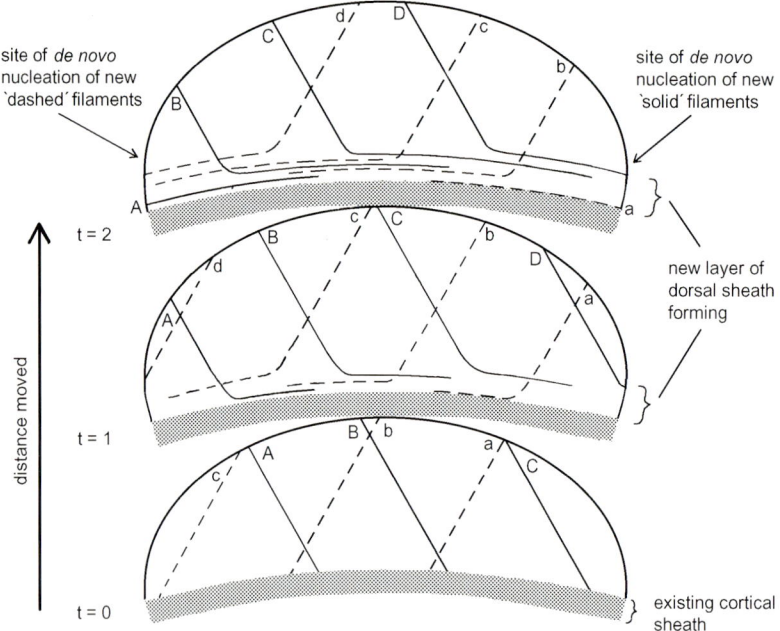

Fig. 1. A model showing a mechanism for the formation of the dorsal sheath from lamellipodial actin filaments. The diagram indicates position of the lamellipodium of a single locomoting fibroblastic cell at three different times (t =0, 1 and 2). The leading edge of the lamellipodium is represented as an arc and the existing dorsal sheath at the proximal edge of the lamellipodium as a shaded area. Three or four filaments from each of the two overlapping sets of actin filaments that constitute the lamellipodial network are represented as solid or dotted lines and are labelled A,B,C etc. or a,b,c etc. respectively. The first formed filaments are labelled 'A' (or 'a'). There would be several hundred such filaments in a fibroblast lamellipodium. The model is based on the assumption that each filament assembles in a straight line, and will therefore meet the leading edge ever further to the right (dotted filaments) or to the left (solid filaments) as the cell moves forwards. Both sets of filaments will then become the new forward edge of the dorsal sheath at the rear of the lamellipodium as this boundary moves forward. Sites of new filament nucleation are indicated.

Filaments at the leading edge do not meet the plasma membrane at right angles, but can be considered as two overlapping sets of filaments, one set making an acute angle (with respect to one side) and the other, an obtuse angle. As these filaments elongate, their distal ends will necessarily be displaced laterally towards the obtuse angle side until they eventually meet the lateral margin of the lamellipodium (Figure 1). So one set of filaments will move to the left margin and the other to the right. This interpretion implies that 'new' filaments are only nucleated at the sides of the lamellipodium to replace those that have reached the lateral margin on the opposite side. At the leading edge, the filaments simply elongate. This model is consistent with the observations by Heath (1993), that at the proximal edge of the lamellipodium, filaments become aligned across the axis of locomotion to become arcs or be incorporated into the dorsal sheath. The filaments would then simply be progressively and continuously laid down from both directions to form the dorsal sheath rather like computer fan-fold paper lays down when lowered onto a pile, but where alternate sheets represent single actin filaments nucleated at opposite sides of the lamellipodium. This further leads to the implication that these laterally aligned filaments will have opposite polarities (i) that could lead to the association of a different set of actin-binding proteins (those associated with contraction?) and that (ii) if these filaments were pulled out from the rearward moving dorsal sheath by attachment to the substratum, they would have the correct relative polarities to form a nascent stress fibre, a process also observed by Heath (this volume). These inferences are consistent with both the data of Svitkina *et al*. (1986) and of those in this volume by Heath. In general, they indicate the possibility that *de novo* nucleation occurs only at the lateral margins, elongation at the leading edge and depolymerisation somewhere at the rear of the cell. These observations appear to be inconsistent with the data from Symons and Mitchison (1991) relating to the proposed exponential loss of filaments across the lamellipodium from quantitative measurement of phalloidin fluorescence which are in contrast to the ultrastructural observations by Small (this volume) that suggest filaments extend much or all of the way across the lamellipodium. They are also hard to reconcile with the observations of Theriot and Mitchison (1991) where the implication from photoactivation recovery is for exchange to occur across the lamellipodium. However, it may be that in different cells, different proportions of filaments assembled at the leading edge are lost across the lamellipodium as they move rearwards and that some filaments can exchange subunits at ends. The significance of the exchange data has yet to be determined.

The exponential disassembly rate observed in all of the three systems under discussion implies either the dissociation of monomers from the ends of a population of filaments with an exponentially increasing length distribution or the stochastic disassembly of a population of filaments with a particular mean length. In both cases, the half-life of a given mass of filaments would be a function of the mean length of that population and the (constant) rate of depolymerisation from whatever end of the filament was the disassembly end. The only available data comes from ultrastructural observation of the Listeria system where there is no evidence that the average filament length (about 0.2 μm) is different at different positions along the tail (Tilney and Portnoy, 1989). If this is so, then the slow (exponential) loss of filaments with time must be due to stochastic loss of individual filaments rather than a continuous depolymerisation of all the filaments.

The overall rate of loss of filament mass is a function of both the rate of disassembly of individual filaments and of the number of filaments depolymerising; or, at a given mass density of filaments, the average filament length. The off-rate for ADP-actin at the pointed end is 0.3 s^{-1} (Pollard, 1986) or put another way, it would take about 2 min to lose half the monomers in a 0.2 μm actin filament (about 70 monomers), rather slow compared to the measured 30 s half-life. If depolymerisation were from the barbed end then the ADP-actin off-rate is about 7 s^{-1} (Pollard, 1986) giving a value of about 5 s for loss of 50% of the filament mass; certainly fast enough to account for the observed values, but necessarily making the large assumption that values measured for rate constants under *in vitro* conditions have relevance in the cell.

Microinjection of fluorescent actin has been used to study the turnover of actin filaments in stress fibres. The original injection experiments suggested a half-lifetime of several minutes from the kinetics of labelling and from FRAP experiments. Measurement of the loss of fluorescence from a small photo-activated region gave a half-lifetime for actin molecules within stress fibres of a non-motile epithelial cell (PtK2) of about 4 minutes and much slower labelling kinetics in fibroblasts than for the lamellipodium actin (Symons and Mitchison, 1991). The lack of clear evidence for spatially restricted incorporation (or loss) of actin from stress fibres leaves both the mode of stress fibre formation and the turnover of filaments in these relatively long-lived structures an open question. The data reported by Heath (this volume) suggest that 'immature' stress fibres may be literally spun out from the filaments of the cortex; in which case they are assembled from filaments already assembled in the lamellipodium. These observations are entirely consistent with the observations of Svitkina et al. (1986) and suggest the possibility that net assembly of actin filaments may occur only at the distal edge of the leading lamellipodium and that all other actin filament architectures, therefore, are derived from these filaments.

The dynamic properties of actin filaments

Actin assembly is accompanied by the stoichiometric hydrolysis of ATP. ATP bound to assembling monomers is hydrolysed within the filament to yield (eventually) an ADP-actin filament. Comparison between the rate of addition of subunits and the rate of hydrolysis of ATP in a population of rapidly assembling filaments shows that hydrolysis lags behind assembly (Korn et al., 1990). The ATP hydrolysis step is thus uncoupled from the assembly step. The ADP.Pi intermediate is also relatively stable and releases phosphate more slowly than the ATP is hydrolysed (Carlier, 1987). Taken together, these observations show that, while the rate of assembly exceeds the rate of hydrolysis and phosphate release, an individual filament will consist of co-linear segments rich in ATP, ADP.Pi or ADP monomers according to the relative times after assembly into the same filament. There is also a difference between the critical concentrations at opposite ends of the same filament due to the different relative rates of monomer exchange at the pointed and barbed ends; the slower rate of addition at the pointed end means that hydrolysis invariably occurs on the terminal monomer before the next subunit adds on whilst the converse is generally true at the barbed end (Coue and Korn, 1986). The rates are some 20-fold faster at the barbed end (Pollard, 1986). The chemical difference at opposite ends confers different equilibrium constants for the various assembly and disassembly reactions and the equilibrium constant is the reciprocal of critical concentration. Thus, since both ends of the same filament must necessarily share the same monomer pool, there will be net loss of monomers at the pointed end and net gain at the barbed end. This process is called head-to-tail assembly or treadmilling (Wegner, 1976).

If actin polymerisation were an equilibrium system, then the filaments would assemble to equilibrium (which would leave the critical concentration of actin unassembled) and remain there, apart from some random monomer exchange at filament ends. To maintain the actin filament system in flux, a chemical switch is required, that allows one state for actin that will assemble to form stable filaments, then to change that state within the filament such that it is now 'unstable' and will disassemble. The lifetime of a filament is then determined by how quickly the switch operates. This is a steady-state system and requires a continuous influx of energy to prevent its eventual collapse into the 'low energy' equilibrium state. The chemical switch for actin is the hydrolysis of ATP.

Actin monomer binds ATP more tightly than ADP and the ATP form therefore prevails so long as ATP is available. ATP-actin has a low critical concentration in the presence of 100 mM salt and magnesium (conditions prevailing within the cell) and will therefore polymerise until the critical concentration, about 0.1 μM under physiological conditions (Rickard and Sheterline, 1986), of monomer remains (assuming no interference from monomer-binding proteins). Once the monomer is incorporated into the polymer (but not during the assembly step itself), there is a finite probability that the ATP will be hydrolysed. The rate constant for hydrolysis has been measured at about 0.02 s^{-1} or a probability of 1 ATP hydrolysed per second per 50 monomers assembled (Korn *et al.*, 1990). The product of hydrolysis is necesarily ADP.Pi-F-actin. The release of phosphate has a lower probability than hydrolysis, with a measured rate constant of about 0.006 s^{-1} (Carlier, 1987) or a probability of 1 Pi released per second per 170 monomers assembled. Thus after a predictable period of time, the filament which assembled as ATP F-actin converts itself to ADP-F-actin via an ADP.Pi-F-actin intermediate. This is the conformational switch. Put simplisticly, ADP-F-actin has a higher critical concentration than ATP-F-actin, so in the competition for monomers, ADP-F-actin disassembles whilst ATP-G-actin assembles. Monomers will necessarily and continuously shift from ends with high critical concentrations to those with low critical concentrations in the cell given that all polymer ends have access to the same monomer pool.

The number of ends with particular critical concentrations, that is the number of ATP, ADP.Pi and ADP, pointed and barbed ends will be determined by a number of factors: by the thermodynamics of the actin polymer system itself and the immediate history of the system, by the relative affinities and concentrations of end-binding proteins for the different types of end, by the action of severing proteins (the gelsolin family) and also by ionic conditions in the cytoplasmic environment. The location of the various classes of end will define where net assembly and disassembly take place in the cell. From the arguments above it is likely that the terminal subunit at pointed ends will not be an ATP-actin, but because of the slow dissociation of phosphate, it may be an ADP.Pi actin. There may be many ADP.Pi pointed ends in a rapidly turning over population implying that situations where treadmilling can occur may be rare in the cell. As observed directly and discussed below, the major difference in critical concentration at opposite ends when the pointed end is terminated by an ADP.Pi monomer is abolished (Rickard and Sheterline, 1986).

It is now clear that the exchange of ATP for ADP on the newly disassembled monomers is another regulatory point controlled by the monomer-binding proteins thymosin and profilin, and possibly by others (Goldschmidt-Clermont *et al.*, 1991). This regulatory point may be the major linkage between the status of cell surface receptors and control of remodelling of the actin cytoskeleton through the availability of monomers 'charged' with ATP.

The conversion of ATP-F-actin to ADP-F-actin is, as discussed above, time dependent, so since older filaments are more likely to be ADP-F-actin, they are more unstable than recently assembled filaments providing an automatic underlying mechanism for the replacement of old filaments by new. This parallels what appears to occur within the lamellipodium and the 'switch' from ATP- through ADP.Pi- to ADP-F-actin may constitute the steps that determine the end of the assembly phase at the distal margin and the stochastic disassembly step that occurs subsequently. There is direct evidence that these three 'forms' of F-actin have different properties. The off-rate for ADP.Pi monomer at both ends is somewhat (about 10-fold) lower than for ADP monomers (Rickard and Sheterline, 1988). It is this effect that makes the largest contribution to the change in critical concentration of actin in the presence of phosphate. There is also preliminary evidence that binding of actin-binding proteins of the spectrin group (in particular α-actinin) has a lower affinity for ADP.Pi-F- actin than for ADP-F-actin (Hendry et al., 1991). This observation, together with the kinetic observations suggest that ADP.Pi-F-actin has a different conformation than ADP-F-actin, an inference that gains weight from direct observation of negatively-stained filaments by electronmicroscopy (Bremer et al., 1991) and studies on the rheological properties of filaments in solution (Janmey et al., 1990). It remains to be discovered whether capping proteins or members of the gelsolin family can also discriminate the different structures of filaments. If they can, this structural switch caused by ATP hydrolysis could contribute to the sequential evolution of actin filament architectures by the progressive sequential interaction with actin-binding proteins, all of which may be competing for a small number of binding sites on the surface of F-actin and which may have different relative affinities for the different structural forms of filament.

The mechanism by which changes in the status of the nucleotide-binding site can lead to kinetic and structural changes of the filament is not understood. However, the nucleotide sits in the cleft dividing the two major domains of the actin molecule and the γ- and β-phosphates hydrogen bond to amino acids in loops on opposite domains (Kabsch et al., 1990). Since it would be expected that the β-phosphate would repel the now hydrolysed phosphate in the γ-position, its presence might disturb the spatial relationship between the two domains. That these two domains might move relative to one another is suggested by the 3-dimensional structure and by the observation that the closely related structure of hexokinase and a heat-shock protein has a quite different orientation of one domain relative to the other (Flaherty et al., 1991). Inorganic phosphate can also bind back to the γ-site of the nucleotide-binding site of ADP-F-actin (Carlier and Pantaloni, 1988) with an affinity (K_d) of about 1 mM, close to the cellular concentration of phosphate. The binding species is the $H_2PO_4^-$ ion whose pKa is around neutrality, thus changes in intracellular pH could modulate the Pi effect (Rickard and Sheterline, 1986) in addition to changes in phosphate concentration from changes in metabolic status.

The uncoupled ATP hydrolysis by actin filaments confers an immensely complex set of possible conformational states on individual actin filaments with both temporal and spatial implications. Such properties allow for equally complex and subtle regulation of the actin filament cytoskeleton.

References

Amos, L.A. and Amos, W.B. (1991) *Molecules of the Cytoskeleton.* London: Macmillan.

Bennett, V. (1985) The membrane skeleton of human erythrocytes and its implications for more complex cells. *Ann. Rev. Biochem.* **54**, 273-???.

Bray, D., Heath, J. and Moss, D. (1986) The membrane-associated "cortex" of animal cells: its structure and mechanical properties. *J. Cell Sci.* **Suppl. 5**, 71-81.

Bremer, A., Sutterlin, R. and Aebi, U. (1991) Stability and structural dynamics of the F-actin filament. *J. Muscle Res. Cell Motil.* **12**, 483.

Bretscher, A. and Weber, K. (1980) *J. Cell Biol.* **86**, 335-340.

Carlier M. F. and Pantaloni D. (1988) Binding of phosphate to F-ADP-actin and role of F-ADP-Pi-actin in ATP-actin polymerisation. *J. Biol. Chem.* **263**, 817-825.

Carlier, M-F. (1987) Measurement of Pi dissociation from actin filaments following ATP hydrolysis using a linked enzyme assay. *Biochem. Biophys. Res. Commun.* **143**, 1069-1075.

Coue, M. and Korn, E.D. (1986). ATP hydrolysis by the gelsolin-actin complex and at the pointed ends of gelsolin capped filaments. *J. Biol. Chem.* **261**, 1588-1593.

De Lanerolle, P., Adelstein, R.S., Feramisco, J.R. and Burridge, K. (1981) Characterisation of antibodies to smooth muscle myosin light chain kinase and their use in localising myosin kinase in non-muscle cells. *Proc. Natl. Acad. Sci. USA* **78**, 4738-4742.

Flaherty, K.M., McKay, D.B. Kabsch, W. and Holmes, K.C. (1991) Similarity of the three-dimensional structures of actin and the ATPase fragment of a 70kDa heat-shock cognate protein. *Proc. Natl. Acad. Sci. USA* **88**, 5041-5045.

Forscher, P. and Smith, S.J. (1989). Action of cytochalasins on the organisation of actin in neuronal growth cones. *J. Cell Biol.* **107**, 597-602.

Fox, S.E.B. and Phillips, D.R. (1983) Polymerisation and organisation of actin filaments within platelets. *Semin. Haematol.* **20**, 243-260.

Gaertner, A. and Wegner, A. (1991) Mechanism of the insertion of actin monomers between the barbed ends of actin filaments and barbed end-bound insertin. *J. Muscle Res. Cell Motil.* **12**, 27-36.

Geiger, B., Tokuyasu, K.T., Dutton, A.M. and Singer, S.J. (1980) Vinculin, an intracellular protein localised at specialised sites where microfilament bundles terminate at the cell membrane. *Proc. Natl. Acad. Sci. USA* **77**, 4127-4131.

Glenney, J.R. and Glenney, P. (1983) Fodrin is the general spectrin-like protein found in most cells, whereas spectrin and the TW protein have restricted distributions. *Cell* **34**, 503-???.

Goldman, R.D., Chojnacki, B. and Yerna, M.J. (1979) Ultrastructure of microfilament bundles in baby hamster kidney (BHK-21) cells. *J. Cell Biol.* **80**, 759-784.

Goldschmidt-Clermont, P., Machesky, L.M., Doberstein, S.K. and Pollard, T.D. (1991) Mechanism of the interaction of human platelet profilin with actin. *J. Cell Biol.* **113**, 1081-1089.

Handel, S.E., Hendry, K.A.K. and Sheterline, P. (1990) Microinjection of covalently cross-linked actin oligomers causes disruption of existing actin filament architecture in PtK2 cells. *J. Cell Sci.* **97**, 325-333.

Hendry, K.A.K., McGregor, A., Critchley, D. and Sheterline, P. (1991) F-actin structure is modulated by inorganic phosphate. *J. Muscle Res. Cell Motil.* **12**, 483.

Herman, L.M. and Pollard, T.D. (1981) Electron microscopic localisation of cytoplasmic myosin with ferritin-labelled antibodies. *J. Cell Biol.* **88**, 346-351.

Howard, T.H. and Meyer, W.H. (1984) Chemotactic peptide modulation of actin assembly and locomotion in neutrophils. *J. Cell Biol.* **98**, 1265-1271.

Isenberg, G., Rathke, P.C., Hulsman, N., Franke, W.W. and Wolfarth-Botterman, K. E. (1976) Cytoplasmic actomyosin fibrils in tissue cultur cells are contractile. *Cell Tiss. Res.* **166**, 427-434.

Janmey P., Hvidt, S., Oster, G.F., Lamb, J., Stossel, P. and Hartwig, J. (1990) Effect of ATP on actin filament stiffness. *Nature* **347**, 95-97.

Kabsch, W., Mannherz, G., Suck, D., Pai, E.F. and Holmes, K.C. (1990) Atomic structure of the actin:DNAase I complex. *Nature* **347**, 37-44.

Korn E. D., Carlier M.-F. and Pantaloni D. (1990) Actin polymerisation and ATP hydrolysis. *Science* **238**, 638-644

Kreis, T.E. and Birchmeier, W. (1980) Stress fibre sarcomeres are contractile. *Cell* **22**, 555-562.

Lassing, I. and Lindberg, U. (1988) Evidence that the phosphatidylinositol pathway is linked to cell motility. *Exp. Cell Res.* **174**, 1-15.

Lindberg, U., Hoglund, A.-S. and Karlsson, R. (1981) On the ultrastructural organisation of the microfilament system and the possible role of profilactin. *Biochimie* **63**, 307-323.

Mooseker, M.S. and Tilney, L.G. (1975) *J. Cell Biol.* **67**, 725-743.

Pollard, T. D. (1986) Rate constants for the reactions of ATP- and ADP-actin with the ends of filaments. *J. Cell Biol.* **103**, 2747-2754.

Rickard, J. E. and Sheterline, P. (1986). Cytoplasmic concentrations of inorganic phosphate affect the affect the critical concentration for actin assembly in the presence of ADP and cytochalisin D *J. Mol. Biol.* **191**, 273-280.

Rickard, J. E. and Sheterline, P. (1988). The effect of ATP removal and inorganic phosphate on length redistribution of sheared actin filament populations: evidence for a mechanism of end-to-end assembly *J. Mol. Biol.* **201**, 675-682.

Safer, D. (1992) The interaction of actin with thymosin b4. *J. Muscle Res. Cell Motil.* **13**, 269-271.

Sanger, J.M. and Sanger, J.W. (1980) Banding and polarity of actin filaments in interphase and cleaving cells. *J. Cell Biol.* **86**, 568-575.

Sheterline, P and Rickard, J.E. (1989) The cortical actin filament network of neutrophil leucocytes during chemotaxis and phagocytosis. In *The Neutrophil: Cellular Biochemistry and Physiology* (ed. M.B. Hallett), pp 141-165. Boca Raton:CRC Press.

Small, J.V., Isenberg, G. and Celis, J.E. (1978) Polarity of actin filaments at the leading edge of cultured cells. *Nature* **272**, 638-639.

Svitkina, T.M., Neyfakh, A.A and Bershadsky, A.D. (1986) Actin cytoskeleton of spread fibroblasts appears to assemble at the cell edges. *J. Cell Sci.* **82**, 235-248.

Symons, S. H. and Mitchison, T. J. (1991) Control of actin polymerisation in Live and permeabilised fibroblasts. *J. Cell Biol.* **114**, 503-513.

Theriot, J. A., Mitchison, T. J., Tilney, L. G. and Portnoy, D. A. (1992) The rate of actin-based motility in Listeria monocytogenes equals the rate of actin polymerisation. *Nature* **357**, 257-260.

Theriot, J.A. and Mitchison, T. J. (1991) Actin microfilament dynamics in locomoting cells. *Nature* **352**, 126-131.

Tilney, L.G. and Portnoy, D.A. (1989) Actin filaments and the growth, movement and spread of the intracellular bacterial parasite, *Listeria monocytogenes*. *J. Cell Biol.* **109**, 1597-1608.

Wang, Y.-L. (1985) Exchange of subunits at the leading edge of living fibroblasts: possible role of treadmilling. *J. Cell Biol.* **101**, 597-602.

Wegner, A. (1976). Head to tail polymerisation of actin. *J. Mol. Biol.* **108**, 139-150.

Yin, H.L., Lida, K. and Janmey, P.A. (1988) Identification of a polyphosphoinositide-modulated domain in gelsolin which binds to the sides of actin filaments. *J. Cell Biol.* **106**, 805-812.

Zigmond, S.H., Otto, J.J. and Bryan, J. (1979) Organisation of myosin in a submembranous sheath in well-spread human fibroblasts. *Exp. Cell Res.* **119**, 205-219.

Printed in Great Britain © The Society of Experimental Biology 1993   353

MECHANISMS OF REGULATION OF PSEUDOPODIAL ACTIVITY BY THE MICROTUBULE SYSTEM

ALEXANDER D. BERSHADSKY[1] and JURI M. VASILIEV[2]

[1]Chemical Immunology Department, Weizmann Institute of Science, Rehovot 76100, Israel.
[2]Cancer Research Center of the Academy of Medical Sciences of Russia, 24 Kashirskoye shosse, 115478 Moscow, Russia

Summary

Polarization of pseudopodial activity may develop spontaneously or be induced by external signals; this polarization is stabilized by cytoskeletal mechanisms. We have studied the mechanisms of microtubule-dependent control of the polarization of pseudopodial activity. Experiments with cultured fibroblasts exposed to drugs specifically inhibiting or enhancing polymerization of microtubules show that an intact microtubule system is essential, not only for restricting pseudopodial activity to certain sites at the cell edge, but also for enhancing this polarized activity. In other experiments, extension of pseudopods and polarization of cultured fibroblasts was enhanced by *N-ras* proto-oncogene over-expression or by phorbol ester induced activation of protein kinase C. This enhancement of polarization was accompanied in both cases by significant activation of the motility of vesicular organelles. Microtubules in the elongated processes of these cells were enriched in detyrosinated (Glu) α tubulin. Colcemid inhibited both organelle motility and cell process extension in this cell system. Intracellular injection of antibody to kinesin, the protein that moves vesicles toward the plus (distal) end of microtubules, mimicked some effects of microtubule-depolymerizing drugs on cell shape and pseudopodial activity. On the basis of these data it is suggested that, at least in fibroblasts, microtubules direct and enhance the outward component of cortical flow essential for pseudopod extension. This control may be associated with the organelle transport function of microtubules. A model of the stabilization of polarization by reorganization of both the actin cortex and the microtubule system is proposed and discussed.

Introduction

Crawling pseudopod-forming cells, such as amoebae, leukocytes, fibroblasts and neurite-extending neurons, are able to orient themselves in response to external signals, e.g., to gradients of chemotactic molecules or substratum-bound matrix molecules. In response to these signals the cells extend and elongate their cytoplasmic processes, eventually reorganizing and orienting the whole cytoskeleton.

The sensitivity to signal gradients may be extremely high. For instance, leukocytes and slime mold amoebae migrate directionally to the source of chemoattractant, provided the

Key words: cell motility, pseudopodia, microtubule regulation.

concentration of this attractant changes by more than 1-2% across the cell's diameter (Devreotes and Zigmond, 1988). The growing axons of retinal cells are deflected and turned around by gradients of substratum bound repellent proteins, provided that their concentration changed by 5% or more across the diameter of a growth cone (Baier and Bonhoeffer, 1992). Segregation of a cell into active, pseudopod-forming and stable zones is the basis of the reorganization response. As suggested before (Vasiliev, 1987; 1991), polarization initiated by gradients of an external signal is then stabilized by global reorganization of the cytoskeleton.

It is worth noting that some cell types can also polarize their pseudopodia in situations when external conditions are apparently isotropic. A number of agents are described which increase significantly the probability of this spontaneous polarization. For example, addition of any chemotactic agent at a uniform concentration (not as a gradient) to spherical leucocytes leads to the development of front-tail polarity within a few minutes (Wilkinson, 1990).

On the basis of different sensitivities to microtubule-specific drugs, one can operationally distinguish two variants of stabilizing cytoskeletal reorganization: microtubule-dependent and microtubule-independent ones. The actin cortex is the main mediator of pseudopodial changes and it is probable that global reorganization of this cortex is, to a large degree, responsible for microtubule- independent stabilization. The microtubule system is not directly involved in the generation of pseudopodia: cells without microtubules can still extend and retract pseudopods. Nevertheless, all processes involving a high degree of polarization of pseudopodia, e.g. formation and growth of neurites or orientation and directional movements of fibroblasts, are microtubule-dependent. The nature of this microtubule control of polarization is the main topic of this article. Most probably, microtubule control is mediated by reorganization of the actin cortex. Therefore we will begin with a short discussion of the role of the cortex in the generation and polarization of pseudopodia.

Local pseudopodial extension and global reorganization of the cortex

Numerous data indicate that extension of pseudopods is a result of local polymerization of actin microfilaments (for a review, see Condeelis, 1992). In particular, the active edge of a fibroblast is the main site of actin polymerization in the cell (Wang, 1985; Svitkina *et al*, 1986; Okabe and Hirokawa, 1989; Symons and Mitchison, 1991). Possibly, polymerization of microfilaments involves activation of some unidentified nucleating sites under the membrane (Symons and Mitchison, 1991). For further development of a pseudopod, cross-linking of polymerized actin filaments by appropriate actin binding proteins seems to be necessary (Cox *et al*, 1992; Cunningham *et al,* 1992). Myosin I, localized in the advancing pseudopods (Fukui *et al*, 1989), may also play an important role in pseudopod extension (Sheetz *et al*, 1992).

Further evolution of the pseudopodial network of microfilaments involves two different types of change: disassembly and/or acquisition of contractility. Disassembly of actin filaments at the base of lamellipodia was inferred by Symons and Mitchison (1991) who noticed that the width of the zone of polymerized actin did not increase in spite of

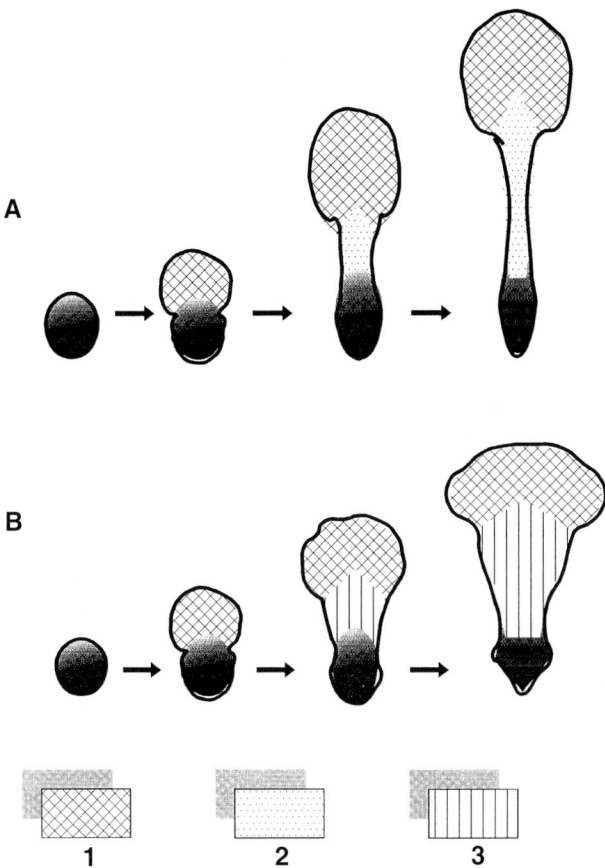

Fig. 1. Highly schematic presentation of two variants of polarization of the actin cytoskeleton. A - "nerve cell" type, with a stalk-like non-contractile process behind the leading lamella. These stalks form due to a partial disassembly of the actin network and usually contain microtubules and intermediate filaments (not shown). B - "fibroblast" type polarization, with a myosin II-containing contractile domain behind the leading lamella. Formation of this domain is associated with the development of cell-substrate contacts. 1 - protrusive domain (actin network); 2 - non-motile, actin-poor domain; 3 - contractile domain (myosin II-containing actin bundles).

continuous addition of new subunits at the anterior edge. An "empty" zone, free from a dense actin network, is often formed immediately adjacent to the base of a lamellopod (Svitkina *et al*, 1984; Rinnerthaler *et al*, 1991); formation of this zone may be a consequence of this disassembly process.

   Acquisition of contractility is probably associated with the penetration of myosin II into the network. Kinetic studies (Conrad *et al*, 1989) have shown that polymerized actin is present in the lamellipodia of fibroblasts from the moment of their extension, while myosin appears several minutes later. We do not know what specific structures are formed by myosin II associated with the pseudopodial actin network. Obviously, acquisition of contractility may lead to retraction of unattached pseudopods. Disassembly alone would lead to thinning of the pseudopod and eventual collapse without retraction (Fig. 1).

The attachment of pseudopods may significantly modify the fate of the pseudopodial network. The surface of pseudopods has special adhesive properties: it can preferentially attach itself to the substratum, other cells or lectin covered beads (Vasiliev *et al*, 1975; Vasiliev, 1982; Kucik *et al*, 1991). Possibly, this facilitated formation of adhesions by the pseudopodial surface is due to the ability of a microfilament network, immediately after its polymerization, to anchor the cytoplasmic domains of membrane proteins. Isometric contraction of the attached networks may lead to their alignment and reorganization into bundles of parallel microfilaments (Fleischer and Wohlfarth-Bottermann, 1975). Simultaneously, the same tension may lead to an increase of contact size, by some unknown mechanism. In this way, initially small cell-substratum contacts may be transformed into large mature focal contacts (Geiger *et al*, 1984; Bershadsky *et al*, 1985). In other words, somewhat paradoxically, contraction may lead not only to retraction of unattached pseudopods, but also to the strengthening of the attachment counteracting this retraction.

Positive external signals, such as chemotactic peptides or growth factors, are known to induce extension of pseudopods in sensitive cells. Some less well studied external agents may act as negative signals preventing pseudopod extension. Paralysis of pseudopodial activity by cell-cell contact, discovered by Abercrombie (Abercrombie, 1970), is the most obvious example of such a negative factor; molecules responsible for this effect are yet to be identified. Recently, a glycoprotein repelling the growth of retinal cell axons has been isolated from the membranes of tectal cells (Stahl *et al*, 1990).

Apparently, both positive and negative agents bind to cellular receptors and activate some signal-processing systems. Some hints about the identity of the systems controlling pseudopodial activity are provided by experiments in which certain components of these systems are selectively activated. Activation of ruffling was induced by microinjection of *ras* protein (Bar-Sagi and Feramisco, 1986). Exposure of cells to phorbol ester (phorbol 12-myristate 13-acetate, PMA), activating protein kinase C, also induced formation of large ruffles in certain cell types (Schliwa *et al*, 1984; Dugina *et al*, 1987). It is probable that signal-processing systems control not only extension of pseudopods but also their further evolution. For instance, large lamellae induced by phorbol ester, in contrast to the lamellipodia of control cells, often do not retract (Danowsky and Harris, 1988; Lyass *et al*, 1988) but undergo collapse starting at their base and progressing to the periphery (Dugina *et al*. 1987). One can suggest that in this system, activation of protein kinase C leads to a decrease of contractility and an enhancement of disassembly.

Almost by definition, each pseudopodial extension leads to segregation of the actin cortex into contracted and extended domains. A series of polarized extensions accompanied by attachment of pseudopods makes this segregation more stable. For instance, the cortex of a substratum-spread polarized fibroblast is segregated into several domains and subdomains: a) leading lamellae with an actin network at the leading edge and a more proximal part without any significant actin structures; b) the proximal contractile cortex with numerous contractile structures such as microfilament bundles and sheaths; c) the tail cortex with several microfilament bundles (Svitkina *et al*. 1984). The cortex of a growing unipolar neuron also can be divided into several well-defined domains: a) the growth cone with distal pseudopods and a central contractile zone; b) the cortex of the shaft consisting of a thin sheath and c) the cortex of the cell body without any prominent specializations.

Segregation of the cortex into contractile and extended domains is a result of the polarization of pseudopodial activity; at the same time, it may stabilize polarization of these activities. The most convincing evidence for the stabilizing role of contractile domains of the cortex is provided by the experiments of Spudich and collaborators with the mutants of *Dictyostelium* amoebae lacking myosin II (for a review, see Spudich, 1989). These mutants, in contrast to wild type amoebae, cannot form a myosin rich tail domain of the cortex and have no stable zones of their surface: they extended pseudopods from all parts of the edge. The general rule, true for most cell types, can be formulated as follows: pseudopods are not induced by external factors in zones having submembraneous contractile cortex structures, especially if these structures are stretched by isometric tension. The other formulation of the same rule is that pseudopods are preferentially extended from previously extended and attached cortex domains. In this way the polarization of pseudopodial activity induced by external factors is stabilized by a reorganization of the cortex.

Signal-induced alterations in pseudopod activity may therefore lead to a general reorganization of the cortex and of the whole cell morphology. For instance, PMA-treated fibroblasts may reversibly acquire neuron-like morphology: they extend long processes with lamellae at the ends and narrow stable proximal shafts (Dugina *et al.* 1987). As suggested above, this change may be due to an enhanced disassembly and decreased contractility of microfilament networks. The exact molecular mechanisms of these shifts remain to be studied; protein kinase C activated by phorbol ester is known to phosphorylate several cytoskeletal proteins including myosin (Ludowyke *et al*, 1989; Papadopoulos and Hall, 1989) and vimentin (Huang *et al*, 1988). Similar alterations of the signal processing systems, similar to those observed in PMA-treated fibroblasts, may be involved in the control of neurite outgrowth during differentiation of neuroblasts.

Segregation of the actin cortex is a highly dynamic process and all the recent evidence supports the old concept of circular cortical flow (Dunn, 1980; for a review, see Bray and White, 1988). Probably, most actin is polymerized in the zone of pseudopodial extension and transferred from there into more proximal domains of the cortex; depolymerization of microfilaments in these domains produces monomers which are transferred back to the active edge and re-polymerized there. It is reasonable to distinguish local and global cortical flow. As indicated by the data of Symons and Mitchison (1991) local flow may circulate between distal and proximal parts of the pseudopodial network at the active edge. Global flow transports parts of the network from the leading lamella into central contractile parts of the cortex and brings monomers back; centripetally moving ruffles and actin arcs are probably visual manifestations of this global flow. Local flow in neurons possibly circulates between peripheral and central parts of the growth cone, while global flow of cortical material circulates between the growth cone and the cell body; this flow probably includes some components of axonal transport. The exact characteristics and mechanisms of cortical flow in various cell types still remain hypothetical.

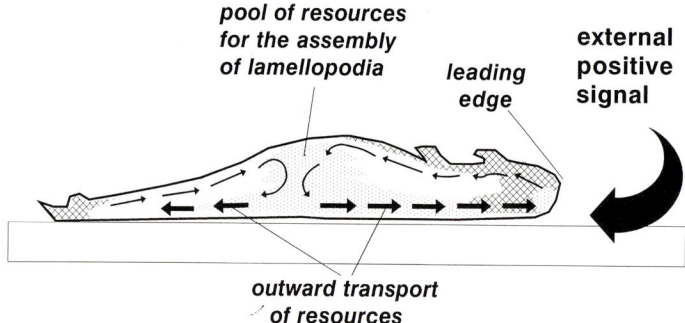

Fig. 2. Circulation of material for the assembly of pseudopods in the course of cortical flow. Pseudopodial activity is polarized because outward flow of material into "successful" pseudopods is stimulated. Thin arrows symbolize the retrograde (centripetal) component of the flow.

It is important to stress that cortical flow is essential for the process of polarization of pseudopodial activity (Fig. 2). In fact, polarization is a consequence of the redistribution of pseudopodial material from regions where pseudopod formation is not appropriately supported by external conditions to regions where it can proceed more successfully. The principle of circulation seems to be more fundamental than the nature of the circulating material and the precise mechanism of circulation. In fact, amoeboid sperm of the nematode, *Ascaris suum*, do not contain actin, but the process of centripetal flow and reassembly of the major sperm protein (MSP) in these cells is apparently similar to the process of actin flow in other crawling cells (Roberts and King, 1991).

What manifestations of polarization are controlled by microtubules? Experiments with microtubule-specific drugs

Alterations of cell shape

As found long ago, fibroblasts treated with drugs which selectively destroy microtubules, such as colchicine, colcemid and nocodazole, lose elongated extensions and cannot polarize their pseudopodia (Vasiliev *et al*, 1970; Gail and Boone, 1971). In recent years, mainly due to technical advances, new important and more detailed information about the effects of these drugs on various aspects of polarization has been obtained. In particular, due to the development of new morphometric indices, namely elongation and dispersion indices, it has become possible to assess quantitatively the degrees of cell elongation and formation of extensions (Dunn and Brown, 1986; Brown *et al* 1989).

Using these methods one can distinguish between microtubule-dependent and microtubule-independent variants of polarization more objectively. For instance, it was found that depolymerization of microtubules by colcemid or nocodazole did not decrease significantly the average elongation and dispersion indices in primary cultures of chick heart fibroblasts grown for 6 h. When the cells are maintained in culture for longer periods, the sensitivity of their polarization is gradually increased and at 48 hr this

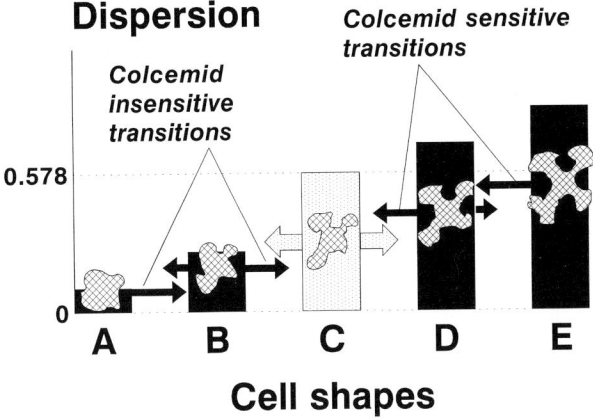

Fig. 3. Sensitivity of cell shape dynamics to microtubule destruction. Outlines corresponding to different stages (A-E) of motile activity of the same cell are drawn. The dispersion values of the outlines, represented by bars, increase from left to right. The arrows attached to the bars symbolize the probabilities of transition of the cell to a state with increased (decreased) dispersion. The greater the dispersion of the cell outline, the lower the probability that it will further increase over the next time period (and vice versa). The dispersion value of 0.578 corresponds to the state of chick heart fibroblasts at which decrease and increase of dispersion are equally probable. Addition of colcemid immediately enhances the tendency of cells with high (above 0.578) dispersion to diminish this value; the tendency of cells with a dispersion value under 0.578 to increase their dispersion is not affected. The scheme is based on the data of I.Karavanova, A.Brown, G.Dunn, A.Bershadsky and J.Vasiliev.

sensitivity becomes high, that is, both indices are significantly decreased in the drug treated cultures. This high sensitivity is retained in secondary cultures. Increase of sensitivity with time is correlated with an increase in the size of spread cells in control cultures (Middleton *et al*, 1989). These results gave reason to suggest that the microtubule system is essential for polarization of larger cells having longer extensions, while shorter extensions may be formed and maintained in the absence of microtubules.

Additional material supporting this suggestion was obtained in experiments studying the effect of colcemid on the dynamics of cell shape changes in individual chick heart fibroblasts in secondary culture (I.D.Karavanova, A.F.Brown, G.A.Dunn, A.D.Bershadsky and J.M.Vasiliev, unpublished). Cells in control cultures underwent dynamic fluctuations of shape (Fig. 3). The types of fluctuations over each 20 min period depended on the value of the dispersion index at the start of the interval: this value tended to diminish in cells with initially high dispersion ("highly polarized cells") whereas it tended to increase in cells with a low dispersion ("weakly polarized cells"). It was found that colcemid during the first 20 min of its action affected only the highly polarized cells, enhancing their tendency to diminish the dispersion, but did not inhibit the increase of dispersion of weakly polarized cells (Fig. 3). In other words, colcemid did not prevent an increase in polarization of the weakly polarized cells but decreased the chance of more highly polarized cells remaining polarized.

In experiments with mouse fibroblasts (Bershadsky *et al*, 1990) it was found that PMA enhances formation of narrow cytoplasmic extensions both in the absence and in the

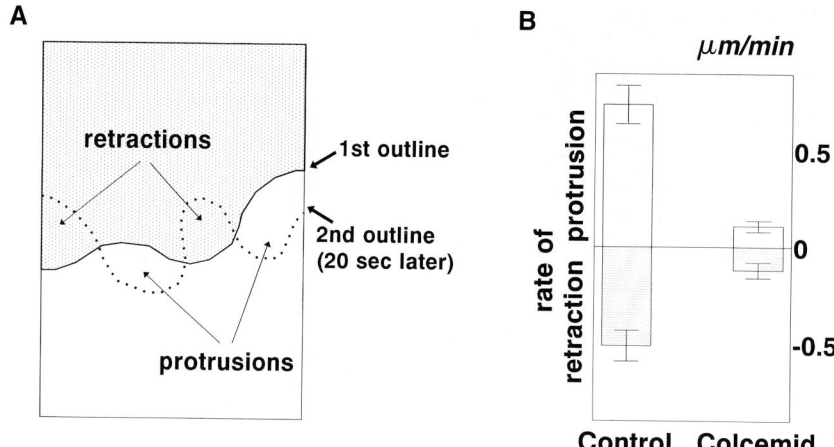

Fig. 4. Inhibition of pseudopodial activity at the leading edge of a migrating fibroblast after destruction of the microtubule system. A - method of evaluation of pseudopodial activity. Definitions of protrusions and retractions are clear from the drawing. The rate of protrusion (retraction) is defined as the total area of protrusion (retraction) per unit of time per unit of the length of the first outline. These rates are measured in μm/min. B - Average rates of protrusion and retraction in control and colcemid-treated human fibroblasts (10 cells of each type were measured). After Bershadsky *et al* (1991).

presence of colcemid. However, in cells without microtubules these extensions developed later, had more bending and thickenings. Thus, microtubules may be essential not only to keep the extensions long but also to keep them straight.

Alterations of distribution and activity of pseudopods

Microtubule-destroying drugs are known to abolish the segregation of the cell edge into active and stable zones: stable zones disappear in the drug-treated cells and the whole of their edge becomes active (Vasiliev *et al*, 1970). The introduction of video microscopy with computer-enhanced contrast made it possible to study alterations of pseudopodial activity in much greater detail. It became possible to assess quantitatively the rate of extension and retraction of individual lamellipodia per unit length of the active edge (Fig. 4A). This assessment was performed in experiments with human fibroblasts moving into a wound (Bershadsky *et al*, 1991) and led to a somewhat unexpected result: both extension and retraction rates at the leading edges of these cells decreased about four fold after 3-4 h of incubation in colcemid-containing medium (Fig. 4B). Thus, microtubules are essential not only for the inhibition of pseudopodial activity at the lateral edges of a polarized cell, they are also necessary for the development of maximal activity at the edge facing the wound.

As expected, re-distribution of pseudopodial activity in colcemid-treated fibroblasts is accompanied by reorganization of the actin cortex. Immunofluorescence microscopy of cells stained with rhodamine-phalloidin for polymerized actin had shown that colcemid induces the disappearance of the large leading lamella extended into the wound. In contrast, the central contractile domain is preserved, so that the ends of stress fibers

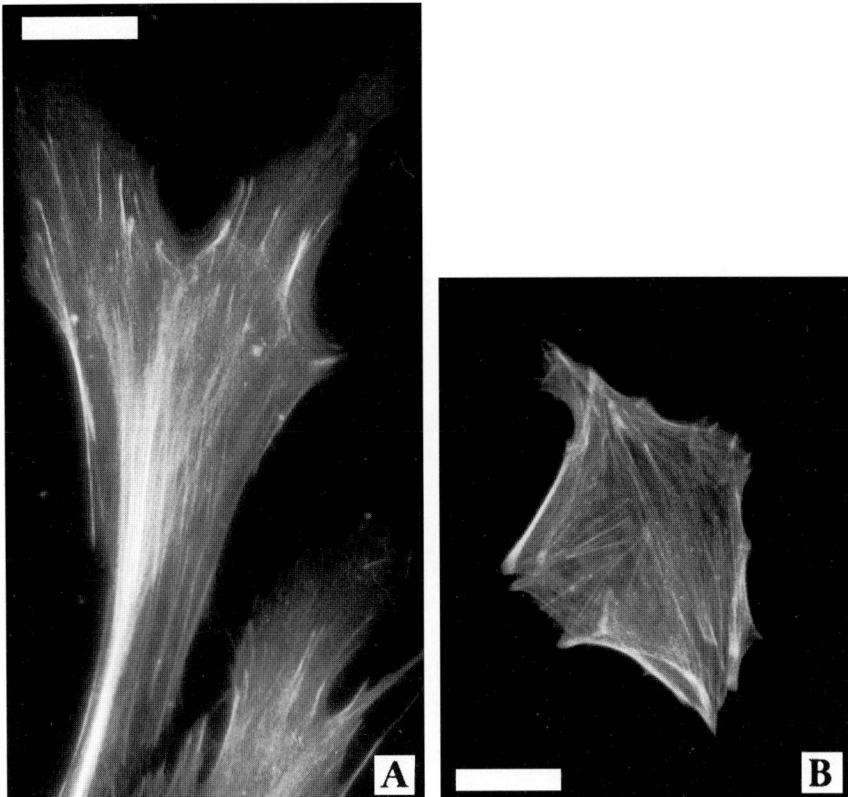

Fig. 5. Actin cytoskeleton of control (A) and colcemid-treated (B) human fibroblasts. Destruction of microtubules leads to a significant reduction of the protrusive domain, while the contractile domain in these cells is well developed. Rhodamine-phalloidin staining. Bars, 20 μm. (Photographs by I.S.Tint).

associated with large mature focal contacts become located just near the leading edge (Fig. 5).

To summarize, these data indicate that the microtubule system is necessary to direct all pseudopodial activity to one part of the edge, the leading edge, and to stimulate activity at that edge. Polarization of pseudopodial activity leads to segregation of the cortex and, more specifically, to formation of a specialized extended domain, the leading lamella. One may suggest that the microtubule system is needed to provide optimal conditions for the global centrifugal flow of cortical material from the central contractile parts of the cortex to the leading edge, in order to meet requirements for a maximal rate of actin polymerization at that edge.

The importance of an intact microtubule system

In cultured fibroblasts, as in many other cells, microtubules form a structurally integrated radial system. The integrity and intactness of this system are essential for the control of the actin cortex structure and the cell shape. This conclusion is supported by two groups of data: by experiments with taxol and with graded concentrations of colcemid.

The effect of taxol at the molecular level is opposite to that of colcemid. This drug stabilizes microtubules in a polymerized state (Schiff and Horwitz, 1980). Due to this effect, taxol promotes polymerization of free microtubules not associated with any organizing centers, so that gradually the system of microtubules radiating from the perinuclear center is replaced by numerous aggregates of free microtubules (DeBrabander *et al*, 1981). This disintegration of the microtubule system was found to be accompanied by the loss of an elongated cell shape and reversal of the polarization of pseudopodia (O.Ivanova, O.Pletyushkina, I.Kaverina and J.M.Vasiliev, unpublished). In other words, two drugs having opposite effects at the molecular level, produced similar loss of polarization at the cellular level.

In a series of unpublished experiments we (I.S.Tint, N.Alieva, A.D.Bershadsky and J.M.Vasiliev) compared the effects of various concentrations of colcemid on human fibroblasts. It was found that significant decreases in the dispersion and elongation indices and in the relative sizes of extended lamellar domains were produced not only by the "standard" concentrations of colcemid (about 500 ng/ml), used for depolymerizing most microtubules, but also by much lower concentrations (15-20 ng/ml). Immunofluorescence examination after staining with anti-tubulin antibodies showed that cells exposed to this low concentration of colcemid retained an almost normal- looking microtubule system (Fig. 6). Thus, the control of polarization is a function of a microtubular system which is highly sensitive even to minor alterations in the normal organization of this system.

How microtubules control polarization. Possible role of microtubule-mediated organelle transport

Morphological observations show that the polarized extension of pseudopods is accompanied by an ordered redistribution of organelles. For instance, video microscopy of the leading lamella of human fibroblasts migrating into a wound reveals numerous worm-like mitochondria oriented in parallel to the direction of cell movement. Often these mitochondria move individually toward the active edge and, more specifically, toward the zones of maximal pseudopodial activity at that edge (Fig. 7). The polarized extension of growth cones was found to include the so-called "engorgement" stage, that is, the periodic displacement of groups of vesicular organelles into the recently attached distal part of the growth cone (Burmeister *et al*, 1991).

As already described, exposure of fibroblasts or epithelial cells to PMA induces formation of long, polarized extensions. Enhanced polarization of cell shape can also be induced by hyperactivation of the expression of *N-ras* protein in cells transfected with a construct carrying a normal *N-ras* gene under the control of a dexamethasone responsive promoter (McKay *et al*, 1986). We (A.Y.Alexandrova, V.B.Dugina, H.Paterson, A.D.Bershadsky and J.M.Vasiliev, in press) have found that in both these systems the increase of polarization of cell shape was correlated with considerable increases in the motility of intracellular vesicular organelles. The average velocity of organelles increased

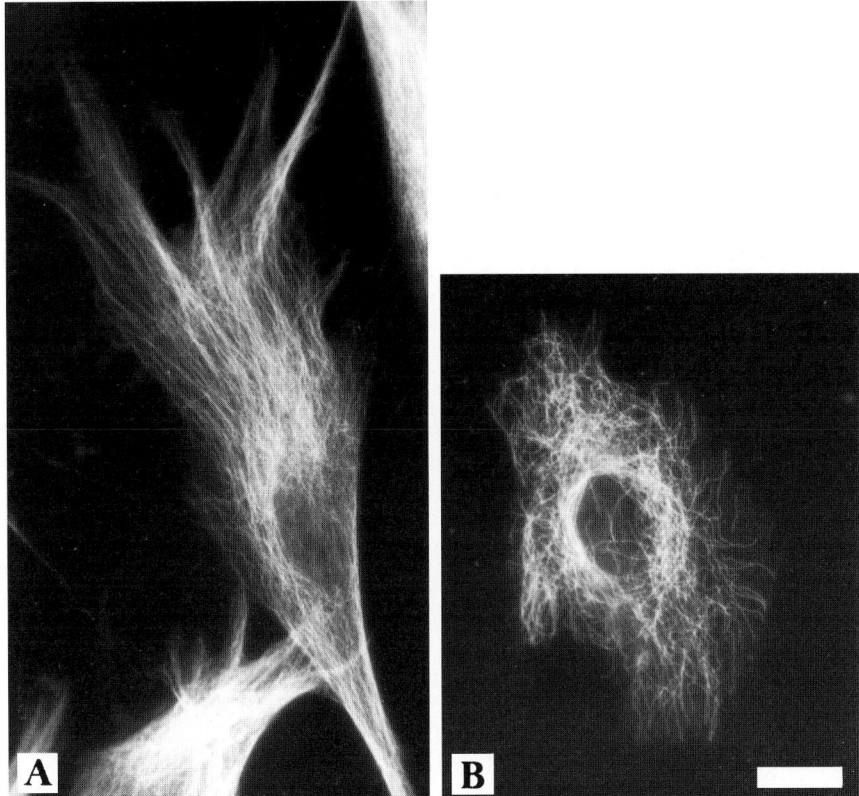

Fig. 6. Subtle changes in the microtubule system lead to a significant alteration of cell shape. Tubulin antibody staining of control (A) and 25 ng/ml colcemid-treated (B) human fibroblasts. This dose of colcemid only slightly diminished the density of the microtubule array and the mean length of the microtubules. At the same time, dispersion and elongation indices fell to the levels characteristic for cells with completely depolymerized microtubules. Bar, 20 μm. (Photographs by I.S.Tint).

8-fold after activation of protein kinase C by PMA. Sphingosine, an inhibitor of protein kinase C, diminished both cell polarization and particle velocity; this agent also prevented the increase in particle motility induced by PMA. In a similar way, dexamethasone treatment leading to high levels of *N-ras* expression increased the organelle motility more than 10-fold (Fig. 8). These data, summarized in Fig. 9, give reason to suggest that microtubule-dependent transport of organelles may play an important role in the development and maintenance of polarized cell shape.

More direct proof of the central role of microtubule-dependent organelle transport in the control of polarization has been obtained recently (A.D.Bershadsky, V.I.Gelfand, F.K.Gyoeva, V.I.Rodionov and J.M.Vasiliev, unpublished) in experiments with intracellular injection of an antibody to kinesin into directionally moving polarized human fibroblasts. Kinesin (Vale *et al*, 1985) is a collective name for a group of closely related "microtubule motor" proteins (for a review, see Endow, 1991). Kinesin is a microtubule-activated ATP-ase (Kuznetsov and Gelfand, 1986) with the ability to induce

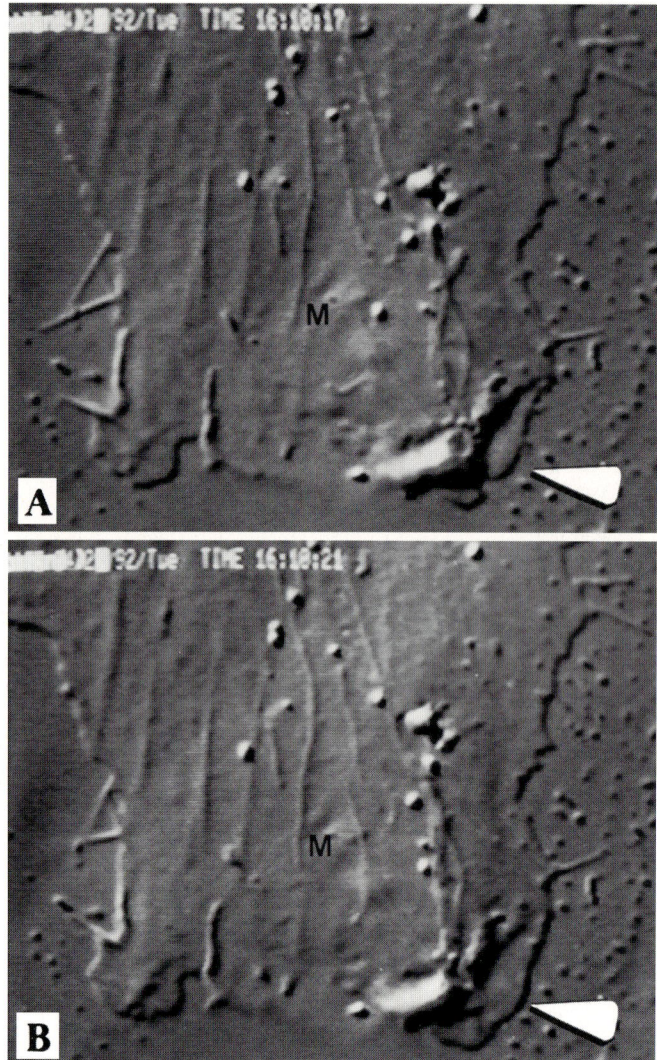

Fig. 7. Active edge of a cultured human fibroblast. A,B - sequence of two photographs from the video monitor; the interval between photographs is 4 sec. Numerous vesicles and organelles, including worm-like mitochondria (M), are seen. Notice that lamellopod protrusion (arrowhead) is developed in close proximity to a group of several mitochondria. Video enhanced DIC microscopy. The width of photographs corresponds to 30 μm. (Photograph by V.I.Rodionov).

gliding of microtubules *in vitro* (Vale *et al*, 1985). Kinesin was shown to be specifically associated with some membrane-bounded organelles including mitochondria and synaptic vesicles (Leopold *et al*, 1992). It is responsible for organelle movements along microtubules from minus to plus ends e.g. from perinuclear centers to the cell periphery.

We used rabbit antibody to kinesin heavy chain (head region); this antibody recognizes kinesin-like proteins from different types of cells. After injection into fish melanocytes

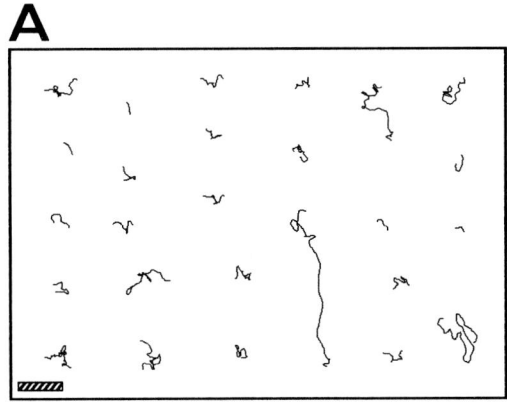

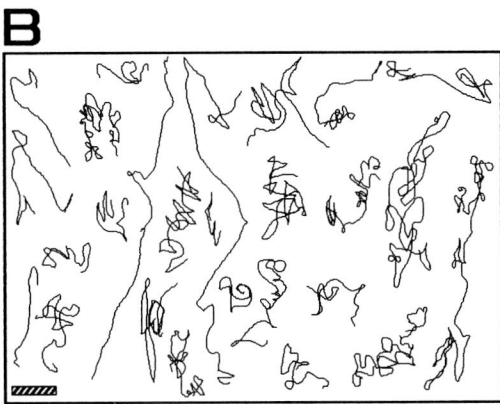

Fig. 8. Effects of increased expression of the normal *N-ras* gene on intracellular particle motility in Rat-2 fibroblasts. A Rat-2-derived fibroblast line that has been transfected with a construct (McKay *et al.* 1986) containing multiple copies of the *N-ras* proto-oncogene under the control of a dexamethasone-sensitive promoter was used. A - collection of particle traces from 3 untreated, control cells, B - traces from 3 dexamethasone-treated cells. Duration of tracing - 2 min; bars - 2 μm. Data of A.Alexandrova, V.Dugina, H.Paterson, A.Bershadsky and J.Vasiliev.

this antibody blocked anterograde pigment particle movement, while their retrograde movement was not disturbed (Rodionov *et al,* 1991). Injection of this antibody into cultured fibroblasts leads to a concentration of mitochondria in the central part of the cell and a clearing of these organelles from the cell periphery (see Fig. 10A).

3 h after injection of these antibodies, human fibroblasts developed certain changes similar to those observed in cells exposed to microtubule-depolymerizing drugs: a) these cells had a less polarized cell shape than controls (Fig. 11) as indicated by significant decreases in elongation and dispersion indices; b) the rate of extension and retraction of pseudopods at the leading edges of these cells migrating into a wound was significantly lower (decreased about twofold) as compared to control non-injected cells or cells injected with pre-immune serum (Fig. 10B). As expected, examination showed that the

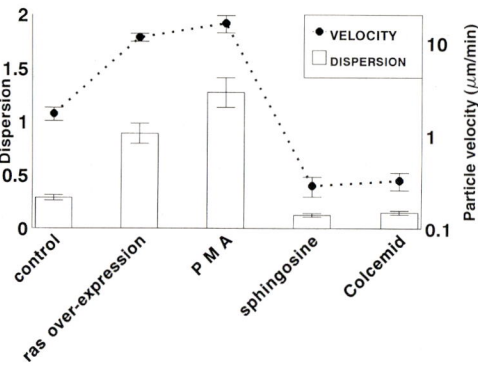

Fig. 9. Relationship between the velocity of intracellular particles and the degree of polarization of cell shape. Average values of particle velocity and average indices of shape dispersion are plotted for several types of treatment of Rat-2 fibroblasts. To estimate particle velocity, 5-15 cells were analyzed for each type of treatment. To calculate dispersion indices, 50 cell outlines were measured in each case.

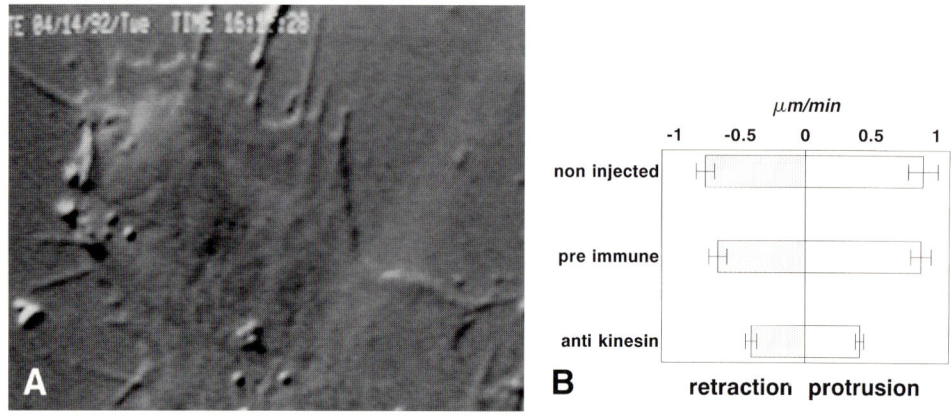

Fig. 10. Modulation of lamellopodial activity of a cultured human fibroblast after microinjection of antibody to kinesin. A - Photograph of the leading edge of a kinesin antibody-injected cell. Notice the complete absence of mitochondria. Video enhanced DIC optics; width of the photograph corresponds to 25 μm. (Photograph by V.I.Rodionov). B Average values of the rates of protrusion and retraction for non-injected, pre-immune antibody-injected and kinesin antibody-injected human fibroblasts. 10 cells of each type were analysed. Data of A.D.Bershadsky, V.I.Gelfand, F.K Gyoeva, V.I.Rodionov and J.M.Vasiliev.

cells injected with anti-kinesin antibody retained a morphologically intact microtubule system. Recently, Ferreira *et al*, (1992) have shown that suppression of kinesin expression in cultured hippocampal neurons using antisense oligonucleotides, decreased the outgrowth of dendrites and axons about two-fold but did not stop this growth completely. These data are not sufficient to prove that all microtubule-dependent regulation of polarization is mediated only by kinesin but they show that kinesin-mediated transport of organelles is at least an important component of this regulation.

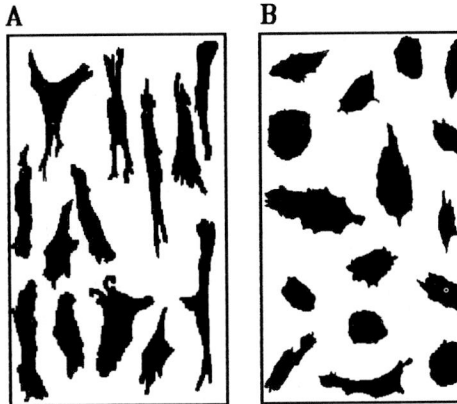

Fig. 11. Outlines of control (A) and kinesin antibody-injected (B) cultured human fibroblasts. Notice the decreased polarization of cells injected with kinesin antibody. Average indices of dispersion and elongation (calculated for two hundred outlines of each type) were significantly lower for the population of kinesin antibody-injected cells. Data of A.D.Bershadsky, V.I.Gelfand, F.K Gyoeva, V.I.Rodionov and J.M.Vasiliev.

What is the specific role of organelle redistribution in the regulation of polarization? Developing the ideas of Abercrombie (Abercrombie, 1970) Singer and collaborators (Rogalski *et al*, 1984; Kupfer *et al*, 1987) suggested that insertion of membrane material into the leading edge may play a critical role in directional motility and that the microtubule system may function as a steering mechanism, directing vesicles from the Golgi apparatus for fusion with the plasma membrane at the leading edge. At present this suggestion remains valid although the existence and role of the directional "membrane flow" is doubtful (see discussion in Heath and Holifield, 1991). Vesicles moving along microtubules can also transfer some proteins that are necessary for the assembly of pseudopods, as well as certain regulatory proteins. Hollenbeck and Bray (1987) described fast moving "parcels" containing cytoskeletal proteins in the growing processes of nerve cells. Fereira *et al* (1992) have found that inhibition of kinesin expression leads to a block in the outward transport of protein GAP-43 and synapsin I. Possibly, the role of the outward movement of mitochondria may be also important. Mitochondria may release ATP, and release or remove ions and other regulatory factors controlling polymerization of actin in certain domains of the cortex and its depolymerization in other domains.

The possible role of intermediate filaments in microtubule-dependent polarization deserves special discussion. It is well known that distribution of vimentin-containing intermediate filaments is determined by the microtubule system (for a review, see Traub, 1985). In particular, depolymerization of microtubules is known to induce perinuclear collapse of the intermediate filament system (Traub, 1985). Kinesin may play an important role in the regulation of intermediate filament distribution by microtubules; intracellular injection of the kinesin antibody was found to produce the collapse of intermediate filaments (Gyoeva and Gelfand, 1991). There is growing evidence that the intermediate filament system is also structurally and functionally connected with the actin

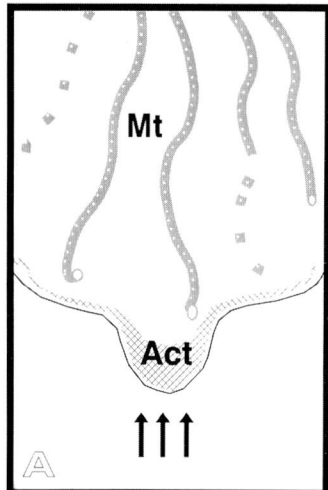
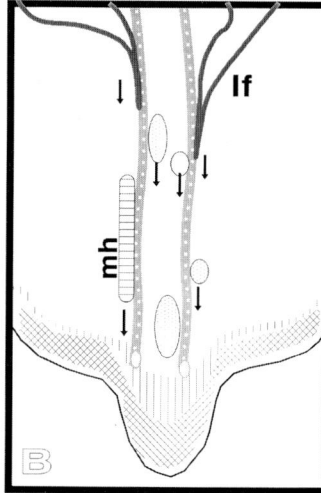

Fig. 12. A scheme summarizing the proposed model of microtubule- dependent polarization of pseudopodia. Three arrows in the drawing "A" symbolize an external signal; Mt - microtubules, "Act" - region where formation of lamellopodia and assembly of the actin network occurs (cross-hatched area). Drawing "B": If - intermediate filaments; circles and ellipses with small arrows - directionally moving organelles and vesicles; mh - mitochondrion.

cortex (Bershadsky *et al*, 1987; Hollenbeck *et al*, 1989; Tint *et al*, 1991b; for a recent review, see Skalli and Goldman, 1991). It is possible that intermediate filaments may act as mediators, functionally connecting the microtubular system with the actin cortex. Of particular interest is a recent finding that protein kinase C may be associated with intermediate filaments (Spudich *et al*, 1992). This association may be related to data showing that PMA-induced extensions of fibroblasts formed in the presence or in the absence of colcemid always contain intermediate filaments. Of course, when extensions are formed in the presence of colcemid, intermediate filaments are not accompanied by microtubules (Bershadsky *et al*, 1990). Possibly, activation of protein kinase C associated with intermediate filaments plays some important role in the formation and stabilization of these extensions. At present, however, we have no definitive evidence for or against the essential role of intermediate filaments in microtubule-dependent or microtubule-independent polarization.

Still another unsolved question is the significance of post-translational modification of tubulin in microtubules. Specific antibodies reveal microtubules with such modifications of tubulin as detyrosination or acetylation. By themselves, these types of modification do not change any known property of microtubules. However, they are usually regarded as markers correlated with some unknown alterations increasing the stability of microtubules. Microtubule systems of several types of polarized cells were found to develop certain zones with microtubules enriched in detyrosinated and/or acetylated tubulin. These zones were developed in fibroblasts when they started to migrate into a wound (Gundersen and Bulinski, 1988), in transfected NIH-3T3 cells after morphological transformation induced by overexpression of *N-ras* (Prescott *et al*, 1989), in the shafts of neurites formed by neural cells (Robson and Burgoyne, 1989; Lim *et al*, 1989) and in the stalks of extensions formed by PMA-treated fibroblasts (Tint *et al*, 1991a). We do not know what the mechanisms of tubulin modification are which accompany increased polarization and what the functional role is of increased stability of microtubules. Possibly, stable microtubules are needed to provide more permanent highways for ordered and rapid translocation of organelles over long distances (see discussion in Gelfand and Bershadsky, 1991).

Conclusions: a model for cytoskeletal reorganization responsible for the polarization of pseudopodia

The facts and considerations presented in this paper give reason to propose a hypothetical model of cytoskeletal mechanisms which stabilize the signal-induced polarization of pseudopodia. In short, we suggest that polarization is stabilized first by reorganization of the actin cortex induced by the extension of pseudopods; in the next stage polarization is enhanced and further stabilized by the microtubule-dependent redistribution of organelles. Let us discuss in more detail the suggested stages of polarization using the cultured fibroblast as a prototype cell.

1) The extracelluar environment induces initial differences in the distribution of pseudopods

External signals acting locally at cell surface receptors and activating signal-processing systems induce the extension of pseudopods. These pseudopods can then form attachments to the substratum. Due to random variations or unequal intensity of signals acting on various zones of the surface and also due to intrinsic randomness of the process of pseudopod assembly itself, the rate and numbers of extensions formed in these zones may be different. Local negative signals, e.g. contact paralysis, may also produce differences. Stability of pseudopod attachments may also be unequal in various zones due to gradients of adhesive proteins on the substratum.

2) Tensions exerted by competing pseudopods lead to reorganization of the cortex

Competition between initial pseudopods promotes further development of polarization. Successful pseudopods, that (for example) have established relatively stable contacts with the substratum, survive, while unsuccessful ones disassemble and retract and their material is adopted by survivors. In the course of this competition, development of tensions that lead to a global reorganization of the actin cytoskeleton plays an important role. When microfilament networks within pseudopods develop contractility, they begin to exert tension on the central cortex, stretching this cortex and aligning microfilaments. The stretched zones between the attached pseudopods become stable: pseudopods cannot be extended perpendicular to the direction of microfilament alignment. It is also possible that when extended pseudopods are much more numerous at one pole of the cell, their tension detaches less numerous pseudopods at the other pole. The "loser" pole then contracts and stops extending pseudopods. Thus, tension-induced reorganization of the actin cortex leads to microtubule-independent stabilization of pseudopodia.

3) Microtubule-dependent redistribution of organelles further stabilizes polarization

The ends of radiating microtubules grow into all parts of the peripheral cytoplasm, including all the extended pseudopods. Naturally, they become more numerous in the "successfully" extended and attached zones, than in the "losing" zones of peripheral cytoplasm. We do not know yet whether any special features of successful pseudopods exist which somehow stabilize the microtubules growing into these pseudopods. One can suggest, for example, that contact of microtubule ends with cytoskeletal specializations corresponding to initial cell-substrate contacts (Rinnerthaler *et al,* 1988) may promote microtubule stabilization.

Microtubules, after extension into a process, start to redistribute organelles between the center and periphery (Fig. 12). These organelles regulate and enhance the global flow of cortex material from the center to the active edges. As we discussed previously, the microtubule-based transport of material for pseudopod formation, certain regulatory substances and energy sources, can be crucial for the promotion of pseudopod formation and their further extension. For instance, organelles may be redistributed in such a way that disassembly of microfilaments is promoted in the central contractile zones, while polymerization of actin is stimulated at the active edge. External signals activating signal-processing systems induce the extension of new pseudopods at the active edge and stimulate organelle motility, thereby continually reinforcing polarization. In particular, these reinforcing signals may induce and enhance differences between pseudopodial activities at the edges of two peripheral extensions, e.g. between the leading edge and the tail edge of an elongated migrating fibroblast.

4) Additional reorganization of the microtubule system reinforces polarization

When polarization is developed, the microtubule system may undergo certain alterations better adapting it to maintain the polarized state of the cortex. One of these changes is reorientation of the microtubule-organizing center to a position near the side of the nucleus facing the leading edge (for a review, see Singer and Kupfer, 1986). Possibly, this displacement is due to the tension exerted by actin structures on the microtubules linked to them directly or indirectly. Another variant of reorganization involves chemical modifications leading to increased stability of certain groups of microtubules (see above). Intermediate filaments associated with microtubules may also become involved in the control of the organization of the cortex.

To summarize, our model postulates the existence of a system of amplifying cytoskeletal mechanisms transforming subtle differences in the intensity of signals at various poles of the cell into more or less stable polarization of cell structures. These mechanisms act at post-translational levels and, of course, their action may be followed or accompanied by re-programming of the synthesis of cytoskeletal and other proteins leading to further alterations of cell shape. Needless to say, there should be considerable differences in the functions of these systems in the cells of various tissue types and these differences are responsible for the differences in morphology and motility between various polarized cells. Nevertheless, the general principles of cytoskeletal stabilization of polarity induced by external signals are probably similar in cells of various types.

References

Abercrombie, M. (1970). Contact inhibition in tissue culture. *In Vitro* **6**, 128-142.

Baier, H. and Bonhoeffer, F. (1992). Axon guidance by gradients of a target-derived component. *Science* **255**, 472-475.

Bar-Sagi, D. and Feramisco, J. R. (1986). Induction of membrane ruffling and fluid-phase pinocytosis in quiescent fibroblasts by ras proteins. *Science* **233**, 1061-1068.

Bershadsky, A. D., Ivanova, O. Y., Lyass, L. A., Pletyushkina, O. Y., Vasiliev, J. M. and Gelfand, I. M. (1990). Cytoskeletal reorganization responsible for the phorbol ester-induced formation of cytoplasmic processes: Possible involvement of intermediate filaments. *Proc. Natl. Acad. Sci. U.S.A.* **87**, 1884-1888.

Bershadsky, A. D., Tint, I. S., Neyfakh, A. A. and Vasiliev, J. M. (1985). Focal contacts of normal and RSV-transformed quail cells. Hypothesis of the transformation-induced deficient maturation of focal contacts. *Exptl. Cell Res.* **158**, 433-444.

Bershadsky, A. D., Tint, I. S., and Svitkina, T. M. (1987). Association of intermediate filaments with vinculin containing adhesion plaques of fibroblasts. *Cell Motil. Cytoskel.* **8**, 274 -283.

Bershadsky, A. D., Vaisberg, E. A. and Vasiliev, J. M. (1991). Pseudopodial activity at the active edge of migrating fibroblast is decreased after drug-induced microtubule depolymerization. *Cell Motil. Cytoskeleton* **19**, 152-158.

Bray, D. and White, J. G. (1988). Cortical flow in animal cells. *Science* **239**, 883-888.

Brown, A. F., Dugina, V., Dunn, G. A. and Vasiliev, J. M. (1989). A quantitative analysis of alterations in the shape of cultured fibroblasts induced by tumour-promoting phorbol ester. *Cell Biol. Intern. Reports* **13**, 357-366.

Burmeister, D. W., Rivas, R. J. and Goldberg, D. J. (1991). Substrate-bound factors stimulate engorgement of growth cone lamellipodia during neurite elongation. *Cell Motil. Cytoskeleton* **19**, 255-268.

Condeelis, J. (1992). Are all pseudopods created equal? *Cell Motil. Cytoskeleton* **22**, 1-6.

Conrad, P. A., Nederlof, M. A., Herman, I. M. and Taylor, D. L. (1989). The correlated distribution of actin, myosin and microtubules at the leading edge of migrating Swiss 3T3 fibroblasts. *Cell Motil. Cytoskeleton* **14**, 527-543.

Cox, D., Condeelis, J., Wessels, D., Soll, D., Kern, H. and Knecht, D. (1992). Target disruptiom of the ABP-120 gene leads to cells with altered motility. *J. Cell Biol.* **116**, 943-955.

Cunningham, C., Gorlin, J., Kwiatkowski, D., Janmey, P. and Stossel, T. (1992). Requirement for actin binding protein for cortical stability and efficient locomotion. *Science* **255**, 325-327.

Danowsky, B. A. and Harris, A. K. (1988). Changes in fibroblast contractility, morphology and adhesion in response to a phorbol ester tumor promoter. *Exp. Cell. Res.* **177**, 47-59.

DeBrabander, M., Geuens, G., Neydens, R., Willebrords, R. and DeMey, J. (1981). Taxol induces the assembly of free microtubules in living cells and blocks the organizing capacity of the centrosomes and kinetochores. *Proc. Natl. Acad. Sci. USA* **78**, 5608-5612.

Devreotes, P. N. and Zigmond, S. H. (1988). Chemotaxis in eukaryotic cells: a focus on leukocytes and Dictyostelium. *Ann. Rev. Cell Biol* . **4**, 649-686.

Dugina, V. B., Svitkina, T. M., Vasiliev, J. M. and Gelfand, I. M. (1987). Special type of morphological reorganization induced by phorbol ester: reversible partition of cell into motile and stable domains. *Proc. Natl. Acad. Sci. USA* **84**, 4122-4125.

Dunn, G. A. (1980). Mechanisms of fibroblast locomotion. In Cell Adhesion and Motility (A. S. G. Curtis and J. D. Pitts, eds.), pp. 409-423. Cambridge: Cambridge University Press.

Dunn, G. A. and Brown, A. F. (1986). Alignment of fibroblasts on grooved surfaces described by a simple geometric transformation. *J. Cell Sci.* **83**, 313-340.

Endow, S. A. (1991): The emerging kinesin family of microtubule motor proteins. *Trends in Biochem. Sci. (TIBS)* **16**, 221-225.

Ferreira, A., Niclas, J., Vale, R. D., Banker, G. and Kosik, K. S. (1992). Suppression of kinesin expression in cultured hippocampal neurons using antisense oligonucleotides. *J. Cell Biol.* **117**, 595-606.

Fleischer, M. and Wohlfarth-Bottermann, K. E. (1975). Correlation between tension, force generation, fibrillogenesis and ultrastructure of cytoplasmic actomyosin during isometric and isotonic contractions of protoplasmic strands. *Cytobiologie* **10**, 339-365.

Fukui, Y., Lynch, T. J., Brazeska, H. and Korn, E. D. (1989). Myosin I is located at the leading edges of locomoting Dictyostelium amoebae. *Nature* **341**, 328-331.

Gail, M. H. and Boone, C. W. (1971). Effect of colcemid on fibroblast motility. *Expl. Cell Res.* **65**, 221-227.

Geiger, B., Avnur, Z., Rinnerthaler, G., Hinssen, H. and Small, V. J. (1984). Microfilament-organizing centers in areas of cell contact: Cytoskeletal interactions during cell attachment and locomotion. *J. Cell Biol.* **99**, 83s-91s.

Gelfand, V. I. and Bershadsky, A. D. (1991). Microtubule dynamics: mechanism, regulation and function. *Ann. Rev. Cell Biol.* **7**, 93-116.

Gundersen, G. G. and Bulinski, J. C. (1988). Selective stabilization of microtubules oriented towards the direction of cell migration. *Proc. Natl. Acad. Sci. USA* **85**, 5946-5950.

Gyoeva, F. K. and Gelfand, V. I. (1991). Coalignement of vimentin intermediate filaments with microtubules depends on kinesin. *Nature* **353**, 445-448.

Heath, J. P. and Holifield, B. F. (1991). Cell locomotion: new research tests old ideas on membrane and cytoskeletal flow. *Cell Motil. Cytoskeleton* **18**, 245-257.

Hollenbeck, P. J., Bershadsky, A. D., Pletjushkina, O. Y., Tint, I. S. and Vasiliev, J. M. (1989). Intermediate filament collapse is an ATP-dependent and actin-dependent process. *J. Cell Science* **92**, 621-631.

Hollenbeck, P. J. and Bray, D. (1987). Rapidly transported organelles containing membrane and cytoskeletal components: their relation to axonal growth. *J. Cell Biol.* **105**, 2827-2835.

Huang, C. K., Devanney, J. F. and Kennedy, S. P. (1988). Vimentin, a cytoskeletal substrate of protein kinase C. *Biochem. Biophys. Res. Commun.* **150**, 1006-1011.

Kucik, D. F., Kuo, S. C., Elson, E. L. and Sheetz, M. P. (1991). Preferential attachment of membrane glycoproteins to the cytoskeleton at the leading edge of lamella. *J. Cell Biol.* **114**, 1029-1036.

Kupfer, A., Kronebush, P. J., Rose, J. K. and Singer, S. J. (1987). A critical role for the polarization of membrane recycling in cell motility. *Cell Motil. Cytoskeleton* **8**, 182-189.

Kuznetsov, S. A. and Gelfand, V. I. (1986). Bovine brain kinesin is a microtubule-activated ATPase. *Proc. Nat. Acad. Sci. USA* **83**, 8530-8534.

Leopold, P. L., McDowall, A. W., Pfister, K. K., Bloom, G. S. and Brady, S. T. (1992). Association of kinesin with characterized membrane-bounded organelles. *Cell Motil. Cytoskeleton* **23**, 19-33.

Lim, S.-S., Sammak, P. J. and Borisy, G. G. (1989). Progressive and spatially differentiated stability of microtubules in developing neuronal cells. *J. Cell Biol.* **109**, 253-263.

Ludowyke, R. I., Peleg, I., Beaven, M. A. and Adelstein, R. S. (1989). Antigen-induced secretion of histamine and the phosphorylation of myosin by protein kinase C in rat basophilic leukemia cells. *J. Biol. Chem.* **259**, 8808-8814.

Lyass, L. A., Bershadsky, A. D., Vasiliev, J. M. and Gelfand, I. M. (1988). Microtubule-dependent effect of phorbol ester on the contractility of cytoskeleton of cultured fibroblast. *Proc. Natl. Acad. Sci. U.S.A.* **85**, 9538-9541.

McKay, I. A., Marshall, C. J., Cales, C., and Hall, A. (1986). Transformation and stimulation of DNA synthesis in NIH-3T3 cells are a titratable function of normal p21N-ras expression. *EMBO J.* **5**, 2617-2621.

Middleton, C. A., Brown, A. F., Brown, R. M., Karavanova, I. D., Roberts, D. J. H. and Vasiliev, J. M. (1989). The polarization of fibroblasts in early primary cultures is independent of microtubule integrity. *J. Cell Sci.* **94**, 25-32.

Okabe, S., and Hirokawa, N. (1989). Incorporation and turnover of biotin-labelled actin microinjected into fibroblastic cells: An immunoelectron microscopic study. *J. Cell Biol.* **109**, 1581 1595.

Papadopoulos, V. and Hall, P. H. (1989). Isolation and characterization of protein kinase C from Y-1 adrenal cell cytoskeleton. *J. Cell Biol.* **108**, 553-567.

Prescott, A. R., Vestberg, M., and Warn, R. M. (1989). Microtubules rich in modified alpha-tubulin characterize the tail processes of motile fibroblasts. *J. Cell Sci.* **94**, 227-236.

Rinnerthaler, G., Geiger, B. and Small, J. V. (1988). Contact formation during fibroblast locomotion: Involvement of membrane ruffles and microtubules. *J.Cell Biol.* **106**, 747-760.

Rinnerthaler, G., Herzog, M., Klappacher, M., Kunka, H. and Small, J. V. (1991). Leading edge movement and ultrastructure in mouse macrophages. *J. Struct. Biol.* **106**, 1-16.

Roberts, T. M. and King, K. L. (1991). Centripetal flow and directed reassembly of the major sperm protein (MSP) cytoskeleton in the amoeboid sperm of the nematode Ascaris suum. *Cell Motil. Cytoskeleton* **20**, 228-241.

Robson, S. and Burgoyne, R. D. (1989). Differential localization of tyrosinated, detyrosinated, and acetylated alpha tubulins in neurites and growth cones of dorsal root ganglion neurones. *Cell Motil. Cytoskeleton* **12**, 273-282.

Rodionov, V. I., Gyoeva, F. K. and Gelfand, V. I. (1991). Kinesin is responsible for centrifugal movement of pigment granules in melanophores. *Proc. Natl. Acad. Sci. USA* **88**, 4956-4960.

Rogalski, A. A., Bergmann, J. E. and Singer, S. J. (1984). Effect of microtubule assembly status on the intracellular processing and surface expression of an integral protein of the plasma membrane. *J. Cell Biol.* **99**, 1101-1109.

Schliwa, M., Nakamura, T., Porter, K. R., and Euteneuer, U. (1984). A tumor promoter induces rapid and coordinated reorganization of actin and vinculin in cultured cells. *J. Cell Biol.* **99**, 1045-1059.

Schiff, P. B. and Horwitz, S. B. (1980). Taxol stabilizes microtubules in mouse fibroblast cells. *Proc. Natl. Acad. Sci. USA* **77**, 1561-1565.

Sheetz, M. P., Wayne, D. B. and Pearlman, A. L. (1992). Extesion of filopodia by motor-dependent actin assembly. *Cell Motil. Cytoskeleton* **22**,160-169.

Singer, S. J. and Kupfer, A. (1986). The directed migration of eukaryotic cells. *Ann. Rev. Cell Biol.* **2**, 337-365.

Skalli, O. and Goldman, R. D. (1991). Recent insight into the assembly, dynamics, and function of intermediate filament networks. *Cell Motil. Cytoskeleton* **19**, 67-79.

Spudich, A., Meyer, T. and Stryer, L. (1992). Association of the beta isoform of protein kinase C with vimentin filaments. *Cell Motil. Cytoskeleton* **22**, 250-256.

Spudich, J. A. (1989). In pursuit of myosin function. *Cell Regulation* **1**, 1-11.

Stahl, B., Muller, B., von Boxberg, Y., Cox, E. C. and Bonhoeffer, F. (1990). Biochemical characterization of a putative guidance molecule in the chick visual system. *Neuron* **5**, 735-743.

Svitkina, T. M., Neyfakh, A. A. and Bershadsky, A. D. (1986). Actin cytoskeleton of spread fibroblasts appears to assemble at the cell edges. *J. Cell Sci.* **82**, 235-248.

Svitkina, T. M., Shevelev, A. A., Bershadsky, A. D. and Gelfand, V. I. (1984). Cytoskeleton of mouse embryo fibroblasts. Electron microscopy of platinum replicas. *Europ. J. Cell Biol.* **34**, 64-74.

Symons, M. H. and Mitchison, T. J. (1991). Control of actin polymerization in live and permeabilized fibroblasts. *J. Cell Biol.* **114**, 503-513.

Tint, I. S., Bershadsky, A. D., Gelfand, I. M., and Vasiliev, J. M. (1991a): Post-translational modification of microtubules is a component of synergic alterations of cytoskeleton leading to formation of cytoplasmic processes in fibroblasts. *Proc. Natl. Acad. Sci USA* **88**, 6318-6322.

Tint, I. S., Hollenbeck, P. J., Verkhovsky, A. B., Surgucheva, I. G. and Bershadsky, A. D. (1991b). Evidence that intermediate filament reorganization is induced by ATP-dependent contraction of the actomyosin cortex in permeabilized fibroblasts. *J. Cell Sci.* **98**, 375-384.

Traub, P. (1985). "Intermediate Filaments". Berlin: SpringerVerlag.

Vale, R. D., Reese, T. S. and Sheetz, M. P. (1985). Identification of a novel force generating protein, kinesin, involved in microtubule-based motility. *Cell* **42**, 39-50.

Vasiliev, J. M. (1982). Pseudopodial attachment reactions. In Cell Behaviour (R.Bellairs, A.Curtis and G.Dunn, eds.), pp. 135-158. Cambridge: Cambridge University Press.

Vasiliev, J. M. (1987). Actin cortex and microtubular system in morphogenesis: cooperation and competition. *J. Cell Sci. Suppl.* **8**, 11-18.

Vasiliev, J. M. (1991). Polarization of pseudopodial activities: cytoskeletal mechanisms. *J. Cell Sci.* **98**, 1-4.

Vasiliev, J. M., Gelfand, I. M., Domnina, L. V., Ivanova, O. Y., Komm, S. G. and Olshevskaja, L. V. (1970). Effect of Colcemid on the locomotory behavior of fibroblasts. *J. Embryol. Exp. Morphol.* **24**, 625-640.

Vasiliev, J. M., Gelfand, I. M., Domnina, L. V., Zakharova, O. S. and Lyubimov, A. V. (1975). Contact inhibition of phagocytosis in epithelial sheets: alteration of cell surface properties induced by cell-cell contacts. *Proc. Natl. Acad. Sci. USA* **72**, 719-722.

Wang, Y.-L. (1985). Exchange of actin subunits at the leading edge of living fibroblast: Possible role of treadmilling. *J. Cell Biol.* **101**, 597-602.

Wilkinson, P. C. (1990). How do leucocytes perceive chemical gradients? *FEMS Microbiol. Immunol.* **64**, 303-312.

Printed in Great Britain © The Society of Experimental Biology 1993   375

MYOSIN FUNCTION IN THE MOTILE BEHAVIOUR OF CELLS

DIETMAR J. MANSTEIN

National Institute for Medical Research, London NW7 1AA, UK

Summary

Cells undergo a wide variety of movements, such as directed locomotion, extension and retraction of cell surface projections, saltatory movement of intracellular particles and cytoplasmic streaming. These events involve changes in the organization and function of cytoskeletal structures that contain actin and myosin. *Dictyostelium* has proven to be a very useful model system for studying these events. Its actomyosin-based motility resembles that of mammalian cells and has been extensively characterized, from the standpoints both of biochemistry and cell biology. Furthermore, the *Dictyostelium* cytoskeleton can be specifically altered using gene-targeting and other molecular genetic approaches.

Introduction

Myosins are complex multidomain proteins that interact with actin filaments and adenosine triphosphate (ATP) to produce mechanical force and displacement. All myosins contain a globular head domain (approx. 90×10^3 M_r) that contains the binding sites for the myosin light chains and has the catalytic and actin binding properties of the myosin molecule. It has been shown that this globular head fragment, which is also referred to as Subfragment 1 or S-1, is sufficient to cause sliding movement of filaments *in vitro* (Toyoshima *et al.*, 1987; Manstein *et al.*, 1989a) and to produce a force *in vitro* that is comparable with that produced by each head of myosin in muscle during isometric contraction (Kishino and Yanagida, 1988). For historical reasons, members of the myosin superfamily have been divided into two operational categories. The first group consist of the conventional myosins, all of which show the same structural pattern. The COOH-terminal halves of two myosin heavy chains associate with each other in an extended α-helical coiled-coil while the NH$_2$-terminal half of the polypeptide chains folds to form the globular head domains, each of which binds two light chains. The second group is far more diverse. Unconventional myosins consist of the generic motor domain, that is common to all actin based motorproteins, attached to a variety of structurally and functionally distinct tail domains and are associated with one or more light chains. Members of this group include single headed as well as double headed myosins (Fig. 1). For a recent review on unconventional myosins see Cheney and Mooseker (1992).

Dictyostelium is an attractive model organism for the study of cytoskeletal proteins and cell motility. This organism resembles higher eukaryotic cells in appearance and motile behaviour and is amenable to classical and molecular genetics. The cell

Key words: myosin, cytokinesis, cell motility, *Dictyostelium*.

physiological function of conventional myosin in *Dictyostelium* has been addressed by applying molecular genetic techniques to alter the expression of the myosin heavy chain gene (*mhc*A) and the gene encoding the essential myosin light chain (EMLC). Overexpression of antisense RNA was used to reduce the expression of the myosin heavy chain (MHC) to less than 1% (Knecht and Loomis, 1987) and the EMLC to less than 0.5% of wild-type levels (Pollenz *et al.*, 1992). The high frequency of homologous recombination events in *Dictyostelium* was exploited in order to generate cell lines that produce a truncated heavy meromyosin fragment (HMM) and less than 0.1% of wild-type levels of intact myosin (De Lozanne and Spudich, 1987). This approach was later extended to eliminate the entire *mhc*A gene, generating cells that are completely devoid of conventional myosin (Manstein *et al.*, 1989b). MHC null cells as well as the other myosin-defective cell lines display the same range of pleiotropic effects on many cellular functions. The fact that three different molecular genetic approaches led to the same phenotype strongly suggested that the observed effects are due solely to the lack of conventional myosin and not to secondary mutations. Definitive genetic proof of this point was achieved by the reintroduction of the cloned gene into the MHC null cells (Egelhoff et al., 1990).

The purpose of this article is to review the work on the cell physiological function of conventional myosin. Special emphasis will be given to recent work using molecular genetics to dissect myosin function in *Dictyostelium discoideum*.

Cleavage furrow formation

The importance of myosin in the division of mitotic cells by cytokinesis has been established by many studies. Microinjection of inhibitory antibodies into starfish blastomeres (Mabuchi and Okuno, 1977) and *Drosophila* embryos (Kiehart et al., 1990) was shown to disrupt cytokinesis. Immunofluorescence studies using *Dictyostelium* (Yumura *et al.*, 1984) and sea urchin blastomeres (Schroeder, 1987) show that myosin concentrates and disperses in the contractile furrow in coincidence with the beginning and end of the furrowing that leads to cell division. In *Dictyostelium*, the myosin that is concentrated in the cortical ectoplasm during interphase becomes dispersed throughout the cytoplasm, in the form of myosin filaments, at late prophase. Most myosin remains localized in the endoplasm during metaphase and anaphase. At the anaphase-telophase transition, myosin accumulates again in the cortical ectoplasm from where it relocates to the equator. Finally, at late telophase, myosin almost exclusively co-localizes with the constricting cleavage furrow (Kitanishi-Yumura and Fukui, 1989). These findings are in good agreement with the contractile ring model which assumes that bipolar myosin filaments pull in both directions on actin filaments that are attached with their barbed ends to the plasma membrane.

Genetic proof that myosin is indeed required for normal cytokinesis has been established by depletion of the MHC from *Dictyostelium* (De Lozanne and Spudich, 1987; Knecht and Loomis, 1987). Phenotypically, the lack of myosin manifests itself in an inability of the cells to grow in suspension culture. Myosin depleted cells become large and multi-nucleated and massive lysis starts to occur after 4 to 5 days. However, when

Actin Based Motors

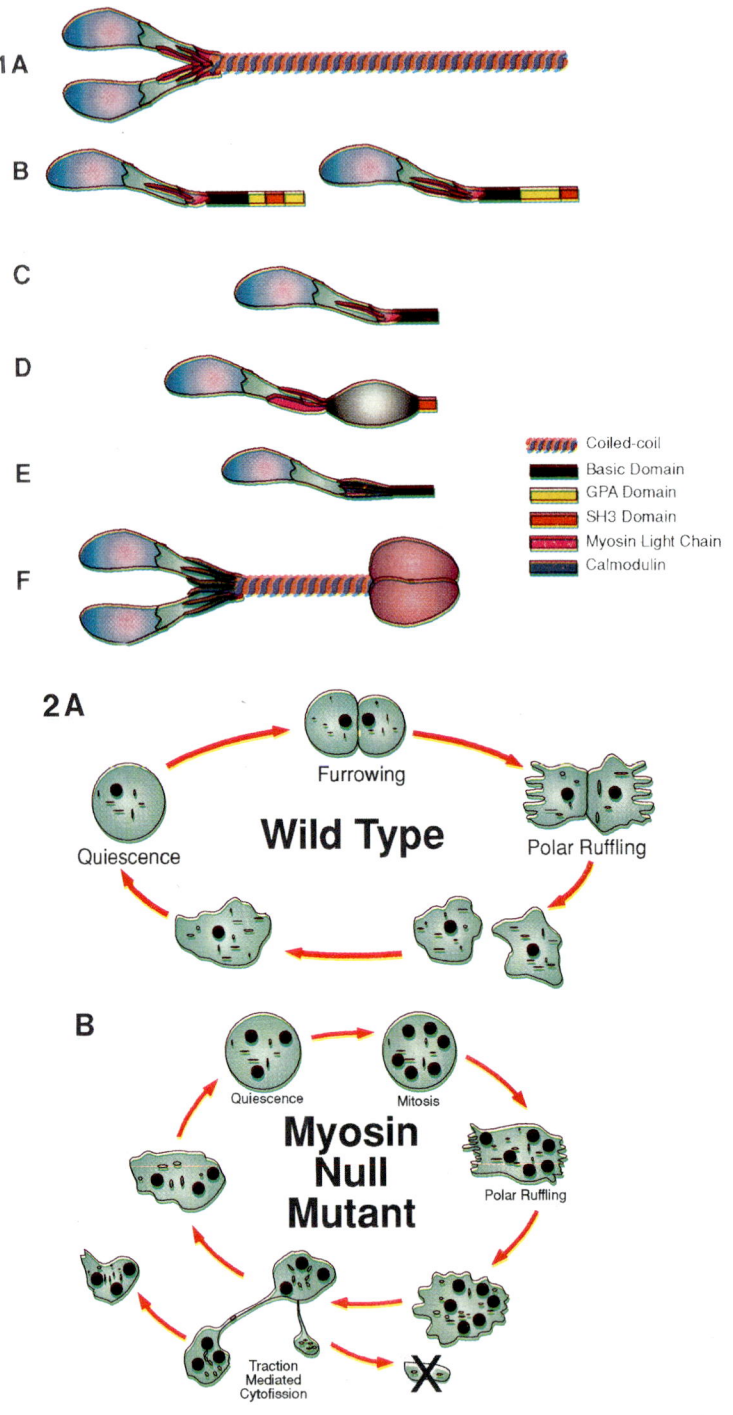

Legend:
- Coiled-coil
- Basic Domain
- GPA Domain
- SH3 Domain
- Myosin Light Chain
- Calmodulin

1A
B
C
D
E
F

2A

Furrowing

Quiescence

Wild Type

Polar Ruffling

B

Quiescence

Mitosis

**Myosin
Null
Mutant**

Polar Ruffling

Traction
Mediated
Cytofission

X

Fig. 1. Schematic representation of actin based motor proteins: (A) Conventional myosin has two globular motor domains and a tail that forms the backbone of bipolar filaments. It consists of two identical heavy chains (185 to 230 kD) and two pairs of light chains (14 to 23 kD). Conventional myosins are abundant in muscle tissue but can be found in most eukaryotic cells. (B) Single-headed protozoan myosins containing a basic domain of ~220 amino acids that confers membrane-binding activity and a ATP-independent actin binding site. This second actin binding site is formed by a ~50 amino acids *src* homology domain (SH-3) that is either situated at the tip of the tail, next to a GPA domain, or within a GPA domain. GPA domains are characterized by their unusual concentration of glycine, proline, and alanine residues. *Dictyostelium* myosins IB, ID and *Acanthamoeba* myosins IA, IB, IC are examples of this class of actin based motors. Acanthamoeba myosins IA, IB, IC are associated with one or two specific light chains. (C) Single-headed protozoan myosins that contain a membrane binding site but lack a second actin-binding site. Examples of this group are *Dictyostelium* IA and IE. (D) The *Acanthamoeba* 177 kD myosin has a unique 800 amino acid tail domain except for the SH-3 domain at the tip of the tail. (E) Brush border myosin has similar to the *Dictyostelium* myosins IA, IE a short tail that contains a membrane-binding site. Brush border myosin is associated with 3 to 4 calmodulin molecules. Calmodulin binds also to at least one of the single-headed *Dictyostelium* myosins (Zhu and Clarke, 1992) and the *Dilute* class of actin based motors. (F) The *Dilute* myosins are dimers with two globular motor domains a central rod-like segment and two terminal globular domains. Members of this class are the mouse *Dilute* protein, p190, a protein isolated from chicken brain, and the yeast proteins *MYO*2 and *MYO*4.

Fig. 2. Schematic comparison of the cell cycle of *Dictyostelium* wild-type and myosin null cells. In contrast to the mitotic cell division normally observed with wild-type cells (A), *Dictyostelium* null mutants undergo amitotic cell divisions by "traction mediated cytoplasmic fission" (B).

returned to a surface these cells can fragment into daughter cells by literally crawling away from each other. This amitotic cell division has been termed 'traction mediated cytoplasmic fission' and allows MHC– cells to grow with doubling times of ~12 hours, only slightly slower than the parental cell line (Spudich, 1989). The efficiency of this process is also reflected in the finding that the majority of cells in mid-log phase cultures are small and hardly distinguishable from wild-type cells. Traction mediated cytofission can also occasionally be observed in surface-attached cultures of wild-type cells and may represent cell division at an earlier evolutionary level.

The creation of *Dictyostelium* MHC null mutants in which the entire coding region of the *mhc*A gene was deleted (Manstein *et al.*, 1989b) makes it possible to map the functional domains of the myosin molecule by introducing altered *mhc*A genes into these cells. Egelhoff *et al.* (1991) have shown that complementation of the null mutants with a MHC that lacks the 34 kD carboxy-terminal portion of the molecule partially restores cellular functions like the ability for growth in suspension. However, the removal of phosphorylation sites needed for the proper regulation of myosin filament disassembly, that are contained within the 34 kD fragment, leads to excessive localisation of the truncated myosin to the cell cortex and other cytoskeletal abnormalities. Further extensions of this approach will undoubtedly help to elucidate other structural features of the MHC that are involved in the regulation of contractile activity, correct localisation, assembly and disassembly during cell division.

Intracellular particle movement

In *Dictyostelium* rapid saltatory movements of small intracellular particles and endocytotic vesicles can be readily observed by phase-contrast microscopy. In cells that are actively migrating most of this movement is directed anteriorly, towards the site of cell expansion. Particle movement is normally restricted to the endoplasm and excluded from the ectoplasmic layer and pseudopods. However, in MHC null cells intracellular particles as well as microtubules are found to invade these actin-rich regions. Wessels and Soll (1990) used video enhanced DIC optics in combination with computer-assisted motion analysis systems (Soll, 1988; Wessels *et al.*, 1989) to compare intracellular particle movement in MHC minus cells and wild-type strain AX4. Their results show that while the proportion of intracellular vesicles and mitochondria is similar in AX4 and MHC null cells, the null cells contain far more immobile particles and particle movement lacks directionality. Although in null cells far fewer particles move at rates >1 μm/sec, the maximum velocity of 3 μm/sec observed for AX4 cells exceeds the maximum speed determined for particle movement in MHC null cells only by a factor of two. Another important difference between AX4 and MHC null cells is displayed when cells are pulsed with 1 μM cAMP. While cAMP has no effect on intracellular particle behaviour in MHC null cells, AX4 cells pulsed with cAMP display the same lack of directionality in particle movement and depressed level of particle velocity that is normally observed in null cells.

These results clearly demonstrate that conventional myosin affects the behaviour of intracellular particles. However, it remains open whether myosin directly interacts with particles or influences their motility by sustaining the general integrity of the cytoskeleton. The observations made after pulsing cells with cAMP suggest a more indirect role of myosin and point at the importance of the molecule in the organisation of the cortical cytoskeleton and in maintaining cell polarity.

Cell motility and chemotaxis

Immunofluorescence studies of *Dictyostelium* amoeba and poly-morphonuclear leukocytes show how myosin may mediate cell polarity in locomoting cells. These cells display a characteristic asymmetric distribution of contractile proteins. Actin and the majority of actin-binding proteins are found throughout the cortex. However, most of the actin is concentrated in the advancing pseudopod, whereas conventional myosin is localized specifically at the posterior end of the cell (Valerius et al., 1981; Stossel et al., 1985; Yumura and Fukui, 1985). For *Dictyostelium*, the correlation between myosin distribution and cell polarity can be further tested by comparing the effects of chemoattractants on wild-type and MHC null cells. Chemotactic cells develop polarity even in the absence of a chemoattractant gradient, however, attractants like cAMP can modulate this inherent polarity. Three different sets of conditions have been used to study the effect of cAMP on *Dictyostelium*. Alternatively, cells were exposed to a non directional pulse of attractant, a stable gradient of attractant, or cAMP was applied locally to individual cells from a micropipette.

It is one particular advantage of *Dictyostelium* as a model system that the response of cells to non-directional pulses of cAMP is so synchronous that it can be followed by biochemical means in large populations of cells (Berlot *et al.*, 1985). Wild-type cells exposed to uniform solutions of cAMP respond by a large recruitment of actin to the insoluble cytoskeleton within 5 seconds of stimulation, concomitant with rounding up of the cell and the formation of surface blebs. Following this initial 'cringe' reaction, cells proceed to extend pseudopodia and ruffles in all directions, indicating a complete loss of cell polarity (Fukui *et al.*, 1990). When null cells are stimulated in the same way, they form surface blebs and become irregular in shape but, unlike wild-type cells, do not round up, suggesting that myosin is required for an active rounding up response. Additional evidence that conventional myosin contributes as force generator in the rounding up response comes from experiments where *Dictyostelium* cells were depleted of ATP by exposing them to millimolar concentrations of sodium azide (Pasternak *et al.*, 1989). In MHC null cells azide has no effect on the morphology and causes only a small increase in stiffness, while in wild-type cells depletion of ATP causes a rigor-like contraction of the cytoskeleton. Cells that contain conventional myosin retract all surface projections, become spherical, and their stiffness increases five-fold. However, it is important to point out that MHC null cells can round up. Spontaneous rounding up of MHC null cells can frequently be observed with cells grown in suspension and on plastic surfaces and is most likely a passive response to the cytoskeletal changes associated with mitotic events (see Fig. 2).

It is well known that *Dictyostelium* amoeba move by pseudopod extension. Directional stimulation of *Dictyostelium* with the chemo-attractant cAMP results in the almost instantaneous formation of one, rarely more, pseudopods and the progressive movement of the cell content into the enlarging projection. As a consequence the whole cell moves forward in the direction of the source of cAMP. In a stable gradient of cAMP, null cells move with just 20% the efficiency of wild-type towards the source of chemoattractant (Wessels *et al.*, 1988). Furthermore, the initial area of new pseudopods and the maximum area of pseudopods are far smaller than those observed for wild-type cells. This finding could be another manifestation of the lack of directed vesicle movement in the null cells, as vesicle movement towards the anterior of the translocating cell may be involved in membrane insertion for the expanding pseudopod.

Local stimulation of null cells with a micropipette extruding cAMP elicits the fast formation of a pseudopod from the near site of the cell, similar to the initial response observed with wild-type cells. However, the movement of cell content into the newly formed pseudopod is slower and the rate of translocation towards the source of cAMP is only about one third the rate of wild-type cells (Gerisch and Manstein, unpublished observations). This observation confirms that the initial rapid protrusion process does not require myosin and suggests that it is caused by directed actin polymerisation, similar to the formation of an acrosomal process by sea cucumber sperm cells (Tilney and Inoue, 1985).

Conclusion

Conventional myosin has been shown to play a critical role in a number of motility processes. In *Dictyostelium*, myosin is required for the capping of cell surface-receptors

(Pasternak *et al.*, 1989), morphogenetic changes associated with development (Manstein *et al.*, 1989b), the furrowing event of cytokinesis, and the generation of cell polarization. However, even the MHC null cells display many forms of movement, including cell migration, and intracellular vesicle movement, albeit at depressed rates, formation of cell surface extensions, and karyokinesis. These observations are generally in good agreement with studies carried out in other biological systems ranging from yeast to leukocytes and as such endorse the value of a model system like *Dictyostelium* that can be simultaneously studied using biochemical, cell biological and molecular genetic approaches.

References

Berlot, C.H., Spudich, J.A., and Devreotes, P.N. (1985). Chemoattractant-elicited increases in myosin phosphorylation in *Dictyostelium*. *Cell* **43**, 307-314.

Cheney, R.E. and Mooseker, M.S. (1992). Unconventional myosins. *Current Opinion in Cell Biology* **4**, 27-35.

De Lozanne, A. and Spudich, J.A. (1987). Disruption of the *Dictyostelium* myosin heavy chain gene by homologous recombination. *Science* **236**,1086-1091.

Egelhoff, T.T., Manstein, D.J., and Spudich J.A. (1990). Complementation of myosin null mutants in *Dictyostelium discoideum* by direct functional Selection." *Dev. Biol.* **137**, 359-367.

Egelhoff, T.T., Brown, S.S., and Spudich, J.A. (1991). Spatial and temporal control of nonmuscle myosin localization: Identification of a domain that is necessary for myosin filament disassembly *in vivo*. *J. Cell Biol.* **112**, 677-688.

Kiehart, D.P., Ketchum, A., Young, P., Lutz, D., Alfenito, M.R., Chang, X.-J., Awobuluyi, M., Pesacreta, T.C., Inoue, S., Stewart, C.T., and Chen, T.-L. (1990). Contractile proteins in *Drosophila* development. *Ann. N. Y. Aca. Sci.* **582**, 233-251.

Kishino, A. and Yanagida, T. (1988). Force measurements by micromanipulation of a single actin filament by glass needles. *Nature* **334**, 74-76.

Kitanishi-Yumura, T. and Fukui, Y. (1989). Actomyosin organization during cytokinesis: Reversible translocation and differential redistribution in *Dictyostelium*. *Cell Motil. Cytoskel.* **12**, 78-89.

Knecht D. and Loomis W. (1987). Antisense RNA inactivation of myosin heavy chain gene expression in *Dictyostelium discoideum*. *Science* **236**, 1081-1086.

Mabuchi, I. and Okuno, M. (1977). The effect of myosin antibody on the division of starfish blastomeres. *J. Cell Biol.* **74**, 251-263.

Manstein, D.J., Ruppel, K.M. and Spudich, J.A. (1989a). Expression and characterization of a functional myosin head fragment in *Dictyostelium discoideum*. *Science* **246**, 656-658.

Manstein, D.J., Titus, M.A., De Lozanne, A., and Spudich, J.A. (1989b). Gene replacement in *Dictyostelium*: generation of myosin null mutants. *EMBO J.* **8**, 923-932.

Pasternak, C., Spudich, J.A., and Elson, E.L. (1989). Capping of surface receptors and concomitant cortical tension are generated by conventional myosin. *Nature* **341**, 549-551.

Pollenz, R.S., Chen, T.-L. L., Trivinos-Lagos, L., and Chisholm, R.L. (1992). The *Dictyostelium* essential light chain is required for myosin function. *Cell* **69**, 951-962.

Schroeder, T.E. (1987). Fourth cleavage of sea urchin blastomeres: Microtubule patterns and myosin localization in equal and unequal cell division. *Dev. Biol.* **124**, 9-22.

Soll, D.R., Voss, E., Varnum-Finney, B., and Wessels, D. (1988). The "dynamic morphology system": a method for quantitating changes in shape, pseudopod formation and motion in normal and mutant amebae of *Dictyostelium discoideum*. *J. Cell. Biochem.* **37**, 177-192.

Spudich, J.A. (1989). In pursuit of myosin function. *Cell Regulation* **1**, 1-11.

Stossel, T.P., Chaponnier, C., Ezzell, R.M., Hartwig, J.H., and Janmey, P.A. (1985). Nonmuscle actin-binding proteins. *Ann. Rev. Cell Biol.* **1**, 353-402.

Tilney, L.G. and Inoue, S. (1985). Acrosomal reaction of the Thyone sperm. III. The relationship between actin assembly and water influx during the extension of the acrsomal process. *J. Cell Biol.* **97**, 416-424.

Toyoshima, Y.Y., Kron, S.J., McNally, E.M., Niebling, K.R., Toyoshima, C. and Spudich, J.A. (1987). Myosin subfragment-1 is sufficient to move actin filaments *in vitro*. *Nature* **328**, 536-539.

Valerius, N.H., Standahl, O., Hartwig, J.H., and Stossel, T.P. (1981). Distribution of actin-binding protein and myosin in polymorphonuclear leukocytes during locomotion and phagocytosis. *Cell* **24**, 195-202.

Wessels, D., Soll, D.R., Knecht, D., Loomis, W.F., De Lozanne, A., and Spudich, J.A. (1988). Cell motility and chemotaxis in *Dictyostelium* amebae lacking myosin heavy chain. *Dev. Biol.* **128**, 164-177.

Wessels, D., Schroeder, N.A., Voss, E., Hall, A.L., Condeelis, J., and Soll, D.R. (1989). cAMP-mediated inhibition of intracellular particle movement and actin reorganization in *Dictyostelium*. *J. Cell Biol.* **109**, 2841-2851.

Wessels, D. and Soll, D.R. (1990). Myosin II heavy chain null mutant of *Dictyostelium* exhibits defective intracellular particle movement. *J. Cell Biol.* **111**, 1137-1148.

Yumura, S., Mori, H., and Fukui, Y. (1984). Localization of actin and myosin for the study of ameboid movement in *Dictyostelium* using improved immunofluorescence. *J. Cell Biol.* **99**, 894-899.

Yumura, S. and Fukui, Y. (1985). Reversible cyclic AMP-dependent change in distribution of myosin thick filaments in *Dictyostelium*. *Nature* **314**, 194-196.

Zhu, Q. and Clarke, M. (1992). Association of calmoduline and an unconventional myosin with the contractile vacuole complex of *Dictyostelium discoideum*. *J. Cell Biol.* **118**, 347-358.

INDEX OF AUTHORS

Index of Authors

INDEX OF SUBJECTS

Index of Subjects

Proud Heart, Fair Lady

Proud Heart, Fair Lady

Elayn Duffy

ROBERT HALE · LONDON

© Elayn Duffy 1998
First published in Great Britain 1998

ISBN 0 7090 6204 4

Robert Hale Limited
Clerkenwell House
Clerkenwell Green
London EC1R 0HT

2 4 6 8 10 9 7 5 3 1

Photoset in North Wales by
Derek Doyle & Associates, Mold, Flintshire.
Printed in Great Britain by
St Edmundsbury Press, Bury St Edmunds, Suffolk.
Bound by WBC Book Manufacturers Limited, Bridgend.

For my Mother without whom so much would have been impossible. And for 'the believers' who had faith in me when I had none. To my agent Janet Glass, Russell, Nat and Ali, Al and Linda, Phillipa, Beverley, Pat, Jackie, Carole, Christopher, Paul, Alex, Roisin, Craig, Julie, Adrian, Noel, Dawn and my darling Martin. Also to remember my Beloved who was my joy and whom I miss every day.

One

Arabella Hastings' voice faltered as she read from the letter held in front of her, and finally stopped. Her suddenly nerveless fingers allowed the letter to fall into her lap. She blinked several times before turning to her sister, to see her own astonishment mirrored on her face.

'But it's too incredible,' Kathryn breathed, first to gain some measure of composure. Arabella's blue eyes filled with tears.

'Bella, Bella, we'll think of something. Don't look so! And please don't turn into a watering-pot on me now! We must think, Bella!' Kate knew that her sister's admittedly lovely head could rarely concentrate for two minutes at a time! But she could not give her attention to a bout of the vapours now, not when she was trying to marshal her own thoughts.

Kathryn's mind reeled as she thought of the letter's news. Her thoughts raced from one speculation to another with little conscious control. She found herself thinking of Aunt Jane, whose reaction, and probably that of any member of polite society, would have been tears of joy; not that Aunt Jane would let herself get red eyes. Her attention to her appearance was matched only by her attention to the business of her neighbours – or her determination in the hunt to find her nieces suitable husbands. A husband was Aunt Jane's barometer of success.

'But what can we do Kate?' demanded Arabella. She tapped the letter in her lap. 'He will expect an answer and Aunt Jane. . . .' Her voice started to wobble dangerously. She took a deep breath and started again. 'It's a very good offer.'

7

'You're not thinking of accepting!' exclaimed Kathryn.

'How can I refuse such a brilliant match for an army captain?' Arabella wailed. Kathryn hurriedly thrust a handkerchief into her hand.

'Aunt Jane will be delighted that you have two suitors. It's a shame you can't work the same magic for me,' she said in an attempt at lightness. She took Arabella's hand and continued in the same vein. 'Let me see the letter; maybe we can write back and say you have smallpox or are becoming a nun!'

Kathryn's sally failed to lift Arabella's spirits. Her fingers plucked at the handkerchief, her eyes not seeing the damage she was inflicting on the delicate lace edging.

Kathryn continued to examine the letter. Her mouth tightened several times and once she exclaimed, 'What intolerable conceit!' She stabbed her finger at the offending phrase. 'Really,' she said, 'he doesn't entertain the thought that you might refuse, even for a moment. Then all this tosh about a promise between our fathers. I wonder if he even knows your Christian name, "Miss Hastings".'

'What does it matter what he calls me if. . . .' Arabella stumbled to a halt, aware of the expression on her sister's face. 'What are you planning, Kate?'

Arabella knew that look. It was the same look that had preceded her sister's short career as a chimney sweep, when she had wedged herself up the drawing-room chimney for three hours, completely ruined the rugs, and sent yet another governess looking for a new position. Arabella knew only too well the havoc Kathryn could cause in one of these moods.

Kathryn please!' She attempted to put as much disapproval into those syllables as Aunt Jane could, but even to her own ears it sounded more like a plea than a rebuke.

Kathryn was almost hugging herself with glee. She began to pace, sketching in the air with her hands as she explained her plan to Arabella.

'Well?' she questioned. 'It solves all our problems and . . .'

'It's fraud,' wailed Arabella, once again reaching for the handkerchief.

'Don't be such a goose, Bella. Look at this.' She pushed the letter under her sister's nose. 'Nowhere does it say it is with Arabella Margaret Elizabeth Hastings that he is proposing to enter into the holy state of matrimony.' Her voice had taken on a pompous cast. 'It would serve him right to be married to Miss Hastings, the younger, rather than to the beauty of the family.' She continued to examine the letter. 'He might be buying a horse, Bella!'

But Arabella had taken umbrage at Kathryn's first words. 'I am not the beauty of the family,' she protested hotly. Kathryn smiled at her sister's defence of her. But, as most of the county would testify, Kathryn could not hold a candle to her sister. But then beauty such as Arabella's was rare. Kathryn once again marvelled that her parents could be responsible for her sister's classical features, luminous skin and deep blue eyes all framed by natural guinea-gold curls. Some susceptible young men from neighbouring estates had acclaimed her as a modern day Helen.

'Now don't fly into the boughs with me!' said Kathryn quickly, and recalled to her sister's mind the current problem. 'It's not really fraud, Bella. He states quite clearly that the agreement was to marry the daughter of Edmond Hastings. This being decided after our father saved his father's life. He probably hasn't even enquired whether there is more than one Miss Hastings.' She looked at her sister's worried expression. 'Listen, Bella. This way you can still marry Tom. I will fulfil his honoured father's wishes and Aunt Jane will congratulate herself on marrying two dowerless girls in a season.'

Arabella, a slight frown creasing her brow, said, 'But he will surely have a description of me.' She gestured at Kathryn's hair. 'Even if he were buying a horse he would notice if it was the wrong colour!'

Kathryn couldn't help laughing. Arabella looked aggrieved. 'There's no need to laugh at me, it was you who mentioned horses!' Before she could continue in this fashion, Kathryn interrupted her.

'If you were already married when this arrived, it would naturally be assumed that the offer was meant for me.'

'But I am not married to Tom and this Viscount whatshisname

will be expecting to marry the daughter of Edmond Hastings, a blonde!' She sat back with the air of someone having won an argument.

'We shall have to be careful, Bella. We can't let Aunt Jane take too close a look at the letter. I shall reply immediately, stating what an honour and so on, and explain that Arabella is a family name for me and I am actually called Kathryn. I shall also tell him that the matter is in my uncle's hands. We shall explain to Aunt Jane that, acknowledging your recent engagement to Captain Tom Ripley, he has naturally addressed his offer to me.'

Kathryn's heart-shaped face was alight with mischief and her green eyes sparked, as she continued to add details to the plan. Arabella continued to bite her lips and eye her sister with a growing feeling of helplessness.

Two

Arabella would have been surprised to know that the Right Honourable the Viscount Philip Devlin was experiencing much the same feeling of impending disaster. He listened to the dry, legal voice of his father's lawyer dash any hope he might have had of extricating himself from his looming matrimonial duty.

'Damn it all, man!' He slapped his riding boot with his whip to underline his profanity. 'It's absolutely medieval to be forced into marriage with a complete stranger.'

Mr King reached for a copy of the will. 'My Lord, there is no . . .'

The viscount interrupted him. 'I know. I know.' He narrowed his sherry-coloured eyes. 'I'm thinking another adviser would sing a different song.'

Mr King straightened in his chair, his voice now holding an unmistakable note of affronted virtue. 'That, of course, is your prerogative. However, any reputable legal adviser would have to agree. There is no way of overturning your late father's will.'

'All right. All right. Don't ruffle your feathers. I don't take being backed into the corner very well, do I?' The viscount bestowed on him a charmingly rueful smile. Mr King permitted himself a small smile in return and murmured something about a galling situation. Philip Devlin strode across the room. Mr King watched his immaculate form, skin-tight, biscuit hessian moulded to muscular thighs and his jacket needed no padding around the shoulders. There was no doubt that he deserved his reputation as the most eligible bachelor in London, not only for his wealth and title but also for his

11

dashing good looks. Unfortunately, the viscount's disposition was not as sunny. Mr King thought of the long-standing feud between father and son, ended only by death. Enough to sour any man, let alone a man with the wealth and influence to indulge any or all of his desires.

'Brandy?' the viscount asked. Mr King declined with a shake of his head and a discreet cough.

'Well, there's no help for it, let me know the worst. This Miss country mouse, is she ugly? What did my august father choose for his only son?' His bantering tone did not quite hide his bitterness.

Mr King felt a great deal of sympathy for the viscount, who would have been much surprised by this attitude. But the puritan nature of an instinctive lawyer recoiled at the excesses that had marked the viscount's career so far.

The viscount regarded Mr King from under lowered brows. His volatile temperament had not reacted well to the posthumous decree of his father. The father, who, in his lifetime, had failed to influence his son in anything, ensured, in death, absolute obedience to his wishes. His bride had been chosen and there was nothing he could do to escape. In order to inherit his beloved Meadowsdene and Kingsgrey Court, his father would dictate from the grave Philip's choice of wife.

In his mind's eye he completed a portrait. Mr King's dry voice supplied the details; blonde hair, twenty years old, daughter of Edmond Hastings, dead last summer from heart trouble. Unkind voices said heart trouble from a bottle. It did not matter if she was beautiful or good-tempered, she would shortly become Viscountess Devlin. Good fortune for a girl with no dowry to speak of, marrying not only into one of the oldest houses in England, but one whose fortune had survived and, indeed, increased over the generations. A slight sneer briefly marred the viscount's handsome face; much good it will do her, he vowed silently.

Many other young ladies and their proud mamas would fruitlessly gnash their teeth at an unknown country mouse carrying off one of the most glittering matrimonial prizes of the day. The viscount's lip curled sardonically. Well, she might find her husband was regarded as a great catch in the eyes of society, but it would only be in society's eyes that she had a husband. He would give her his name and his financial support and she would enable him to claim his inheritance; a fair trade. He would live in his London house, he decided. Apartments would be made ready for her at Meadowsdene or Kingsgrey, she could choose which she preferred, he decided magnanimously. He expected that his new wife would be suitably grateful for his generosity, but beyond that she did not enter his plans.

News of his impending nuptials had indeed caused great fuss in the Hastings household, but it was not the joyful gratitude the viscount expected.

In the drawing-room, Kathryn and Arabella waited for their brother to appear. Arabella started to ruin yet another handkerchief, fretful at the delay. But when the door opened, instead of the brother they were expecting, Tom Ripley erupted into the room. Tall and handsome in his newly won captain's uniform, the decorations for his heroic part in the Battle of Waterloo still pinned to his coat, he moved directly to Arabella and possessed himself of her hand.

'Darling! I mean, Miss Hastings, is it true?' he said, his eyes searching her face.

'Tom! What are you doing here? I thought you were not returning for at least another week.'

'I came as soon as I got your sister's note.'

Arabella turned accusingly to her sister. 'Kathryn! How could you?'

Kathryn looked appealingly at Tom. 'If you will only listen to me, I can solve all our problems!' Her sweeping statement drew a frown from her sister's beau. But before he could remonstrate with her, Arabella retired behind her badly treated handkerchief wailing, 'I can't let you sacrifice yourself for me.'

Kathryn silently condemned her sister's taste in literature and laughed at her dramatic pronouncement.

'Sacrifice!' she laughed. 'I shall be a viscountess and terribly grand. You have two offers for your hand and I have none. Do you condemn me to a life as a maiden aunt, dependent on Aunt Jane for my keep?' Her eyes sparkled and the afternoon sun threw a halo of light behind her. She regarded them expectantly. Arabella seemed about to argue the number of her sister's suitors but allowed herself to be silenced by Tom.

'What do you mean, Kathryn? I cannot condone one of your wild schemes and I will not allow you to entangle Arabella.' In the heat of the moment he forgot to use her formal name. Kathryn sat down and smiled demurely at Tom.

'Do you want to marry my sister?'

'More than you can know. But what has this to do with your emergency? And what devilment are you hatching?'

'If you follow my plan, Tom, you can marry Arabella, I can escape Aunt Jane, Jess can go to university and we are all happy.'

'Explain yourself,' said Tom. 'You're not making any sense.'

'Why, Tom, can't you guess? The emergency is the odious viscount and his offer. He must get himself a Hastings for a wife in order to get the late viscount's worldly goods. But who is to say which Hastings he marries? It's not as if he had a preference! He's never even seen us!'

'But doesn't he know who he's marrying?'

'Not personally,' said Kathryn, with a patience she did not feel. 'That's the point.'

Tom frowned in perplexity. 'Even so, he will have been told that his fiancée is tall and blonde, which you are not.'

'Exactly what I was saying. It's as if we can read each other's thoughts,' said Arabella, gazing at Tom adoringly.

Kathryn sighed despairingly and raised her voice slightly to ensure their attention. If falling in love meant that you became incapable of holding a normal conversation or a thought in your head for longer than a minute, she could well do without it.

'I shall colour my hair and wear large heels and by the time he notices, it will be too late.'

Arabella opened her mouth to argue, but just then there was a commotion in the hall, and a few moments later Aunt Jane swept into the room.

Her ensemble was a masterpiece in lilac. A colour she considered eminently suitable for a matron, and one she also believed highly flattering to her fading blonde looks. She arranged herself on the settle with a great deal of ceremony. A screen was precisely placed following her direction and, not until she was entirely satisfied with its positioning, did she remove her gloves and turn her attention to her nieces. Her sharp eyes took in every detail of their attire, noting the careful repairs and ingenious tricks used to conceal the fact that their gowns were at least three years old. She inclined her head to acknowledge Tom's stammered greeting and hurried bow.

'I should not feel surprised to find you here, but I cannot say I approve of your running tame in the house.'

Her remarks were interrupted by Kathryn. 'Aunt Jane, we have good news.'

'Really my dear, you must learn to curb your tongue; it is most unattractive for a gently brought-up young lady to be overly pert. How you ever expect to catch a husband with some prospects, I declare I don't know.'

Before she could enlarge on this Kathryn announced, 'We shall soon be hearing wedding bells, Aunt.'

'If you mean that Mr Latimer has been prevailed upon to allow Arabella and her soldier to set a date for the wedding, it is hardly news my dear. I hope that you and Jessamy will finally come to me. It will be a bit of a squeeze, but as I said to Mr Latimer, you are family. It will mean being cramped and economies will have to be made to send Jessamy to university, but we shall contrive. I am sure you will be a great support to your uncle and I. Particularly with your uncle's health, and I, too, am not a well woman. Not that I parade my suffering as some are wont to do.' She sniffed disparagingly. 'Too solicitous to others and not the same for myself.'

Knowing her aunt could continue in this vein indefinitely, Kathryn took a deep breath, repressing a shudder at the picture of the future being painted, and declared, 'Maybe we should make it a double wedding.'

Aunt Jane stopped in mid-breath and went quite pink. It did not take long for her to recover her tongue.

'Kathryn, tell me instantly what scheme of yours this is! Thank goodness Mr Latimer isn't here. A shock like this could make him ill for days.'

Both her nieces hid their grins; they were quite sure that their uncle's heart condition was dependent on how great was his need for escape from his determined and highly voluable wife.

'My smelling salts, I need my smelling salts.' Arabella retrieved the bottle from her aunt's reticule. With these gripped tightly and with an expression of great trepidation, she faced her niece.

'There isn't any need for such drama, Aunt Jane,' said Kathryn. 'In fact,I believe we should celebrate.' At this she gave the bell a sharp tug, answered some moments later by the butler.

'Champagne please, Peters. The best we have.' She ignored his raised eyebrows and continued with a hint of a smile, 'And glasses for all of us.'

Kathryn faced her family and, with an air of triumph, declared, 'I am pleased to announce the engagement of Miss Arabella Hastings to Captain Thomas Ripley and also that of Miss Kathryn Hastings to the Right Honourable Viscount Philip Devlin.'

The result of her words exceeded that of her wildest expectations. Arabella closed her eyes and gripped convulsively on Tom's hand; the returning Peters nearly dropped the tray – the severe look he gave her did nothing to abate Kathryn's enjoyment of the consternation she had caused. Aunt Jane announced that she was going to faint, then immediately gave lie to this and demanded to know exactly what was going on, else she would succumb to a violent bout of hysterics.

Kathryn passed glasses of champagne to her dumbfounded audience. 'Fortification,' she explained – with a smile. She then seated herself next to Aunt Jane and began the story she had described to Arabella.

'As I'm sure you recall Father telling us as children, he saved the late Viscount Devlin from . . .'

'I never did believe that tale,' Aunt Jane declared roundly. 'Footpads indeed!'

Before she could continue, Kathryn went swiftly on. 'That's as may be, Aunt Jane. We all thought the same until this letter arrived.' She picked up the letter and began to read. '*After the*

death of the viscount's father some weeks ago, it was found that his
will contained a clause stating that, due to a great service rendered
to him by our father, a marriage between the two families was
agreed. In order for Viscount Philip Devlin to inherit his estates he
must marry the daughter of Edmond Hastings.' Here she stopped,
wondering if Aunt Jane had noticed the omissions. In fact, Aunt
Jane recovered with remarkable speed, started to swell and her
face became wreathed in smiles.

'Congratulations! My dear, dear girls. What luck! What good
fortune! At last your father has done something for his family.'
Kathryn wondered irreverently whether he had to be dead to do
it. Aunt Jane beamed at the room.

'I don't usually agree with drinking in the day, so injurious to
those like myself with a delicate constitution, but a celebration is
definitely called for.' With this she presented Tom with her now
empty glass.

Aunt Jane departed in a flurry of kisses and promises, to begin her
journey back to Mr Latimer, and her campaign to inform most of
the county as fast as possible of her nieces' coup.

As soon as she was gone, the last member of the Hastings family
poked his head around the door.

'Has the old dragon gone?' he asked. 'Am I safe?'

'Yes,' Kathryn answered. 'But you missed Aunt Jane's moment
of supreme triumph,' she added.

'What! You mean Lady Rushfort has finally declared Aunt Jane
to be the better whist player? Or maybe the Prince Regent has
invited her to the palace? Or maybe . . .'

Before he could continue his flights of fancy Kathryn laughingly
interrupted him,

'Come in and hear the news. I expect you will be delighted too!'

Jessamy Hastings entered the room, smiling broadly. At eigh-
teen, he was slight for his age and his face still showed signs of his
delicate health. He made his way towards the window-seat,
slightly dragging his left leg. No one offered to help him, as they
knew he hated being treated as a cripple; one reason for his avoiding Aunt Jane. Kathryn joined him and showed him the letter.

'Well!' he said finally. 'The old man was telling the truth. And Bella will be a viscountess if I know Aunt Jane. Not delighting me so far, Kate, I must say.'

'No,' said Kathryn. 'I am afraid that I will be the aristocrat in the family and shall lord it over you all.' Jess looked at his sister, noting her bright eyes and air of triumph.

'Come on, Hellkat, you'd better tell me what you've done.'

'The viscount needs to marry Miss Hastings in order to inherit. It makes no odds which Miss Hastings. So I will dye my hair; I've already written to the lawyer with an explanation of the change in name. Arabella marries Tom; you go to university without worrying about the cost; and I escape from Aunt Jane. We all live happily ever after!'

Jess did not return her smile, his eyes searching her face.

'Katie,' he said, using his childhood name for her. 'What do you think the viscount will do when he discovers the deception?'

Kathryn shrugged. 'Why should he care? All he wants is his inheritance – either of us will do for that.'

Jess turned to Tom. 'Couldn't you stop her?' Tom looked uncomfortable.

'I tried, but you know how she is. . . .'

'And now that Aunt Jane has got hold of the news there's no stopping it!' Jess finished.

Kathryn rounded on her brother. 'But it's a brilliant plan. It solves all our problems. Why on earth aren't you happy?'

'Katie,' said Jess, 'can't you see – you solve everyone's problems except your own. Do you think you and this viscount will live happily ever after?'

Three

Kathryn took the towel from her head with a feeling of dread. Although her mood had improved, she could not stem a growing feeling of apprehension. What met her eyes did nothing at all to alleviate this feeling. Gone were her coppery curls, instead she had a lifeless yellow mop. Some of her hair seemed to have broken off and was very thin in places. Arabella stared at the hair which remained in the towel, blinking back her tears.

'Now I've got the same colouring as you, I can wear that blue gown of yours I like so much.' Kathryn chattered on, having no awareness of what she was saying. Her eyes remained fixed on her reflection. The harsh yellow leached colour from her face and her skin looked positively pasty, far away from her usual healthy complexion. She poked her tongue out at her reflection and deliberately turned her back on the mirror. Her smile didn't waver as she faced Arabella.

'Well, twins we are not. But, my love, it will do – until after the wedding at any rate.'

Arabella gazed at her sister, chewing her lower lip. 'Kate, I am not sure that. . . .' Unable to continue, she waved her hand expressively at her sister.

'Don't lose heart, Bella,' Kathryn said bracingly. 'Everything will turn out fine, you'll see. You and Tom will live happily ever after. Just like we used to plan when we were children.' She clasped her sister's hand. 'Bella, you're half frozen, come closer to the fire. Tom will never forgive me if you catch so much as a cold.' She settled her sister in a chair close to the fire and disposed herself on a cushion at her feet.

'You're trying to bamboozle me, madam.' Her smile robbed the words of any sting. 'Soon you'll be married to Tom with babies of your own. Then you won't have such fidgets about me.'

Arabella laughed. 'Fidgets over you! Like the time you nearly broke your neck following Father in the hunt, and the time you fell into the coke and when you ate those flowers and we thought they had poisoned you. . . .'

'Honestly, Bella. . . .' Before Kathryn could continue, Arabella's tears spilled over.

'What will happen to you when we are not together? Will this viscount take care of you? What if he's unkind and if. . . .' Bella dissolved into incoherent tears.

'Bella, please don't. Everything will be fine. Why, I'll be his saviour; he must marry me to inherit his father's fortune. So I will be the gift horse, so to speak. Anyway, I can always inflict myself on my sister and brother-in-law for as long as they let me.'

'You will always have a home with us.' Arabella mopped her face vigorously. 'Tom and I can never repay the sacrifice.' She looked on the verge of tears again.

'Fiddle,' said Kathryn roundly. 'I shall have a large house or two and a wardrobe of such immense proportions that you shall never see me in the same dress twice.' Arabella managed a watery smile at her sister's banter, but before she could reply Jess entered the room.

'I don't like your hair, Katie,' he said at last. Kathryn wanted to cry. But she was determined that they should never ever suffer for her rash decision. Calling up her most brilliant smile, she ignored the growing knot of apprehension in her chest.

'Would you deny me my adventure, Jess? Sentence me to a life with Aunt Jane?' She gazed at her brother. He avoided her eyes and embarked on what was obviously a prepared speech.

'Maybe if I could find a post after university we could live together. That way you would not have to suffer the dragon for too long.' He cleared his throat nervously. 'Of course, with my leg, I don't know what work. The civil service maybe. I would do anything to spare you hardship.'

Kathryn's eyes filled with tears, but to shed them now would stretch his new manly dignity too far. Swallowing her tears, Kathryn could not resist hugging him tightly.

'Whatever happens, I know you will be there to look after me.'

Arabella joined them and for a moment they were all linked. 'After all,' said Kathryn, 'there is nothing we Hastingses cannot overcome.'

Kathryn recalled these words, and the naïve confidence with which they were said, and wished herself back as hard as she could. However, no fairy godmother appeared to grant her wish. She waited in the draughty church porch for her cue to walk down the aisle. She wondered how far she would get if she just started running. Mr Latimer, Aunt Jane's patient spouse, eyed her nervously, smiling in what he hoped was a reassuring, fatherly way every time he thought she was looking at him.

Suddenly the door opened and the organ wheezed into life. Right on cue, Aunt Jane started to sniffle into her handkerchief. But Kathryn knew that no red eyes would be allowed on today of all days. Aunt Jane had driven everyone in the neighbourhood to distraction on the matter of her gown. No one escaped the monologues on the virtue of silk over satin, or the latest fashionable bonnets from France. The result of her labours with bales and bales of emerald-striped silk reminded Kathryn strongly of a galleon under full sail.

Such details fled from Kathryn's mind as she and her uncle reached the altar rail. As her uncle stood back, she found herself for the first time face to face with her groom. His gaze was so intense that she thought for a moment that he could see into her very soul. For some reason this made her blush vividly. She was horribly aware of her red face and straw-like hair. The sugar-pink roses in her hair and gown only served to make her look more sallow and jaundiced than ever. She attempted to smile at the man she was about to promise the rest of her life to; but no answering smile lifted his brow and his expression became one of arrogant disdain as he turned towards the vicar.

This fluffy-haired octogenarian, who had married the viscount's parents, beamed sunnily on them all. The sort of vicar any young girl would imagine at her wedding, Kathryn thought wryly. The congregation settled and the service began.

Aunt Jane, whose eyes showed not a trace of red, frowned slightly at Kathryn's back. She normally showed such good taste. What had possessed the girl to dye her hair? Aunt Jane mentally sniffed; had they sought her advice she would never have chosen pink, an insipid sugary colour she had never liked. However, despite the dress, her niece would soon be the Viscountess Devlin. As she thought this, she sat a little straighter and smoothed her skirt. Well, there would be clothes enough in the future, she thought, and started to day-dream pleasantly of the conversations that could begin with the phrase, 'My niece, the Viscountess Devlin. . . .'

Kathryn's mouth became very dry and a rushing sound filled her ears. For an awful moment she thought she might faint for the first time in her life. Her voice quavered as she made her wedding vows. The strange figure at her side made his vows in a deep confident voice, and the touch of his hand was cool and steady. When the rings were exchanged, Kathryn was miserably aware of her own clammy, shaking hands. In what seemed an amazingly short time the ceremony reached its end. Kathryn's heart started to beat wildly as her husband lifted her veil for the traditional kiss. There was no warmth in his eyes as, dispassionately, he kissed her cheek.

The service had gone without hitch. Kathryn had half expected . . . what had she expected? Someone to dramatically denounce her? No fear of that now. In a shorter time than Kathryn had thought possible, she had changed from Miss Kathryn Hastings with no fortune to recommend her, to a peeress married to one of the richest men in the country, with one of the oldest titles. The change in her status was immediately apparent at the wedding breakfast. Whereas once her opinion was only of interest to herself, now she found that even Aunt Jane hung on her every comment.

The breakfast continued with no opportunity for the bride and groom to have any private conversation. Kathryn watched her husband – how odd it was to use that word for this stranger –

conducting himself with punctilious politeness. But only once did his smile reach his eyes. This was at the entrance of a straight-backed old lady in a black dress. Although she held an ebony cane, the viscount insisted that she take his arm and solicitously he guided her to a chair next to the fire. The lady, she was informed by a rotund partner of Mr King, was his nurse and first governess. The viscount treated her as an honoured family member. Kathryn wondered whether this tall, arrogant man would treat her with the same courtesy. Her eyes roamed around the room with its delicate straw-gold walls and perfectly toned carpets. The viscount's choice, wondered Kathryn? Magnificent, but tasteful. If only her hair had turned nearly as nice a colour.

Of the few guests present, Kathryn could only put a name to two of the faces, and one of them was Aunt Jane. Kathryn fully expected her aunt to swoon from pride. It was no wonder there was no sign of her uncle. She hoped her aunt wasn't being too voluble on the fact that the Hastings family had seen two weddings in recent weeks. Thinking about how different Arabella's wedding was caused a lump in Kathryn's throat. Arabella, in their mother's wedding dress, had been married in the small village church, surrounded by people she had known all her life. Aunt Jane interrupted Kathryn's thoughts, and for once she was grateful.

'Well, my dear, or should I say Your Ladyship?!' she tittered girlishly. 'That I should see the day! I always thought it would be your sister, what with her looks.' Before Aunt Jane could pursue this thought, Kathryn hurriedly distracted her.

'Don't you think it is a very small wedding for such an important man, Aunt?' Any sarcasm, was lost on Aunt Jane.

'Of course His Lordship is still officially in mourning.' Kathryn noticed Aunt Jane had no trouble using his title. 'So it would not be fitting to have a society wedding,' Aunt Jane continued wistfully. 'These people are family retainers, His Lordship's godfather is the gentleman with the stick, just leaving. The rest are witnesses needed legally for the inheritance.' She waved a hand dismissively. 'Of course, when you go to London you will move in the highest circles. With the cream of the ton. . . .' Before Aunt Jane could continue with her ecstatic monologue, Kathryn's attention was claimed by the viscount.

'Madam.' He offered her his arm. Kathryn swallowed nervously as he led her to the centre of the room. She felt the weight of many eyes upon her and suddenly her heart was pounding furiously. In order to hide this weakness, she set her mouth firmly and, with an intense concentration, gripped the glass of champagne that the viscount gave her. Such was her single-mindedness, she missed the opening remarks of a speech the viscount was giving to the wedding party. His voice sounded weedy and distant, compared to the beating of her own heart pounding in her ears. Dimly she realized a toast was being made, and tried to stretch her mouth into a smile, but the effort was too much. She swallowed her champagne, hoping that no one could see how shaky the hand was that held the glass.

To Kathryn's relief, the party seemed to be breaking up. She smiled automatically, relief lending it sincerity, at the departing guests. When the last of them had gone she sank gratefully into the nearest armchair. Her relaxed attitude did not last as she realized she was left with one stranger – her husband.

Four

The viscount found the wedding almost as much strain as his wife. He deplored the necessity of publicly fulfilling his despised father's wishes. Only his great love for his family seat of Kingsgrey, and an awareness of what was due to his ancient and honourable name, had constrained him over the last few weeks from tearing up the will and wishing it all to the devil. He did not look on his wife with any joy, but regarded her as a slightly tedious encumbrance and somewhat distastefully, as the means of his public humiliation. Very soon the whole matter could be relegated to the background of his life. He was not a cruel man and would have been astonished if anyone had accused him of such, but he did not see his bride as having any personality of her own. As far as he was concerned, the purpose of the wedding had been fulfilled and now he wished her well. She would certainly not interfere with him and, as long as she conducted herself as a lady of quality, he would not interfere with her. Her agreement to his plan was not in question; the viscount felt he was being more than reasonable and fully expected her grateful acquiescence and thanks.

Kathryn surreptitiously examined the viscount's face as he prepared to tell her of his plans for her future. His face revealed nothing of his inner thoughts. His hair was brushed in the latest Corinthian style; Kathryn had seen Jess try to emulate this with no great success. His most striking feature, Kathryn decided, was his eyes, large and the colour of conkers or sweet sherry. His gaze seemed to look right through you, often with an arrogant stare. Kathryn could quite believe his reputation as a ruthless deflator of the pretentious.

Kathryn, however, did not fail to notice the humorous quirk of his mouth or . . . her train of thought was interrupted by the viscount.

'I hope you found everything to your satisfaction, madam?' he said formally. He stood in front of the fireplace coolly gazing at her.

'Indeed. Everyone has been most kind,' Kathryn stammered. She swallowed and managed to speak in something resembling her own voice. 'I thank you for your consideration.' Then, after a pause, 'Sir.' He would not accuse her of having manners less formal than his own.

He bowed at her words and added in a dry tone, 'I am glad the efforts of my staff met with your approval.' Kathryn stiffened slightly, so he was making it clear that there had been no effort as far as he was concerned. He moved towards the door.

'If you will be so kind as to join me in the library, there are some matters I must discuss with you before my return to London.' He did not offer Kathryn his arm but opened the door and waited for her to pass through. His manners, thought Kathryn tartly, left a lot to be desired. As she passed through the door she glanced into the viscount's face; his eyes flicked past her with complete indifference as he then strode ahead of her to open the library door.

Kathryn wondered what was so important that it would not wait till morning. The strain of the day had given her a headache and her new shoes pinched cruelly at her heels. She felt sure she would get a blister. Oh! How she longed for the peace of her room. It was this thought that jerked Kathryn fully awake once more. She hoped fervently that the viscount had arranged separate suites to allow her to become accustomed to her new married status. As if he had read the perturbation in her eyes, the viscount smiled cynically.

'Allow me to allay your fears, madam; I am not in the habit of forcing my attentions on blushing maidens, unless invited.' Kathryn blushed, furiously trying not to. The viscount seemed amused by her reaction and continued, 'Nor shall I inflict my presence on you longer than is strictly necessary.' Nor mine on you, thought Kathryn.

Abruptly he turned his back on her and strode towards the desk. 'I have arranged a quarterly allowance to be paid to you, however is convenient. The household expenses are dealt with by my man, Richard Spires. I am sure you will find me generous in this area.' The cynical look had returned to his face. 'Naturally, there are both horses and carriages available to you should you wish to shop in Hadfield. You may order the household as you wish. Mrs Bentham, the housekeeper, will introduce you to the staff in the morning. I am sure you will be able to make yourself a comfortable home here.'

Kathryn watched him as he spoke. He did not pause in his speech, nor did he seem to expect any reply. Kathryn's temper started to rise as she noted the arrogant lift of his chin and the sardonic twist of the lips at the word comfortable. Doubtless he thinks I have come from a hovel and will grovel with everlasting gratitude because he throws me a few crusts, Kathryn thought.

The viscount continued, oblivious to her reaction, 'You may have family to visit, of course, but please let Richard know your plans.'

'Will I not be able to inform you myself, My Lord?'

The viscount lifted his eye glass and viewed her sardonically.

'I, madam? I will remain in town until the end of the season. What my plans will be after that I don't know. But, rest assured, I will inform you of my movements should I feel it necessary.'

Kathryn's eyes started to sparkle with indignation. The arrogance of the man. She had just drawn breath to voice her complete lack of interest in him or his plans. 'If you think that I will spend my time thinking of you . . .' she began. But he interrupted her smoothly.

'Madam, do not be alarmed. I will endeavour most successfully not to think of you, as you doubtless will not think of me.' He opened his snuff box with a practised flick and offered it to Kathryn who, still trying to think of a scathing reply to his last comment, ignored the proffered box.

'I blend my own you know. A secret mixture known only to myself. I fancy that you will not find a more subtle blend,' he said conversationally. Kathryn could not believe the effrontery of the man. Now he was striking up the merest commonplace as if she were a boring guest at her own wedding.

Kathryn ignored his comments with the contempt she felt they deserved. 'Do I understand this correctly? That you intend to leave me here while you continue to live in London? We will not have the same household?'

'Such concern of your wifely duty,' the viscount drawled and, with a pronounced sneer, added, 'I should consider myself a lucky man.' Before Kathryn could respond in kind the viscount continued in a slightly lecturing tone, 'You will find that among members of the ton, this sort of arrangement is quite unexceptional. To those of breeding, a marriage in name only is the most civilized arrangement – providing there are no problems regarding the succession. And I feel I have been more than generous in my terms.'

'Doubtless you consider yourself included in the ton,' she began.

The viscount just gazed at Kathryn. With her lank, yellow hair which had started to come loose from her head-dress, her pale face appeared even more colourless as the sugar-pink in her dress was reflected. The viscount bowed ironically.

'That, madam, I leave entirely to your judgement. I will wish you goodnight as I leave for London within the hour.' At that he strode out of the room without giving her another glance.

Kathryn's immediate reaction was to follow and inform him as to exactly what she thought about his 'generous terms', but a timid knock on the door stayed her. The door opened a couple of inches; all Kathryn could see was the top of a mob cap.

'Is it all right to come in . . . er . . . madam?' a breathless voice asked nervously.

'Of course,' said Kathryn. At this the mob cap was followed by a neatly turned-out maid.

'Begging your pardon, madam, but Mrs Bentham thought you might like to see your suite now?' The maid, obviously obeying her instructions, bobbed nervously at Kathryn.

'Thank you.' As she said this Kathryn realized how tired she was. Waves of weariness washed over her. Seeing the girl's anxiety, Kathryn forced herself to smile. Trying to put her at ease, she said,

'I feel so foolish not knowing the way to my own room.'

'Just you follow me, Mi'Lady, if there is anything I can do, I mean. . . .' She stammered to a halt. Kathryn attempted another smile.

'I'm sure you'll look after me very well.' Probably better than my so-called husband, Kathryn thought. But she was too tired to continue that train of thought.

On the way upstairs, Kathryn discovered that her guide was called Rebecca and that normally she was the upstairs maid, and that her ambition was to be a proper lady's-maid. As they reached the top of the stairs, Kathryn asked about the portraits that hung there and on into what looked in the shadows like a gallery. Rebecca claimed no knowledge of them, beyond the fact that they were all viscounts and they were all dead. Mr Spires, however, she was informed, could recite all their histories, ' 'im being a university man'.

'Your rooms are in the old castle,' Rebecca informed her, as she pushed open a door. Kathryn gathered that the current manor house had been built on the remains of the original fortified castle. The room, being part of the west tower, was circular. Two huge sets of windows allowed her a view of the long, maple tree-lined drive on her right and over the formal gardens on to dimly outlined hills on her left. Heavy velvet curtains were drawn quickly by Rebecca. A large, stone-clad fireplace dominated the room and added to the gothic feel. Finally the bed, a large four poster hung in dark blue and silver, glimmered richly in the candle-light. A thick Turkish rug on the floor reflected the same blue as the bed.

'Your sitting-room can be reached by a private staircase here.' Rebecca pushed at the panelling between the bed and the fireplace and a cleverly concealed door sprang open, revealing spiral stairs down. Kathryn was fascinated and determined to explore the whole house thoroughly, to discover its other secrets. She would certainly have time, she thought wryly. A small door, almost hidden by the dressing-table, led to a dressing-room and, luxury of luxuries, a bath. Kathryn gazed at this longingly.

Rebecca, seeing the direction of her gaze, asked, 'Would you care for a bath, madam?'

'Oh no!' said Kathryn. 'It's much too late for all that trouble.'

'No trouble, Mi'Lady. I pull the water up directly from the kitchen.' She crossed to what Kathryn thought was a cupboard and opened the doors to reveal a simple pulley system that allowed water to be poured into a bucket in the pump room next to the kitchen and raised to the dressing-room.

'It used to be a chimney flue,' Rebecca said, as if that explained everything. But Kathryn was not in the mood to question her. A bath, however obtained, was most welcome. She did not hesitate for a moment.

'Yes, please,' she said. This time she did not have to force a smile. Immediately, Rebecca rang a bell down the flue and lowered a bucket. 'Mrs Bentham said you would fancy a bath. The water should be good and hot by now.'

Thank you Mrs Bentham, thought Kathryn gratefully to her unknown benefactor. Kathryn returned to the dressing-table and started to remove her head-dress. Rebecca bustled up behind her and began to remove the pins and flowers at the back. Kathryn's brittle hair was vigorously brushed but it still remained dull and lifeless. Kathryn sighed at her reflection.

'It is a shame it's so dry,' Rebecca remarked shyly.

'It's the colour,' Kathryn answered candidly. Rebecca stared thoughtfully for a moment longer, then bustled off to attend to the bath.

Later when Kathryn was preparing to retire, the question hovering on Rebecca's lips was finally asked.

'What colour was your hair, madam?' Suddenly conscious that this was not the most tactful enquiry, she flushed and fluttered in her confusion. Kathryn, not the slightest bit dismayed, reached for the locket at her throat. Inside rested a lock of hair and a miniature of a woman, not a great beauty certainly, but she had red-gold hair and a pair of laughing eyes. Kathryn held the locket so that Rebecca could see the contents clearly.

'My hair was a little darker than this.'

'She looks like a noble lady, Madam,' Rebecca said.

'Yes,' said Kathryn, gently touching the face with a caressing forefinger. As she closed the locket Rebecca smiled at her shyly.

'We'll make up a rinse, madam,' she promised, 'and we'll set it to rights in no time,' she finished optimistically. Her country-girl face glowed with importance as she tidied the room. Her soft heart had been touched by Kathryn's fragile dignity and her pale, tired face. Nor had the shadows beneath her eyes escaped notice. But, as she told her mother before going to sleep, 'She's a real lady, no matter what her hair is like; even Pendleton – a notorious high stickler – cannot find fault with her manner.' She paused. 'It is strange though, her hair I mean. I wonder why she would do such a thing.'

Her mother had no patience with her daughter's wonderings and told her not to bother her head about the doings of the gentry. 'Not for the likes of us to judge,' she said with finality, and blew out the candle.

Five

Kathryn slept late the next morning and for a few moments before she was fully awake, she could almost believe she was still at home. This agreeable day-dream did not last long, and Kathryn was soon gazing around the strange room which was now hers. She resolutely thrust her growing feeling of homesickness to one side and told herself that she would be the greatest goose ever to pine for her old home when she was surrounded by such luxury. Interrupting her thoughts, Rebecca slipped into the room.

' 'Morning, ma'am,' she trilled, as she bobbed a quick curtsy. 'Did you sleep well?' she enquired, as she patted the pillows into position and then presented Kathryn with her cup of chocolate.

'Mrs Bentham asks that you meet her and the rest of the staff this morning and allow her to show you round. So you can order things as you wish.' Kathryn listened to this in growing alarm.

'I shall be happy to be introduced to both Mrs Bentham and the staff. But as to ordering things or changing Mrs Bentham's arrangements, I would consider that a great impertinence as I have scarcely been here for a day.'

'Well, Mrs Bentham will be relieved to hear that, madam,' replied Rebecca frankly. 'She being very fussy about her arrangements besides having been here since she was a girl.'

'I'm sure that Mrs Bentham and I shall deal extremely well.' Kathryn was rather looking forward to meeting her. Dealing with old family retainers was something she was used to at least. She grimaced as she sipped her chocolate.

'Would you prefer tea, ma'am?' Rebecca asked anxiously. 'It

won't take a minute for me to fetch you some. . . .' Kathryn smiled at her fussing.

'This will do for today, but if you're to remain as my maid, please bear in mind that tea is the only civilized drink at this time of day.' Kathryn said this with deliberate nonchalance and waited for Rebecca's reaction.

'Me, madam – your maid! Oh madam!' Rebecca's face went from red to white and back again. 'Oh madam! I'll learn ever so quick, you'll see. Thank you, ma'am.'

Kathryn's spirits were lifted by the obvious delight and happiness Rebecca radiated as she fussed around self-importantly.

'What will you be wearing today, madam?'

Kathryn thought of her meagre wardrobe and was tempted to tell Rebecca there was not much choice. But, to respect Rebecca's new-found dignity, she gave instructions for her brown worsted dress to be set out. Hard wearing and no fear of damage, as she fully intended to explore her new home as soon as she had the opportunity.

Rebecca did as she was bid and it was not more than half an hour later that Kathryn entered the parlour where breakfast was laid out. As she entered the room, a young man jumped to his feet. He stepped towards her and bowed over her hand.

'Richard Spires. His Lordship's secretary, at your service, madam.'

A pair of hazel eyes regarded her and Kathryn thought she detected a friendly twinkle there. She smiled to acknowledge the introduction and thanked Pendleton for her tea. Nothing was said until he finally retired.

'You should be honoured. Pendleton does not normally demean himself by serving breakfast. Usually the under butler is the best we can expect at this time of day.' He smiled engagingly as he chattered on, obviously trying to put her at ease. He had an open and boyish countenance, and, judging from the remains on his plate, as big an appetite as Jess had ever displayed. But while there were lines of pain around her brother's eyes and mouth, this man's face showed only laughter and a healthy interest in the outdoors. She dragged herself back from her contemplation of his face in order to reply to his questions.

'Yes indeed, my rooms are very luxurious and in the old castle so I'm told. Rebecca seemed to think that you could tell me about the pictures on the stairs, a sort of introduction to my new family. . . .' Her voice trailed off.

'Of course,' he said hurriedly, to cover the sudden halt in conversation. 'I would be happy to show you round and tell you something of the history of the Devlins, I mean your family . . . your new family.' He stopped, embarrassed. Kathryn felt a tug at her heart-strings, his blushing made him look no older than Jess and she felt a rush of sympathy.

'Well, I shall start my day with Mrs Bentham. Perhaps I can accept your invitation for tomorrow?' He agreed with alacrity and opened the door for her to pass through. Kathryn vaguely hoped that Jess's manners would be as pretty.

Mrs Bentham was a formidable matron, with hair that was pulled back into a bun from which no stray curl escaped. Her sharp eyes missed no corner where dust might lurk, nor did they miss any broom not wielded with sufficient energy. Mrs Bentham introduced herself and her sharp eyes examined Kathryn with none of the courtesy of her voice. The under butler and maids were introduced. Kathryn gave up trying to remember their names.

This was followed by a complete tour of the linen rooms, airing cupboards and closets; she counted sheets, bottles of preserves and pickles and exclaimed over the amount of apples in the store-room. By the time Kathryn had completed the grand tour of Kingsgrey Court, lunch had been missed and the best part of the day had gone.

Mrs Bentham had thawed noticeably during the day. Kathryn's lady-like manners, knowledge of housekeeping and pretty deference to Mrs Bentham's judgement had gone a long way to reconciling that formidable lady to a new mistress. Kathryn's firm refusal to accept the keys of the household heralded the final accolade of approval from Mrs Bentham.

'Would you care for a cup of tea, madam, in the privacy of my own parlour?'

Kathryn was ushered with all formality into an overcrowded parlour. A fire blazed merrily in the grate as Kathryn moved carefully to a chair. Caution was necessary, as every available surface was densely packed with tiny glass and china figures.

'Oh! What a lovely collection!' Kathryn exclaimed. Mrs Bentham bridled with pleasure.

'Many years it's taken me to collect these and I wouldn't let any of those silly girls near 'em. Oh no, I dust these myself twice a week, with a wash in soda once a month.'

'They certainly reflect your good care,' said Kathryn, unable to tear her gaze away from a particularly grotesque figure of a begging dog, with huge painted eyes. How it was possible to concentrate on anything with those huge soulful eyes staring at you, Kathryn could not imagine.

Tea was made with all due ceremony and Kathryn sipped gratefully at her cup of inky brew. Mrs Bentham relaxed, in what Kathryn was sure she considered her kingdom.

Kathryn speculated as to what Mrs Bentham thought of her employer's attitude to his wife. It certainly would not have escaped anyone's notice that His Lordship had left the house within hours of the wedding. Whatever Mrs Bentham and the staff conjectured, Kathryn was relieved that none of them asked for an explanation. She looked at Mrs Bentham over the tea things.

'How long have you worked for the viscount?' she asked, as innocently as she could.

'The old viscount hired me, it must be nigh on thirty years ago now. Seen a few changes in my time.' She shook her head and sighed. Kathryn waited and her patience was rewarded.

'The old viscount, now that was His Lordship's grandfather, now there was a one! 'Tis said he risked his whole fortune on one roll of the dice.' She sniffed disparagingly. 'Addicted to gaming and,' she paused, for a meaningful look, 'other vices.'

Burying her smile in her teacup, Kathryn wondered whether or not he had won. If he were anything like his grandson, Kathryn hoped rather waspishly that he had lost. Thinking of the current viscount, Kathryn's smile vanished.

Mrs Bentham's sharp eyes did not miss Kathryn's expression. Behind her crusty exterior she had a good deal of sympathy for his 'little bride'. As she had told Pendleton, when they had shared their usual sherry the previous evening, 'No good will come of it. He's not one to be told.'

Pendleton nodded at her words. 'If only Her Ladyship. . . .' He sighed and sipped at his sherry, as they both pondered the current state of affairs.

Bringing Mrs Bentham back to the present, Kathryn said in the hope of hearing more, 'You must have known the viscountess.' It did not escape the housekeeper's notice that she did not refer to herself as the viscountess.

'Oh! She was a beauty. Men wrote poetry about her, never have you seen the like.' Mrs Bentham's expression grew dreamy. 'But. . . .' She drew herself up hurriedly, recalling her dignity and did not finish the sentence. Kathryn sensed a mystery, but did not press her to continue. She accepted her dismissal from Mrs Bentham's parlour a few minutes later with gracious thanks.

Six

The next morning Kathryn finally found her way to the library. Not yet familiar with her new home, she had taken a wrong turn. Richard Spires was already there when she arrived and leapt to his feet as she entered the room.

'I am sorry to have kept you waiting, Mr Spires,' Kathryn said, 'I found myself quite lost. It would have taken me longer, only luckily I found one of the housemaids to give me direction.' She seated herself and gestured to him to do the same.

'How remiss of me!' he exclaimed. 'I should have thought to escort you myself. I remember when I arrived it took me a se'night before I no longer needed directions!' he added, with an attempt at comedy. 'You must allow me to offer my services as guide – at least for meal-times, else you might starve.'

Kathryn thanked him for his offer, but said she did not wish him to get into trouble with the viscount for ignoring his duties. Richard cheerfully tried to allay her fears.

'Indeed, I find time hanging on my hands. His Lordship does not attend the house as often as it would suit me. My aim is to enter politics,' he explained. Not wishing to discuss the viscount, Kathryn returned to the subject of his house.

Richard rattled on agreeably about the features of the house and its owners. Finishing, 'Viscount Edward Devlin was famous for his parsimony.' Suddenly he smiled. 'I must be boring you, prosing on like this. Whatever will you think of me?'

'Oh, it's fascinating,' Kathryn assured him. 'Is there a picture of this nip-farthing gentleman? I wonder if his features show his nature or if it was disguised by a handsome face,' she mused.

'You shall be able to judge for yourself as his likeness is in the gallery.'

'Do continue, Mr Spires,' Kathryn encouraged. 'Your history is most informative.' His enthusiasm again recalled Jess to her mind, but she pushed that thought firmly away. If she thought of how lonely and confused she had become in the last few days, she was afraid she would start to cry, never to stop.

Richard Spires saw no hint of her thoughts in the interested countenance she showed to him, and continued his story.

'Finally there is the most modern part of the house, making a third wing and completing a horseshoe shape around a central courtyard. This contains the master suite, complete with its own dining-room and servants' quarters. It was built by the viscount's grandfather.'

'The one who risked everything on one roll of the dice?' Kathryn asked.

'Yes,' he laughed. 'Legend has it that he had to throw a seven with his last roll in order to win. Before doing that he raised the bet, risking the whole of the Devlin estates. When he won, he restored the family fortune. His opponent killed himself two days later, unable to face the shame of losing the ancestral lands. Some ugly stories of the time hinted at cheating, but nothing came of it and the viscount died peacefully in his bed.'

Kathryn found herself listening to him with increasing interest. His honest enthusiasm was infectious. 'How old is the oldest part of the house?' she asked.

'The original manor was built around 1260, but I doubt whether any of that remains. Maybe the dungeons and part of the—'

'Dungeons!' Kathryn interrupted him.

'Certainly, they are closed up now, of course, except for certain areas used for storage. The Devlins used to be the only law in these parts. From feudal times the Lord of the Manor had the power of life and death. For many years the Devlins were absolute rulers of the surrounding country.'

Kathryn thought that this probably explained the arrogance of the current title-holder. He would be perfect in the part of a despotic overlord.

Richard continued, oblivious to her train of thought, 'The Devlin family have passed down this property and its responsibilities since 1376. Of course, the current viscount has no son as yet.' He stopped, suddenly becoming conscious of who he was speaking to. Blushing, he mumbled sheepishly, 'Of course, you already know that.'

Kathryn pretended not to notice his blushes, knowing that to acknowledge his predicament would only increase his embarrassment. 'That is the most amazing achievement surely. With wars and disease it is a wonder that any descendants of the original family survive,' Kathryn commented.

'Oh, I agree,' said Richard, the enthusiastic note back in his voice. 'Really a most astonishing feat. You can see in the gallery how strong the family resemblance is throughout the generations. A family tree to be proud of.'

Kathryn finished her tea, much revived by the liquid which she enjoyed scalding hot.

'If you have finished your tea, I will be glad to show you the picture gallery.' He grinned boyishly. 'Although you must be wishing an end to my chatter.'

'Not at all,' Kathryn laughed. 'I have a younger brother, Mr Spires. I fear my sensibilities have been blunted by him. As long as you do not subject me to hunting stories, or the merits of rival prizefighters, or the intricacies of the mathematical tie, I shall feel nothing but gratitude, I assure you.'

Richard Spires laughed. 'The viscount is the expert in that area. A veritable Tulip of the Ton!' Kathryn's curiosity about the viscount had now reached considerable proportions. Luckily, Richard needed little encouragement to talk of his employer. Admiration shone from his eyes as he informed Kathryn of his patron's horses, art collection, his fame as a leading member of the Four Horse Club and the Corinthian set. Kathryn could not reconcile this paragon, this nonpareil, to the arrogant thoughtless selfish. . . . She dragged her thoughts back to the present.

The tide of praise had stemmed and Richard was pointing out

interesting features on the main staircase. 'Built in 1540 and an outstanding example of its type.'

They reached the first portrait together. It showed a striking woman with dark hair and flashing eyes. She wore an old fashioned riding habit, its masculine lines accentuating her elegant features. Kathryn looked closely at the expression, the slight frown between the brows and the petulant tilt to the mouth. She looked spoilt, decided Kathryn, not even the beauty of her face could hide her nature.

Richard told Kathryn she was the viscount's aunt. She had married his father's younger brother when he was serving in Spain in the diplomatic service. On his death she had arrived in England with her daughter.

'Unfortunately, the poor lady did not survive the English climate and she succumbed to influenza.' Richard's tone of voice led Kathryn to believe that a romantic heart beat in his breast.

'Her daughter was cared for by the viscount's father until six months before he died, when she married. A close neighbour, so she is still a frequent visitor here.' Kathryn wondered if this woman was the recipient of Richard's romantic leanings. Unrequited, of course, as all the best romances are.

'Ah. Here is the Viscount Thomas Devlin. Founder of the family fortunes.'

Kathryn studied the picture critically. The viscount had a plump face and a smug expression. His most striking feature was his sherry-coloured eyes, given prominence by thick black eyebrows. Looking at the rest of the portraits, Kathryn realized this was a family feature. She thought of the current viscount's face, her husband's face – such thinking did not come easily. His eloquent eyes capable of implacable icy stares or sparkling welcome. Kathryn found herself wondering how the viscount would be remembered by his descendants at Kingsgrey. If, of course, there were any. Kathryn did not pursue this line of thought further.

Kathryn and Richard continued to tour the rest of the gallery, watched by generations of Devlin eyes, the men with their proud demeanour and arrogant stares. Kathryn thought they would be glad the current viscount was continuing the tradition.

There were fewer pictures of the female members of the family, Kathryn noticed. Doubtless due to the fact that they were only important for breeding purposes: to continue the unbroken male line.

One picture, however, instantly caught Kathryn's eye. The sketch was mounted in a simple gilt frame. It showed a dark-haired girl in a party frock. Probably her first grown-up dress Kathryn thought. Her face was alight with expectation, her wide-open eyes and slightly parted lips giving the feeling of a rose starting to unfurl with the promise of summer glory. While she stared, Richard continued with his history.

'The viscount's mother at her coming-out ball, the year before she was married. There is another in the dining-room, started a few years later, which she never lived to see completed. But this one I like to think captures her essence. How she'd like to be remembered.'

Kathryn turned to him questioningly.

'A tragic accident,' Richard explained, 'a riding accident, with her son still in short coats.'

'His father did not remarry?' Kathryn asked.

'No. No.' Richard shook his head. 'Never had the heart, it's said. Left his son here while he travelled and never mentioned her name again.' Once again Kathryn thought she detected the romantic in him. 'Guinivere,' he said softly.

How sad, thought Kathryn. No comfort for the father in the son? Kathryn put this train of thought to one side with a promise to return to it later in the privacy of her own room.

'No tour would be complete without mentioning the ghost of Kingsgrey Court,' Richard said, emerging from his reverie.

'Ghost?' asked Kathryn intrigued. 'Am I to be haunted by a grisly spectre rattling his chains in the midnight hour?' Richard laughed at her ghoulishness.

'If you listen to some of the tales the servants tell, you would think so,' he replied. 'The story goes that the daughter of the fifth viscount, the gambler, was promised in marriage to some ageing rogue in order to pay her father's gambling debts. When she failed to change his mind, she tried to flee to her sweetheart. She knew of an old smugglers' tunnel, running from the dungeons here to the woods about a quarter of a mile away.'

A cloud covered the face of the sun and the shadows in the gallery multiplied. Richard, standing in the window casement, was no more than a silhouette to Kathryn, his voice eerily disembodied. His tone grew more hushed as he continued, 'Her father, suspecting her plan, trapped her in the dungeons. Unable to reach the tunnel and knowing she had no other avenue of escape, she hanged herself. The next day her sweetheart, hearing of the impending marriage, brought the viscount cash in order to buy her freedom, but he was too late. The servants say she searches for her love, crying piteously for him to save her.'

Kathryn swallowed. A woman sacrificed to the Devlin pride. With a shudder she hoped her own fate would be a happier one.

Seven

Kathryn awoke to see bright sunshine flooding the room and her spirits lifted at this turn in the weather. She gave Rebecca a cheery 'Good morning' as she sipped her tea. A note sat on her tray; she picked it up with a questioning glance at Rebecca.

'It's from Miss DeWinter, ma'am.' For a moment Kathryn had no idea to whom she was referring. Then it came back to her. The lady at the wedding who was, by all accounts, the nearest thing the viscount had to a mother. Rebecca chattered on while she prepared Kathryn's bath.

'She were 'is Lordship's first nurse. The viscount's father wanted to get rid of her but Her Ladyship would have none of it. Me mum says that she would allow no man to rule her. That's what killed her really.'

'Why? What do you mean, Rebecca?'

'It was his horse you see. He had forbidden her to ride it. But that just spurred her on, Mum says. Anyway the horse bolted and they brought her poor broken body back in a wagon.' Rebecca seemed to relish this vision. 'They say the viscount was never the same again. He went away and there was only her you see, Miss DeWinter, to look after His Lordship.'

'How do you know all that?' Kathryn asked, fascinated by Rebecca's story.

'Born 'ere you see, ma'am. I mean, me mum and me grandma. It's like it's the history of our own family see?' She finished filling the bath. 'Mum always says I rattle on too much, ma'am, I hope you'll excuse me if. . . .' She stammered to a halt biting her lip, obviously afraid that she had blundered.

'Don't worry, Rebecca. As a stranger here I think I need all the help I can get.'

Rebecca bustled off, relieved that she had not ruined her chance to be a lady's-maid. Tranquillity was restored with her departure and Kathryn opened the note.

Written in spiky italics was an invitation to tea signed by Miss DeWinter. Kathryn felt a lively curiosity rise in her. She remembered the viscount's attitude to her at the wedding. Anyone who could evoke a human response from that supremely arrogant, high-handed, so-called husband of hers was definitely worthy of further investigation. Kathryn rose and headed for her bath.

Some half-hour later, she emerged from her room wearing a charming morning dress of French muslin. Her hair was gathered at the base of her neck hiding the worst of the damage.

When she arrived at the breakfast parlour Richard was just finishing, but he allowed her to pour him another cup of coffee as an excuse to linger. He sipped at it while she made her breakfast with a little toast and fruit, ignoring the ham, bacon, kidneys and scrambled eggs also on offer.

The weather outside promised another balmy autumn day, and Kathryn announced her intention of exploring the garden and cutting some late blooms for the house.

'I agree,' said Richard. 'It would be a shame to waste the day indoors, but I wonder if you could spare me half an hour this morning? His Lordship left instructions for certain monies to be forwarded to you. I am expected in London early next week. So, if it is convenient, I would like to complete these matters as soon as possible.'

Kathryn was quite taken aback when he mentioned his return to town. So she was to remain here alone, bereft of company, while His Lordship caroused in London with no more regard for her than an old boot. Kathryn crumbled the remaining toast on her plate. Her voice, however, showed nothing of her anger. Only her nearest family would have noticed the look in her eyes that denoted a temper that had quite rightly earned her the nickname 'Hellkat'.

'Of course. Shall we retire to the study?' Kathryn rose and led the way, trying to keep a check on her temper. She told herself that Richard did not deserve to be the whipping boy for his graceless master. Although, Kathryn thought, he did show questionable taste in employers. Pushing these thoughts to one side, Kathryn sat down and forced herself to concentrate on the matter in hand.

'Firstly, there is the question of your allowance,' Richard said, handing her a sheet of paper. 'This amount will be paid to you each quarter either in a bank or in cash. There is a strong box containing estate papers you could use for safe-keeping.' Kathryn was for a moment quite speechless. The amount named in the agreement she held seemed to her an enormous sum.

'Naturally as the mistress of Kingsgrey you must order the house as you please, and the estate manager shall deal with those costs.' Richard consulted a list on the desk. 'His Lordship would appreciate notice of your plans; to have house guests or go visiting yourself. Apart from this—' Here Richard stalled. Kathryn raised an interrogative eyebrow.

'Yes,' she prompted.

Richard flushed. 'What I mean to say is—' He swallowed and his eyes filled up with sympathy. 'This is what His Lordship desired me to say.' He took a deep breath. 'If these arrangements meet with your approval, he will not interfere with your "amusements", providing they are conducted with decorum.' He stopped, gazing miserably at the papers in front of him. 'I'm sure that . . . I mean His Lordship's meaning was. . . .'

'Yes, I gather his meaning,' said Kathryn drily. 'You may tell His Lordship that I have no intention of taking any interest in him, his amusements or his likes and dislikes! In fact, it is a great relief that I will not be expected to forgo the pleasure of my own company for the dubious pleasure of his!'

The viscount raised an eyebrow when Richard reached this part of the story. His secretary studiously avoided his eyes.

'I take it you have some sympathy with the lady's point of view? Well, she may start to interest me, this unknown bride of mine. She certainly made an impression on you.' His Lordship gazed meditatively into the fire.

Richard continued, 'The Lady Kathryn is most well thought of. Mrs Bentham says she is most genteel and has a nicety of taste which . . .'

'Yes. Yes. I recollect she is a paragon of virtue and quite a hit with all of you.' He stretched out one foot in front of him. 'What do you think?' He inspected his spotless boot. 'Wilby has tried a secret recipe for a new shine. I really think this is superior to his best.'

'You are choosing to make fun of me, My Lord!'

'Don't ruffle your feathers at me,' the viscount laughed. 'I can see that I shall have to watch my step, else all my family retainers will be ousting me in favour of Her Ladyship.' Richard grinned at the viscount's mock horror, knowing full well that at least two family retainers would be separated from him only by death, and even that was not certain: Wilby, his valet, and Underwood, his groom. Each regarded the viscount as his sole responsibility and a constant state of warfare existed between the two. But woe betide any stranger who interfered, theirs was very much a private war.

His Lordship helped himself to a large draught of porter, a supposed cure for the fog of brandy he had consumed the night before. He then turned his attention to the invitations that littered the mantelpiece. Even though the season had only just begun, the viscount could spend his whole day moving from one function to another. From breakfast in the park to midnight card parties, his days could be filled with ceaseless pleasure. It caused a great many matrons much gnashing of teeth that the viscount did not choose to spend his days so. In their eyes he spent too much of his money gaming and too much of his time with actresses. But if he did deign to appear, very late but very charming, the delighted hostess would sing his praises for days to come, especially to their less-honoured cronies.

A constant round of activity allowed him to put his wife out of his mind most successfully. In fact, he discovered he could forget he was married for nearly a whole day at a time, if he really tried.

There were, however, occasions when he found it very convenient to recall that he was now a married man. He eyed a pink envelope, it shrieked the scent of violets across the room at His Lordship's tender head. He opened it with a sigh and, raising his eyeglass, glanced quickly at the letter. Really, Charlotte was too demanding. Only last month he had bought her two carriage horses, perfectly matched, high steppers for her to cut a dash with in the park, as she was a poor rider. Now she had her eye on some rather fine china figurines for her drawing-room. The viscount made a decision; she could have the figurines as a farewell gift, somewhat sooner than the lady fondly imagined. He penned a suitable note, instructed Richard to purchase the figurines and arrange for both to be delivered. Richard made no comment on the viscount's instructions. However, His Lordship was quite aware of his disapproval and had to hide his smile at Richard's puritanical expression.

'You agree with my decision?'

'Most definitely, My Lord. The money squandered on such a female! Not a woman, for no true woman would behave in such a manner.'

No, thought the viscount, you would not see the charm of such a dazzling piece of muslin. A diamond of the first water, but too rapacious, even for him.

'I hear the Earl of Osterley has been most particular in his attentions of late,' Richard continued. 'With his reputation, it is no more than she deserves.'

The much maligned earl was a notorious carouser and womanizer, but not the demon suggested by the tone of Richard's voice.

'Poor Charlotte,' murmured the viscount, 'to deserve such a fate. Really you are too hard.'

Richard snorted but refused to rise to the bait. The viscount had no regrets, a pleasant interlude now over in a civilized fashion. But he was left feeling restless, he needed a change of scene. A knock, shortly followed by the butler, interrupted his thoughts.

'Lord Avon,' he announced. A young man with a mop of sandy hair followed him into the room. Dressed for riding, he sported a vivid waistcoat and pale-yellow pantaloons.

'Lord Devlin, I am as parched as a desert,' he said by way of greeting.

'Bring some claret please, Grimble,' said the viscount. 'We'll go on with this later, Richard.' He gestured his friend to a chair, which he took, with a groan. Grimble left the claret and the viscount served them both.

'Going it rather heavy aren't you, dear fellow?' the viscount asked, seeing his friend's pale face. Bloodshot eyes focused on the viscount with difficulty.

'Doing it rather too brown, Devlin. I seem to recall your presence last night. Damn it, man! At least have the grace to pretend you are suffering instead of looking so smug.' He held out his glass for a refill.

'Been thinking of leaving town for a while,' said the viscount.

'Why?' asked Avon baldly. 'Not for the good of your health if today is anything to go by. Season's just starting.'

'I feel the need for some air. Been thinking of going down to Leicestershire.'

'Would have thought you'd go to Kingsgrey, what with. . . .' Avon buried his head in his glass when he saw the expression on the viscount's face.

'No. There is nothing that would take me there.' There was a note of finality in his voice. Avon decided definitely he would not bring up the subject a second time. 'Going to try those chestnuts I bought at Tattersalls last week.' Avon offered no comment and concentrated on finishing his drink. At his encouragement the viscount opened a second bottle.

'Why not ask a couple of the chaps? Make a party of it?'

His Lordship shrugged. 'As you like. We'll leave on Friday then. If your nags can get you there!'

'Oh! I do believe I hear a wager coming on! Let's make it interesting. One hundred guineas to the first man at the door!' The viscount did not blanch at this enormous sum.

'Done!' He raised his glass and drank deeply.

Eight

Kathryn thoroughly enjoyed her tea with Miss DeWinter. Even though the lady was tied to her room more and more with arthritis, she had a lively mind and an entertaining manner.

Kathryn was especially interested in her stories of the viscount. Although it was a little difficult to get rid of the image of the viscount as an insufferable pig, she felt her heart go out to the solemn little boy described by Miss DeWinter. The boy adored his mother and, while devastated by her loss, discovered he had also lost a father.

'The old viscount could not bear the boy near him after the accident. Reminded him too much of her, I think,' said Miss DeWinter, her eyes misting over with memories. 'He was too young to understand of course, my Philip. Then his father kept going away, for months and months, once even for a whole year. He just cut him out of his life. Philip started to blame his father for the accident. When his father wanted his comfort, he had none to give.' She sighed. 'Too late. I think those are the saddest words in the English language. What a wealth of opportunity is hidden behind those words.' She fell into silent contemplation.

Kathryn sipped her tea thoughtfully. The old lady's words had given her much to ponder on. She could see that being forced by his father into a marriage not of his own choice must have been anathema to him. But that still did not give him the right to make her suffer for it. She brooded on this for a while, but was brought out of this reverie by Miss DeWinter.

'You must forgive me. I must be boring you, going on like this.'

49

The old lady's discerning eyes were as bright as buttons as they regarded her. Kathryn was willing to bet that there was not much which escaped Miss DeWinter's eyes and astute mind.

'You must forgive me if I have. His Lordship sees to all my material needs and doctoring,' – Kathryn thought she relished using his title – 'but I do miss company.'

Kathryn knew exactly what she meant and said, 'Now that we are neighbours, you'll probably be wishing me away by the end of the month.'

'You are a sweet girl,' said Miss DeWinter and then almost to herself, 'What a mull he's made of everything.'

Startled Kathryn said, 'Why, whatever do you mean?'

'Why did you dye your hair?' Miss DeWinter asked abruptly.

Kathryn flushed and did not answer immediately. She wasn't prepared with a flippant answer. Despite herself, a single tear coursed down her face. Miss DeWinter squeezed her hand sympathetically. Suddenly Kathryn's tears welled up and she sobbed out the whole story on Miss DeWinter's compassionate shoulder: the letter, the success of her plan, the attitude of the viscount and how miserable she was.

Quite exhausted at the upset, she dried her tears and sat quietly as Miss DeWinter ordered more tea. When they were settled, the old lady started to speak.

'The reason I'm telling you this is to make you see that it's not you he is trying to hurt. There was very little communication between father and son. In recent years none at all. Philip had an income left to him by his mother, so his father could not exercise any influence. When the will was read, it was like a red rag to a bull. It's not you he fights so hard against. It's just that you are the physical reminder that his father controlled him from the grave.' Her face was solemn. 'I tried, but he was such a sensitive little boy.' She shook her head sadly. Kathryn had trouble seeing the viscount as sensitive. Miss DeWinter was arguing his case very well, but Kathryn still had reservations.

Miss DeWinter, seeing the expression on Kathryn's face, patted her hand. 'Be patient, my dear,' she advised. 'Tell Rebecca to call here tomorrow I'll give her instructions for a rinse for your hair.'

Kathryn thanked her for all her kindness and left feeling that she had at least one friend at Kingsgrey.

Slightly comforted, Kathryn retired and was asleep as soon as her head hit the pillow. If her dreams were disturbed by a small, motherless boy with black eyebrows and striking eyes, she kept the knowledge to herself.

Kathryn received her first guests as mistress of Kingsgrey – Lord and Lady Courtney. Any nerves Kathryn might have had were instantly dispelled by Emma Courtney, née Devlin. A creamy-skinned, dark-haired beauty, she showed a great resemblance to the woman whose portrait Kathryn had seen in the gallery.

As soon as Emma saw Kathryn, she advanced to clasp her hand smilingly.

'I hope you'll forgive us just barging in. But, as I said to William, we're family now, we're positively supposed to barge.' While she chattered on, her eyes stared at Kathryn in frank appraisal.

Her husband, Lord William Courtney, was a large, slow-moving man with a pleasant, open countenance, some years older than Emma. He smiled at his petite wife and said, 'Don't let her frighten you, she's been eaten up with curiosity for weeks.'

His wife pouted adorably. 'You're giving away all my secrets, William.'

He laughed. A deep rumbling sound. 'Now, before I get myself into trouble, I will leave. But I hope to see you soon at Evesham.' With a last cheery wave he left the two women alone.

'Now we can be cosy,' said Emma. 'He was under strict orders not to leave me on my own here, in case you turned out to be a dragon,' she added candidly. 'So now you have to put up with me for at least half an hour!' Kathryn laughed at this artless speech and felt her heart warm to Emma. It seemed she would like at least one member of the Devlin family.

Pendleton entered the room with a tray. 'Some refreshments, madam.' Kathryn smiled at him.

'Thank you,' she said. Pendleton left the room enjoying the afterglow of her smile.

Emma, noticing Pendleton's manner, was somewhat startled,

but surprisingly kept her own counsel.

'It'll be so lovely having you next door. I declare I have been bored, bored, bored.' She smiled her lovely smile and continued, 'I hope we will be friends, you know, but I am afraid I have a confession to make.' She raised her pansy eyes to Kathryn's. 'Yesterday, while riding, I stopped in to see Miss DeWinter. She was sure we would deal extremely together.'

Kathryn was not surprised at her use of fashionable slang, as she could see from Emma's dress that she was in the very pink of fashion. Kathryn poured two glasses of ratafia and passed one to her unexpected guest.

'You will think me a perfect block, but you are the viscount's cousin?'

'Oh! Very much younger cousin, please!' said Emma, with an impish smile. Kathryn could not help smiling back. 'As William said, you must visit Evesham soon, to get to know your new family.'

'I'm sure they are just as curious,' said Kathryn astutely.

Emma laughed. 'You are quite right, of course. But I see that our fears have been quite groundless. How foolish we must seem.' She smiled disarmingly. 'I have made a terrible hash of this, haven't I?' She lowered her eyes. 'Will you ever forgive me?' The sidelong glance that accompanied this was full of artful coquetry.

Kathryn laughed out loud for the first time in weeks. 'I can see that I shall have no defences left.'

Emma gave an appreciative chuckle. 'Miss DeWinter said I should like you and I can see that she is right, as usual. We shall be great friends in no time.'

This prediction was fulfilled in an amazingly short time, it seemed to Kathryn. She missed Arabella and Jess so much and Emma had a charm that was hard to resist. Unlike her cousin, Kathryn thought wryly.

The ministrations of Rebecca and Miss DeWinter's herbs had restored the condition of her hair and dulled the yellow colour to a rather pleasant shade of coppery blonde only slightly lighter than her natural colour.

One morning when Kathryn sat with Emma in her small sitting-room, Emma was insisting that Kathryn replenish her wardrobe.

'Really, first you refuse all my invitations to Evesham, refuse to let me write to Philip, and now you won't even go shopping with me!'

Kathryn laughed at her tragic face. 'Well, I suppose I do need some more gowns and a riding habit maybe,' she said musingly.

'We shall go to Madame Zoe in Bond Street.'

'No!' exclaimed Kathryn sharply.

'My dear, why ever not? She may cost a little more, but even with the pittance William gives me for clothes, I go to the best. It is false economy to do otherwise,' she added mendaciously.

Kathryn smiled reluctantly at her banbury story. Far from keeping her on a tight purse string, William fulfilled her smallest whim. No expense was spared in dressing his wife from the tip of her adorable curls down to the elegantly shod little feet.

'I think we should go into Canterbury to shop,' said Kathryn.

'Why?' said Emma in honest amazement. 'Surely Philip is not such a nip-farthing that he grudges you some decent gowns? If that is so then he is beyond the pale and I shall tell him so myself! No matter what you may say about writing!'

Kathryn had not described her one and only interview with the viscount to Emma. Miss DeWinter had obviously kept her confidence. Kathryn's excuses were starting to wear thin; Emma was her friend after all, she deserved to know the truth. There was nothing to do but tell Emma the whole story.

'Why, of all the infamous rogues! No wonder you have been shutting yourself up here,' Emma exclaimed. By the end of the tale Emma wholeheartedly agreed with Kathryn on the subject of the viscount's behaviour.

'Philip really does deserve a set down for his behaviour, really, I had no idea he was such a boor.' Her eyes started to flash militantly. 'And what,' she demanded of Kathryn, 'are you going to do about it?'

Nine

Kathryn stared at her friend in stunned silence for some moments. Images from the last few weeks flashed through her mind; Arabella, the letter, her hair, the wedding, and her brave words to Jess. She could almost hear him. 'Do you think you and this viscount will live happily ever after?' Maybe she ought to be grateful that she had succeeded in her deceptions so far with no disasters. And yet . . . Kathryn's thoughts became more and more muddled.

'That's all very well Emma, but. . . .' She didn't finish her sentence; instead she avoided Emma's eyes and gazed down at her serviceable grey gown.

'I have an idea of such genius that you will be blinded by my brilliance!' Emma announced.

Kathryn smiled at her friend's theatricals. 'Without a doubt,' she retorted. 'You are well known for the sharpness of your wits!'

'Now, now,' said Emma, 'be nice. You haven't even heard my solution.' She settled back in her chair with the air of a smug kitten and smilingly said, 'We shall go to London. William will open the town house as a base for our operations.' Kathryn was somewhat startled at her choice of words. 'I have been reading about the Duke of Wellington's victories,' Emma said, by way of an explanation. 'You will become Kathryn Chambers, an old schoolfriend of mine from Hertfordshire, I think.' She tapped her chin thoughtfully. 'I'm sure that you can engage his interest, and then' – she snapped her fingers – 'poof. You can puncture his consequence simply by revealing who you are.' She smiled at Kathryn expectantly.

'What if he recognizes me?' Kathryn said. 'And what if I can't engage his interest?'

'I've thought of all that,' Emma said loftily. 'If he recognizes you, he will send you back to Kingsgrey. Which, to be frank, is exactly the situation you are in now.'

'And what if he doesn't fall madly in love with me? Just like that.' Kathryn snapped her fingers to underline her point.

'That's the beauty of my plan,' Emma said, in a superior tone. 'People only have to *think* that you have attached his interest. His reputation, you see, is very important to him.' For a moment her eyes misted over with memory. Suddenly she smiled. 'He was always unbearably smug. Anyway, how could you not attach his interest? Your hair is much improved; with new clothes and the attentions of a hairdresser of the first consequence, you cannot fail.' She dimpled wickedly. 'Especially as you have someone of my intellectual stature to help you!'

Kathryn laughed, but she was still dubious about the strategy Emma was proposing.

'Write to Philip,' Emma continued persuasively. 'Tell him you are going to visit your family. Then William will arrange to open our town house and by the time the season gets into full swing we shall be settled.'

'I don't know,' said Kathryn doubtfully, shaking her head.

Emma took Kathryn by the hand. 'Let's go to town and have some fun. You can return whenever you wish.' With that she led Kathryn to the desk and watched her write.

If Kathryn had any doubts about their plan, they were dispelled by the reply she received to her letter. As the viscount was in Leicestershire, Richard had replied that she should delay her plans until his return, when his permission could be sought.

Kathryn snorted. Permission! Honestly, anyone would think she was his prisoner, not his wife. Well, he would learn that she could not be treated this way. Emma was quite right if she didn't assert herself now, this arrogant viscount would continue his despicable behaviour. Really, the odious man deserved no consideration. She crumpled the letter and tossed it on the fire, watching it disintegrate with a great deal of satisfaction. Kathryn yanked the bell. It would not take long to pack, then they would see that Kathryn Hastings was no man's lackey.

The bell was answered by Pendleton. Kathryn immediately informed him that she was going to visit her sister. With his beady eyes fixed on her, Kathryn felt sure he knew she was bamboozling him. She covered her nerves well, and was grateful that her voice remained firm.

'I don't know the exact date of my return,' said Kathryn, avoiding Pendleton's eyes. 'I will send a message.' She smiled with relief when he did not argue with her. 'Will you ask Rebecca to attend me? I have a lot to do.' With that she left the room and almost skipped up the stairs, her mind humming with plans.

Kathryn reached her room and, a few seconds later, Rebecca arrived a little breathlessly.

'Madam. Pendleton . . . I mean Mr Pendleton said that you were going away!'

'Quite right,' said Kathryn. Rebecca's face fell ludicrously. 'Why, whatever is the matter?' Kathryn asked.

'I was wondering, ma'am, if you would be needing me . . . I mean, whether I. . . .' She stammered to a halt.

'Of course I will need you, Rebecca! In fact I have a very great favour to ask of you.'

'Of me! Oh, madam! If there is anything I can do, I would be happy to. Truly!'

Kathryn smiled. 'It is a little complicated. I need you to keep a secret. I am not going to my sister's, I am going to London with Lady Courtney. While I'm there I will be known as Miss Kathryn Chambers. No one must know who I am. Not even the viscount. I would like him to get to know me as Miss Chambers, before he knows that I am his wife.' Kathryn waited for Rebecca's reaction.

Rebecca's romantic heart had been sorely puzzled by the viscount's marriage arrangements. Already she was devoted to Kathryn, not only because of her promotion, but because her fanciful nature had already cast Kathryn as a lady in distress.

'Madam, may the Devil take me tongue if I should ever breathe a word. I swear it, ma'am!'

'Thank you,' said Kathryn. 'I knew I could rely on you.' Rebecca went quite pink with pleasure at the compliment. 'Now, we must pack!' With this, Kathryn opened a drawer and turfed it on to the bed.

'I'll deal with that, My Lady,' Rebecca said, seeing that the amount of work would be double with Kathryn helping.

'Maybe you're right,' Kathryn agreed. 'I shall see if Miss DeWinter can see me. I must tell her of my plans.' Feeling better than she had for weeks, Kathryn bustled off, leaving Rebecca alone to pack, much to her relief.

Kathryn found Miss DeWinter's apartments, only getting lost twice, and tapped on the door.

'Come in.' Miss DeWinter beamed with pleasure when she saw who her visitor was. 'What a lovely surprise. A visitor for tea.' She gestured Kathryn to a seat, her bright eyes noting Kathryn's obvious excitement. 'You look like you've had good news,' she probed.

Kathryn took a deep breath. 'I've made a decision.' She bit her lip, her eyes on Miss DeWinter's face. 'I'm going to London.'

'Let me ring for some tea and then you can tell me all about it.' This was accomplished with the minimum of fuss, and Kathryn fortified herself with a few sips of the scalding brew before continuing her story.

'So,' said Miss DeWinter, 'what has made you decide to go to London? And do I sense Emma's influence at all?' With a smile she continued, 'That little minx winds us all around her pretty little fingers.'

Kathryn laughed at her description and answered with perfect honesty, 'Partly, but mostly because of me.' She shifted uncomfortably, trying hard to recognize the truth of her motives. 'Before I accepted Arabella's offer of marriage, I thought I was to be sentenced to my Aunt Jane! Ugh!' Kathryn still shuddered at the idea. 'Now it seems I have exchanged one prison for another.'

'So what is the answer?' Miss DeWinter asked.

'I will go to London and attempt to gain the viscount's interest. To prove to him that I will not be ignored!' Her hands gripped the cup more tightly and Miss DeWinter feared for the china.

'Even if it doesn't help matters, they can't get any worse.' Miss

DeWinter still made no reply. 'I left my safe life behind me looking for something more.' Kathryn struggled to make Miss
DeWinter understand. 'Maybe I made a mistake, but now I've
started, I must try. Then, if I remain here the rest of my days. . . .'
Her voice trailed off.

'It may well be safer, my dear,' said Miss DeWinter softly.

'I'd rather have an hour of wonderful than none at all. This may
be my last chance; they may even be the last memories I can make.'
Kathryn swallowed and managed a slight laugh. 'Whatever would
Emma think if she could hear me now? She feels that her cousin
should be taught a lesson.'

'Indeed. She has attempted that more than a few times, but with
limited success. You must not let her beguile you against your
better judgement.'

Kathryn sighed. 'So you don't think it's a good idea?' she asked
glumly.

'No. No. I'm not saying that exactly, but—' Miss DeWinter
paused searching for the right words. 'Be careful you are not so
caught up in your own devious plans that you are ensnared in a
trap of your own making.'

A little baffled by this remark, Kathryn was just about to ask for
clarification, when her eye caught sight of the ormolu clock.

'Heavens is that the time! I must fly. Emma will take me to
Evesham to spend the night there, then on to London.' Impulsively
she bent and kissed the old lady's cheek. 'Wish me luck?' she
asked.

'Oh, my dear, of course. I shall remember you in my prayers
every day. Will you write and let me know how you go on?'

After Kathryn had given her promise, she left Miss DeWinter to
her nap, not giving the old lady's words another thought.

Kathryn had been behaving with a confidence she didn't feel, and
her relief at seeing Emma was instant. Her airy confidence allayed
Kathryn's fears.

'William,' Emma said, 'is already in town, making suitable
arrangements.'

Kathryn was astonished. But Emma just dimpled wickedly and

denied that she'd had any trouble persuading him to help them. She had even forced him to invite his Aunt Sybil – a confirmed man-hater – to stay, in order to lend them the respectability of a chaperon. Kathryn looked at her in awe.

'How did you manage to arrange that? It hasn't been two days since I decided to go to London with you!'

'I put arrangements in hand as soon as you wrote to Philip,' Emma replied, gazing at her friend in mock severity. 'You must realize, Kathryn, that it is sometimes necessary to make a push for what you want. All it takes is a little gumption.' This was a new side to the frivolous and pleasure-loving Emma. Kathryn kissed her cheek.

'How will I ever thank you?'

'Save your thanks,' Emma laughed. 'I would have been moped to death without you.' Then, turning her mind to practical matters, she asked, 'Now, where is Rebecca? And we also need to speak to Mrs Bentham.'

Kathryn watched her organizing the household with the military precision of an old campaigner. Before she had a chance to protest, she was on her way to Evesham, from where they would drive to London.

Ten

Kathryn was amazed by the number of boxes and trunks Emma had brought to London. A seemingly endless line followed her.

'Just you wait,' Emma chortled. 'I defy you to criticize until we see how many trunks, let alone boxes, you bring back from London!' Kathryn remembered the bankroll hidden in her meagre jewellery.

'Do you think William would hold my money for me?' she asked tentatively. 'Would that be the proper thing to do?'

'Surely you can't be worried about money!' Emma looked at Kathryn in amazement. 'Philip is one of the richest members of the ton. Not as much as the Golden Ball perhaps,' she said judiciously, 'but certainly able to dress you suitably.'

Kathryn flushed. 'But if I am not the Viscountess Devlin I cannot draw on my allowance. . . .' She halted in embarrassment.

'Don't fret,' Emma comforted her. 'We shall think of something before the bailiffs get you.'

In the face of this airy confidence Kathryn could not help but be reassured. William gladly took her bankroll and agreed to keep strict account for her.

Once the tiresome difficulties of settling down, unpacking and finding the hundred things that vanish during a move, had been dealt with, Kathryn and Emma sallied forth to Bond Street.

'I hope no one will remember you in that pelisse,' Emma said. 'I cannot wait for you to see Madame Zoe.' She eyed Kathryn across the carriage. 'Maybe a visit to the pantheon. I have already made an appointment with Monsieur Anton. He shall wait on us

60

later this afternoon.'

Kathryn made no demur at this plan. She was looking forward to abandoning herself to the frivolous, but exhilarating pursuit of shopping.

Emma showed herself to be a veteran campaigner. She and Madame Zoe hummed and hawed over her, discussing colours, pleats, flounces, half capes, gloves and French knots. Kathryn was moved to protest at the number of gowns they felt she required. Four morning dresses, two ball gowns, two riding habits, the list went on and on. When Kathryn thought of the cost of all her finery, she was very much afraid that she had spent most of her allowance already.

Emma scoffed at her, deciding on several gowns for herself, and saying that she could pay next quarter. And that she shouldn't let it worry her.

'For no one appreciates a girl, especially a single girl, who is a penny-pincher. Gentlemen reserve financial issues to themselves.'

'You sound like Aunt Jane!' Kathryn retorted.

'So long as I don't resemble your Aunt Jane, I shall take that as a compliment!' Kathryn showed her the very tip of her tongue. Emma gurgled, 'How you have forgotten your manners! Anyway for my part I am quite happy to let William deal with the money and the bills.'

'You're just interested in spending it.'

'Now who sounds like Aunt Jane?' Emma replied with a toss of her head. Both ladies dissolved into giggles.

Kathryn recovered herself first, struck by a dress that Madame had just brought out. Made from layers and layers of palely tinted superfine chiffon, the skirt fanned out creating a subtle rainbow. The bodice, made of delicate ivory silk, was also covered by a layer of chiffon which was then gathered, slightly off the shoulders, leaving the arms bare.

Madame Zoe had calculated correctly; as soon as Kathryn saw the dress she knew she must have it.

'It's beautiful, Madame,' she breathed.

'You will look like an angel,' Madame agreed. 'Sometimes you must inspire adoration as well as passion in the hearts of men,' she continued. 'It could be ready for you next week.' Kathryn resolutely ignored the voice in her mind shouting about the cost, and agreed.

During that week there was a constant stream of visitors at the Courtney residence in Cavendish Square. Hairdressers, seam-stresses, dancing masters came and went until Kathryn's head whirled. The hairdresser was Kathryn's particular favourite. Monsieur Anton was a small, dapper Frenchman. He danced around Kathryn humming and hawing at the state of her hair.

He cut the front short, teasing it into capricious curls, left longer at the back he dressed it high on her head. This added to her height and seemed to crown her in a wave of titian hair, more beautiful than any jewels. Finally Emma pronounced her ready to meet society.

First, William, Emma and Kathryn went over the story that explained Kathryn's presence. Aunt Sybil had been unable to oblige them as her sister's children had all contracted measles and she felt it her duty 'to succour them in their hour of need'.

Emma said that phrase made her feel sick and why must such stupid people insist on having four children! Not one of them had had the grace to remain healthy when they needed Aunt Sybil.

Kathryn had to laugh at her friend's uniquely selfish point of view. But they still had the chaperon problem to solve. Finally, William was persuaded to invite his Aunt Agatha to fill the breach. William seemed to have a plethora of aunts. A good job too, Kathryn thought, as Emma pronounced this aunt to be the laziest and fattest woman she had ever met. But, she owned, also one of the most good-natured. The problem seemed admirably solved.

William went to his club for dinner, leaving the two ladies to enjoy a comfortable coze. They agreed to the name Kathryn Chambers, also that when Aunt Agatha arrived, she would be admitted into their confidence. Emma was right when she predicted that, as long as it did not interfere with her comforts, Aunt Agatha would cause no problem.

They also decided on the story that Kathryn and Emma had attended the same school, but that Kathryn had lived quietly in the country since – with her grandmother, Emma decreed. There was something infinitely respectable about having a grandmother, she explained. This would be her first trip to London, something at least she wouldn't have to pretend.

'Would anyone be likely to recognize you?' Emma asked. Kathryn thought for a moment.

'No, we have no acquaintance in town. We lived most quietly. No one will recognize this Cinderella!' Suddenly another thought struck her. 'Will I be able to visit Jess at Cambridge?'

Emma tutted. 'Will you put your family out of your mind for a few minutes? The plan was to teach your high-handed husband a lesson.' She gazed at Kathryn, her pansy eyes troubled. 'Have you decided to see the plan through?' Receiving no reply, she continued: 'I think it's a fine idea to bring Philip down a peg or two. He was always the most arrogant of men, albeit with cause. His behaviour to you has, of course, been inexcusable.' She struggled to continue.

Kathryn stared into the roaring library fire. Her freshly cut hair was curled and dressed in the latest fashion and in her figured muslin with French knots she looked a completely new person. Her confidence had grown, with Emma's help, and she was sure that whatever happened she would at least have a few weeks of fun and frivolity to remember. However, her anger against the viscount had not abated one jot in the intervening time, Kathryn told herself as she turned to Emma.

'I intend to teach that odious cousin of yours a lesson he will never forget!' She started to chuckle, 'Imagine, when he discovers that it is his despised wife, not a beautiful stranger with whom his name is linked.' Kathryn reflected on this picture with a good deal of relish. Various day-dreams of the viscount writhing in humiliation as she watched afforded her a great deal of satisfaction.

Revenge she thought would be very sweet.

Leicestershire, His Lordship thought, must have the highest rainfall in the country. For a week the weather had remained stubbornly wet. Sheets of rain poured from the sky, showing no sign of letting up.

Avon thrust a glass of claret into the viscount's hand. 'Damn fine stuff this, Devlin. And thank God for it,' he smiled.

'Little else to do. My luck has finally deserted me,' the viscount replied. This statement was met with a snort.

'Never known a fellow as lucky as you, Devlin, 'pon my word!' Avon once again drank deeply. 'Look at this race now. Driving like the Devil himself, and no harm come to you. A charmed life is what you have. Damned if it isn't!'

The viscount remained silent as he replenished his guest's glass. The door opened to admit Sir Rupert Roscommon.

'Well, that's the last of them off. Damned weather. Can't hunt in this mud river.' He dragged a hand through his already disordered locks. 'Is that the claret?' He made a beeline for the decanter. Baronet Rupert Roscommon cultivated a Byronic air and his clothes reflected this artistic unconcern. 'Wonderful stuff,' he said appreciatively. 'Have you been hiding it here?'

'And deny my friends a drink? What sort of block do you take me for?'

'A block with a first-rate cellar, of course!' Rupert quipped. 'So, tomorrow the last of the carousers return to town. Shame about the hunting. But I did win a tidy sum from Wilerby at faro.' He grinned wolfishly. 'Going to Rutland's bash then?'

'So, what's your interest in Rutland?' Avon chortled. 'As if we didn't know.'

Rupert smiled. 'The Divine Dinah.' He sighed. 'My latest poem, don't you know. I will present it to her at the ball.'

'Wasting your time there, dear boy,' said Avon. 'Still got an earl. Old Rutland's a better catch than a mere baronet.'

Rupert continued to smile. 'She will not be able to resist my latest poem. I have called it "Ode to Her Smile".' The viscount and Avon exchanged a significant glance. Rupert didn't notice, a rather vacuous expression on his face showed that his thoughts weren't with them.

'The fair Dinah will throw him for the earl,' Avon said to the viscount in a low tone.

'She's not that mercenary, old thing,' he replied.

'Mother is,' said Avon, as if that clinched the argument. The viscount was unable to disagree. He sighed at the thought of his friend's hopeless love life.

'God spare me from that,' he muttered, as he poured them all another drink.

'Of course, you have never been in love,' said Rupert as if that explained everything.

'Love! Hah!' the viscount snorted. 'Love is for women. There is nothing in love that a sensible man cannot deal with.' Once again the decanter was passed round.

'You, of course, are a sensible man!' Rupert retorted.

'Sensible enough to know that love is something that has no control in my life. I control it!' The viscount waggled a finger at him.

'Doesn't it take two?' Avon asked. 'Not that I'm any expert,' he added hurriedly.

The viscount regarded him balefully over the rim of his glass. A sneer settled on his face, and with a disparaging twist of his mouth he replied, 'Pshaw! Two! Once I express an interest in a woman there is no question that she will,' – he paused – 'shall we say, accede to my wishes.' He grinned lecherously.

'So, because you are in control, any woman will succumb to your charms?' Rupert questioned with seeming innocence.

'Oh ho!' Avon cried. 'Do I hear a wager coming on?'

'Why not?' said the viscount. 'I'm sure you will benefit from my example!'

'Lord, you're a smug devil, 'pon my soul you are!' Avon exclaimed.

Rupert, with a dangerous twinkle in his eye said, 'Right, I'll wager my winnings from Wilerby. One hundred guineas that you will not succeed with a woman of my choosing!'

'Done,' the viscount cried. He got up with difficulty and negotiated his way to the bell. 'I think this calls for another bottle. The memory of it may ease the pain of your losing!'

Eleven

Emma was sitting in the drawing-room with Kathryn, who was struggling with the intricacies of the cotillion, as she heard the story of the race. Kathryn kept losing time as she reacted angrily to the story. If it had been anyone else she would have admired a driver of such nerve and skill. But it was her husband. No, bridegroom!

'And now all the old tabbies are agog to know what made him run so far so fast!'

A week after his wedding. Making it public that he wanted nothing to do with his new bride. Kathryn felt the humiliation sting and a mortified flush mounted in her cheeks. She thrust out her chin and said, 'Well, shortly they will have the opportunity to judge for themselves.'

Emma watched Kathryn's face and said with satisfaction, 'I knew this would not deter you! And I must say that even William admits. . . .' She stopped, flustered. Kathryn grinned at her.

'It's nice to know that I have such loyal friends. If it wasn't for you I would be a virtual prisoner at Kingsgrey.'

'Tosh,' said Emma roundly. 'You would have contrived I'm sure, and I can't let you deny me the opportunity of finally seeing Philip get the set down he so richly deserves!' Kathryn had to laugh at the fierce expression on her face. 'When we were children, rather I was the child and he a young man, everyone seemed to think a match between us was a good idea. No matter how insufferable I made myself, they carried on. And believe me, I can make myself pretty insufferable! Yet he remained unbearably smug!'

But before Kathryn could comment they were interrupted. The footman entered and proffered a silver tray to Emma; on it rested an elegant calling card. As soon as Emma recognized it, she ordered him to bring the guests up immediately.

A formidable matron in a remarkable hat, containing no less than three ostrich feathers, swept into the room. She was followed by a nervous-looking young gentleman in an eye-catching waist-coat which combined blue and yellow stripes to the most startling effect. Emma swept forward to welcome them.

'Lady Etherington! How charming you look in that bonnet! And I see you have brought Harold, such a treat!' She gave him her hand. 'How do you do? Please allow me to present an old school-friend of mine, Kathryn Chambers. Kathryn, this is Lady Etherington, whom I'm sure you'll have read about in the society pages of *The Times*, and her son, Harold.'

Introductions completed, Lady Etherington was seated with all due ceremony, Harold hovered nervously around her.

'Do sit down, Harold,' she said in majestic tones. Harold perched himself on the edge of the chair she had indicated. Seeing her wish obeyed, she ignored him and turned her full attention to Kathryn and Emma.

Lady Etherington noted Kathryn's new finery and had priced its cost with great accuracy. She enquired about Kathryn's family. On being told she was an orphan, with only a grandmother as her clos-est family, she set about discovering whether she was wealthy enough to consider for her Harold.

Kathryn stuck to the tale she and Emma had agreed. Lady Etherington found her background unexceptional, and her fortune not great enough to render her eligible for her only son. Having dealt with this, she then proceeded to bring Emma up to date with the latest gossip. Kathryn listened, not with a great deal of atten-tion, as she did not know the people of whom they spoke. However, when the viscount's name was mentioned, her flagging interest revived.

'Shocking behaviour! Racing curricles on the public roads,' Lady Etherington continued blightingly. 'However, the gentlemen seem to see this childishness as of the highest importance.

Gambling and low-class habits, I say.'

Harold looked as though he'd like to protest at this point, but on catching his mother's eye he subsided without comment.

'It's not as if Devlin needs the money,' she continued. 'Probably trying to forget his father's will,' she finished with an air of triumph. Lady Etherington lowered her voice conspiratorially. 'I hear she's staying at Kingsgrey. Alone.' She paused, but receiving no comment she probed, 'Have you met your new cousin yet, my dear?' The innocence of her tone was not reflected in her face as her bright eyes focused hopefully on Emma.

'No. Not yet,' said Emma. 'Our plans brought us to town before we could become acquainted.'

'Is it true that she is, how shall one put it, shunning society because of a' – her voice dropped to a hushed whisper – 'deformity?' Her florid face moved closer to Emma, her nose almost twitching with curiosity. Kathryn had trouble suppressing a laugh, wondering what deformity she had in mind.

Emma looked up to meet Lady Etherington's avid gaze, widening her eyes into a guileless expression, and asked, 'Whatever do you mean, Lady Etherington?'

'I only ask you, my dear, in order to warn you. Of course, I don't believe such tittle-tattle myself. But there are folk who are spreading such a wicked story.' Her face had resumed its normal expression of rigid disapproval. 'However, I'm sure you'll understand that my interest in this is almost maternal, as your own mother isn't here to guide you, I felt it my duty to warn you.' She refreshed herself with a sip of tea. 'Madness in the family I hear. No one has seen anything of the Hastings family for years now. As the new viscountess and wife of the biggest catch in the marriage mart, why else would he not bring her to town?'

This spurious piece of logic nearly had Kathryn in whoops, her next words, however, replaced the mirth with indignation.

'Certain gentlemen, I hear, have been placing wagers on whether she is mad or deformed, or both!' Her face registered a little distaste. 'I'm glad to say that my Harold is not among that set. He shows no such vices that would sadden a mother's heart. Indeed, I am the luckiest of women.'

Harold definitely was not the luckiest of sons, thought Kathryn vindictively.

Emma, however, answered with perfect aplomb. 'I can't say. Pray, what is your opinion, madam?'

Lady Etherington considered for a moment before speaking. 'I think that it must be madness, or that she is so unprepossessing that society holds no interest for her.'

'Maybe I should ask Philip,' Emma said lightly, but with a dangerous look in her eye. Lady Etherington failed to see this and forced a high-pitched giggle.

'My dear! What a thought! Can you imagine!' No one obviously could as she did not get a reply. More than a little put out at her failure to elicit any juicy tidbits and realizing that she had probably gone too far in her speculations, Lady Etherington took her leave, with the faithful Harold shadowing her devotedly.

As soon as they were out of earshot, both girls dissolved into helpless giggles. When they had recovered somewhat, Kathryn said, 'They are placing wagers on the nature of my *disability*! The impudence!' But her sense of humour did not allow her to feel angry for long. She turned to Emma and pulled a comic face.

'Whatever do you think she meant by unprepossessing?' She screwed up her face to make it as ugly as possible, crossing her eyes at the same time. Emma once again dissolved into helpless laughter. Kathryn continued, 'A wooden leg, perhaps, or maybe I'm terribly fat? Although,' she remarked waspishly, 'I should have trouble matching her.' A wicked grin twitched at the sides of her mouth. 'I wonder if she has to rely on Harold to help her upstairs? That's probably why he is too tired to talk!'

Arabella Ripley, née Hastings, fluttered around her bedroom in an agony of indecision. Her agitated pacing was interrupted by a firm knock on the door.

'C - come in.' She quavered. A white-haired gentleman with a broad country face entered, smiling reassuringly.

'Doctor James at your service, as always. Now what has got you so agitated? Hmm? I came as soon as I got your note.' He moved stiffly toward her. Really, thought Araballa, he must be well over seventy by now, but it never occurred to her to call someone else.

It was about an hour later that the doctor left Arabella alone once more. She stared in her glass for a long time, trying to see if she looked any different. Her eyes were bright, her face flushed and her breathing uneven; she looked supremely lovely. Her eyes paid no account to her features and she continued to regard her reflection intently. Finally, taking a deep breath she left her room to seek her husband.

She found him reading the monthly accounts with a worried frown. As she entered, his countenance cleared miraculously and he smiled dotingly at her.

'Well, this is a lovely surprise, my dear. Do you think you could find your way to balancing these accounts for me? I fear I have made a terrible hash of them.'

Taking him by the arm she led him to the settle and made him sit, a serious expression in her sky-blue eyes and a solemn cast to her mouth.

'Is everything all right, my dear? Has something happened?'

She gestured him to silence, took another deep breath and began. 'Doctor James has just left. . . .'

'Are you ill? Darling, why didn't you tell me? I shall send to London immediately.' Watching him, her expression changed, her fears evaporated, and a slight smile played across the corners of her mouth.

'Well, I think I could do with Fanny's help, but not yet.'

Her husband gazed at her in confusion. 'But Fanny is the midwife. . . .' Realization galloped across his face and dawned in his eyes. 'I . . . I . . . mean . . . Oh Lord!' He leapt to his feet, tears starting in his eyes. 'Oh my precious life. How happy you have made me.' With infinite care he placed his arms about her and together they sat on the settle. 'You must rest. My mother craved pickled onions,' he added inanely. 'But anything your heart desires shall be provided, even if I must send to the moon for it.'

Arabella found tears starting in her eyes.

'No. No tears, my dear. You have made me the happiest man on earth and I shall spend the rest of my days ensuring that you never have to shed another tear again.' Tenderly, as if holding a delicate blossom, he cupped her face in his hands and lovingly brushed her lips with his.

Twelve

As soon as the polite world learned that the Courtneys were back in town, invitations to dinner, balloon ascensions, picnics, balls, the theatre and breakfasts started to arrive. Kathryn was amazed by the social whirl Emma was planning.

'Let me decide on the invitations,' she said one morning to Kathryn. 'You concentrate on enjoying yourself and our plan.' As she spoke, she deftly arranged the morning post into three piles; bills, invitations, accepted and refused. Emma planned an assault on the polite world that a general would envy. Everyone who was anyone would notice Kathryn, whose humour and intelligence, not to mention her good looks, would make her the most sought-after belle of the season. Emma was sure that her friend would take the ton by storm.

That afternoon, Kathryn returned to the house alone, after riding with Emma in Rotten Row at the fashionable hour. Emma was now visiting William's mother. Kathryn entered the drawing-room to find the viscount awaiting her. Instantly he rose to his feet at her entrance. Despite her surprise, she couldn't help but notice the fall of his cravat, his perfect top boots and how his hair curled in at his neck. She decided on a plan. Evincing surprise at finding him in the drawing-room and adopting her most disdainful attitude, she addressed him. 'Are you waiting for Lady Courtney? Have you a message for her? Well, you may certainly leave it with me and return to your work.' She treated him to her best impression of Lady Etherington's haughty glance.

His sherry eyes flashed at her tone but before he could reply,

Kathryn swept on, thoroughly enjoying the opportunity to give him a taste of his own medicine.

'The footman should not have shown you in here, it is a family room, not a reception room.' She continued to eye him imperiously. 'Please state your business. I cannot dally here on your account!'

The viscount, who was used to hostesses welcoming him with open arms, had never been spoken to this way before. Choosing to see the funny side, he smiled down on her. She was dressed in a dark-blue riding habit, trimmed in Hussar style with a high peaked hat. The blue made her hair the colour of red-gold and highlighted her unusual green eyes. She made a charming picture.

Seeing him smile at her, his sherry eyes alight with warmth, Kathryn felt herself respond to his charm. Pulling herself together sharply, she awaited his reply.

The viscount, thinking she was delightful and looking forward to her sweet apologies when she discovered her mistake, bowed over her hand and introduced himself. To his surprise, his identity did not cause the lady's expression to alter one whit. She gazed at him, unmoved, noticing with satisfaction his astonishment at her attitude.

'How very surprising,' she said, 'a viscount who is mistaken for a messenger. I must beware that I do not suppose another messenger boy for a viscount.' Kathryn took off her gloves. 'Well, the manners maketh the man.' With this cryptic utterance she seated herself, and with the obvious air of starting again, gestured him to be seated and asked him if he wanted any refreshment.

The viscount, who was totally baffled by this behaviour, asked, 'And may I know whom I have the honour of addressing?'

Kathryn eyed him with no answering smile. 'I am Miss Chambers,' she said formally, adding no extra information.

'Charmed,' said the viscount through gritted teeth. Her attitude was beginning to rankle. Normally he was received with exquisite courtesy, his hosts fully aware of the honour he was doing them. Not that he was high in the instep, he would tell himself, and he was definitely no snob. But he could administer the most crushing snub to an encroaching mushroom, and a glance through his quizzing glass could depress the most determined social climber. Never before had he been treated as if he were of no more consequence than the merest macaroni. Kathryn noticed with satisfaction that his confusion was turning to indignation.

With a slightly bored air she ordered tea and turned to the viscount with an expression of polite interest. The viscount, who had often worn such an air when he wished to be elsewhere, decided that the joke had gone on long enough. He opened his mouth to deliver a speech that would correct Miss Chambers' attitude towards him but he was balked in this by the arrival of Emma and Lady Etherington followed by the dutiful Harold. Such company did nothing to sweeten the viscount's mood.

'Oh, Philip! What a lovely surprise! I see you have met my great friend Kathryn Chambers.' Emma did not fail to note the expression on the viscount's face and burned with curiosity to know what had passed between them.

Kathryn rose and giving the viscount her sweetest smile, excused herself. Offering Harold her seat she said in a satisfied voice, 'Now you two gentlemen must have a great deal in common! I'm sure you won't miss me!' She walked gracefully to the door. Smiling at Emma, she demanded a little petulantly, 'You should have warned me about your relatives you know!' On this note she departed.

Emma peeked at the viscount under her lashes; his expression, as she later told Kathryn, was one of baffled fury. A second later, however, it was gone, but not forgotten, Emma was sure.

About a week following this incident, Emma and Kathryn were preparing to spend the evening at Almack's. William awaited the ladies in the library. His wait was interrupted by Avon and Rupert.

'Just popped in on our way to Brooks's,' Avon enunciated very carefully. Rupert tutted in a superior fashion.

'Bit over the boughs,' he said confidingly. William eyed them with disapproval.

'I am escorting my wife and Miss Chambers to Almack's tonight.' The traitorous thought of joining them instead pushed firmly away.

'Miss Chambers eh? The filly that gave Devlin a set down t'other day?' Avon asked.

'What's this?' Rupert demanded. Avon described with great glee the encounter between Kathryn and the viscount. Rupert chuckled triumphantly.

'You know what this means, don't you? Almack's it is! Send a note to Devlin!'

'What are you talking about?' William said worriedly. 'You'll never get into Almack's in pantaloons and what's it got to do with Devlin?'

Rupert smirked at the two faces turned questioningly towards him. 'Miss Chambers is my choice; I think my money's safe there.'

Realisation dawned on Avon's face, then he broke into a gusty laugh.

'What are you two raving about?' William had been listening to them in growing confusion. Rupert was concentrating on filling his brandy glass, so it was Avon that explained that maybe, for once, the Right Honourable Viscount Philip Devlin's famous luck had deserted him.

'That's the worst idea I've ever heard,' William protested. 'Absolutely! Quite beyond the pale, by God!' He fell silent as he realized that fate had played into Kathryn's hands. Suddenly he felt that his involvement was too much strain entirely. Avon and Rupert continued to argue the morals of the wager.

'Look,' said William desperately, 'let's meet at White's tomorrow. We need a clear head to discuss this.'

Avon eyed him with as much dignity as he could muster.

'What difference will it make to the girl? She obviously has no *tendre* for Devlin, so no damage will be done to her.'

'What about him?' said William desperately. This was met with gales of laughter.

'He's even married! So what could possibly go wrong?' Avon continued.

William finally convinced his visitors that a clear head on the morrow would be more helpful to their discussions. So they set off into the night, arguing as to where to stop for dinner.

William poured himself a well-earned brandy, his mind reeling.

As his wife entered the room, his anxious expression softened. She was dressed in yards and yards of ivory silk with a sparkle in the weave, so she looked as though she were surrounded by stars.

'I must say you become more lovely with every passing day.'

She pirouetted in front of him. 'Thank you, kind sir.' He bent forward to receive an airy kiss on the cheek. 'The bills will arrive shortly you know.' She grinned impishly at him.

'Do you intend to beggar me?' he asked, kissing her fingers. His good mood did not last and he found himself confessing to Emma the conversation prior to her entrance.

Emma listened to his tale, her colour rising and her eyes flashing. 'Oh! How infamous! He deserves everything. I shall not feel a moment's guilt. Not one!' She stamped a small foot for emphasis.

'What has happened to put you in such a temper, Emma?' Kathryn asked from the doorway. The effect she had on them was most gratifying. Kathryn's mirror had told her a transformation had taken place, and their eyes told her that it was for the better. Dressed in white, suitable for her 'unmarried' status, her dress was stunningly simple, straight classical lines showed her figure to an advantage. Her hair was dressed à la Sappho, dotted with small white flowers and one red-gold ringlet resting on one shoulder. As she walked, her underskirt of frothy lace was visible through cunningly hidden splits in the seams of the lower part of her skirt. This gave her the effect of walking on billowing clouds. Emma was the first to speak.

'You will break hearts tonight,' she prophesied. Kathryn blushed becomingly at her praise, quite forgetting her question.

'I am the most fortunate of men,' William said with a bow. 'I will be the envy of every man in England!'

Indeed, Kathryn did create quite a stir on her arrival at Almack's. As she was a little older than most of the debutantes, and used to managing her own home and speaking to strangers, she was not a tongue-tied partner. Her humour and intelligence as well as her pretty manners soon endeared her to all. Even the strict matrons of Almack's pronounced her 'a most charming girl'.

Within half an hour of arriving, Kathryn's dance card was full. As she was swept into quadrilles, country dances and waltzes, she was profoundly grateful to Emma for insisting on dancing lessons. Several of her partners asked permission to call on her and Kathryn had the satisfaction of knowing she was quite a hit.

When she joined Emma and William for supper, no less than three gentlemen begged for the honour of filling her plate. Kathryn was thoroughly enjoying herself and Emma laughingly teased her that she would become a rival to the 'Divine Dinah'. Kathryn had been introduced to this famous beauty and had duly admired her perfect guinea curls, rose-bud mouth and porcelain complexion. Surprisingly, Kathryn had found her to be a most unaffected girl, not in the least conceited about her looks. As she said to Emma, she looked forward to continuing the acquaintance.

It wasn't until the early hours of the morning that they returned to Cavendish Square. Kathryn was sure that she wouldn't sleep a wink after all the excitement, but as soon as her head hit the pillow, she was asleep and dreaming of waltzing in the arms of a tall, dark stranger.

Thirteen

The following morning, Kathryn had the luxury of breakfast in bed. Her bedroom caught the morning sun, making the pink and white room a blaze of colour. Kathryn was enjoying her dish of tea and the sound of starlings quarrelling over the crumbs she provided, when she was interrupted by Emma. Seeing that she had something of importance to tell her, Kathryn dismissed Rebecca. Emma, still labouring under extreme indignation, could not sit still while she related the story told to her by William.

'I feel quite ashamed of his behaviour.'

Kathryn's eyes hardened as she heard the details of the wager. Emma finally ran out of names to call the viscount and continued, 'But it does gain us one advantage; no matter how you treat him, he will return in the hope of winning his bet.' She then subsided into seething silence.

Kathryn continued to sip her tea, her face not showing her inner turmoil. Really! First, he did not have the grace to ask for her hand himself – the fact that this had worked to her advantage was not the issue, thought Kathryn angrily – then he treated her as if she was no more than luggage to be dumped out of the way when she was no longer needed. On top of that, she was a laughing stock, due to his famous race a mere few days after the wedding. Now he was wagering with his cronies that his famous charm could make a fool of her. It was beyond enough. Emma was anxiously watching her friend's face, trying to gauge her expression.

'I feel there can be only one response to this,' Kathryn announced, putting aside her tea. 'I feel a very expensive trip to

Bond Street coming on. Shall we take the carriage? In case I cannot manage all my parcels!' she dimpled wickedly. 'Emma,' she said, throwing back the coverlet and reaching for the bell, 'I intend to be the most delectable wager the viscount has ever entered and little does he know it will be totally at his expense!'

Her aim was certainly achieved, if her success at the Earl of Rutland's masked ball was anything to go by.

William and Emma were dressed as Pierrot and Pierrette, while Kathryn had chosen, on the advice of Madame Zoe, to go as Circe, the mythical sorceress who had enchanted men with her beauty and invited them to their doom. Once under her spell the love-lorn men turned into pigs. Kathryn was pleased at the symbolism of her costume.

A simple green shift was belted in woven gold; her hair was a cascade of red-gold ringlets spun with pearls and small blue flowers which spilled over her bare shoulder. On the other, the chiffon was caught in a gold brooch in the shape of an owl. Her underskirt, also in gold, was heavily embroidered with gold and silver thread. The skirt of the shift flowed only to her knees, allowing the underskirt to glitter as she moved. She held a gold-painted staff, with the head of a pig. Her gold mask made her eyes almost cat-like. Kathryn's blood trilled through her veins and, in a fanciful moment, she wondered if some of Circe's power might be with her tonight. She hoped the enchantress would wish her well at least; after all, their intentions were the same.

Kathryn would not have been human if she had not enjoyed the stir of excitement that her entrance created. Eyes followed her every movement, admiring eyes, jealous eyes, and the eyes of her husband.

Rupert was chuckling to himself. The viscount eyed him sourly. 'I take it you have found an impregnable female?' If he was honest, the bet did not seem as good an idea as it had in Leicestershire. But he knew that unless he could think of a very good reason not to continue, he had no choice. The viscount was dressed as a pirate. Long boots and a sword at his side, he sported a black cloak with scarlet lining, his mask no more than a band over his eyes.

Finally Rupert exclaimed, 'Ah. If I'm not mistaken, your prize has just come into view.'

The viscount followed his gaze. Kathryn was dancing with a Roman soldier. The viscount watched appreciatively until the end of the dance, when she rejoined Emma and William.

'God's teeth, Rupert! That is Miss Chambers! I didn't know icebergs were included!'

'Too much for you, dear boy?' Rupert drawled. 'I understand she has already had the dubious pleasure of meeting you.'

The viscount remained silent, his thoughts racing. Of all the women, he had to choose one the very sight of whom irritated him. Rupert was watching him expectantly.

'How long until we settle the bet?'

'Two months,' Rupert replied promptly. The viscount thoughtfully swirled the remains of the drink around his glass.

'Very well,' he replied and, swallowing the remains of his drink, straightened his shoulders and muttering, 'Once more into the fray,' made his way directly to where Kathryn was seated.

Once again, he found himself gazing into Kathryn's coolly appraising eyes. Pushing away his misgivings, he bowed extravagantly over her hand.

'I must compliment you on your costume, madam. You will have every man here tonight dancing to your tune!'

'Yes?' replied Kathryn coolly. 'You are probably right. There are some men here who would definitely make very good pigs!'

For a moment Philip was stunned at the unexpectedness of her reply. Suddenly she found his eyes twinkling down at her as he laughed.

'I have often made that observation myself, you know.' A friendly smile followed.

'Indeed,' Kathryn said frigidly, forcing herself not to respond to his charm. She turned back to Emma, showing him an indifferent shoulder. Despite this distinct lack of encouragement, the viscount persevered.

'May I have the honour of a dance, madam? Even if I do run the risk of becoming even more of a swine than I already am.' Kathryn deliberately ignored this sally and consulted her dance card.

'I would be pleased to dance one of the country sets with Your Lordship.' Kathryn knew the viscount never danced country sets. His eyes flashed at this deliberate snub.

'What, no waltz for one who would risk even his humanity for one dance?'

'The amount of risk is dependent surely on the amount of humanity there was to start with.' Kathryn's heart started to beat a little faster, she was really being quite abominable to him. 'I'm sorry, My Lord,' she said, sounding far from penitent, 'all my waltzes are promised.'

The viscount, who had confidently expected her to accommodate her dance card around his wishes, did not enjoy the unaccustomed feeling of being thwarted. He made his excuses with as much grace as he could muster.

'Round one to the lady, I think!' Rupert said, watching the viscount retire, somewhat baffled, from the lady's company.

'But not the war,' Avon replied.

'Not yet,' Rupert chuckled.

Kathryn was well satisfied, everything was going according to plan, so far. But a small worm of guilt had started work on her conscience. He had looked so, so shocked when she had rebuffed him. And when his eyes were sparkling down on her, just for a moment, she wished. . . . There her thoughts just became more and more muddled. To regain her composure, she sat for a while in the conservatory.

Suddenly she heard voices behind the screen. Immediately she recognised the voice of the viscount, but she had no idea as to the identity of the second voice.

'So you are married I hear, to a great country beauty.' It was a female voice, husky and as smooth as velvet.

'Hah!' the viscount exploded. 'Yes, a great beauty for some incestuous village yokels. The Viscountess Devlin. Beauty enough to sour milk. A sad little donkey of a woman. For all I know she is still picking the hay seeds from her teeth and dirt from her nails!'

Kathryn's face burned. He would pay for that, she promised herself, this conceited, self-satisfied, arrogant, insolent, overweening. . . . Here, due to her gentle upbringing, she could think of no epithet vile enough for him. She swept back to the ball-room and proceeded to have a wonderful time; she was quite the belle of the ball. The viscount could not have reached her if he'd tried, because of the ranks of her admirers. So, with a gleeful Rupert in tow, Philip left early, to assuage his battered vanity by winning an obscene amount of money at Brooks's. Although Lady Luck was smiling on his cards, he could not banish from his mind the memory of a pair of sea-green eyes gazing fearlessly back at him.

Kathryn was up early the following morning as she had not slept at all well and, despite her success the previous evening, she was not in the best of moods. Feeling the need for exercise, she walked to Hookhams Lending Library to wander for an hour amongst its shelves looking for something that appealed to her. Her concentration had not been good of late; her mind would wander and she would find herself staring at the same page for hours and have no idea what was on it.

Dressed in a dark-blue wool gown with a matching bonnet framing her lovely face, Kathryn stepped out of the library into the bright autumn sunlight, and, not minding where she was going, suddenly collided with a dapper young gentleman walking in the opposite direction. She was about to apologise to him, when she suddenly froze, dropping her books once again.

'Lady Kathryn! Why, whatever are you doing here?' The kindly face of Richard Spires swam back into her vision. 'Are you all right? Here, lean on my arm.' Solicitously he supported her. Kathryn stared at him aghast.

'What are you doing here?' she stammered.

'I just asked you that! Where are you staying? Does His Lordship know you are in town?' Realizing that they were attracting attention standing in the middle of a busy thoroughfare, Kathryn pulled him along with her as she started to move.

'Not here. I'm staying at Cavendish Square. Return with me and I will explain everything.' Obediently Richard followed her. Now that the initial shock was over, he noticed her becoming new haircut and admired the trim figure she cut in her dark-blue walking dress and matching bonnet.

Kathryn had never been so grateful to see anything as she was to see the front door. Quickly she hustled him inside. Emma was in the parlour. When she saw Richard she almost dropped the vase of flowers she was holding. Flowers filled the whole room, mute testimony to a successful campaign to make her mark on society and on the viscount. Richard was almost as amazed to see Emma. He looked from one to another in shock.

'Richard, do sit down and stop acting like such a nodcock!' Kathryn said.

'Where on earth did you find him!' Emma almost wailed.

'Outside Hookhams.' Kathryn untied her bonnet and sank wearily into a chair. Richard was amazed at the change in her and couldn't take his eyes from her. A fact that was lost on Kathryn, but not on Emma.

'What are you doing here? Did the viscount send for you?' Richard questioned her.

'Richard, can I rely on you?' She fixed her luminous eyes on his. 'If you betray me I am quite undone!'

Richard's romantic heart skipped a beat. A lady in distress and such a pretty one.

'Will you keep a secret from the viscount, Richard? For me?'

Richard flushed slightly; he was certainly not used to beautiful women treating him as though he were their only hope. Kathryn watched his face closely as she continued.

'I wanted His Lordship to get to know me as Kathryn rather than as the wife he doesn't want. At least, we might get on together as strangers. He doesn't write them off as quickly as he did his wife. Then, even if he sends me back to the country for ever, I'll have tried, and I'll have a few memories to keep me company.' Kathryn resisted the urge to plead with him.

'I have no wish to cause you any grief, My Lady. If it is your wish, then I will mention nothing to the viscount.'

A beautiful smile bloomed on Kathryn's face. 'Oh thank you,' she cried, 'I should have known I could rely on you.'

'I hope you will allow me to call on you and see how you go on,

and, of course, to offer you any help that I can.' Kathryn continued to beam at him, and Richard, a slightly dazed expression on
his face, smiled back.

Fourteen

Philip Devlin was not the only one facing set-backs in his romantic aspirations. Sir Rupert Roscommon had not met with the response he expected when he handed Dinah his poem, 'Ode to Her Smile'. Her mama had firmly insisted that she return it unread, as it was not suitable for an unmarried girl to receive such gifts. This indomitable lady continued that, of course, once Dinah was married it would be for her husband to decide. Rupert's volatile nature had been hurled into despair by this. He had not left his lodgings for days and, during that time, he had consumed copious amounts of claret and brandy.

The viscount eyed his friend in exasperation. Trying to control his temper, he stalked to the window and by dint of his broad shoulders finally let in some fresh air.

Rupert blearily opened half an eye, to watch his friend's efforts. He was sprawled over a table, oblivious to the parchments under his hands. He started to giggle.

'What a fine nursemaid you make, my friend! Do you wish to clean them while you're there? Then, if I feel the urge to dash my body to the ground, I shall reflect that despite everything, at least I left my room with clean windows!' The viscount glared at him sourly but made no reply.

Rupert brought a glass to his lips, discovering that it was empty, swore and hunted furiously for a full bottle in the debris that surrounded him.

'If you think this will impress the dragon lady mother of your beloved, I fear you are very much mistaken. All you are doing is convincing her you are a bad lot and that she was right in the first place.'

'It doesn't matter. All is lost. My life is over!'

'You sound like a twopenny drama,' the viscount said dismissively.

'Yes. Two pennies is about right. How can I have the temerity to think that a goddess like Dinah would consider a man like me, with nothing to recommend him but his loving heart!'

The viscount poured himself a drink. If he had to listen to much more of this, the drunker he got the better.

'I don't wonder at her conclusion if you think so little of yourself! Why not leave the decision to Dinah? If she is as wonderful as you say. . . .' Philip let the sentence hang.

'How can I expect her to incur the wrath of her family, and the censure of society in order to marry me?'

'The only way is to confound the old tabbies and prove them wrong by behaving in exactly the opposite way to what they expect! Become the perfect suitor.'

'It will do no good. I am not an earl, and my fortune is no more than modest, even with my expectations from my godmother.'

'So, they are right, these old biddies, you are spineless and you display a remarkable lack of breeding!'

Rupert, astounded by his words, immediately leapt to defend himself.

'I'll have you know that there were Roscommons who accompanied the Conqueror and my family signed the Magna Carta! I count among my ancestors some of the noblest families in England.' He took a deep breath. 'And if you think. . . .'

The viscount, unable to contain himself any longer, let out a burst of laughter. Rupert, realizing he had been manipulated, smiled rather shamefacedly.

Characteristically he ran a hand through his disordered locks.

'You really think it's worth a try?'

The viscount shrugged. 'After all, what have you got to lose?'

Admitting the truth of this, Rupert eyed his friend and asked, 'When did you become such an expert?'

'If you had taken your face out of the bottle for a few moments you would have realized that yourself.'

'Is it all so hard and fast with you? Black and white? Will nothing break through your reserve?'

'If you mean will I ever make such a cake of myself over a woman, the answer is, most definitely not. I do not allow my emotions to rule my life. And I cannot see what you gain in doing so.'

'All right, enough of the lecturing. As you are so insufferably righteous, you can buy this poor emotional wreck a hearty breakfast.'

'I'll do better buying you a watch and chain,' the viscount retorted, 'as it is now dinner-time!'

Unfazed, Rupert grabbed his coat. 'Good. In that case I can start with breakfast and work my way round to the splendid dinner I'm sure you will supply!'

Emma and Kathryn were seated in the drawing-room awaiting the arrival of Aunt Agatha.

'There's no need to watch the road, you know,' said Emma. 'It will be at least half an hour after she arrives that she will finally make it in here.' At Kathryn's puzzled glance she explained. 'We're on the first floor. Aunt Agatha is not greatly enthusiastic about stairs, although she did say in her letter that she had lost some weight. She has discovered a wonderful new doctor, and a miracle diet. Something to do with vinegar.' Emma frowned at Kathryn's giggles. 'Aunt Agatha is an authority on diets as you will discover!' she finished in a dire voice.

Kathryn continued to arrange some yellow roses in a bowl, a gift from one of her many admirers.

'Who are they from?' asked Emma.

'Your beloved cousin Philip Devlin,' Kathryn replied, a trifle smug. 'Rather unoriginal, don't you think?'

'But rather splendid for the time of year,' Emma pointed out. 'Here, these are for you.' Emma handed Kathryn a pile of letters. Kathryn opened the first and blanched slightly at its contents.

'How can I have spent that much at Madame Zoe's?' she said in horror. She opened the next to find that it was another bill, and the next, and the next. 'Whatever can I do?' she said in alarm.

Emma pooh-poohed her concern. 'If you wish, William will pay them immediately and you can repay him next quarter. Although no one pays their dress-maker after the first bill.' She considered a second. 'And I'm sure they don't expect you to.' Kathryn reflected ruefully that it was only the well-heeled who would make such statements.

Emma, seeing the look on her face, said bracingly, 'Come, put them on the fire and forget them.' Before she could continue, a rumpus erupted on the street below. 'Quick.' She gestured Kathryn to her side.

A carriage had drawn up and orders could be heard given in an increasingly frantic female voice. A large two-step mounting block was brought round to the carriage door. The door then opened, a pudgy hand was extended and was taken by the largest and newest footman. The carriage started to rock, slightly at first but with growing impetus. All that could be seen of the occupant was a large hat, liberally strewn with silk flowers, which was pushing out of the door. The footman grabbed the other hand, and exerted every ounce of his strength and with one final heave Aunt Agatha's shoulders were freed from the door.

Kathryn and Emma could not contain their laughter, tears running down their cheeks and their sides aching. Aunt Agatha's voice could now be heard more clearly giving conflicting commands.

'How on earth does she get in the coach?' Kathryn gasped.

'Maybe they should sit her on top!' Emma suggested, dissolving into fresh gales of laughter.

'They should harness her up, and give those poor horses a rest!' By the time they had recovered some composure, Aunt Agatha had been deposited on the street.

'But this isn't the end of it,' chortled Emma. 'She still has to get up the stairs.'

'Maybe we should get Harold to give her a helping hand?' suggested Kathryn. 'After all, he has been in training with his mother.'

At last Aunt Agatha reached the drawing-room, leaning heavily

on the arm of a copiously sweating footman. Once her bulk had
been arranged on the sofa, she needed a few minutes to catch her
breath, and some hefty sniffs at her sal volatile. Now Kathryn had
time to realize how vast Aunt Agatha was; she had never seen
anyone that overweight before. Her corpulent figure was truly
awe-inspiring as was her predilection for girlish bows witnessed by
the prodigious number strewn around her person. Fascinated,
Kathryn could hardly take her eyes off her. Tea arrived and at last
Aunt Agatha recovered enough to converse with them.

'Oh! My dears. What a journey! I declare I am quite exhausted.'

'You're looking very well, Aunt,' Emma protested.

'No thanks to that upstart doctor! I declare I've been half dead
with hunger! Then he tells me that I am still eating too much! A
morsel of toast in the morning, a light lunch, only five courses at
dinner, and no account taken of my constitution. There is an
imbalance which means I gain weight from next to nothing to eat,
I assure you.'

'I thought the vinegar had helped, Aunt. Although I find it a
most revolting idea I must admit,' Emma said.

'Well, now I have the answer. I have been told by the highest
authority that a pound of apples before every meal is the answer.
Something to do with the acid. Anyway, it stops the body making
fat,' she nodded sagely.

'I should think it would make you very ill,' Emma commented
frankly.

'No, my dear, you don't understand. This is scientifically
worked out.'

Kathryn avoided Emma's eyes, afraid that she would start
giggling again.

'You should rest, Aunt. Will you be joining us tonight? We are
invited to a small card party.'

Aunt Agatha's expression lifted. 'Cards and a light supper
sounds most agreeable. I shall rest now so as to be at my best.' She
attempted to stand up. After much heaving she finally managed to
stand unaided. Luckily she had been given a room on the same
floor.

'Just in case all those stairs give her a stroke!' Emma remarked.

'How on earth did she get to be that size?' Kathryn asked incredulously.

'You won't ask that once you have dined with her!'

Kathryn laughed.

It was this image the viscount was presented with as he stood on the threshold of the room. His heart caught for a second as he watched Kathryn, sitting in the window. The afternoon sun had turned her hair to liquid gold and, for a moment, she was surrounded by a halo of light, making her into a creature too beautiful to be completely real. As soon as she became aware of his presence, the laughter fled from her face leaving a peculiarly closed look in its place. Philip wondered what she was scared of, under what trouble she was labouring. Somewhat to his own surprise, he found himself wishing he could banish that look from her face forever. He dismissed the thought quickly, before he acknowledged that he would sacrifice himself and large chunks of the human race just to spare her a moment's pain.

Kathryn, seeing him enter, made no move to welcome him, and responded with only the briefest of nods to his most winsome smile. The viscount refused to be deterred.

'Philip!' cried Emma. 'You just missed Aunt Agatha.'

'Your Aunt Agatha,' he corrected. 'I am no relative to that battleship of a woman. You should watch William, in case it runs in the family!'

'Have you just come to criticize my family-in-law,' asked Emma, 'or have you come to make yourself useful?'

The viscount seated himself and regarded Emma through his eyeglass, but before he could answer Kathryn interrupted.

'Why Emma! What a foolish notion. Your estimable cousin is useful just by being no more than himself. A true original. What more than what we see before us, could we possibly ask?' Her tone was nowhere near as innocent as her eyes, and she returned the viscount's sharp look blandly. Emma smothered a laugh.

'Fie!' she retorted. 'Philip is a lot more useful than he looks.'

'Shall I leave you to a discussion of my many virtues? Or would you rather hear my reason for honouring you with a visit?' Somehow he felt his sense of humour slipping.

Both ladies rearranged their features into expressions of polite interest. 'Please,' Kathryn said, 'we are agog.'

'I have decided to allow you, Emma, to hold a ball at Devlin House this season. At least, I hope you will consider being my hostess. Whatever arrangements you wish, at my expense of course.'

Emma stared at him in amazement. The viscount felt rather smug at finally seeming to render them speechless. Emma stared at him suspiciously.

'You said that never in a thousand years would you allow me to fill your house with buffoons, stuffing themselves at your expense and quaffing your expensive wines!'

'I changed my mind,' Philip said carelessly. 'If of course you don't wish to be mi'hostess I'm sure I can find someone to oblige.'

This deliberate barb made Emma exclaim, 'Don't you dare!' Excitement had lent her cheeks colour, and her brain could almost be seen feverishly making plans.

Annoyed at the viscount's success in manipulating Emma, Kathryn decided it was time she entered the fray.

'Oh dear! I have the most shocking memory. Please forgive my ignorance.' Her innocent voice seemed guileless. 'But why do you need Emma as a hostess?' She raised her eyebrows questioningly. 'Surely your wife will perform that role?' Kathryn's heart was beating so fast that she was surprised that they could not hear it. Emma blanched slightly as she waited for the viscount's reply.

'My wife is indisposed.' Philip's voice was tight with rage, his nostrils pinched white. He hadn't realized that she knew he was married.

'How disappointing,' Kathryn said, with great sympathy. 'You must be longing to introduce her to your friends and family. I wish her a speedy recovery and look forward to meeting her.' With that Kathryn swept across the room, hiding her shaking hands and resisting an impulse to run. The viscount's expression was volcanic. Kathryn wanted to be as far away as possible when he erupted.

Philip stared after her in bafflement. Never before had he been treated in such a fashion. This chit of a girl dared treat him as if he was of no more importance than the merest moonling. As his first flash of anger evaporated, he was surprised by his dismay at her behaviour. She knew of the existence of his wife! He felt an urge to try and explain to her that it wasn't his fault, that if it had been his choice. . . . He drew his thoughts in sharply and he dragged himself back into the present to find Emma regarding him curiously.

'Are you all right, Philip?' she asked. 'Does that expression mean that you have changed your mind?'

'No. No. Of course not. I had not realized that my marital status was common gossip!'

'But Philip, why should you keep your marriage a secret?' At this, she told Kathryn, his face looked as though he were smelling old milk. In fact, he was so wrapped up in his own thoughts that before he could realize it, he had agreed to accompany them to Lady Felton's card party that evening. Even with Aunt Agatha!

'I have never seen him behave like that,' she said musingly, 'and I'd have bet a fortune that he would never accompany Aunt Agatha anywhere. Let alone to a fashionable squeeze.'

Kathryn shrugged with as much carelessness as she could manage, but avoided meeting Emma's eyes.

'So far so good,' she said, with a lightness she was far from feeling.

'Yes,' said Emma doubtfully. 'Only. . . .'

'Only what?'

'I have never seen Philip so vulnerable before.'

'Hardly vulnerable,' said Kathryn. 'He's as vulnerable as a wild boar!'

'No. I mean I've never seen him concerned with what was said about him before.'

'Nonsense. All he's interested in is winning his bet,' Kathryn said, much more firmly than she felt. Emma smiled but said nothing. Feeling the silence, Kathryn said, 'So how do you intend to make this ball the most spectacular of the season?'

Emma smiled broadly. 'By spending the most vulgar amount of Philip's money, of course!'

Fifteen

Aunt Agatha was dressed in black with silver bows. Kathryn wondered idly how many there were in all, or how long it might take to count them. She too had chosen silver for this night, silver and white highlighting her flame-like hair. As straight as a candle and dressed in moonlight and starshine. The viscount had manoeuvred arrangements so Aunt Agatha, William and Emma travelled in one carriage, while he and Kathryn had another to themselves. Kathryn regarded his profile from under lowered lashes, his strong chin and slightly aquiline nose. Feeling her scrutiny he smiled down at her.

'Emma will be calling me a wretch. What space will be left them I can hardly imagine!'

'You deserve much worse!' Kathryn laughed. 'Poor Emma's dress will be quite crushed.'

'Not to mention the poor horses,' the viscount added. Kathryn giggled and visibly relaxed. Hoping to keep her in this cordial frame of mind the viscount continued, 'Doubtless she will have her revenge. Once, after some childish prank of mine, she filled my riding boots with porridge and sewed up the arms of all my jackets!'

'A bad woman to cross,' said Kathryn. 'It must run in the family.'

'Aha! I see that my dismal reputation has preceded me. How can I prove myself worthy of a smile or a waltz?' He pulled a face and Kathryn continued to giggle. Just as he thought he detected a softening in her attitude, the carriage lurched to a sudden halt and raised voices could be heard.

'What the dev— I mean what is the meaning of this?' The viscount thrust his head out of the carriage window.

'Sorry, Mi'lord. This ruffian blocks the road.' A horse was desperately trying to pull a cart loaded high with timber. His bones showed pathetically through his shabby coat. Kathryn could see the whites of his eyes as the horse strained to obey his master. A small boy stood on the side of the road, tears streaming down his face, his wide eyes fixed with horror on the scene before him.

'You worthless old nag! Not worth the feed! It's the knackers for you. You lazy sod!' Each of these threats was punctuated with liberal use of his master's whip.

Kathryn peering through the coach window was horrified. Without a second's thought she jumped from the carriage and bore down on the whip-bearing driver.

'Stop that this instant!' she roared. Stunned, the driver turned expecting to face a formidable matron, instead he was faced by a diminutive girl dressed in spangled silk. A light drizzle was falling and drops caught in Kathryn's hair, reflecting the light, sparkling like diamonds. 'How dare you treat that poor creature in such a fashion? It is you who should be whipped!'

The carter just laughed and raised his whip again. Infuriated, Kathryn prepared to wrestle the whip from his grasp. But before she could blink the viscount appeared from nowhere and, grabbing the astonished man by the throat, pinned him against the side of the cart. Kathryn watched wide-eyed in amazement. The child ran to the horse as he stood panting weakly, bloody weals rising already on his back.

'I think you should apologize to the lady, don't you?' The viscount's grip on the carter tightened perceptibly. 'Or shall I horse whip you as the lady desires and you so richly deserve?'

'Now, now sir. I didn't mean no 'arm. No offence meant, Your Lordship. A man has to earn his crust, you know.' A wheedling tone entered his voice. 'It's hard, young master, for an honest man.' He eyed the viscount surreptitiously.

'If there were not a lady present I'd see you got what you deserved. You rogue.' While they were talking, Kathryn had turned towards the horse left trembling between the shafts, and its young friend. Still absolutely outraged, she turned to the driver once again, her eyes flashing. She picked up the whip. Even the viscount felt a moment's trepidation on seeing the expression on her face. Her eyes glittered, the only sign of life in her stony face.

'Would you like a taste of your own medicine, you scurvy dog! Do you beat women and children as well as dumb animals, who have no protection from scum like you?' She paused for breath and Philip could see a pulse beating wildly in her throat. She turned her brilliant eyes to the viscount searching his face. The viscount glanced towards the child whose huge eyes were still filled with tears and horror. Feeling the spotlight on him, Philip rose to the challenge of what to do next magnificently.

'Damn it all!' he said through clenched teeth, as his fist slammed satisfyingly on to the wretched driver's chin.

'Bravo!' shouted Kathryn, totally carried away. The small child stopped crying and even the horse looked somehow perkier.

'Peters! Take this cur to the Watch and see him soundly whipped.' He gestured to his groom and driver, who obediently leapt forward. Freed of his prisoner, he turned towards the horse. Kathryn watched in admiration as he dealt with everything with such assurance, with such competence. You could feel safe with such a man, Kathryn thought, no matter what happened. She turned her attention to the small boy.

The viscount led the horse from between the traces. Once the load was removed from his back he moved with surprising liveliness. The viscount rubbed his nose and mumbled some reassuring nonsense. Just then the second coach party arrived on the scene. William came hurrying towards them, his face puckered with anxiety.

'Whatever is amiss?' he asked, his eyes taking in the chaos around them that Kathryn had hardly even noticed.

'Nothing that you need worry about. I have just had the pleasure of purchasing a horse.' He tossed a guinea to the other groom and gave orders for the man to be paid and the horse taken to his stables. He then turned to Kathryn who was holding the hand of a small, grubby boy.

'My Lord, he is from the poorhouse. That wretch has treated

him terribly.' Her green eyes turned to him in absolute faith that
he would know what to do. Swallowing a sigh he consigned the
boy and the horse to Peters' care with orders to send them both to
Kingsgrey. As soon as possible, he added with feeling.

With no further fuss, Kathryn found herself back in the carriage
and once again alone with the viscount.

'I don't quite know what to say,' Kathryn said shyly.

'And you're not usually lost for words!' They both laughed.
Suddenly Kathryn was serious once again.

'You were absolutely magnificent,' she said earnestly. The
viscount felt a warm glow at her words and started to thoroughly
enjoy her admiration. So far in their acquaintanceship this was a
novel sensation. Kathryn saw the self-satisfaction settle on his face
and resolved to prick that little balloon straight away, just in case
he thought he had succeeded in thawing her.

'You have made a friend for life, you know.' Deliberately she
made her voice simper to him. The coach pulled up at Lady
Felton's residence and Philip, thinking that he finally had the
perfect opportunity, bent his face close to hers.

'And what do friends for life receive?' he said softly, wondering
whether or not to kiss her. Suddenly she opened the door and
whisked away.

'I can't imagine what you would want to receive from a horse!'
she laughed, 'but I'm sure you'll think of something. You have had
such magnificent instincts in this matter so far!' With that she took
Emma's arm and prepared to make her entrance.

The viscount was speechless. He simply stared at what was now
an empty space.

'I say, old fellow. Whatever is the matter?' William asked, concerned at Philip's expression. Suddenly Philip slapped William's
shoulder boisterously.

'I have been snubbed! Quite deservedly I may add,' he said,
laughing at the confused look on William's face. 'I actually think
there is a chink in that armour after all. I'll give Rupert a run for
his money yet!'

Lady Felton's horse-like face was astonished. What a coup!
Viscount Devlin arriving early at her party. Her complexion took

on a red tone which clashed with her emerald-green silk dress. Fair
or grey-haired women should never wear green, Kathryn thought.
It made Lady Felton look positively jaundiced. The hostess was
quite overcome by the honour bestowed on her by the viscount's
more than punctual arrival. She fluttered over him with a fawning
smile on her face. Kathryn watched her distastefully. Really if
that's how people spoke to him, no wonder his consequence was
so inflated.

There were three gaming-rooms and a cold collation was laid
out in the dining-room. Kathryn, who had never gambled in her
life before, watched wide-eyed at various tables, wondering if she
dare try a hand herself. Suddenly she found the viscount at her
elbow.

'Why don't we try *rouge et noir*?' he suggested, taking her arm.

'I'm not sure I'll be good at it. It all seems very fast and for so
much money!' Kathryn confided, enjoying the security of his arm
in surroundings so new to her. The viscount was relishing his role
as her champion once more.

The evening was a great success. Not only was Philip making
some headway in his bet with Rupert but, he realized with some
surprise, he was also thoroughly enjoying the company of this
most singular of women. He had to admit to himself he had never
met anyone quite like her before.

As he placed bets to her commands, he teased her that she
would soon be a most hardened gamester.

Kathryn retorted, 'I see that you, of course, know exactly what
a hardened gamester is like!'

Philip laughed at her sparring and, without actually noticing,
she started to respond to his charm.

Normally the viscount would never have been seen betting on a
little ball and a wheel; his usual territory would have been the deep
play on the faro table, but Philip didn't notice any desire to leave
her side. Somewhat ruefully, he realized he was behaving as the
captivated, rather than captivating.

Others had also noticed his unusual behaviour. Gossip spread,
flowing from group to group throughout the party. Many inquisi-
tive eyes watched them, not all of them with malice, but all of

them with astonishment.

A woman dressed in red and black, highlighting her dark beauty, watched the pair with increasing outrage.

'Who is that?' she demanded of the cavalier at her side. Languidly he raised his eyeglass in the direction she indicated.

'Why, your replacement by the looks of things, mi'dear,' he drawled. 'Poor Charlotte, have you lost your viscount? I warned you not to play fast and loose. He's not one who'll put up with your games. Now a much younger girl has replaced you in his affections.' Ignoring her livid expression, he continued, 'Friend of the Courtneys I hear. Nice gel, so I'm told.'

Through clenched teeth she turned on her hard-hearted companion. 'Take me home. Now!'

'What, afraid to enter the lists, my dear? Well, at your time of life it is to be expected. Can you lean on me or would you prefer one of the footmen?' Her reply was to almost drag him up the stairs. They headed for the doors, one of them chuckling the whole time.

Kathryn was in fine fettle. Her pile of chips was steadily growing bigger.

'How much have I won, My Lord?'

'I should say around one hundred pounds.'

Kathryn stared at him in amazement. 'Are you sure?'

The viscount laughed at her incredulous expression.

'Now that you have made your fortune, can I interest you in an ice or glass of punch?' Kathryn readily agreed, more than a little shaken at the amount of money she could have lost. When the viscount presented her with a pile of notes, however, she could not contain her glee.

Kathryn and Philip strolled on the balcony. It was now a fine autumn evening with a hint of bonfires in the air. She was magnificent. Her spangled silk dress clothed her in radiance and the stars in the sky could not compete with the stars in her eyes. Philip was bewitched.

Suddenly, aware of his scrutiny, Kathryn turned towards him. Before she could protest he swept her into his arms, claiming her lips and leaving her breathless.

Unbeknown to them they were watched by several interested pairs of eyes. Rupert, who had been scrutinizing the viscount's progress all evening, watched wistfully.

'And what of the little wife?' Avon's voice startled Rupert from his reverie.

'You don't mean Philip's serious!' Rupert's eyes widened in surprise. 'No. It is the bet. His pride won't let him lose, that's all.' He didn't sound convinced, and regarded Avon with a troubled expression.

'No. He will win the bet, I'm sure. Only—'

Avon paused reflectively. 'Only what will he lose to do so? And, more to the point, can he afford it?'

Sixteen

Kathryn paced the room restlessly. Meanwhile Emma sat at her desk surrounded by lists. Lists of flowers, chairs, people, tables, wine, champagne, everything that Emma simply had to have in order to throw the ball of the season. Kathryn could not concentrate. Ever since Lady Felton's card party, she hadn't been able to settle to anything. Emma had a fair idea of what the problem was, but sensibly she kept her own counsel.

'I cannot make up my mind!' Emma said in exasperation. 'You must help me, otherwise I shall be grey by the time I have worked my fingers to the bone to ensure that Philip throws *the* ball of the season!' Kathryn smiled. There was nothing on earth that would have stopped Emma organizing Philip's ball. But it was a real problem finding an original idea or theme, everything had been done at least twice.

'We've racked our brains already. There cannot be anything that has not been done.'

Emma glared at her friend. 'We must! I will not have it said that it was just another squeeze.' She nibbled her thumbnail in concentration.

'What about mythology? Or astrology? Or nursery rhymes? Or dress as your husband?' Kathryn succeeded in making Emma laugh. 'Flowers maybe or a country fair? What about characters from the Arabian Nights?' Emma sat up at this. Kathryn continued, 'Or a slave market? What about one of the seven virtues or seven sins?'

'Aunt Agatha would never forgive me! Unless she went as gluttony, we would be the laughing stock,' Emma exclaimed.

'We could always just write all of them down on a sash or something and wind it round her!'

'Do you think seven would be enough?' Both girls collapsed into giggles. The door suddenly opened and the topic of conversation entered.

'I hope I'm not interrupting you.' Aunt Agatha sailed across the room, rather like a long procession of one person, Kathryn thought.

'Oh no! Of course not. I wish you would help us decide on a theme for Philip's ball.' Aunt Agatha settled in her chair, which emitted several worrying creaks.

'I always find the simpler the idea, the better it works,' she started.

Emma interrupted her. 'What we need is an idea no one has used before.'

Aunt Agatha blew out her cheeks. 'Well, I'm told that King Arthur's round table will hang at Westminster soon. They cannot say if it is the original, of course, but it would make a pretty party.'

'Why! Aunt Agatha. I do believe that to be a tremendous idea!' Emma started to enlarge on the theme. 'I can decorate the ballroom as if it is Merlin's cave, and have the sword in the stone, and I shall build a dragon of flowers!' Both Kathryn and Aunt Agatha looked askance; neither thought this a good idea, but prudently refrained from saying so. Suddenly Kathryn had an inspiration.

'What about making the theme the sea? We could make an underground grotto in the ballroom and we'll make the dining-room into King Neptune's hall.'

'We could make designs from shells and flowers for the tables,' Emma said, with growing enthusiasm. 'I'll have a waterfall in the conservatory, a better one than that cheapskate Maria Henshaw, and I'll colour the champagne and the punch.' She started to scribble frantically in her notebook. 'Blackbeard's treasure will be next to the waterfall, with a few pirate skeletons and a . . . a shipwreck!'

Kathryn wondered whether she should have remained silent. Emma's plans were becoming more and more lavish. Telling herself firmly that it was the viscount's problem, she said, 'I think

I'll change my library books. Is there anything I can get you?' Both Emma and her aunt were deep in a discussion on the rival merits of Vales or Kentons for the catering, so hardly noticed her leave.

Kathryn walked thoughtfully to her room. She had not seen the viscount since the Feltons' card party. When he kissed her it was as if her whole world paused for that moment. She had never been kissed before and it left her with a welter of conflicting emotions.

A walk would do her good, she decided, help clear her mind. She pulled on a frivolous bonnet of pink ruffles and swansdown feathers. Looking at her reflection she wondered at the fact that she still looked the same as before. Much like her sister, she expected to see a different person in the mirror.

As she collected her library books from the parlour, a footman knocked apologetically on the door.

'Lord Etherington begs leave to wait on you, madam.' Kathryn was somewhat startled.

'Is Lady Etherington with him?'

'No, madam.'

Harold on his own! Kathryn could hardly credit it.

'Lady Courtney will be glad to receive him I'm sure.'

The footman cleared his throat apologetically.

'He asked for you, madam.'

Puzzled and a little exasperated, Kathryn agreed to receive him.

Harold arrived at the door of the parlour and, on finding Kathryn alone, launched into what seemed to be a much-practised speech.

'Apologies for this untimely interruption, madam. I can see you are just about to go out, but I wonder . . . I would like to invite you to a picnic I am organizing for tomorrow. Three carriages and riders to go to Kew, where we shall lunch on the lake. I hope you find this plan agreeable, and I would very much like you to . . . I . . . I mean, you would honour me by accepting.'

Kathryn listened to his speech in growing astonishment. Surely Lady Etherington had not let him off her apron strings?

'Why Harold, who else will be in the party?' she asked, hedging for time.

'I have invited your good self first, so I cannot definitely say. I have given it much thought, and I am sure you will find my choice of companions most edifying.' Kathryn was not sure that she knew what that meant.

'Will you be asking Lady Courtney and her aunt?'

'If you wish it, naturally I shall extend my invitation to these ladies.' His face had taken on a complacent air. Kathryn was also sure that he had started preening slightly.

'Shall I see your dear mother tomorrow?'

Harold licked his lips nervously, and the smug air was replaced by one of definite tension.

'Unfortunately, my mother has been called away most suddenly, so I shall be hosting this party myself.' The last words were said with such firmness to show the timid nature of the rest of this speech.

Rather intrigued by the incipient rebellion she sensed, Kathryn agreed. A pleased flush mounted his cheeks at her words and he excused himself to wait on Emma and Aunt Agatha.

Mulling over to herself the incredible sight of Harold doing anything without his mother, Kathryn made her way to the library. Once there she could find nothing to tempt her fancy and left empty-handed.

On her return, she found the viscount taking tea with Emma. They had been reading through the reams of lists Emma had written to organize the ball.

'Come and have some tea to warm you up,' Emma said. Kathryn had not seen the viscount since the card party and, quite illogically, she was pleased to see him.

'I cannot decide what to do for the best,' Emma said in exasperation.

'Just because I will not let you dye my champagne blue, or force my guests to eat seafood when it is out of season, or let you burn the house down with a dragon, or flood it with an artificial lake!'

'All right. All right. I admit they might have been a little extreme, but, my clever cousin, what are you suggesting?!' Kathryn sipped her tea and watched the exchange with interest.

'Why not,' he paused for dramatic effect, 'as it is Hallowe'en. . . .'

'Oh not a Hallowe'en party, Philip! That's hardly an original idea,' Emma interrupted crossly.

'If you will let me finish what I am saying,' Philip continued, waiting for their attention. 'What I propose is that the ball is held in Hell.' They stared at each other in silence for a moment.

'People could come as devils or sirens, or even,' Emma's voice grew more and more excited, 'people that should be in Hell past and present.'

Kathryn had to admit that it was a very good idea and wished she could add to it. Emma continued to chatter as she started making fresh plans. The viscount turned to Kathryn.

'What do you think, have I made myself useful at last?'

Kathryn smiled at him, almost forgetting herself for a moment. Calmly she placed her teacup on the tray.

'Indeed, I should have expected such a thought from you.' Turning to Emma she continued, 'I suggest that you pave the way to Hell with good intentions. Down the entrance stairs, between the ballroom and gaming tables, on and around the supper tables perhaps?'

'What an excellent idea!' Emma said, chuckling to herself. 'Oh! This is famous, we shall throw the ball of the season.' Once again she started to fill sheets of paper. The viscount watched her fondly.

Kathryn was surprised yet again by this complex and fascinating man. His sincere regard for Emma and Miss DeWinter was a complete opposite side of the man she thought she hated. Her thoughts were interrupted.

'I wonder if you would care to visit your protégés. They are still at my stable here. If you wish it, I can send the horse to a happy retirement at my estate at Kingsgrey with his young protector as a groom.'

Kathryn thought how far away Kingsgrey seemed now.

'I would love to see them before they go,' she replied.

'Indeed. You will hardly recognize either of them. By the by, have you thought of a name for your first horse?'

'I shall have to wait for inspiration to strike, I'm afraid. I fear I may have to ask for help.'

'I shall always help you,' Philip said quietly. 'Never fear, you can always trust me.' Startled, Kathryn looked at him wondering whether she had heard him correctly. A little surprised at himself and more than a little embarrassed, he turned back to Emma and the planning continued apace.

'I saw Harold without his mother!' said Emma. 'I can't say I fancy the party though. Even if it is the one and only time poor Harold will be let off her apron strings. It would be best for me to concentrate on the ball, I think.'

'Oh Emma! I have already accepted! You can't sentence me to Harold with only Aunt Agatha's support!' The expression of dismay on Kathryn's face made both her companions laugh. 'It is not amusing! I don't even know who else will be in the party.'

'I shall be happy to escort you and Aunt Agatha and I shall also endeavour to stop Harold boring you to death.' Kathryn listened to this in amazement. Her wide eyes searched his face, maybe he was joking after all. 'Unless of course you prefer his company to mine?'

Before Kathryn could reply Emma said, 'What an excellent plan. Now I shall have no need to feel guilty!' Kathryn and Philip's eyes met and they both smiled in silent communication. Emma, oblivious to this, continued, 'You can tell me what I missed, so remember everything. I'm sure Lady Etherington will also be interested.'

It was a soft autumnal day, the leaves on the trees blazed gold and red, the last dazzling display before the rigours of winter. Two carriages and three riders set out. Kathryn was in the first carriage which she shared with Dinah and Mr Parsons; behind them were the two Misses Drake who shared with Aunt Agatha. Harold, the viscount and Sir Rupert rode close to the carriages.

Kathryn enjoyed renewing her acquaintanceship with Dinah, who reminded her of Arabella. She was glad to be spared the company of either of the Misses Drake, two lemon-faced young ladies of no great fortune, or intelligence. Harold informed her that they were his cousins, which did not surprise Kathryn at all.

Mr Parsons was a delightful old man, unable to ride because of his gout, but his discomfort did not stop him regaling his companions with amusing stories, most drawn from his time at the court of George III, whose peculiarities caused both commoners and the Prince Regent much amusement.

Harold was most particular in his attentions to Kathryn, pulling his mount close to her side of the carriage. On the other side Rupert was unable to take his eyes from Dinah. Cut out, His Lordship had to put up with the gushing and positively mawkish behaviour of the Misses Drake, which did nothing to improve his humour. Harold had not been best pleased with the inclusion of the viscount in his party, and was determined to keep Kathryn's attention to himself. Kathryn was horrified when Harold's intentions were apparent. The whole day would be spent listening to Harold prosing on, Kathryn thought she would expire from boredom. Luckily, Dinah was no more enamoured of her company, so the two girls determinedly spoke to each other, turning a shoulder on the men.

Aunt Agatha watched this behaviour with amusement; when she chuckled both her chins wobbled.

'In my day the men had a little more gumption, and the women more spirit,' she sniffed, and fixed her beady eyes on the girls in front of her.

'Surely every young lady should obey the wishes of her parents?' Miss Drake the elder asked.

'Only if they have to,' Aunt Agatha replied.

Seeing themselves effectively dismissed, Rupert drew up to ride with the viscount and Harold went ahead to lead the party, much to everyone's relief. Rupert was scowling darkly at Harold's back.

'Has our host displeased you?' the viscount asked lightly.

'How has he managed to put together such an ill-suited party? I fear today will be a great trial for us both,' he said pessimistically.

'I would have thought this a great opportunity to get your Dinah alone.'

'What do you mean?' Rupert demanded, giving the viscount his full attention.

'Aunt Agatha will make no objection, and there can't be anyone here that she would prefer surely? I'm certain you can manage to lose yourselves in the gardens.'

'As you intend to do, no doubt!' retorted Rupert.

Kathryn unfurled her parasol, in the palest of pinks to flatter her complexion. Her dress was in a slightly darker shade, trimmed in matching ribbons. Even sitting next to the Divine Dinah the most celebrated beauty of the day, did not detract from her loveliness. In fact, Dinah's colouring proved to be the perfect foil for Kathryn's auburn flame beauty.

'Have you visited Kew before?' Dinah asked Kathryn.

'No, this will be my first visit,' Kathryn replied. 'Lord Etherington has told me about the oldest tree in England.' She tried to sound enthusiastic and smiled ruefully at Dinah. 'I must confess gardens have never really fascinated me.'

Dinah laughed. 'I know nothing about them myself.'

They soon arrived at their destination. Luckily, it was easier for Aunt Agatha to get out of a landau with the top down than it was a closed carriage and she negotiated her descent with no problems.

Harold started busily to organize them but Aunt Agatha demanded refreshment and insisted on going straight to the picnic site. The unfortunate Harold tried to insist that the party stayed together, but with no success. As the host, he escorted Aunt Agatha and his cousins to the picnic site, while the viscount and Rupert, seizing the chance, escorted the remaining ladies around the botanical garden.

'I'll catch up with you, never fear,' Harold said. Kathryn saw this as a threat rather than a promise, especially as his words were accompanied by a knowing smile. It was with a sense of great relief that she saw him leave. Now, however, she had another problem, spending time alone with the viscount, but neither girl would let the other out of her sight as it was unthinkable for a girl of respectable reputation to be alone with a man.

Rupert and Philip, however, managed to arrange two quite separate conversations without too much trouble.

The viscount and Kathryn were seated on a bench overlooking a pond. Kathryn was very tense and uncharacteristically quiet. Philip tried to amuse her with several anecdotes, but she continued to be distracted. Philip narrowed his sherry eyes, staring into the distance, unable to believe that she would spurn his company for that of Harold. Still, he was somewhat put out by his reception.

Nor was Rupert's reception quite what he would have wished. Down on one knee, hording a single rose, he attempted to propose to Dinah, who, on hearing his first phrase – one he had worked on so hard too – burst into tears and refused to speak to him. Drawing up the shreds of his dignity he bowed to the still sobbing Dinah.

'I am sorry to have caused you such distress, madam. I shall remove myself from your company and you will not be troubled by me again.' With that he turned smartly and left her. Dinah's sobs became even more distressed.

Kathryn went to her side. Glaring at the viscount she said, 'Your friend has a lot to answer for!' The viscount was about to protest, but Kathryn swept on, 'At least go away and give this poor girl some privacy! Unless Sir Rupert intends to do any more harm!' With that she led Dinah away.

A short time later they were joined by Harold, a little pink from the unusual activity of running in order to catch up. Yet again his plan to get Kathryn alone was foiled. Kathryn insisted that she could not leave Dinah's side, glad that she had such a good excuse.

It was a rather subdued party after that, and not all Harold's attempts at humour and jollity could change it. The Misses Drake seemed to be mesmerized by the amount of food Aunt Agatha consumed and could hardly take their eyes off her. Kathryn could feel a headache coming on, and wished that she had never come.

The viscount and Sir Rupert found themselves staring at the two ladies and falling into brooding silences, the viscount behaving with only a little more dignity than Rupert, a fact that had not escaped his attention, although his companions for the moment seemed oblivious to his nightmare.

Mr Parsons, subdued with the wine and the late autumn sunshine, dozed off. Not to be outdone, Aunt Agatha followed suit and emitted small snores. which jostled the crumbs on her gown.

Harold suggested that they might enjoy a trip on the boating lake. The ladies disposed themselves in punts. The sweating Harold claimed Kathryn to himself and Sir Rupert and the viscount escorted Dinah. After exclaiming dutifully on his non-existent skill with a pole, or stick or whatever it was called, Kathryn felt she had performed above and beyond the call of duty.

The afternoon grew chilly and finally Harold had to admit defeat and turned back towards the jetty. Dinah and her two companions stepped safely on to dry land. The viscount stretched out a steadying hand to Kathryn, which was firmly ignored.

Somehow Harold had got his pole stuck and, unable to free it, the punt moved and left him clinging like a small despairing squirrel to the pole before landing in the water with a majestic splash. The water was only chest deep and he arose spluttering and gasping, picking weeds off his clothes. The two Misses Drake hurried to the water's edge, all concern. Harold just wanted the afternoon to be over, a sentiment echoed by nearly all his guests.

After what seemed like hours, they were finally back in the landau, Aunt Agatha taking one double seat and Kathryn sharing the other with a still slightly weepy Dinah.

'It is a great shame that you do not favour Sir Rupert's suit.'

'I thought you understood.' Keeping a wary eye on the sleeping Aunt Agatha, she said, 'My mama would never never allow. . . .' Her eyes again filled with tears. Kathryn squeezed her hand encouragingly.

'Have you explained this to him?' she asked.

Dinah shook her head. 'But I was hoping that maybe you would pass on a letter to him?' She gazed at Kathryn hopefully.

Inwardly, Kathryn sighed, not wanting to be involved in a closet romance when she already had enough problems of her own. Instead she said, 'What an excellent idea! So dry your eyes and compose yourself. I will give you what help I can.'

As soon as they reached the outskirts of town, Rupert took his leave, not looking any of them quite in the eye. The viscount, thinking that he should keep a close watch on his friend, also bade the group farewell. He did so a little reluctantly, as he hadn't had a chance to discover why Kathryn was so displeased with him.

At last they were home. Aunt Agatha refreshed from her nap started to describe the food Harold had provided to Emma. Kathryn, pleading a headache, went straight to bed.

Seventeen

Kathryn slept late the next morning. She had been awake for hours trying to find what to do for the best. Rebecca brought in her breakfast, along with several letters. Kathryn sipped her tea and ignored the letters.

'How are you enjoying London, Rebecca?' she asked.

'Ooh madam, I never thought I would be living in London. I'll have so much to tell mi'mam when we go back. She's never gone no further than Canterbury. I bought her a shawl and mi'sister a picture of Saint Paul.' Kathryn could not imagine what she would do with that.

'So you're not homesick then?'

'Why, we've only been here six weeks! I haven't even been to Bedlam to see those poor, mad, slobbering creatures yet.' She regarded Kathryn anxiously. 'Are you sure you are all right, ma'am? You be looking a bit peaky.'

Kathryn just smiled and said she would ring when she wished to dress, then sat back and allowed her chaotic thoughts free rein. She thought of Philip, on the one hand kind, competent, amusing, handsome. But on the other hand, as Rebecca's mum would say, handsome is as handsome does. Kathryn had to admit though, his attitude did seem to be thawing and the over-weaning arrogance was seen less and less.

Her mind approached the incident of the kiss, and shied away from examining her own feelings and concentrated on how the kiss fitted into her plan for revenge. Kathryn sighed; somehow her anger at the viscount's insults was not as ready as it once was, and,

if she were honest with herself, she did not really think the plan
was such a good idea after all. But what could she do about it?

In frustration she turned to her letters. A bill was first. Typical,
Kathryn thought, maybe I should spend the rest of the day in bed.
She opened it impatiently and gasped in horror. She couldn't have
spent that much! Her hands started to shake; to pay this she would
have to take the money from not the next quarter but the one after
that. She would have to send the clothes back she decided. But
both of the gowns had been worn, and the shoes with the sparkling
heels had been caught in the rain and were completely ruined.
Agitatedly, Kathryn rang the bell. Once she was dressed she would
seek out William and ask his advice. Oh! How could she have been
so stupid! She hurriedly chose a plain muslin dress in white with
figured arms and a high neck. This accentuated her high cheek-
bones and emerald eyes. Unaware and for the moment careless of
her looks, she ran downstairs seeking William.

Her enquiries were not successful, Lord Courtney had left the
house early and was not expected back for some time. Kathryn
ground her teeth in frustration, she wanted this dealt with now!

'Pardon me, ma'am. A Mister Richard Spires awaits you in the
drawing-room. He said he was a friend of yours.' The butler's tone
showed that he doubted this fact. Kathryn's thoughts were not
complimentary to Richard as she made her way to the drawing-
room.

As she entered the room she did not see how Richard's face lit
up at the sight of her, or that his hand was shaking as he greeted
her. Seeing the signs of a frown on her face, he kept hold of her
hand and asked, 'Is everything all right? Can I be of service in any
way?' Usually Kathryn would never have dreamt of confiding in
Richard, but with her emotions in a turmoil, she could not resist
the offer of a friendly shoulder. So she thrust the bill into his hands
and admitted she had vastly overspent. To her chagrin she found
herself sniffing, handkerchief in hand, exactly the way Emma did.

Richard read the letter and asked her, 'How much do you owe
altogether?'

'Everything else is paid,' said Kathryn indignantly. 'That's why I
can't pay this one, and William has already advanced me next

quarter's money, and I thought . . . I thought. . . .' She retired into the handkerchief once more.

'Do not concern yourself! It breaks my heart to see you in such distress. I will pay this bill, and William, from the Kingsgrey account, which is yours to spend as you choose as its mistress.'

Kathryn gazed at him in wonder. 'Do you mean it? Oh! Thank you. Are you sure it will be all right? It seems a bit strange to be taking money off His Lordship for my bills when. . . .' Her voice trailed off. Richard, enjoying the feeling of being her saviour, smiled expansively.

'I assure you that there is nothing for you to worry about.' Relief washed over Kathryn at his words. With almost a joyous heart she offered him some tea.

Once it had been poured Richard, deciding he might not get as good an opportunity again, broached the reason for his call.

'I wonder if you would care to visit some of the sights of London with me?' His mouth had gone quite dry. He rushed on before she could reply, 'I know that people who live here often find the sights boring so I. . . .' He swallowed nervously.

Kathryn was rather touched, reminded once again of Jess. So between that and the gratitude she felt at that moment, she readily agreed. A date was made for a visit to the British Museum and Richard left with a satisfied glow, looking forward to their next meeting, his romantic heart beating a little quicker.

William tapped lightly on his wife's bedroom door.

'Come in,' she trilled, a smile blossoming on her face when she saw her husband. 'I hope you haven't come to tell me off,' she said. 'I have just received a new bonnet, which drives all thought of money matters out of mind!' She placed an adorable confection of straw and coquettish feathers on her curls.

'In the presence of such loveliness how could I think of anything so grubby as money?' The worried look, however, did not leave his face.

'Are you really angry with me?' Emma said, in a small voice.

'Oh, no, my love.' He held out his arms to her and she settled on his knee, but still he did not smile.

'Whatever is the matter?' she cried.

William sighed. 'I must admit I am a little worried about Kathryn.'

'What do you mean?'

'Well. Philip does seem quite taken with her, and she with him. But this plan to humiliate him. . . .'

He stopped uncomfortably.

'What of his plan? The bet? Have you forgotten that? What do you expect her to do, stop just because he asked her to dance?' Emma was growing quite heated. 'Honestly, all you men stick together. The instant a woman stands up for herself and gives as good as she gets, you think there's something wrong.' She climbed off his knee and, with cold dignity, informed him that she was just about to have a bath. William did not protest at his dismissal. As he knew Emma in one of these moods, he resolved to keep out of her way for a while.

Emma was almost as surprised as her husband at her reaction. But William's ill-timed thoughts had coincided with Emma's honest amazement at Kathryn's success. And she felt very proud of Kathryn's beating the viscount on his own terms. Crossly she started to pin up her hair. Maybe it was time for her to stop relying on her feminine charm and wiles to get her own way, and to start using Kathryn as an example for her behaviour.

But then she had to admit that William did not deserve the same treatment as her odious cousin. He would not dream of behaving in the same way. She started to feel a little guilty. Then she remembered that the bill for a court dress, in Spanish lace, had not yet arrived. Perhaps she should delay any reconciliation until then. Satisfied with this plan, she began her bath.

Kathryn had not seen anything of Philip for several days. Not that she missed him, she told herself, only that this was a set-back to her plan. Maybe he had already won his bet. This thought caused her much pain, but she refused to examine why. She could settle at nothing and roamed the house restlessly. Emma tried to interest her in plans for the ball, but to no avail.

One morning she was sitting in the parlour flicking through *The*

Ladies Journal in a desultory fashion, when Dinah Neville was announced. She entered, wearing a deep-red coat trimmed in dark fur. A matching cossack style hat was perched on top of her golden curls. Kathryn was forced to admit that she was in the best of looks, and felt positively colourless beside her.

'I cannot stay long,' Dinah said breathlessly, 'Mama is at Hookhams and I must meet her there directly.' She glanced over her shoulder, as if she expected her mother to appear at any minute. Kathryn stifled an impulse to tell her to grow up and stop being such a goose. Instead she smiled reassuringly.

'Have you the letter for me? I will be sure to see it reaches Sir Rupert immediately.'

'Oh, you are so kind. I have asked him to wear a blue domino at Ambassador Barcombe's masque. Mama cannot protest if she does not know who he is.' This logic seemed a little flawed to Kathryn, but she did not have the energy to pursue the matter. Instead she accepted the billet and once again swore eternal secrecy. Then in a swirl of red Dinah was gone.

Kathryn sat and penned a note to Sir Rupert inviting him to tea the following afternoon. She wished her problems could be solved as simply.

Unable to contain herself, she embarked on her own and then Emma's mending but sewing gave her too much opportunity for her thoughts to wander. The viscount as a hero, as a boy, as an arrogant husband. Her emotions swung like a pendulum. When had a man plagued a woman so! Gaining no conclusions from her fevered thoughts, and having then given herself a headache, she retired early to bed.

The next day Kathryn rose early for her trip to the British Museum with Richard. Determined to put Philip out of her mind, she wore a most becoming walking dress in emerald green trimmed in pale-gold sarcenet, giving a golden tone to her green eyes and contrasting her magnificent Titian hair.

Her first visitor of the day, however, was not Richard as expected, but the man who had been haunting her thoughts for days. He was dressed in superfine biscuit pantaloons and a pair of dazzlingly shiny hessians. He also sported a high-waisted driving coat with row upon row of capes with mother-of-pearl buttons. York tan gloves and a curly rimmed beaver completed his toilette.

Kathryn's mouth went dry, and she could think of absolutely nothing to say. Suddenly she was conscious that she was grinning inanely at him. To try and hide her pleasure at seeing him, her expression immediately changed.

'I thought you were Mr Spires,' she said. 'I certainly was not expecting such eminent company.' Her face was coolly interested, none of her pleasure at seeing him reflected.

Philip was somewhat taken aback. For one wild moment when she was smiling at him . . . he wished. . . . He could not take that thought further.

Suddenly, feeling rather gauche he said, 'I was wondering whether you would like to drive with me in the park. It bodes to be a fine day for driving. I have two new greys to try. Can I tempt you?'

Every fibre in Kathryn's body pulled her towards accepting his invitation and wishing Richard to the devil. She was surprised the viscount couldn't tell. Before she could choose one way or the other, Richard entered. Looking from one to the other, he swallowed nervously.

'My Lord! Excuse me I. . . .'

Kathryn's decision was made. 'Ah Richard!' she cried. 'I am quite ready as you can see. You cannot say that it is my fault if we are tardy!' She pulled on her gloves. 'I am sorry, My Lord, but as you can see I have a prior engagement.'

'I am sure Richard will excuse you.' The viscount looked at him pointedly.

'Of course, My Lord.' He bowed preparing to leave.

What a bully! Kathryn thought. Pulling rank on Richard!

'I can assure you that I have no wish to change our arrangement, Richard,' she said briskly. 'Shall we go?' She glanced once at the viscount's face, a mask of fury, and feeling herself ridiculously close to tears, almost ran from the room.

Richard had to trot to keep up with her. Not quite as insensitive as Kathryn imagined, he offered no comment until they reached the entrance to the museum. By this time Kathryn had her emotions under control, and could face the rest of the day with some semblance of equanimity.

'I would not have minded deferring our visit had you business with His Lordship.'

'No. No. I wouldn't hear of it,' Kathryn lied. 'He only came to see if I were free to drive with him in the park.'

'Drive with him!' Richard's amazement was evident.

'Yes. Why? You make it sound incredible.'

'Well, you see, it is,' Richard tried to explain. 'He never takes up females. Never!' Kathryn felt like crying again. No wonder he had expected her to change her plans. She couldn't let him think she was at his disposal, but he obviously felt it was a special occasion.

Kathryn's heart fell to her boots. Her enthusiasm for the day's pleasure, including the marbles recently brought by Lord Elgin, could not have been lower. What she felt like doing was going to bed and crying for at least a week.

'You look very lovely, I am not surprised that His Lordship was willing to break the rule of a lifetime. For you are an exceptional lady.' He blushed vividly.

Kathryn began to feel alarmed, surely he didn't think that she had chosen him over the viscount. It sounded suspiciously like the beginning of a speech to make an offer. He of all people knew that she was married!

But Richard, guide book in hand, droned on and on.

'Have you ever considered teaching?' Kathryn asked, with a slight edge to her voice.

'My tutors at Cambridge suggested that career, but I felt politics was my particular future. Hence my job with the viscount.'

'What is the viscount like to work for?' Kathryn enquired. Anything to get away from the lecture, she told herself, not that she was really interested in anything to do with that man.

'His mind is first rate. However his concentration does not match this, and anything that does not interest him will not get done. Merry making seems to be his prime occupation. But he has been more than kind to me,' he added, in case she thought him ungrateful.

Kathryn, tired of him, the museum and the viscount, suggested they finish the tour. It had been a long day.

Kathryn arrived home feeling as if she had walked from London to Brighton. Emma was still writing lists for the ball but, when she saw Kathryn, all this was forgotten.

'You look exhausted! Here, sit, I'll pour you a glass of ratafia. Whatever have you been doing?' Kathryn sank gratefully into a chair and undid her bonnet.

'The British Museum with Richard Spires.' She sipped her drink. Emma eyed her closely.

'Are you telling me that it's true?'

'What?'

'That you turned down Philip in order to visit all those mummies and bits of pots?'

'If you mean did I have a prior engagement, when your cousin asked me to take a drive with him, then yes, that's true.' Tears welled up in her eyes. 'I have spent the whole day wishing I could have said yes, but . . . but I couldn't let him think . . . I mean the bet and my plan.' She wiped her eyes. 'I don't know what's the matter with me.' Emma joined her on the settle.

'Do you think maybe your feelings for my cousin have undergone a change?' she asked tentatively.

'Emma. What are you suggesting?'

'Well, you seem so well suited,' she started lamely.

'Let us be clear; after the way your cousin has treated me, there is no force on earth that could even make me vaguely like him. Let alone what you are suggesting.'

When she was in bed that night, unable to sleep, she tried to convince herself that she had been telling the truth.

Eighteen

William peeped around his wife's bedroom door.

'Is it safe for me to enter?' he said. 'I come bearing gifts.'

Emma was trying to decide what jewels would go best with her gown of cream watered silk. 'What do you suggest?' She held out one necklace of rubies and a cameo locket.

'Neither,' said her doting husband, and handed her a slim, dark-blue box. 'So, I am forgiven?' Emma was already excitedly opening her gift.

'Of course, my love. How could I stay angry with you?' She raised an innocent pair of eyes to his.

'You minx,' he laughed. 'Do you like it?'

Emma's mouth fell open as she gazed at a ravishing diamond necklace made from six large diamonds surrounded by eight smaller ones, linked by delicate filigree in white gold.

'I had it specially made. A perfect setting for my perfect rose.'

Emma threw her arms around him. 'If you merely bought it for my forgiveness you must return it.' Though it almost broke her heart to say it. 'It is a gift worthy of a better occasion.'

When she was with her husband her mercenary little heart thawed. A beautiful, exotic, social butterfly, she always fluttered back to her rock-like husband. 'You are much too good to me,' she continued. 'I am such a crosspatch!'

'What, remorse for your cruel treatment of me!' William laughed. 'The day you accepted my offer you made me the luckiest man alive. I could never believe you preferred me to Devlin!' He clasped her to him, gently, unwilling to crush her dress. 'I wish

I could clothe you in jewels, even a diamond nose pin to prove my devotion.' Emma laughed and dropped a fleeting kiss on his cheek. 'Anyway,' he said, releasing her, 'have you forgotten already? Our wedding anniversary is not that far away.'

'I know,' she said indignantly, 'but it is still a month or more away!'

'I couldn't resist giving it to you as soon as I got it.' He placed the necklace around her long white neck and battled with the clasp. 'Do you really like it?' he asked anxiously.

'It is perfect,' said Emma. 'I shall be the envy of every woman there!' This thought made her smile even brighter and she descended the stairs on her husband's arm in high good humour.

Kathryn awaited them in the library. William had taken a box at the highly fashionable Drury Lane Theatre. Kathryn's gown was simple and uncluttered white velvet, elegantly tailored. Framed by a small ruff, her long white neck looked almost too delicate to support such a wealth of hair. Her brilliant green eyes flashed, adding fire to her icy sparkle.

Even though it was Kathryn's first visit to the theatre, she could not feel excited, try as she might. The viscount had been conspicuous by his absence in the days following her ill-fated visit to the British Museum. She wondered if he too would be at the opening tonight. Maybe she would have the chance to speak to him, to try and explain. Kathryn duly admired the necklace and envied Emma her happy marriage. Something that she would never know, especially now, she thought.

There were crowds surrounding the theatre. The Prince Regent was due to arrive and it was rumoured that Princess Caroline, too, would make an appearance. The scene at Leicester Square was chaotic and Kathryn's eyes were dazzled by the gowns and jewels, even on the men.

Finally they reached their box. From this safe haven they could observe the rest of the audience. Emma pointed out the famous names, Beau Brummel, Lord Byron and the Duke of York. Kathryn could scarcely believe that she was in such exalted company. Aunt Jane would be in heaven, she thought, and tried to be a little more enthusiastic.

In an opposite box, a lady dressed in crimson and gold sat staring at Kathryn intensely.

'Who is that?' Kathryn asked William. 'Do we know her?'

William following her gaze muttered, 'I don't think you have met. That is Charlotte Mountjoy, with her, the Earl of Osterley.' As he spoke they were joined by another man. Kathryn could not see his face at first, but the lady was delighted to see him by the welcome she gave. The earl, however, did not look so pleased and soon removed himself. Kathryn then recognized the other occupant, who leant so solicitously over the lady's hand and stole a kiss behind her fan, Philip Devlin.

Kathryn felt a cold knife stab her heart as she recognized him. There was no mistaking him, and she knew he must be able to see her. Digging her fingernails into her hand, she turned and smiled brilliantly at William as she accepted a glass of champagne. Soon they were joined in the box by Sir Rupert Roscommon, who devoted his whole attention to Kathryn. She wished she could be flattered, but his sole topic of conversation was Dinah. He apologized profusely for not replying to her note. When she told him of the letter, he looked completely dumbstruck. Suddenly he raised her hand extravagantly to his lips and promised her anything she desired. Kathryn had to laugh, looking, just as she intended the viscount to see, as if she was having the time of her life.

Philip watched her gaiety through narrowed eyes and flirted even more outrageously with Charlotte.

'So you say you missed me?' she asked him archly. He kissed the palm of her hand and whispered in her ear.

'What do I have to do to prove it to you?' His lips lingered by her ear and his breath tickled her neck.

Kathryn, seemingly unaware of what was taking place in the opposite box, gave the impression that she was the belle of the ball, and enjoying every minute of it. Which was as well, as she thought the effort would kill her. Keeping a brilliant smile on her face, talking and laughing, she wondered if the night would never end. Her performance certainly convinced the viscount.

Charlotte noticed the looks he kept darting at the box opposite, but held her tongue; she would deal with that little madam when the time came. Right now, she set out to charm him back into her arms, and no prissy schoolgirl was going to stop her.

Finally the curtain came down, and the night was nearly over. Kathryn sank back into the coach seat, feeling as if she were at least eighty-three. She said nothing for the whole journey home and on reaching the house, went sadly to bed.

The viscount and Charlotte went on to supper in a private room at The Golden Lion. Charlotte looked like a cat that had just caught a canary.

'Who was the girl in white with your cousin?' she asked. The viscount who was deep into his cups, smiled at her blearily.

'That's a secret,' he smiled.

'Not from me, surely?' She put her hand on his. 'When we are such good friends!' With the tip of one finger she caressed his cheek.

'Charlotte, can you keep a secret?' He attempted to sit up. 'That girl is one hundred guineas to me.' He laughed bitterly.

'What do you mean?'

'It's a wager, to get her to thaw or not. Two months up in a couple of weeks. Can't lose mi'money!' Charlotte stored away this nugget of information for more thought later. For the moment, she had her hands full with a drunken and amorous viscount. Fortunately, it was not long before he slipped into unconsciousness.

As she made her way home, her agile mind wondered the best way to utilize the information she had gleaned. A malicious smile crossed her face. Her coachman thought she was the most beautiful woman he had ever seen, unaware of the thoughts behind her loveliness.

Philip opened one eye blearily and a face swam in and out of his vision. Wilby, his valet, flung open the blue and gold drapes, and late morning sunshine flooded the room.

Groaning he lifted his head from the pillow, as gently as possible, and eyed his faithful retainer with as much disgust as he could manage.

His shaky hand received a glass of milky fluid and with a

grimace, the viscount started to drink.

'All of it,' his dour companion growled, waiting for the empty glass.

'Whatever has put you in such a humour?' said the viscount to an unyielding back.

'What humour would Your Lordship prefer?' Knowing that there would be no peace until this grumpy retainer had had his say, Philip prepared himself as best he could.

'I could remind you that it took two footmen to carry you to bed, and three to keep you there! Wanting to fight the Devil and all his minions! The mirror in the hall is broken and one of the youngsters doesn't think his mother would allow him to stay in such a household!'

Inwardly, the viscount cringed at the recitation of his crimes. 'Had a bit of a blue, did I?' A telling stare was his reply. Unable to strain his throbbing mind any longer, he staggered into his bath, hoping that the cure would soon start to do its work.

Finally, resplendent in a startling dressing gown with embroidered peacocks, he relaxed with a mug of porter and cast a cursory glance at his letters. His concentration, however, failed miserably at this task and he threw them back on the tray with a disgusted expression and went down to the library.

His thoughts turned, as they so often did, to Kathryn. So beautiful, yet there was something, a challenge in her eyes, her refusal to admit the undeniable. What was the undeniable? For the first time he looked honestly at his feelings. The thought of spending the rest of his life without Kathryn, of not having her beside him in everything he was and everything he did, was enough to make him mad.

As the impact of this revelation washed over him, a certainty grew within him. Unless he acted very soon, she would be lost to him forever. An icy fear bit at his heart. For the first time in his adult life he was afraid. A voice interrupted his thoughts.

'Are you recovered enough for company?' Avon waved his hat round the door. 'Wilby warned me that I could be risking life and limb.' A cautious head followed the hat.

'Where does he get the gall to discuss me with you, I should like to know!' This strenuous speech threatened the return of his headache, and Avon entered without further comment, bringing with him the brandy.

Having settled himself with the decanter at his elbow, he turned to his friend. 'Bit over the boughs, mi'boy?' he asked discreetly.

'If you have been talking to that disloyal cur who calls himself my valet, doubtless you know more about my behaviour than I do!' Deserting the porter he gestured to Avon to pour him a brandy. Downing it quickly for courage, he said, 'I don't think I can continue with the bet. I will pay Rupert and there the matter can end.' Avon made no comment, gazing at the brandy in his glass. 'Say something! Even if only I told you so! You were quite right.' A small smile crossed his face. 'But you're not allowed to tell anyone I said that.'

'And Miss Chambers?' Avon risked asking.

The viscount was silent. A ready reply rose to his lips but remained unuttered. He avoided his friend's gaze. Suddenly he sighed.

'What shall I do, Harry?' Avon was surprised at the use of that childish name, lost at the public school of his youth for the identifiable title he bore.

'I think I'm in love with her. She haunts me. . . .' He gripped his glass tightly. 'Without her my life might as well end.'

Avon finished his brandy and poured himself a generous replacement. The viscount refused. 'Aren't you going to say I told you so?'

Nineteen

Emma chewed her thumbnail. Piles of paper surrounded her, the preparations for the ball were on the same scale as those for the war, William remarked frequently. But Emma was not thinking about the ball. She was genuinely fond of her cousin, despite all her words. As for Kathryn, she was like the sister she always wanted.

But Emma was worried. Their plan had seemed like such a good idea. But not now. Emma could see that both Kathryn and Philip should stop the game, and realize that there was something more important, the fact that they loved each other.

Emma's mind searched for a way that she could facilitate this happening. Her thumbnail was suffering much damage as a result.

William interrupted her thoughts. Emma curled up on his knee and tucked her head on his shoulder. Once settled she sighed contentedly.

'I wish I knew what to do! There must be a way . . .' she began. William placed a finger over her lips.

'We cannot interfere. Kathryn would never forgive you! As for Philip!' William shuddered.

'But they are throwing it all away!' she protested. William released her grip on his neck cloth with difficulty. It would never be the same again he thought ruefully.

'Maybe the ball will give them a chance to clear things up.' He hoped he sounded more optimistic than he felt. His wife said nothing, her agile mind busy with possible plans.

'We must think of a stunning costume for Kathryn, one that no

124

man can resist.' She cast a sidelong glance at him, 'Except you, of course.' He laughed and kissed the top of her head.

'We also have to give them the chance to be alone together,' she mused. 'I wonder if, accidentally of course, they could be locked in the library or something?' She raised her pansy eyes towards her husband. 'What do you think?'

'I think we have to let them get on with it. We cannot interfere.'

'But . . .' she started.

'No,' he reiterated. 'They must find their own way to happiness.'

Dinah picked at the food in front of her. Her mother watched her, eyes glinting, as she saved up her comments for when they were alone.

'Are you ill, child?' her father asked in concern. His wife glared at him. Paying no attention, Lord Neville continued, 'Are you determined to fade away? You have not touched your food for days! Do I call the doctor? Are you ailing?' His voice was growing peevish. 'I declare, it's enough to put me off mi'own dinner. Dashed if it ain't.' His portly frame showed how rare this was. He returned his attention to the duck in front of him and continued to munch. Married for many years, his customary haven, his club, was beckoning. Most domestic crises were dealt with by his wife.

How they had produced a child of such surpassing loveliness was a mystery. But it was this child who had his heart. The apple of his eye, Dinah was responsible for his discomfort. He did not want to disagree publicly with his wife – in fact, he always agreed with his wife, privately and publicly. Now, however, his allegiance was being tested. His need for a peaceful life with his spouse, coupled with the lure of his club, offered an escape from the type of family row he abhorred. His thoughts were clearly written on his face. Unable to stand it, his recalcitrant daughter fled from the table in tears.

As the rest of her family stared after her in astonishment, her mother pursed her lips in anger. An uneasy silence reigned. Lady Neville's face promised consequences and even Lord Neville held his tongue.

At last the meal was over. Lord Neville had no intention of staying in the house a moment longer than was necessary. His club had a passable port, and his wife could not delay him in his quest, but on behalf of his daughter, he would sacrifice an early escape to get to the bottom of the situation.

Usually unwilling to gainsay his wife in anything, his digestion could no longer stand the tension surrounding meal-times with the family. Driven to either eating every meal at his club or staying at home and facing his wife, reluctantly he decided, he must make an effort for his daughter. He requested his wife to join him in the study and sent for Dinah to join them there.

He sipped a brandy to steady his nerves, promising himself that as soon as he could he would renew his acquaintance with the whole bottle.

'I have had enough of this!' he announced. 'A man's home cannot be disrupted this way. It is not the role of the family to add to the problems of the master of the house, but to ease them. I will have peace in my own home!' He reached for his brandy glass. Dinah started to sob, much to her father's disgust. Ignoring her in the hope she would stop, he turned to his wife.

'Do not bother yourself, my dear!' Lady Neville smiled at her husband, who narrowed his eyes suspiciously. 'I will deal with this small domestic matter.' She turned to her daughter, her mouth tight with fury. 'Our daughter will respect my wishes!'

Lord Neville had had enough. His dinner ruined, and now his wife speaking to him as if his opinion was of no consequence. It was time to put his foot down. Quaffing the rest of the brandy he puffed out his chest.

'Enough!' he slammed the glass to the table. 'I will deal with *our* daughter!' He poured himself another brandy. 'So kindly be seated.'

Lady Neville shrank back to her seat. Never had he spoken to her this way before. Never!

'What ails you, girl? Come now, there is nothing to be afraid of.'

Dinah was amazed at the change in her father and, even under her mother's glacial gaze, sobbed out her desire to marry Sir Rupert Roscommon rather than the Earl of Rutland. Even though he could not stand to see his daughter unhappy, he was not at all sure that Roscommon was the man to care for his treasure.

'She is turning down a good match,' Lady Neville burst out, unable to keep silent any longer. 'For what? A handsome mountebank, living on his expectations for years! And a mere baronet!'

'Silence!' Lord Neville roared.

Even Dinah stopped crying and gazed at her father in shock. Their instant obedience gave him the confidence to continue. If he had guessed how easy it was, he would have done it years ago. He poured himself another brandy and realized he had no idea which course to follow. To give his blessing or not.

Remembering his own father he said, 'I will think on it.' Holding up his hand to stem their protests, he continued, 'Until then I will have peace! I will give you my decision when I see fit!'

This was a phrase he had hated his father using, but now it fitted his needs exactly. With a final glare at his audience he finished his drink and stalked from the room.

For the second time in four months Mr King had the honour of advising Viscount Philip Devlin. This time, however, it was a very different viscount sitting before him.

Gone was the insolent, angry young man, gone was the bitter veteran of many emotional wars. In their place was a man of purpose, poise and quiet confidence. Mr King allowed himself a moment to speculate as to the cause of this. He was ill-prepared for the viscount's next comment.

'Mr King, I must have a divorce,' the viscount said baldly. Mr King forgot himself so far as to drop his pen. 'As soon as possible,' he added.

'May I be so bold as to ask why, My Lord?' Before he could give an answer Mr King continued, 'Your pardon, let me ring for some refreshment.' Decanters were delivered by a dusty-looking clerk who watched in amazement as Mr King helped himself to a sherry, unknown in the forty years he had worked there. The viscount refused all drinks. 'I wish to marry someone else. I believe I can get an annulment for non-consummation.' His mouth curled distastefully at these words. Mr King sipped again at his sherry, having difficulty believing he was not dreaming.

'There is a possibility that your inheritance would be at risk if you insist upon this course of action.' The sun illuminated the dry office of the lawyer, motes of dust swimming in a golden stream. Mr King watched them distractedly as he waited for a reply.

'I rely on you to minimize that risk as far as possible. But I must make this clear: I will suffer no impediment. My divorce must be settled as quickly as possible. Not even the loss of my estates will stop this matter moving with speed. I must be a free man!'

Horrified, Mr King reached for another drink. This was contrary to every lawyer's instinct he possessed. Pursue at all costs! No loss too much! What was the world coming to? He could hardly believe his ears.

'My Lord, am I to take it that you do not care about your lands?'

'Nothing will delay this most urgent matter. I will have my freedom, and soon, no matter what the cost.' The viscount spoke quietly, but implacably. Mr King shivered slightly.

'Naturally we will do our best, but a matter of such delicacy. . . . May one ask what the lady concerned has said?'

The viscount stared at him blankly for a moment.

'When I tell her I expect she will agree to my plans.' He has not even told his wife, Mr King thought incredulously.

'Your wife's co-operation would ease matters somewhat,' he said delicately.

The viscount, who had thought he meant Kathryn, hurriedly said, 'Very well. I will assure her consent, in writing. You will have the document by the end of the week. You may inform me of your progress then.' With that he was gone.

Mr King sat unmoving for several minutes. Such extraordinary behaviour. Gossip had reached his ears that the viscount had been seen much in the company of one particular lady. But that he would lose his title and lands for a pretty face! No, Mr King was sure that there was more to this than met the eye. Naturally he would not stoop so low as to gossip on the viscount's affairs, but he determined to enquire discreetly into his recent activities, in the hope of discovering what had precipitated such an enormous change.

He shook his head, thinking to himself that love was a hard taskmaster, but so was the viscount. Mr King started to consider the best way forward.

Twenty

Kathryn was feeling rather low. Philip had been noticeable by his absence. She had neither seen or heard of him since that disastrous night at the theatre. Kathryn's feigned nonchalance hid a far from indifferent heart, thinking that now he must have won his bet and he no longer needed to pretend any preference for her. Emma noticed her heavy eyes and distracted air but said nothing, thinking that she would discuss it when she was ready, and nothing would be gained by pushing her.

Kathryn had thrown herself into preparations for the ball. Hand-painted gold signs were made for the flight of stairs leading down to the ballroom, each quoting a good intention. Walls were draped in red and gold and a supper of red food was planned. If the food was not already red then it would be dyed, Emma decided. You could not make a dinner of red cabbage, beetroot, tomatoes and pomegranates.

Emma had also ordered a huge devilish face to be cast in plaster, which would be painted red and hung overlooking the dancing. Oil would be poured into the fountain so real flames would burn. Manacles and cages would be hung from pillars and walls, warped mirrors would be used to make the ballroom look larger and to reflect back the guests' costumes. She had also hired a small troupe of players to dress as furies and devils and perform suitably hellish tricks. And she had finally found a fire-eater. She was determined that this should be the ball of the season, even if it was the death of her. William thought it more likely that the fire-eater would be the death of all of them.

'What do you think if the flowers at the top of the stairs are white until halfway down they become red?'

'Good idea.' said Kathryn absently.

'Have you decided what to wear yet?' Emma continued. 'I must admit I do fancy myself as Lucrezia Borgia or maybe Morgan le Fay.' Kathryn made no reply. 'Are you listening to me?' she asked sharply.

'Oh yes. Yes. You were talking of flowers.'

Emma tutted in frustration. 'You are miles away. Is there anything wrong?'

'No, nothing wrong. I was just thinking that perhaps your cousin has already won his bet and my chance will have passed.' She gave Emma a tight smile. 'If that is the case I must be thinking of going home.' Emma stared at her aghast.

'You don't mean it. The season's in full swing! Even if the bet is over he will still pay court to you. After all, you are the only woman to have been asked to drive with him.'

Kathryn smiled sadly. 'Except I didn't have the sense to accept his offer and now he has done with me.'

Emma became quite concerned. This was most unlike Kathryn's usual bubbly spirits. 'The ball will give you the perfect opportunity to exact your revenge.' She paused and then rushed on, 'Or whatever you decide is the best thing to do.'

Kathryn glanced at her suspiciously. 'What do you mean?'

'I just thought that perhaps you and Philip might have something else apart from your revenge and the bet. I was hoping. . . .' She trailed off.

Kathryn laughed in a brittle fashion. 'Whatever gave you a foolish idea like that? I'm sure I don't know what you mean.' She tossed her head. 'Your cousin has absolutely no interest in me beyond that stupid bet! Look at how he behaved at the theatre, with that . . . that woman.' A lump started in her throat. 'Oh Emma,' she said, in a broken little voice, 'whatever am I going to do?' A lone tear trickled down one cheek.

'Oh no! Don't cry! Why, I'm sure you're quite wrong. Philip has never behaved like this before. I'm sure that he does care for you,' she finished lamely.

'Of course; I'm sure that he threw himself at that woman because he wanted to be with me!' Kathryn sniffed. 'I won't let it spoil your ball though. I'll return to Kingsgrey soon and hope to hide my identity. It should not be that difficult as he has no interest in me as Kathryn Chambers or as his wife.'

Emma's heart went out to her, but sympathy she felt would not be welcome. Kathryn set her face determinedly. 'I don't want to talk about it again. Least said soonest mended after all.' If only that were true Kathryn thought. If only!

Charlotte Mountjoy threw a perfume bottle at the wall. It shattered, oozing the sweet smell of violets. It was just too frustrating, she wanted to scream with fury. She lifted another bottle, considering its destruction. Her mercenary little soul, however, noted that it was the last bottle of her favourite orchid essence, and it was returned to the table.

No, she would not waste her perfume, she would turn her energies towards revenge. How dare he treat her as a convenience? After all his attentions she had not heard from him in over a week. The heavily scented note delivered by her footman had been returned unopened. He would not get away with it she swore, no man treated her like this.

She began to pace, swirling her skirts around the room as she turned one way then another. What about that chit of a girl he'd been running all over town with? She admitted that it was difficult for someone like her to cause trouble to one of the viscount's wealth and standing, but the girl was a different matter altogether.

Charlotte sat in front of her mirror, tweaking at her hair, finding no pleasure in her creamy complexion and large brown eyes fringed by ridiculously long lashes. That mouse of a girl was going to discover that the Viscount Philip Devlin was using her to win a bet. And not only that, but he had married not more than four months ago. That was something else to add to the account. Charlotte had had quite a fancy to become a viscountess. At one time it had seemed possible. After all, wasn't she lovely enough? She posed showing a ravishing profile.

'Planning a portrait, mi'dear?' Lord Osterley stood at the door. Seeing the remains of her temper tantrum, he raised an eyebrow. 'Well, sweeting. What has got you in such temper?' He wrinkled his nose. 'I never have liked violets. I shall give you lilies.'

'So you have come to gloat! I suppose you have spread it all over town.' She flung down her brush. 'That miserable cur! I shall have my revenge you know!'

'You are beautiful when you are angry. I almost find you irresistible when you are in this mood.' His mocking tone brought a flush to her cheeks.

'I will not be laughed at!' Charlotte said with gritted teeth.

'Viscount beware, eh?' He opened his snuff box and helped himself to a generous pinch. 'So what will you do? Give him the lecture of his life? Or merely try to frighten him into submission? You have the temper of a child, my dear.' Another bottle of perfume launched itself over the room.

'Are you so set on your viscount that you cannot look elsewhere?' His eyes for once were serious. Charlotte, totally intent on her own mood, did not reply. Osterley's mouth tightened as he watched her. Dressed in rose damask, her hair à la Sappho, cheeks flushed, lips parted, eyes blazing, she looked magnificent. But the mind of his lady was on another man.

'Have you had invitations to the Devlin ball?' she asked suddenly.

Resigned to her mood, Osterley answered, 'You wish me to take you? Hardly good taste, my dear!'

'I will be masked, you need not worry that the viscount will count you a worthwhile opponent!'

Suddenly he could bear it no more. That she should make such efforts for him! That she should ask his help! His voice, louder than he intended broke into her reverie.

'And what will I get for my part in this charade?'

His brutal question surprised her. Impatient, wrapped in her own thoughts, she answered, 'Whatever you wish. You have received cards?' She looked at him anxiously. 'You will take me, Harry?' She raised tear-filled eyes to his.

'Yes. Yes. I shall dance to your tune as always.' Charlotte kissed his cheek and, unaware of the longing on his face, started to describe her revenge.

Twenty-One

The Viscount Philip Devlin consigned another ball of paper into the fire. He chewed the end of his quill in frustration. No matter how hard he tried, he could not seem to find the words. Maybe it would be better to face her and ask for a divorce. Philip, however, was loath to leave London. He still had to speak to Kathryn; after all it was for her he needed a divorce – if she would have him.

Trying to arrange his affairs, before declaring himself to Kathryn, seemed to be the best plan. He had deliberately avoided her company thinking to free himself of his love for her. But not seeing her proved that, far from ridding himself of her, he could not live without her.

Once again he started to write. Finally he settled for a business-like tone, informing his wife of his wish for a divorce as soon as possible. He offered no explanation, saying simply that it was not a fault with her and further details could be discussed face to face. He read his scrawled lines once more, cursing his lack of words. This onerous task was interrupted by Avon.

'Hello! I hope I'm not disturbing your muse. Rupert led you to try versifying?'

'I don't think there's such a word!'

'Well there ought to be, by Gad!' Avon seated himself in front of the fire, stretching out his long legs and admiring the polish on his boot.

'I hear Emma is about to bankrupt you with this ball.'

'Believe me, that is the least of my problems at the moment,' Philip replied ruefully. 'In fact, I may not have the funds nor the house to entertain for much longer, so she can do her worst.'

'Whatever do you mean? You can't be telling me that you are short of a nugget or two!'

'I may be left with my grandfather's inheritance only. Not to be sniffed at, but hardly on the same scale as the Devlin inheritance.'

'What are you talking about? Should I call a quack? Or maybe you're a bit bosky?' He watched his friend anxiously.

Philip laughed. 'My dear friend, once again you now have cause and full permission to say I told you so.' Avon was still perplexed. Philip poured them both a drink.

'God's teeth man! Will you tell me what lunacy you are spouting?' He took his drink from a serious-faced viscount.

'You'll probably need it,' the viscount observed. 'I have instructed my lawyers to file for divorce as quickly as possible. I have decided to ask for the honour of the hand of Kathryn Chambers in marriage.'

Avon threw his brandy back in one, and refilled his glass. Philip watched him with bitter humour.

'So, now you know you were quite right when you said that I was a smug devil.' Avon was thunderstruck, unable to utter a word. 'Can't you manage at least one I told you so? I did after all warn you that I am in love.'

Avon just stared at him.

'Have you told her? Your wife, I mean.'

'You find me in the act of writing to beg my freedom.' He smiled sardonically and tossed down his brandy.

'You mean to say that in the course of the bet you really fell in love? And you'll give up all this?' Avon waved his hand expressively. 'Have you asked her yet? I mean, Miss Chambers? I thought you meant to speak to her at the ball.'

'I don't want to frighten her. I was hoping to approach her as a free man. But now I don't know if even that will be enough. I have made poor work of wooing, I have to admit.'

'For heaven's sake, man! What are you thinking of? Throwing everything away for a pretty face?' He started to splutter. 'Have you no pride, no sense of duty?' He looked into Philip's eyes, 'Have you gone mad?'

'Easy. Easy,' the viscount said. 'Believe me when I say that I have never been more serious in my life. I have battled many hours. Drunk and sober.' He raised his glass as if for a toast. 'To the man who is in control.'

Avon sank into a chair, groping for the bottle. He poured large measures into both their glasses. Philip started to speak quietly, almost to himself.

'The thought of spending the rest of my life without her at my side is unbearable. Without her, I care nothing for my place in society or estates, nor even my life. So I will gladly give all I possess, every last shilling of it, everything I am. Because without her I am dust.' He hung his head in his hand. 'May God forgive me, but there is no life for me without Kathryn, no joy except from her hand.'

Never before had Avon seen Philip like this. He could not help a feeling of wonder at the change in him.

'What will you do?'

'Try,' he replied shortly. 'Try to convince her to wait for a fool of a man to get a divorce. Try to convince a stranger to give me my heart's desire.' He turned to Avon. 'Will you disown me when I am a penniless wretch?'

'I am honoured that you have confided in me,' Avon said with honesty, 'I will do anything I can to aid you.' Embarrassed by his emotion, he added, 'A toast then. To love. May it conquer all.'

If only, thought Philip, if only!

Sir Rupert Roscommon fidgeted anxiously with his mask. Borrowed from a smaller friend, his blue domino was somewhat tight. Trust her to choose blue; he felt conspicuous in his borrowed finery. Ambassador Barcombe's party was a little flat, just another squeeze Rupert thought, except that tonight Dinah, his muse, his inspiration, had agreed to meet him at a secluded seat in the conservatory. Punctuality was, unfortunately, not one of her virtues.

Suddenly a pink domino appeared. Rupert's heart soared, she had kept her promise. Also she wore the roses he'd sent her.

Rupert felt himself the happiest man on earth.

'Rupert,' a soft voice called.

'I am here, my love.' Taking her hand he kissed it passionately. 'Next to you where I always wish to be.' Dinah ignored his flowery speech and tugged him back into the shadows.

'Hush. Quietly, in case you are heard. If Mama finds out that you are in the blue domino I don't know what she'll do!' Her voice trembled expressively.

'Has she bullied you terribly?'

'No. No, nothing of moment. But I have good news! Father has actually told Mama that he would decide, should you offer for me.' She blushed at her words.

'Offer for you! I would do anything for you.' He took a deep breath to launch into another one of his speeches, but his muse firmly interrupted him.

'Father must start to like you. If we can think of a way for him to be indebted to you he could not refuse. Depend on it!'

'I will prove to him that I am worthy of such a jewel. Have no fear, my love I will find a way!' His nervous sweetheart pulled away from him.

'I must fly. Shall I see you at the library on Thursday?'

'Death will not stop me,' he declared extravagantly to Dinah's back, as she slipped away into the ballroom.

Rupert sighed blissfully. Soon all barriers would be overcome. He felt a rush of gratitude towards Kathryn without whom this and the other clandestine meetings would never have taken place. This was causing him some pangs of guilt. The wager now seemed childish and, even worse, rather spiteful.

How could he accept her help in attaining his heart's desire, when he knew he did not deserve such kindness? He decided to tell Devlin the wager was finished and, if necessary, pay the one hundred guineas. Conscience appeased, he returned to the ballroom to spend the rest of the evening surreptitiously watching his love, while working on a new problem: how to gain the gratitude of Lord Neville. He decided he would ask Kathryn's advice. She had managed the impossible so far, he felt sure she would not fail him now.

*

Emma prodded at the dish in front of her. Somehow dyed red savouries and pastries just did not look appetizing.

'Maybe I should just stick to the red punch and leave the food alone,' she said to Kathryn.

'No one really eats at these affairs,' Kathryn pointed out. 'Do you?'

'No,' agreed Emma, 'and people will remember the effect not the taste.' She nibbled at a pastry. 'Actually they taste better than they look.'

'Thank God for that!' Kathryn laughed. Their giggles were interrupted by the announcement of a guest. Mr Spires requested the honour of waiting on them. Kathryn pulled a face, but Emma agreed. It would be interesting to know what Philip had been up to lately.

Richard bounded in looking as fresh faced as a schoolboy, Emma noted.

'I have come to offer my apologies that I have not been able to visit since our trip to the museum. I have had my nose close to the grindstone. His Lordship has been settling many matters of late.' He smiled at Kathryn, looking to all the world like a friendly puppy, making it impossible to snub him, Kathryn thought in exasperation.

'I hope that you will save me at least one dance at the Devlin ball. That is if you have one free?' He looked nervously at Kathryn, who murmured a suitable reply.

'I have watched the preparations with great interest. Such an original idea, Lady Courtney. Not one has ever seen the like!' Emma preened at his compliments.

'It has been a great deal of work. I hope the result will be worth it. Do you like the idea of paving the way in with good intentions? And the fire-eater?'

They entered into a discussion comparing her ideas to the most recent entertainments offered by and to the ton. Kathryn made no comment, wondering what 'matters' the viscount had been settling. The same thought had obviously occurred to Emma.

'So, what has my cousin been doing that makes you too busy to call?'

'He has been in the family archives settling investment and estate matters, almost as if he were planning a trip. But he has said nothing to me. I just follow instructions you know.' Emma's mind was working furiously. The family archives! Whatever was he up to?

Richard's mind was taken up with Kathryn and he answered Emma's questions with only half of his attention. How lovely she looked, he thought. Dressed in primrose morning dress with French knots and delicate lace at her throat and cuffs, he thought that she was as bright as a ray of sunshine and as lovely as the dawn.

He had dressed himself with great care. His coat was by Weston and the most expensive item he owned; his cravat tied into a 'mathematical' a fearsomely difficult feat that had taken him some time. A fresh flower pinned to his lapel completed his toilette. Kathryn, however, paid no heed.

'I will be visiting Kingsgrey at the end of next week, probably just after the ball,' he continued. 'There are some papers the viscount needs. I must confess that he has been somewhat mysterious.' Kathryn's blood ran cold and her mind was instantly seized with horror. Did this mean he would tell the viscount of her deception? Or would he continue the charade? Her thoughts must have been written across her face.

'Do not fret, My Lady, your secret is safe with me.' He squared his shoulders in what was probably supposed to be a manly pose, thought Kathryn spitefully.

'Naturally, I cannot lie to him.' Neither Kathryn nor Emma thought that there was anything natural about it! He continued, 'And I shall have no need to!' Kathryn wished he would get on with it and suppressed an urge to shake him. 'I shall simply say that you are visiting your sister and I did not see you.' He beamed beatifically at them, obviously expecting their compliments on his logic. Kathryn said nothing, she merely looked at him.

'That is truly clever!' Emma said, to fill the silence. 'You are a truly chivalrous man,' she continued, more than a little desperately. Kathryn remained silent. Casting about for something else to say, she happily thought of Miss DeWinter.

'I hope you will be so kind as to take some letters for me? It will be wonderful not to have to cross my lines, I can tell her all our news.' Kathryn managed a smile of agreement.

Richard still feeling heroic and magnanimous appeared not to notice Kathryn's ominous silence. Tea was served and Emma continued a stream of inane chatter.

'What costume have you chosen for our ball?' Emma asked.

'What about a snake?' Kathryn suggested, 'The one in the Garden of Eden. Or as Judas?' Emma nearly choked on her tea.

'You have such original ideas!' Richard said with admiration. 'I'm afraid that my costume is somewhat predictable. I will be a demon.'

'I'm sure you'll make a marvellous demon!' Emma flattered. Now it was Kathryn's turn to hide her giggles.

'I shall do my best to be suitably devilish!'

Kathryn thought that she would explode if she could not laugh soon. Luckily, Richard seemed to be about to take his leave of them. He could not go fast enough as far as Kathryn was concerned. As soon as they heard his footsteps on the stairs, they both burst into gales of laughter. Tears of merriment starting in their eyes, it took them several minutes to compose themselves.

'I hope none of the servants heard us. They will think we have run mad!' said Emma, mopping her eyes. 'You were positively wicked to bait the poor boy so! And he so fond of you!'

Kathryn pulled an eloquent face. 'His self-congratulation was too pompous for words,' she declared. 'Anyway, I think he would make an excellent snake.' Again they started to giggle. Kathryn had paid absolutely no attention to his attire and not much attention to his conversation, unless he mentioned the topic constantly on her mind: Philip Devlin.

'It would serve you right if he did tell. You should be ashamed of yourself,' Emma said sternly, doing her best to look suitably cross.

'Oh, really! And what about "I'm sure you'll make a marvellous demon"?' She fluttered her eyelashes. Both girls again dissolved once more into helpless giggles.

Twenty-Two

Miss DeWinter sat by the window watching the gardener and his boy sweep up the leaves, preparing for a bonfire. She sighed, old bones and winter were definitely not compatible she thought. A rap on the door interrupted her thoughts.

'A letter for you, ma'am,' the most junior footman announced.

'Thank you, Luke. Would you ask cook to heat some of my cure for me? My old bones creak in this weather.' She smiled at him. Lord, he did resemble his father. She could remember the day he started at the Hall, and the day they carried him out feet first. She tried to shake off her sombre mood and eagerly examined her mail.

It was a letter from Philip! Such a treat! She broke the seal, amazed at the number of pages he had filled. He had not written at such length since he had first started at public school. She frowned at his writing, her eyes were not what they used to be, but somehow she did not want to ask anyone else to read it for her.

When she had finished she was glad of that decision. What a jumble! Philip on the verge of losing everything, in order to marry a woman who was already his wife! Really she was getting a little too old for this. Just then her cure arrived on a tray also bearing a light luncheon designed to please an old lady's finicky appetite. Scrambled eggs, a little ham and rice pudding with a generous dollop of jam. Usually her favourite dessert, on this occasion it failed to tempt her.

Her first impulse was to tell Philip the truth. The thought that he would lose his beloved family home was too terrible to contem-

143

plate. But that would not solve the personal problems involved. Miss DeWinter racked her brains seeking the best solution. Round and round her thoughts went, yet the best course of action still eluded her.

Unable to sit still any longer, she pulled on a shawl and made her unsteady way to the rose garden, the favourite retreat of Philip's mother in times of confusion. Maybe there she would find her inspiration.

Sheltered from the wind and in the path of fitful sunshine she took a seat. Leaning forward on her stick and letting her mind wander she sat for some minutes. She found herself remembering Philip as a child, a lonely boy, closed as a nut. But his heart once given was ever true. She thought of the time he had searched for hours in the snow for his lost dog. After almost freezing to death, boy and dog returned safely home. Of the time when a huge thunderstorm had woken the house and, too scared to sleep alone, he had crept into her bedroom. When he thought her asleep, he had settled on the floor beside her bed, slipping his cold hand into hers for reassurance.

She thought also of Kathryn. Such a slight frame to bear the whole responsibility for her family. Such a wild spirit to be caged. Somehow these two were drawn together, by fate and by choice. And yet foolish pride threatened to keep them apart.

Miss DeWinter considered her choices. She could reveal the deception to Philip who, through misplaced pride, would probably refuse to have anything more to do with Kathryn. She could hide it from him and hope that their feelings would bring them together before any blame could be shared. But, in the meantime, Philip might lose everything chasing an illusion. She reread his letter.

In it he spoke of how this love had crept up on him, that the tables had been turned on him by his own arrogance, and how he had never been happier. He asked her to wish him joy, and despite her fears, realize that Kathryn was his destiny.

As always he asked for her advice; how should he approach his wife? A stranger. Should he beg, bribe, blackmail, or bash her into giving him his heart's desire? It was ironic, he admitted, that his whole happiness, once again, rested in the lap of an unknown woman. First for his inheritance, now for his love.

Nowhere did his letter say if he had revealed his feelings to his lady (and his wife, if he only knew!). Miss DeWinter could not believe that Kathryn would knowingly allow him to lose his home in order to continue her planned revenge. Although from the sound of Emma's recent letters it seemed likely that Philip's feelings were reciprocated by Kathryn, so why did the deception continue? Unless Kathryn was convinced that he did not love her. Why she should miss something so obvious was beyond her. Oh! What a tangle!

Miss DeWinter, after much deliberation, decided to write to Kathryn. She would explain what Philip was planning and ask her to intervene before things went too far. How, she did not specify. She hoped she was not too late! If only, thought Miss DeWinter, if only!

Jess thought his journey by stage would probably rattle the teeth out of his head. He clenched them grimly and wondered for the third time in the past hour, how much longer he would be made to suffer this. A particularly violent bounce awoke the man sitting opposite. Although Jess was grateful that the snoring had stopped, he now ran the risk of having to talk to him. In an effort to avoid this he closed his own eyes, feigning sleep. But he could not keep it up, as being unable to see in the lurching carriage made him feel as though he were seasick.

Instead, he looked out at the countryside, not going by speedily enough as far as he was concerned. Snow-coated hedgerows and bare trees. The temperature had steadily dropped the further north from Cambridge they travelled. Jess was looking forward to getting home. Exciting as university was, he could not deny that he had been homesick.

Arabella had written to him with her good news and also with the request that he be the godfather. Jess felt nearly overwhelmed at this offer. He intended to take his duties as uncle and godfather very seriously. He wondered if he would see Kathryn or whether her plans for the viscount meant that she could not come home for Christmas.

The lurching coach came to a sudden stop. Much to Jess's relief he abandoned the snoring man and struggled with his trunk.

'Hey there!' a cheery voice hailed him. Jess turned to see Thomas striding towards him. 'I brought the gig over for us, quicker than the cart. They'll take your trunk.' He beamed with pleasure. 'Here, whatever am I thinking of? You must be frozen! Let's get a drink inside you!'

Jess, unable to get a word in edgeways, just smiled and allowed himself to be dragged into the tavern.

It was some hours later that they returned home, later even than the cart. Arabella, resting in the library, heard them long before she saw them. After a lot of banging about in the hall sheepishly they entered to greet her.

'I see you two have been warding off the chill. It's mighty cold I understand!' Jess bent to kiss her and breathed a heady mixture of wine over her.

'It's wonderful to be home!' he said, as he sank gratefully into a chair.

'Yes,' Arabella said drily. 'I'm sure you are quite exhausted all the same, why not have a rest before dinner. Then you can tell us all your news.'

He grinned at her. 'I consider myself dismissed.' He struggled to rise. 'You know Thomas,' he said, meandering to the door, 'I think I'm slightly bosky!' He fumbled at the door and headed off in the direction of his room.

'I'll get cook to set dinner back an hour,' Arabella said to Thomas, 'and maybe you should also get some rest. I hope your headache will teach you not to lead my brother astray again!' To take the sting out of her words she blew him a kiss as he tottered from the room.

Dinner was duly served an hour late to allow for the recovery of both men, who appeared in a remarkably good mood. Arabella was the pale and shaky member of the party while they attacked their meal with gusto.

'Will Kathryn be here for Christmas?' Jess asked.

'Oh I do hope so. I have missed her so terribly!' Arabella replied. 'I worry about her in London, with that man!'

'I feel more sympathy for "that man". After all, he has had no experience with Hellkat!'

Arabella played with her pheasant, eating no more than a morsel. Thomas watched her with concern.

'Are you well, my love? Shall I send for Doctor James?'

'No. No. I am just in the doldrums. Nothing to concern the doctor with. Remember what he said? Women are made to have babies and a sickly start is Nature's way of making sure that the mother takes good care of herself.' Arabella wished she could believe him. No woman would ever have babies if for three-quarters of a year she felt like this. The soufflé made her feel sick, and the game pie was revolting. As for the trout, she could not even bear to smell them.

'Shall I ask Cook to prepare you something else? What would tempt you?' Thomas dragged himself away from his dinner to ask.

'Please stop fussing, Thomas. I am looking forward to Jess's news and I do not need you to nanny me!' she snapped. Almost instantly she regretted her harsh words. Promising herself that she would apologize later, she turned to her brother.

'I don't think you have missed us at all! Your letters are so full of your adventures. Have you made nice friends?'

'Now who's fussing?' Jess asked, and they all laughed, easing an atmosphere which had started to build. 'I have been asked by Jamie, James Fakenham that is, to pootle up to his place before term starts. Get in some hunting. The rest of the chaps will probably be there.' He eyed his sister warily. 'Unless of course you object?'

'Of course not! Have I become such a witch? As long as you do not kill yourself and that you promise not to push yourself too hard.' She looked as though she might continue, but then smiled and added, 'Oh dear. Thomas will desert me as a nag.' She smiled at her husband.

'Have you told Kathryn your good news? I would have thought she would have been here hot foot!'

'I have written that Doctor James is not yet absolutely certain. I wouldn't want her to change her plans for a false alarm.' She avoided her brother's eyes. 'Anyway, she'll probably be here for Yuletide.'

'How is the dragon? Our beloved aunt?' Jess asked. 'I half expected her to greet me.' He closed his eyes. 'I shudder at the thought.' He opened his eyes and attacked his second helping of dressed crab.

'Not that it affects your appetite,' Arabella noticed drily. 'She means well.'

'Oh yes! Does she think I crippled my brain not my leg?' Arabella had to admit that even she sometimes found Aunt Jane's behaviour galling in the extreme.

'Maybe while I'm in such a crabby mood I ought to speak to her? Vent my spleen so to speak. I can always blame my condition, after all,' she said.

Jess laughed. 'I would like to see that! But you must not excite yourself, and certainly not on my account!'

'Yes sir!' Arabella said. 'You and Thomas are acting like nurse-maids or guard dogs!'

'Well, as father and godfather that is our responsibility. So now, my girl, you will do as you're told!' Jess stood and walked towards her. With almost no sign of a limp she thought with relief. He served her some creamed potatoes and a little more pheasant. 'Now eat!' he ordered. 'Unless I must feed you?' Laughing she picked up her fork and began to eat.

Kathryn and Dinah strolled around Hookhams Library, pretending interest in the shelves while they talked.

'So now I must persuade Papa. But how? He must give his consent. He must!' Dinah sounded as though another crying bout was imminent.

'Now. Now. Don't upset yourself,' Kathryn said desperately, checking that no one was paying them any notice. 'We must think!' Dinah sniffed and obediently wiped her face.

'Does your father have any special interests? Is there anything that Rupert Roscommon can do for him?'

'I have racked my brains,' admitted Dinah, 'but can think of nothing.'

Kathryn sighed. Didn't she have enough problems of her own? She ought to let them sort it out for themselves, having nothing to do with it. But she could not resist the appeal in Dinah's blue eyes.

'I don't suppose Rupert can suddenly make himself rich?' Kathryn asked lightly.

Dinah, however, answered in perfect seriousness. 'He has expectations from his godmother and another uncle, but nothing to speak of.' Dinah peeked out at Kathryn under her furred hat. Rather like a timid squirrel Kathryn thought. Then she thought what impudence! Pronouncing that the beauty of the season looked like a squirrel!

'You must try and talk to your father. I am persuaded that he will not deny you when he realizes your heart is truly engaged.'

'But what shall I say? When he shouts at me I get so muddled.' She grasped Kathryn's arm. 'You will help me?' she implored.

'I will see what I can think up,' Kathryn assured her.

Lady Neville approached them, her sharp eyes noticing that their heads were close together in conversation.

'Now, what is it that you girls talk of so secretively?' Her smile did not quite reach her eyes. She stared at her daughter.

'Dinah was telling me about her costume for the Devlin ball,' Kathryn answered smoothly. Really the woman was enough to give you goosebumps, Kathryn thought.

'Have you chosen, madam?' she asked.

'I will dress as a witch most probably, but such mummings are for you young people, not ladies of my station.' She inclined her head graciously. 'Now we must take our leave. Come, Dinah,' she ordered peremptorily. Dinah gave Kathryn one last agonized glance before being led away.

Kathryn turned her attention back to the shelves. Witch was right! What an odious woman; she could understand why Dinah and her father needed help. But what could she do?

No scheme came to mind. Only Dinah talking to her father. But if her mother found out, she would be sure to stop it. Kathryn did not know if she could be much help. She and Rebecca walked slowly back to the house, but no flash of inspiration came.

As they reached the house, Kathryn was convinced that it was the viscount's carriage outside and hurried her steps. But she was doomed to disappointment. It was Lady Etherington and Harold.

This is definitely not my day, thought Kathryn, I should have stayed in bed.

Lady Etherington had heard of their little jaunt and Harold's preference for one of the guests. Determined to put a stop to what she regarded as an unfortunate connection, she had come to ensure that this was fully understood. Harold was soon to be sent to Europe on a short tour to avoid any complications.

Kathryn realizing the lady's intention began to reassure her. But before she could speak, an imp of mischief overtook her and those words remained unsaid.

'How lovely to see you again! I must thank you again for our day out. Such a treat.' Kathryn turned her most dazzling smile on Harold, who smiled back fatuously. Looking from one to the other, Lady Etherington hurriedly intervened.

'I was just saying to my dear Emma, I shall miss the man of the house sorely while he is in Europe.'

'The man of the house?' Kathryn echoed. 'Oh! You mean Harold is going away!' She turned her large emerald eyes to his. 'Indeed, I am sure we all will miss him.' She sighed. 'Especially one who has been lucky enough to spend a little time in his company.' Lady Etherington's colour started to rise.

'Harold has to sow his wild oats now. Though I'm sure, in the future, he will find a suitable wife, that time is not now.' She glared, daring anyone to contradict her. Harold paid no attention to her at all.

'I must say that I will miss my special friends very much,' Harold stated. Feeling that she had gone a little too far, and boggling at the thought of Harold's wild oats, Kathryn turned to Lady Etherington.

'Did you not consider joining the trip, My Lady?'

Lady Etherington was glaring at Harold. 'I can see that my son has been sadly lacking his mother's good advice, so perhaps that is a good idea. Maybe a trip to his uncle in Scotland, where I would be close to hand, of course.'

Kathryn thought that either sounded hellish, and Harold looked more than a little perturbed. Kathryn took pity on them.

'I, too, will return home soon. Probably soon after the ball.' She rose and smiled at the relief evident on Lady Etherington's face. 'So we will save our goodbyes until then.' Excusing herself she retired to her room to try and find a solution, not only to Dinah's problem but her own.

Twenty-Three

Kathryn had spent many hours trying to decide on a costume. The white dress seen at Madame Mirelle's would be a perfect choice, but what character from hell should she be? In that dress she looked like an angel, hardly suitable. She spread the dress on the bed and racked her brains. Perhaps she should find another dress. a little more suited to the theme, but it was probably the only chance she would have to wear this particular dress. And somehow she felt it would be lucky.

Kathryn entered the drawing-room to see if Emma had any ideas. Aunt Agatha, however, was there alone.

'You look a little down, my girl. Anything I can do?' Kathryn sat down opposite her. Today Aunt Agatha was dressed in a blue ensemble, with only a sprinkling of her favourite decoration, bows. She rather resembled a wedding cake, thought Kathryn. But the face that topped it was genuinely concerned, and the twinkling eyes kind.

'I was thinking about my costume for Emma's ball. I wanted to wear a dress from Madame Mirelle but it's white, and I cannot find a character to fit it. So I must forgo the dress. I wanted to pick someone else's brain. I am foxed.' She smiled resignedly.

'Well, I can't advise you until I've seen the frock.'

'I'll bring it down,' Kathryn offered.

'No. No. I'll come up.' After a gargantuan effort she left the chair and Kathryn followed her from the room. Kathryn felt a little guilty; she shouldn't have dragged her up the stairs. But Aunt Agatha surprised her, she negotiated the stairs with amazing agility.

Aunt Agatha must have noticed her surprise.

'I find it easier to move around when I actually want to get somewhere,' she said blandly. Kathryn smiled in appreciation; who would have guessed it?

'I see what you mean. It is indeed a splendid gown, perfect for the ball.'

'But I cannot honestly think of a character to match it. So I will have to think of something else.'

'Now, now. Wait awhile. Let's think. Who would be in Hell! Pandora, perhaps, the nosy one with the box.' Kathryn snorted with laughter and thought that she must tell Jess.

'Someone classical or biblical? What about the recording angel, could she visit Hell?' Aunt Agatha continued to muse. 'Or a Roman goddess. Would there be pagans in Hell?'

'I don't know,' Kathryn admitted. 'If it was before Christianity would it count and send them to Hell?'

'We could discuss theology all night but it won't get you in that dress to the ball.' A slight frown denoted her mental effort. 'Someone who might be visiting Hell.' She tapped her teeth. 'Could you be "hope" turning her back?'

'A bit tenuous,' Kathryn answered.

'Yes What about a fury? Torturing men with what they can't have?' Kathryn shook her head. 'Fetch me my comfit box, will you? It's in my room,' Aunt Agatha asked. 'I always think better with a sweet in my mouth.' Kathryn hurried off obediently.

Aunt Agatha offered her a toffee. Kathryn accepted, willing to see if anything helped.

'What about the daughter of Death? Or what about the one who was made into stars? You know, a consolation or something.'

'Constellation,' Kathryn laughingly corrected.

'Well, whoever. But she won't do if I can't remember her name. What about Helen of Troy? Not only is she pagan, but she committed adultery, and sent hundreds of warriors to their deaths!'

'Perhaps.' Kathryn was not overly enthusiastic.

'What was his name who went to Hell to get his wife? Orpheus? Does that sound right? Could you be his wife? Wait a moment though.' Her face showed intense concentration. 'Mother Nature

had a daughter and she married the Lord of Hell. And something
to do with springtime. Where is that book on Greek myths?' She
fretted. 'Look in the library, child. I'll try and remember her
dratted name!'

Kathryn followed her instructions and returned to her room
with a volume of *Greek Myths and Legends*, not more than fifteen
minutes later. Aunt Agatha pounced on this, thumbing eagerly
through the index.

'Ha!' she cried. 'I have it! Perfect!' she said, with a good deal of
satisfaction. 'Ring the bell and get Loretta, my abigail, to bring me
my jewel box. The big one mind!' Not understanding any of this,
Kathryn did as she was told.

'Will you please tell me?' she said finally. 'I don't think I can
restrain myself much longer!'

Aunt Agatha smirked in Cheshire-cat style.

'Demeter, Earth Goddess had one daughter, Persephone. She
married the Lord of Hell. While she is with him the earth goddess
sleeps, giving us autumn and winter. When her daughter is
returned to her, she wakes, giving us spring and summer. So,
Persephone is the goddess of spring, who visits Hell for six months
every year!'

Kathryn was speechless. Aunt Agatha was right. It was
absolutely perfect! An incredulous smile spread over her face.
Impulsively, she kissed Aunt Agatha soundly on the cheek. She
went a pleased shade of pink at Kathryn's delight. Kathryn took
the book and read the legend herself.

Meanwhile, Loretta delivered the large jewel box to her mistress
and was shooed out of the room. Aunt Agatha opened it and
rummaged energetically.

'Sit here, child.' She gestured to the seat she had recently
vacated in front of the dressing-table mirror. Puzzled, Kathryn
seated herself. From the box Aunt Agatha produced a tiara. Light
and delicate, it traced flowers of opals, pearls and diamonds. It
framed Kathryn's face perfectly. Aunt Agatha beamed in satisfaction.

'Perfect,' she breathed. 'You will be the goddess of spring, visiting her husband Lord of Hell, and mixing with denizens of that

dread place.' She sounded like she was reciting poetry. There was a lot more to Aunt Agatha than met the eye, Kathryn realized.

'Thank you,' Kathryn said. 'You have performed a miracle. As for the jewels, I will take great care of them until they are returned.'

'No. No. They are jewels for someone as young and lovely as yourself. I have no daughters so I wish you to keep it.'

Kathryn gasped at her generosity. 'I couldn't possibly,' she began.

'Pass it to your daughters with my blessing. It truly pleases me to imagine the pretty young faces it will adorn.'

Kathryn squeezed her hand. 'I shall be proud to wear your gift. And proud to tell my daughters of their Aunt Agatha.' Eyes shining, she turned back to her reflection. With luck and her friends, she prayed she could mount a successful assault on His Lordship's heart, not realizing that it was already hers.

Arabella groaned, the pains in her side and back were growing worse. Nothing seemed to help. Thomas stared anxiously down at her.

'I am going for Doctor James.' He held up his hand to stop any protests. 'This is not normal and he is going to come and examine you.'

Arabella's memories of the competent, kindly doctor of her youth were no longer reassuring, she had to admit.

'In fact,' Thomas continued, 'I shall get Doctor Fletcher instead.' Taking his wife's hand he tried to smile reassuringly. 'I shall return shortly. Never fear.' With that he ran for his horse.

Jess paced restlessly in the room below, wishing that there was something he could do. Struck by a thought, he rang the bell. Peters answered promptly.

'Have you sent a carriage for Fanny?' Jess demanded.

'Indeed sir. It left not more than half an hour ago,' Peters answered. 'Is the mistress still bad?' he asked. Jess nodded, not trusting his voice. 'I know I speak for all the staff, sir, when I say how sorry we are and we all pray for her recovery.'

'Thank you, Peters. All our prayers are with her.' Suddenly there was a commotion at the door. To Jess's horror he recognized the voice of his Aunt Jane. Peters tactfully withdrew.

'Where is she?' Aunt Jane demanded of Jess. 'Take me to her instantly!'

'Aunt Jane, I think you should wait for the doctor,' he began.

'Really, and I suppose you know what to do at the sickbed?' she asked sharply. 'I'm sorry, Jessamy,' she said, 'I shouldn't have said that.' Jess could not believe his ears. 'But hurry now and take me to her.' She waved him along.

This was not the Aunt Jane Jess was used to. Patting his arm, she left him at Arabella's door and said, 'I'm sure everything will be all right. I won't let anything happen to her.' Jess felt quite choked and reassured. Nothing would stop Aunt Jane after all. He returned to the library to resume his lonely vigil.

Aunt Jane entered the sickroom and started to give orders. Water was to be boiled. Soothing draughts of tea and other infusions were to be brought. A fresh nightgown, smoothed pillows and a cooling cloth on her forehead, relieved some of Arabella's pains. But what relieved her most of all, was that Aunt Jane was in charge. Never again, she vowed, would she hear a word against her.

As the pains grew worse she held her aunt's hand in an ever-tightening grip. But not for one moment did Aunt Jane stop the soothing flow of words and reassurances. Even though Arabella's grip was almost certainly painful, Aunt Jane paid it no attention.

Fanny finally came and was spirited upstairs to Arabella's room. Jess continued his solitary vigil in the library, able to do nothing more constructive than search the night through a window for a sign of the doctor and Thomas.

Finally they arrived. Thomas dragging the doctor and almost forcing him to run up the stairs. The doctor was admitted, but the door was firmly closed in Thomas's face. He, too, was sent to the library to fret and fume.

The minutes had stretched to over an hour by the time the doctor joined them.

'Is she going to be all right?' Thomas demanded, as soon as the door opened. Doctor Fletcher was immediately followed by Aunt Jane.

'How is she?' Thomas demanded. 'What is going on?'

Aunt Jane seated herself by the fire. For the first time in his life, Jess saw his Aunt Jane act like an old woman. A cold hand clenched his throat and he expected the worst.

'There is a chance, a good chance that your wife will be fine, Captain Ripley,' the doctor began. Thomas opened his mouth as if to interrupt, but Aunt Jane shushed him with a raised hand and a frown. The doctor continued, 'I'm afraid, though, that there is not much hope for the unborn babe. If I had been called sooner. . . .' The sentence was left hanging. Thomas groaned and hung his head in his hands.

'She insisted on Doctor James,' he said brokenly. 'If anything happens to her I swear I'll. . . .'

'Now. Now,' Aunt Jane interrupted. 'This is not helping Arabella.' Thomas subsided, for now.

'I am afraid that Mrs Ripley has lost a lot of blood and there may be internal complications. If she survives, the next few days will be crucial. I am sorry I cannot give you better news. Only that I will do my damnedest to save her life.' Aunt Jane did not even protest at his language, and he slipped from the room, leaving the family alone.

For the longest time, Jess thought, nothing was said. Each one of them was wrapped in their own emotions. Finally Aunt Jane broke the silence.

'Thomas, I feel that Kathryn ought to be here. Is there anyone you can send?' Her voice broke slightly. 'I feel I would be of most use upstairs.' Jess was amazed to see tears pouring down the dragon's face, and before he thought about it, he had crossed the room and put a comforting arm around her.

'Where's the Aunt Jane that nothing can stop?' he asked quietly. She muffled a sob in his shoulder. Jess could think of nothing to do except . . . he offered her his handkerchief.

'Oh, you're always the thoughtful one,' she said and, patting his hand, continued, 'My dear, you must be strong, strong enough for all of us.' He followed her gaze to Thomas, still absolutely motionless.

'Is there anything he could do?' Aunt Jane asked. She's asking me! Really thought Jess, today his Aunt Jane had been a revelation. He would never feel the same about her again. Smiling, in what he hoped was a reassuring fashion, he turned to Thomas.

'Arabella would want Kathryn to be here. What do you say, Tom?' The only reply he got was a muffled sob. Jess raised his voice slightly. 'Captain Ripley!' That got his attention. He raised his head, his eyes revealing him to be a man in torment. Aunt Jane went to him and clasped his hands.

'We must send for Kathryn,' she said. 'Jess can ride.' She eyed him in concern. 'You will accomplish nothing by making yourself ill.' A touch of the old dragon there, thought Jess. 'You must be strong for Arabella's sake.'

This seemed to penetrate the fog in Thomas's brain. He began to pull himself together. As Jess reminded him, he was a soldier after all. Both Jess and Aunt Jane felt that a crisis had been passed and overcome. Relief flooded their faces.

'I will bring Kathryn home,' Thomas said quietly. 'I must do something. I feel so helpless.' He paused, bringing himself under control. 'I will take the coach and four. Even so, it will take at least a day and a half to get there and the same back.' He stood purposefully. 'I must leave immediately.' He strode away to send orders to the stables.

'He will feel better for doing something,' Jess said.

'Yes,' Aunt Jane agreed. 'A good idea.' Jess basked in her approval. 'We all depend on your good sense you know, as the head of the family now.'

Jess could not believe his ears. Head of the family! Aunt Jane calling him head of the family!

'I'm sure you'll make us Hastings proud.' She had to stand on tiptoe to kiss his cheek. 'I will be needed upstairs.' She moved to the door, leaving Jess alone with his feelings.

He stood in front of the fire, feeling strangely removed from his surroundings. He felt the weight of his ancestry settle on his shoulders. This must be what they call coming of age, he thought.

His new responsibilities were heavy, but not uncomfortably so. They fitted him snugly, and he promised to bear them with honour. Especially now, when the family was facing one of its darkest hours. Oh Arabella! he thought, why had this day only come with such a tragic turn of fate?

Twenty-Four

Kathryn fiddled with her dress as Rebecca buttoned her up. The changes wrought by Madame Mirelle in order to give the outfit a more classically Greek feel had delighted her. One side was gathered on her shoulder, leaving the arm bare. The other shoulder was also uncovered, a winding gold-coloured bracelet snaked up her arm towards it. A gold rope crossed over her front emphasising her tiny waist. Layered chiffon created a shifting rainbow of colour around her. The costume was crowned with her Titian hair, a myriad of curls woven with pearls and tiny white blooms, the tiara giving her stars among them. One smooth curl lay on her shoulder.

Rebecca sat back on her heels after fitting soft embroidered slippers on Kathryn's feet. Her face bright with admiration.

'Oh, madam! You look beautiful. I declare I have never seen anything so lovely; you ought to be painted!' Kathryn laughed. Her mirror showed her that she was in the best of looks. Her heart fluttered as she thought of the evening ahead, adding colour to her cheeks and sparkle to her eyes.

A cough attracted her attention. William and Emma were standing in the doorway. Emma was dressed as Lucrezia Borgia in cloth of gold and black velvet. Her hair was twirled into a complex knot, pinned with jet clips and she held a black mask. William sported a cardinal's red, and a false moustache.

'Cardinal Richelieu at your service,' he smiled.

'We have a present for you,' Emma said gleefully. William advanced purposefully towards her and presented her with a golden parcel. Kathryn pulled excitedly at the wrappings. When she saw what they contained she gasped. A serpent with emerald eyes flashed at her, his body in four coils was decorated with complex carving finishing in a ruby tail.

Emma watched her face anxiously. Kathryn immediately replaced the brass snake on her arm.

'Thank you! I don't know what to say. All your gifts.' She felt quite choked. 'I will never forget what you have done for me.' Kissing William and Emma, she swallowed her tears.

Tonight. If everything did not come right, tonight would be her last night as Kathryn Chambers. Tonight she would find out if she had made any impression on the heart of her husband. Not for her revenge, not for her pride, but because she loved him. Kathryn finally admitted the truth to herself. A truth that had been increasingly obvious to those around her, and, though little did she know it, also to those around her husband.

Kathryn followed her friends downstairs for a light dinner before they left for the Devlin house in Grosvenor Square. They had to be the first, in order to greet their guests, so they were leaving at an unfashionably early hour.

Kathryn could hardly swallow a morsel. Aunt Agatha frowned at her as she played with her food.

'For Heaven's sake, girl! Eat something unless you faint dead away and spend the evening unconscious on the settle.' Kathryn laughed and obediently finished her salmon mousse. Aunt Agatha's costume was as one of the horsemen of the apocalypse.

'I am "Famine" you know. So you must listen to me.' Her eyes sparkled as she placidly finished her fish and started on the lamb cutlets.

Emma grew increasingly restless, her nerves beginning to show. She stared at the clock significantly.

'All right. All right,' said her long-suffering husband. 'We will leave soon. It still lacks three hours until the show begins.'

'I just want to check those last-minute details,' Emma replied. 'Like the punch, the pink champagne, the flowers, the fountain of fire. . . .'

'You need not bother your head over the drinks, my dear.

Devlin will see that the wine is excellent.'

'And if he sees one problem he will never let me forget it! Please, William, hurry up! Before Philip spoils my ball!'

'I give up! Come ladies, we shall follow this worrywart at a more decent pace.'

Emma sprang from the carriage as soon as they had stopped, and dashed up the stairs. Before she reached the door it was opened by Philip.

'I wondered how much longer you would last before you came to check on me,' he greeted her, but his eyes slid past her to the other occupants of the coach. His face lit up as Kathryn stepped down. The mist blurred with the outline of her dress, and she looked as if she had stepped straight from his dreams to his sight.

Tonight. Tonight he would explain it all to her. His wife, his plan, his love. Suddenly he felt as nervous as a boy, and his mouth went dry. None of his emotions showed on his face, and Kathryn's eager eyes could see no pleasure in him at the sight of her.

Determined to be optimistic, Kathryn did not allow his cool reception to upset her. She took the steps slowly, savouring this moment to remember, a moment of stillness before all the excitement started. She found Aunt Agatha next to her and she smiled nervously.

'Faint heart never won fair lady,' Aunt Agatha said, 'I, however, prefer, If you do not ask you do not receive.' She sailed magnificently past Kathryn and entered the house.

The effect on first entering the ballroom was really quite spectacular, thought Kathryn. The steps swept down to the dance floor, each painted with a good intention 'I shall not envy', 'I will stop gambling', 'I shall not drink', glared up at you in gold letters. Overhead, a satanic mask brooded and wall coverings shimmered in the candle light. The fire-eater was demonstrating his art to Emma. The place had a wonderfully gothic air to it. Kathryn had to admit it, Emma had worked a miracle.

'Do you like it?' a voice said behind her ear. Startled, Kathryn turned to find Philip at her elbow. He was dressed as Doctor Dee in a black and red Elizabethan costume. He looked suitably satanic Kathryn thought.

'Yes, I do. Emma deserves the greatest admiration. She has created the ball of the season. You must be very proud.'

'Allow me to be your escort through Hades.' He took her arm and led her down the stairs.

'I must compliment you on your costume once again. Little did I think to see the goddess of spring at my ball. You suit the part very well; I can see you spreading hope and love in your wake with just your smile.' Kathryn nearly laughed at this obvious flattery.

'I'm not very good at it, am I?' He smiled disarmingly. Kathryn found herself grinning back at him. Before anything more could be said, they were pounced upon by Emma.

'What do you think?' She gazed at them worriedly. 'What do you think of the fountain of fire?' Before they could answer she rushed on, 'Did you notice the flowers?' Philip stopped her by placing his hands on her shoulders.

'Emma, you have done the impossible! And created the most original ball ever. It will be the envy of this season and any other!'

'Such praise!' Emma laughed. 'I'm glad you like it, Philip.'

'I'm very proud of you.' He kissed Emma's cheek and she blushed vividly, for the first time in Kathryn's memory. The moment was interrupted by one of the caterers and Emma rushed off to deal with him. Kathryn started her job of distributing the place cards in the correct places for the midnight supper. She did not see Philip again until he, Emma and William formed a reception line, for the incoming guests. Before the first guest arrived, the viscount left the line to speak to Kathryn.

'I hope that you have placed me next to you at dinner. There is something of particular importance that I must discuss with you.' Before Kathryn could reply he returned to his post.

Soon the guests filled the ballroom, admiring the fountain of fire and praising Emma's originality. Kathryn created quite a sensation and she was the most courted woman at the ball. Not that she cared. She was aware only of Philip, his every move noted, every smile scrutinized. Impatiently she bided her time until their discussion could take place.

Richard Spires was one of a legion of admirers who surrounded her, one of the lucky ones who was granted a dance. As he led her on to the floor for a quadrille he felt proud enough to burst. He tried to portray something of this to Kathryn, but she refused to take him seriously, treating him like a younger brother. When he returned her to her waiting admirers he wondered if she would ever take him seriously.

To Kathryn, the evening was something from a childhood dream of being the Belle of the Ball. Compliments flowed like champagne and she wallowed in the attentions of her court, attentions she was used to Arabella receiving, not herself.

The viscount was also aware of Kathryn's every move. He twitched in jealousy every time she rose to dance with another man. And had to stop himself from physically separating her from some of her more ardent swains. At last he claimed her for his promised waltz. He took her reverently into his arms and they twirled around the dance floor, saying nothing, gazing into each other's eyes.

Their behaviour had not gone unnoticed, and several tongues started wagging. Charlotte Mountjoy watched this display with growing fury. Her gown portrayed her as Jezebel of Old Testament fame. Her faithful companion watched as she started to rage. For once Lord Osterley felt her inattention keenly. Her every thought was of the viscount. The earl wondered if there was any limit to her vanity, or to his patience.

Another couple attracting unwanted attention was Sir Rupert Roscommon and Dinah. Under her mother's glacial eyes they had danced together and tried the punch. But Dinah was shaking so hard that she could barely sip at it. Her costume of a black cat emphasised her pale face and Rupert's temper flared as he watched her shaking hands. He wished he could take her away from that harpy of a mother. In his hands she would be happy forever, he swore to himself, dreaming of their future together.

Emma was lapping up praise for the ball of the season. She grinned smugly at her husband as they danced.

'This has made the old tabbies sit up a bit. A hostess to be reckoned with!' she chortled.

'Now, my dear, there's no cause to be smug.'

'It's my party, I can be smug if I like!' She poked the very tip of her tongue at him. 'It is positively compulsory for those clever enough to throw the ball of the year, you know!'

'I see you shall become insufferable and I shall be forced to introduce desperate measures!'

Emma giggled at his serious expression. 'Such as?' she asked archly.

'You will soon find out,' he promised. But her attention had now shifted to Kathryn and Philip. Knowing his butterfly of a wife, this sudden change of subject did not worry him unduly.

'Do you think he will say something? Has he mentioned anything to you?' Emma questioned her husband. 'If they do not do something soon, I may be forced to take them in hand.' Her husband was horrified by this remark.

'That Philip would not suffer! You will only anger him by interfering. Besides,' he added, 'from the looks of them they are well on the way to an understanding.' Emma tutted at her husband impatiently.

'I'm perfectly sure even Philip can manage to tell Kathryn that he loves her, but how is Kathryn to tell him that they are already married?' William had to admit she had a point.

'Surely, once they realize that they love each other, that will not matter.' It sounded rather lame even to his own ears.

'If you think that Philip will greet the news that he has spent these last weeks wooing his wife, in front of every member of the ton, with equanimity, you are very much mistaken!'

William sighed. She was absolutely right. His anxious look returned.

The fire-eater was a great success, as was the fountain of fire. Red devils waltzed with a host of furies. Some of the more original costumes included Richard III, Oliver Cromwell (which Kathryn thought was a bit harsh), Cleopatra and the snake from the Garden of Eden.

Dinah and Kathryn sat watching the dancers as their would-be partners fetched them some punch.

'Is there something wrong?' Kathryn asked her. Seeing Dinah's shaking hand, she clasped it impulsively in her reassuring grip. Dinah seemed to be on the verge of tears.

'You are too kind! It is only that I. . . .' Her lip started to tremble. 'Papa still has not decided, and Mama is so . . . so. . . .' Kathryn squeezed her hand sympathetically.

'I think that Rupert Roscommon should approach your father and answer any doubts that he has of his expectations or anything else. Your father may appreciate a brave man.'

'If only there were something that would bring Rupert to his attention. He knows and likes the Earl of Rutland and I'm afraid that . . . that. . . .' This time the tears spilled over.

'It is the only advice I can offer,' Kathryn said frankly. 'Unless you elope?' Dinah gazed at her in horror. Seeing her expression she continued, 'You must at least make a push, Dinah! I doubt that being a watering-pot helps matters any.'

'I'm not like you!' Dinah wailed. 'You aren't afraid of anything and Mama says that I am afraid of everything!'

Kathryn tried to speak bracingly. 'Well, which is the stronger? Your fear, or your love for Rupert?' Dinah was greatly struck by this. 'I have put Rupert and your father on the same table. I hope that helps.'

'You are too kind, too good,' Dinah gushed. 'I will do my best.' She pressed her lips together, in a not altogether successful attempt to stop them trembling.

Kathryn was swept off by yet another admirer, leaving Dinah alone, but not for long as the Earl of Rutland bore down on her. Rather than waiting like a frightened rabbit as was her wont, she rose and went in search of her father, her expression suddenly reminiscent of her mother's determined glare.

Charlotte had bells around her ankles, which tinkled musically as she moved. She had tried to manoeuvre a private word with Philip, but to no avail. This set-back had not improved her temper and she plotted her revenge furiously. Lord Osterley was not his usual cynical self; Charlotte's determined pursuit of the viscount made him realize that he wanted all her energies concentrated on him. As this was glaringly not the case, he felt his temper wearing thin.

'He might not speak to me, but I'll wager she will!' She smiled at the scene in her mind's eye. 'Yes. Yes, that will do very well.' She turned her brilliant eyes back to the earl. 'Shall we dance?'

The clock struck midnight and a hellish feast was served. Kathryn had the viscount to her right and Aunt Agatha to her left. As expected, she concentrated on her food to the exclusion of all else, leaving Kathryn and Philip free to talk.

Now they had the opportunity to speak, the viscount found himself strangely tongue-tied. She was so lovely, his throat constricted as he sought the words to explain his feelings for her. Kathryn's stomach lurched and she was too nervous to speak for fear of her voice shaking. She looked at the surrounding tables wishing he would say something. A loud burst of laughter attracted Kathryn's attention, a group in the corner was laughing immoderately. Probably at some joke of Emma's, Kathryn thought. The viscount interrupted her thoughts.

'I wonder if you would consent to walk with me after dinner? There is a fine view of the fountain from the conservatory window.'

'This matter you wish to discuss with me, is of a private nature?' Kathryn fished.

'Have you tried the lobster?' he asked smoothly. 'A speciality of my chef.' Frustrated, Kathryn waited impatiently, dinner seemed to go on forever.

Charlotte flirted with her fan, and from behind its lace imparted the latest gossip to yet another guest. The rumour, once started, flowed across the room, accompanied by several barks of laughter. Philip and Kathryn saw nothing of this, engrossed as they were. But they would, Charlotte would see to that. Finally the dinner meandered to a close, leaving Kathryn and Philip to pursue privacy in the conservatory.

Twenty-Five

Dinah smiled hopefully at her father, who was starting to feel positively harassed.

'My dear,' he said, 'I have only your best interests at heart. I do not know that he can support you. That you might be better off with a slightly older man of assured fortune, than this poet of yours, must be considered.' He felt his daughter's eyes boring into him on one side and the icy glare of his wife on the other. Had a man ever been so afflicted? In desperation he headed into the gaming-room.

Sir Rupert Roscommon had already sought sanctuary there, at the faro bank with a bottle of brandy. Lord Neville needed no second invitation to join him. They settled over their cards; Rupert held the bank and had won a sizeable stack of coins piled in front of him. The two other players, a rakish pair, had been betting heavily.

Lord Neville continued to play as the stakes grew higher. Finally out of coins, Lord Neville said, 'I will give you a paper, you'll take mi'note.'

'Cash only, I'm afraid, we leave London tonight,' said one of them smoothly.

Lord Neville flushed. 'Are you suggesting that my note is not good?' Anger drove good English from his mind.

'No insult intended, Mi'Lord. But we'll see the colour of your coin or not at all.' Before Lord Neville could reply to this calumny, Rupert entered the fray.

'As you do not have the good manners to take the note of a

gentleman, I, as the bank, will cover Lord Neville's bets. I trust that you have no problem with that?'

Lord Neville eyed him approvingly; the young fella had a sensible head on him, he'd give him that. Turning his attention back to the cards he proceeded to trounce his opponents in fine style. His temper steadily improved and when the other gentlemen had to yield their places, as no one would take their paper, he was feeling remarkably pleased with himself and the world.

Maybe he should consider his daughter's words again, Lord Neville thought. His run of luck continued and he continued to gaze upon Rupert sunnily.

Kathryn and the viscount sat in the conservatory, ostensibly admiring the fountain of fire. Before Philip could broach the subject closest to his heart, another group of revellers joined them. Charlotte carefully orchestrated their entrance. Ensuring that she could be overheard, Charlotte began to speak in carrying accents.

'I assure you, I had it from Philip himself! He's been chasing the chit for some fool bet with Roscommon. Can you imagine!' Male laughter followed. 'The silly girl thinks she has made a catch! What simpletons these country girls are.'

Kathryn froze as she heard a voice telling an appreciative audience that Philip's only interest in her was due to the bet. She could hear them laughing.

Kathryn felt as though every drop of blood had drained from her body. He had told that woman! Joked with her about it! She turned to an aghast viscount.

'I hear now what it was that you so urgently wished to discuss with me!' She stood staring him in the eyes. 'I hope the stakes were worth it.' With that she turned and ran to the door, straight past a gloating Charlotte and her audience. Blinded by tears, Kathryn's only thought was to leave as fast as possible. Ignoring the surprised faces, she fled through the ballroom and on. to the front door.

'Please call Lady Courtney's coach.' A surprised footman did as he was bid and Kathryn vanished into the night.

Appalled by this turn of events, the viscount tried to catch up with Kathryn. But the gossiping tongues had eagerly noted

Kathryn's flight with a great deal of interest, as the story of the bet spread. Bright eyes waited avidly for the next happening. Really, Lady Emma did throw the most interesting parties.

Charlotte gloated at the success of her plan, most satisfying. Lord Osterley watched her beautiful face covering the malice in her thoughts.

'I will take my leave of you,' he said suddenly. Charlotte immediately gave him her full attention.

'If you find it too boring here we could stop at Oscar's for supper?'

'I think not.' He had never used that tone of voice to her before. Placing her hand on his arm she tried laughingly to placate him.

'I declare you are miffed with me! Have I been ignoring you?' Her bantering tone and flirtatious eyes elicited no response from him.

'I have had enough, Charlotte. Enough of watching you throw yourself at another man. Enough of your petty little schemes.' His tone was contemptuous. 'I have wasted enough time on you. You have shown only too clearly that you do not reciprocate my feelings. So I wish you goodbye.' Turning his back on her he walked away.

'What do you mean *goodbye*?' Charlotte's poise was showing several cracks as she followed him. Once again the party-goers were treated to a first-hand experience of another juicy titbit. Unable to delay him, Charlotte found herself trotting after him in an effort to gain his attention.

Emma watched her with narrowed eyes. Whatever had prevailed upon Kathryn to make such an exhibition of herself, Emma was sure that Charlotte was to blame. Emma longed to go after Kathryn, she had looked so upset, but she could not draw any more attention to the incident. If the hostess were to leave early as well! Balked from following her friend, Emma made a bee-line for the viscount. He was also suffering from a desire to follow Kathryn, and his dislike of being food for the gossips. He saw Charlotte and Osterley leave and promised himself that Charlotte had not heard the last of this. Emma materialized at his side.

'Dance with me,' she ordered abruptly. They took to the floor for the cotillion. As the dance progressed, Emma hissed, 'What happened? How could you let her make such a cake of herself in public?'

'She overheard Charlotte recounting a story for the amusement of her companions. A story that I told her, about a bet. A bet that involved Kathryn and whether or not she could be made to view me in a more kindly light, shall we say?' His mouth was tight and his voice hard. Emma was so amazed that she almost came to a complete halt and stopped the dance. The viscount assumed it was his admission that caused this. Actually, it was shock at the fact that Charlotte knew about the bet, and that it was Philip who had told her!

'Don't you start!' The viscount pulled her after him. 'For Heaven's sake concentrate!' Emma pulled herself together.

'You told Charlotte Mountjoy that you were pursuing Kathryn as part of a bet?' Emma spoke as if she could not believe her ears.

'You don't understand!'

'Well, on that point at least you are correct,' Emma said, and retired into a pointed and angry silence.

'Whatever am I to do, Em?'

'Well, you have won your bet. Presumably you should spend your winnings. On some friends perhaps, if you have any left!'

The dance ended and Philip took her arm and they strolled towards the fountain.

'You don't understand, Em. I'm in love with her.' Emma just looked at him. 'I've made such a hash of things, but I don't know what I'll do if I lose her.' Emma started to thaw slightly at his words. 'It's so strange. She is my first thought on waking and my last thought on sleeping. I cannot imagine any part of my life without her.' They sat watching the leaping flames.

'I have started divorce proceedings. Mr King hopes that it will not mean disinheriting me, but that is of no importance.'

'No importance!' Emma yelped. 'Have you gone mad?'

'No, I think I may just be coming to my senses. What good is my inheritance if I cannot share it with her?'

Emma's mind was working furiously. She could not stand by and let this happen. For a moment she opened her mouth to tell him the truth, then closed it quickly. Instead she said, 'Let me speak to her, try and explain things.' A thought occurred to her. 'How exactly did Charlotte Mountjoy know of this?'

'I was drunk; you remember that night at the theatre? I did not have such a good recollection of the night's events, but that I do remember.' He grasped her hand. 'Emma, I swear that there is nothing between me and Charlotte, since before Kathryn ever came into my life. You must believe me!'

'It doesn't really matter what I think, does it?'

Philip's eyes searched her face. 'Please help me, Em.'

Emma was confused, she had never seen Philip like this before. Finally she relented. 'I will do what I can,' she promised, and hope flared in his eyes.

'I shall send Underwood with a note,' Philip planned, 'asking her to speak to you.'

'Let's get this night over, then we can sort out this sorry mess,' Emma said drily. More of a mess than Philip could even imagine, she was certain.

Dinah and her mother searched for Lord Neville, Dinah at least coupling it with a search for Sir Rupert Roscommon. Finally they came upon them in the gaming-room. The first bottle of brandy had been replaced twice and both the men were a little bleary-eyed.

'I see you have been most thorough in your examination of Dinah's poet, as a possible suitor!'

'Do you see what I mean, lad? At you all the time. Man needs his peace!'

Lord Neville gazed owlishly at the outraged countenance of his wife. Rupert refilled their glasses and both men tossed them straight down.

'A man must be master in his own home,' Lord Neville continued. 'A woman's place is—' He stopped, forgetting what he was going to say. 'Well, it's . . . it's. . . .'

'Wherever the man says it is!' Rupert finished for him.

'Exactly what I was meaning to say! You have a good head on you, lad.'

'Not only for the brandy!' said Lady Neville, with a voice like cracking ice. 'If you have finished your cosy chat, your daughter and your wife wish to return home. If it is not too much trouble?' Lord Neville scratched his cheek as he mulled over her words.

'Yes, let's send the womenfolk home! Take the carriage and return home. I shall join you there later.' He turned his back on his nonplussed wife and smiled broadly at Rupert.

Lady Neville nearly succumbed to hysterics then and there. What had happened to the manageable man she had married? She distinctly felt a migraine coming on, and was led by her concerned daughter to the coach and home.

'Fancy a nightcap, old boy? They have a wonderful port at my club, you know. Fancy a snifter? After all you might, only might mind, become my son-in-law. Unless, of course, I have managed to show you the error of your ways?' He eyed Rupert hopefully. 'No? Oh well. Perhaps if I told you about my wedding day.' He began a long and involved anecdote as, arms around one another they staggered off towards St James's and White's club for gentlemen.

Kathryn sat back in the carriage. On the way there the mist had made the town fairy-like, now it made it cold and damp, and even a little sinister, she thought. Her mind was like a frightened horse; any thought that recalled tonight's humiliation caused it to buck and veer off in any direction. Kathryn found herself thinking of a rhyme she had learnt as a child. *Georgie porgie, pudding and pie, kissed the girls and made them cry.* . . . She could not remember the rest, something about running away.

This idea stuck in her mind. To have to face everyone after fleeing the ball in tears and the thought of seeing the viscount again was too mortifying for words. Kathryn's mind chased round and round, whatever was she to do? Maybe the best idea would be to go home as soon as possible, tomorrow perhaps. She could start packing as soon as she got back.

The thought of him brought with it, not memories of Kingsgrey, but of her real home, with Jess and Arabella. A wave of homesickness flooded through her. That was where she would have a chance to sort out her feelings. And, more to the point, avoid any meeting with her 'husband'.

A feeling of betrayal grew within Kathryn, so much for thinking there was more to it than that. She had been a silly little fool. Who was she to disrupt his well-ordered life? The Mountjoy woman had been right, she was a simpleton if she believed that he loved her, when it was glaringly apparent that she had been no more than an amusement to him and his cronies.

Never had she been so relieved to hear the front door close behind her and she ran straight up the stairs to her room. Rebecca was not there, to her relief as she needed to be alone. She sank her head in the pillows and cried as if she would never stop.

Just when she realized that she loved Philip, she did not know. But she had been more than a fool to imagine her feelings returned. And yet, he had been so caring, so open. What did that matter now, she told herself angrily, the whole time he had been laughing at her, while in the arms of his mistress. She was no more than a joke.

Her pride told her that she would have to pull herself together, at least until she got home. She splashed cold water on her face and wondered if she dare go downstairs and get herself a brandy. She had heard that it was good for shock, and she felt she needed all the help she could get.

Before she could decide, there was a tentative knock on her bedroom door. Checking in the mirror to make sure that she did not look too ravaged, she took a deep breath and opened the door, expecting to see Rebecca or Emma. Instead a nervous footman stood before her.

'I'm sorry to disturb you, miss. But Captain Thomas Ripley insisted that you were informed of his arrival.' He gave her a worried look. 'I hope that was right, miss.'

Kathryn answered automatically, her thoughts racing ahead. Why was Tom here? Had there been an accident? She ran downstairs to the parlour where he was waiting.

'Tom! Whatever is wrong? Is it Jess? Arabella?' The questions tumbled from her lips. Seeing his face, she cried, 'You look exhausted! Sit down. I'll fetch the brandy.' She returned moments later bearing the brandy and two glasses. Pouring huge measures, she handed him his drink. She sipped at hers, only just managing not to choke. How could people enjoy drinking this? she thought to herself. It tastes like acid. But Tom's drink seemed to revive him and a little colour entered his grey cheeks.

'I have come to bring you home.' Kathryn heard these ominous words, her heart racing. 'Arabella is ill. A problem with the baby.'

'The baby!' exploded Kathryn. 'Why didn't she tell me?'

Tom waved her to silence. 'Doctor James was no help. Doctor Fletcher says that there is little hope for the babe, but he hopes to be able to save Arabella.' He spoke as if from rote. Kathryn could hardly believe she was hearing this. She must go to Arabella before . . . oh God! Before it was too late.

Kathryn rang the bell energetically. The young footman answered her call. Nervously he looked from one to the other.

'Tell Rebecca to pack me a portmanteau, I leave in half an hour. Send the coachman to me and ask Cook to pack a basket of cold food for the journey.' He stared at her as she listed her orders. 'Well? Run then!' she finished and turned her attention back to Thomas.

'You are in no fit state to drive. Lord Courtney's man will take a turn while you rest. That way we will not have to stop on the road; we will change horses as necessary. Do you have enough money?' Thomas was too exhausted to do more than nod, glad that he did not have to spare any energy for thinking, as staying awake was strain enough at the moment.

Kathryn continued to give orders crisply and the household hummed to her command. While these preparations were made, she dashed off a quick note to Emma. Promising to write a full account of the evening's happenings, she simply told her that a family emergency needed her immediate presence. She told Emma not to worry and to hold her tongue until they met again.

Giving the note to the butler to deliver, she buttoned up her warmest pelisse and added a scarf and gloves. It looked like being a long, cold night. Thomas still looked half dazed, Kathryn insisted that he put on his military top coat, else he'd catch his death.

The basket was packed into the coach and Kathryn and Thomas, well bundled in blankets, prepared to set off. In a few hours,

Thomas would take over the driving and, swopping in this fashion they should finish the journey by the same time two days hence. The horses stamped, impatient to be off, and the sleeping city seemed to pay them no heed. One person, however, was paying quite close attention to what was occurring, as he knew he would be expected to make a full account of it.

Twenty-Six

Charlotte Mountjoy was a worried woman. The Earl of Osterley actually seemed serious in his wish to be rid of her. He did not respond to her most winsome smiles or her sharpest questions. Finally, unable to endure any more, she burst into tears. This did gain a reaction from him. Silently he passed her his handkerchief. Genuine tears started to run down her face, and forgetful of her looks for once, she sobbed like a child.

Osterley escorted her into the house, sat her down and gave her a brandy. After choking it down, her sobs abated somewhat. Red-eyed, she gazed up at him.

'Why are you so angry with me? You've never spoken to me that way before! What did I do?' she wailed.

Osterley watched her, as always with wonder at her beauty. As he had discovered to his cost, this beauty appeared to be only skin deep.

'I cannot accept the other men in your life Charlotte, and I know that you will not hold yourself only unto me. So, our ways must part.' He rose as if to leave.

'No. No. Don't. I don't understand. As you know the only thing I really ever had was you. Philip was a childish game and a childish revenge. I can understand you not liking me for that.' All artifice gone from her face, she stood in front of him, gazing at him squarely.

'You are the only worthwhile thing in my life. My only treasured friend.' She touched his face lightly. 'You who know the worst of me, also know the best. Will you give me another chance?' Tears filled her eyes once again.

With a hoarse cry, he took her into his arms, kissing her passion-ately. Joy spread through her body, she had not lost him. The moments passed. When Charlotte opened her eyes it was to see the earl kneeling at her feet. Still a little light headed, she did not real-ize his intentions immediately. His words left her dumbfounded.

'Charlotte, will you do me the honour of becoming my wife?'

'Marry you? You want me to marry you?'

'And be true!' he added, not entirely humorously.

Charlotte felt as though her heart would burst. Somehow she had never expected to be married. 'I would be honoured beyond words to become your wife,' she whispered As she said these words, it was as if the years peeled away and left her as innocent as she had been at sixteen.

'There was, is, and will be only one man in my heart.' The bargain was sealed with the first of many kisses, all of which would leave her a little trembly about the knees.

Emma stared at the note in disbelief. A family emergency? Or a desperate effort to run away? She pulled off her head-dress – it was giving her a headache, and she could not think when she had a headache. She read the note again. What was she going to tell Philip? Maybe, if she showed him the note, he could draw his own conclusions. Please God he would delay the divorce until Emma could speak to Kathryn.

A rumpus in the hall heralded the arrival of Philip. He stopped when he saw her face.

'What? What has happened? Emma! For God's sake talk to me!' Dumbly she handed him the letter. The viscount had to read it twice.

'She can't have got my note! What emergency? You know her family, where has she gone?' Emma could not bring herself to look her cousin in the eye. As luck would have it, Underwood tapped on the door.

'Aha! Just the man I wanted to see,' said the viscount. 'What of your errand?'

'Well, sir, by the time I got 'ere the lady were leaving. With an officer, going by his coat. They had a coach and four. leavin' in a mighty hurry.' The viscount and Emma exchanged a bewildered look.

'Miss Chambers left the house?'

'With luggage,' Underwood interrupted.

'With luggage,' Emma repeated. 'In a coach and four, accompanied by an army officer.' She thought a moment. 'Which direction did they take? Did you remark it?'

'They took the north road, as far as I could see, madam.'

Emma forced her numb lips to smile. 'Thank you, Underwood.' He retired, curiosity unsatisfied.

Philip Devlin stared through the window into the mists instantly thinking the worst: Kathryn had eloped; she had fled; fled into the arms of a dashing officer. Kathryn had eloped with someone else because of what he had done. His face paled, as these thoughts raced around and around. He had to find her. He turned to Emma.

'Where do her family live? It may be true, after all.'

Emma did not know what to say.

'I'm sure she will write to me soon. Since her grandmother died I am not sure that I know where her family is. Philip, she has not eloped.'

'How do you know?'

Emma longed to say, because she is already married! But she could not betray her friend.

'Please, Philip. I know she has not eloped.' Emma tried desperately to convince him.

'So who is the "military gentleman" then? Oh no! I can see that I have driven the woman I love into running away with some dashing captain!' He started to pace the room. 'But I will find her. Just to give her a choice. I cannot let my stupidity ruin her life.'

Emma offered a silent prayer. Please God, let this turn out well. Please let me make the right decision.

'Underwood will continue the search. I will go to Kingsgrey. My divorce must proceed as quickly as possible. I will speak to my wife, then I will join the search for Kathryn.' His shoulders seemed to bow at the enormity of his task, but he straightened and with all the arrogance Emma knew so well, said, 'I am sure I can convince her that I am a more suitable husband than a country captain.'

Emma almost told him the truth then and there. It was only through the most stringent self-control that she managed to hold her tongue. Luckily Philip was distracted by William's entrance.

'Have you questioned the servants?' the viscount demanded.

'They can tell us nothing beyond the fact that they took enough food for two days. Also that my coachman went with them, presumably to relieve the other driver. They must have been planning to go quite a way.' Philip's knuckles whitened as he clenched his fists. William scratched his head.

'What can we do?' His assumption that they were all in this together touched Philip and he was profoundly grateful for their support.

'I will go to Kingsgrey to persuade my wife to divorce me.' His lips twisted disdainfully. 'I am sure I can make it a profitable undertaking for her.'

'What about Kathryn?' William asked. 'If you leave it that long, she could go anywhere and you might never find her!' This thought caused an unfamiliar pain to blossom in Philip.

'It is useless if I find her and have nothing to offer her. Especially if Gretna is their destination. If he has offered her marriage, how could I offer her anything less?' He swallowed and continued, 'Her reputation can only be saved, after this foolish action, if she marries the man she is with. And I intend to be that man.'

William's conscience started to bother him. The viscount was going through such pain, risking his fortune, his reputation and, most surprisingly of all, his dignity. All for a woman he was already married to! If Emma was not going to say anything, William thought in horror, he was going to have to. But before he could speak, Emma jumped up, to his great relief.

'Philip, you must believe me, Kathryn has not eloped!'

Philip smiled sadly at her. 'I know you are trying to stop me feeling guilty. But it is my fault she has undertaken this desperate course of action, I cannot let her ruin her life because of my stupidity.' His voice grew hoarse, and his fists clenched. Emma and

William stared at each other with expressions of distress. How much longer could they hold their tongues?

Sir Rupert Roscommon opened one eye, blearily trying to bring his surroundings into focus. The green hangings around his bed swam sickeningly. His head felt as if he had been kicked repeatedly by several vicious horses. He tried to remember if horses had been present the previous night. As memory returned to him, he groaned, remembering the amount of brandy he had consumed. His groan was echoed by an unseen companion, hidden from his view by the back of the settle. Rupert remembered hazily that he had a guest. He tried to speak, but somehow the words did not reach his mouth. His head fell back to the pillow; he would try again in a minute.

His valet entered bearing a tray and two glasses. Placing these on the table he flung open the drapes.

'Good afternoon, sir.' Rupert ignored this pointed comment and mutely held out his hand for the glass of Doctor James's patented cure. Sipping slowly at its sweetness, the room came back into focus and this time stayed there.

'I will return with some tea, sir.' His disapproving back retired from the room.

Rupert tottered over to the chair by the fire. His guest still lay on the settle, his cure untouched. Rupert lifted it and pressed it upon him.

'Drink this. It will help.' The prone figure grasped the glass as a drowning man would grasp a line. Tea was served and Rupert waved away his valet's suggestion of dressing so the two of them could recover in peace.

Finally, the figure of Lord Neville straightened and sat upright, his wig askew and his eyes only half open. At least he is still alive, Rupert thought, that's the first time he has moved in the last hour.

Both men regarded each other a little sheepishly, the instant friendship of the night before evaporating in the cold light of day.

'I suppose we were a bit over the boughs last night,' Lord Neville observed. 'Do you remember anything about a peacock feather?'

'No. Why?' Rupert replied, puzzled. Wordlessly Lord Neville produced a feather from his pocket. Rupert started to chuckle and soon they were both roaring with laughter.

'Ooh. My head,' Rupert groaned. 'Stop making me laugh! I am not a well man!' Their hilarity abated and the atmosphere lost its edge of embarrassment.

'Shall I ring for some food?' Rupert asked.

'And more tea.'

When they had slaked their appetite for tea, muffins with jam, and savoury pastries, they felt nearly themselves again.

'What time is it?' Neville suddenly demanded.

'After three,' answered Rupert, squinting at the clock. Lord Neville uttered a groan. Nothing to do with the brandy this time, but at the thought of the reception awaiting him at home. Would he never be free of women?

'Everything all right, old chap?' Rupert was concerned, maybe the carousing last night had been too much for him. Trust his luck, he'd probably killed his future father-in-law, trying to persuade him to agree the match. Rupert nearly groaned again himself. He eyed Lord Neville solicitously.

'I have a fine full-bodied claret, help build you up before you go.' He rang the bell energetically. His valet answered.

'The good claret, please Charles,' Rupert ordered. 'Nothing like the hair of the dog that bit you to set a man up. Why, by the time you choose to return home you will be in fine fettle, and certainly not in the mood for any female reproaches.'

Rupert felt that if the worm turned, it could only be to his advantage but in encouraging the spirit of rebellion, he would portray himself as a less than ideal son-in-law. His head still felt too fuzzy to think this out. Gratefully, he accepted a glass of claret from Charles, in the hope that it would help. Lord Neville also accepted a glass with much the same wish.

They sat companionably in silence for a while, each involved with his own thoughts. Lord Neville imagined the performance when he got home, his lady wife was not noted for her docile temperament, and experience had taught him what to expect. He sighed.

Rupert's thoughts were just as depressing. Despite their current camaraderie, Rupert was sure that it would not be enough to stand Lady Neville. Just thinking of her made Rupert shudder and he sought solace in his glass, as did Lord Neville. The uniting club to which all men belonged could, for the moment, provide them with comfort, but they both knew that this respite would be brief.

Another bottle was broached, and the mood soon lightened. The fire crackled merrily and rain spattered against the window; both men felt entirely comfortable.

'Maybe you should delay your departure? The weather is rather inclement, and I would be glad of your company in opening another bottle.' That ought to show which way the wind blew, Rupert thought.

Lord Neville mulled over this invitation. He was in no hurry to go home, where he would find no peace. On the other hand he had only a passing acquaintance with Rupert, for all he was a suitor for his daughter's hand. His love of comfort, however, won, and, as he toasted his toes, he looked forward to his dinner. Maybe this young chap was not as bad as he had thought, his manners were pretty enough and his cellar not at all bad. These things must also be taken into account, he thought mendaciously to himself. He revelled in his unaccustomed freedom.

Rupert was wondering how to broach the subject of Dinah. Lord Neville was obviously comfortable and this might be the best chance he would get. He took a deep breath, but before he could speak there was a sharp rap on the door.

'Come in!' Rupert said, annoyed at the delay.

'A letter, sir.' Charles presented it on a silver tray. 'Delivered by hand.'

Rupert took and opened the letter with bad grace, chafing at the postponement of his speech. He had thought of several most uplifting paragraphs designed to move the hardest of hearts. Lord Neville did not realize how near he had been to hearing a speech that might have put him in mind of a Sunday sermon. It might even have been enough to drive him back to the arms of his acid-tongued wife.

Rupert read the letter with growing incredulity. By all the gods!

It was fantastic! Headache forgotten, he leapt to his feet.

'My dear Lord Neville!' Rupert could hardly hold still long enough to talk.

'Whatever is it?' Lord Neville demanded, watching Rupert caper around him. Had he lost his mind? Or was the claret stronger than he had imagined?

'My dear Lord Neville!' Rupert panted at him.

'Yes. Yes, you have already said that!' Really the boy must be mad!

'I have the greatest pleasure in asking you for the hand of your daughter Dinah in marriage!' He clutched the letter to his chest. 'As a concerned father you naturally want to know my financial situation. Let me assure you that I can support Dinah most comfortably, as I have just inherited twenty thousand a year from my wonderful dead godmother!'

'My dear boy!' He shook Rupert's hand energetically. 'Let me be the first to congratulate you!' Twenty thousand a year! Neville thought, suddenly regarding Rupert in a different light. His mood of male solidarity, however, had not quite dissipated.

'Are you sure? As a man of independent means, substantial independent means, are you sure you want to get married? To our daughter? My dear wife's daughter?'

Rupert looked at him with stars in his eyes; at this moment the world was to him a beautiful place. Ignoring Lord Neville's implied warning, he replied, 'More than anything on this earth!' Seeing that there was no persuading him to hold on to his bachelor life a little longer, Lord Neville agreed.

'Let us go immediately! To tell her this wonderful news.' Rupert started to fling off his dressing-gown.

Resignedly, Lord Neville prepared himself to face the bleak outdoors. He was not particularly enamoured with the bleak indoors he was going to either.

Suddenly a thought occurred to him. His wife could hardly look upon a suitor of twenty thousand a year, who was longing to marry Dinah, as an unsuitable connection. And he would be responsible for bringing him up to scratch. Suddenly the trip home looked rather rosier.

Taking a flask of brandy with them, against the cold, they finally presented themselves at the Neville residence. Blowing clouds of air before them they mounted the stairs and assaulted the bell. Their arrival had obviously been spied, as both Lady Neville and her daughter awaited them in the hall, their expressions remarkably similar. No more than expected, Lord Neville thought. He decided that the best form of defence was attack.

'Let us adjourn to the library, Wife. I must inform you of my plans.' Astounded for a moment, Lady Neville could not say a word. This situation, unfortunately, would not last long, Lord Neville knew. So he moved quickly.

'You and Dinah may retire to the drawing-room.' He waved Rupert in the right direction. Rupert, smiling fatuously, was marched off by Dinah, with an expression more than vaguely reminiscent of her mother on her face, and some very similar sentiments behind it, which would be aired.

Lord and Lady Neville withdrew to the library. Lady Neville prepared to deliver the speech she had been working on since the ball. Lord Neville, however, intended not only to have the last word, but also the first.

'I have one comment to make, then I am sure that you will relish the chance to talk to me.' *That* he knew from experience.

'I have given Sir Rupert Roscommon permission to pay court to Dinah.' Lady Neville looked as though she were going to interrupt, but Lord Neville hurried on, amazed that she kept silent. 'So there are two reasons to celebrate.' He paused, starting to enjoy himself.

What is the old fool talking about? Lady Neville thought impatiently. But she was glad that she held her tongue when she heard what he had to say.

'Sir Rupert is now the beneficiary of a fortune that gives him twenty thousand a year. This "handsome mountebank" as you call him, is one of the catches of the season. I secured his forgiveness over your attitude and, I am glad to say, he has very kindly accepted my apologies. I expect that he and Dinah will have an announcement to make shortly. That might be a good time for you to add your apologies to mine.'

Lady Neville could not believe her ears. Twenty thousand! A satisfied smirk crossed her face. What a coup! She then heard her husband's last words. Apologize! Her face became very red. Lord Neville wondered whether she would succumb to apoplexy or just a fit of the vapours.

As Rupert later described to Avon, the apology was a masterpiece given in a high-pitched voice, from a glassy expression. If he had not known better, Rupert laughed, he would have suspected that she, not Lord Neville, had used the flask of brandy.

Twenty-Seven

Emma had hardly slept a wink since the ball. She fretted at the lack of news from Kathryn, her heart racing at every knock on the door. William worried with her, but, as she said, Philip was the only family she had. Worry had robbed Emma of her gloss and her effervescence. She would soon be worn to a shadow, William thought, if she continued in this way. Yet it had only been three days.

Emma fidgeted around the room. Her restlessness allowing her to settle at nothing. When a discreet knock was heard at the drawing-room door, Emma nearly jumped out of her skin.

'A letter, My Lord.' Barely had the butler said the words than Emma was upon him, snatching the letter from the tray. Maddeningly it defied her attempts to open it, she ground her teeth in frustration. Finally she spread it before her.

Emma's first reaction was one of disappointment. The signature was not Kathryn's but Miss DeWinter's. Emma scanned the neat lines of italics in amazement. Philip had told her of his plans! Emma frowned, and glanced at the opening. It was addressed to Kathryn. Miss DeWinter was warning her of Philip's intentions and telling her to return to Kingsgrey as soon as possible. She rubbed a tired hand across her forehead, she had been hoping that Kathryn had fled to Kingsgrey. But she had never seen this letter. A family emergency? What had Kathryn meant? Emma racked her brains.

Unnoticed, Aunt Agatha had followed the butler into the room. Seating herself by the fire she regarded the pair of them with almost as resigned an expression.

'Any news?'

'No. In fact I think this is where I am supposed to say "The plot thickens".' Emma attempted to laugh, but she sounded closer to tears.

Aunt Agatha took pity on her.

'If it's Kathryn you want to get a message to, I can furnish you with her whereabouts.' Her announcement fell like a stone and Emma and William gazed at her stupefied. Suddenly William burst out laughing.

'Have you taken up witchcraft, Aunt? Or have you received a message from the spirit world?'

Aunt Agatha regarded him with stony eyes.

'Well, if you don't need my help. I'm sure you have already solved the problem yourselves.' She made as if to leave. Emma gestured at William commandingly.

'I'm sorry, Aunt Agatha. I'm a boor. What is it you can tell us?'

For a moment she considered him from under lowered brows, then finally she started to speak.

'While you have been chewing at your fingernails and wearing the carpets with your pacing, I have been applying my mind to the problem. Kathryn has gone to her sister and brother-in-law. There is a problem with her babe. It was her brother-in-law who came to take her, a captain of some division or other.' She sat back, enjoying the amazement on their faces.

'How . . . how?' Emma gasped. A satisfied smirk appeared on Aunt Agatha's face. William simply goggled at her in amazement. Aunt Agatha let them stew for a while, enjoying their bafflement.

'I spoke to her abigail, of course. I remembered Kathryn mentioning that she had brought her from Kingsgrey, and how devoted she was as a result of her promotion. Well, I thought if anyone knew what was happening, it would be her. And I was right.' She smiled at them, gratified, she did not like to be thought of as predictable. 'The poor girl was too terrified to say anything to anyone.'

Emma rushed across the room and kissed her soundly on both cheeks.

'You are a wonder!' she announced, her old energy returning to her in an instant. Resilient she certainly was, thought William, as she gave him her orders.

Kathryn had never been so tired in her life and she felt as though her teeth were being rattled from her head as she was pitched round yet another corner. They had been travelling for nearly two days. Two days of watching Tom shrink before her very eyes as exhaustion and fear took their toll.

Half of Kathryn wanted to be home now. The other wanted the journey to last forever, dreading what they would find at its end.

Finally the road started to look familiar. As Kathryn watched the passing countryside, dusk was descending into night. A full moon rose and bathed the world in its bright silvery light. But neither Kathryn or Tom had eyes for the beauty of the scenery.

Tom was insisting that he should take the reins despite his obvious exhaustion. His reason: to reduce any delay as he knew the way, but this cut no ice with the coachman. Finally Tom had to settle for taking a seat on the box and guiding him from there.

At last the coach gave a final lurch and halted. Kathryn swung open the door and not waiting for any assistance, jumped and started up the steps, but she was not quicker than Tom who raced through the door first.

Jess met them in the hall, relief etched on his face. He hugged Kathryn to him for a long moment, then led them both into the drawing-room where Aunt Jane sat.

Kathryn was shocked at her appearance, she looked deflated, thinner, and, Kathryn noted with horror, there was a definite tremble in her Aunt's hands.

'What news?' croaked a hoarse Tom.

'The doctor is still with her. He said that the crisis will come tonight. He has managed to stop the bleeding, but her fever still rages.' Aunt Jane spoke jerkily and Kathryn realized that she was trying to stem her tears.

'But the babe is lost.' Her aunt's voice was no more than a whisper yet the words thundered through Kathryn's head. She went swiftly to her aunt's side and clasping her cold, shaky hands, tried to give comfort where there was none to be had.

At her words, Kathryn shrugged off her cape and made for the stairs. Tom had collapsed in a chair and Jess started to remove his boots. At Kathryn's glance he paused, saying, 'I'll take care of him. You go on.'

Kathryn climbed the stairs feeling a cold hand constrict her throat and her heart like lead inside her. Taking a deep breath she entered Arabella's room.

The new doctor, Kathryn could not remember his name, was sitting on the edge of Arabella's bed. Shirt sleeves rolled up, cravat and waistcoat missing, sweat plastering his brow. He looked as though he had been in a fight. At the sight of her he rose.

Kathryn forced her suddenly leaden legs towards the bed where Arabella lay. She would have hardly recognised her. Her skin was slightly yellow, her usual milk-and-roses complexion replaced by a waxy pallor. Her hair was braided but strands clung to her forehead and neck. Her breath came in short gasps and she tossed and muttered.

The doctor led her towards the fire.

'Viscountess? I am Doctor Fletcher.' Kathryn did not even notice his use of her title. 'I am sorry that we meet under such circumstances.' Kathryn dragged her eyes away from the bed and tried to concentrate on what he was saying.

'If your sister's fever breaks tonight then I think all will be well, if not. . . .' His sentence petered to a stop.

'I want to help,' Kathryn said suddenly. The doctor eyed her appraisingly.

'Very well. But if at any time you feel you cannot continue, you will inform me immediately. Understand?' Kathryn nodded. The doctor's expression softened. 'Why don't you sit with her, while I check on the captain, else we'll have two patients.'

At this he left Kathryn and Arabella alone. Pulling a chair next to the bed, Kathryn sat and clasped her sister's burning hand.

'Bella?' she said softly. 'Bella, it's me, Kathryn.' Anxiously she searched her sister's face for any response. After a couple of seconds Arabella's eyelids fluttered and her eyes opened. For a moment they looked at Kathryn, then fluttered closed once more.

Kathryn found herself gripping her sister's hand tightly as though she were trying to will strength into her body. No further response was forthcoming but Kathryn did not let go.

The next few hours were spent obeying the terse instructions of Doctor Fletcher. Fetching, carrying, trying to get Arabella to drink the draughts that had been prepared, but Arabella's fever grew worse. She rolled and flung out her arms in her delirium. After some hours, Aunt Jane relieved her and Kathryn tried, with no success, to snatch some sleep.

Doctor Fletcher worked indefatigably and Kathryn admired his dedication and determination. If anyone could wrestle with death and win, it would be him, Kathryn thought. Such fanciful thoughts did not seem out of place in Arabella's sickroom that night.

Dawn had started to stretch bright fingers over the sky when Doctor Fletcher called Kathryn sharply. She ran to the bed fearing the worst.

Doctor Fletcher smiled at her, a beaming smile. Kathryn, in blossoming hope, turned to the patient. Arabella was still pale, but the waxy pallor had disappeared and her breathing had steadied as if she were now only asleep.

Kathryn turned to the doctor, happiness setting her green eyes aglow.

'Come,' said the doctor. 'Your family must be told the good news.' Fanny remained with Arabella as they made their way downstairs.

Tom was stretched out on a chair, Aunt Jane was dozing on the settle and Jess was standing by the window not seeing the dazzling dawn before his eyes. As they entered, the three of them jerked towards the door, both Aunt Jane and Tom immediately fully awake. Three pairs of eyes bored into them. Kathryn gave them a trembling smile.

'The crisis has passed and with careful nursing our patient will return to full health,' Doctor Fletcher announced. Aunt Jane burst into tears. She rummaged for her handkerchief.

'Oh, how foolish of me.' She mopped vigorously at her face. 'Only, the relief. I can't tell you how. . . .' She hid behind the handkerchief sobbing. No lectures on red eyes now. Kathryn sat and put an arm round her, feeling a bit teary herself.

Tom was on his feet, his haggard face alight and his eyes bright. 'Can I see her?'

The doctor smiled, and led him from the room.

'Aunt, let me take you upstairs to bed, you must get some rest.'

'Take me!' said Aunt Jane, with a flash of her usual spirit. 'Certainly not!' But she smiled and dropped a kiss on Kathryn's head to rob her words of any sting.

Kathryn turned to Jess. 'Funny, I never thought I would see Aunt Jane. . . .'

Jess laughed at the expression on her face.

'Yes, she's been full of surprises recently.' He was about to tell her of their conversation when Arabella was taken ill, but saw Kathryn's eyelids start to droop. The story could wait.

'Do I have to send you to bed as well?' He pulled her arm and led her towards the door. As he pushed her through it he said, 'Sleep well, Hellkat. Welcome home.'

Kathryn managed not to cry until she had closed the door of her room.

Twenty-Eight

Over the next two days, Arabella continued to improve. Tom spent every possible moment with the patient, driving the doctor and his helpers to distraction. Through everything Kathryn maintained a tranquil calm which nothing seemed to pierce. It was as if her worry over Arabella had used up all her energy.

Jess finally managed to pin her down in the library, determined to discover what was amiss.

'So, how progresses your revenge on the viscount?' he hazarded. Kathryn gazed at him with huge green eyes which suddenly filled with tears. Jess was absolutely horrified at the effect of his words. He led her to a chair, patting her hand sympathetically, at a loss for what else to do. Kathryn started to sob.

'What has the wretch done to you? I promise you he will pay for it. Kat? It cannot be so terrible,' Jess cried. Kathryn regained a small measure of control.

'It's not him. Only . . . oh Jess! I have made such a fool of myself!' A flash of her old spirit tinged these words. 'I thought I was so clever. I thought that he really cared for me. And then. . . .' She choked into incoherent sobs.

'And then what?' Jess asked.

'I . . . I started to mind about him, what he thought, how he felt. I thought he felt the same, but I was just a joke for his cronies and his mistress and who knows who else!'

Jess was left somewhat confused by this outburst. Seeing Kathryn in tears for the first time in years thoroughly upset him.

'Calmly, Kat. Tell me the whole. You arrived in London. . . .'

A short while later, Kathryn had managed to impart most of her story, choking back the sobs which constantly threatened to overcome her.

'And now you say you love him?' Jess was incredulous.

Kathryn laughed harshly. 'Yes. How stupid can you get?' And with that she once again dissolved into tears. Jess had no idea how to deal with weeping women. He decided to fetch her a glass of water and hurried from the room.

There was a commotion in the hall. A filthy, dishevelled messenger was thrusting a package at the footman, insisting that it was urgent and must be given to the Viscountess Devlin at once.

Curiously, Jess accepted the package and retraced his steps to the library. He was relieved to find that Kathryn had recovered her equilibrium without the aid of a glass of water.

'This has just arrived for you. Viscountess.' Kathryn looked up horrified when she heard the use of her title. She took the packet in a trembling hand and slowly undid it.

Two letters fell into her lap. First, Kathryn read Emma's note, dated two days before. As she read, her eyes widened in amazement. Jess watched her with concern as the letter fell from her nerveless fingers. He retrieved it from the floor.

'May I?' He took Kathryn's silence as consent and read the letter. In it Emma described the events after Kathryn's flight. Her description of the viscount's behaviour surprised Jess especially.

'It appears you left too soon, Hellkat. Seems you did snare your viscount after all. Now what's the matter?' Kathryn was reading the second letter forwarded by Emma. Her face drained of colour; even her lips were pale.

'Oh no! He can't mean it!' she muttered.

'What are you talking about?' Jess cried in frustration. Kathryn handed him the other letter, written by Miss DeWinter. Jess read it quickly.

'But surely this is good news. He loves you and when he discovers that he will not lose his inheritance because he is already married to you, he'll be pleased.' He watched the expression on Kathryn's face. 'Won't he?'

Kathryn laughed mirthlessly.

'You don't understand. He hated his father. He hates his wife because his father chose her. Once he discovers that I am the woman he was obliged to marry for no other reason than his despised father's whim. . . .' Her voice tailed off.

'I know it's hard, Kat, but he deserves to know. He is risking everything, you must tell him before things go too far.'

Kathryn nodded slowly. Jess was absolutely right of course, Philip had every right to know.

'Your friend says that the viscount will arrive at Kingsgrey tomorrow evening. That means we will be able to beat him there at any rate.'

'We?'

'You don't think I am going to let you go alone, do you? At least I will be there to support you, no matter how this turns out.'

Kathryn smiled at him gratefully; his presence would be welcome. Her thoughts returned to Philip. She could hardly believe it. He loved her! If only it could stay that way. But Kathryn had little hope of this, it was only she thought, that she might die from the pain of losing him.

The viscount had received no news of Kathryn, although his enquiries were wide-searching and very expensive. He left no stone unturned, but it seemed as though she had vanished into thin air.

Fretting at the lack of action, the viscount's first recourse was to the brandy bottle, but no amount of liquor gave him any respite from the promptings of his heart.

Even in his self-enforced solitude, some gossip still reached his ears. The planned wedding of Charlotte and the Earl of Osterley had caused tongues to wag, and rumour had it that the earl had bested the viscount in fixing her interest.

Whereas once he would have turned up his nose at such news, now he only felt a deep envy and renewed his efforts to find his love. His friends found him distracted and positively anti-social. He had refused an invitation to a select party to celebrate the engagement of Sir Rupert Roscommon to the Honourable Dinah Neville. Selfishly, he could not bear to see others enjoying the happiness that was so nearly his, that he might never now experience.

Was he the only man in London not on the verge of marrying his sweetheart?

Thinking of marriage reminded him of his wife. Which led him to cursing his father for having caused him all this difficulty in the first place.

Finally he could bear it no longer. With a sudden burst of energy, he made his preparations for his visit to Kingsgrey. Unable to bear inactivity any longer he left London a day earlier than originally planned.

When Kathryn and Jess arrived at Kingsgrey they were met by Miss DeWinter's personal maid, who insisted that Kathryn visit her mistress immediately before even removing her bonnet.

Puzzled, Kathryn and Jess obediently followed the maid's trotting form. Moving at high speed they reached Miss DeWinter's rooms and were ushered in.

'Oh! At last. I have been so worried.' Miss DeWinter did indeed look extremely agitated.

'Whatever has happened?' Kathryn asked. She could feel the cold hand of fear clutch at her throat. What could possibly go wrong now?

'Philip is already here!' Miss DeWinter exclaimed. Kathryn's knees gave way and she almost fell into the nearest chair.

'Now, my dear, don't panic. I have told him you were visiting your sister's sickbed. Which is actually true.'

'I can't face him! Not yet.' Kathryn looked round desperately for Jess.

'Oh. I must apologize for my manners. This young gentleman is . . ?' Miss DeWinter asked.

Kathryn was not capable of performing the introduction, so Jess had to introduce himself.

'I am Kathryn's brother, Jess.' He limped over to take her hand. Miss DeWinter appeared quite taken aback by this information, and examined the young man in front of her with a great deal of attention. Jess smiled, slightly embarrassed by her scrutiny.

Kathryn was staring into space, inhabiting her own thoughts and totally unaware of the effect meeting Jess had had on Miss DeWinter.

'Why don't you retire to your bed, exhausted by the strain of the last few days? That wouldn't be lying either, would it?'

'I can see that I have met an exceptional dissembler. I shall have to watch myself around you ladies!' Jess's attempt to lighten the situation fell somewhat flat. Miss DeWinter's attention was firmly centred on Kathryn and had missed his remark completely.

'Go to bed, child. I will tell Philip that you are not to be disturbed, tonight at least. I don't think any force on earth could delay Philip longer than that I am afraid.' She gazed at Kathryn with concern. 'Courage, my dear. Take tonight and prepare for the morrow. It may not be as bad as you think.' A thoughtful expression was on her face and her mouth was set in a determined line.

Kathryn allowed herself to be led to bed and slipped beneath the covers with a huge sense of relief, almost as if her blankets were her only protection against the world. The stress she had been through in the last few days meant that she fell asleep almost at once.

When Kathryn was settled, Jess felt at a bit of a loss. He found Miss DeWinter regarding him with her button eyes.

'Do you have any suggestions for me? Should I retire to bed as well?' Miss DeWinter took his arm and led him from the room.

'No. No. I think I should introduce you to Philip.' She smiled up at him. 'It could be important.' Confused, Jess allowed himself to be ushered from the room.

The viscount bounded up the steps to the door. Having seen the coach he knew that his wife had returned and was desperate to face her as soon as possible. His plans were frustrated, however, by Miss DeWinter.

'Ah! Philip my dear. Would you join me in the library; there is someone I want you to meet.'

'I thought I would speak directly to my wife.' Miss DeWinter tottered slightly and instantly Philip was at her side. 'You should be resting; let me take you to your room. Do you need the doctor?'

Miss DeWinter smiled at his concern and leant on him a little more than she needed to.

'Let me introduce you to our guest. I fear your lady wife cannot be bothered tonight.'

He guided her into the library.

Jess stood by a roaring fire, gazing into the flames, wondering how he was going to deal with the man who had caused his beloved sister such heartache.

Miss DeWinter performed the introductions and the two men eyed each other warily across the length of the room. Having seated herself, Miss DeWinter ordered refreshments and dinner for two.

'I hope you will not mind my not joining you, but I feel the need of an early night.' Both men started to fuss round her and somehow, before they knew it, they were settled in front of the fire, enjoying a pre-dinner drink.

'How were those horses, my dear? Were they worth such an exorbitant amount of money?' Miss DeWinter asked. At once Jess gave the conversation his full attention. Horses, as Miss DeWinter had guessed, were a favourite topic. Within half an hour, the two were totally engrossed in the relative merits of the viscount's hunters.

When dinner was announced, Miss DeWinter excused herself, in the knowledge that the two men, despite their disparate ages and experience, were well on the way to an understanding. Their hunting stories became more and more outrageous and, she guessed, would continued well into the night.

Although she was weary and she knew that she would spend most of the night planning, Miss DeWinter took time to look in on Kathryn.

Kathryn was wide awake and worrying. She shrank back under the blankets at Miss DeWinter's knock, fearing that the viscount had not been put off. When she saw Miss DeWinter she smiled in relief.

'I just came in to wish you good night, my dear.' She dropped a kiss on Kathryn's brow. 'It will be all right, you know. I shall give you all the help I can. Try not to worry.'

'How is Jess?' Kathryn asked.

'Oh. I left him with Philip recounting some very tall hunting stories. No need to worry about them.'

Kathryn tried to be comforted. However her worrying could not overcome her physical exhaustion and within five minutes of Miss DeWinter's departure she was sound asleep.

The viscount and Jess retired to the library. With glasses at their side they relaxed, at ease with each other. The tension of the last few days seemed to flow from them and, for the first time since Kathryn's disappearance, the viscount was able to relax. However, the knowledge of what he was going to do to Jess's sister began to make him feel uncomfortable. After another couple of brandies he could stand it no longer.

'I am sorry that we have met under such circumstances.' The viscount fiddled with his glass nervously. 'I am afraid that what I have to discuss with your sister will cause her great upset.'

Jess wished he had not brought the subject up. It made him feel very small to lie to a man for whom, despite the fact that they had just met, he felt a great liking. He mumbled a reply into his glass. The viscount continued, not noticing his distress.

'I have made a terrible botch of these matters, I fear.' He poured himself another drink. 'I only hope that you will both forgive me.'

Jess was seriously tempted to tell him the whole story then and there, but with great difficulty held his tongue. Luckily the viscount was distracted and did not continue this train of thought.

'I hope you will not find this impertinent, but does your leg cause you any problems when you hunt?'

Unusually Jess had no difficulty in discussing his lameness with the viscount.

'Actually I feel happier riding a horse than I do walking. The horse at least does not treat me like a cripple. All men are equal on a horse.' Jess finished his drink and continued, 'At least then I do not have to suffer the pity of others; they cannot see my leg when I ride.'

The viscount nodded. 'Yes. It is terrible to be treated differently to everyone else.' The viscount sighed. 'How did it happen? An accident?'

'A childhood illness. I was too young even to remember that once I had two good legs like everyone else.' The viscount heard the pain in Jess's voice and turned the subject to less distressing avenues.

It was late when they eventually retired and, as Philip slowly dropped into slumber, he mused that Jess was exactly the sort of brother he would have wished for. His last thought before sleeping was regret that this likeable, intelligent young man should have had to suffer such a cruel turn of fate.

Twenty-Nine

Kathryn awoke the next day to a feeling of impending doom. Her hands shook as she drank her morning tea. The contortions of her stomach made her excuses not to eat breakfast in the parlour quite true.

She sat and stared at her reflection in the mirror on her dressing-table. A light tap on the door caused her to leap to her feet in horror. Jess poked his head around the door.

'Kat. You look like you are about to face the gallows.' He crossed the room and embraced her. He could feel her trembling and he cast about desperately for something bracing and reassuring to say.

'Courage, my dear. It will be all right. You'll see,' he added lamely. Kathryn saw his worried face and made a determined effort to pull herself together. She tried a smile.

'At this rate I shall drive us both into a spasm. Is he. . . ? Is he waiting for me?'

'He told me to ask you to meet him in the library in half an hour.'

'He asked you!'

Jess blushed slightly, feeling guilty for consorting with the enemy.

'Yes. Well, you see yesterday evening we had dinner and a few drinks and I found him. . . .' His voice trailed to a halt. Kathryn immediately regretted her sharp retort.

'Yes, he is good company. I'm glad that you like him.' She smiled. 'If only. . . .' Her eyes filled with tears. Jess's worried face swam before her. Getting herself in hand she tried to be reassuring.

'I'm sure it will work itself out and anyway I will be fine. How can I not be, with my loving family behind me? Don't worry, I will be fine.'

Jess said, 'He seems a reasonable man. He will understand, I'm sure.' He felt his words inadequate, but could think of nothing to add. Kathryn smiled and dropped a kiss on his cheek. In a strange way, by reassuring Jess she had reassured herself and she now felt almost capable of facing her husband.

Kathryn squeezed Jess's hand and then made her way to the library.

She was, of course, nearly fifteen minutes early and she spent the time rehearsing what she would say, but the speeches she created in her head became confused. Maybe she should just write him a note and leave it in her stead. She could say she was in the garden, then if he wanted to see her or if he didn't. . . . But didn't he deserve an explanation face to face? After all she had done to him it was the least she could do.

Kathryn gazed sightlessly at the garden wishing with all her heart for the happy ending she had so blithely expected what seemed like a lifetime ago.

The click of the door closing caused her to jump and slowly she turned around.

The viscount, not expecting to see her already waiting for him, also jumped a little in surprise. But that was nothing to his amazement at seeing it was Kathryn there.

His first impulse was to take her in his arms and never let her go. Reality, however, intruded and a hundred questions rose to his lips, and for a full minute as confusion reigned he could not say a word. Finally, in a voice he would never have recognized as his own, he managed to stutter, 'Kathryn! I don't understand. What are you doing here? I would have searched the world for you, only to find you on my doorstep!'

Kathryn drank in the sight of him, willing herself to remember his every detail. There were shadows beneath his eyes, and lines of stress had etched themselves around his mouth. Kathryn hated herself for being the cause of them.

Their eyes locked and in that moment Philip knew something was very wrong.

'What does this mean?' he asked, almost whispering in trepidation. 'Where is my wife?'

Kathryn could find no voice to reply. She simply held up her left hand to allow him to see the crested ring he had placed on her finger.

Philip blanched and his lips became pinched.

'You are my wife?!' Realization hit him like a cannon ball and it took all his pride not to sink into the nearest chair and allow his bewilderment free reign. Kathryn thought her heart would break as she watched him reel in shock.

'I think you had better explain.' His voice sounded distant, cold and his expression was stony.

Kathryn pressed her trembling hands together and willed herself not to cry.

'After the wedding, I was angry, humiliated. You hurt me and I wanted to hurt you.' Philip snorted at this and curled his lip in a disdainful fashion, but offered no comment.

'I went to London with Emma, at first for revenge . . . but then. . . .' Philip made as though to leave the room.

'Wait. Philip, let me explain. Please. If you then want me to go,' – Kathryn swallowed hard – 'I will. But please, five minutes is all I ask.'

Philip turned towards her. His back straight and all his defences in place. So distant, Kathryn thought. How will I ever make him understand? Her words started to come in a great rush.

'Then, I thought that maybe we could get to know each other as people, not as the woman you were forced to marry and the man who humiliated her. A clean slate. I thought I had nothing to lose. You seemed to hate me anyway so I thought it could not get any worse.' Kathryn felt the pleading note in her voice but could not stop it. 'I never for one moment thought it would come to this.'

'No, you didn't think at all. Was playing with my emotions fun? Did you laugh, My Lady Viscountess? Was this game of yours amusing?' His bitterness stung Kathryn like a whip and her temper began to rise.

'What of your wager? Putting money on gaining my interest! What of that, My Lord? Did you think of the effect on me? No, of course not. One set of rules for me and another for you?' Kathryn's eyes flashed green fire and her hands had somehow planted themselves on her hips.

'I was going to give up everything for you,' Philip roared, 'because I thought I loved you! Now . . . now I discover that I was married to you all along. How you must have laughed. What a royal fool I have been.'

'No!' Kathryn cried. 'You don't understand, I, too, fell in love. Then I had to endure your mistress explaining to me that I am no more than a wager. Did you both find me amusing? And you say *I* was laughing!'

'You lied to me. Ever since I have known you. Lies, all of it lies!' His voice had risen to a shout.

'Not all lies. Philip, please believe me, the feeling was mutual. I would have told you after the ball but. . . .'

'Oh, just in time for me not to lose my estates. How kind of you. I should be grateful, I suppose.' Philip started to pace. Kathryn gripped the back of the chair, her knees weak.

'Feelings? What do you know of my feelings? You made all these arrangements without so much as a word to me. Doubtless you expected me to fall into your arms like an over-ripe plum!'

No longer shouting, Philip answered in such a cold, dead voice that Kathryn would rather he were shouting at her. 'I wanted the woman I loved to love me in return. To be able to love me with no impediments, no obstacles.' A bitter smile twisted his mouth. 'What a foolish expectation from the woman my father chose for me. He thought to land me for life with a callous harpy and it seems he succeeded. And the fact that I thought myself in love with you, must make his revenge complete.'

During his speech Kathryn's face crumpled and tears started to wend their way down her cheeks. At his final words she uttered a small wordless cry and fled from the room. In the hall she almost collided with Miss DeWinter, but such was her heartache and misery that her only thought was for escape. Leaving Miss DeWinter behind her she fled into the garden.

Rather than following her, Miss DeWinter stepped into the library. The viscount had sunk into a chair, his head held in his hands. No movement showed that he was even aware of her presence. Miss DeWinter briskly rang the bell and told a nervous maid to ask Mr Hastings to join them.

Finally the viscount spoke. 'You knew, didn't you? Yet you said nothing. You betrayed me too.'

'Don't judge me yet, Philip. Have I ever, in your memory, done anything to hurt you?' When he didn't answer she continued, 'So, give me a few minutes of your time now. Then, if you still feel I have betrayed you when I have finished, I will leave your house forever and you need trouble yourself over me no more.'

The viscount raised his eyes to hers. They were the eyes of a man in Hell, whose soul was being torn from him. His eyes told her that at this time he would welcome death, as life no longer held any joy for him.

Jess arrived and somewhat warily entered the room. His eyes took in the scene and instantly he wished himself a hundred miles away. He could not imagine why his presence was required.

'Please, Jess, sit down; you, too, have a right to hear this.'

'Do you need an audience to humiliate me further? Why not bring the servants in? Let everyone in on the joke that I was on the verge of giving up everything in order to marry a woman who was already my wife. My father chose well; she is a heartless harpy.'

Jess opened his mouth to protest but was silenced by a gesture from Miss DeWinter.

'Philip, don't compound your foolishness. Listen to me. You have fallen in love with a warm, wonderful girl who would like nothing on this earth better than to spend the rest of her life loving you.' Philip looked as though he were going to protest, but Miss DeWinter swept on regardless. 'She wasn't chosen by your father in order for him to revenge himself on you, nor to have the posthumous pleasure of finally having you obey his wishes. He chose Kathryn because of Jess.'

Both men gaped at her in amazement. 'Before you cart me off

to the asylum, at least let me finish. The damage to Jess's leg was not caused by a childhood illness, but by your father. An accident while he was carousing. Your father, Philip, and yours, Jess, were somewhat the worse for wear when a race was suggested. It was during that race that a small boy was caught under the wheels of the leading coach. Your father's coach, Philip, and the small boy was Jess. The will was his effort at atonement.'

Stupefaction was written on both men's faces. Miss DeWinter allowed herself a minute of smugness at the effect of her words.

'Does Kathryn know?' Jess's words broke the heavy silence.

'No,' said Miss DeWinter baldly. 'She, too, thinks she was chosen by the old viscount as a punishment for his son.'

Philip's face was haggard and his hands shook.

'Dear God. What have I done?'

In a sudden burst of energy he leapt to his feet. Miss DeWinter, smiling slightly, gestured towards the garden.

'You will probably find her in the rose garden.'

The viscount hurried to the tall library windows. A thought suddenly struck him and he stopped abruptly.

'Jess, please don't think too harshly of me. I will do my best to make things right. For both of you.'

Jess nodded, still too stunned to speak. Then he was gone, leaving Jess and Miss DeWinter alone. Miss DeWinter sighed; she was unaccustomed to all the excitement.

'If you would be so kind as to escort me to my room I will be happy to answer your questions on the way, but I am afraid that I really must rest.'

Jess solicitously offered his arm on which she leant heavily.

'Maybe you should take up fairy-godmothering. I was beginning to think no magic on earth would solve this sorry tangle.'

Miss DeWinter laughed. 'There is only one magic and that is love. Let us hope that now it will run true.'

The viscount erupted from the library moving at a run. Never had the rose garden seemed so distant. Finally he spotted the tell-tale flutter of Kathryn's skirts. She was seated on the bench where he was told his mother had spent many hours.

At the sound of his approach, Kathryn leapt to her feet. Her face was puffy from her tears but her eyes were clear, emerald green flashing at him as she prepared to continue their quarrel.

Before she could say anything to Philip he took her hand and started to speak.

'I have come to beg your forgiveness. I do not know whether you will find it in your heart to forgive me. But I swear that, if you let me, I will spend every day of the rest of my life trying to make you happy.' He gazed at an astounded Kathryn.

Clasping her little hand tightly he continued, 'If my behaviour has put paid to any feeling you had for me, I will warm my heart at the flame of your memory and think myself fortunate. But if you think you could love me too, I swear you will never regret it. I would sacrifice myself and the world to spare you one moment's pain. Everything I am or ever will be is yours.'

Suddenly he dropped to one knee pressing her hand to his heart. 'I offer you this heart a little battered and unaccustomed to use, but for what it is worth, it will be true and will love you to the end of time and beyond.' His eyes desperately searched her face for his answer.

Kathryn thought she should pinch herself in case she was dreaming, but she didn't, as this was a dream she wanted to continue for the rest of her life.

'Why?' she breathed. 'Are you ill?'

'If love can be called a sickness of the heart than I pray I will never recover. I am also sick with fear that you will spurn me.'

'Oh never. Never.'

At her words, Philip leapt to his feet and gathered her into his arms.

'Oh my love. I will cherish every day with you and know myself to be in paradise. For even Heaven holds no joy for me unless you are there.'

Holding her so tightly he could feel the fluttering of her heart he gazed down at her and finally claimed her lips, vowing to himself that while he drew breath he would never let them escape from him again.